SLOPE STABILITY AND STABILIZATION METHODS

SLOPE STABILITY AND STABILIZATION METHODS

Second Edition

LEE W. ABRAMSON
Hatch Mott MacDonald
Millburn, New Jersey

THOMAS S. LEE
Parsons, Brinckerhoff, Quade & Douglas
San Francisco, California

SUNIL SHARMA
University of Idaho
Moscow, Idaho

GLENN M. BOYCE
Parsons, Brinckerhoff, Quade & Douglas
San Francisco, California

WILEY

Slope Stability and Stabilization Methods
Second Edition

Authorized reprint by Wiley India Pvt. Ltd., 4436/7, Ansari Road, Daryaganj, New Delhi - 110002.

Reprint: 2019

Printed at: Sai Printo Pack Pvt. Ltd.

ISBN: 978-81-265-4749-4

CONTENTS

PREFACE

The successful analysis and stabilization of soil and rock slopes requires an in-depth understanding of several subjects and concepts. Until the first edition of this book had been published, never had one volume assembled all the information needed to provide adequate treatment of each required area. This new edition includes an introduction to the general concepts used in slope stability studies, a discussion of the geologic features that usually give slopes their "personality," treatment of the groundwater and seepage issues that frequently cause slope stability problems, investigative and testing techniques for developing an appropriate subsurface ground model for evaluation, slope stability analysis procedures ranging from simple rules of thumb and design charts to the more complex rigorous limit equilibrium methods, stabilization design methods, and issues related to design, construction, and long-term monitoring of slopes with geotechnical instrumentation and the like. The book has been written by practitioners who, through their knowledge and experience, have assembled the background information, theory, analytical methods, design and construction approaches, and practical examples necessary for one to carry out a complete slope stability project. This book is intended to be a reference for practicing geotechnical engineers and geologists and a textbook for graduate courses in geotechnical engineering and geology. The few subjects not covered in detail are supplemented by ample references for additional information.

Retaining the best features of the previous edition, this new book has been completely updated to address the latest trends and methodology in the field. Features include:

- All-new chapters on shallow failures and stability of landfill slopes
- New material on probabilistic stability analysis, cost analysis of stabilization

alternatives, and state-of-the-art techniques in time-domain reflectometry to help engineers plan and model new designs
- Tested and FHWA-approved procedures for the geotechnical stage of highway, tunnel, and bridge projects
- Sound guidance for geotechnical stage design and planning for virtually all types of construction projects

LEE ABRAMSON
TOM LEE
SUNIL SHARMA
GLENN BOYCE

ACKNOWLEDGMENTS

This book is based on a manual and seminars developed by Parsons, Brinckerhoff, Quade & Douglas, Inc., for the United States Federal Highway Administration. Many individuals, too many to name here, participated and assisted in that project. The authors wish to thank their colleagues at Parsons Brinckerhoff, the Federal Highway Administration, several State Departments of Transportation, the United States Forest Service, the United States Army Corps of Engineers, several contractors and manufacturers, and others who shared their knowledge and expertise with us.

The San Francisco office of Parsons Brinckerhoff and the University of Idaho at Moscow provided clerical, graphical, and computer support. The authors are very grateful for these institutions' contributions to this effort. Special thanks go to Peter Settle, Mitchell Fong, Jennie Fong, Galen Samuelson Nagle, Ken Burke, Arlene Brown, Stephanie Cunningham, Lou Omania, and all those not mentioned by name.

Finally, our families deserve a great deal of credit for allowing us to turn our homes into libraries and word-processing centers and for supporting our long hours away from home when necessary. In gratitude for their love and support, we dedicate this book to Meryl, Deena, Ariel, and Gavrila Abramson; Marilyn, Jason, and Jana Lee; and Eleanor Bried, Berrin, and Julian Boyce.

ABOUT THE AUTHORS

LEE WAYNE ABRAMSON is a Vice President and Eastern Regional Manager of Hatch Mott MacDonald. He has nearly 25 years of experience in geotechnical engineering including the planning, design, and construction of underground structures, foundations, retaining walls, embankments, and slopes for transit, highway, power, nuclear waste, water, wastewater, railroad, commercial, military, mining, and research projects. Lee is a licensed engineer in Wisconsin, New York, Georgia, Washington, Oregon, and California. He has authored several papers related to geotechnical engineering as well as a *Handbook on Ground Control and Improvement*, published by John Wiley & Sons. Lee has Bachelor and Masters degrees from the University of Illinois–Champaign Urbana. He has also taken courses at the Georgia Institute of Technology and the University of California–Berkeley. Lee lives in Millburn, New Jersey, with his wife, Meryl, and three daughters, Deena, Ariel, and Gavrila.

THOMAS SIK-CHUEN LEE is a Senior Professional Associate of Parsons, Brinckerhoff, Quade & Douglas, Inc. He has over 22 years of experience in geotechnical engineering related to slopes, retaining structures, foundations for transit, highway, commercial, and residential projects. His past experience includes six years as a geotechnical engineer for the Hong Kong Geotechnical Control Office, and 17 years in private consulting firms in the United States. In Hong Kong, Tom was involved in engineering design and stability studies for soil and rock slopes and retaining structures under the Landslide Prevention Measures Program run by the Hong Kong government. In the United States, Tom has served owners, engineers, contractors, and attorneys as an expert witness on the subject of landslide investigations and stabilization, forensic engineering, and interpretation of anticipated geotechnical conditions. Tom has also served as a board consultant for public agencies. Tom is a geotechnical engineer licensed in California and a professional engineer in New York and New Jersey. He has authored and coauthored several papers related to slope stability and remediation. Tom has a Bachelor of Science degree from Brown University and a Master of Science degree from the Massachusetts Institute of Technology. Tom lives in Union City, California, with his wife, Marilyn, and two children, Jason and Jana.

SUNIL SHARMA is a Professor of Civil Engineering at the University of Idaho, Moscow, Idaho, where he teaches courses in Earthquake Engineering, Soil Dynamics, and Geotechnical Engineering. His research interests and publications for the past 20 years have been directed toward earthquake engineering, slope stability, foundation engineering, numerical methods, groundwater and seepage, and computer software design and development. Dr. Sharma is also the author of the popular slope stability computer program XSTABL, which is used by many highway agencies and consultants. Sunil has a B.Sc. (Hons) degree in Civil Engineering from the University of Leeds, England, and Masters and Ph.D. degrees from Purdue University in Indiana. He is a licensed professional engineer in Idaho.

GLENN M. BOYCE is a Senior Professional Associate with Parsons, Brinckerhoff, Quade & Douglas, Inc. Glenn has 18 years of experience in geotechnical engineering including the design of slopes, embankments, excavations, and tunnels for highway, transit, and pipeline projects. A handful of Glenn's project assignments have included design and construction of major highway systems through difficult terrain. He is a licensed civil engineer in California and Arizona. Glenn has Bachelor and Masters degrees from Drexel University in Philadelphia and a Ph.D. from the University of California–Berkeley. Glenn lives in Piedmont, California, with Eleanor, his wife, and their two children, Berrin and Julian.

CHAPTER 1

GENERAL SLOPE STABILITY CONCEPTS

1.1 INTRODUCTION

The evolution of slope stability analyses in geotechnical engineering has followed closely the developments in soil and rock mechanics as a whole. Slopes either occur naturally or are engineered by humans. Slope stability problems have been faced throughout history when men and women or nature has disrupted the delicate balance of natural soil slopes. Furthermore, the increasing demand for engineered cut and fill slopes on construction projects has only increased the need to understand analytical methods, investigative tools, and stabilization methods to solve slope stability problems. Slope stabilization methods involve specialty construction techniques that must be understood and modeled in realistic ways.

An understanding of geology, hydrology, and soil properties is central to applying slope stability principles properly. Analyses must be based upon a model that accurately represents site subsurface conditions, ground behavior, and applied loads. Judgments regarding acceptable risk or safety factors must be made to assess the results of analyses.

The authors have recognized a need for consistent understanding and application of slope stability analyses for construction and remediation projects across the United States and abroad. These analyses are generally carried out at the beginning, and sometimes throughout the life, of projects during planning, design, construction, improvement, rehabilitation, and maintenance. Planners, engineers, geologists, contractors, technicians, and maintenance workers become involved in this process.

This book provides the general background information required for slope stability analyses, suitable methods of analysis with and without the use of computers, and examples of common stability problems and stabilization methods for cuts and fills. This body of information encompasses general slope stability concepts, engineering geology principles, groundwater conditions, geologic site explorations, soil and rock testing and interpretation, slope stability concepts, stabilization methods, instrumentation and monitoring, design documents, and construction inspection.

Detailed discussions about methods used in slope stability analyses are given, including the ordinary method of slices, simplified Janbu method, simplified Bishop method, Spencer's method, other limit equilibrium methods, numerical methods, total stress analysis, effective stress analysis, and the use of computer programs to solve problems. This book is intended for individuals who deal with slope stability problems, including most geotechnical engineers and geologists who have an understanding of geotechnical engineering principles and practice.

1.2 AIMS OF SLOPE STABILITY ANALYSIS

In most applications, the primary purpose of slope stability analysis is to contribute to the safe and economic design of excavations, embankments, earth dams, landfills, and spoil heaps. Slope stability evaluations are concerned with identifying critical geological, material, environmental, and economic parameters that will affect the project, as well as understanding the nature, magnitude, and frequency of potential slope problems. When dealing with slopes in general and slope stability analysis in particular, previous geological and geotechnical experience in an area is valuable.

The aims of slope stability analyses are

(1) To understand the development and form of natural slopes and the processes responsible for different natural features.

(2) To assess the stability of slopes under short-term (often during construction) and long-term conditions.

(3) To assess the possibility of landslides involving natural or existing engineered slopes.

(4) To analyze landslides and to understand failure mechanisms and the influence of environmental factors.

(5) To enable the redesign of failed slopes and the planning and design of preventive and remedial measures, where necessary.

(6) To study the effect of seismic loadings on slopes and embankments.

The analysis of slopes takes into account a variety of factors relating to topography, geology, and material properties, often relating to whether the slope was naturally formed or engineered.

1.3 NATURAL SLOPES

Many projects intersect ridges and valleys, and these landscape features can be prone to slope stability problems. Natural slopes that have been stable for many years may suddenly fail because of changes in topography, seismicity, groundwater flows, loss of strength, stress changes, and weathering. Generally, these failures are not understood well because little study is made until the failure makes it necessary. In many instances, significant uncertainty exists about the stability of a natural slope. This has been emphasized by Peck (1967), who said:

> Our chances for prediction of the stability of a natural slope are perhaps best if the area under study is an old slide zone which has been studied previously and may be reactivated by some human operations such as excavating into the toe of the slope. On the other hand, our chances are perhaps worst if the mechanism triggering the landslide is (1) at a random not previously studied location and (2) a matter of probability such as the occurrence of an earthquake.

Knowing that old slip surfaces exist in a natural slope makes it easier to understand and predict the slope's behavior. Such slip surfaces often result from previous landslides or tectonic activities. The slip surfaces may also be caused by other processes, including valley rebound, glacial shove, and glacial phenomena such as solifluction and nonuniform swelling of clays and clay–shales. The shearing strength along these slip surfaces is often very low because prior movement has caused slide resistance to peak and gradually reduce to residual values. It is not always easy to recognize landslide areas (while postglacial slides are readily identified, preglacial surfaces may lie buried beneath glacial sediments). However, once presheared strata have been located, evaluation of stability can be made with confidence.

The role of progressive failure in problems associated with natural slopes has been recognized more and more as time goes on. The materials most likely to exhibit progressive failure are clays and shales possessing chemical bonds that have been gradually disintegrated by weathering. Weathering releases much of the energy stored in these bonds (Bjerrum, 1966). Our understanding of landslides involving clay and shale slopes and seams has increased largely due to the original work by Bishop (1966), Bjerrum (1966), and Skempton (1964).

1.4 ENGINEERED SLOPES

Engineered slopes may be considered in three main categories: embankments, cut slopes, and retaining walls.

1.4.1 Embankments and Fills

Fill slopes involving compacted soils include highway and railway embankments, landfills, earth dams, and levees. The engineering properties of materials used in these structures are controlled by the borrow source grain size distribution, the methods of construction, and the degree of compaction. In general, embankment slopes are designed using shear strength parameters obtained from tests on samples of the proposed material compacted to the design density. The stability analyses of embankments and fills do not usually involve the same difficulties and uncertainties as natural slopes and cuts because borrow materials are preselected and processed. Because fills are generally built up in layers, analyses are required for all steps in the life of the project including:

(1) All phases of construction
(2) The end of construction
(3) The long-term condition
(4) Natural disturbances such as flooding and earthquakes
(5) Rapid drawdown (for water-retaining structures like earth dams)

Debris and waste landfills differ from other engineering slopes in that they typically have an extreme variety of material types, sizes, and characteristics. These materials are extremely difficult to characterize and analyze.

Constructed fills have been used since antiquity with varying degrees of success and failure. In ancient times, they were used to construct earth fill dams for storing irrigation water. One of the oldest recorded earth fill dams is the dam completed in Ceylon in the year 504 BC, which was 11 miles long, 70 feet high, and contained about 17 million cubic yards of embankment (Schuyler, 1905).

It is well known and documented that compaction of soils increases their strength. Tools and methods for compacting soils were developed long before the principles of compaction were discovered in the 1930s. For a long period of time, before the building of the first road compaction roller in the 1860s, cattle, sheep, and goats were used to compact soils. For example, in the United States, the 85-foot high Santa Fe water supply dam of New Mexico was compacted by 115 goats in 1893 (Highway Research Board, 1960).

Although mechanical equipment has been used to compact soil since the late 1860s, the engineering literature prior to the 1930s gave no evidence that anyone had established the relationships between moisture content, unit weight, and the compaction effort, relationships that are now documented as the fundamental principles of soil compaction. Between the 1930s and 1940s, the principles of compaction were widely known and discussed among engineers. The "Proctor curve" was a result of these studies. Following the work by Proctor, numerous investigations and reports were prepared to increase knowledge of compaction principles, which, through modifications and upgrading, resulted in widely used compaction testing standards.

Today, as in the past, earth materials continue to be used for embankment fills and backfill behind retaining structures because of their widespread availability and relative economy. Backfill is compacted in lifts that vary from 6 inches to 3 feet, depending on the types of soils and proposed use. Different types of compactors, ranging from large sheep's foot and vibratory rollers to small hand-operated tampers, have been developed and used to compact soils.

Embankment fills generally consist of:

(1) Cohesionless soils (sands and gravels)
(2) Cohesive soils (silts and clays)
(3) A mixture of cohesionless and cohesive soils, gravels, and cobbles (herein called earth–rock mixtures).

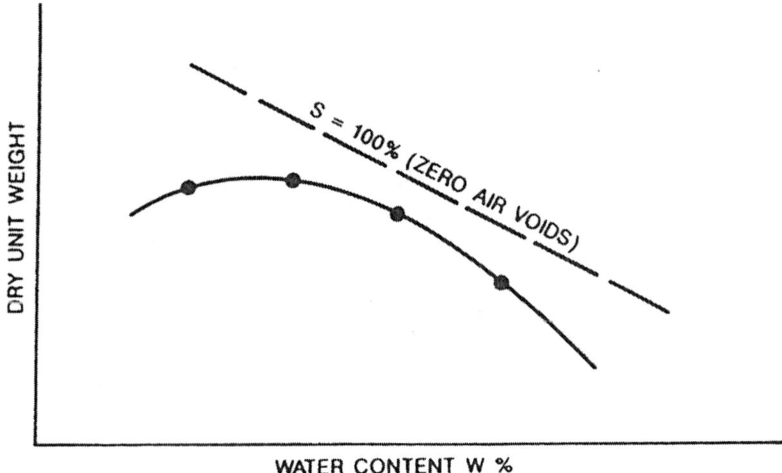

Figure 1.1 The moisture–density relationship (Proctor curve).

Organic soils, soft clays, and silts are usually avoided. The range of particle sizes of embankment fills is governed, for economic purposes, by the availability of the materials from nearby borrow areas.

The density of the materials after being excavated from borrow areas is usually very low and can be increased by compaction with mechanical equipment. In general, when the moisture content of the compacted soil is increased, the density will increase under a given compaction effort, until a peak or maximum density is achieved at a particular optimum moisture content. Thereafter, the density decreases as the moisture content is increased. The variations of moisture contents with density of the compacted soil is generally plotted in curves (Proctor curves) similar to Figure 1.1. The point of 100 percent saturation is called the saturation line, which is never reached since some air (pore space) always remains trapped in the soil material.

Cohesionless and cohesive soils behave differently when being compacted and have different compaction curves under the same compaction effort. Engineering characteristics of fills are discussed in the following sections. Typical engineering properties for compacted soils include maximum dry unit weight (standard compaction), optimum moisture content, typical strength, and permeability characteristics.

In general, soil shear strength varies with soil type and compaction conditions. Samples compacted dry of optimum moisture content appear stronger and more stable than those compacted wet of optimum moisture content. Increasing the compaction effort on soil reduces the permeability by reducing the amount of void space between the soil particles. When soil is compacted dry of optimum moisture content, the permeability is increased with an increase in water content. There is a slight decrease in permeability if the water content exceeds the optimum value.

Cohesionless Fills Cohesionless soils generally consist of relatively clean sands and gravels that remain pervious when compacted. These soils are represented by

Unified Classification System soil groups SW. SP. GW, GP, and boundary groups of any of these.

During the compaction process, compacted cohesionless soils are not affected significantly by water content because they are relatively pervious. The Proctor curve, as depicted in the dry density versus optimum moisture content graph, is usually round. Hence, their compactness is usually evaluated based on their relative density, as introduced by Terzaghi (1925) and defined as

$$D_r = \frac{e_{max} - e}{e_{max} - e_{min}} \qquad \text{(Eq. 1-1)}$$

where D_r = relative density expressed as a percentage
e_{max} = void ratio of the soil in its loosest state
e = void ratio of the soil being tested
e_{min} = void ratio of the soil in its densest state

Since void ratio is related to dry density for a given specific gravity, Equation 1-1 can be written as

$$D_r = \frac{\gamma_{d\max}(\gamma_d - \gamma_{d\min})}{\gamma_d(\gamma_{d\max} - \gamma_{d\min})} \qquad \text{(Eq. 1-2)}$$

where $\gamma_{d\max}$ = dry density of the soil in its densest state
$\gamma_{d\min}$ = dry density of the soil in its loosest state
γ_d = dry density of the soil being tested

The loose or dense state of cohesionless soils is usually judged by relative density as defined by Terzaghi. Typical ranges of relative density of sand are

$0 < D_r < \frac{1}{3}$ Loose sand
$\frac{1}{3} < D_r < \frac{2}{3}$ Medium dense sand
$\frac{2}{3} < D_r < 1$ Dense sand

Terzaghi further defined compactibility as

$$F = \frac{e_{max} - e_{min}}{e_{min}} \qquad \text{(Eq. 1-3)}$$

where F is very large for well-graded cohesionless soil such as SW and GW soils. Table 1.1 presents a list of compactibility values (F) for cohesionless soils.

Cohesive Fills Cohesive soils consist of those that contain sufficient quantities of silt and clay to render the soil mass relatively impermeable when properly compacted. Unlike compacted cohesionless soils, whose physical properties are generally improved by compaction to the maximum dry unit density, the physical properties of cohesive soils are not necessarily improved by compaction to a maximum unit

TABLE 1.1 List of Compactibility Factor F

Classification	γ_{min}	γ_{max}	e_{min}	e_{max}	Maximum Size	D_{10}	C_u	C_c	F
SP-SM	90	108	0.54	0.84	#16	0.058	6.1	2.2	0.555
SM	75	97	0.83	1.36	$\frac{3}{4}''$	0.0065	31.0	5.5	0.638
SP	92	112	0.48	0.80	#4	0.15	3.0	0.93	0.667
SP	93	113	0.46	0.77	$1\frac{1}{2}''$	0.16	2.4	0.92	0.674
SP	95	116	0.43	0.74	#4	0.30	3.7	1.0	0.721
SP-SM	92	113	0.46	0.80	$\frac{3}{4}''$	0.08	3.0	0.88	0.739
SP	85	107	0.54	0.94	#30	0.10	2.3	1.3	0.740
SP	97	118	0.40	0.70	$1\frac{1}{2}''$	0.11	3.2	1.2	0.750
SP	99	120	0.38	0.67	$1\frac{1}{2}''$	1.8	4.4	0.76	0.763
SM-ML	83	108	0.62	1.11	#4	0.012	8.3	1.5	0.790
SP-SM	79	103	0.60	1.08	#30	0.09	2.4	1.5	0.800
SP	103	124	0.33	0.60	$\frac{3}{8}''$	0.17	5.0	0.75	0.818
SM	105	126	0.31	0.57	5″	0.02	350.0	0.30	0.838
SP-SM	87	112	0.48	0.90	#4	0.08	3.0	1.3	0.875
SM	82	108	0.54	1.02	#16	0.023	6.5	1.4	0.889
SW-SM	95	119	0.39	0.74	3″	0.05	10.0	1.4	0.897
SP	98	122	0.36	0.69	#4	0.37	5.1	1.2	0.917
SW-SM	98	125	0.34	0.71	3″	0.07	6.8	1.0	1.088
SP-SM	97	124	0.33	0.70	$\frac{3}{4}''$	0.10	5.0	1.4	1.121
SP-SM	84	115	0.44	0.97	$1\frac{1}{2}''$	0.085	4.7	1.4	1.205
SP-SM	94	123	0.34	0.76	$1\frac{1}{2}''$	0.12	4.4	1.3	1.235
SM	99	128	0.31	0.70	3″	0.02	240.0	1.8	1.258
SP-SM	80	114	0.44	1.06	#16	0.07	3.7	1.6	1.409
SW-SM	80	116	0.42	1.07	$1\frac{1}{2}''$	0.074	6.6	2.4	1.547
SM	83	120	0.38	0.99	#4	0.015	26.0	6.1	1.605
SM	102	134	0.23	0.62	$\frac{3}{4}''$	0.01	120.0	1.9	1.695
GN-GM	113	127	0.31	0.47	3″	0.14	86.0	1.2	0.517
GP-GM	112	129	0.32	0.52	3″	0.03	200.0	0.50	0.625
GW-GM	116	133	0.26	0.44	5″	0.17	171.0	2.2	0.692
GP-GM	110	128	0.30	0.51	3″	0.11	191.0	15.0	0.700
GP-GM	117	133	0.24	0.41	5″	0.125	160.0	4.0	0.708
GW-GP	111	130	0.27	0.49	3″	0.20	105.0	7.5	0.815
GP	116	134	0.23	0.43	5″	0.27	111.0	6.2	0.870
GW	119	139	0.24	0.45	3″	0.51	45.0	2.2	0.875
GW	120	139	0.20	0.39	3″	0.45	51.0	1.6	0.950
GW	119	139	0.21	0.41	3″	0.18	94.0	1.1	0.952
GW	111	132	0.25	0.49	3″	2.9	9.7	1.8	0.960
GP	115	136	0.22	0.44	5″	0.38	29.0	0.61	1.000
GP	114	135	0.22	0.45	3″	2.0	11.0	0.77	1.045
GW-GM	121	141	0.19	0.39	3″	0.30	77.0	2.3	1.052
GM	122	141	0.17	0.36	$1\frac{1}{2}''$	0.025	381.0	3.0	1.118
GW-GM	114	137	0.21	0.45	3″	0.60	16.0	1.2	1.143
GW	112	138	0.20	0.48	3″	2.0	12.0	1.3	1.400
GW	109	137	0.21	0.52	3″	2.0	14.0	2.6	1.476

(Continued)

TABLE 1.1 *(Continued)*

Classification	γ_{min}	γ_{max}	e_{min}	e_{max}	Maximum Size	D_{10}	C_u	C_c	F
GP	114	140	0.18	0.45	3″	1.7	10.0	0.76	1.500
GM	101	132	0.25	0.64	$1\frac{1}{2}$″	0.03	260.0	12.0	1.560
GW-GM	111	139	0.19	0.49	3″	1.8	13.0	2.3	1.578
GP	115	142	0.17	0.44	3″	0.31	87.0	8.2	1.588
GW	123	146	0.13	0.34	3″	0.21	124.0	1.1	1.615
GW-GM	110	139	0.19	0.50	5″	0.42	43.0	2.1	1.631
GW-GM	115	142	0.17	0.45	3″	0.15	133.0	1.1	1.647
GP-GM	112	140	0.18	0.48	3″	0.42	26.0	4.2	1.667
GW-GM	112	140	0.18	0.48	5″	0.25	56.0	1.0	1.667
GW-GM	114	142	0.16	0.45	3″	1.2	15.0	1.7	1.812
GP	112	141	0.17	0.48	3″	1.4	7.1	0.73	1.823
GW-GM	118	147	0.12	0.40	3″	1.3	19.0	1.1	2.333

Source: Hilf (1991).

density. For example, Figure 1.2 indicates that the strength of compacted silty clay decreases with increasing molding water content (Seed and Chan, 1959).

Whether compacted cohesive soils should be placed dry or wet of optimum for the same density depends on the type of construction. During construction of a fill slope, data (Lee and Haley, 1968) indicate that it is better to place cohesive soils dry of optimum. Such data are presented in Figures 1.3 and 1.4. Nevertheless, for earth dam construction, there is an increasing tendency to compact the cores of earth dams on wet of the optimum moisture content to minimize crack development and subsequent formation of seepage channels. A balance must be struck between the resultant lower strength and potential pore pressure problems caused by using a higher initial water content.

By changing the moisture content of compacted clays, a pronounced change in their engineering properties results. The effects of compaction dry and wet of optimum on the shear strength, permeability, compressibility, and structure of cohesive soils are shown in Table 1.2.

Earth–Rock Mixtures There has been a considerable increase in the usage of earth–rock mixtures in the fills of high embankments over the last 40 years. Such soils are heterogeneous mixtures of particles that may range in size from large boulders to clay. The mixing of the larger size particles enhances the workability of the soil in the field and increases the overall strength of the soil.

Past research studies by the Bureau of Reclamation (Holtz and Gibbs, 1956) suggested that the strength of earth–rock mix fill depends on the amount of rock to be mixed with the in situ soils. The strength will increase with the amount of rock until some threshold percentage, for instance about 50 to 68 percent for sand–gravel mixtures, is reached. Further increase in gravel contents produces little to no increase in strength. Variations in friction angle with gravel contents of earth–rock mixtures for coarse-grained soils are shown in Figure 1.5.

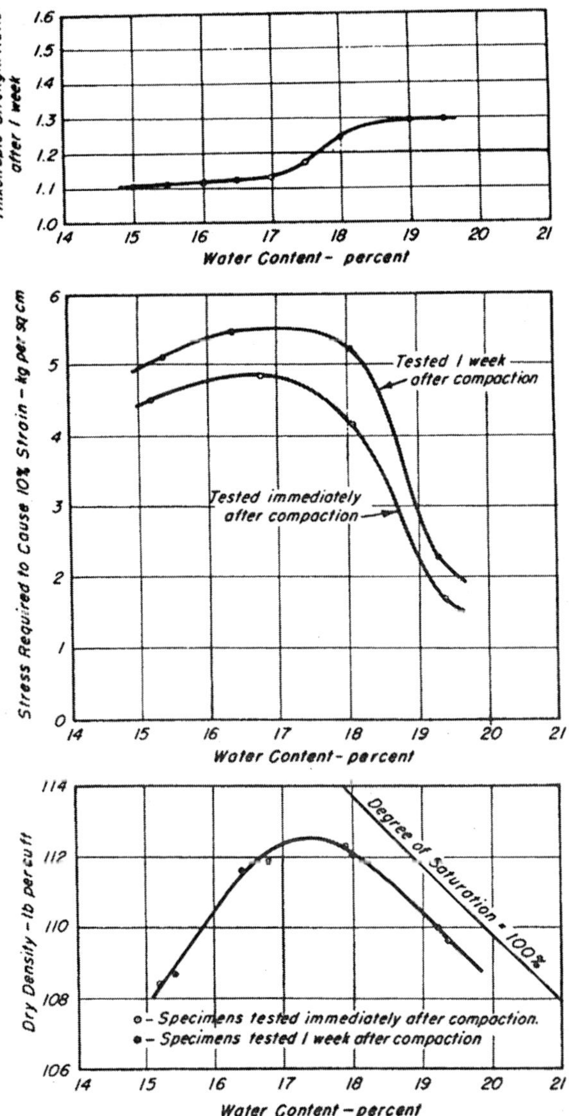

Figure 1.2 Strength of compacted clay versus moisture content. (From Seed and Chan, 1959, reproduced by permission of ASCE.)

Embankments on Weak Foundations Embankments are sometimes built on weak foundation materials. Sinking, spreading, and piping failures may occur irrespective of the stability of the new overlying embankment material. Consideration of the internal stability of an embankment-foundation system, rather than just the embankment, may be necessary. A simple rule of thumb based on bearing capacity theory can be used to make a preliminary estimate of the factor of safety against

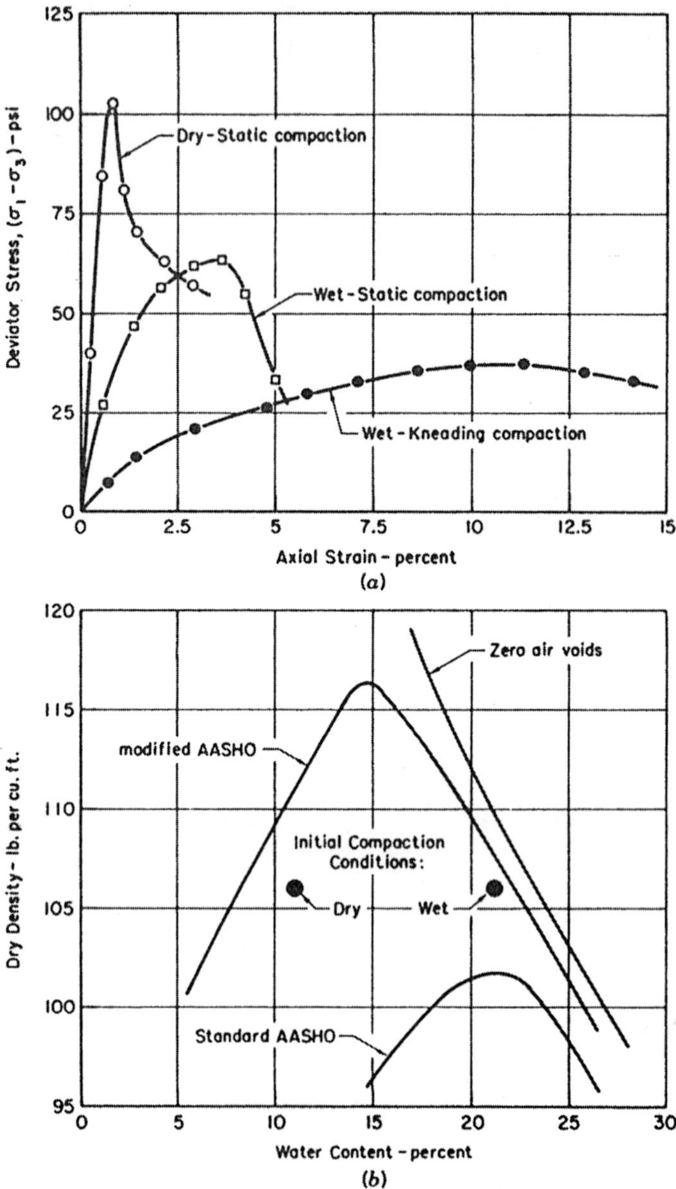

Figure 1.3 Compaction and unconfined compression characteristics of Higgins clay. (*a*) Unconfirmed compression tests. (*b*) Compaction curves. (From Lee and Haley, 1968, reproduced by permission of ASCE.)

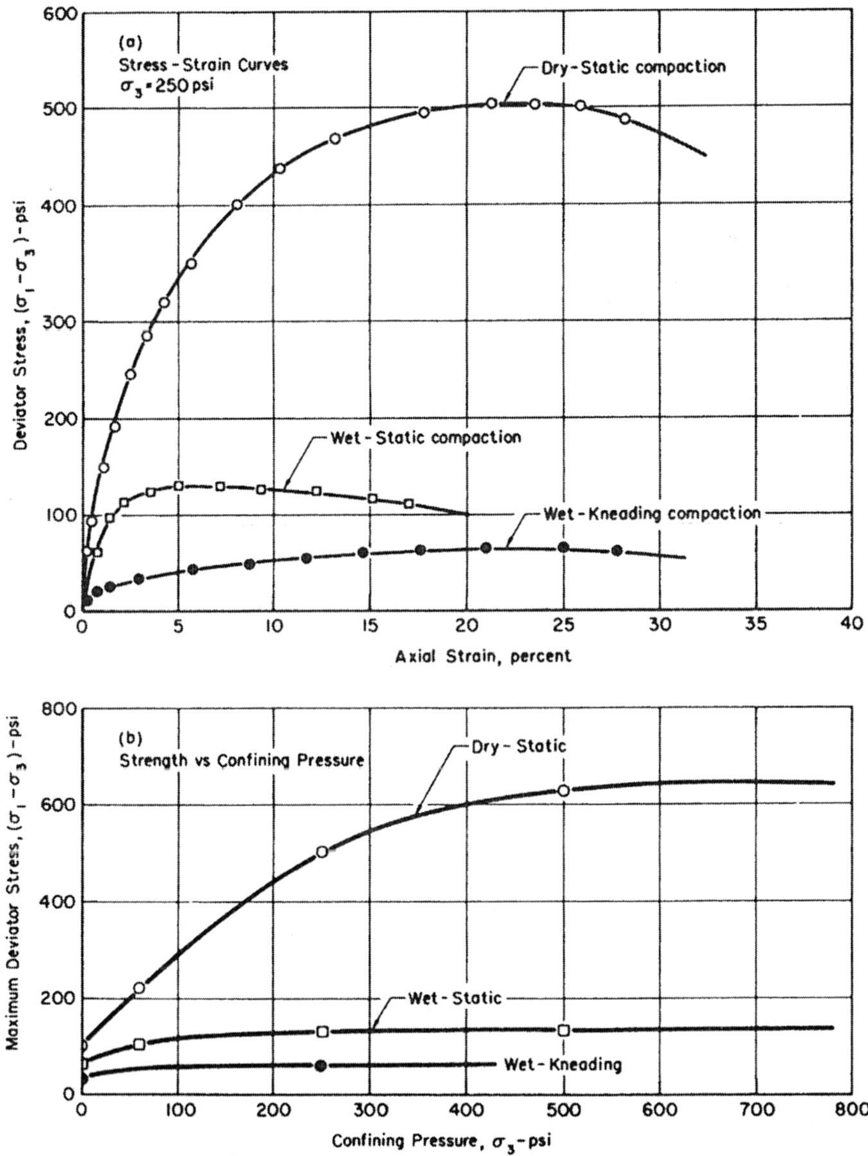

Figure 1.4 Unconsolidated-undrained tests on compacted Higgins clay. (*a*) Stress–strain curves = 250 psi. (*b*) Strength versus confining pressure. (From Lee and Haley, 1968, reproduced by permission of ASCE.)

TABLE 1.2 Comparison between Dry of Optimum and Wet of Optimum Compaction of Cohesive Soils

Physical Properties	Effects of Compaction
	Shear Strength
As molded	
Undrained	Dry side much higher
Drained	Dry side somewhat higher
After saturation	
Undrained	Dry side somewhat higher if swelling prevented; wet side can be higher if swelling permitted
Drained	Dry side about the same or slightly higher
Pore-water pressure at failure	Wet side higher
Stress–strain modulus	Dry side much greater
Sensitivity	Dry side more apt to be sensitive
	Permeability
Magnitude	Dry side more permeable
Permanence	Dry side permeability reduced much more by permeation
	Compressibility
Magnitude	Wet side more compressible in low stress range, dry side in high stress range
Rate	Dry side consolidates more rapidly
	Structure
Particle arrangement	Dry side more random
Water deficiency	Dry side more deficient, therefore more water imbibed, more swell, lower pore pressure
Permanence	Dry-side structure more sensitive to change

Source: Table extracted from *Soil Mechanics*, by Lambe and Whitman (1969).

circular arc failure for an embankment built over a clay foundation. The rule is (Cheney and Chassie, 1982)

$$\text{FOS} = \frac{6c}{\gamma_{\text{fill}} \times H_{\text{fill}}} \qquad \text{(Eq. 1-4)}$$

where FOS = factor of safety

c = cohesion of foundation clay (pounds per square foot)

γ_{fill} = unit weight of embankment fill (pounds per square foot)

H_{fill} = height of embankment fill (feet)

The factor of safety computed using this rule serves only as a rough preliminary estimate of the stability of an embankment over a clay foundation and should *not* be

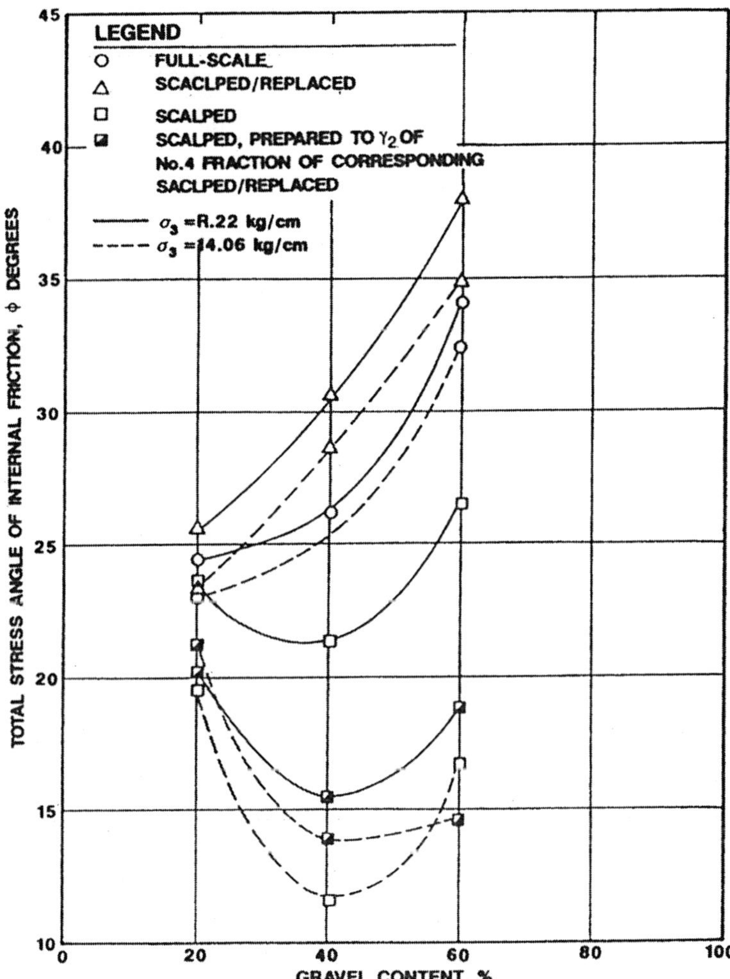

Figure 1.5 Total stress angle of internal friction versus gravel contents. (Donaghe and Torrey, 1985.)

used for final design. The simple equation does not take into consideration factors such as fill strength, strain incompatibility between embankment fill and the underlying foundation soils, and fill slope angle. In addition, it does not identify the location of a critical failure surface. If the factor of safety using the rule-of-thumb equation is less than 2.5, a more sophisticated stability analysis is required (Cheney and Chassie, 1982).

Figure 1.6 shows the variations in safety factor, strength, pore pressures, load, and shear stresses with time for an embankment constructed over a clay deposit. Over time, the excess pore pressure in the clay foundation diminishes, the shear strength of the clay increases, and the factor of safety for slope failure increases.

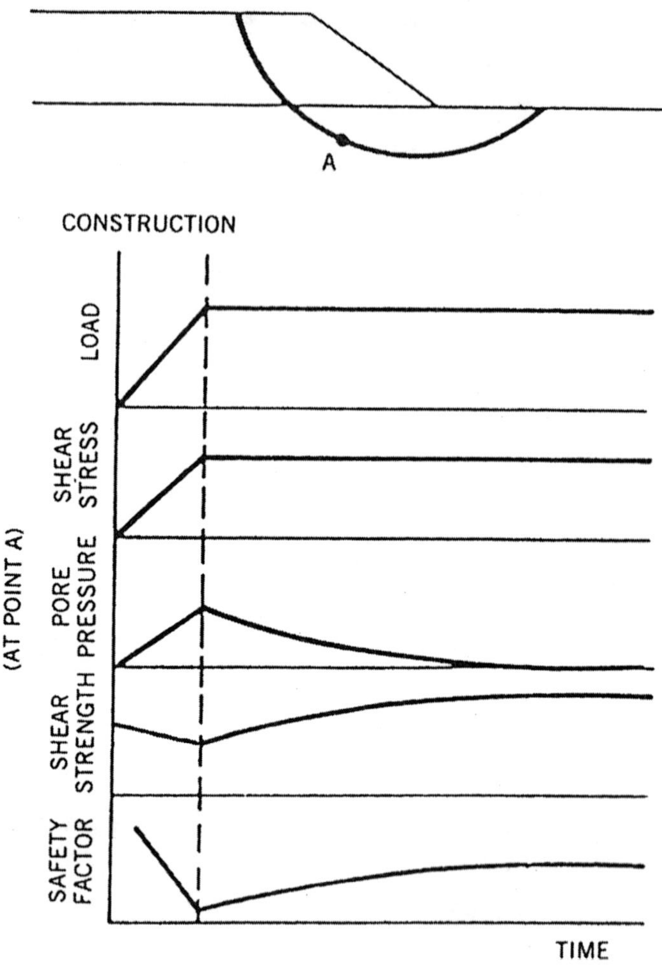

Figure 1.6 Stability conditions for an embankment slope over a clay foundation. (From Bishop and Bjerrum, 1960, reproduced by permission of ASCE.)

Embankment fills over soft clay foundations are frequently stronger and stiffer than their foundations. This leads to the possibility that the embankment will crack as the foundation deforms and settles under its own weight and to the possibility of progressive failure because of stress–strain incompatibility between the embankment and its foundation. Design charts developed by Chirapuntu and Duncan (1977), using finite element method analyses, depict the effects of cracking and progressive failure on the stability of embankments on soft foundations. These charts may be used as a supplement to conventional stability analyses. The use of geosynthetic reinforcement in the fill may prevent the initiation of cracking and subsequent failure in these cases. Alternatively, it may be necessary to remove the soft foundation materials or locate the fill at another site.

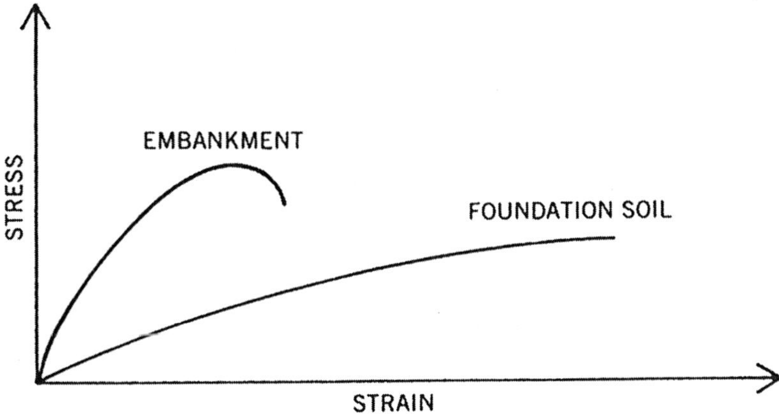

Figure 1.7 Example of stress–strain incompatibility.

Peak strengths of the embankment and the foundation soils cannot be mobilized simultaneously because of stress–strain incompatibility (Figure 1.7). Hence, a stability analysis performed using peak strengths of soils would overestimate the factor of safety. Many engineers perform stability analyses using soil strengths that are smaller than the peak values to allow for possible progressive failure.

Shale Embankments Embankments constructed of shale materials often have slope stability and settlement problems. According to DiMillio and Strohm (1981), the underlying causes of shale fill slope failures and excessive settlement frequently appear to be:

(1) Deterioration or softening of certain shales over time after construction
(2) Inadequate compaction of the shale fill
(3) Saturation of the shale fill

These types of failures have been found to be typical in many areas from the Appalachian region to the Pacific coast. In general, severe problems with shales in embankments are found in states east of the Mississippi River rather than west of the river (DiMillio and Strohm, 1981). Embankments can use fill originating from shale formations successfully if the borrow source is not particularly prone to long-term decomposition and if adequate compaction and drainage are required. In addition, shale embankments should be keyed into any sloping surfaces by using benches and installing drainage measures to intercept subsurface water that may enter the foundation area. Guidelines for design and construction of shale embankments have been established by Strohm et al. (1978).

1.4.2 Cut Slopes

Shallow and deep cuts are important features in any civil engineering project. The aim in a slope design is to determine a height and inclination that is economical and that will remain stable for a reasonable life span. The design is influenced by the purposes of the cut, geological conditions, in situ material properties, seepage pressures, construction methods, and the potential occurrence of natural phenomena such as heavy precipitation, flooding, erosion, freezing, and earthquakes.

Steep cuts often are necessary because of right-of-way and property line constraints. The design must consider measures that will prevent immediate and sudden failure as well as protect the slope over the long term, unless the slope is cut for temporary reasons only. In some situations, cut stability at the end of construction may be a critical design consideration. Conversely, cut slopes, although stable in the short term, can fail many years later without much warning.

To a certain degree, the steepness of a cut slope is a matter of judgment not related to technical factors. Flat cut slopes, which may be stable for an indefinite period, are often uneconomical and impractical. Slopes that are too steep may remain stable only for a short period of time. A failure may pose a danger to life and property at a later date. Failures could involve tremendous inconvenience and the expense of repairs, maintenance, and stabilization measures.

Figure 1.8 shows the general variations of factor of safety, strength, excess pore pressure, load, and shear stresses over time for a clay cut slope. The initial shear strength is equal to the undrained shear strength on the assumption that no drainage occurs during construction. In contrast to embankment slopes, the pore pressure within the cut increases over time. This increase is accompanied by a swelling of the clay, which results in reduced shear strength. Thus the factor of safety decreases over time until an unstable condition is reached. This, for the most part, explains why clayey cut slopes sometimes fail a long time after initial excavation.

For cuts in overconsolidated clays, the in situ shear strength is a direct function of the maximum past overburden pressure. The higher the maximum past overburden pressure, the greater the shear strength. However, if the clay is subjected to long-term unloading conditions (permanent cuts), the strength of the clay no longer depends on the prior loading. The strength of a cut slope will decrease with time. The loss in strength is attributed to reduction of negative pore pressure after excavation. This loss in strength has been observed to be a time-dependent function related to the rate of dissipation of negative pore pressure.

In practice, the loss in strength after cuts are made is not easily determined. According to McGuffey (1982), the time dependency of clay cut slope failures can be hypothesized to be a function of the Terzaghi hydrodynamic lag model. The estimated time to failure can be expressed as (McGuffey, 1982)

$$t = \frac{h^2 T_{90}}{C_v} \qquad \text{(Eq. 1-5)}$$

where t = time to failure

h = average distance from the slope face to the depth of the maximum negative pore pressure

T_{90} = time factor for 90% consolidation = 0.848

C_v = coefficient of consolidation (square feet per day)

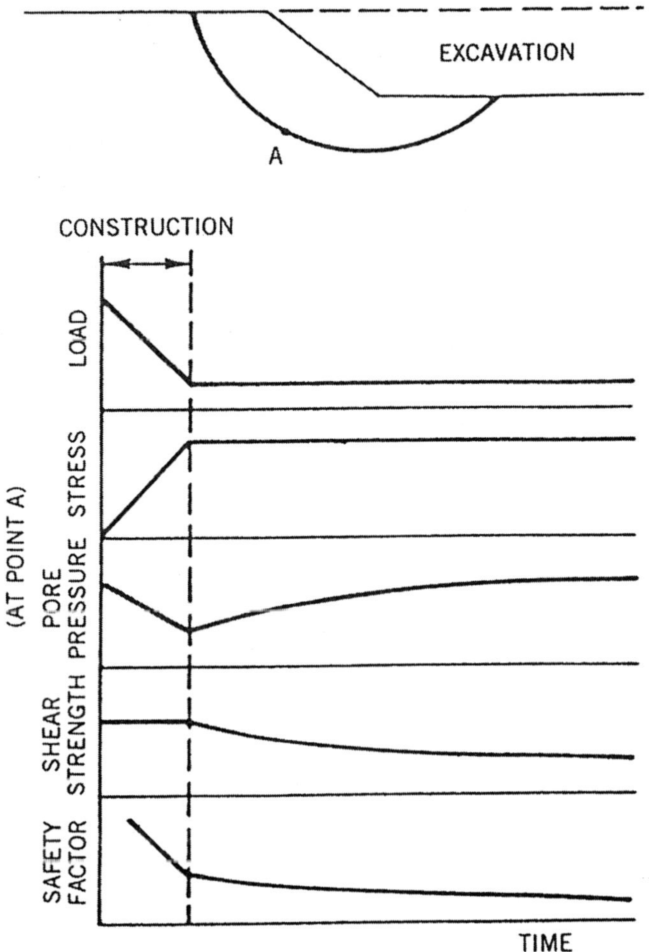

Figure 1.8 Stability conditions for a cut slope. (From Bishop and Bjerrum, 1960, reproduced by permission of ASCE.)

This model was used with some success by McGuffey (1982) to determine and back-analyze the time for stress release leading to slope failures in clay cuts in New York.

Long-term cut slope stability is also dependent on seepage forces and, therefore, on the ultimate groundwater level in the slope. After excavation, the free-water surface will usually drop slowly to a stable zone at a variable depth below the new cut surface. This drawdown usually occurs rapidly in cut slopes made in sand but is usually much slower in clay cut slopes. Although typical rates and shapes of groundwater drawdown curves have been proposed for cut slopes, none has proved useful for correctly predicting the time or rate of drawdown of preconsolidated clays. The main obstacle to such prediction comes from the difficulty in correctly modeling the recharge of the area in the vicinity of the cut slope.

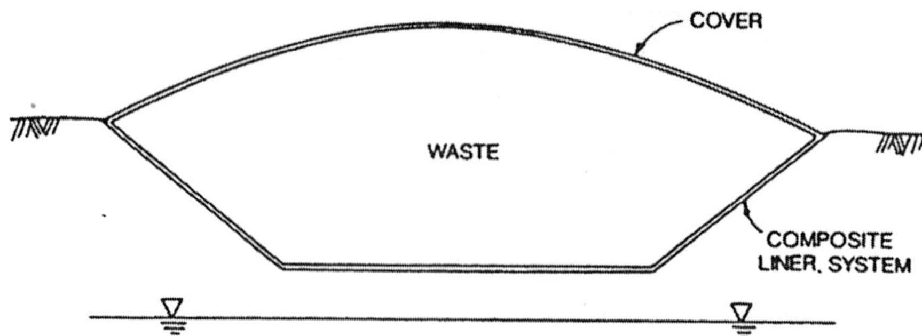

Figure 1.9 Cut and fill slopes used for landfills. (From Mitchell and Mitchell, 1992, reproduced by permission of ASCE.)

1.4.3 Landfills

Landfills are a special case where both cut and fill slopes are involved (Figure 1.9) and where the fill materials are much less than optimum. To make matters worse, except for very old landfills, zones of clay barriers and, more recently, geosynthetic barriers (Figure 1.10) are placed between the fill materials and the natural ground, creating an extra zone that must be characterized and analyzed with respect to short-term and long-term stability. Also, the plethora of environmental regulations that are imposed on existing and new landfills places an extremely heavy burden on the engineers and geologists to accurately characterize and analyze the short-term and long-term behavior of landfills for the life of the project and beyond.

Landfills may contain organic materials, tree limbs, refuse, and a variety of debris that are commonly dumped, pushed, and spread by bulldozers, and then compacted by refuse compactors. Compaction of landfills is somewhat different from the compaction of soils, particularly with respect to crushing. Compaction crushes (collapses) hollow particles, such as drums, cartons, pipes, and appliances, and brings the crushed particles closer together. It may be expected that landfills, compacted at the top only, will be relatively loose. Landfill materials are commonly soft, and large voids can be encountered.

The evaluation of landfill slope stability is similar to the analysis of other types of slope stability problems. Selection of proper values for the strengths of the waste and foundation materials, and of proper shearing resistances along the interfaces within the liner and cover systems, is the most critical part of any stability study. The greatest difficulties and uncertainties are associated with evaluation of the strength and stress–strain properties of the liner system materials and interfaces, and of the waste fill.

Very little is known about the geotechnical engineering properties of landfills. The paucity of data results in part from the difficulties in credible sampling and testing of refuse. This difficulty is further compounded by the fact that refuse composition and properties are likely to change erratically within a landfill and are also likely to decompose with time.

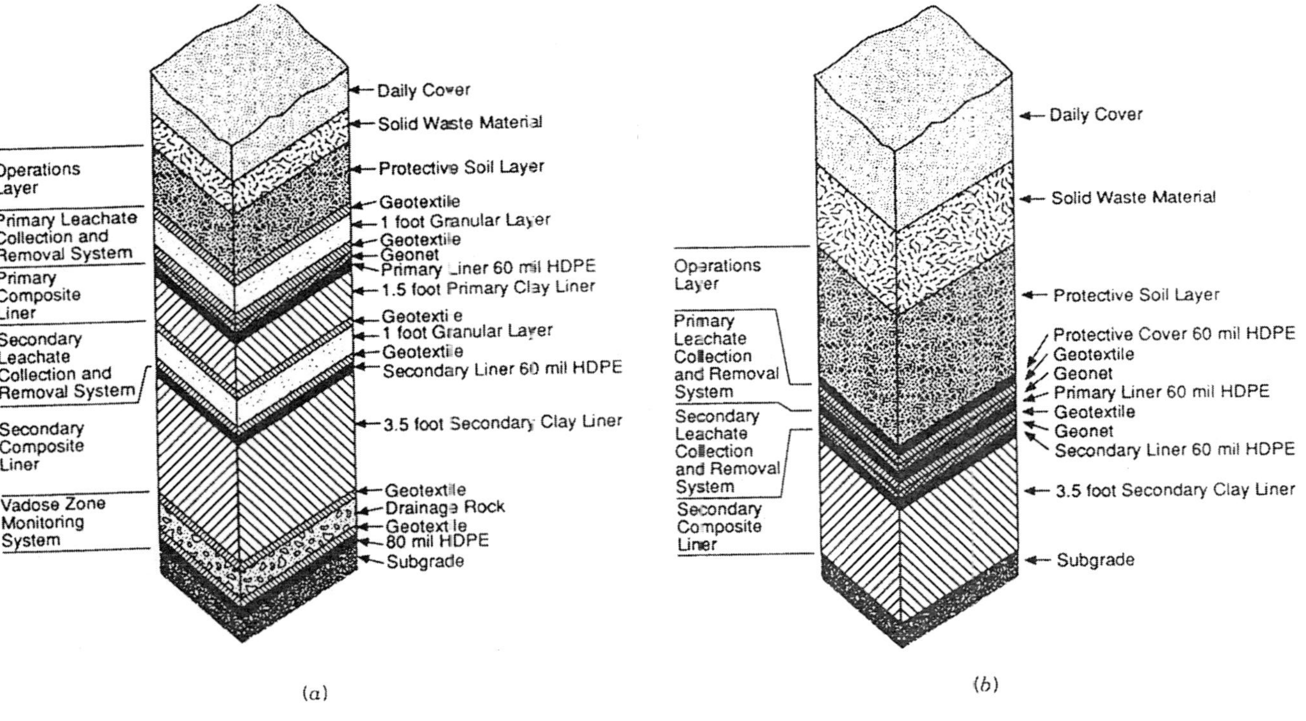

Figure 1.10 Geosynthetic barriers used for landfills. (*a*) Base liner. (*b*) Side slope liner. (From Byrne et al., 1992, reproduced by permission of ASCE.)

The unit weight of landfill is highly dependent on a number of factors, including initial composition, compactive effort, decomposition, settlement, and moisture content. Although the density of the landfill will typically increase with depth, considerable variability should be expected within relatively short distances (Landva and Clark, 1987).

The moisture content of landfill materials is also dependent on a number of interrelated factors, such as the initial composition, landfill operating procedures, the effectiveness of any leachate collection and removal systems, the amount of moisture generated by biological processes within the landfill, and the amount of moisture removed with landfill gas (Mitchell and Mitchell, 1992). The water content of the landfills in the United States ranges from 10 to 50 percent.

Shear strengths of the landfills may be estimated by means of:

(1) Laboratory testing
(2) Back-calculation from field tests and operational records
(3) In situ testing

Laboratory samples are usually reconstituted from landfills before they are tested. Direct shear tests have been commonly used to determine shear strength parameters of the landfill materials. Back-calculation of an existing landfill based on field load tests also can be made to estimate the shear strengths of the landfill (Converse et al., 1975). However, the back-calculated strengths are usually conservative by an unknown amount because the back-calculation assumes failure of the slope (that is, a factor of safety equal to 1). Vane shear and standard penetration tests have been used to estimate the shear strength of refuse in a landfill near Los Angeles, California (Earth Technology Corporation, 1988). The shear strength data obtained by these in situ testings may not be representative of the actual conditions because both the vane shear device and the standard penetration sampler are small compared with the inclusions (for example, tires, wood, carpet) that make up the landfill.

True cohesion or bonding between particles is unlikely in landfills. However, there may be a significant cohesion intercept that results from interlocking and overlapping of the landfill constituents. This interpretation is supported by some laboratory test results and the common observation that vertical cuts in landfills can stand, unsupported, to considerable heights (Mitchell and Mitchell, 1992).

Alshunnar (1992) proposed the following design considerations for landfill slopes:

(1) Groundwater conditions before and after construction of the landfill
(2) Subsurface conditions
(3) Construction sequence
(4) Adjacent site conditions and history
(5) Site topography
(6) External loads such as from construction equipment, stockpiles, earthquakes, and so on
(7) Liner geometry and configuration
(8) Filling sequence

It is extremely difficult to reconcile the uncertainties implicit in landfill site characterizations (e.g., landfill material types and characteristics, future land use at and around the site, the probability of severe natural phenomena like earthquakes, tsunamis, sinkholes, etc.) with the guarantees that regulatory agencies require of operators and owners related to siting, operations, and closure. Seed and Bonaparte (1992) stated:

> There are some considerable uncertainties with respect to material properties and dynamic response characteristics associated with both: (a) waste fill masses, and (b) base liner systems and final cover systems. As a result, current analysis and design methods are generally based on a sequence of conservative assumptions at various stages of the analyses, with the resulting cumulative level of conservatism generally selected so as to offset the possible impact(s) of uncertainties at each stage of analysis and design. In the opinion of the authors, the compounding of conservative assumptions, both in terms of property/parameter selection and analysis/evaluation methodology, at multiple stages during the overall analysis and design process should provide a conservative final result when carefully implemented.

An example of how modern regulations and construction methods of landfills surpassed analytical characterization and design methods is the case of the Kettleman Hills Landfill B-19 Phase IA failure. On March 19, 1988, there was a failure in Landfill B-19, Phase IA at the Kettleman Hills Class I hazardous waste treatment, storage, disposal facility (Byrne et al., 1992). Approximately 580,000 cubic yards of waste and other material had been placed to a height of 90 feet above the base at the time of the failure (Figure 1.11). The entire mass slid a horizontal distance of about 35 feet toward the southeast, and vertical slumps of up to 14 feet along the sideslopes of the landfill were observed after the failure (Figure 1.12). Initial study of the failure indicated that it had most likely occurred within the liner system and identified both geomembrane/clay and various synthetic/synthetic interfaces as candidate sliding surfaces. The characteristics of the landfill base and sideslope liners are shown in Figure 1.13. Further study, testing, and two- and three-dimensional slope stability analyses after the failed waste and liner materials were excavated concluded that

(1) The mechanism of failure consisted of slip along multiple interfaces within the landfill liner system.

(2) Low liner interface strengths (residual friction angles as low as 8 degrees) relative to the constructed geometry of the landfill were clearly the underlying cause of the failure.

(3) The predominant surface of sliding during the failure appeared to be the geomembrane/clay interface of the secondary liner system that apparently behaved in an essentially undrained mode during the approximately one year of waste loading prior to failure.

(4) Undrained shear strength testing of the clay/geomembrane interface indicated that the shear strength was sensitive to the as-placed moisture and density conditions of the clay.

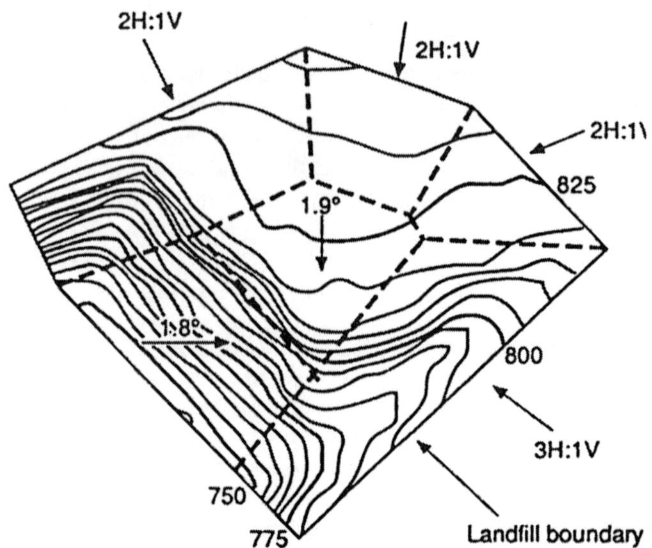

Figure 1.11 Kettleman Hills landfill layout. (From Byrne et al., 1992, reproduced by permission of ASCE.)

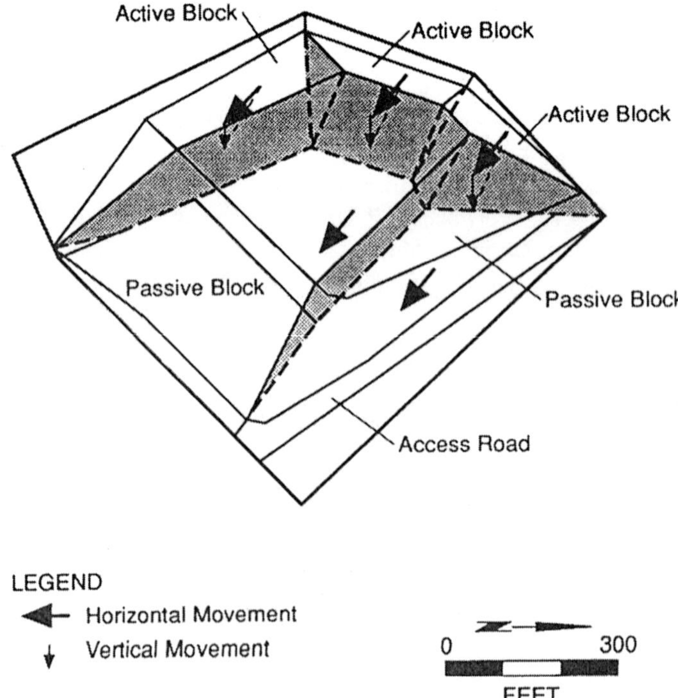

Figure 1.12 Slope failure at Kettleman Hills Landfill. (From Byrne et al., 1992, reproduced by permission of ASCE.)

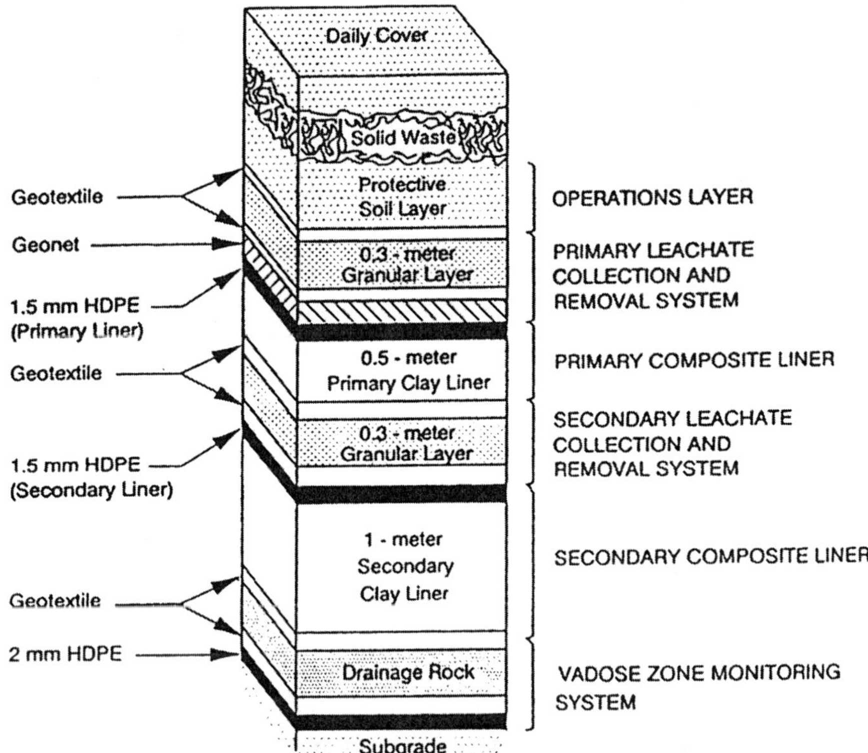

Figure 1.13 Kettleman Hills landfill liner characteristics. (From Byrne et al., 1992, reproduced by permission of ASCE.)

(5) The calculated factor of safety using a three-dimensional model and residual shear strengths was 0.85. This was consistent with the observed occurrence of large displacements following failure initiation and the attainment of residual strength conditions over the entire slip surface.

(6) The failure demonstrated that specifications for the placement of liner clay must focus not only on achieving specific permeability requirements, but also on developing liner shear strengths that are adequate to support both the interim and final geometric configurations of the landfill.

It should be noted here that three-dimensional analysis was required for this case because of the complex geometry and difficulty in selecting a typical two-dimensional section to analyze. However, as Duncan (1992) states, "The factor of safety calculated using 3D analyses will always be greater than, or equal to, the factor of safety calculated using 2D analyses." This is true for all cuts and fills, not just landfills.

1.4.4 Retaining Structures

Retaining structures are frequently used to support stable or unstable earth masses. The different types of retaining structures, as shown in Figure 1.14, are:

(1) Gravity walls (e.g., masonry, concrete, cantilever, or crib walls)
(2) Tieback or soil-nailed walls
(3) Soldier pile and wooden lagging or sheet pile walls
(4) Mechanically stabilized embankments including geosynthetic and geogrid reinforced walls

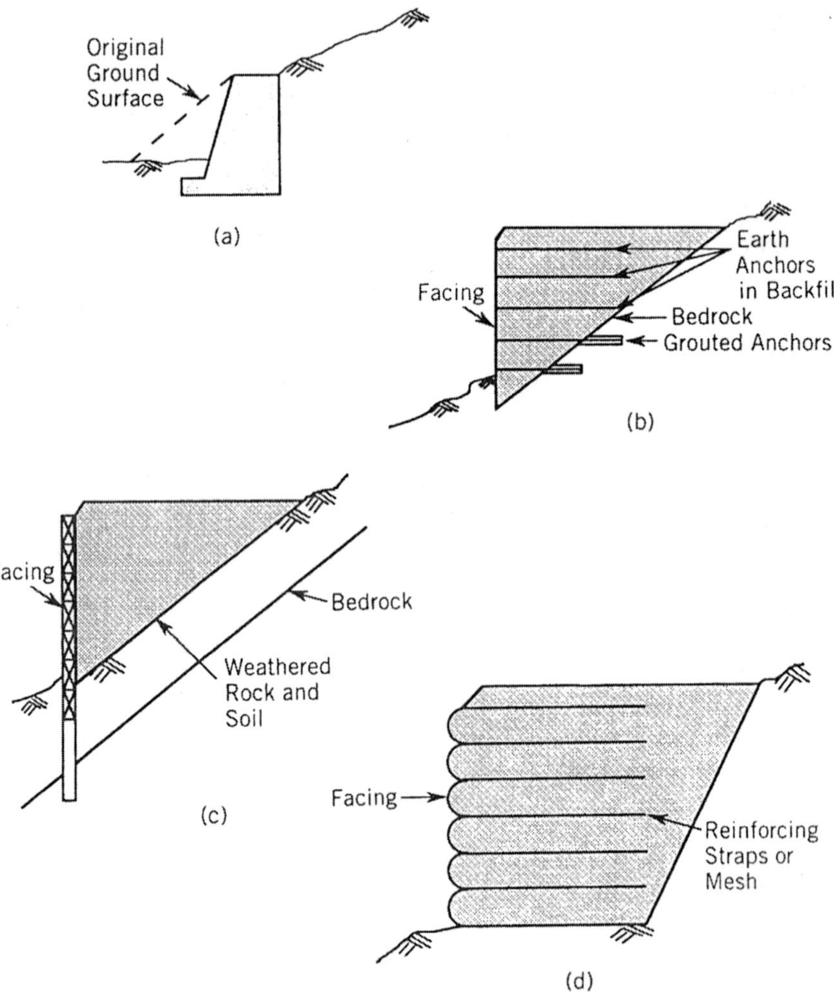

Figure 1.14 Types of retaining structures. (*a*) Gravity retaining wall. (*b*) Tieback retaining wall. (*c*) Sheet pile cantilever wall or soldier pile. (*d*) Mechanically stabilized embankment.

Retaining structures are used in seven principal ways, as shown in Figure 1.15. The design of retaining structures requires three primary considerations:

(1) External stability of the soil behind and below the structure
(2) Internal stability of the retained backfill
(3) Structural strength of retaining wall members

1.5 LANDSLIDES

When a slope fails, it is often called a landslide or a slope failure. Several classification methods and systems have been proposed for landslides. The one adopted in this book and most consistently around the world is the one proposed by the International Association of Engineering Geologists (IAEG) Commission on Landslides.

1.5.1 Features and Dimensions of Landslides

Typical features of a landslide are shown schematically in Figure 1.16. The observable features of a landslide are

(1) *Crown* The practically undisclosed material above the main scarp
(2) *Main Scarp* A steep surface on the undisturbed ground at the upper edge of the landslide
(3) *Top* The highest point of contact between the displaced material and main scarp
(4) *Head* The upper parts of the landslide between the displaced material and main scarp
(5) *Minor Scarp* A steep surface on the displaced material produced by differential movements
(6) *Main Body* The part of the displaced material that overlies the surface of rupture
(7) *Foot* The portion of the landslide that has moved beyond the toe
(8) *Tip* The point on the toe farthest from the top
(9) *Toe* The lower margin of the displaced material
(10) *Surface of Rupture* The surface that forms the lower boundary of the displaced material
(11) *Toe of Surface of Rupture* The intersection between the lower part of the surface of rupture and the original ground surface
(12) *Surface of Separation* The original ground surface now overlain by the foot of the landslide
(13) *Displaced Material* Material displaced from its original position by landslide movement

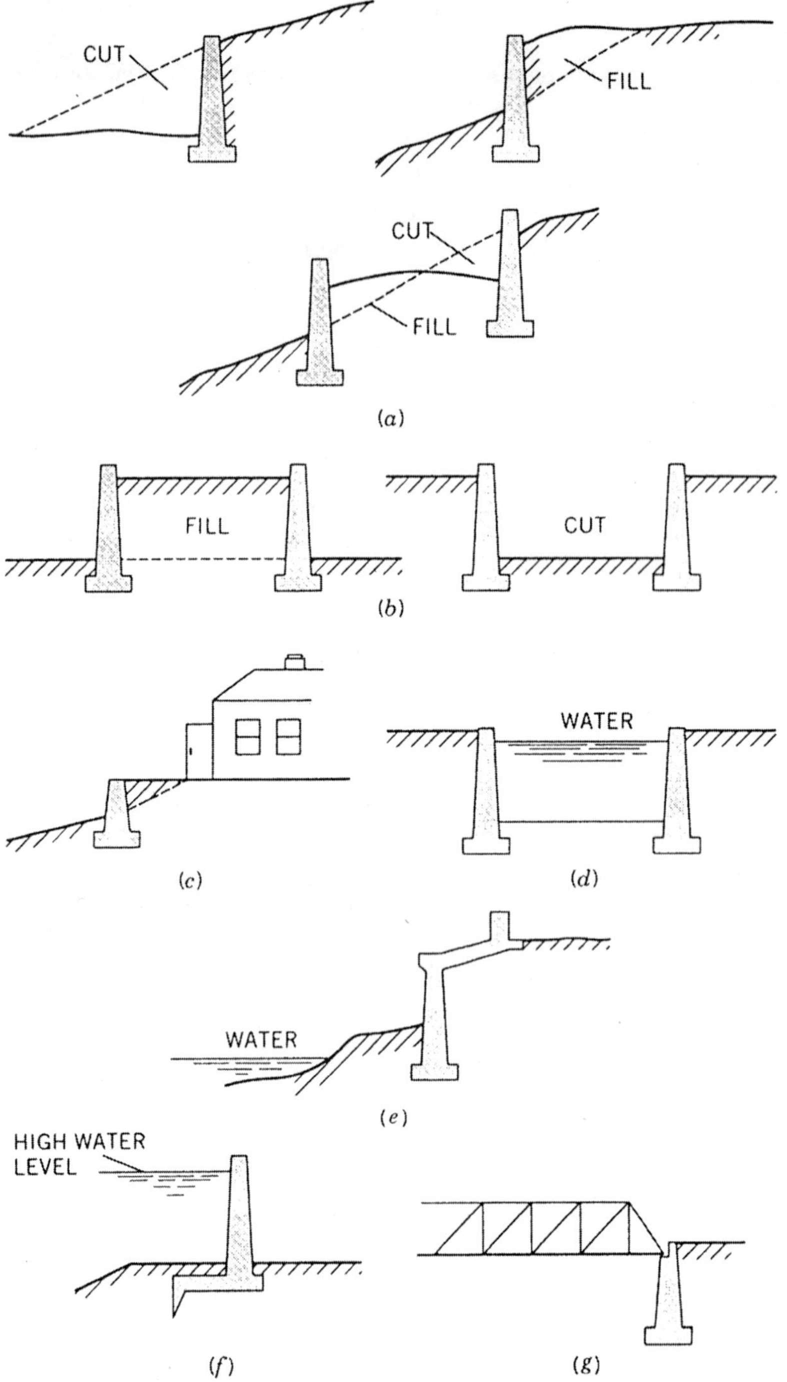

Figure 1.15 Common uses of retaining structures. (*a*) Cuts and fills. (*b*) Retained cuts and fills. (*c*) Site levelling. (*d*) Canals. (*e*) Waterfront protection. (*f*) Water retention. (*g*) Bridge abutments.

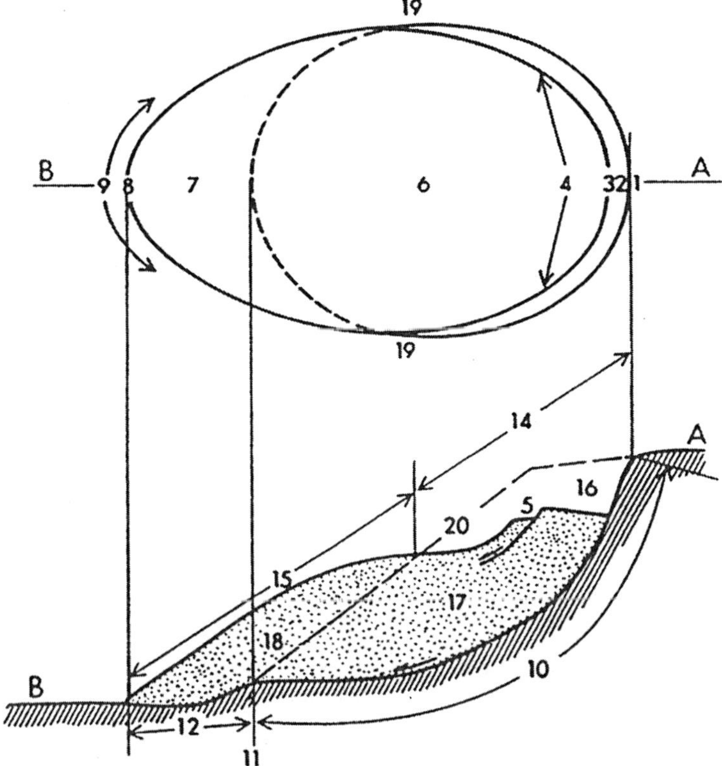

Figure 1.16 Landslide features. The upper portion of the figure is a plan of a typical landslide. The dashed line is the trace of the rupture surface on the original ground surface. In the section of the lower portion of the figure, crosshatching indicates undisturbed ground, and stippling shows the extent of the displaced material. (Cruden and Varnes, 1992.)

(14) *Zone of Depletion* The area within which the displaced material lies below the original ground surface

(15) *Zone of Accumulation* The area within which the displaced material lies above the original ground surface

(16) *Depletion* The volume bounded by the main scarp, the depleted mass, and the original ground surface

(17) *Depleted Mass* The volume of displaced material that overlies the rupture surface but underlies the original ground surface

(18) *Accumulation* The volume of the displaced material that lies above the original ground surface

(19) *Flank* The undisclosed material adjacent to the sides of the rupture surface

(20) *Original Ground Surface* The surface of the slope that existed before the landslide took place

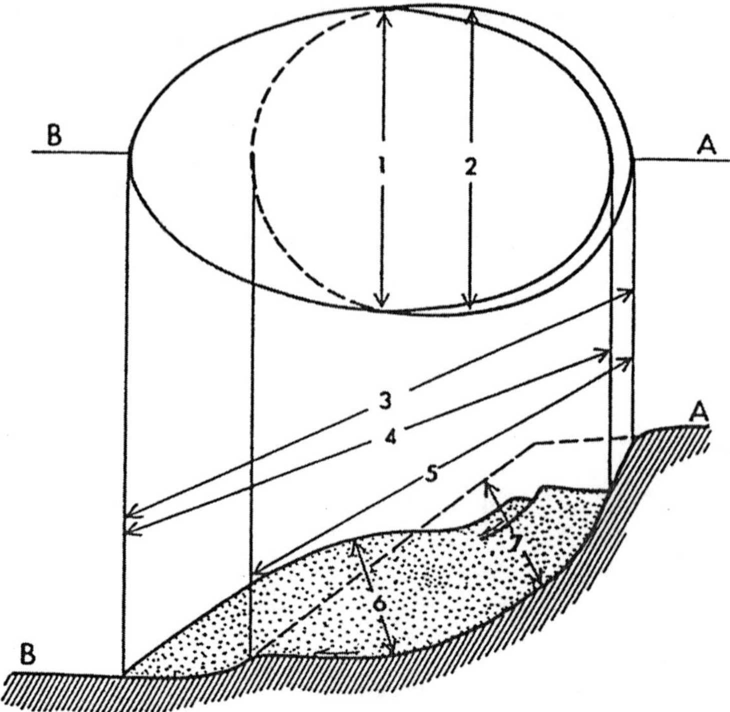

Figure 1.17 Landslide dimensions. In the section, crosshatching indicates undisturbed ground and the broken line is the original ground surface. The dashed line in the plan is the trace of the rupture surface on the original ground surface. (Cruden and Varnes, 1992.)

Similarly, the IAEG proposed standardized typical dimensional variables for landslides as shown in Figure 1.17 and discussed below.

(1) *Width of Displaced Mass,* W_d The maximum breadth of the displaced mass perpendicular to the length, L_d

(2) *Width of the Rupture Surface,* W_r The maximum width between the flanks of the landslide, perpendicular to the length, L_r

(3) *Total Length,* L The minimum distance from the tip of the landslide to its crown

(4) *Length of Displaced Mass,* L_d The minimum distance from tip to the top

(5) *Length of the Rupture Surface,* L_r The minimum distance from the toe of the surface of rupture to the crown

(6) *Depth of the Displaced Mass,* D_d The maximum depth of the displaced mass, measured perpendicular to the plane containing W_d and L_d

(7) *Depth of the Rupture Surface,* D_r The maximum depth of the rupture surface below the original ground surface measured perpendicular to the plane containing W_r and L_r

TABLE 1.3 Velocity Class

Class	Description	Velocity (mm/sec.)
7	Extremely rapid	5×10^3
6	Very rapid	50
5	Rapid	0.5
4	Moderate	5×10^{-3}
3	Slow	50×10^{-6}
2	Very slow	0.5×10^{-6}
1	Extremely slow	

Source: Cruden and Varnes (1992).

1.5.2 Landslide Rates and Types of Movements

It is important to distinguish types of landslides according to the rate of slope movement. Rates of movement range from less than 6 inches per year to more than 5 feet per second according to Cruden and Varnes (1992). The rate of movement can be expressed in multiples of 100, as shown in Table 1.3.

The kinematics of landslides (i.e., how movement is characterized throughout the displaced mass) constitutes another way of classifying landslides. After Cruden and Varnes (1992), there are five kinematically distinct types of landslide movements (Figure 1.18):

(1) Falling
(2) Toppling
(3) Sliding
(4) Spreading
(5) Flowing

Each type of landslide has a number of common modes. Falling and toppling are features frequently associated with rock slopes, whereas the latter three are related to soil slopes.

A slide is a downslope movement of a soil mass occurring dominantly on surfaces of rupture or relatively thin zones of intense shear strain (Cruden and Varnes, 1992). Movement is usually progressive from an area of local failure. The first overt signs of ground movement are usually cracks in the original ground surface along which the main scarp of the slide will form. The slide may be translational or rotational, or a combination of both, which is called a compound slide.

Translational slides often involve movement along marked discontinuities or planes of weakness, including previously existing failure planes. In clay soils, translational slides take place along saturated sand or silt seams, particularly where these zones of weakness dip roughly parallel to the existing slope. Rotational slips have a failure surface that is concave upwards and often occur within an intact soil mass. Classic, purely rotational, slope failures (slumps) most commonly occur in relatively homogeneous materials, such as those found in constructed fills and embankments. In addition,

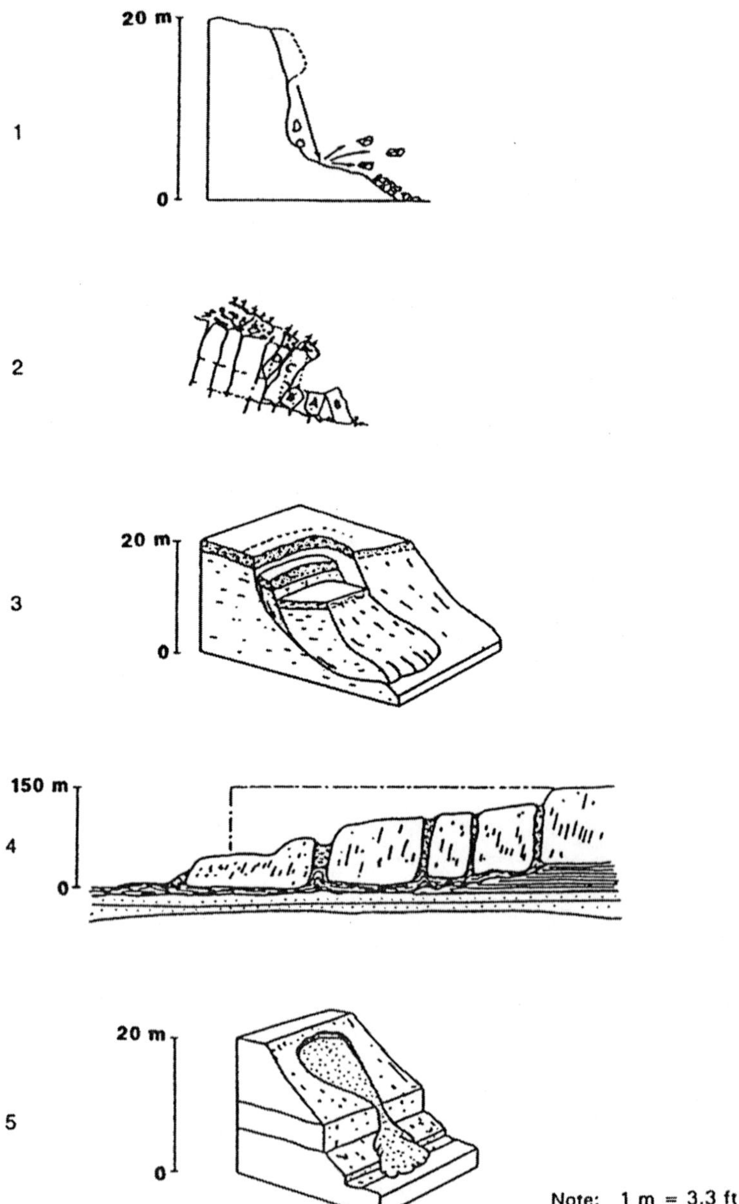

Figure 1.18 Types of landslides. (1) Fall. (2) Topple. (3) Slide. (4) Spread. (5) Flow. Broken lines indicates the original ground surfaces. Arrows show portion of the trajectories of individual particles of the displaced mass. (Cruden and Varnes, 1992.)

slumps may degenerate to flows, as do the translational slides, if dilated slide debris becomes saturated.

Spread is defined by Cruden and Varnes (1992) as an extension of a soil mass combined with a general subsidence of the fractured mass into softer underlying material. The rupture surface is not a surface of intense shear. Spreads may result from liquefaction of granular deposits or failure of weak cohesive soils in a slope (Schuster and Fleming, 1982). They commonly occur on shallow slopes.

A flow is a spatially continuous movement in which surfaces of shear are short-lived, closely spaced, and usually not preserved. The distribution of velocities in the displacing mass resembles that of a viscous liquid. The lower boundary of the displaced mass may be a surface along which appreciable differential movement has taken place or it may be a thick zone of distributed shear. Slides may turn gradually into flows with changes in water content, mobility, and evolution of movement. As the displaced material loses strength and gains water or encounters steeper slopes, debris slides may become rapid debris flows. There are five modes of flows: skin flows, lahars (debris flow from volcanoes), channelized flows, open-slope debris flows, and debris avalanches.

Slopes also can undergo creep movements that are too slow to be detected, for example, less than 1 inch per year. Creep can be categorized as continuous or of a seasonal nature. Continuous creep often occurs under the action of low shear stress and may continue for long periods without resulting in a failure. Thus creep may be a factor in progressive failure. Continuous creep is common in overloaded argillaceous (clayey) soils and rocks.

Solifluction movements (Figure 1.19) are a form of seasonal creep that occur in glacial areas because of annual freeze-thaw cycles. Their influence generally extends to shallow depths. However, it must be noted that areas that have no present-day glacial activities may have had such processes in the past. Previous glacial activity is sometimes evidenced by zones of weakness in the ground that are reactivated by new construction or natural events. These zones may be located below the ground surface because geologic or engineered processes can cause changes in topography.

Landslides occur in fill slopes as well as in natural cut slopes. The degree of compaction of fill slopes may have considerable influence on the speed of movement. The most remarkable example of such movements is the failure of slopes at Sau Mau Ping, Hong Kong, in 1976, which resulted in the loss of many lives. The lack of compaction resulted in a dramatic failure because the loose fill was fully saturated and became "quick."

Post-failure movements often occur on existing slip surfaces at rates that may vary from 8 inches to 20 feet per year (Skempton and Hutchinson, 1969). These movements may be caused by changes in pore water pressure or by other types of external disturbance. In some soils, the movement during initial failure is such as to bring the slipped mass into a more stable position, and there may be no post-failure movements. Mud flow movements may occur where the initial slide has changed the soil into a broken and softened mass.

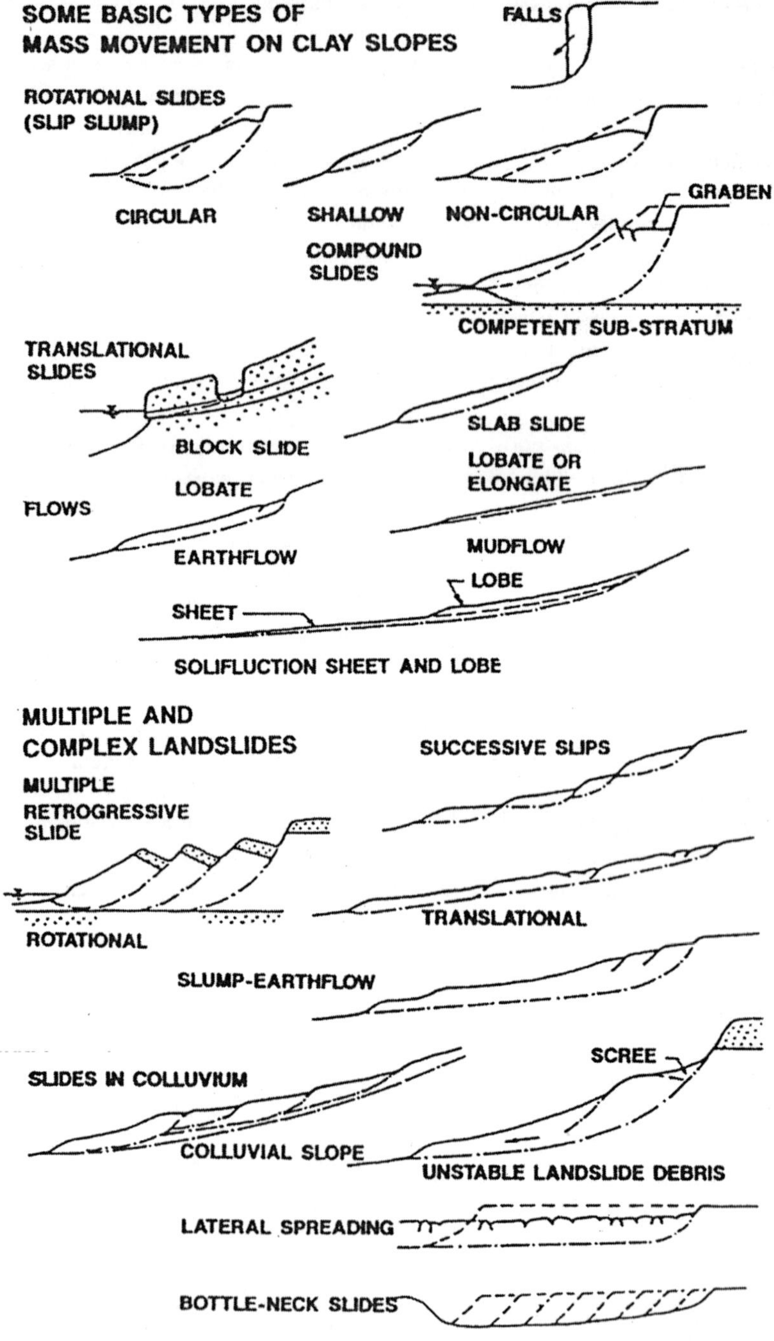

Figure 1.19 Types of movement in clay slopes. (Skempton and Hutchinson, 1969.)

1.6 FACTORS CONTRIBUTING TO SLOPE FAILURES

Slope failures are often caused by processes that increase shear stresses or decrease shear strengths of the soil mass. Processes that most commonly cause an increase in the shear stresses acting on slopes are listed in Table 1.4. Processes that most commonly cause a decrease in the shear strength of slope materials are listed in Table 1.5.

Residual soil and weathered bedrock can be weakened by preexisting discontinuities such as faults, bedding surfaces, foliations, cleavages, sheared zones, relict

TABLE 1.4 Factors That Cause Increases in Shear Stresses in Slopes

(1) Removal of support
- A. Erosion
  1. By streams and rivers
  2. By glaciers
  3. By action of waves or marine currents
  4. By successive wetting and drying (e.g., winds, freezing)
- B. Natural slope movements (e.g., falls, slides, settlements)
- C. Human activity
  1. Cuts and excavations
  2. Removal of retaining walls or sheet piles
  3. Drawdown of bodies of water (e.g., lakes, lagoons)

(2) Overloading
- A. By natural causes
  1. Weight of precipitation (e.g., rains, snow)
  2. Accumulation of materials because of past landslides
- B. By human activity
  1. Construction of fill
  2. Buildings and other overloads at the crest
  3. Water leakage in culverts, water pipes, and sewers

(3) Transitory effects (e.g., earthquakes)

(4) Removal of underlying materials that provided support
- A. By rivers or seas
- B. By weathering
- C. By underground erosion due to seepage (piping), solvent agents, etc.
- D. By human activity (excavation or mining)
- E. By loss of strength of the underlying material

(5) Increase in lateral pressure
- A. By water in cracks and fissures
- B. By freezing of the water in the cracks
- C. By expansion of clays

Source: From Highway Research Board (1978).

TABLE 1.5 Factors That Cause Reduced Shear Strength in Slopes

(1) Factors inherent in the nature of the materials
- A. Composition
- B. Structure
- C. Secondary or inherited structures
- D. Stratification

(2) Changes caused by weathering and physiochemical activity
- A. Wetting and drying processes
- B. Hydration
- C. Removal of cementing agents

(3) Effect of pore pressures

(4) Changes in structure
- A. Stress release
- B. Structural degradation

Source: From Highway Research Board (1978).

joints, and soil dikes. Relict joints and structures in residual soils often lose strength when saturated. Slickensided seams or weak dikes may also preexist in residual soil and weathered rock slopes. Faults, bedding surfaces, cleavages, and foliations have more influence on rock slope stability than soil slope stability.

1.7 BASIC CONCEPTS APPLIED TO SLOPE STABILITY

The discovery of the principle of effective stress by Terzaghi in 1920s marks the beginning of modern soil mechanics. This concept is very relevant to problems associated with slope stability. Consider three principal stresses, σ_1, σ_2, and σ_3, at any point in a saturated soil mass and let u be the pore water pressure at that point. Changes in the total principal stresses caused by a change in the pore water pressure u (also called the neutral stress) have practically no influence on the volume change or on the stress conditions for failure. Compression, distortion, and a change of shearing resistance result exclusively from changes in the effective stresses, σ_1', σ_2' and σ_3', which are defined as

$$\sigma_1' = \sigma_1 - u, \qquad \sigma_2' = \sigma_2 - u, \qquad \text{and} \qquad \sigma_3' = \sigma_3 - u \qquad \text{(Eq. 1-6)}$$

Therefore, changes in u lead to changes in effective stresses.

Slope materials have a tendency to slide due to shearing stresses created in the soil by gravitational and other forces (e.g., water flow, tectonic stresses, seismic activity). This tendency is resisted by the shear strength of slope materials expressed by Mohr–Coulomb theory (Figure 1.20) as

$$S = c + \sigma_n \tan\phi \qquad \text{(Eq. 1-7)}$$

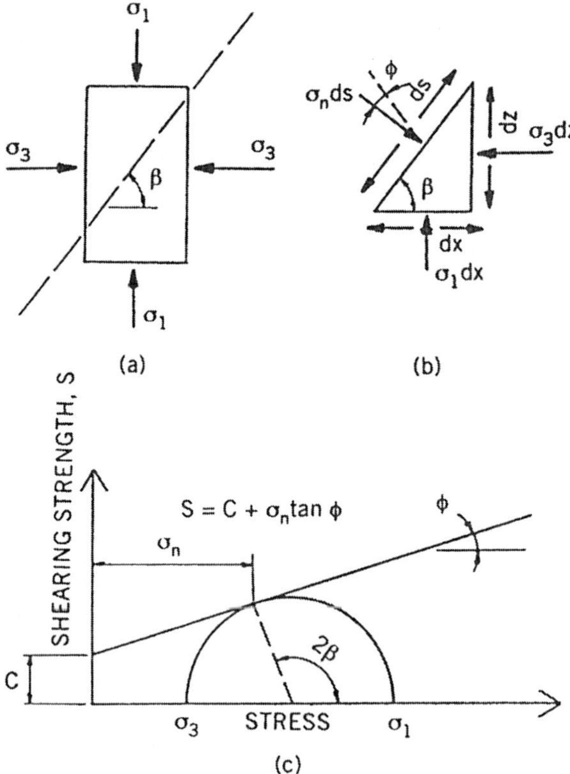

Figure 1.20 Mohr–Coulomb envelope. (*a*) Soil element. (*b*) Stress vectors. (*c*) Shear strength envelope.

where S = total shear strength of the soil
c = total cohesion of soil
σ_n = total normal stress
ϕ = total angle of internal friction

In terms of effective stresses,

$$S' = c' + (\sigma_n - u)\tan\sigma' \qquad \text{(Eq. 1-8)}$$

where S' = drained shear strength of the soil
c' = effective cohesion
σ_n = normal stress
u = pore water pressure
ϕ' = angle of internal friction in terms of effective stress

Practically all slope stability analyses are based on the concept of limit equilibrium (Figure 1.21). In the equation

$$\tau = \frac{S}{\text{FOS}} \qquad \text{(Eq. 1-9)}$$

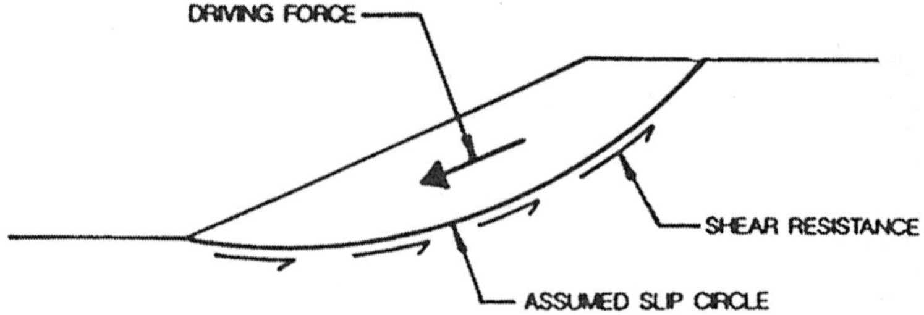

Figure 1.21 Geometry used in slip circle analysis.

where τ = shearing stress along the assumed failure surface
S = shear strength of the soil
FOS = factor of safety

A state of limit equilibrium is assumed to exist when the shearing resistance along an assumed failure surface (slip circle) equals the shear strength of the soil, or in other words, when FOS equals unity. The determination of $c - \phi$ shear strength values is made by field investigations and laboratory testing.

1.8 TYPICAL INPUT DATA FOR SLOPE STABILITY ANALYSES

1.8.1 Geologic Conditions

To perform slope stability analyses, it is essential to understand the geology of the site. Basic geological features that could affect the stability of slopes include

(1) Slope material fabric such as mineral types
(2) Mineral orientations and stratification
(3) Discontinuities and bedding planes resulting from faults and folds, schistosity, cleavages, and fissures
(4) Geological anomalies, such as soft seams, bedrock contact, and previously sheared zones
(5) Degree of weathering
(6) Groundwater
(7) History of previous landslides
(8) In situ stresses

These geological features should be considered when undertaking a slope stability analysis.

1.8.2 Site Topography

Site topography is an overt clue to past landslides and potential instability. For example, the head region of a landslide may be recognized by the presence of slump

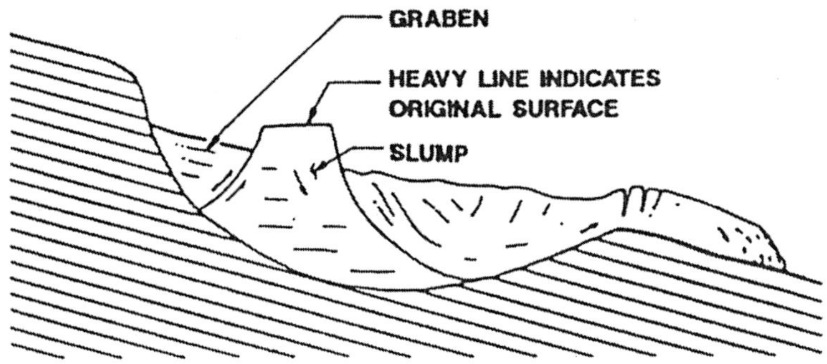

Figure 1.22 Graben on slump slide. (Schuster and Krizek, 1978.)

grabens that have rotated, as shown in Figure 1.22. Trees tend to be tilted downhill rather than uphill as they near the head of a scarp as in Figure 1.23. For identification of landslides and areas of potential instability, more detail than that usually shown on topographic maps is often required. Interpretations by topographers who are not specifically looking for landslide features can obscure the special geometric or topographic forms that diagnose landslides. Therefore, special mapping using additional detailed site reconnaissance and aerial photographic mapping should be used to supplement topographic information.

The nature and origin of land forms, together with the process of land form development, play an important role in understanding the relationship of slope form to stability. Slopes similar in inclinations, materials, and geology may behave differently, depending on their topographic aspects, which may control moisture, seepage, and pore pressures. Surface drainage patterns may be a clue about the nature of underlying soils.

1.8.3 Possible Effects of Proposed Construction

During design of a project that involves cut or fill slopes that could be prone to failure, one way to avoid stability problems is by changing the project site. When this

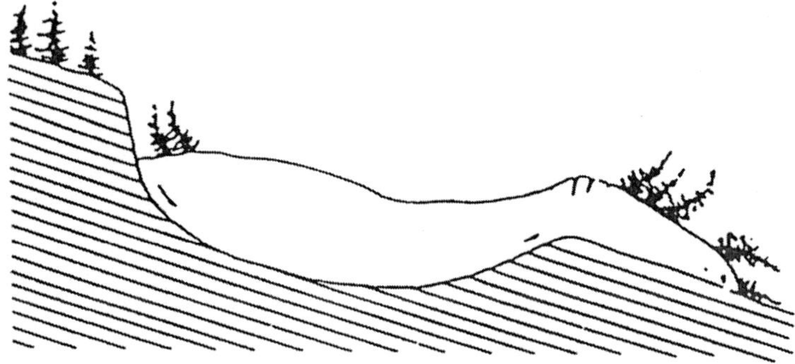

Figure 1.23 General orientation of trees on slump landslide. (Schuster and Krizek, 1978.)

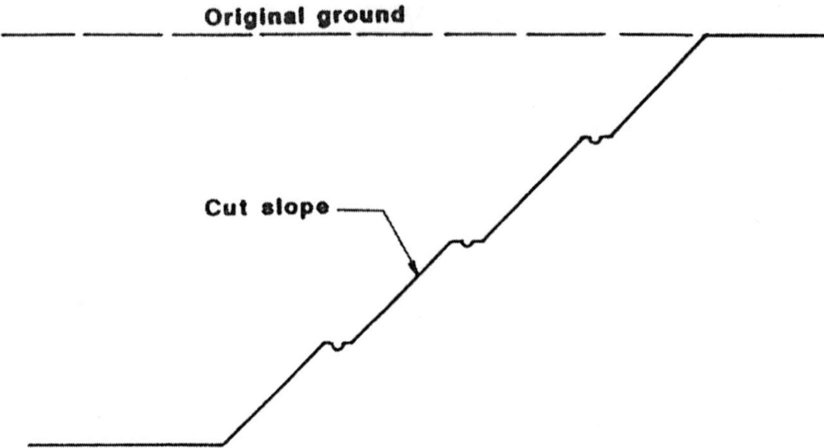

Figure 1.24 Provision of berms and ditches on a large cut slope.

is not possible or practical because of economic or other reasons, the designers must determine where to form the slopes at the site and what slope angles will provide the most stability and least maintenance.

Slope cuts release residual horizontal stresses and cause expansion of the slope. It is essential not to undermine the toe of the cut slope but to maintain the slope as "loaded" as possible. The main causes of cut slope instability are overcutting the toe or oversteepening the slope angle.

When designing a slope, it should be designed for minimum exposure to precipitation. The smaller the volume excavated for a cut and the steeper the slope, the smaller will be both the exposed slope face and the amount of precipitation received by the cut surface. Berms with ditches on the uphill side of the cut may be required to collect and eliminate runoff water (Figure 1.24). Often a steep slope will prove successful (if the ground is strong enough) where a complicated slope with flatter angles might fail, because with the steep slope, the amount of slope face exposed to precipitation is minimized. Failures of flat cut slopes in loess, where the rain dissolves the natural cementing agents of the soil, are an example.

Construction of embankment slopes, using local experience and practice, generally does not pose stability problems if competent foundation soils are present. Embankments with steep slopes concentrate stresses at the toe. Settlement is encouraged in this area and can lead to instability. The toes of very flat slopes are not as intensively loaded.

Drainage patterns of existing terrain may be altered as a result of proposed construction. Sometimes, fill slopes are constructed in valleys that may be part of stream systems. In such cases, adequate drainage measures must be devised to accommodate subsurface flows. Where cuts are made in a ridge or valley, the long-term groundwater flow patterns set up may cause changes detrimental to the stability of the newly constructed slopes or of the existing in situ slopes that were stable prior to construction.

Construction methods are an important aspect in maintaining slope stability. For example, a cut should be excavated so that rain and spring water can be drained easily. Cuts should be made stratum by stratum to allow the groundwater table to lower evenly, thus avoiding the development of large localized zones of water pressure in undrained slopes having strata of varying rates of drainage. The construction of embankment fills over soft clay foundations should be carried out in stages to minimize local failures and global instability brought on by excessive pore water pressure buildup.

Sometimes for expedience, slopes are cut initially steeper than the final design ones. The slopes are then trimmed to the designed angle after mass excavation. This practice is considered to be inappropriate in most cases. When this practice is used, cracks and fissures may open during mass excavation that will degrade slope continuity and future slope stability performance.

1.8.4 Material Properties

Soils are particulate materials in which the void spaces between particles may be filled with liquid (water), or gas (air), or both. Fundamental understanding of soil behavior requires a knowledge of mineralogy. In general, cohesionless soils consist of minerals, such as quartz, feldspar, and mica. Quartz, which occurs in abundance, has a very stable crystal structure. Feldspars are weak and less stable and occur less abundantly in soil than in igneous rocks. Mica occurs in stacks of layers bonded together weakly, and its presence increases compressibility under loading, as well as swelling upon unloading. The clay minerals are silicates of calcium and aluminum. Water is absorbed strongly by the clay minerals. This type of bonding between water and clay minerals plays an important role in slope stability.

The properties of natural soils depend on the proportions of different sizes of fine-grained and coarse-grained materials and also on their mineralogy. In addition, the properties also depend on the process of formation, density, and stress history. For example, the angle of shearing resistance of talus and colluvial material may range from 20 to 50 degrees, depending on the gravel fraction. The values of ϕ for residual soils may range from 20 to 40 degrees, depending on the clay fraction, water content, and relict rock structure. As clay fraction increases, the ϕ value generally decreases. Of the clay minerals, montmorillonite gives the lowest ϕ values; illite and kaolinite give higher values. Tables 1.6 to 1.8 summarize the engineering properties of typical soils.

The properties of compacted soils used in the construction of embankments depend to a significant extent on the borrow material characteristics, methods of compaction, compaction effort, and moisture content at which they are compacted. The proportion of different soil sizes and the moisture content greatly influence the efficiency with which the soils can be compacted and the resulting mechanical properties.

The properties of residual soils are quite different from sedimentary soils. The difference is that residual soils are formed by the in situ physical and chemical weathering of underlying rock, while sedimentary soils are formed by a process of erosion and transportation followed by deposition and consolidation under their own weight. The latter may undergo further alteration after deposition through processes such as leaching, secondary consolidation, and thixotropic effects (Bjerrum, 1966).

TABLE 1.6 Void Ratio, Water Content, and Unit Weight of Typical Soils[a]

Soil	Particle Size and Gradation				Voids[b]			Voids[b]		Unit Weight[c] (lb/ft³)						
	Approximate Size Range (mm)			Approximate Range Uniform	Void Ratio			Porosity (%)		Dry Weight			Wet Weight		Submerged Weight	
	D_{max}	D_{min}	Approximate D_{10} (mm)	Cofficient C_u	e_{max}, loose	e_{cr}	e_{min}, dense	D_{max} loose	D_{min} dense	Min., loose	100% Mod. AASHO	Max., dense	Min., loose	Max., dense	Min., loose	Max., dense
Granular Materials																
Uniform materials																
Equal spheres (theoretical values)	—	—	—	1.0	0.92	—	0.35	47.6	26	—	—	—	—	—	—	—
Standrad Ottawa sand	0.84	0.59	0.67	1.1	0.80	0.75	0.50	44	33	92	—	110	93	131	57	69
Clean uniform sand (fine or medium)	—	—	—	1.2–2.0	1.0	0.80	0.40	50	29	83	115	118	84	136	52	73
Uniform, inorganic silt	0.05	0.005	0.012	1.2–2.0	1.1	—	0.40	52	29	80	—	118	81	136	51	73
Well-Graded Materials																
Silty Sand	2.0	0.005	0.02	5–10	0.90	—	0.30	47	23	87	122	127	88	142	54	79
Clean fine to course sand	2.0	0.05	0.09	4–6	0.95	0.70	0.20	49	17	85	132	138	86	148	53	86
Micasceous sand	—	—	—	—	1.2	—	0.40	55	29	76	—	120	77	138	48	76
Silty sand and gravel	100	0.005	0.02	15–300	0.85	—	0.14	46	12	89	—	146[d]	90	155[d]	56	92

							Mixed Soils									
Sandy or silty clay	2.0	0.001	0.003	10–30	1.8	—	0.25	64	20	60	130	135	100	147	38	85
Skip-graded silty clay with stones or rock fragments	250	0.001	—	—	1.0	—	0.20	50	17	84	—	140	115	151	53	89
Well-graded gravel, sand silt, and clay mixture	250	0.001	0.002	25–1000	0.70	—	0.13	41	11	100	140	148[e]	125	156[e]	62	94
							Clay Soils									
Clay (30–50% clay sizes)	0.05	0.5μ	0.001	—	2.4	—	0.50	71	33	50	105	112	94	133	31	71
Colloidal caly (−0.002 mm: 50%)	0.01	10Å	—	—	12	—	0.60	92	37	13	90	106	71	128	8	66
							Organic Soils									
Organic silt	—	—	—	—	3.0	—	0.55	75	35	40	—	110	87	131	25	69
Organic clay (30–50% clay sizes)	—	—	—	—	4.4	—	0.70	81	41	30	—	100	81	125	18	62

Source: Terzaghi and Peck (1967).

[a] General note: Tabulation is based on $G = 2.65$ for granular soil, $G = 2.7$ for clays, and $G = 2.6$ for organic soils.

[b] Granular materials may reach e_{max} when dry or only slightly moist. Clays reach e_{max} only when fully saturated.

[c] Granular materials reach minimum unit weight when at e_{max} and with hygroscopic moisture only. The unit submerged weight of any saturated soil is the unit weight minus the unit weight of water.

[d] Applicable for very compact glacial till. Unusually high unit weight values for tills are sometimes due to not only an extremely compact condition but to unusually high specific gravivty values.

[e] Applicable for hardpan.

TABLE 1.7 Approximate Unconfined Compressive Strength of Clay in Terms of Consistency Values

Consistency	q_u (tsf)	Remarks on Field Identification
Very soft	0.25	Exudes between fingers when squeezed in hand
Soft	0.25–0.5	Molded by light finger pressure
Medium	0.5–1.0	Molded by strong finger pressure
Stiff	1.0–2.0	Indented by thumb
Very stiff	2.0–4.0	Indented by thumbnail
Hard	>4.0	Difficult to indent by thumbnail

Source: Terzaghi and Peck (1967).

Therefore, it is widely believed that a direct application of conventional soil mechanics concepts to residual soils is likely to cause misleading conclusions about the properties of at least some of the residual soils (Wesley, 1974; 1990). This view is best illustrated by Vaughan's quotation (1985):

> The development of the "classical" concepts of soil mechanics . . . has been based almost exclusively on the investigation of sedimentary deposits of unweathered soil. These concepts have been found almost universally inapplicable to the behavior of residual soil, and misleading if inadvertently applied.

The material properties of intact rock are largely dependent on mineralogy, hardness, and unconfined compressive strength. However, in most rock slope stability problems, it is the rock mass properties that are important. The rock mass properties are largely dependent on the frequency, orientation, and shear strength of the rock joints, foliations, bedding planes, shears, and faults.

1.8.5 Shear Strength

The stability of a slope cannot be analyzed without knowledge of appropriate shear strength values regardless of the method of limit equilibrium analysis used. The two

TABLE 1.8 Angle of Internal Friction Values for Cohesionless Soils[a]

	ϕ Values (degrees) When		
Description	Loose	Medium	Dense
Nonplastic silt	26–30	28–32	30–34
Uniform fine to medium sand	26–30	30–34	32–36
Well-graded sand	30–40	34–40	38–46
Sand and gravel	32–36	36–42	40–48

Source: Hough (1969), and Lambe and Whitman (1969).

[a]*Note:* Within each range, assign lower values if particles are well-rounded or if there is significant soft shale or mica content; assign higher values for hard, angular particles. Use lower values for high normal pressures (>4000 psf) than for moderate normal pressures (1000 to 4,000 psf).

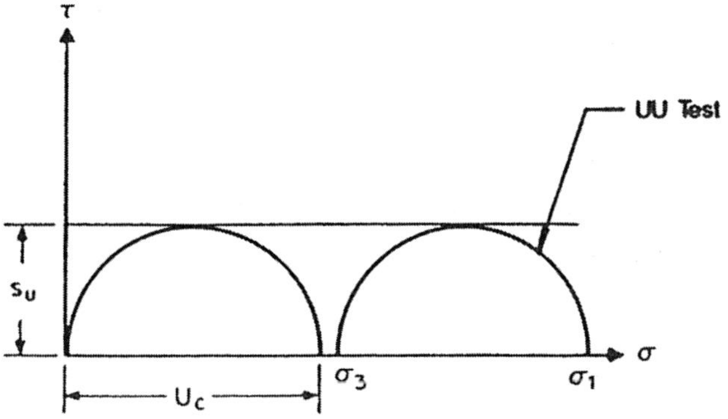

Figure 1.25 $\phi = 0$ strength envelope.

types of shear strength used in stability analyses are the undrained shear strength, c_u, and the drained shear strength, c'. Undrained shear strength is used in total stress analyses, whereas drained shear strength is used in effective stress analyses.

Undrained Shear Strength Natural deposits of saturated clay are frequently loaded or unloaded rapidly relative to the rate at which consolidation or drainage can occur. An ideal undrained condition may be assumed for such circumstances. The water content and the volume of the clay remain constant during the undrained loading and excess pore pressures are generated. The shear strength for such conditions is defined as the undrained shear strength, c_u.

If the undrained behavior of saturated clays is analyzed in terms of total stresses, the evaluation of pore water pressures is unnecessary. Under this situation, the $\phi = 0$ method of analysis is assumed and the undrained shear strength, c_u or u_c, is equal to the cohesion value of the Mohr–Coulomb envelope for total stresses (Figure 1.25). For these assumptions, the undrained strength of a saturated clay is not affected by changes in confining pressures as long as the water content does not change.

For normally consolidated fine-grained soils, there is a unique relationship between water content, the logarithm of undrained strength, and the logarithm of the major principal consolidation stress. A relationship exists between the ratio of undrained strength and the effective consolidation pressure with the plasticity index, PI (Terzaghi and Peck, 1967):

$$\frac{c_u}{c'} = 0.11 + 0.0037\,\text{PI} \qquad \text{(Eq. 1-10)}$$

where c_u = undrained strength (pounds per square foot)
c' = effective consolidation pressure (pounds per square foot)
PI = plasticity index (percent)

For overconsolidated clays, the three-way relation between water content, consolidation pressure, and undrained strength is not unique, as it is influenced by the degree of overconsolidation. However, it has been shown that the normalized undrained strength can be estimated from

$$\frac{(c_u/c_0')oc}{(c_u/c_0')nc} = (\text{OCR})^{0.8} \qquad \text{(Eq. 1-11)}$$

where OCR $=$ overconsolidation ratio $= c_{\max}/c_0'$
$c_{\max}$ $=$ maximum past pressure
c_0' $=$ existing overburden pressure

Drained Shear Strength Based on the effective stress principle, the maximum resistance to shear on any plane in the soil is a function not of the total normal stress acting on the plane (Equation 1-7), but of the difference between the total normal stress and the pore pressure (Equation 1-8). This equation is illustrated in Figure 1.26.

In general, the pore pressure consists of the water pressure within the soil voids (pores), u, and the pore pressure change, Δu (excess pore pressure or pore pressure deficiency), as induced by a change in loading. The pore pressure related to the groundwater level is hydrostatic if the water is not moving and is equal to

$$u = \gamma_w z_w \qquad \text{(Eq. 1-12)}$$

where γ_w $=$ unit weight of water
z_w $=$ vertical depth below the groundwater table

If the groundwater is moving, the pressure head (h_u) can be found from a piezometer or a flow net (Figure 1.27). In this case, the pore pressure related to the flowing

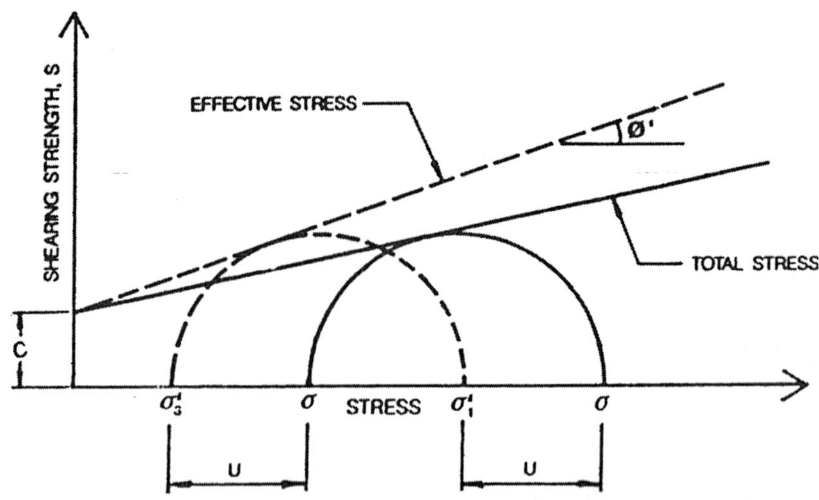

Figure 1.26 Effective stress and total stress envelopes.

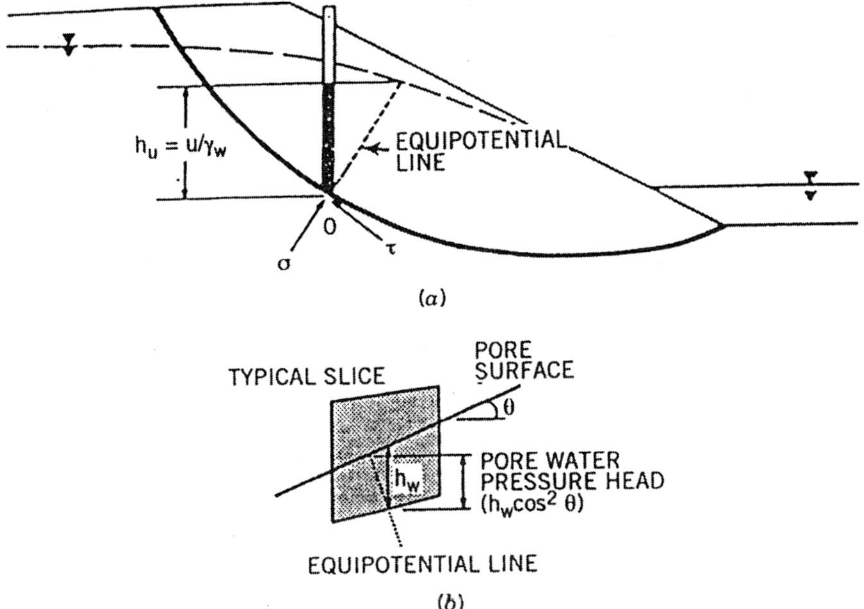

Figure 1.27 (*a*) Measurement of pore pressure. (*b*) Calculation of pore water pressure head from phreatic line.

groundwater can he computed as

$$u = \gamma_w h_u \tag{Eq. 1-13}$$

where h_u = piezometric head.

The phreatic surface is straight but inclined at angle θ with the horizontal, the piezometric head $h_u = h_w \cos^2 \theta$, and h_w is the vertical distance to the phreatic surface (Figure 1.27).

When soil is either loaded by building an embankment or unloaded by excavation, the soil volume change will result in a pore pressure change, Δu. This change in pore pressure may increase or decrease with time, depending on the type of soil and the type of stresses involved. Under the fully drained (long-term) conditions, Δu dissipates, and thus $\Delta u = 0$. For partially drained or undrained conditions, the evaluation of u depends on the relative rate of loading as compared to the rate of drainage within the soil.

The magnitude of pore pressure change, which develops as a result of stress changes in undrained soil, was proposed by Skempton (1954) as

$$\Delta u = B[\Delta\sigma_3 + A(\Delta\sigma_1 - \Delta\sigma_3)] \tag{Eq. 1-14}$$

where $\Delta\sigma_1$ = change in major principal stress
$\Delta\sigma_3$ = change in minor principal stress
A and B = pore pressure parameters

For fully saturated soils, B is equal to 1, whereas for partially saturated soils, B lies between 0 and 1, depending on the degree of saturation and the compressibility of the soil skeleton. A varies with the magnitude of shear strain, initial density, and overconsolidation ratio of the soil. A is positive for those soils that tend to compress when sheared (for example, loose sand and normally consolidated clay) and negative for those soils that tend to dilate when sheared (for example, dense sand and overconsolidated clay).

An important concept exists such that any change in effective stress would change the density at failure, as well as changing the shear strength. Similarly, any action that changes the density at failure must produce a change in shear strength. For the ultimate condition where the strength of normally and overconsolidated soils approaches residual strength (Figure 1.28), there is a unique relationship among effective stress, shear strength, and water content, such that knowledge of any of these three quantities specifies the other two quantities. At the peak resistance, this three-way relation is not unique, as it is influenced by the degree of overconsolidation (Figure 1.29). But this relationship still helps to visualize the effect of changes in effective stress and density upon strength (Lambe and Whitman, 1969). Other factors affecting the drained strength of a soil sample are

(1) Sample disturbance
(2) Rate of testing

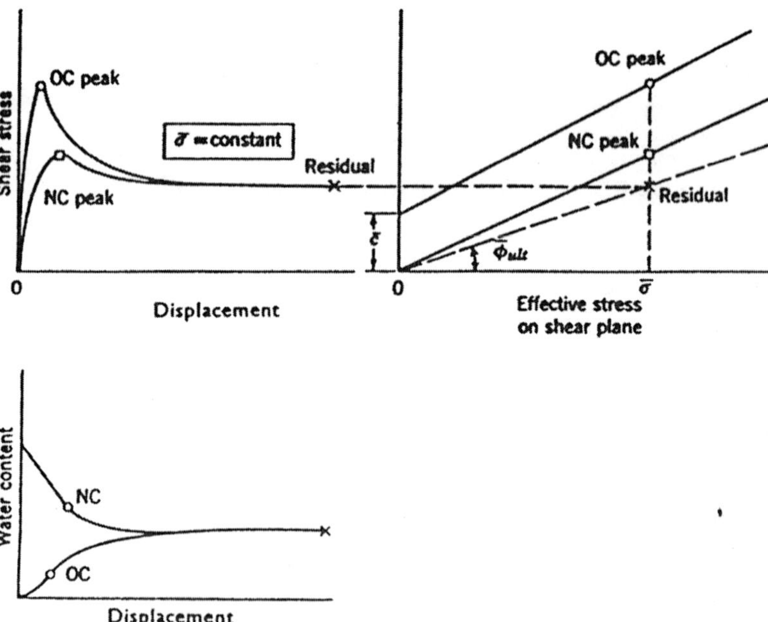

Figure 1.28 Relationship between peak and ultimate conditions. (Lambe and Whitman, 1969.)

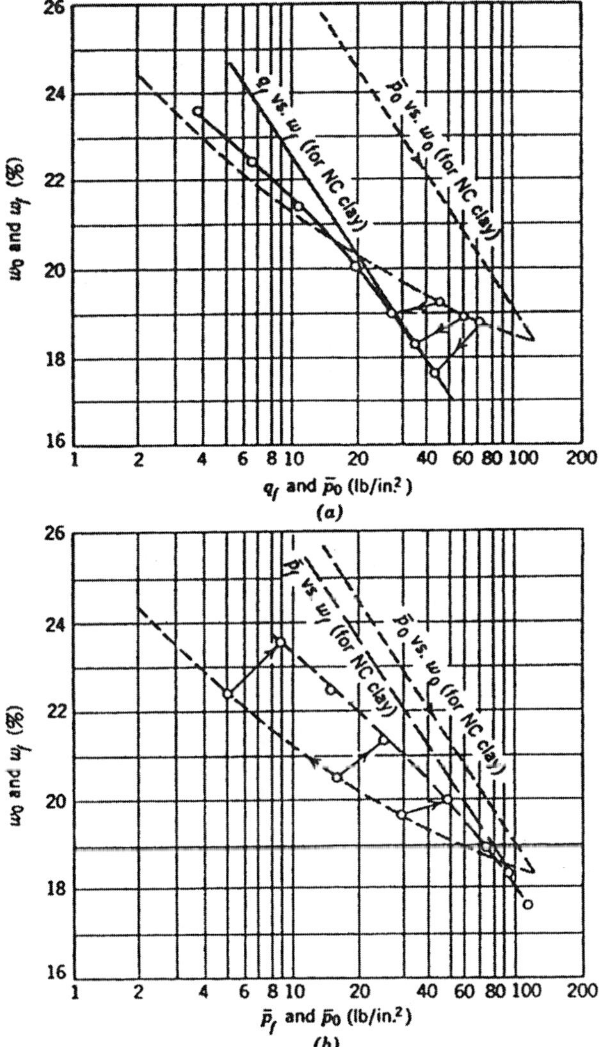

Figure 1.29 Stress-volume relations for overconsolidated Weald clay. $\sigma'_m = 120$ psi. (Lambe and Whitman, 1969.) (*a*) Normal and shear stresses. (*b*) Initial and final stresses.

(3) Anisotropy

(4) Creep

Relationships between Undrained and Drained Strength of Cohesive Soils The undrained strength of a saturated soil is unaffected by changes in total stress unless accompanied by a change of water content (or a change in volume). Such conditions are prevalent immediately after loading or unloading of saturated

TABLE 1.9 Critical Conditions for the Stability of Saturated Clays

	Soil Type	
	Soft (NC) Clay	Stiff (Highly OC) Clay
	Foundation Loading	
Critical conditions	Unconsolidated undrained (UU) case (no drainage)	Probably UU case but check consolidated drained (CD) case (drainage with equilibrium pore pressures)
Remarks	Use $\phi = 0$, $c = \tau_{ff}$ with appropriate corrections	Stability usually not a major problem
	Excavation or Natural Slope	
Critical conditions	Could be either UU or CD case	CD case (complete drainage)
Remarks	If soil is very sensitive, it may change from drained to undrained conditions	Use effective stress analysis with equilibrium pore pressures; if clay is fissured, c' and perhaps ϕ' may decrease with time

soft clays with very low permeability. Therefore, such a soil behaves as if it has a value of $\phi = 0$, that is, the total stress Mohr–Coulomb envelope is a horizontal line. The $\phi = 0$ concept should be used for stability problems in normally consolidated and slightly overconsolidated soils.

However, if the overconsolidation ratio (OCR) is between or higher than 4 to 8, the volume of the soils tends to increase significantly during shearing with a decrease of pore water pressure. Thus the undrained strength exceeds the drained strength, but high negative pore pressures tend to draw water into the soil with consequent swelling and reduction of strength. Therefore, the undrained strength cannot be relied upon, and its use in stability analyses would lead to unsafe results. On the other hand, for normally loaded, lightly overconsolidated soils, the tendency to decrease in volume implies that strength would increase with time, and thus use of undrained strengths would lead to results on the safe side. According to Terzaghi and Peck (1967), the $\phi = 0$ concept should not be used for clays with overconsolidated ratios higher than about 2 to 4. Table 1.9 summarizes the critical shear strength and drainage conditions for stability analyses.

1.8.6 Groundwater Conditions

Besides gravity, groundwater is the most important factor in slope stability. Groundwater can affect slope stability in five ways:

(1) Reduces strength

(2) Changes the mineral constituents through chemical alteration and solution

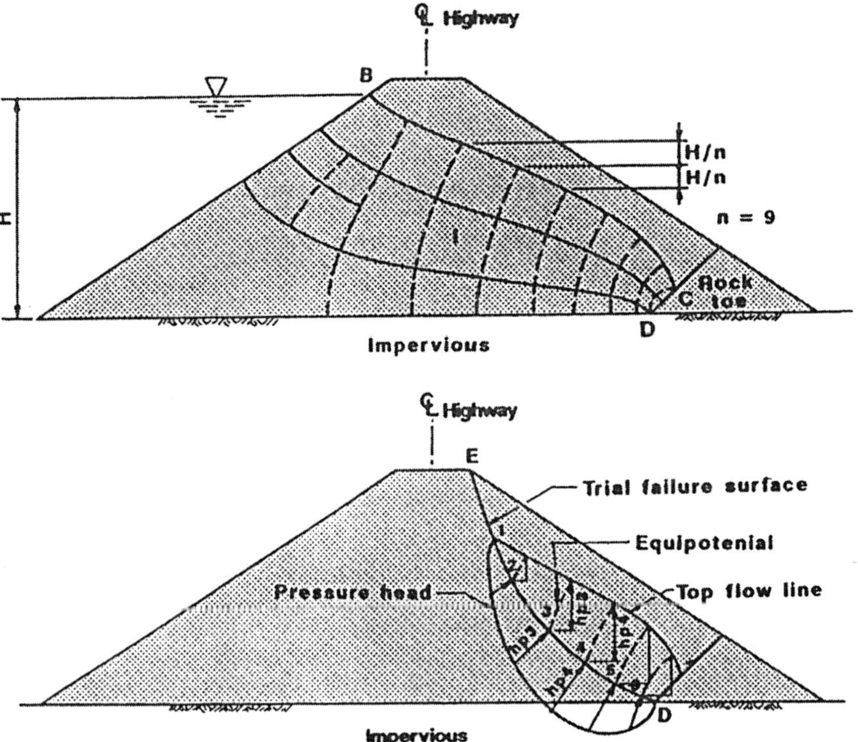

Figure 1.30 Flow through highway embankment.

(3) Changes the bulk density
(4) Generates pore pressures
(5) Causes erosion

The determination of pore water pressure is facilitated by the theory of seepage based on Darcy's law of flow and hydraulic gradient. Closed-form, numerical, and graphical solutions have been developed for seepage problems and reference may be made to textbooks by Cedergren (1967), Harr (1962), and Lambe and Whitman (1969). The graphical flow-net approach is often very valuable in practical application concerning slopes (Figure 1.30).

1.8.7 Seismicity

The release of energy from earthquakes sends seismic acceleration waves traveling through the ground. This type of transient dynamic loading instantaneously increases the shear stresses in a slope and decreases the volume of voids within the materials of the slope, leading to an increase in the pressure of fluids (water) in pores and fractures. Thus shear forces increase and the frictional forces that resist them decrease. Other factors that affect the response of slopes during earthquakes include

(1) Magnitude of the seismic accelerations
(2) Earthquake duration
(3) Dynamic strength characteristics of the materials affected
(4) Dimensions of the slope

After an earthquake a slope may be stronger, weaker, or the same as before.

As mentioned earlier, earthquakes can increase shear stresses and reduce shear strength by increasing pore water pressures. Liquefaction of sand and silt lenses during earthquakes can cause progressive failure of a slope. Seed (1966) has discussed many examples that show how minor geological details have been extremely important in the development of landslides caused by earthquakes.

Earthquake-triggered landslides in the United States occur most frequently in Alaska and California. They can range from small debris flows to rock slides of great size and depth. Several methods of analysis have been developed to study the stability of slopes during earthquakes. The most common ones include the pseudostatic limit equilibrium method and New mark's slope displacement method. The pseudostatic limit equilibrium method is a modification of the conventional limiting equilibrium analysis to include seismic forces that are assumed to be proportional to the weight of the potential sliding mass times a seismic coefficient A expressed in terms of the acceleration of the underlying earth expressed as a function of gravity (Figure 1.31).

Newmark's displacement method is a method that considers the displacements of an embankment during an earthquake. It is a combination of conventional pseudostatic procedures with basic dynamic considerations involving the ground motion.

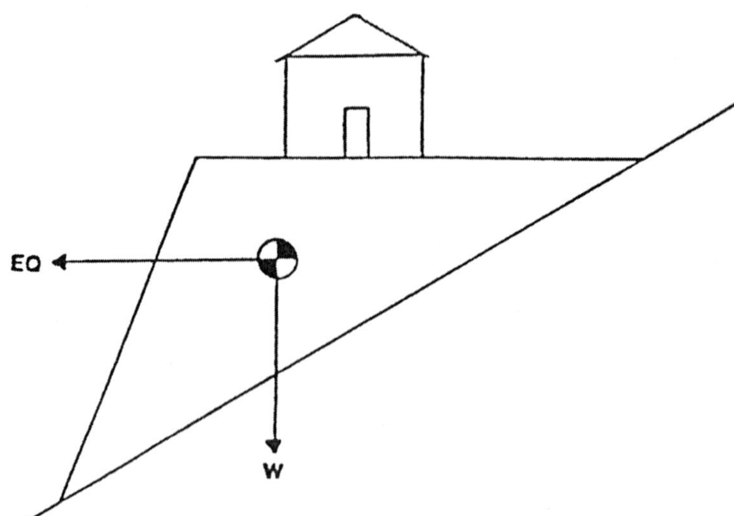

Figure 1.31 Forces of earthquake and gravity acting on slope.

1.9 SUBSURFACE MODEL AND BACK-ANALYSIS FOR SLOPE STABILITY ANALYSES

For any slope stability analyses, details of topography, geology, shear strength, groundwater conditions, external loadings, and plan curvature of slope (three-dimensional effect) must be known and interpreted judiciously to obtain the most representative subsurface model for analyses.

Because of the difficulties inherent in the classical design approach to slopes, back-analysis of a slope failure often provides valuable information for future design purposes. This can only be meaningful, however, in circumstances where the majority of factors that contributed to the failure can be evaluated. Results of the back-analysis calculations should provide an unambiguous measure of the shear strength at failure. All too often, back-analyses are conducted without adequate information being available on the mode of failure or on the pore pressure that existed at the time of failure.

With the aid of back-analysis methods, relevant shear strength parameters can be obtained that otherwise would not be obtained through conventional laboratory testing. However, it must be remembered that these parameters are applicable only to a specific geological formation in one geographical location, as they are products of the topography, geology, and climate of that location.

Back-analysis is a useful procedure for developing an analytic model of a failed or failing slope. This model unsually consists of five components (Filz et al., 1992):

(1) Landslide geometry including the ground surface, slip surface, and material boundary locations
(2) Pore water pressures on the sliding surface at time of failure. These are necessary for effective stress analysis
(3) External loads acting on the slope at the time of failure
(4) Unit weights of the materials involved in the landslide
(5) Strength of materials along the failure surface

Often, the first four components of the model can be evaluated with reasonable accuracy based on field and laboratory investigations. Back-analysis is often used to establish the fifth component of the model, that is, the soil strengths, on the assumption that the factor of safety was equal to 1 at the time of failure. Because of large deformations, residual strengths are often in effect along the existing failure surfaces, and the material strengths can be characterized by values of effective stress, residual friction angle, and the effective cohesion intercept.

The above model is applicable to a single material along the sliding surface. In the case where there are two or more materials along the sliding surface, back-analysis, such as the one mentioned above, does not result in a unique residual frictional angle and cohesion value for the materials. In such circumstances, the back-analysis procedure proposed by Filz et al. (1992) would include four steps:

(1) Laboratory test results and/or correlations with index properties (Figure 1.32) are used to establish trial values of shear strength along the failure surface.

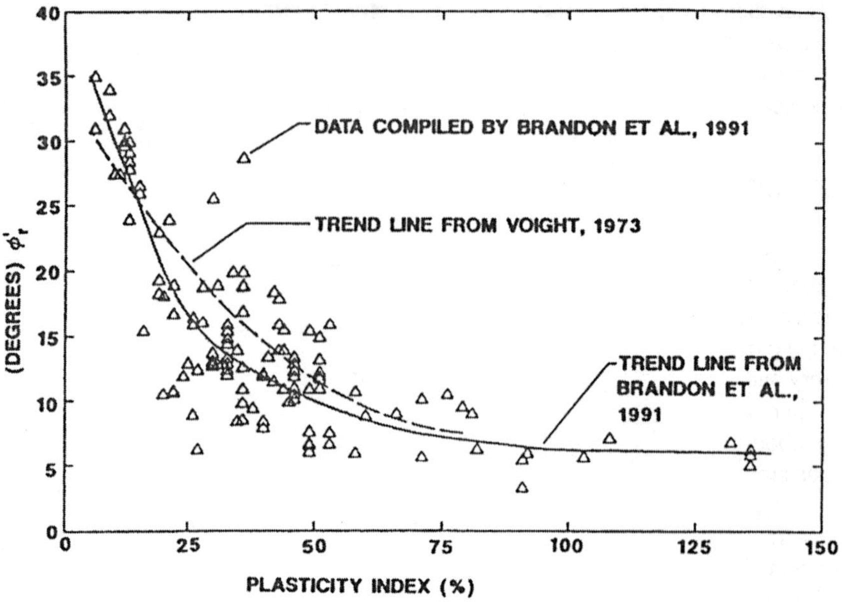

Figure 1.32 Plasticity index residual friction angle relationship. (Voight, 1973, and Brandon et al., 1991.)

Often to simplify the analysis, c' is assumed equal to zero and only the residual friction angle is used.

(2) A stability analysis is performed using slope geometry, groundwater levels, and external loading conditions at the time of failure. The analysis yields a factor of safety, FOS, that corresponds to the trial strengths from Step 1.

(3) The trial strengths from Step 1 are adjusted using the safety factor computed in Step 2, according to the following formulas:

$$\phi_r'(\text{adjusted}) = \tan - 1\frac{\tan\{\phi_r(\text{trial})\}}{\text{FOS}} \quad \text{(Eq. 1-15)}$$

$$c_r'(\text{adjusted}) = \frac{c_r(\text{trial})}{\text{FOS}} \quad \text{(Eq. 1-16)}$$

If extensive local experience with a particular material is available, that experience, as well as Equations 1-15 and 1-16, could be used to adjust the strength of some materials more than others. Note that, in general, ϕ_r' is fixed at a certain value and c_r' is varied based upon local experience to expedite the time involved in the back-analysis.

(4) The results of Step 3 can be verified by reanalyzing the slide using the newly calculated strengths. The final back-calculated strengths that produce a safety factor equal to unity are appropriate for the existing sliding surface, where the shear strength has been reduced to residual.

Note that confidence in this back-analysis is increased when the same set of strengths results in a value of safety factor close to unity for several different cross sections through the landslide and when the back-calculated strengths are in reasonable agreement with laboratory tests and correlations with index properties.

1.10 CONCLUSIONS

Problems associated with the stability of natural slopes are very different from those of engineered slopes. Cuts and embankments therefore should be dealt with as a totally different entity. The main differences are (1) the nature of the materials involved, and (2) the influence of human activities and the environment depending on geologic history and climatic conditions that prevail or will prevail in the future.

When designing slopes, engineers must obtain sufficient information and knowledge about the nature and homogeneity of the component materials. Although numerous methods of analysis are available, there is no general method of analysis that can be applied to all slopes because the state of internal stress in the ground and the stress–strain relationship within a slope before or at the time of failure cannot be determined with absolute reliability. Almost all of the methods in use today are limit-equilibrium analysis methods, which do not truly simulate the failure mechanism of the slope. This is because all limit-equilibrium methods assume an incipient failure at a factor of safety of one, and they neglect the stress-deformation increments or decrements within the slope itself. Nevertheless, limit equilibrium analysis methods are widely used as a tool to design slopes.

Because of the inherent weaknesses of the limit-equilibrium methods, design of slopes has to rely upon experience of the engineers as well as the general policy established by the organization responsible for the project design.

REFERENCES

Alshunnar, I., 1992. "The Design of Landfill Slopes," *Proceedings of a Specialty Conference on Stability Performance of Slopes and Embankments—II*, Geotechnical Special Publication No. 31, ASCE, pp. 1232–1243, June.

Bishop, A. W., 1966. "The Strength of Soils as Engineering Materials," *Geotechnique*, Vol. 16, Nos. 1–4, pp. 91–130.

Bishop, A. W., and L. Bjerrum, 1960. "The Relevance of the Triaxial Test to the Solution of Stability Problems," *Proceedings of the Research Conference on Shear Strength of Cohesive Soils*, Boulder, Colorado, ASCE, June.

Bjerrum, L., 1966. "Mechanism of Progressive Failure in Slopes of Overconsolidated Plastic Clay Shales," Preprint, *ASCE Structural Engineering Conference*, Miami.

Brandon, T. L., G. M., Filz, and J. M. Duncan, 1991. "Review of Landslide Investigation, Phase I—Part B, Olmstead Locks and Dam," A Report to the Louisville District of the U.S. Army Corps of Engineers, June.

Byrne, R. J., J. Kendall, and S. Brown, 1992. "Cause and Mechanism of Failure, Kettleman Hills Landfill B-19, Phase IA," *Proceedings of a Specialty Conference on Stability Performance of*

Slopes and Embankments—II, Geotechnical Special Publication No. 31, ASCE, pp. 1188–1215, June.

Cedergren, H. R., 1967. *Seepage, Drainage, and Flow Nets*. New York: Wiley.

Cheney, R. S., and R. G. Chassie, 1982. *Soils and Foundations Workshop Manual*. Washington, DC: Federal Highway Administration, November.

Chirapuntu, S., and J. M. Duncan, 1977, "Cracking and Progressive Failure of Embankments on Soft Clay Foundations," *International Symposium of Soft Clay*, Bangkok, Thailand, July.

Converse, Davis and Associates, 1975. "Slope Stability Investigation, Proposed Final Slope Adjacent to the Pomona Freeway," Report prepared for Operating Industries Inc., Monterey Park, California.

Cruden, D. M., and D. J. Varnes, 1992. "Landslide Types and Processes," *Landslides: Investigation and Mitigation*. Washington, DC: Transportation Research Board, National Academy of Sciences, Chapter 3, July.

DiMillio, A. F., and W. E. Strohm, Jr., 1981. "Technical Guidelines for the Design and Construction of Shale Embankments," *Transportation Research Record 790—Shale and Swelling Soils*. Washington, DC: National Research Council.

Donaghe, R. T., and V. H. Torrey, 1985. "Strength and Deformation Properties of Earth-Rock Mixtures," Technical Report GL-85-9, U. S. Army Corps of Engineers Waterways Experiment Station, Vicksburg, Mississippi, p. 54.

Duncan, J. M., 1992. "State-of-the-Art: Static Stability and Deformation Analysis," *Proceedings of a Specialty Conference on Stability Performance of Slopes and Embankments—II*, Geotechnical Special Publication No. 31, ASCE, pp. 222–266, June.

Earth Technology Corporation, 1988. "Instability of Landfill Slope, Puente Hills Landfill, Los Angeles County, California," Report Submitted to Los Angeles County Sanitation District.

Filz, G. M., T. L. Brandon, and J. M. Duncan, 1992, "Back Analysis of the Olmstead Landslide Using Anisotropic Strengths," *Transportation Research Board, 71st Annual Meeting*, Washington, DC, January, 12–16.

Harr, M. E., 1962. *Groundwater and Seepage*. New York: McGraw-Hill.

Highway Research Board, 1960. "Factors That Influence Field Compaction of Soils," *Special Report No. 272*, A. W. Johnson and J. R. Sallberg, Eds., Washington, DC.

Highway Research Board, 1978. "Landslides and Engineering Practice," *Committee on Landslide Investigation, Special Report No.* 29, E. B. Eckel, Ed., Washington DC.

Hilf, J. W., 1991. "Compacted Fill," *Foundation Engineering Handbook*, H. Y. Fang, Ed. New York: Van Nostrand-Reinhold, Chapter 7.

Holtz, W. G., and H. J. Gibbs, 1956. "Triaxial Shear Tests on Previous Gravelly Soils," *Journal of the Soil Mechanics and Foundations Division*, ASCE, Vol. 82, No. SM1, pp. 1–22.

Hough, B. K., 1969. *Basic Soils Engineering*, New York: Ronald Press.

Lambe, T. W., and R. V. Whitman, 1969. *Soil Mechanics*. New York: Wiley.

Landva, A. O., and J. J. Clark, 1987. "Geotechnical Testing of Waste Fill," Report Submitted to the Institution of Civil Engineers, London.

Lee, K. L., and S. C. Haley, 1968. "Strength of Compacted Clay at High Pressure," *Journal of the Soil Mechanics and Foundations Division*, ASCE, Vol. 94, No. SM6, pp. 1303–1332.

McGuffey, V. C., 1982. "Design of Cut Slopes in Overconsolidated Clays," *Transportation Research Board No. 873—Overconsolidated Clays: Shales*. Washington DC: National Research Council.

Mitchell, R. A., and J. K. Mitchell, 1992. "Stability Evaluation of Waste Landfills," *Proceedings of a Specialty Conference on Stability Performance of Slopes and Embankments—II*, Geotechnical Special Publication No. 31, ASCE, pp. 1152–1187, June.

Peck, R. B., 1967. "Stability of Natural Slopes," *Proceedings: ASCE*, Vol. 93, No. SM4, pp. 403–417.

Schuster, R. L., and R. W. Fleming, 1982. "Geologic Aspects of Landslide Control Using Walls," *Proceedings of the Application of Walls to Landslide Control Problems*, R. B. Reeves, Ed., ASCE National Convention, Las Vegas, Nevada, April.

Schuster, R. L., and R. J. Krizek, 1978. *Landslides Analysis and Control*, Special Report 176. Washington, DC: Transportation Research Board, National Academy of Sciences.

Schuyler, J. D., 1905. *Reservoirs for Irrigation*. New York: Wiley.

Seed, H. B., 1966. "A Method for the Earthquake Resistant Design of Earth Dams," *Journal of the Mechanics and Foundations Division*, ASCE, No. SM1, pp. 13–41.

Seed, R., and R. Bonaparte, 1992. "Seismic Analysis and Design of Lined Waste Fills: Current Practice," *Proceedings of a Specialty Conference on Stability Performance of Slopes and Embankments—II*, Geotechnical Special Publication No. 31, ASCE, pp. 1521–1543, June.

Seed, H. B., and C. K. Chan, 1959. "Structure and Strength Characteristics of Compacted Clays," *Journal of the Soil Mechanics and Foundations Division*, ASCE, Vol. 85, No. SM5, pp. 87–128.

Skempton, A. W., 1954. "Soils," *Building Materials, Their Elasticity and Inelasticity*, M. Reiner, Ed. Amsterdam: North-Holland, Chapter X.

Skempton, A. W., 1964. "Long-Term Stability of Clay Slopes," *Geotechnique*, Vol. 14, No. 2, pp. 77–101.

Skempton, A. W., and J. N. J. Hutchinson, 1969. "Stability of Natural Slopes and Embankment Foundations," *7th International Conference on Soil Mechanics and Foundation Engineering*, Mexico City, State of the Art Volume, pp. 291–340.

Strohm, W. E., G. H. Bragg, and T. W. Ziegler, 1978. *Design and Construction of Compacted Shale Embankments: Volume 5, Technical Guidelines*, Federal Highway Administration, U.S. Department of Transportation, Department FHWA-RD-78-141, December.

Taylor, D. W., 1948. *Fundamentals of Soil Mechanics*. New York: Wiley.

Terzaghi, K., 1925. *Erdbaumechanik auf bodenphysikalischer.* Grundlage, Vienna: Deuticke.

Terzaghi, K., and R. B. Peck, 1967. *Soil Mechanics in Engineering Practice.* New York: Wiley.

Vaughan, P. R., 1985. "Mechanical and Hydraulic Properties of In Situ Residual Soils," *Proceedings: 1st International Conference on Geomechanics in Tropical Lateritic and Saprolitic Soils*, Brazil, pp. 1–33.

Voight, B., 1973. "Correlation Between Atterberg Plasticity Limits and Residual Shear Strength of Natural Soils," *Geotechnique*, Vol. 23, No. 2.

Wesley, L. D., 1974. "Tjipanundjang Dam in West Java, Indonesia," *Journal of the Geotechnical Engineering Division*, ASCE, Vol. 100, No. GT5, May, pp. 503–521.

Wesley, L. D., 1990. "Influence of Structure and Composition on Residual Soils," *Journal of the Geotechnical Engineering Division*, ASCE, Vol. 116, No. 4, April, pp. 559–603.

CHAPTER 2

ENGINEERING GEOLOGY PRINCIPLES

2.1 INTRODUCTION

Understanding the geology of a region is of paramount importance in tackling problems associated with slopes and slope development. Local geological details, such as (1) geometry of the subsurface, (2) soil properties, and (3) groundwater (which are sometimes called the "three-rules" of slope stability), have a considerable influence on the performance of individual slopes. Thus slope stability evaluation is an interdisciplinary endeavor requiring concepts and knowledge from engineering geology and soil/rock mechanics. Any slope stability method of analysis must give due consideration to significant geological features. Awareness of geology is necessary for appropriate idealization of ground conditions and the subsequent development of realistic geotechnical models.

The "three-rules" of slope stability is to be addressed at length in the subsequent chapters. This chapter focuses on the principles of engineering geology and discusses geology-related landslide prone occurrences, which is followed by an in-depth overview of fundamentals of landslides. Chapter 3 will address the principle of groundwater in slope stability, and Chapter 5 will discuss various methods of obtaining soil strength parameters to be used in slope stability analyses.

2.2 TYPES AND CHARACTERISTICS OF GEOLOGIC SOIL DEPOSITS

Soils are unconsolidated sediments that either are transported to their present place by water, glacier, and air or are formed in place from earlier deposits of sediments and from local bedrock (residual soil). The different transporting agents have different sedimentation characteristics and affect the properties of their soils in different ways. Soils must be recognized in terms of the means of their transportation as well as manner of deposition. Different types of soils include (1) alluvial deposits (by water), (2) glacial deposits (by glaciers), (3) eolian deposits (by wind), (4) alteration (residual) deposits, (5) colluvial/talus deposits (by gravity), (6) marine deposits, and (7) melanges. Each of these soil types have unique engineering characteristics that allow prediction of anticipated behavior of a particular formation.

A classification of soils by origin and mode of occurrence is given in Table 2.1. Included in this table are the depositional environment, the occurrence either as deposited or as subsequently modified, and the typical material associated with the formation. A general distribution of soils in the United States, classified by origin, is illustrated on Figure 2.1. The nomenclature for Figure 2.1 is given on Table 2.1.

2.2.1 Alluvial Deposits

Alluvial deposits were transported by running water and settled out when the speed of water flow was no longer sufficient to carry them. Deposits formed in river valleys are called fluvial, and those formed in lakes are called lacustrine. The deposits are generally of relatively narrow particle size range regardless of whether they consist of cobbles and gravels from rushing rivers and creeks, sands from moderately moving rivers, or clays from sluggish rivers, or from precipitation water moving in sheets down the sides of gentle slopes. Since river flow and location vary considerably over time, multiple zones of varying grain size are frequently encountered within one deposit. These soils do not exhibit distinct horizontal strata and are usually unconsolidated unless subject to removal of overburden. These characteristics provide a favorable situation for landslides. Alluvial soils are usually found over the western part of the United States, and in the southwest United States. General engineering properties for specific example locations are listed in Table 2.2.

River of fluvial deposits are transported by river flow. The deposits may range in size from boulders to colloidal clay, depending on the speed of river flow and the consequent carrying capacity. At a particular location, the particle size range is usually very narrow. Fluvial deposits are typically stratified and extremely variable with frequent interbedding. Permeability in the horizontal direction is often greater than in the vertical. Unless subject to fill placement, removal of overburden, or desiccation, the deposits are normally consolidated. Clays tend to be soft and sands tend to be loose to medium dense. Because of their variability in compositions and engineering properties, fluvial deposits have high susceptibility to slides if unsupported.

Lake or lacustrine deposits are usually the fine-grained materials deposited on lake bottoms. They may contain appreciable amounts of organic matter and also of fragments of shells and skeletons from aquatic animals. If exposed in valley walls or cuts, land areas that were former lake beds can be expected to present slope problems. Slides of considerable magnitude have occurred in lake clays under each of the following circumstances: (1) where lake clays are interbedded with or are overlain by granular deposits, and (2) where lake clays overlie bedrock at shallow

TABLE 2.1 Soil Classification by Origins and Distribution of Principal Soil Deposits in the United States

Origin of Principal Soil Deposits	Symbol for Area in Figure 2.1	Physiographic Province	Physiographic Features	Characteristic Soil Deposits
Alluvial	A1	Coastal plain	Terraced or belted coastal plain with submerged border on Atlantic; marine plain with sinks, swamps, and sand hills in Florida	Marine and continental alluvium thickening seaward; organic soils on coast; broad clay belts west of Mississippi; calcareous sediments on soft and cavitated limestone in Florida
Alluvial	A2	Mississippi alluvial plain	River floodplain and delta	Recent alluvium, fine grained and organic in low areas, overlying clays of coastal plain
Alluvial	A3	High Plains section of Great Plains province	Broad intervalley remnants of smooth fluvial plains	Outwash mantle of silt, sand, silty clay, lesser gravels, underlain by soft shale, sandstone, and marls
Alluvial	A4	Basin and range province	Isolated ranges of dissected block mountains separated by desert plains	Desert plains formed principally of alluvial fans of coarse-grained soils merging to playa lake deposits; numerous nonsoil areas
Alluvial	A5	Major lakes of basin and range province	Intermontane Pleistocene lakes in Utah and Nevada. Salton Basin in California	Lacustrine silts and clays with beach sands on periphery; widespread sand areas in Salton basin
Alluvial	A6	Valleys and basins of Pacific border province	Intermontane lowlands, Central Valley, Los Angeles Basin, Willamette valley	Soils weathered in place from metamorphic and intrusive rocks (except red shale and sandstone in New Jersey); generally more clayey at surface
Residual	R1	Piedmont province	Dissected peneplain with moderate relief. Ridges on stronger rocks	Soil weathered in place from metamorphic and intrusive rocks (except red shale and sandstone in New Jersey). Generally more clayey at surface
Residual	R2	Valley and ridge province	Folded strong and weak strata forming successive ridges and valleys	Soils in valleys weathered from shale, sandstone, and limestone; soil thin or absent on ridges
Residual	R3	Interior low plateaus and Appalachian plateaus	Mature, dissected plateaus of moderate relief	Soils weathered in place from shale, sandstone, and limestone
Residual	R4	Ozark plateau, Ouachita province,	Plateaus and plains of moderate relief, folded	Soils weathered in place from sandstone and limestone

TABLE 2.1 *(Continued)*

Origin of Principal Soil Deposits	Symbol for Area in Figure 2.1	Physiographic Province	Physiographic Features	Characteristic Soil Deposits
		portions of Great Plains and central lowland, Wisconsin driftless section	strong and weak strata in Arkansas	predominantly, and shales secondarily; numerous nonsoil areas in Arkansas
Residual	R5	Northern and western sections of Great Plains province	Old plateau, terrace lands, and Rocky Mountain piedmont	Soils weathered in place from shale, sandstone, and limestone including areas of clay-shales in Montana, South Dakota, Colorado
Residual	R6	Wyoming basin	Elevated plains	Soils weathered in place from shale, sandstone, and limestone
Residual	R7	Colorado plateaus	Dissected plateau of strong relief	Soils weathered in place from sandstone primarily, shale and limestone secondarily
Residual	R8	Columbia plateaus and Pacific border province	High plateaus and piedmont	Soils weathered from extrusive rocks in Columbia plateaus and from shale and sandstone on Pacific border. Includes area of volcanic ash and pumice in central Oregon
Loessial	L1	Portion of coastal plain	Steep bluffs on west limit with incised drainage	30 to 100 ft of loessial silt and sand overlying coastal plain alluvium; loess cover thins eastward
Loessial	L2	Southwest section of central lowland; portions of Great Plains	Broad intervalley remnants of smooth plains	Loessial silty clay, silt, silty fine sand with clayey binder in western areas, calcareous binder in eastern areas
Loessial	L3	Snake River plain of Columbia plateaus	Young lava plateau	Relatively thin cover of loessial silty fine sand overlying fresh lava flows
Loessial	L4	Walla Walla plateau of Columbia plateaus	Rolling plateau with young incised valleys	Loessial silt as thick as 75 ft overlying basalt; incised valleys floored with coarse-grained alluvium
Glacial	G1	New England province	Low peneplain maturely eroded and glaciated	Generally glacial till overlying metamorphic and intrusive rocks, frequent and irregular outcrops; coarse, stratified drift in upper drainage systems; varved silt and clay deposits at

(Continued)

TABLE 2.1 *(Continued)*

Origin of Principal Soil Deposits	Symbol for Area in Figure 2.1	Physiographic Province	Physiographic Features	Characteristic Soil Deposits
				Portland. Boston, New York, Connecticut River Valley, Hackensack area
Glacial	G2	Northern section of Appalachian plateau, Northern section of Central lowland	Mature glaciated plateau in northeast, young till plains in western areas	Generally glacial till overlying sedimentary rocks; coarse stratified drift in drainage system, numerous swamps and marshes in north central section; varved silt and clay deposits at Cleveland, Toledo, Detroit, Chicago, northwestern Minnesota
Glacial	G3	Areas in southern central lowland	Dissected old till plains	Old glacial drift, sorted and unsorted, deeply weathered, overlying sedimentary rocks
Glacial	G4	Western area of northern Rocky Mountains	Deeply dissected mountain uplands with intermontane basins extensively glaciated	Varved clay, silt, and sand in intermontane basins, overlain in part by coarse-grained glacial outwash
Glacial	G5	Puget trough of Pacific border province	River valley system, drowned and glacial	Variety of glacial deposits, generally stratified, ranging from clayey silt to very coarse outwash
Glacial	G6	Alaska peninsula	Folded mountain chains of great relief with intermontane basins extensively glaciated	In valleys and coastal areas widespread deposits of stratifed outwash, moraines, and till; numerous nonsoil areas
		Hawaiian Island group	Coral islands on the west, volcanic islands on the east	Coral islands generally have sand cover; volcanic ash, pumice, and tuff overlie lava flows and cones on volcanic islands; in some areas volcanic deposits are deeply weathered
Nonsoil areas		Principal mountain masses	Mountains, canyons, scablands, badlands	Locations in which soil cover is very thin or has little engineering significance because of rough topography or exposed rock

Source: Hunt (1984a) adapted from *Design Manual DM-7* (1971).

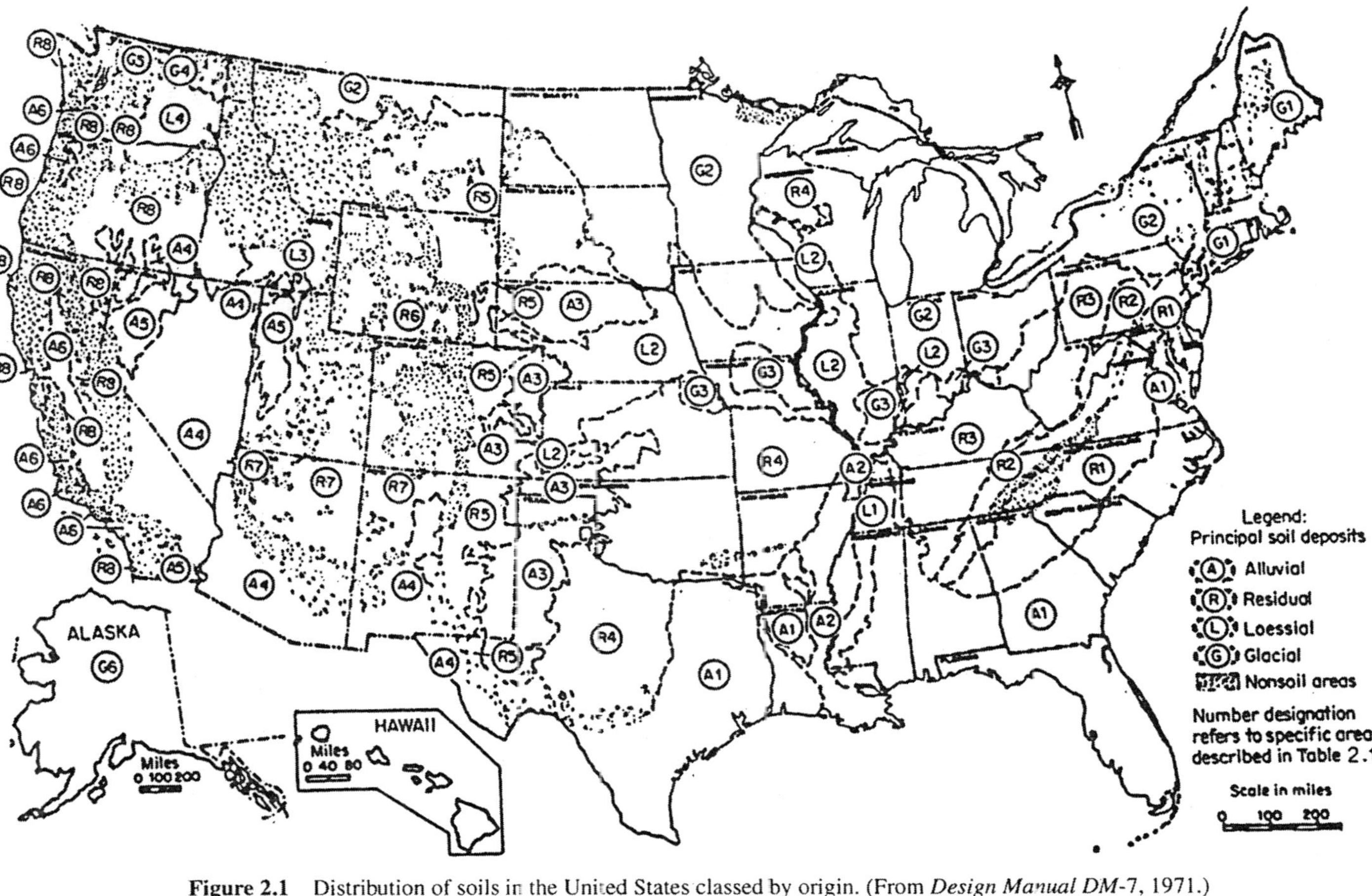

Figure 2.1 Distribution of soils in the United States classed by origin. (From *Design Manual DM-7*, 1971.)

TABLE 2.2 Strength and Sensitivity of Cohesive Soils

Material	Type[a]	Location	γ_d (g/cm^3)	w (%)	LL (%)	PI (%)	S_u (kg/cm^2)	$\bar{c}$ (kg/cm^2)	$\bar{\phi}$	Remarks
Clay Shales (Weathered)										
Carlisle (Cret.)	CH	Nebraska	1.48	18		90		0.5	45	ϕ extremely
Bearpaw (Cret.)	CH	Montana	1.44	32	130			0.35	15	variable
Pierre (Cret.)		South Dakota	1.47	28				0.9	12	
Cucaracha (Cret.)	CH	Panama Canal		12	80	45				$\phi_r = 10°$
Pepper (Cret.)	CH	Waco, Texas		17	80	58		0.4	17	$\phi_r = 7°$
Bear Paw (Cret.)	CH	Saskatchewan		32	115	92		0.4	20	$\phi_r = 8°$
Modelo (Tert.)	CH	Los Angeles	1.44	29	66	31		1.6	22	Intact specimen
Modelo (Tert.)	CH	Los Angeles	1.44	29	66	31		0.32	27	Shear zone
Martinez (Tert.)	CH	Los Angeles	1.66	22	62	38		0.25	26	Shear zone
(Eocene)	CH	Menlo Park, CA	1.65	30	60	50		Free swell 100%: $P = 10$ kg/cm^2		
Residual Soils										
Gneiss	CL	Brazil: buried	1.29	38	40	16		0	40	$e_0 = 1.23$
Gneiss	ML	Brazil: slopes	1.34	22	40	8		0.39	19	c, ϕ–
Gneiss	ML	Brazil: slopes	1.34		40	8		0.28	21	unsoaked
Colluvium										
From shales	CL	West Virginia		28	48	25		0.28	28	$\phi_r = 16°$
From gneiss	CL	Brazil	1.10	26	40	16		0.2	31	$\phi_r = 12°$

					Alluvium					
Back swamp	OH	Louisiana	0.57	140	120	85	0.15			
Back swamp	OH	Louisiana	1.0	85	85	50	0.1			
Back seamp	MH	Georgia	0.96	54	61	22	0.3			$e_0 = 1.7$
Lacustrine	CL	Great Salt Lake	0.78	50	45	20	0.34			
Lacustrine	CL	Canada	1.11	62	33	15	0.25			
Lacustrine (volcanic)	CH	Mexico City	0.29	300	410	260	0.4			$e_0 = 7, S_r = 13$
Estuarine	CH	Thames River	0.78	90	115	85	0.15			
Estuarine	CH	Bangkok		130	118	75	0.05			
Estuarine	MH	Maine		80	60	30	0.2			
					Marine Soils (Other than Estuarine)					
Offshore	MH	Santa Barbara, CA	0.83	80	83	44	0.15			$e_0 = 2.28$
Offshore	CH	New Jersey		95	95	60	0.65			
Offshore	CH	San Diego	0.58	125	111	64	0.1			Depth = 2 m
Offshore	CH	Gulf of Maine	0.58	163	124	78	0.05			
Coastal plain	CH	Texas (Beaumont)	1.39	29	81	55	1.0	0.2	16	$\phi_r = 14, e_0 = 0.8$
Coastal plain	CH	London	1.50	25	80	55	2.0			
					Loess					
Silty	ML	Nebraska-Kansas	1.23	9	30	8		0.6	32	Natural $w\%$
Silty	ML	Nebraska-Kansas	1.23	(35)	30	8		0	23	Prewetted
Clayey	CL	Nebraska-Kansas	1.25	9	37	17		2.0	30	Natural $w\%$

(Continued)

TABLE 2.2 *(Continued)*

Material	Type[a]	Location	γ_d (g/cm^3)	w (%)	LL (%)	PI (%)	S_u (kg/cm^2)	$\bar{c}$ (kg/cm^2)	$\bar{\phi}$	Remarks
				Glacial Soils						
Till	CL	Chicago	2.12	23	37	21	3.5			$e_0 = 0.6$ (OC)
Lacustrine (varved)	CL	Chicago	1.69	22	30	15	1.0			$e_0 = 1.2$ (NC)
Lacustrine (varved)	CL	Chicago		24	30	13	0.1			
Lacustrine (varved)	CH	Chicago	1.18	50	54	30	0.1			$S_t = 4$
Lacustrine (varved)	CH	Ohio	0.96	46	58	31	0.6			$e_0 = 1.3$ (clay)
Lacustrine (varved)	CH	Detroit	1.20	46	55	30	0.8			$e_0 = 1.25$ (clay)
Lacustrine (varved)	CH	New York City		46	62	34	1.0			$S_t = 3$
Lacustrine (varved)	CL	Boston	1.35	38	50	26	0.8			$\phi_r = 3°$
Marine[b]	CH	Seattle		30	55	22			30	$S_t = 128$
Marine[b]	CH	Canada-Leda clay	0.89	80	60	32	0.5			$S_t = 7$
Marine[b]	CL	Norway	1.34	40	38	15	0.13			$S_t = 75$
	CL	Norway	1.29	43	28	15	0.05			

Source: Hunt, 1984.

[a]Permission being sought.

[b]Permission being sought.

depth and the base level of erosion of the area is generally lowered (Transportation Research Board, 1978).

2.2.2 Glacial Deposits

Glacial deposits were transported by glaciers, whose action resembles a giant bulldozer. Materials are both pushed forward with droppings on the side and crushed when overridden by the glacier. Warm temperatures provide general thawing of the glaciers, which stops their forward movement and permits settling out or further movement of suspended rock particles by flowing water. Glacial deposits may vary in grain size composition from boulders to clays. Glaciers often produce a disordered landscape, with inhibited drainage and the development of bogs. These soils are commonly found in the northeastern and eastern United States.

There are four types of glacial deposits, namely (1) glacial drift, (2) till, (3) glaciofluvial deposits, and (4) glaciolacustrine deposits. Glacial drift is rock debris that has been transported by glaciers and deposited either directly from the ice or from the meltwater. Till is unsorted, unstratified, unconsolidated, heterogeneous material deposited directly from the ice (ice has no sorting power) and generally consists of clay, silt, sand-gravel, and boulders, interbedded in varying proportions. Till is usually dense to very dense and is of high strength and low compressibility. Standard Penetration Test (SPT) values tend to be high because of boulders and gravels. The high values are not a reliable indicator of dense till.

Glaciofluvial deposits are materials moved by glaciers and subsequently sorted and deposited by streams flowing from the melted ice. The deposits are usually stratified, and they may occur in the form of outwash plains, deltas, and terraces, and they tend to vary from loose to medium compact.

Glaciolacustrine deposits consist of lacustrine soils derived from outwashes that are deposited in glacial lakes formed during previous periods of glaciation. The typical infolding of glacial lakes consists of varved clays (or alternating thin layers of clay and silt) with occasional sand seams or partings, as shown in Figure 2.2. The varves can range from several millimeters to several centimeters in thickness as found in Connecticut and New Jersey. Varved clays are either normally or overly consolidated, depending on the process of deposition. As for engineering properties, glaciolacustrine deposits have low activity and low sensitivity. Some data on the strength and sensitivity of varved clays from a number of U.S. cities are listed in Table 2.2.

Glacial deposits vary greatly in terms of composition and engineering properties. Soils for earthwork and foundation engineering should be thoroughly explored and their properties tested as comprehensively as possible.

2.2.3 Eolian Deposits

Eolian deposits are transported by winds. They range from sand dunes to loess deposits whose particles are predominantly of silt size with a certain amount of fine sand and aggregated clay particles. Typical engineering properties are shown in Table 2.2.

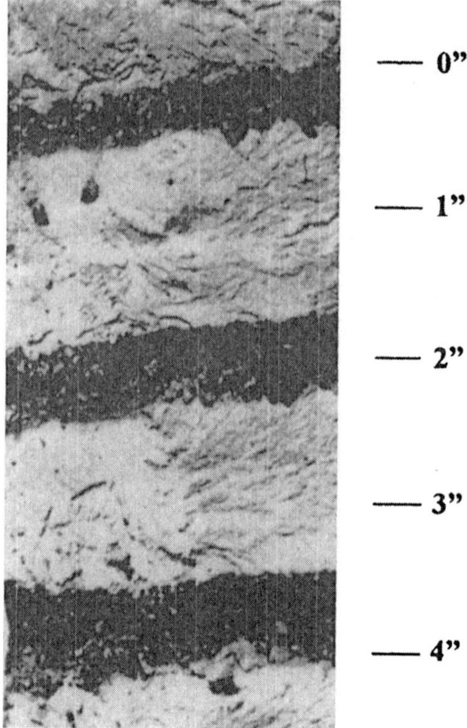

Figure 2.2 Sample of varved clayey silt.

Dune sand deposits are recognizable as low elongated or crescent-shaped hills, with a flat slope windward and steep slope leeward of the prevailing winds. Usually, these deposits have very little vegetation cover. The material is very rich in quartz and its characteristics typically consist of a limited grain size range, usually fine- or medium-grained sand; no cohesive strength; moderately high permeability; and moderate compressibility. In the United States, dunes are common to Nebraska, Kansas, Iowa, Mississippi, Indiana, and Idaho.

Loess deposits cover extensive areas in the temperate zone plains regions. Loess consists mainly of angular particles of silt and/or fine sand. Although of low density, the naturally dry loessial soils have a fairly high strength because of the clay binder. However, they are easily eroded when close to the groundwater table. When the soil is flooded or rained on, collapse of the soil structure and large settlements can occur. Protection of the foot of a cut against saturation during heavy rainstorms is therefore important. Otherwise, landslides may occur as triggered by erosion at the toe of the slope. Cuts in loess are often set back more than usual from the roadway to prevent damage or interruption of traffic due to local failure.

Slope failures in loess (e.g., bank erosion) are attributed to the removal of the binding material by seepage below the groundwater table. The binding material, frequently calcium carbonate, can be dissolved within a few weeks or months. When this happens, the loess assumes the character of a supersaturated rock flour that flows like molasses (Terzaghi, 1950). Consequently, subsurface tunnels are formed within the soil mass, which eventually lead to ultimate collapse of overburden soils (Figure 2.3) For engineering work in urban development on loess slopes, it is important to know the potential and possible behavior of the upper 5 to 10 feet of the soil because disruption to utilities, slabs, pavements, shallow foundations, and so on, can occur.

Eolian deposits are normally regarded with suspicion, especially as foundations for earth embankments. Such deposits should be avoided if practicable to do so; however, they can be used when properly explored and evaluated. Information on the in-place density of eolian soils is of vital importance in planning their usefulness for foundations of structures (Bureau of Reclamation, 1968).

2.2.4 Residual Deposits

Residual deposits are formed in place by mechanical and chemical weathering of their parental bedrocks. Figure 2.4 depicts a typical form of the deep tropical weathering profile (Little, 1969). Some of the residual soil engineering properties are shown in Table 2.2. These soils are found over much of the eastern part of the United States east of the Appalachian Mountains, in the southeastern United States, and in other tropical areas such as Hawaii, Guam, and Asia.

Laterites and saprolites are two types of residual soils. The laterites are typically bright red to reddish brown soils, which are formed initially by weathering of igneous rocks with subsequent leaching and chemical erosion due to high temperature and rainfall. The colloidal silica is leached downward, leaving behind aluminum and iron, which become highly oxidized (hence the red color) and are relatively insoluble in the high pH environment (greater than 7). Well-developed alteration soils are generally porous and relatively incompressible. Saprolites are zones consisting of completely weathered or highly weathered bedrock that contain soil-like materials but retain the original relic rock structure.

The type of parent bedrock has a very pronounced influence on the character of the residual soils. The rock type should always be determined in assembling data for the appraisal of a residual soil deposit.

The stability of slopes in residual soils is very difficult to predict on the basis of field or laboratory tests. The engineering properties of such soils can appreciably differ from those of sedimentary soils with the same composition and grain size distribution. Their properties can vary considerably laterally and with depth due to differential weathering patterns. A residual soil deposit will frequently contain unweathered boulders (corestones) and unweathered rock layers.

Landslides are common in residual soils, particularly during periods of intense rainfall. The rainfall-induced landslides are usually shallow in nature. The well-developed internal drainage of alteration and allophanic soils is conducive to water infiltration, subsequent reduction in pore-water tension, and consequent

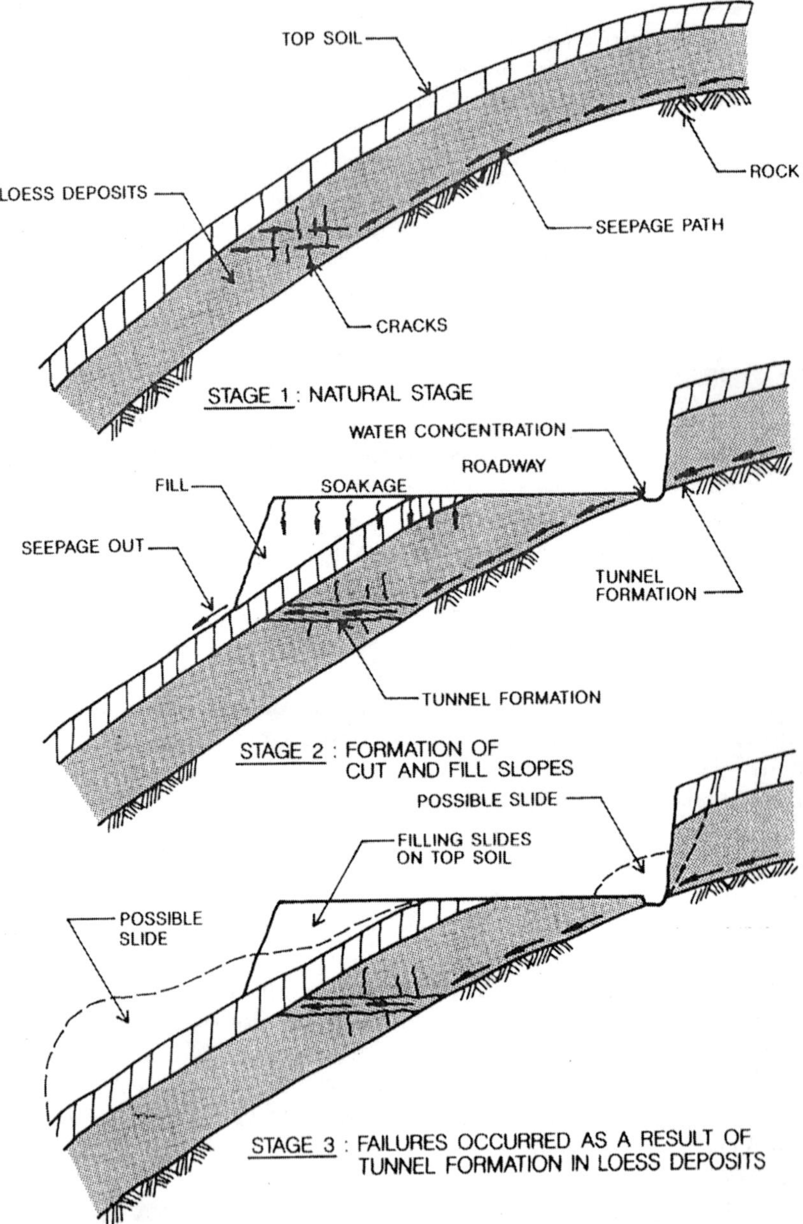

Figure 2.3 Schematic diagrams showing landslides as a result of tunnel formation in loess deposits. (Evans, 1977.)

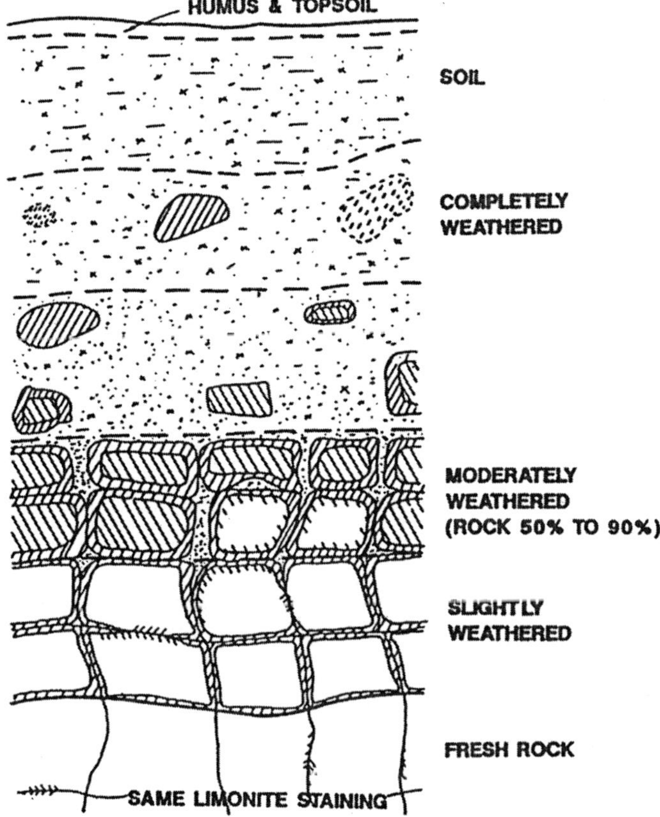

Figure 2.4 Schematic diagram of typical residual soil profile. (Little, 1969.)

sliding. The role of soil suction in these rainfall-induced landslides is still under examination.

2.2.5 Colluvial/Talus Deposits

Colluvial/talus deposits are residual soils formed by weathering of parent soil and rock materials that moved downslope by gravity on steep slopes (Figure 2.5). These materials are easily identified on air photos as bare slopes in mountainous areas, but they are not obvious on vegetated lower slopes.

Colluvial/talus deposits are often loose and unconsolidated. An unstable condition can exist when colluvial/talus deposits rest on slopes, and further slope movements are likely in such instances. Slope movements before total failures range from the barely perceptible movements of creep to the more discernible movements of several inches per week (Hunt, 1984a). The natural causes of these movements are weathering, rainfall, snow and ice melt, earthquake-induced vibrations, and changing water levels as a result of floods or tides. Cuts made in colluvial or talus slopes are expected to

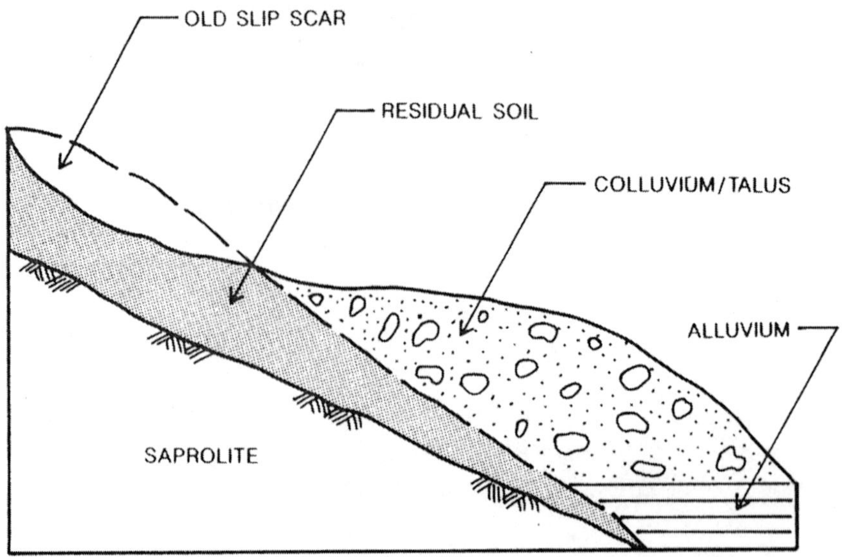

Figure 2.5 Colluvial/talus deposits moving downslope.

become less stable with time and usually lead to failure unless retained or removed. Table 2.2 shows the engineering properties of some of the colluvial deposits found in West Virginia and Brazil.

2.2.6 Marine Deposits

Marine deposits originate from two general sources: (1) terrestrial sediments from rivers, glaciers, wind action, and slope failures along the shoreline, and (2) marine deposition from organic and inorganic remains of dead marine life and by precipitation from oversaturated solutions.

Marine deposits consist of sands, silts, and clays. Marine sands are normally composed of quartz grains, which are hard and virtually indestructible, although the deposit may be compressible because of the loose arrangement of the quartz particles. Marine clays are usually normally consolidated and soft in nature, but at depths below the sea floor of about 300 feet or more, they are often stiff to very stiff in consistency. Table 2.2 shows the engineering properties of some of the marine deposits in the United States and elsewhere. Marine clays tend to flocculate and settle quickly to the bottom, and hence tend to be devoid of laminations and stratifications.

Marine clays from glacial runoff deposited in marine estuaries along coastlines are called glacial-marine clays. They are often of high sensitivity as a result of leaching. Sensitivity tends to increase with time as groundwater continues to leach the salt, slowly weakening the deposit until it can no longer retain its natural slope and a failure results often in the form of a flow.

Highway embankments are sometimes constructed over marine deposits. Care must be exercised to avoid formation of mud waves during construction. Both short- and long-term stability analyses of embankments constructed on marine clays should be carried out to evaluate the embankment stability.

2.2.7 Melanges

Melange is a French word for "mixture" and is defined in the Dictionary of Geological Terms by Bates and Jackson (1984) as

> A mappable body of rock that includes blocks of all sizes, both exotic and native, embedded in a fragmented and generally sheared matrix.

In other words, melanges are rock masses composed of competent rock blocks of varying sizes, embodied in a weaker soil matrix. Medley and Goodman (1994) have coined the term "bimrocks" for such block-in-matrix rocks.

Melanges are found in the mountains of over 60 countries around the world. They are exemplified by the Franciscan Assemblage (the Franciscan)—a regional scale jumble in northern California—as well as by the argille scagliose of the Italian Apennines and the extensive melanges of Turkey and Iran. The blocks of melanges in northern California usually vary from being lenticular to tabular. The ratio of maximum axial dimension to minimum block size is about 2 to 3 (Medley and Goodman, 1994).

Because of their heterogeneity and complex nature, melanges are difficult geotechnical materials with which to deal in analyzing slope stability. For example, it is very difficult, if not impossible, to obtain an "undisturbed" melange sample from conventional drilling. This is because the drilling resistances of the harder and softer materials are so different that the harder materials may gouge into the weaker materials, resulting in significant sample disturbance. In most cases, the weaker materials are lost during sampling. Even if one were able to retrieve an undisturbed sample, it would be improbable that the sample would be representative of the melange mass of interest. Hence, large in situ test samples are the most common approach to finding and testing a representative sample of melange.

When working with melanges, it is a general practice to assume the strength of the weaker matrix. But this may be overly conservative, as it neglects the possibility that the blocks strengthen and stiffen a melange mass. Studies by Volpe et al. (1991) suggested that the strength of a melange mass could be represented by the weighted average of the strength of the weaker matrix and stronger blocks based on their volumetric proportions. However, no theoretical basis was provided for this strength model. Recent studies on physical model melanges by Lindquist (1994) indicate that the strength and deformation properties of melanges are enhanced by increasing the block proportions. This might explain why some of the "melange" slopes can stand at very steep angles (more than 45°) without suffering any instability.

Figure 2.6 Terracettes on hilltop. (Beavis, 1985.)

2.2.8 Other Types of Deposits

Other types of deposits include volcanic clays and soils that are derived from the deposition of new minerals within a primary soil formation as a result of hardening. Three broad groups are considered:

- Volcanic clays
- Duricrusts, where the primary formation is hardened by the inclusion of iron, aluminium, carbonate, or silica
- Permafrost, where the formation is hardened ice

Volcanic Clays Volcanic clays are deposited as a result of volcanic ash and dust being thrown into the atmosphere during volcanic eruptions. The ash and dust are often altered by weathering processes into montmorillonite clay, which is found in most states west of the Mississippi, as well as in Tennessee, Kentucky, and Alabama.

Ash and dust produced by the volcanic activities often modify the landscape. Potentially unstable hill slopes are marked by terracettes (Figure 2.6), by mudflows (Figure 2.7), and by landslides. Any construction activity on these slopes will tend to lead to renewed movements. Areas of so-called reactive soils are occupied by soils that are expansive and unsaturated in arid and semiarid areas. The soils tend to expand upon wetting and shrink upon drying, thus setting up process-induced stresses within the soil mass. These stresses may produce damage to, or even failure of, the slopes.

Duricrusts Duricrusts are highly indurated zones within a soil formation, often of rocklike consistency, and include either laterite ironstone or ferrocrete (iron-rich), bauxite (alumina-rich), calcite or caliche (lime-rich), and silcrete (silica-rich)

As discussed earlier, laterites extend over very large areas in tropical regions. In the advanced state of induration, laterites are extremely stable and durable, resist chemical change, and will not soften when wet. Ironstone is a form of limonite found in some

Figure 2.7 Mudflow on hilltop. (Beavis, 1985.)

coastal plain formations as discontinuous beds usually from 3 to 7 feet thick. In the Atlantic Highlands of New Jersey, it caps the hills, making them very resistant to erosion. Bauxite is an alumina-rich variety of alteration soil with a smaller area and wider distribution than the iron forms, and is of much less engineering significance. Caliche (calcite) is common in hot and semiarid regions such as Texas, Arizona, and New Mexico. It is usually variable in form, hard and rocklike in many areas and soft in other areas. Therefore, it is a highly questionable material with respect to slope stability.

Permafrost Permafrost is the soil that remains permanently frozen. The soil formation is hardened by ice. The unconfined compressive strength of ice commonly varies from 21 to 76 tons per square foot, depending on the ice temperature (Terzaghi, 1952). Ice has a tendency to creep, especially under high loads (i.e., greater than 2 tons per square foot).

Solifluction is the term used to describe downslope movements resulting from the freezing and thawing of silty soils. This phenomenon is most common between the southern boundary of seasonal frost (the 5°C mean annual temperature isotherm) and the southern boundary of the permafrost region. At the foot of slopes subject to solifluction, the soil strata may be intricately folded to a depth of more than 10 feet (Hunt, 1984a).

2.3 TYPES AND CHARACTERISTICS OF ROCKS

Rocks are made up of small crystalline units known as minerals. They form the greater part of the crust of the Earth. The engineering properties of rocks vary depending on their origins. Three broad rock groups are present, based on their origins:

(1) Sedimentary rocks are mainly formed from the breakdown products of older rocks, the fragments of which have been sorted by water or wind and built up into deposits of sediment (e. g., shale, sandstone). Some sedimentary rocks are formed by chemical deposition.

(2) Igneous rocks are derived from hot materials that originated below the Earth's surface and solidified at or near the surface (e.g., granite, basalt).

(3) Metamorphic rocks are derived from earlier igneous or sedimentary rocks through transformation by means of heat or pressure so as to acquire conspicuous new characteristics (e.g., slate, schist, gneiss).

Rock has no formal classification equivalent to that for soils. Bedding, faulting, jointing, and weathering in rock have a pronounced effect on its behaviors. Listed below are common rocks that have an engineering significance on slope stability.

2.3.1 Shales

Two types of shales are known to exist: the laminated shale, which are fine-grained sedimentary rocks having a fissle, laminated, or thinly stratified structure, and non-laminated argillaceous shales, which are formed by the solidification of clays and silts through the process of compaction, or compaction and cementation. Depending on the character and degree of solidification, shales vary widely in their behavior to exposure and to stress and strain, and there is no sharp demarcation line between the compaction type and the cemented type. Those that have been formed by compaction alone revert to the original muds from which they were formed when subjected to drying and wetting. The rate at which they undergo this disintegration is a function of the extent of drying, the degree of compaction, and the nature and fineness of the constituent particles (Burwell and Moneymaker, 1950).

Low-grade clay-shales of the compaction type may completely disintegrate after several cycles of drying and wetting. On the other hand, high-grade, coarser-textured, compaction shales of the coal measures of Pennsylvania, West Virginia, and Ohio may require many cycles of drying and wetting before total disintegration. It is therefore important to recognize the types of shale at hand, especially their character and degree of solidification, before making any engineering assessment of slopes composed of compaction type shales.

Some shales have very high strengths compared to good concrete and are of low permeability, both of which would enhance the slope stability. Other shales behave more like soils than rocks, and introduce some of the most troublesome problems in engineering constructions—problems of consolidation under load, of rebound upon unloading, of rapid deterioration under alternate drying and wetting, of slides from subvertical valley walls and cut slopes, and of shear and sliding failures.

The well-cemented shales are usually regarded as being stable, competent rocks from a slope stability point of view. The problems that may arise from the well-cemented shales are conditions where the shale strikes with the course of the river and dips toward the river in the cut slope. Under such conditions, weak bedding planes may cause slides if the dips are flatter than the valley walls, or if construction cuts have slopes steeper than the dip. In addition, there is always a possibility that the development of hydrostatic pressure along the bedding planes may endanger the slope near the river. The problem becomes more acute if the cuts are in tilted formation and in formations containing bentonitic materials.

There are no precise methods to analyze the stability of shale cuts on the basis of test data. Most generally, the problem of design becomes one of thorough exploration and comparison with existing slopes in similar materials. Frequent helpful information can be obtained by observing and studying the effect of stream trenching, seepage, and existing loads on the local materials. Sometimes, existing highway or railroad cuts can shed some light on the problem.

2.3.2 Sandstones

Sandstones are classified according to the size of their granular constituents as fine-, medium-, or coarse-grained, and according to the nature of their cementing or matrix materials as siliceous, calcareous, ferruginous, or argillaceous sandstones. Like shales, sandstones have a wide range of strength and durability characteristics, although their granular nature, high angle of internal friction, and their cemented condition eliminate many of the problems associated with shales. The problems related to slaking are rare in sandstones. However, sandstones are commonly interbedded with shales, and this association often affects their engineering properties. In some cases, it even accentuates the undesirable properties of the shale by permitting access of water to the shale-sandstone contacts. Contact seepage may lubricate weak shales and cause slides in many natural and cut slopes. It may also build up excess hydrostatic pressures along the relatively impervious beds of shales and appreciably reduce its resistance to sliding.

A major problem in slope stability associated with sandstones results from the fact that they are cut by many joint fractures, particularly if they have been subjected to sliding. Even virtually horizontally bedded sandstones are troublesome because of this condition, especially near valley walls where the influence of weathering, creeping, and elastic rebound of associated shales results in the opening of fractures and subsequent deep-seated movements toward the river.

2.3.3 Limestones and Related Carbonate Rocks

Limestones free from solution cavities offer good materials for slopes. However, bedding planes separated by layers of clay or soft shale, and those inclined downstream may, under certain conditions, often serve as sliding planes and result in slope failures. Fault zones, particularly large ones, in limestones involve problems of watertightness, shearing strength, and stability. All limestones in humid regions are to some degree cavernous, regardless of the degree of their chemical purity, the complexity of geologic structures, or the position of the groundwater table. Solution cavities often present sliding problems.

2.3.4 Igneous Rocks

Igneous rocks such as granite, dolerite, and gabbro, when unweathered, are often very strong and exhibit extremely high shear strength for any engineering construction purposes. Hydrothermally altered rock, however, may be soft and weak even though it is unweathered. Igneous rocks, if hydrothermally altered along shear zones, would pose sliding problems for slopes. Joints, faults, and shear zones often permit the entry of water and air into the rock mass, and facilitate weathering within the resistant rock mass. When weathering becomes so advanced, sliding along the weathered zones would ensue.

2.3.5 Pyrocrastic Volcanic Rocks

Pyrocrastic volcanic rocks were formed from the fragmentary materials blown from the vents of volcanoes. Like shales and sandstones, they offer a variety of strength, permeability, and behavior under conditions of exposure. Like the shales, their engineering properties depend on the degree of solidification. Many agglomerates are so well cemented that they are a good slope-forming material. The major problems of these rocks is the rapid disintegration upon exposure to drying and wetting. It is, therefore, essential to protect these materials from serious deterioration upon exposure during construction. From the viewpoint of sliding tendencies, tuffs are among the worst offenders. The clay mineral montmorillonite is not an uncommon constituent of tuffs. Its presence should always be regarded as a danger signal. Slopes that contain tuffaceous beds daylighting toward the river are usually of questionable stability and must be thoroughly investigated for potential slides.

2.3.6 Metamorphic Rocks

Metamorphic rocks are formed as a result of the metamorphism of igneous and sedimentary rocks. Through the process of alteration, these rocks have undergone textural and mineralogical changes, or both, so that their primary characters are altered or even lost. Consequently, metamorphic rocks exhibit a wide range of engineering characteristics. The most common groups are schist, gneiss, slate, phyllite, and marble.

The metamorphic rocks vary considerably in terms of slope stability. Foliation in gneiss or schist renders the rock weaker and more susceptible to weathering and decay. Some metamorphic rocks are relatively incompetent after being deformed by faulting and folding. These structures, along with joints, facilitate rapid decay into weak rocks. Rocks such as quartzite and gneiss are relatively strong if they are not deformed or deeply decayed. Some mica schists are very weak physically. Upon weathering, they become slippery and fail under light loads. Marble, the metamorphic derivative of limestone, has the same advantages and disadvantages as other carbonate rocks in terms of slope stability.

2.4 GEOLOGICAL FEATURES ASSOCIATED WITH SLOPES

Before any method of slope stability analysis is to be performed, due consideration must be given to significant geologic features associated with the slope under investigation. Sometimes, when it is not economically feasible to conduct detailed investigation, simple methods of analysis may suffice in the light of geotechnical as well as engineering geological experience and precedent in the area. Geologic awareness is thus of great value in planning site investigation and laboratory testing programs. The role of geology in relation to slope stability has been emphasized by Terzaghi (1950), Terzaghi and Peck (1967), and others.

Basic geologic features associated with slope stability are (1) soil/rock fabric, (2) geological structures, (3) discontinuities, (4) groundwater, (5) ground stresses, (6) weathering, (7) preexisting landslide activities, (8) clay mineralogy, and (9) seismic effects.

2.4.1 Soil/Rock Fabric

Soil and rock have their own individual fabric. Mineral fabric may be sufficiently developed in some rocks to influence their engineering characteristics and properties. Examples of this are the fabrics of schists, slates, shales, and laminated clays, which can cause marked anisotropy in their strength and deformation characteristics. In some cases, the decay of the mineral fabric could result in complete loss of strength upon disturbance (e.g., quick clays). Both micro and macro fabric (e.g., major joints and bedding) are important features related to slopes. Details such as mineral orientation, stratification, fracture, faulting, shear zones, and joints must be gathered and assembled for a meaningful slope stability model to be developed.

2.4.2 Geological Structures

Geological structures of the slope-forming materials are a dominant feature in the slope behavior. For example, the succession, thickness, and attitude of beds are of direct relevance to consideration of potential instability in sedimentary rocks. These structures play an important role in understanding slope development processes, formation of valleys, ridges, escarpments, and the development of residual soils, talus, and colluvial deposits. Other major and minor structural discontinuities, such as faults, folds, and joints, must also be carefully studied and mapped. Their engineering significances are often discussed in great details in books on engineering geology. In order to predict slope stability accurately, it is essential to recognize features such as a sequence of weak and strong beds, thin marker beds, old failure surfaces, fault zones, and hydro geological effects.

Sequence of Weak and Strong Beds Weak rocks like shales and claystones, which may be slickensided due to alteration, are often critical in the development of slope instability.

Thin Marker Beds Such beds may consist of coal, bentonite, or clay seams, or carbonaceous shale. These beds are easily missed during routine investigations.

Old Failure Surfaces or Shear Planes These surfaces and planes often lead to slope instability if reactivated by weathering process, human activities such as overfilling, or undercutting by rivers and streams.

Fault-Related Geologic Features Features such as the presence of ground-up materials near fault zones are often remolded, which may result in less resistance to sliding because of loss in shear strength. Groundwater is attracted to a fault zone due to the greater conductivity of the fractured and loosened rock to be found in the fault zone. Faults can act as conduits for flow of water, which explains why rocks adjacent to them are often found to be hydrothermally altered. Replacement of original minerals by clays, zeolites, and silica or calcite, as well as precipitation of these minerals in void spaces, grossly changes the character of the rocks near the fault zones, as a result of which stability problems would ensue.

In the case of soluble rocks, a fault usually localizes solution in them, leading to planar caverns developed along the fault zone. Caverns may develop along faults even in nonsoluble rocks as a result of washing out of gouge and crushed rock, and the opening of extension fractures oblique to the fault plane as a by-product of movement along the fault (Goodman, 1992).

Hydrogeologic Features Such a feature might be a competent water bearing stratum, for example, sandstone, limestone, gravel, or sand, overlying a much less permeable cohesive soil or a highly weathered shale. The excess water in the previous stratum contributes to slope failure in a number of ways—by creating high pore pressures in the materials forming the slope, by a washout of fines at the interface of the two layers of contrasting permeability, and by softening the underlying cohesive soils that make up the lower slope (Watson, 1984)

2.4.3 Discontinuities

A steep hard rock slope can be stable with a height of thousands of feet if it is free of discontinuities. The importance of discontinuities, water pressures (so called cleft pressures) in discontinuities, and water seepage forces within a rock mass must not be overlooked (Terzaghi, 1960). Assuming no seepage forces, the orientation of a slope with respect to significant discontinuities would affect the potential relative movements. Figure 2.8 shows possible relationships between slope stability and the discontinuities (often measured in terms of strike or dip directions and dip) as discussed by Terzaghi (1962). The angle of friction along discontinuities depends on factors such as roughness, waiveness, weathered seams, infills, and continuity of joints. Gouge material separating two rock surfaces may have very low strength to resist potential instability. The degree to which infill material controls stability depends on its thickness, extent, strength, and the proportion of the slip surface that passes through it.

2.4.4 Groundwater

The way groundwater flows, its pressure and gradient at any point within a slope depend on the local geology. Water plays a very important role in slope stability, which will be discussed at length in subsequent chapters. In brief terms, water can influence the strength of slope-forming materials by (1) chemical and hydrothermal alteration and solution, (2) increase in pore water pressures, and subsequent decrease in shear strength, (3) reduction of apparent cohesion due to capillary forces (soil suction) upon saturation, and (4) softening of stiff fissured clays and shales.

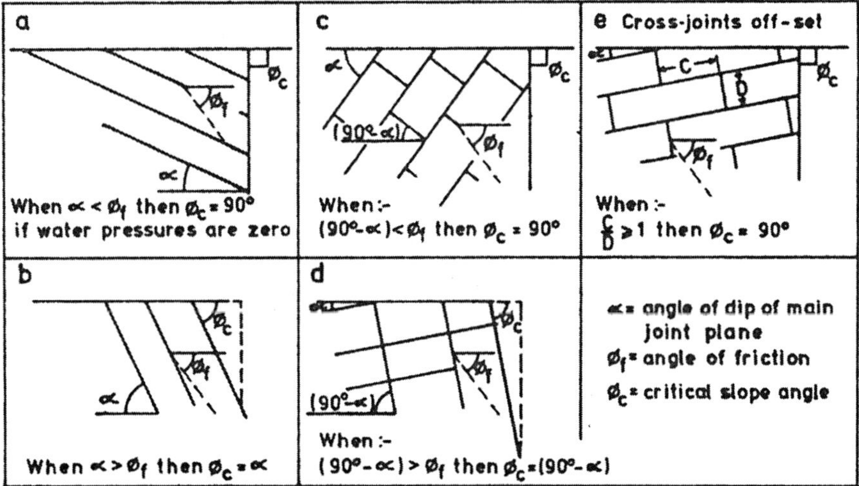

Figure 2.8 Effect of structure on rock slope stability. (From Terzaghi, 1962, reproduced by permission of the Institute of Civil Engineers.)

2.4.5 Ground Stresses

All slope-forming materials are subject to initial stresses as a result of gravitational loading, tectonic activity, weathering, erosion, and other processes. Stresses produced by these processes are embodied in the materials themselves, remaining there after the stimulus that generated them has been removed. For this reason, they are called residual stresses.

Many structural features are produced by stress relief. Stress release activity is an important feature in many rock formations. High lateral stresses have played a crucial role in initiating landslides in overconsolidated clays and clay-shales. The importance of the overconsolidated clays and clay-shales in landslides has been discussed at length by Bjerrum (1967), and that of hard rocks by Bjerrum and Jorstad (1968).

The stress history of slope-forming materials is very important and is often used to differentiate normally consolidated soils (i.e., normally loaded soils that have never been subject to an effective overburden stress greater than their present overburden pressure) from overconsolidated soils (i.e., soils that have been precompressed by an overburden effective stress greater than their present one). There are significant differences in terms of engineering properties between these two types of soils. This aspect will be discussed in detail in the subsequent chapters.

2.4.6 Weathering

There are two types of weathering—one is the chemical weathering due to chemical changes; the other is the mechanical weathering as a result of wind, temperature

TABLE 2.3 Common Chemical Processes

Process	Description
Solution	Dissociation of minerals into ions, greatly aided by the presence of CO_2 in the soil profile, which forms carbonic acid (H_2CO_3) with percolating rainwater
Oxidation	The combination of oxygen with a mineral to form oxides and hydroxides or any other reaction in which the oxidation number of the oxidized elements is increased
Reduction	The release of oxygen from a mineral to its surrounding environment: ions leave the mineral structure as the oxidation number of the reduced elements is decreased
Hydration	Absorption of water molecules into the mineral structure; note that this normally results in expansion, some clays expand as much as 60%, and by admitting water hasten the processes of solution, oxidation, reduction and hydrolysis
Hydrolysis	Hydrogen ions in percolating water replace mineral cations: no oxidation–reduction occurs
Leaching	The migration of ions produced by the above processes; note that the mobility of ions depends upon their ionic potential: Ca, Mg, Na, K are easily leached by moving water, Fe is more resistant, Si is difficult to leach, and Al is almost immobile
Cation exchange	Absorption onto the surface of negatively charged clay of positively charged cations in solution, especially Ca, H, K, Mg

Source: Blythe and de Freitas (1984).

changes, freeze-thaw cycles, and erosion by streams and rivers. The processes involved in chemical weathering and mechanical weathering are listed, respectively, in Tables 2.3 and 2.4. The rate of chemical weathering ranges from a few days to many years and can affect both the short- and long-term stability of slopes (Blyth and de Freitas, 1984). On the other hand, mechanical weathering may take years before it has any adverse effect on slopes.

Chemical weathering is the breakdown of minerals into new compounds by the action of chemical agents; acids in the air, in rain, and in river water. Destruction of diagenetic bonding by chemical weathering in overconsolidated clays and clay-shales has been discussed at length by Bjerrum (1967) in his classic state-of-the-art paper on progressive failure in slope in overconsolidated plastic clays and clay-shales. Weathering of marl could increase its liquid limit and natural moisture content, and decrease its bulk density, permeability, and shear strength (Chandler, 1969). Chemical weathering depends not only on the slope materials, but also on other factors such as climate, drainage characteristics, and geological details.

Mechanical weathering is a process by which rock is broken down into smaller fragments as a result of energy developed by physical forces, for example, freeze-thaw cycles and temperature changes. When water freezes in a fractured rock, enormous energy will build up as a result of expansion of the frozen water, which may split off

TABLE 2.4 The Processes of Mechanical Weathering

Process	Description
Mechanical unloading	Vertical expansion due to the reduction of vertical load by erosion; this will open existing fractures and may permit the creation of new fractures
Mechanical loading	Impact on rock, and abrasion, by sand and silt size windborne particles in deserts; impact on soil and weak rocks by rain drops during intense rainfall storms
Thermal loading	Expansion by the freezing of water in pores and fractures in cold regions, or by the heating of rocks in hot regions; contraction by the cooling of rocks and soils in cold regions
Wetting and drying	Expansion and contraction associated with the repeated absorption and loss of water molecules from mineral surfaces and structures: (see Hydration in Table 2.3)
Crystallization	Expansion of pores and fissures by crystallization within them of minerals that were originally in solution; note that expansion is only severe when crystallization occurs within a confined space
Pnuematic loading	The repeated loading by waves of air trapped at the head of fractures exposed in the wave zone of a sea cliff

Source: Blyth and de Freitas (1984).

pieces of rock. Forest or brush fires may generate heat great enough to break up rocks. Plants can also play a role in mechanical weathering. The roots of trees and shrubs in rock crevices sometimes exert sufficient pressure to dislodge previously loosened fragments of rock (Figure 2.9), much as tree roots heave and crack sidewalk pavement.

Weathering has a pronounced effect on the permeability of the slope-forming materials. Relict geological structures, such as joints, bedding planes, and faults, affect in varying degrees the weathering profiles in slope stability problems concerning residual soils, for example, laterites and saprolites in Guam and Hawaii. Sequences of varying weathering zones (as regards strength and permeability) can develop in igneous and metamorphic rocks. A sequence of a low-strength, low-permeability zone overlying a high-permeability zone often leads to slope instability. Table 2.5 gives a description of a generalized profile for igneous and metamorphic rocks (Deere and Patton, 1971)

2.4.7 Preexisting Landslide Activities

Slopes in preexisting landslide areas are often prone to failures, which will be discussed further in Section 2.5. Knowledge of local geology is very valuable in understanding both recent and ancient landslide activities. Slope-forming materials in these areas are usually ground up and consist of varying percentages by weight of

Figure 2.9 A white birch tree growing in a crevice pries a large block from a low rock cliff in Hermosa Park, Colorado. (After Leet et al., 1978, from U.S. Geological Survey.)

clays, silts, sands, gravels, cobbles, and boulders. The materials are often remolded and take up a lot of moisture as a result of remolding. Shear strengths usually reach their residual values.

Thin lenses of clays or marls are often found to have served as sliding planes for preexisting landslides. These thin layers can be viewed as a transformation of the parent material as a result of hydrochemical alteration. An example of this is the Cortes de Pallas landslide in Spain, where a thin marl layer triggered the reactivation of a large prehistoric landslide (Alonso and Lloret, 1993). The failure mechanism was postulated to be due to the removal of the dolomite by dissolution and transportation by percolating waters. Given the high permeability of the fractured limestone and the imperviousness of the underlying marl layer, it is probable that the water had flowed downwards along the limestone and marl interface, where slightly acidic rain water (pH = 5.7) may have dissolved carbonate minerals before reaching a geotechnical equilibrium with the surrounding materials.

2.4.8 Clay Mineralogy

The fine particles in soils are composed of clay minerals that exhibit plasticity when mixed with water. Clay minerals are silicates of calcium and aluminium. Water is strongly absorbed by clay minerals. The structure and behavior of adsorbed water in clay minerals are different from those of normal water, and it plays an important role in soil behavior.

TABLE 2.5 Description of a Weathering Profile for Igneous and Metamorphic Rocks

Zone and Description	Relative Permeability	Relative Strength
	Residual Soil	
IA — top soil, roots, organic material, zone of leaching and eluviation, may be porous	Medium to high	Low to medium
IB — clay-enriched, also accumulation of Fe, Al, and Si	Low	Commonly low, high if cemented
IC — relict rock structures retained, silty grading in sandy material, less than 10% corestones, often micaceous	Medium	Low to medium, relict structures very significant
	Weathered Rock	
IIA — transition zone, highly variable, soil-like to rocklike, fines common, coarse sand to fine sand, spheroidal weathering common	High	Medium to low where weak structures and relict structures are present
IIB — partly weathered rock, rocklike soft to hard rock, joints stained to altered, some alteration to feldspars and micas	Medium to high	Medium to high
	Unweathered Rock	
No iron stains to trace along joints—no alteration of feldspars and micas	Low to medium	Very high

Source: Deere and Patton (1971).

The most common clay minerals are illite, kaolinite, and montmorillonite (sometimes called smectites). Of these minerals, montmorillonite has the highest swelling potential and is the most troublesome in terms of slope stability. Due to weathering, chemical and hydrothermal alteration, or both, clay minerals are often found in the joints, shears, and faults in both argillaceous and nonargillaceous rocks. Therefore, clay minerals have a significant influence on the behavior of rock masses. An example of this is bentonite seams found in competent rock. Bentonite is a clay derived from the alteration of volcanic dust and ash deposits and is mainly composed of montmorillonite. Owing to the capacity of this material to adsorb water within the crystal lattice, bentonite swells enormously on the addition of water. Because of its high swelling potential, bentonite is considered one of the problem soils attributable to many landslides.

As for soils, mineralogy has a strong influence on Atterberg limits, which are often used in slope stability problems to identify and classify soils, and to determine their shear strengths through empirical correlations.

2.4.9 Seismic Effects

Many major and minor landslides have occurred during earthquakes in the past. Geological features, whether major or minor, have a significant influence on slope stability during earthquakes. Earthquakes result in an increase of shear stresses and a reduction of shear strength by increasing pore pressures. Liquefaction of small saturated sand and silt lenses within a slope can result in progressive failures of materials that may be relatively insensitive to seismic disturbance. Examples of how minor geological details play an important role in the development of landslides during earthquakes have been given by Seed (1968). This topic is considered further in Section 2.5.

2.5 LANDSLIDES

A landslide is a mass downward movement of either rock or unconsolidated material. The movement is caused by gravity acting upon materials that are in an unstable state of equilibrium. Movement may be initiated by any change in conditions that upsets the temporary conditions. Three general types of landslides are most commonly encountered in highway and civil engineering works. They are movements involving surficial material, movements involving deep-seated soft soils, and movements involving rock strata. The conditions favoring movements are (1) changes in groundwater conditions; (2) presence of clay or shale that softens when wet, (3) structure, and (4) topography. All types of landslides are dependent upon local geology; hence engineers are interested in the detailed geology of the areas where the slides are likely to occur. Such data can enable them to avoid locations were landslides are possible.

Slides, due to changes in groundwater conditions, occur mostly in wet weather or during snow melts. Underground water softens clay and weathered shale, thus providing lubricating sliding surfaces, and at the same time increases the weight of the material affected. In addition to clay and shale, minerals such as talc, serpentine, and anhydrite provide lubrication. Intense rainfall can result in removal of vegetative cover and erosion of the toe of a slope.

Stratified beds of sand or gravel overlying clay, or of sandstone overlying shale with an appreciable dip daylighting any slopes, could introduce a condition favorable to seepage and sliding on the lubricated beds. Jointing in bedrocks is productive of slides during periods of heavy rainfall or freezing/thawing. Renewed movements on old fault planes are common. Metamorphic rocks such as slates and schists, and sedimentary rocks such as shale and serpentine are the most likely to give trouble. As gravity is the fundamental cause of landslides, topography of great relief is the most subject to slides. Even in areas of low relief, slides can be a problem.

2.5.1 Landslide-Prone Occurrences

The surface of the Earth is constantly being changed by natural forces and human activities. All earth and rock sloping grounds are susceptible to landsliding under severe conditions. Landslides can occur in almost any landform if the conditions are right (e.g., steep slopes, high moisture level, no vegetation cover). On the other hand, they may not occur on the most landslide-prone terrain if certain conditions are not present (e.g., clay shales on flat slopes with low moisture levels). Table 2.6 provides a key to landforms and their susceptibility to landslides, but it is not meant to be all inclusive.

Although landslide-prone areas can be identified through air photos, many of them are too small to be readily detected in small-scale photography. It is therefore necessary to examine and to locate those areas that are conducive to landslides. Typical vulnerable spots are discussed below.

Valley Stress Relief Induced Landslides Naturally eroded slopes created by movement of glaciers often produce V-shaped valleys, such as shown in Figure 2.10. The oversteepened sides of such a valley became overstressed when the glaciers melted and the lateral support provided by the glaciers was gradually removed. As a result, strain energy was released following removal of large overburden soils during previous geological periods, and systems of shear zones subsequently developed. After this stress relief, the valley sides were then subjected to a slow and cumulative process of deterioration and destruction through weathering and through development of pore water pressure at the time of heavy rainfall. Eventually the walls fail, and the failure often occurs on well-defined shear planes or relict joints that developed as a result of stress relief.

The phenomenon of valley stress relief and its geotechnical and geological implications have been emphasized by Ferguson (1967 and 1974), Matheson (1972), and Matheson and Thomson (1973).

Landslide-Triggered Ponds One striking hydrologic phenomenon within a landslide is the many new ponds that could be created on the landslide surface. An example of this is the Manti landslide in Utah (Williams, 1988), where numerous ponds of varying dimensions surfaced within a span of about 4 years (Figure 2.11). Post-slide ponds tend to have distinctly different shapes (more irregular), whereas the pre-slide ponds are more subround to round. Measurements of pond migration on the landslide are normally made directly on time-sequential aerial photographs. The measurements are made either (1) from an object of fixed position on stable terrains outside the landslide, in line with the direction of slide movement, or (2) a baseline of fixed position drawn between two such fixed objects and about perpendicular to the slide direction.

Landslides Caused by Bank Erosion Banks undercut by streams or rivers are subject to erosion by running water. Where the banks are made of soil or unconsolidated material, the weakest and the most favorable slide-prone position is often

TABLE 2.6 Key to Landforms and Their Susceptibility to Landslides

Topography	Landform or Geologic Materials	Landslide Potential[a]
I. Level terrain		
A. Not elevated	Floodplain	3
B. Elevated		
1. Uniform tones	Terrace, lake bed	2
2. Surface irregularities, sharp cliff	Basaltic plateau	1
3. Interbedded–porous over impervious layers	Lake bed, coastal plain, sedimentary plateau	1
II. Hilly terrain		
A. Surface drainage not well integrated		
1. Disconnected drainage	Limestone	3
2. Deranged drainage, overlapping hills, associated with lakes and swamps (glaciated areas only)	Moraine	2
B. Surface drainage well integrated		
1. Parallel ridges		
a. Parallel drainage, dark tones	Basaltic hills	1
b. Trellis drainage, ridge-and-valley topography, banded hills	Tilted sedimentary rocks	2
c. Pinnate drainage, vertical-sided gullies	Loess	2
2. Branching ridges, hilltops at common elevation		
a. Pinnate drainage, vertical-sided gullies	Loess	2
b. Dendritic drainage		
(1) Banding on slope	Flat-lying sedimentary rocks	2
(2) No banding on slope		
(a) Moderated to highly dissected ridges, uniform slopes	Clay shale	1
(b) Low ridges, associated with coastal features	Dissected coastal plain	1
(c) Winding ridges connecting conical hills, sparse vegetation	Serpentinite	1

TABLE 2.6 *(Continued)*

Topography	Landform or Geologic Materials	Landslide Potential[a]
3. Random ridges or hills		
a. Dendritic drainage		
(1) Low, rounded hills, meandering streams	Clay shale	1
(2) Winding ridges connecting conical hills, sparse vegetation	Serpentinite	1
(3) Massive, uniform rounded to A-shaped hills	Granite	2
(4) Bumpy topography (glaciated areas only)	Moraine	2
III. Level to hilly, transitional terrain		
A. Steep slopes	Talus, colluvium	1
B. Moderate to flat slopes	Fan, delta	3
C. Hummocky slopes with scarp at head	Old slide	1

Source: Extracted from Transportation Research Board (1978) by Schuster and Krizek.

[a] 1 = susceptible to landslides; 2 = susceptible to landslides under certain conditions; and 3 = not susceptible to landslides except in vulnerable locations.

located at the point of maximum curvature of the stream (Figure 2.12). Here, the toe of the soil slope is under constant erosion by the running water, resulting in undermining of the toe and an ensuing failure. However, in areas of rock outcrops, the maximum stream curvature is often occupied by hard rock and the weak spots are to be found on both sides adjacent to that section.

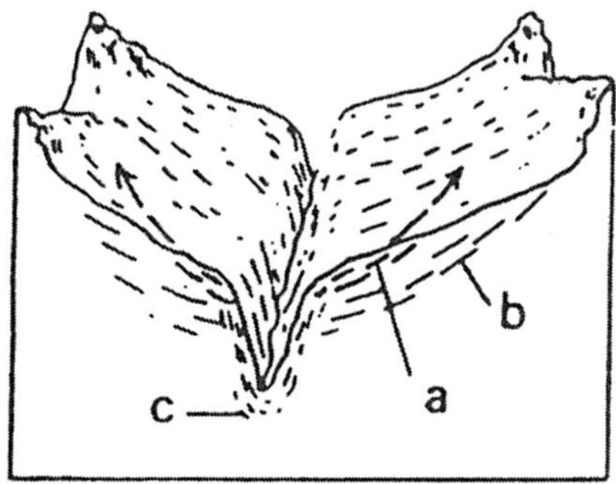

Figure 2.10 Sketch of the gorge as a result of stress relief. a = joints associated with old glacial valley; b = opening associated with bedding plane; c = joints associated with younger river valley. (Blyth and de Freitas, 1984.)

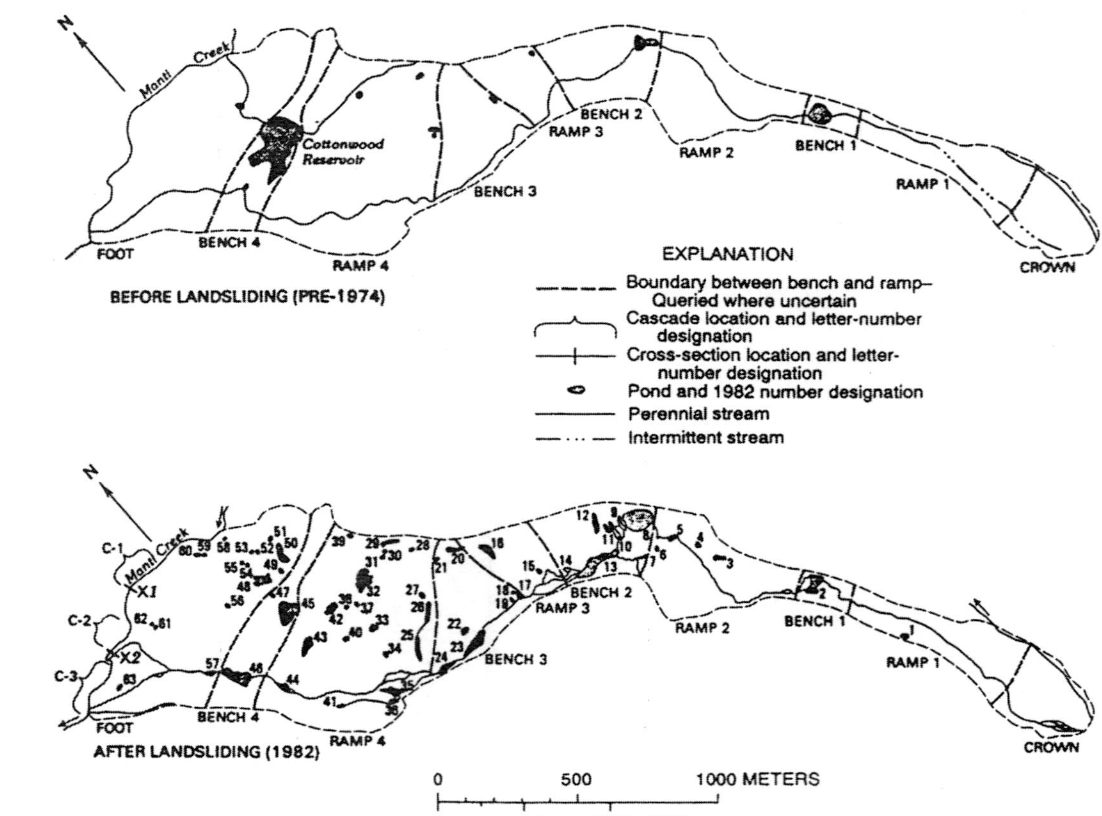

Figure 2.11 Outline and principal hydrologic features of the Manti landslide, Utah. (Williams, 1988.)

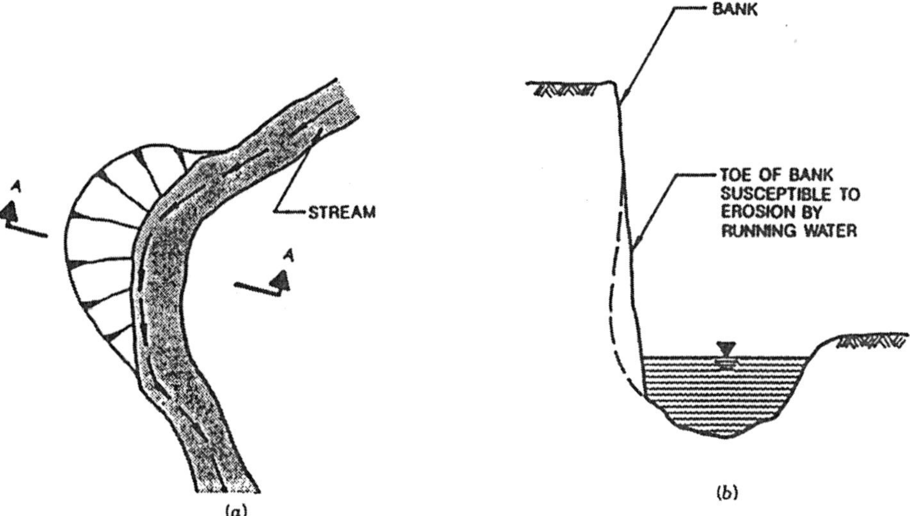

Figure 2.12 Bank erosion by running water. (*a*) Plan View. Most favorable slide-prone position is located at point of maximum stream curvature. (*b*) Section *A–A*

Bank erosion is a common phenomenon in loess deposits, the processes of which include:

Subsurface seepage with tunnel formation and ultimate collapse into gullies
Surface soil creep on steeper slopes, which generates slips and shallow slides.

All these erosive actions are time dependent and influenced by factors such as climate, topography, stratigraphy, physical properties, chemical properties, soil mineralogy, and human activities. For engineering evaluation, particle size distribution and density are two properties that can be measured relatively easily and may in some cases have an influence on erosion.

Weathering Induced Landslides Depending on climatic factors, drainage characteristics, and the nature of residual soils, weathering may occur at a rate rapid enough to be of concern in the design of slope work. Weathering of soil and rock destroys bonds and reduces shear strength. Bjerrum (1967) suggested that weathering of overconsolidated clays and shales increases their recoverable strain energy and consequently their capacity for progressive failure. This occurs due to the destruction by weathering of diagenetic bonding in these materials. Weathering may also be accelerated by slope disturbance and by exposure to atmospheric and other agencies such as stream action.

The weathering profile of a slope has an important influence on its strength characteristics. Table 2.7 gives the shear strength of common weathered igneous, metamorphic, sedimentary rocks, and minerals (Deere and Patton, (1971). Relict geological

TABLE 2.7 Shear Strength of Residual Soils, Weathered Rocks, and Related Minerals

Soil/Rock/Mineral Type	Degree of Weathering	Strength Parameters	
		kg/cm²	Degrees
	Igneous Rocks		
Granite	Partly weathered (Zone IIB)		$\phi_r = 26°–33°$
Granite	Relatively sound (Zone III)		$\phi_r = 29°–32°$
Quartz diorite	Decomposed; sandy, silty	$c = 0.1$	$\phi = 30° +$
Diorite	Weathered	$c = 0.3$	$\phi = 22°$
Rhyolite	Decomposed		$\phi' = 30°$
	Metamorphic Rocks		
Gneiss (micaceous)	Decomposed (Zone IB)	$c = 0.3–0.6$	$\phi = 23°–37°$
Gneiss	Decomposed (Zone IC)		$\phi = 18.5°$
Gneiss	Decomposed (fault zone)	$c = 1.5$	$\phi = 27°$
	Much decomposed	$c = 4.0$	$\phi = 29°$
	Medium decomposed	$c = 8.5$	$\phi = 35°$
	Unweathered	$c = 12.5$	$\phi = 60°$
Schist	Weathered (mica-schist soil)		$\phi = 24.5°$
	Partly weathered	$c = 0.7$	$\phi = 35°$
Schist	Weathered		$\phi = 26°–30°$
Phyllite	Residual soil (Zone IC)	$c = 0$	$\phi = 18°–24°$
	Sedimentary Rocks		
London clay	Weathered (brown)	$c' = 1.2$	$\phi' = 19°–22°$ $\phi_r = 14°$
	Unweathered	$c' = 0.9–1.8$	$\phi' = 23°–30°$ $\phi_r = 15°$
Keuper Marl	Highly weathered	$c' \leq 0.1$	$\phi' = 25°–32°$ $\phi_r = 18°–24°$
	Moderately weathered	$c' \leq 0.1$	$\phi' = 32°–42°$ $\phi_r = 22°–29°$
	Unweathered	$c' \leq 0.3$	$\phi' \geq 40°$ $\phi_r = 23°–32°$
Shale	Shear zones		$\phi = 10°–20°$
	Minerals		
Kaolinite	Minerals common in residual soils and rocks		$\phi_r = 12°–22°$
Illite			$\phi_r = 6.5°–11.5°$
Montmorillonite			$\phi_r = 4°–11°$

Source: Modified from Deere and Patton (1971).

structures, such as joints, bedding planes, and shear zones, influence the nature of the weathering profile. The importance of the weathering profile in slope stability problems was discussed by Deere and Patton (1971), who showed a critical sequence of weathering zones can develop in igneous and metamorphic rocks. For example, a sequence of a low-strength, high-permeability zone overlying a low-permeability zone often leads to instability. The instability is attributed to the buildup of a perched water table during heavy rainfalls, causing a reduction in soil strength and an increase in seepage force in the upper zone.

Drainage- and Seepage-Related Landslides Water contributes greatly to many landslides. Careful examination of existing drainage lines and potential change of drainage routes to the spot under scrutiny should be made. Such drainage may appear on the surface or may go underground and reappear as seepage water that may cause damage to slopes.

Seepage is likely to occur in areas below ponded depressions, reservoirs, irrigation canals, and diverted surface channels. Such circumstances are sometimes overlooked on the ground because water sources may be far above the area under investigation, but they often become obvious in air photos. Leaking utilities are another potential source of excess water.

Areas below diverted surface drainage need particular attention. It has been proven repeatedly, through extensive field experience, that the lower part of an interstream divide, through which surface water seeps from the higher stream bed to the lower one, is the most dangerous section. The seepage often causes erosion at toes of slopes, and undermines the toes of slopes, resulting in ultimate failures.

Landslides Caused by Construction Work done by engineers and contractors can often be the cause of serious slope stability problems. The following construction processes are the main causes of stability problems:

- Modification of the natural conditions of seepage due to fills, ditches, or excavations
- Overloading of weak strata or stratification planes due to fill or a new structure
- Removal, by cutting, of a thin stratum of permeable material that acts as a natural draining blanket of soft clay
- Detrimental increase in seepage pressures or orientation of seepage forces when changes occur in the direction of seepage, as a result of cuts, fills, or other adjacent construction
- Exposure of hard fissured clays to air or water, due to cuts
- Removal of existing slope toes for construction of retaining structures
- Possible leakage of water in pipes and sewers on crest of slopes
- Removal of natural slope vegetation by construction of access roads for equipment and trucks without provisions for adequate drainage systems.

Earthquake-Induced Landslides Loose, saturated sands are particularly vulnerable to liquefaction during earthquakes, leading to flow slides or unstable

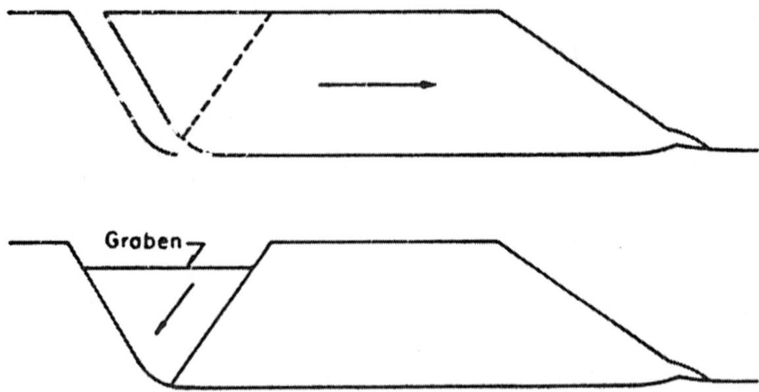

Figure 2.13 Mechanism of graben formation due to sliding on horizontal layer. (Seed, 1970a.)

foundation conditions for overlying sloping deposits. Case histories indicate that during earthquakes, banks of well-compacted fill constructed over weak foundations are more prone to complete failures or severe slumping than are those founded on firm foundation. An excellent example is provided by the behavior of a 4-foot high and 1,500-foot long fill section of the north-south highway between Puerto Varas and Puerto Montt, Chile, which was constructed over a swampy area of ground and surfaced with an 8-inch thick concrete pavement (Seed, 1970a). During the earthquake of 1960, the fill collapsed completely, with the concrete pavement slabs being deposited at the level of the swamp.

Such failures may be characterized by lateral spreading of the base of the fill, and where movement is less severe, the sliding leads to severe longitudinal cracking of the fill. Examples of these failures were found in many sections of highway fills in the Alaskan and Niigata Earthquakes of 1964 (Seed, 1970a).

Thin lenses of loose saturated silts and sands may cause an overlying sloping soil mass to slide laterally along the liquefied layer during earthquakes. As shown in Figure 2.13, a zone of soil at the back end of the sliding mass sinks into the vacant space formed as the mass translates, resulting in a depressed zone known as graben. Buildings in the graben are normally subjected to large differential settlements. Major slide movements can also occur in clay deposits during earthquakes. However, clay deposits often contain sand lenses, and liquefaction of these lenses may well contribute significantly to the slide development in such cases. An example of a large earthquake-induced landslide occurred along the coastline of the Turnagain Heights area of Anchorage during the Alaskan Earthquake in 1964 (Seed and Wilson, 1967).

Fault-Related Landslides Faults are ruptures in rock masses that are associated with differential displacements of opposite sides of the resultant fractures or fracture zones. These deformation movements may not only be limited to slipping along one or many fault surfaces, but may also be associated with distortions, shattering, or crushing of the rock material between the fault surfaces. The latter could give rise to (1) large fragments in a matrix of ground-up material called a fault breccia, or (2) finely ground material called fault gouge—a common type of which with a clay base (often produced by mineral alteration) is called a clay gouge. These zones will frequently be zones of weakness and sometimes trigger and cause slope instability.

The simplest result of faulting is simple fracture. Fractures are often open to a greater or less degree (fissures), and may allow an active seepage or even a rather free flow of water through them. This, in turn, may give access to mineraliferous waters or mineral-bearing solutions, which might form deposits in the fissures, or might invade and modify the composition of the country rock (metasomatic replacement). This action may improve or cause deterioration of the rock from the standpoint of its physical conditions. The results of alteration produced by meteoric water are those typical of weathering, especially solution and decomposition. Rocks with a soluble natural cement may become greatly weakened and made permeable by its solution, and may even be reduced to a mass of loose sand or a mass of impure residual clay. An example of this is sulfides, especially pyrite and marcasite, deposited in or about fault zones, which, when reached by oxidizing waters, may give rise to the production of sulfuric acid, which may attack and alter or weaken the surrounding rock (Lee and Brandon, 1995). Serious problems may arise if slopes are cut into or underlain by any of these post-mineralization faults, and slope instability could ensue.

The surfaces of fault movement may, especially in softer, nonbrittle rocks such as shale and serpentine, become more and more numerous until the mass within the fault zone is dominated by movement surfaces. Separable pieces of the material are usually lenticular and commonly only a fraction of an inch in length, but sometimes up to an inch or more. On removal from the mass, many of the fragments are found to be covered by smooth, slickensided surfaces, which are often culprits for slope instability.

Faults can be recognized by scarps or offsets of strata. But sometimes, it may be difficult to identify them by surface studies. Recognition of faults is, however, not the topic of this chapter, but can be found in engineering geology textbooks (e.g., Goodman, 1992) and publications (e.g., Louderback, 1950).

2.5.2 Fundamentals of Landslides

Introduction A landslide, by definition, implies that movement has taken place. The cause and nature of a landslide are usually invisible to us because they are buried deep beneath the surface or because they are masked by numerous different geologic deposits and groundwater systems. To accurately predict the nature, shape, and causes of a landslide is not an easy task, but it is possible if the investigator is thorough.

Landslides can be identified by the use and interpretation of air photos, other remote-sensing systems such as infrared imagery, satellite imagery, and so on, as well as by field investigations (Chapter 4). For identification of landslides, it is essential to determine the types and causes of slides, so that preventive or corrective action can be taken. Section 2.5.3 focuses on a group of clues, guides, and rules of thumb assembled to help investigators identify features usually associated with landslides.

Types and Mechanisms of Landslides The nature and type of an active landslide must be identified before an evaluation can be made concerning the risk from the slide, or to design corrective treatments. There is an accepted nomenclature for landslides, which has been discussed in Chapter 1. This nomenclature will help in communications with others.

The major factors affecting landslide performance are: (1) the driving forces causing movements, for instance, the weight of the soil system; (2) the seepage forces in the slope; (3) the slope of the failure plane; (4) the strength on the failure plane; and (5) the strength reduction on the failure plane from hydrostatic pressures. The first three factors are usually called the causative forces, whereas the last two factors are the resisting forces.

Causative Forces The weight of the soil system is usually known or can be estimated closely enough for landslide investigation. Seepage forces from internal groundwater flow are the most difficult items to identify and quantify in landslides. The water pressures that build up in any permeable layer or fissure behind the landslide could create additional force, causing failure.

Effect of seismic loading in a landslide is also very difficult to quantify, just like the water pressures. Sufficient background data, if available, can estimate the degree of risk based upon expected earthquake activity.

The major variable that defines the causative force is the angle of the potential sliding surfaces. The steeper the angle, the greater the likelihood of landslide. These sliding surfaces are very difficult to determine reliably from field investigations.

Resisting Forces The primary resistance to landslide movements is the shearing resistance along the failure surface. The shearing resistance along the failure is related to the drained friction angle of the soil on the failure plane. Many landslides are caused by a thin, nearly undetectable layer of clay or bentonite (with drained friction angles from 10 to 25 degrees) found in an otherwise strong material with drained friction angles of 30 to 40 degrees.

The resistance to failure can also be reduced by the increase of water pressures on the failure plane, which has the effect of reducing the frictional resistance by reducing the effective normal forces on the failure plane.

Guides to Landslide Investigations Landslide investigations are usually costly. A deep drill hole (say 150 feet deep) can cost over $10,000 (in 2000) to obtain, and it often does not yield all the information needed. In this regard, it does make sense to carry out preliminary analyses to look for possible controlling features of landslides before starting a field investigation program. Once the features that control the slide are identified, we can use them to design an exploration program.

Mathematical modeling of alternative hypotheses is essential to a complete landslide investigation. Since the factors controlling the performance of landslide can seldom be identified, data gathering must be repeatedly tested against the mathematical models. With the advance of computer technology, hundreds of mathematical models can be evaluated at a cost much less than one subsurface exploration or test pit.

One way to help identify the accuracy of field information and the selected failure model is to run mathematical analyses assigning limiting values to the variables (McGuffey, 1991).

First, test the model by adopting drained friction angles at upper and lower bounds of the material that was encountered in the preliminary inspection. The drained friction angle controlling the failure strengths in the landslide are usually directly related to the plasticity of the fine portion of the material, and, therefore, can be estimated quite accurately if the plasticity index of the material controlling the failure is known. After the drained friction angle is selected, then test the analysis using parameters for the worst soil found in the general area of the slide.

Second, use the information from the subsurface exploration program to estimate upper and lower limits for potential failure slope angles. Limitations of the present exploration technology often restrict our ability to positively define the controlling failure planes. Therefore, the profile information should be tested for the most likely failure model, including other pertinent information available. Examples of surprises are a bentonite seam between competent sound rock layers, or a water-bearing gravel seam in an otherwise extremely hard to very stiff impermeable clay. As a guide, it is often imperative to look first for the obvious, and then, if the information used in the mathematical model does not produce meaningful results compatible to the obvious, look for the exception.

Landslides are often caused by very thin layers of problem soils (e.g., bentonite seams, gouge zones, slickensided surfaces), which cannot be easily identified with confidence during exploration. Therefore, the lower bound of drained friction angle must sometimes be used. For example, in New York, the lower bound value of drained friction angle is usually about 20 degrees (McGuffey, 1991).

Water pressure is an important parameter to be included in a mathematical model. It is very difficult to quantify, as it varies with time and depends on the nature of the soil system. The likely piezometric condition based on the exploration data should first be incorporated into the mathematical model, which should then be continuously reevaluated and revised, based upon the new pieces of information obtained.

Often it is the case that slope movements are associated with maximum hydrostatic pressures in the soil system during heavy rainstorms. Hence, records such as rainfall versus movements are useful to quantify maximum hydrostatic pressures. The physical slide and the mathematical model selected must respond in the same manner to changes in rainfall or groundwater that have been measured, or else there is an improper variable within the model.

Observation of seepage in slopes in the field often gives a clue to the water pressure condition in the slope. If there are springs discharging out of the slope, it can be assumed that there are no artesian pressures in the system.

2.5.3 Useful Clues to Landslide Investigations and Identifications

Landslide Investigations

Depth of Failure Surface The depth of the failure surface below the crest of the slope is usually equal to the distance from the crest of the slope back to the furthest

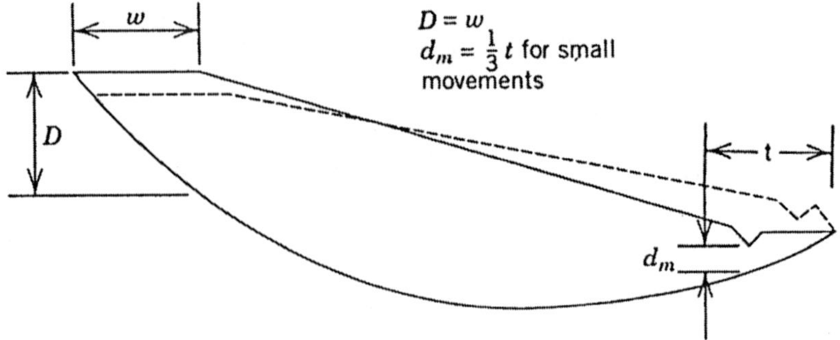

Figure 2.14 Typical depth to failure surface. (McGuffey, 1991.)

shear crack. For failures beyond the toe of a slope, the depth of the failure plane at the toe is usually about one-third of the distance from the toe to the edge of the mud wave (Figure 2.14). If the mud wave exits on a continuing slope, the outlet of the failure surface is usually near the top of the visible mud wave.

The depth of the failure plane may be less than 10 feet if the deep rooted trees in the sliding mass tend to tilt downslope (Figure 2.15).

Breaks in buried utilities, such as culverts and sewer pipes, can give a direct visual identification of where the failure surface exists, and sometimes of how much movement has occurred across the failure surface.

Limits of Landslides The exact limits of a landslide are not easy to identify unless there is a 2- to 3-foot vertical drop on the slope crest. Telephone poles or electric lines tend to tilt in a slide, which often causes tension or sagging of wires between telephone or electric poles. Therefore, the limit of a slide can be roughly identified based on the amount of tension in the wires compared to the average tension between

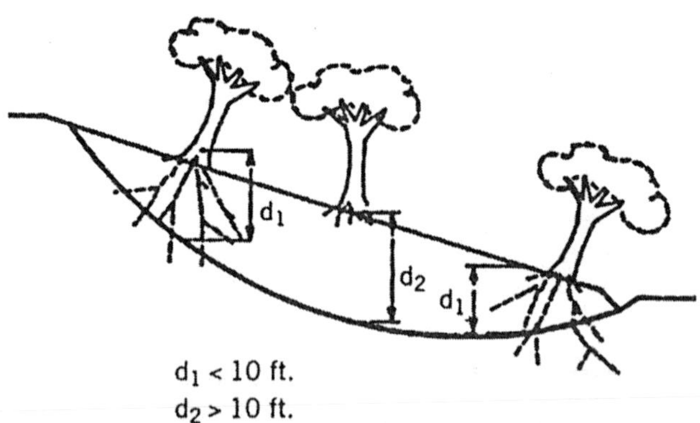

Figure 2.15 Depth of failure surface estimated from trees with deep roots. (McGuffey, 1991.)

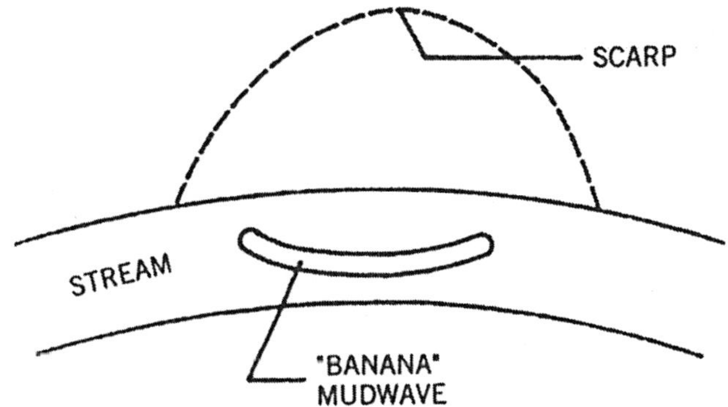

Figure 2.16 Banana-shaped island in a river or stream. (McGuffey, 1991.)

adjacent telephone poles or electric lines. Leaning trees are also often a sign of surface movements. Depth of failure surface may be estimated from trees with deep roots (Figure 2.15)

Man-made features, such as catchpits, masonry walls, guard rails, and so on, are usually built in specific geometric shapes (straight or curved). Any deviations from these shapes would indicate that there has been a differential movement from which limits of a landslide can be determined.

The toe of a slide often shows up in a stream. Extra care must be exercised to identify the failure limits when they are below water. When a stream is undercutting the toe of a landslide, the mud wave that results could create a reverse banana-shaped island in the stream (Figure 2.16). This can be easily identified by the currents, which will show different surface wave patterns over the island, even though the island itself may not be visible. In addition, the change of normal stream flow patterns is often the result of a mud wave shoving the stream into a different position.

Lateral limits of a landslide can sometimes be determined by the uprooting of trees due to landslides. This can be done by comparing the normal tree growth pattern observed on an unfailed slope to the pattern in a landslide area.

Surface drainage in a landslide area is often changed in direction as a result of the slide loosening the normal protective vegetation or stone cover, thus allowing accelerated erosion. By tracing the new direction of the surface drainage, it is possible to determine the approximate limits of the landslide.

Landslide Investigation Resources The landslide investigation process is often unique and site-specific. Resources that can help in landslide investigations include U.S. Geological Survey topographic maps, and air photos from various planning, agriculture, or military groups taken at various times starting just prior to the World War II and continuing to the present time. These resources will be discussed in Chapter 4.

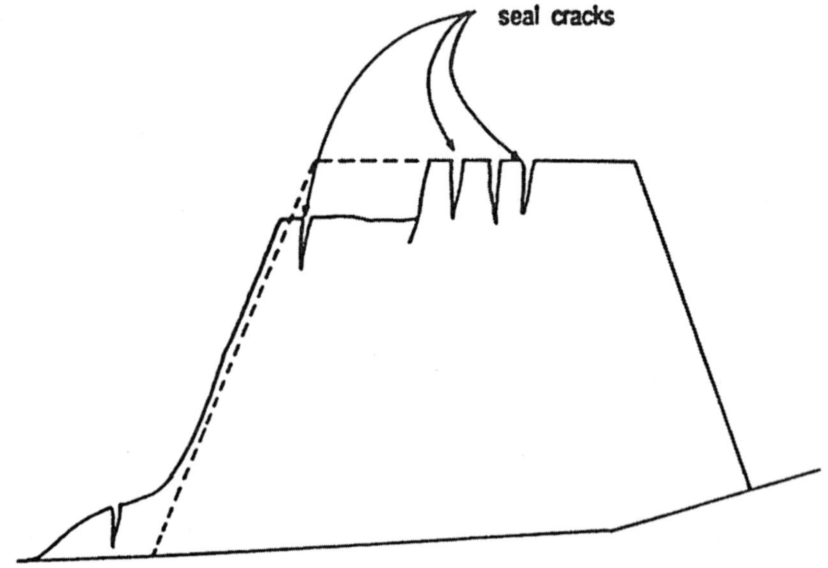

Figure 2.17 Development of escarpment in or above a roadway.

Conclusion The investigation processes for landslides are unique as they are site-specific. Much wasted effort can be saved by following an orderly and well-planned process of analysis and field investigation. Asking good questions and reading the clues that mother nature has provided will save much effort. Start with the simplest, most obvious model based upon field observations, common sense, and sometimes self-intuition before investigating more complex models. Always go back to compare the final answers against the simplest model.

Landslide Identification Quite often, the first visible sign of movement adjacent to a roadway is recorded by settlements of the roadway, a bulge of pavement, and formation of cracks. These manifestations are followed by the presence of hummocky ground whose characteristics are inconsistent with those of the general regional slopes and the presence of a scarp surface (sometimes not very distinct as it is covered with vegetation). The older the landslide is, the more established the drainage and vegetation become on the slide mass. The drainage and vegetation thus help in determining the relative age and stability of the slide.

Landslides in soil slopes are often rotational or noncircular slumps, which can be readily identified from surface indications. The head region of a slump is characterized by steep scarps and by offsets between separate blocks of material (Figure 2.17). If the landslide is active or has been active recently, the scarp is bare of vegetation and may be marked by striations or grooves that indicate the direction of movement. Seeps, springs, and marshy conditions commonly mark the foot and toe of a slump. Moreover, trees tend to be tilted downhill rather than uphill as they near the head.

Old landslides that have been covered by glacial till or other more recent sediments are the most difficult to identify. One who knows the recent geologic history of the region intimately, however, may be able to make some educated guesses as to the probable existence of such slides and even as to where they are most likely to be found. The chances of identifying these old landslides could be increased if borings, test pits, and instrumentations have been performed in the areas.

One point worth mentioning is that even if a landslide has not been identified, it is still necessary to determine whether the ground to be disturbed by the proposed construction will prove reasonably stable. However, no one is able to guarantee the stability of all slopes, no matter how thorough the investigations and how careful the design of the slopes for any construction projects. As a general rule, the amount of investigation and study that is warranted is closely related to the landslide susceptibility of the surrounding area, and also to the degree of damage that might be expected to occur to persons or installations should a slide occur. That is to say, the more serious the consequences of a landslide, the more detailed the search for potential slides should be.

REFERENCES

Alonso, E. E., and A. Lloret, 1993. "The Landslide of Cortes de Pallas, Spain," *Geotechnique*, Vol. 43, No. 4, pp. 507–521.

Bates, R. L., and J. A. Jackson, 1984. *Dictionary of Geological Terms*. New York: Doubleday.

Beavis, F. C., 1985. *Engineering Geology*. London: Blackwell.

Bjerrum, L., 1967. "Progressive Failure in Slope in Overconsolidated Plastic Clays and Clay-Shales," Terzaghi Lecture, *Journal of Soil Mechanics Foundation Division*, ASCE, Vol. 93, No. SM5, pp. 3–49.

Bjerrum, L., and F. A. Jorstad, 1968. "Stability of Rock Slopes in Norway," Norwegian Geotechnical Institute, Publication No. 79.

Blyth, F. G. H., and M. H. de Freitas, 1984. *A Geology for Engineers*, 7th ed. London: English Language Book Society.

Bureau of Reclamation, 1968. *Earth Manual*. Denver: U.S. Bureau of Reclamation.

Burwell, E. B., and B. C. Moneymaker, 1950. "Geology in Dam Construction, Application of Geology to Engineering Practice," *Geological Society of America, Berkey Volume*, S. Paige, Ed., Harvard Soil Mechanics Series, pp. 11–43.

Chandler, R. J., 1969. "The Effect of Weathering on the Shear Strength Properties of Keuper Marl," *Geotechnique*, Vol. 19, pp. 321–334.

Chowdhury, R. N., 1978. *Slope Analysis*. Amsterdam: Elsevier.

Deere, D. U., and F. D. Patton, 1971. "Slope Stability in Residual Soils," *Proc. 4th Pan. Am. Conf. on Soil Mechanics and Foundation Engineering*, State of the Art Report, Vol. 1, pp. 88–171.

Design Manual DM-7, 1971. *Soil Mechanics, Foundations, and Earth Structures*. Alexandria, Virginia: Naval Facilities Engineering Command.

Evans, G. L., 1977. "Erosion Tests on Loess Silt—Banks Peninsula, New Zealand," *Proceedings of 9th International Conference SMFE*, Tokyo.

Ferguson, H. E., 1967. "Valley Stress Relief in the Allegheny Plateau," *Bulletin Association of Engineering Geology*, Vol. 14, No. 1, pp. 63–71.

Ferguson, H. E., 1974. "Geologic Observation and Geotechnical Effects of Valley Stress Relief in Allegheny Plateau," Preprint paper presented to ASCE National Meeting on Water Resources Engineering, Los Angeles.

Goodman, R. E., 1992. *Engineering Geology—Rock in Engineering Construction*. New York: Wiley.

Hunt, R. E., 1984a. *Geotechnical Engineering Investigation Manual*. New York: McGraw-Hill.

Hunt, R. E., 1984b. *Geotechnical Engineering Techniques and Practices*. New York: McGraw-Hill.

Lee, T. S., and S. Brandon, 1995. "The Hoover Slides in Provo Canyon Utah," 31st Symposium on Engineering Geology and Geotechnical Engineering, Logan, Utah, pp. 244–259.

Leet, D. L., S. Jusaon, and M. Kauffman, 1978. *Physical Geology*. Englewood Cliffs, New Jersey: Prentice-Hall.

Lindquist, E. S., 1994. "The Mechanical Properties of a Physical Model Melange," *Proceedings of the 7th Congress of the International Association of Engineering Geologists*, Lisbon, Portugal, Sept. 5–9.

Little, A. L., 1969. "The Engineering Classification of Residual Tropical Soils," *Proceedings of the Specialty Session on the Engineering Properties of Alteration Soils*, Vol. 1, 7th International Conference on Soil Mechanics and Foundation Engineering, Mexico City, pp. 1–10.

Louderback, G. D., 1950. "Faults and Engineering Geology," *Geological Society of America, Berkey Volume*, S. Paige, Ed., Harvard Soil Mechanics Series, pp. 125–150.

Metheson, D. S., 1972. *Geotechnical Implication of Valley Rebound*," Ph.D. Thesis, University of Alberta, Edmonton, Canada.

Matheson, D. S., and S. Thomson, 1973. "Geological Implication of Valley Rebound," *Canadian Journal of Earth Sciences*, Vol. 10, pp. 961–978.

McGuffey, V. C., 1991. "Clues to Landslide Identification and Investigation, Geologic Complexities in the Highway Environment," *Proceedings of the 42nd Annual Highway Geology Symposium*, R. H. Fickies, Ed., Albany, New York, pp. 187–192, May.

Medley, E. W., and R. E. Goodman, 1994. "Estimating the Volumetric Block Proportions of Melanges and Similar Block-In Matrix Rocks (Bimrocks)," *Proceedings of the 1st North American Rock Mechanics Symposium*, Totterdam, Balkema.

Seed, H. B., 1968. "Landslides During Earthquakes due to Soil Liquefaction," Terzaghi Lecture, *Journal of the Soil Mechanics and Foundations Division*, ASCE, Vol. 92, No. SM1, pp. 13–41.

Seed, H. B., 1970a. "Soil Problems and Soil Behavior," *Earthquake Engineering*, R. L. Wiegel, Ed. Englewood Cliffs, New Jersey: Prentice-Hall, Chapter 10.

Seed, H. B., 1970b. "Earth Slope Stability During Earthquakes," *Earthquake Engineering*, R. L. Wiegel, Ed. Englewood, Cliffs, New Jersey: Prentice-Hall, Chapter 15.

Seed, H. B., and S. D. Wilson, 1967. "The Turnagain Heights Landslide in Anchorage, Alaska," *Journal of the Soil Mechanics and Foundations Division*, ASCE, Vol. 93, No. SM4.

Terzaghi, K., 1950, "Mechanics of Landslides, Application of Geology to Engineering Practice," *Geological Society of America, Berkey Volume*, S. Paige, Ed., Harvard Soil Mechanics Series, pp. 83–123.

Terzaghi, K., 1952. "Permafrost," *Journal of Boston Society of Civil Engineers*, January.

Terzaghi, K., 1960. *From Theory to Practice in Soil Mechanics*. New York: Wiley.

Terzaghi, K., 1962. "Stability of Steep Slopes in Hard Unweathered Rock," *Geotechnique*, Vol. 12, pp. 251–270.

Terzaghi, K., and R. B. Peck, 1967. *Soil Mechanics in Engineering Practice*. New York: Wiley.

Transportation Research Board, 1978. *Landslides Analysis and Control*. Special Report 176, R. L. Schuster, and R. J. Krizek, Eds., National Academy of Science, Washington, D.C., p. 234.

Volpe, R. L., C. S. Ahlgren, and R. E. Goodman, 1991. "Selection of Engineering Properties for Geologically Variable Foundations," *Proceedings of the 17th International Congress on Large Dams*, Paris, pp. 1087–1101.

Watson, I., 1984. "Hydrogeologic Control and Statistical Prediction of Active Mass Movement," *Bulletin of the Association of Engineering Geologists*, Vol. XXI, No. 4, pp. 479–494.

Williams, G. P., 1988. "Stream-Channel Changes and Pond Formation at the 1974–76 Manti Landslide, Utah," *U.S. Geological Survey Professional Paper 1311*. Washington, DC: USGS, U.S. Government Printing Office, pp. 45–69.

CHAPTER 3

GROUNDWATER CONDITIONS

3.1 INTRODUCTION

Water is the most important factor in most slope stability problems. Knowledge of groundwater conditions is essential for the analysis and design of slopes. This chapter describes the groundwater flow in soils, and the methods whereby the influence of rainfall on groundwater conditions can be assessed.

The basic soil–water model that exists in nature is described before discussing the groundwater flow in soils. The pressure in a static body of water (Figure 3.1*a*) has a triangular distribution with a magnitude $\gamma_w z$, which is called hydrostatic pressure. This hydrostatic pressure distribution also exists in the pore water surrounding the soil particles in Figure 3.1*b* or at a depth of H_w below the groundwater level in Figure 3.2.

The water zone in the soil mass that has a water table may be divided into a saturated zone below the water table, and a capillary zone above the water table (Figure 3.2). Above the water table, the air voids increase as the distance from the water table increases. The capillary zones are unsaturated. The water in these zones is held in place by capillary attraction and exerts relatively large stabilizing forces on the structure of soil (which is called negative pore pressure or soil suction). The capillary zones could reach a considerable height above the water table in fine-grained soils.

The subsurface water consists of saturated and unsaturated zones, each of which is divided into several components, as given in Table 3.1. Water affects the stability of slopes by

- Generating pore pressures, both positive and negative, which alter stress conditions

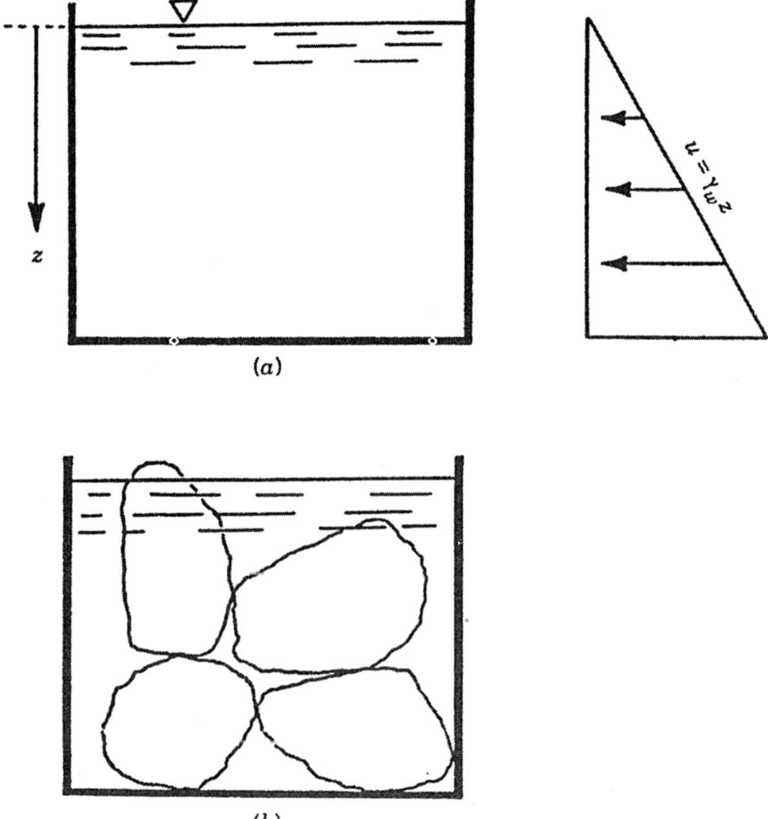

Figure 3.1 Hydrostatic pressure. (*a*) Water. (*b*) Water surrounding soil particles. (Dunn et al., 1980.)

- Changing the bulk density of the material forming the slope
- Developing both internal and external erosions
- Changing the mineral constituents of the materials forming the slopes

3.2 REVIEW OF GROUNDWATER FUNDAMENTALS

Groundwater in its natural state is invariably moving. This movement is governed by established hydraulic principles. Section 1.7 discusses the concept of the effective stress principle developed by Terzaghi, which serves as the basic concept of groundwater fundamentals that are applied to soil mechanics in slopes. This section presents a general review of movement of groundwater and its principles of mechanics.

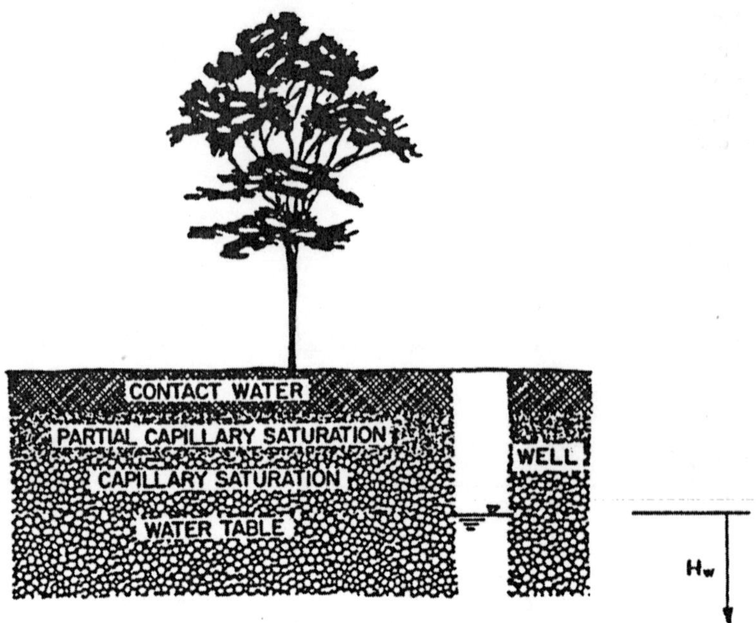

Figure 3.2 Capillary water system. (Dunn et al., 1980.)

3.2.1 Movement of Groundwater

The flow of groundwater is usually very slow and is generally laminar. Turbulent flow of groundwater may occur in large underground passageways formed in such rocks as cavernous limestone or in gravelly soils. In laminar flow, the groundwater near the walls of interstices is presumably held motionless by the molecular attraction of the

TABLE 3.1 Divisions of Subsurface Water

Zones	Water	Process	Division	Pressure
Unsaturated	Hygroscopic	Infiltration	Discontinuous capillary saturation	Gas phase = atmospheric
	Pellicular		Semicontinuous capillary saturation	Liquid phase < atmospheric
	Capillary		Continuous capillary saturation	< Atmospheric
		Water Table		
Saturated	Groundwater (Phreatic zone)	Percolation	Unconfined ground water	> Atmospheric

Source: From Institution of Civil Engineers (1976), reproduced by permission.

walls. Water particles farther from the walls would move more rapidly in smooth, threadlike patterns, for the resistance to motion decreases toward the center of an opening. The most rapid flow is reached at the very center.

The energy that causes groundwater to flow is derived from gravity. Gravity draws water downward to the water table; from there it flows through the ground to a point of discharge in a stream, lake, or spring. Just as surface water needs a slope to flow on, so must there be a slope (or gradient) for the flow of groundwater. This is the slope of water table—the hydraulic gradient—which is measured as the ratio of the vertical distance between the point of intake to the point of discharge (a distance called head) to the length of flow from the two points. This relationship regarding the rate of movement was developed by a French engineer, Henri Darcy, in 1856; it states that

$$v = k(h/l) \tag{Eq. 3-1}$$

where v = the velocity
h = the head
l = the length of flow
k = a coefficient that depends on the permeability of the material, the acceleration of gravity, and the viscosity of water

Equation 3-1 is the famous Darcy's law, which has been used in the study of groundwater hydraulics for many years. It is, however, worth noting that Darcy's law is only valid for laminar flow. Fortunately, as mentioned before, most natural underground flows are laminar in nature so that Darcy's law is usually applicable.

Groundwater moves downward from the water table in broad looping curves toward some effective discharge agency, such as a stream, lake, or spring (Figure 3.3*a* and 3.3*b*). This curving path is a compromise between the force of gravity and the tendency of water to flow laterally in the direction of the slope of the water table. This tendency toward lateral flow is actually the result of the movement of water toward an area of lower pressure, or a point of discharge, for example, the stream channel in Figure 3.3*a*. In other words, the resulting movement is neither directly downward nor directly toward the point of discharge, but is rather along a curving path to the discharge point.

In general, groundwater moves very slowly in the soil mass. This is because the water must travel through small, confined passages if it is to move at all. It is worth considering the porosity and permeability of soils and rocks as in the following.

Porosity The porosity of a rock or soil is measured by the percentage of its total volume that is occupied by voids or interstices. The more porous the rock or soil, the greater the amount of open space it contains. Porosity differs from one material to another. A sand deposit composed of rounded quartz grains with fairly uniform size has a high porosity. But if minerals enter into the deposits and cement the grains into a sandstone, the porosity is reduced by an amount equal to the volume of the cementing agent. A dense massive rock, such as granite, may become porous as a result of fracturing. On the other hand, a soluble mass rock, such as limestone, may have its original planes of weakness enlarged by solution.

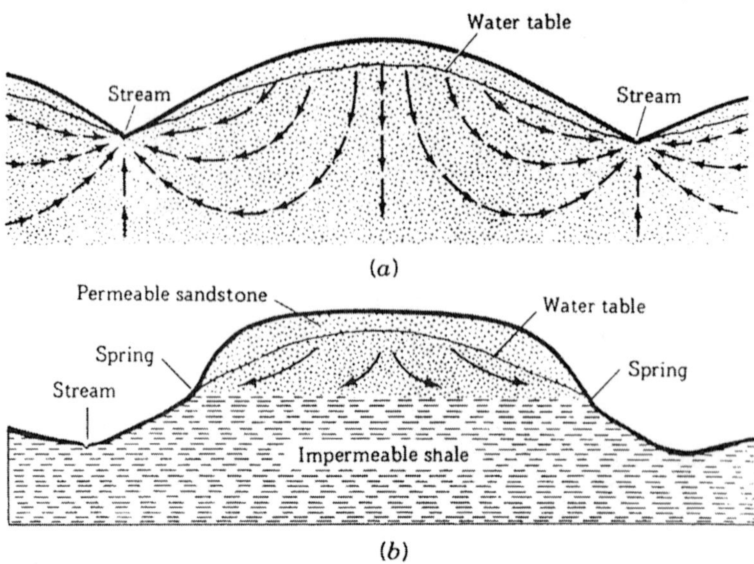

Figure 3.3 (*a*) Flow of groundwater through uniformly permeable material. Note the broadly looping paths of the water flow that converge toward the outlet and may approach from below. (*b*) Springs result from where the water table intersects the surface at the contact of the impermeable shale and more permeable sandstone. (Leet et al., 1978.)

Permeability The ability to transmit groundwater is termed permeability. The rate at which a rock or soil transmits water depends not only on its total porosity, but also on the size of the interconnections between its openings. An example of this is that water passes more readily through the sand than through clay simply because the molecular attraction on the water is much stronger in the tiny openings of the clay than in the sand. It must be noted that no matter how large the interstices of a material are, there must be connections between them if water is to pass through. If they are not interconnected, the material is impermeable, for example, fresh solid rocks with no joints or clays.

3.2.2 Principles of Groundwater Mechanics

A soil mass when fully saturated is composed of two distinct phases: the soil skeleton and the water-filled pores between the soil particles. As shown in Figure 3.4*a*, any stresses (the total stress, σ) imposed on such a soil will be sustained by the soil skeleton (the effective stress, σ') and the pore water (the pore pressure, u). The total stress is equal to the total force per unit area acting perpendicular to the plane. The pore water pressure can be determined from the groundwater conditions, which are discussed in the subsequent sections.

Seepage flows in soil slopes are often encountered, and slope stability analyses must include these flowing conditions. To illustrate the application of the effective

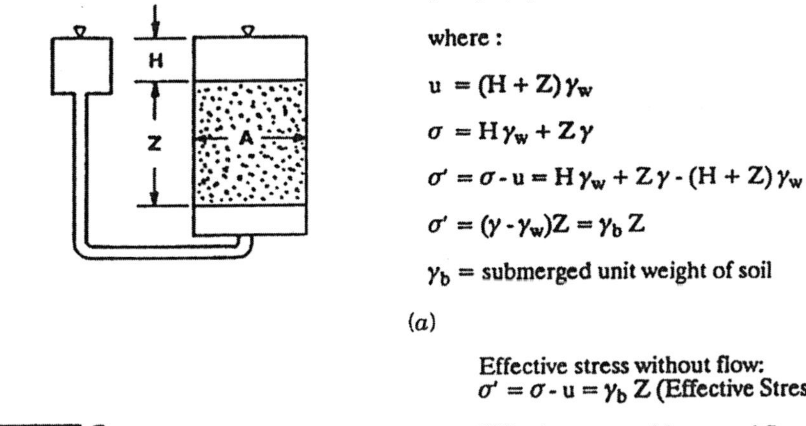

(a)

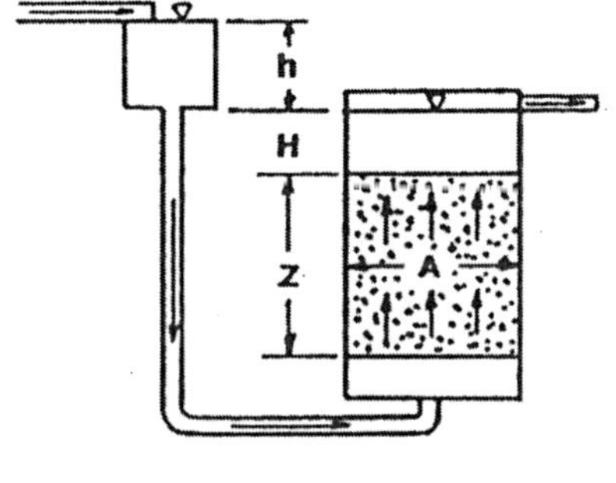

Effective stress without flow:
$\sigma' = \sigma - u = \gamma_b Z$ **(Effective Stress)**

Effective stress with upward flow:
$\sigma' = \gamma_b Z - \Delta u = \gamma_b Z - \gamma_w h$

Hydraulic gradient $i = h / Z$
where h **= differential head**

Therefore, $\sigma' = \gamma_b Z - i \gamma_w Z$

Seepage pressure: $i \gamma_w Z$
Seepage force: $i \gamma_w Z A$
where A = Area vertical to waterflow

(b)

Figure 3.4 Review of groundwater fundamentals. (*a*) Effective stress principle. Total stress = effective stress + neutral stress. In terms of pressure: total pressure = soil skeleton pressure + water pressure. (*b*) Seepage force results from a change in the neutral stress due to groundwater flow.

stress principle in these cases, it is essential to understand the basic principle of water forces that develop in the soil mass. Figure 3.4*b* shows a setup in which water is flowing vertically upward through a soil mass under a total head gradient, i, which is equal to h/Z. The seepage force per unit volume is equal to the total head gradient i times the unit weight of the water, that is, $i(\gamma_w)$. In an isotropic soil, the seepage force always acts in the direction of flow.

Seepage force acts in the direction of flow and tends to move the soil grains in that direction (piping). Piping will occur at a hydraulic gradient greater than the critical hydraulic gradient, i_c, a condition where the effective stress becomes zero:

$$\sigma' = \gamma_b Z - i_c \gamma_w Z = 0 \qquad \text{(Eq. 3-2)}$$

$$i_c \gamma_w Z = \gamma_b Z \qquad \text{(Eq. 3-3)}$$

$$i_c = \gamma_b / \gamma_w \qquad \text{(Eq. 3-4)}$$

where σ' = effective stress
γ_b = submerged unit weight of soil
γ_w = unit weight of water
Z = soil depth
i_c = critical hydraulic gradient

In other words, the critical hydraulic gradient is equal to the ratio of the submerged unit weight of soil and the unit weight of water. For practical purposes, i_c is close to 1.

3.3 SITE CONDITIONS

3.3.1 Groundwater Levels

Groundwater is derived from many sources but primarily originates from rainfall and melting snow. Some water infiltrates into the ground and percolates downwards to the saturated zone at depth, while some water moves over the surface as surface runoff. Groundwater in the saturated zone moves toward rivers, lakes, and seas, where it evaporates and returns to the land as clouds of water vapor, which precipitates as rain and snow. This circulation of water is often known as the hydrological cycle, as indicated in Figure 3.5.

Groundwater levels are rarely static and vary with the rate of recharge or discharge of the groundwater. A uniform distribution of rain or snow melt rarely occurs; thus some areas receive more recharge than others. In such cases, a rise in the groundwater level will usually be greater in some places than others. Fluctuations vary with geology, topography, and proximity to local centers of discharge, such as springs, rivers, and dams that store water or pump water from the ground. Such fluctuations should be studied with observation wells or piezometers and the data presented as maps of groundwater level change (Figure 3.6).

3.3.2 Zones

As mentioned earlier, rainfall or snowmelt on the ground surface will infiltrate into the subsurface materials, which are differentiated into unsaturated and saturated zones. The unsaturated zone is often located above the main groundwater table (phreatic surface) with voids partially filled with water. This zone is sometimes called the zone of aeration, and it extends from the ground surface down through the major root zone (Figure 3.7). Its thickness varies with the soil types and vegetation. Within this zone, the spaces between particles are filled partly with water and partly with air. Molecular attraction is exerted on the water by the soil or rock, and the attraction is also exerted by the water particles on one another (Figure 3.8).

The saturated zone is within the main groundwater regime with voids completely filled with water. Perched groundwater can create saturated zones within unsaturated zones. Different modes of groundwater flow develop in the unsaturated and saturated zones, and affect the stability of slopes.

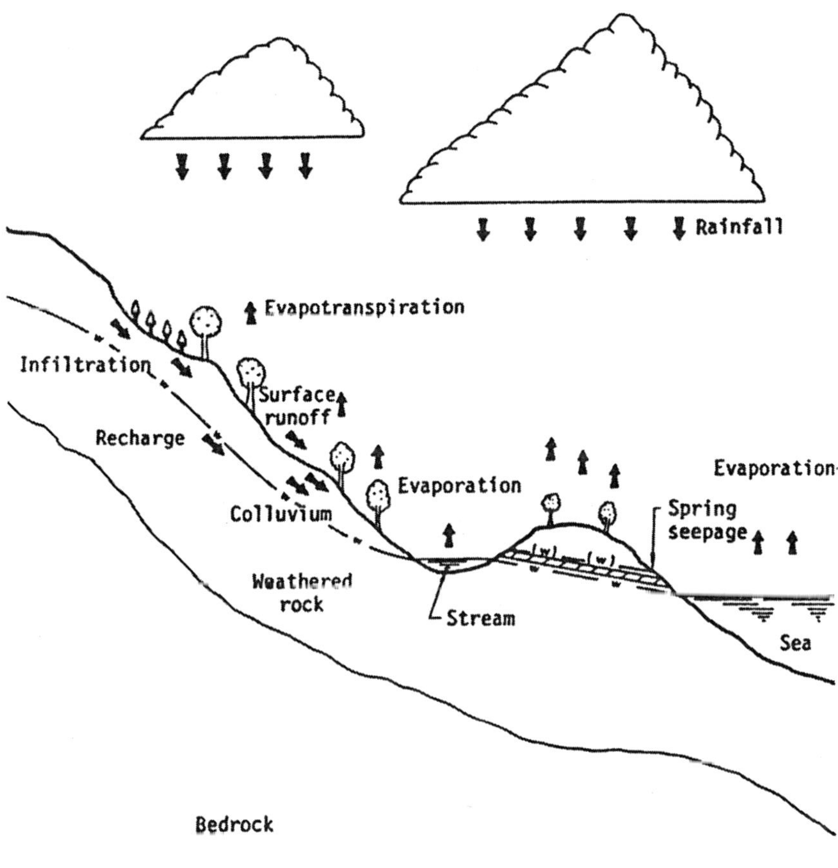

Figure 3.5 Simplified representation of the hydrological cycle. (Geotechnical Control Office, 1984.)

Water penetrates the ground surface and moves downward toward the water table. To percolate through the unsaturated zone, the water flow has to satisfy the capillary requirements of the unsaturated soil profiles, which depend on the dryness of the mineral surfaces surrounding the voids. A measure of these requirements is the moisture potential, which is expressed either as the vacuum necessary to balance soil suction or the logarithm of the head of water in centimeters equivalent to the pressure difference between this vacuum and atmospheric pressure (note that such measurements or analysis are not made in practice). Dry fine-grained soils such as clay have a high

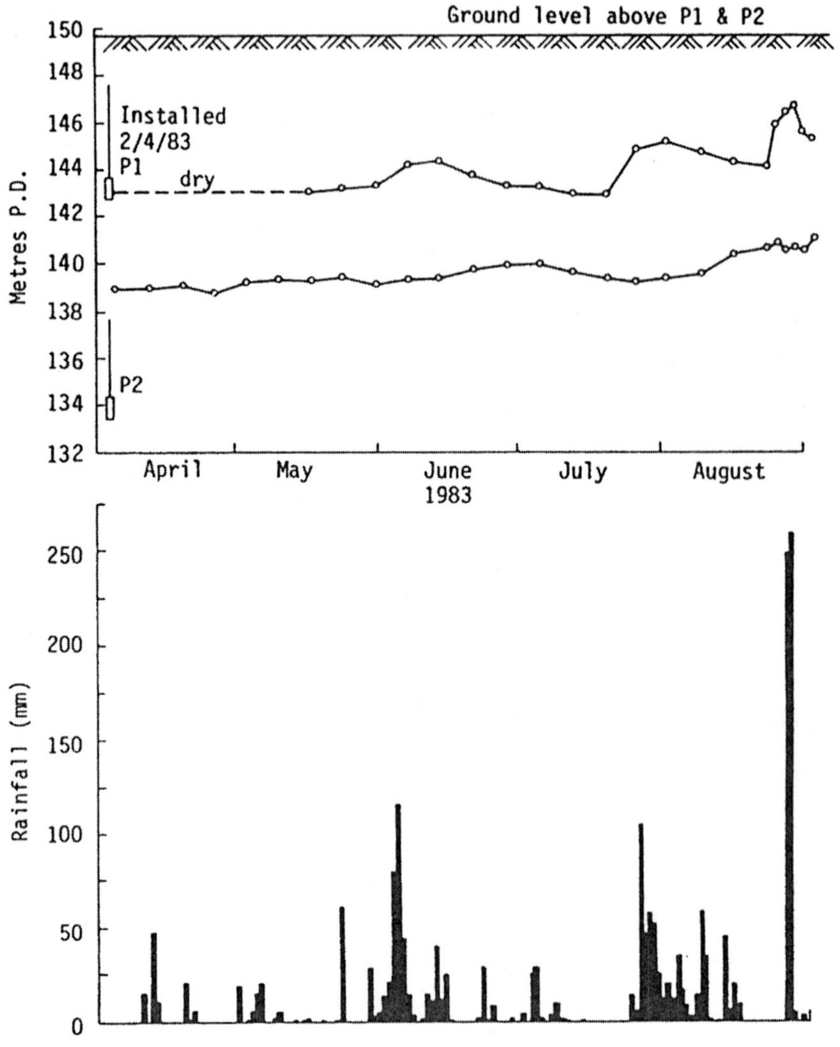

Figure 3.6 Fluctuation of groundwater with periods of rainfall. (Geotechnical Control Office, 1984.)

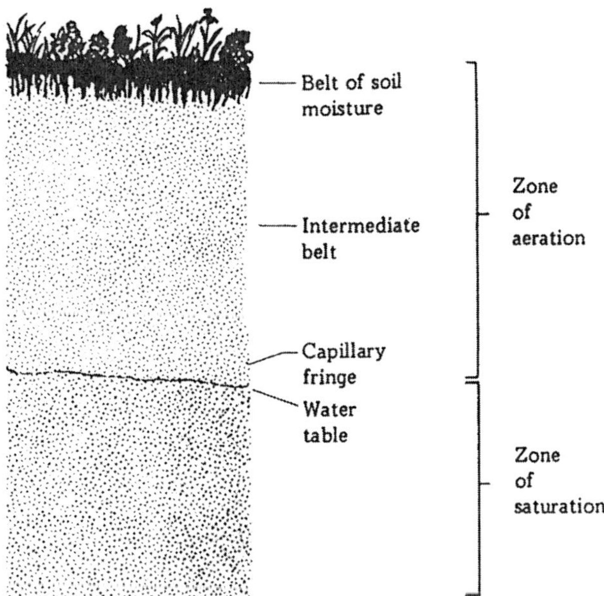

Figure 3.7 Underground water's two major zones: zone of aeration and zone of saturation. The water table marks the upper surface of the zone of saturation. (Leet et al., 1978.)

potential that is satisfied only by a considerable amount of water. When no additional water can be held by the soil against the pull of gravity, the water will drain to deeper levels until it is absorbed by other dry zones or intersects the saturated zone where the groundwater table is located.

3.3.3 Aquifers

Soils and rocks that transmit water with ease through their pores and fractures, respectively, are called aquifers. Typical aquifers are composed of gravel, sand, sandstone,

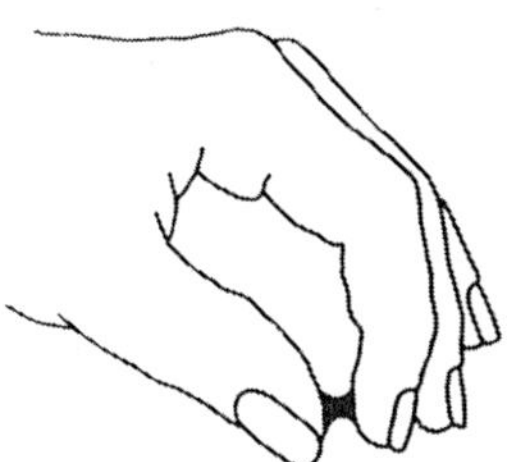

Figure 3.8 A drop of water held between two fingers illustrates the molecular attraction that prevents downward movement. Water is similarly suspended within the pore spaces of the zone of aeration. (Leet et al., 1978.)

TABLE 3.2 Typical Void Sizes for Soil and Associated Permeabilities

Material	Void Size (cm)	Permeability (cm/sec)
Clay	$<10^{-4}-10^{-3}$	$<10^{-6}$
Silt	$10^{-3}-10^{-2}$	$10^{-6}-10^{-4}$
Sand	$10^{-2}-10^{-1}$	$10^{-4}-10$
Gravel	$10^{-1}+$	$10-10^{+2}$

limestone, and fractured volcanic, igneous, and metamorphic rocks. Table 3.2 lists the void sizes that may be expected in these materials and their likely permeability.

Potential aquifers can be identified from detailed records of exploratory borings, well logs, and geologic reconnaissance. The absolute permeability of the stratum does not determine whether it is an aquifer. Instead, its relative permeability compared to the strata above and particularly the strata below is more significant. For example, a stratum of silty fine sand could be an aquifer if it were confined between clay strata, but would be an aquiclude (see Section 3.3.4) when confined between two strata of coarse clean gravelly sand (Figure 3.9).

Aquifers are either confined, if between two impermeable strata, or unconfined, if exposed or not overlain by a less permeable layer. In the former case, the aquifer is fully saturated and the piezometric (hydraulic) head is above, or concident with, the lower boundary of the confining medium (BHC and BHB_3 in Figure 3.10). In the unconfined situation, groundwater does not fully occupy the potential aquifer, and a free water surface (i.e., the water table) exists within the aquifer.

3.3.4 Aquicludes

Aquicludes consist of strata or discontinuities that are sufficiently less pervious than the adjoining strata that they constitute barriers to groundwater. For example, silt or clay washing into a crack in the ground can produce a clastic dike or aquiclude that will block the flow in a sandy seam. Typical aquicludes are clay, shale, and unfractured igneous and metamorphic rocks.

3.3.5 Perched Water

A perched water table is one that is sustained above an aquiclude or an impermeable stratum such as a clayey layer. Perched water may be transient, rapidly developing in response to heavy rainfall and dissipating quickly, or permanent, in response to seasonal variations in rainfall levels.

Like the groundwater table, perched water can be monitored by means of piezometers. They are often first recognized during exploratory borings when water is encountered above the permanent groundwater table (e.g., BHA in Figure 3.10). Shallow failures of the slopes are sometimes associated with the rise of perched water over an impermeable layer.

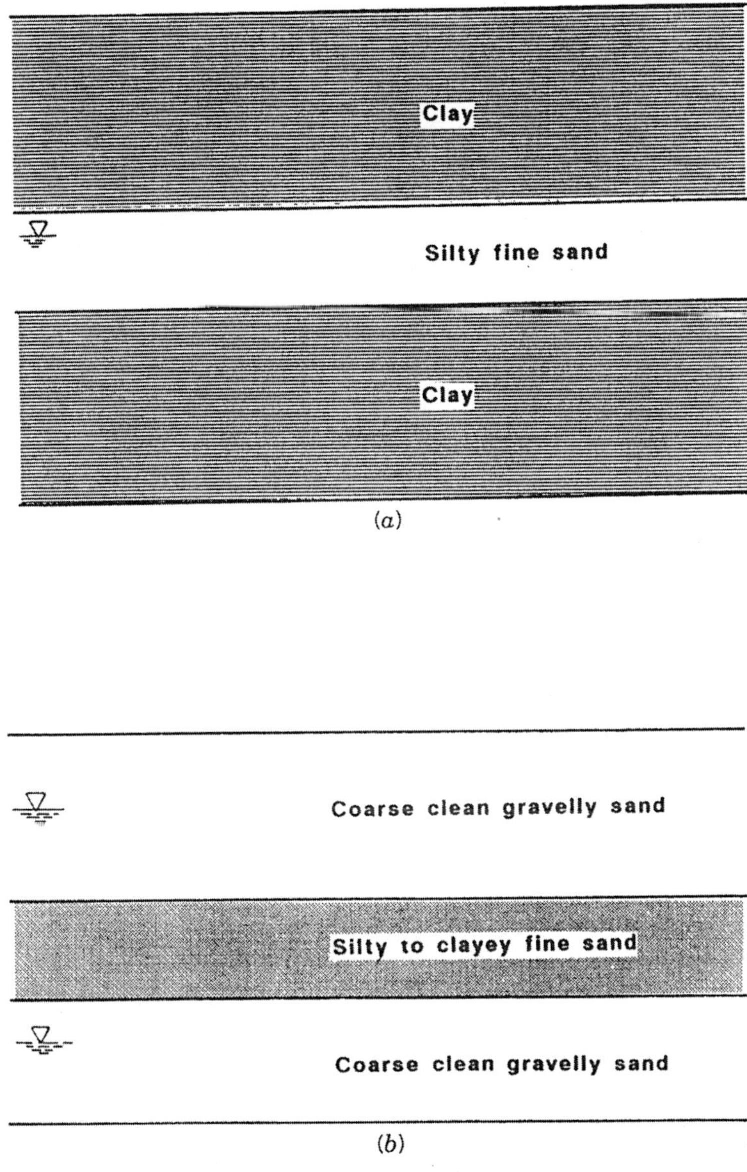

Figure 3.9 Definitions of an aquifer and an aquiclude based on soil permeability characteristics. (*a*) The silty fine sand sandwiched between two clayey layers may be an aquifer. (*b*) The silty fine sand sandwiched between two clean gravelly coarse sand layers may be an aquiclude.

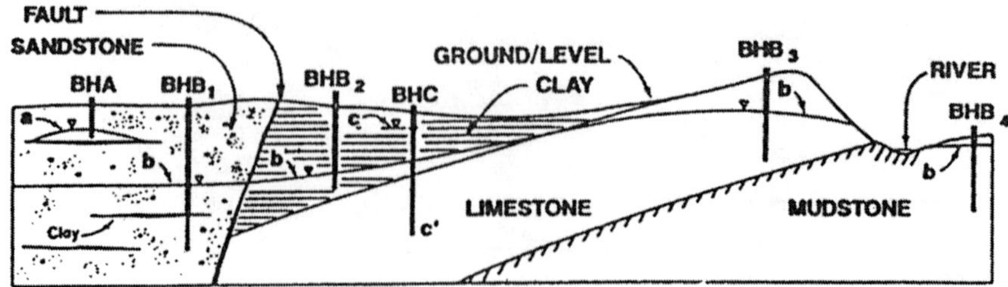

Figure 3.10 A major sandstone aquifer containing clay lenses is faulted against a limestone aquifer that is confined by clay and underlain by mudstone. Note: water tables can exist in aquicludes, as shown in the clay and mudstone. BHA, BHB, and BHC stand for boreholes A, B, and C, respectively. (Blyth and de Freitas, 1984.)

3.3.6 Artesian Water

Artesian water is derived from an artesian aquifer in which the piezometric head of the water pressure is higher than the upper surface of the aquifer, but the water is confined by an overlying aquiclude. In other words, if piezometers are installed in the artesian aquifer, the water in the piezometer tubes would rise to greater elevations from the deeper artesian strata than from strata nearer the ground surface (e.g., the hydrostatic head at point *c* in boring BHC of Figure 3.10). The presence of artesian water should not be overlooked and should be accounted for in any slope stability analysis.

3.3.7 Springs

A spring is a concentrated discharge of groundwater appearing at the ground surface as a current of flowing water. Springs occur in many different forms and have been classified as to cause, rock structure, discharge, temperature, and variability. All springs originate from either gravitational or nongravitational forces. The gravitational springs result from water flowing under hydrostatic pressure, for example, depression springs, contact springs, artesian springs, and solution tubular springs (Figure 3.11). Such springs can have an adverse effect on soil and rock slope stability and should not be overlooked when they are encountered.

Nongravitational springs include volcanic springs associated with volcanic rocks, and fissure springs, which result from fractures extending to great depths in the earth's crust. Such springs are usually thermal.

The discharge of a spring depends on the area contributing recharge to the aquifer and the rate of recharge. Most springs fluctuate in their rate of discharge. Fluctuations are in response to variations in rate of recharge with periods ranging from minutes to years, depending on hydrogeologic conditions. Springs range from intermittent flows that disappear when the water table recedes during a dry season through tiny trickles to an effluence of 3.8 billion liters daily—the abundant discharge of springs along a 16-kilometer stretch of Fall River in California (Leet et al., 1978).

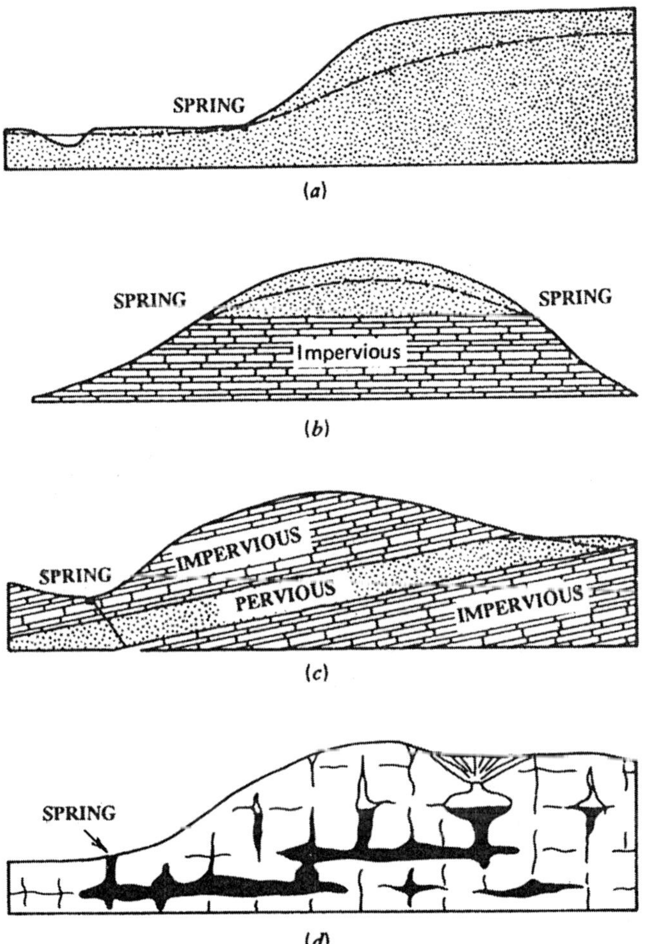

Figure 3.11 Types of gravity springs. (*a*) Depression spring. (*b*) Contact spring. (*c*) Artesian spring. (*d*) Solution tubular spring. (After Bryan; copyright ©1919 by the University of Chicago Press.)

Discharge of springs can be measured in terms of magnitude. Table 3.3 shows there are eight different magnitudes, based on the mean discharge. Large-magnitude springs occur primarily in volcanic and limestone terranes.

3.4 TYPES OF GROUNDWATER FLOW

3.4.1 Runoff

Runoff is that proportion of rainfall or snowmelt that flows from a catchment area into streams, lakes, or seas. It consists of surface runoff and groundwater runoff, where

TABLE 3.3 Classification of Springs According to Magnitude of Discharge

Magnitude	Mean Discharge
First	>100 ft^3/sec
Second	10 to 100 ft^3/sec
Third	1 to 10 ft^3/sec
Fourth	100 gal/min to 1 ft^3/sec
Fifth	10 to 100 gal/min
Sixth	1 to 10 gal/min
Seventh	1 pt/min to 1 gal/min
Eighth	Less than 1 pt/min

Source: Bryan (1919), courtesy of U.S. Geodesic Survey.

groundwater runoff is derived from rainfall or snowmelt that infiltrates into soil down to the water table (Section 3.4.2).

Surface runoff from a catchment area depends on some of the following factors:

- Rainfall intensity
- Area and shape of the catchment area
- Steepness and length of the slopes being drained
- Nature and extent of vegetation or cultivation
- Condition of the surface and nature of the subsurface soils

Surface runoff can be determined by referencing unit hydrographs. In some countries, the method of using unit hydrographs for design of drainage systems for small catchments may not offer any substantial advantage over design methods using empirical equations to represent the complex relationship between rainfall and peak surface runoff. The rational method is commonly adopted, as it is both simple and straightforward to use and, for relatively small catchments, usually yields satisfactory results (Geotechnical Control Office, 1979).

The rational method often used is

$$Q = KiA \qquad \text{(Eq. 3-5)}$$

where Q = maximum runoff in cubic length per time
i = design mean intensity of rainfall in length per time, which is dependent upon the time of concentration
A = area of catchment in square length
K = surface runoff coefficient

The surface runoff coefficient cannot be determined precisely. The recommended value of K for slope drainage is 1 (Geotechnical Control Office, 1984). While some allowance is made for silting by using this value, the drainage system should be designed to minimize siltation and prevent debris from causing blockage. If drainage paths become clogged, higher levels of flooding and infiltration may trigger slope instability.

Visual observation of surface runoff during periods of intense rainfall is desirable. Evidence of sheet runoff sometimes can be found by mud lines and debris that becomes lodged in tall grass and shrubs. Experience shows that the contributions of sheet runoff and intermittent streams to groundwater and pore pressures leading to slope instability are often overlooked in investigations. Securing evidence of surface water and groundwater is a major objective of reconnaissance. Where they are observed in the field, they should be addressed by implementing requisite erosion-protection measures and/or drainage provisions.

Hints of excessive soil moisture can be found in the character of the vegetation or wildlife in the vicinity of the suspect instability area. For example, cattails do not ordinarily grow in low dry areas, and bullfrogs require water for their reproduction and life cycle. Locations of permanent and intermittent surface flows should be recorded on topographic maps. Attention should be paid to the relation of the changes in these water courses to the continuing changes in the area topography.

Rainfall-induced earth movements are governed by the relief surface. For example, earthflows generally head toward large basins in the upper part of the slope; slope debris and weathering material will accumulate in such a basin. Heavy rainfall may trigger the movement of this loose soil mass, resulting in a narrow flow that follows an erosion gully or the channel of a brook, to form a loaf-shaped bulge at the foot of the slope. Another example of the influence of rainfall on earth movement is the vulnerability of concave slopes or valleylike landforms to gully erosions where flows of water are concentrated during heavy rainfalls. Shallow sloughing is common, especially during intense rainstorms in slopes composed of residual soils (Brand, 1982; Lumb, 1975). For shallow sloughing, positive flow regimes with seepage are more or less parallel to the slope face behind an advancing wetting front. Sloughing in residual soils may be attributed to the daylighting relict joints or structures, which lose strength upon saturation by rainwater infiltration.

3.4.2 Infiltration

Infiltration through an unsaturated zone is vertical and causes no positive pore pressures (Section 3.7.1). If the infiltrating rainfall or snowmelt, during its descent, encounters a material of lower permeability, flow will be impeded if the permeability of this lower zone is less than the rate of infiltration. Under this situation, a perched water table will form on the surface of the impermeable zone, and a lateral flow will take place along the upper surface of the impermeable zone (Figure 3.12). Below the impermeable zone, the infiltration rate will be reduced to the value of the permeability of the zone.

When the infiltrating rainfall meets the groundwater table (phreatic surface), most of the vertical component of flow will be destroyed and the lateral flow in the general direction of groundwater flow takes place. Under these circumstances, the groundwater table rises by an amount equal to the depth of saturation caused by the descending

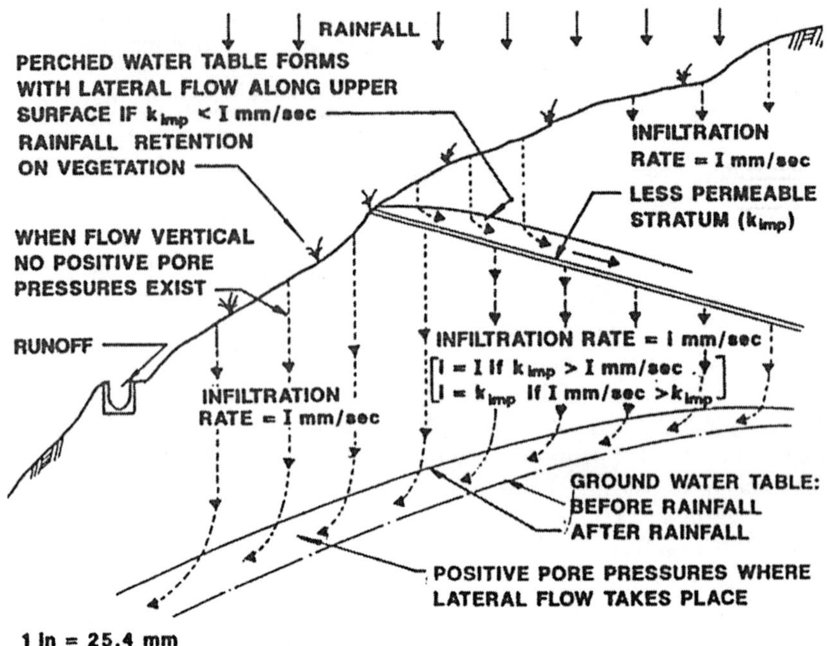

Figure 3.12 Modes of groundwater flow.

zone of infiltration and will be less than the depth of this zone. Within this zone, lateral flow takes place and positive pore pressures exist.

These two modes of groundwater flow affect slope stability by different mechanisms. Above the phreatic surface, the infiltrating rainfall raises the degree of saturation of the soil, which reduces the negative pore pressure and thus the shear strength. As lateral flow develops, pore pressures increase, and as a result effective stresses and shear strength are reduced. The increase of positive pore pressure occurs when the infiltrating rainfall forms a perched water table or has caused a rise in the groundwater table.

Deposits of gravel and sand are able to infiltrate water without difficulty, whereas clay-rich mantles retard the ingress of water and characteristically remain wet after periods of rainfall. Vegetation protects the delicate porous structure of many superficial deposits, especially the crumb-structure of topsoil. Ground covered by vegetation has more uniform infiltration than bare ground.

Infiltration ceases once the voids within the ground are full of water. If the rate at which water is supplied to the surface exceeds the rate at which it can percolate, the infiltration capacity will diminish.

3.4.3 Regional Flow

Regional flow is characterized and often dictated by the local geological boundaries that define the volume of an aquifer, and by hydrological boundaries that define the

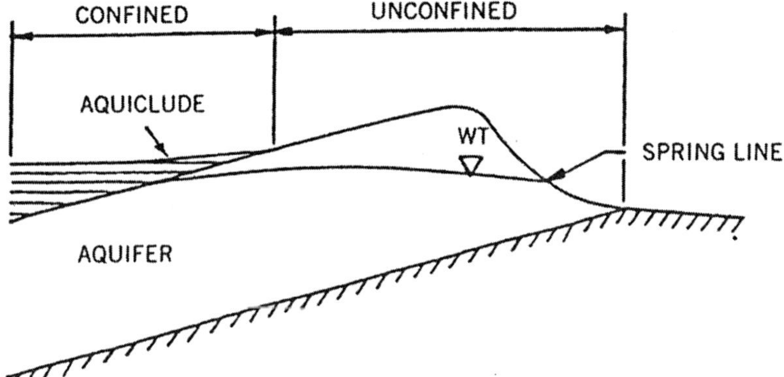

Figure 3.13 Unconfined outcrop of scarp face with dip slope confined beneath aquiclude.

volume of water stored within it. Common geological boundaries are stratification of aquifers and of aquifers against aquicludes (Figure 3.13), the termination of aquifers by faults (Figure 3.14), and unconformable and igneous intrusions (Figure 3.15). Hydrological boundaries include the spring lines (Figure 3.13), coastal and shore lines (Figure 3.16), and river, lake, and reservoir levels (Figure 3.17), which usually fluctuate in elevation and are dynamic.

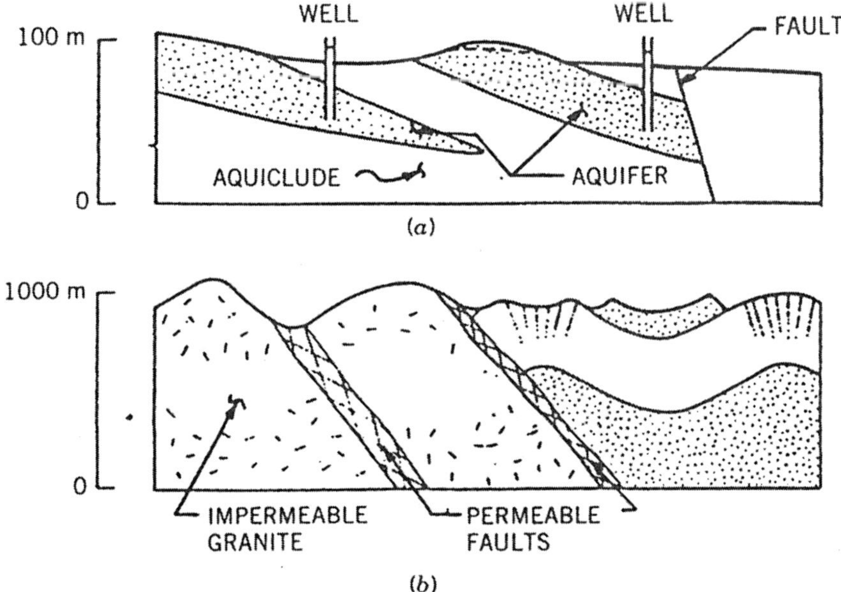

Figure 3.14 Characteristic aquifers. (*a*) As strata. (*b*) As zones associated with faulting and tensile fracturing on fold crest. (Blyth and de Freitas, 1984.)

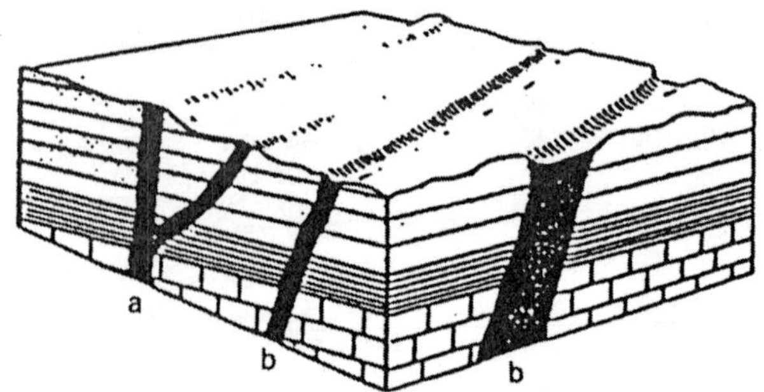

Figure 3.15 Dykes. (*a*) More resistant to weathering. (*b*) Less resistant than the country rock. (Blylth and de Freitas, 1984.)

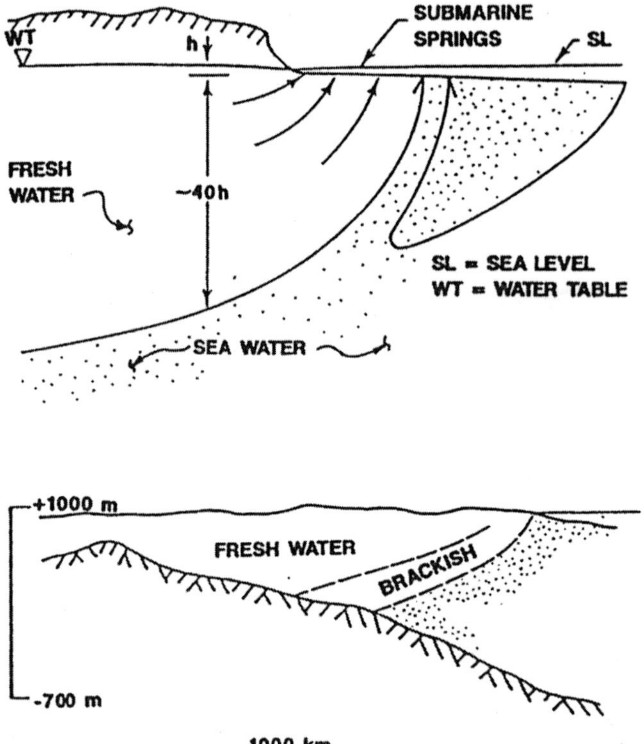

Figure 3.16 Two examples of the relationships between seawater and freshwater at a coast. (Blyth and de Freitas, 1984.)

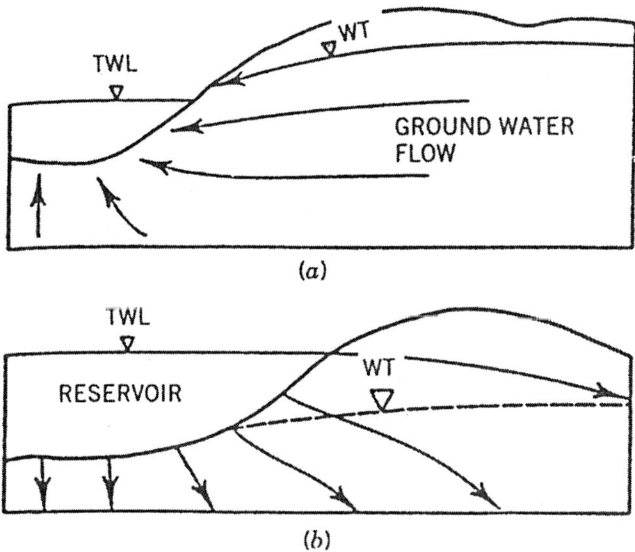

Figure 3.17 Watertight and leaky reservoirs. (*a*) The total head of water in the ground exceeds that in the reservoir and there is no leakage of reservoir water. (*b*) The reverse situation, resulting in leakage. WT = water table; TWL = top water level. (Blyth and de Freitas, 1984.)

Geological boundaries tend not to change with time; however mining or massive earth excavations could disrupt the regional flow. Groundwater that is disrupted in this fashion usually has little chance to recover its original hydraulic characteristics.

Groundwater moves at various speeds through the ground depending on its flow path. Near-surface flows move the fastest and normally supply most of the water that discharges at springs. As the groundwater levels fluctuate, the springs may migrate up a topographic slope in response to periods of recharge in wet weather and recede down the slope in periods of dry weather. The velocity with which groundwater circulates in the ground gradually decreases with depth, and the movement of the deep groundwater may be extremely slow.

3.5 FLUCTUATION OF GROUNDWATER LEVELS

3.5.1 Rainfall

The association between rainfall and slope problems has been discussed in Section 3.4.1. The knowledge of the fluctuation of groundwater table due to the rainfall is of paramount importance in analyzing the stability of slopes.

How much rainfall permeates into a slope depends on the same factors as the surface runoff (Section 3.4.1). Because these factors vary considerably between slopes, it is not possible to draw general relationships between rainfall and groundwater response with accuracy. Ideally, the groundwater conditions assumed in any slope stability analysis should be those observed in the field. However, for very rare rainfall events, this is seldom possible. To tackle this problem, an automatic groundwater monitoring system may be devised and implemented. The Forest Service of the United States has developed instrumentation for monitoring groundwater under inaccessible mountainous terrains and extreme weather conditions (Prellwitz and Babbitt, 1984). The system, powered by rechargeable batteries, stores groundwater data on solid-state integrated-circuit storage modules that are read directly into a host computer for data processing, and hence can be operated unattended.

Alternatively, the critical groundwater table can sometimes be monitored by means of Halcrow buckets installed in open standpipes or piezometers. Halcrow buckets are small buckets (about 1-inch long and 0.2- to 0.3-inch wide) made of fiberglass and connected together in 5-foot-intervals (note: intervals can be varied to any length) by nylon strings (Figure 3.18 and Figure 3.44*b*). The bucket has a small hole in its central portion. Groundwater that rises will enter the hole and fill the bucket. The bucket is

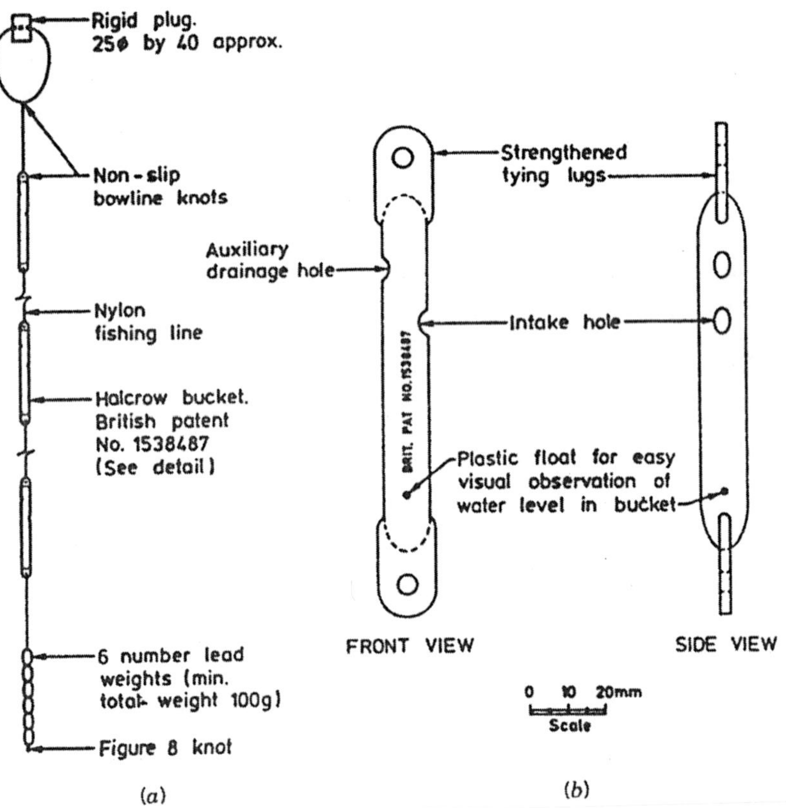

Figure 3.18 Halcrow buckets. (*a*) Assembled bucket string. (*b*) Bucket detail. Notes: (1) not to scale; (2) assembled bucket string must fit into 7.5-millimeter standpipe without sticking. (Geotechnical Control Office, 1987.)

designed in such a way that water cannot move out of the bucket upon entering. When the groundwater table recedes, the water in the higher buckets still remains in place. The groundwater table can therefore be monitored any time before or after heavy rainfalls. The system is very inexpensive and has received wide practical application in Europe and southeast Asia.

Piezometers and open standpipes may be observed closely over at least one wet season and the groundwater response at the site correlated to rainfall. In most cases, this will require the installation of rain gages at, or near, the site and frequent observation of piezometers. Then determination of a site-specific trend between rainfall and groundwater response may be possible.

High groundwater levels often lead to slope failures in natural areas where the groundwater reservoir is recharged solely by infiltration of rain water. The relation between rainfall and slope failure has been well documented in the literature (e.g., Briggs et al., 1975; Crozier and Eyles, 1980; Eyles, 1979; Eyles et al., 1978; Krohn and Slosson, 1976; Nilson and Turner, 1975; Nilson et al., 1976; Radbruch-Hall et al., 1982; Reid et al., 1988; Slosson and Krohn, 1979, 1982). Rainfall is known to have induced major slope failures in southern California in 1973, 1978, and 1980.

As the incidence of slope failures is closely related to periods of intense rainfall, attempts have been made to correlate between the slope failures and antecedent and proximate rainfall (Brand, 1982; Lumb, 1962, 1975). A study by Lumb (1975) showed that there is a definite correlation between slope failures and the intensity of rainfall. Figure 3.19 shows the results in terms of the 24-hour rainfall on the day of landslides and the cumulative rainfall over the previous 15 days. Based on the existing data, it is possible to predict the probability of the future occurrence of landslides.

Many investigations have been launched in the past decade to quantify the relation between precipitation, pore pressure change, and slope failures, particularly in residual soils. Comprehensive investigations in Hong Kong ultimately showed that rainfall history was not particularly important (Brand, 1989). In Hong Kong, if the 24-hour rainfall is high, landslides will occur irrespective of the magnitude of antecedent rainfall. If the 24-hour rainfall is very low, landslides will not occur whatever the magnitude of the antecedent rainfall (Brand, 1989).

3.5.2 Floods

When rain falls at a rate that exceeds the infiltration capacity of the ground, flooding occurs, resulting in ponding of water. Ponding of water in the vicinity of a slope reduces the available strength of the slope since ponded water eventually seeps into the slope and increases pore pressures on potential sliding surfaces. The toes of highway fill embankments are often highly vulnerable to ponding. High concentration of stresses occurs in the toe area, and anything that disturbs that area, including excess water, may cause slope instability. Ponding saturates the soils and increases the driving forces of the slope, possibly above the resistance of the slope of failure.

Areas subject to frequent flooding are prone to instability. One instability mechanism in flooded areas is erosion from rapidly flowing water. A second mechanism is

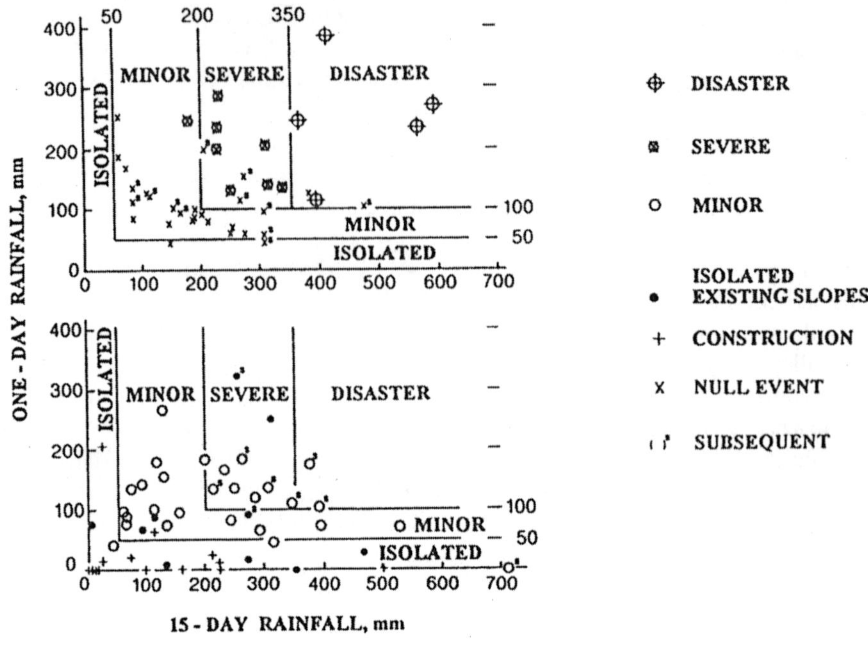

Note: 1 inch = 25.4 mm

Disastrous Event:	Territory-wide damage with more than 50 individual slips recorded in one day
Severe Event:	Wide-spread damage with between 10 and 50 slips in one day
Minor Event:	Localized damage with less than 10 slips in one day
Isolated Event:	A single individual slip

Figure 3.19 Correlation between slope failures and intensity of rainfall. (Lumb, 1975; from Brand, 1982, reproduced by permission of ASCE.)

from a condition known as "sudden drawdown," which most often relates to dikes, dams, and reservoirs, as discussed in Section 3.5.4.

3.5.3 Snowmelt

For slopes, water from snowmelt either infiltrates the ground if the infiltration capacity of the ground is higher than the infiltration rate, or continues to travel downslope as surface runoff. Effects of the snowmelt on the stability of the slope are often dependent upon the layout and configuration of the slope, the material of which the slope is composed, the location of the groundwater table, and the type of slope face protection. Sometimes the ground underneath a snow bank may still be frozen even though the snow has begun to melt on top. In that case, the effects of snowmelt on the slope may be evident a fair distance downslope of the snow bank because the water is not able to infiltrate the frozen ground.

3.5.4 Sudden Drawdown

Sudden drawdown is a rapid lowering in the level of water standing against a slope. This particular situation is commonly encountered in the design of bridge approach embankments at river crossings. In such a case, rapid drawdown occurs when the river falls following a flood. Other examples of sudden drawdown include the lowering of a reservoir adjacent to the upstream slope of an earth dam, lowering of the water level next to a natural slope, and a drop in the sea level next to a slope. Most of the highway slope failures due to sudden drawdown occur at stream crossings.

Failures due to sudden drawdown often occur in embankment slopes composed of clayey materials in which the excess pore water pressures do not have enough time to dissipate, thereby reducing the overall shear strength of the clay materials. Undrained shear strength is lower than drained shear strength. Also, the water acts as a stabilizing pressure against the slope face when in place. If the water level against the slope face is suddenly drawn down, this stabilizing pressure is removed suddenly, but the driving forces within the slope are relieved much more slowly, creating an unbalanced condition.

Figure 3.20 illustrates the three stages necessary for the development of a sudden drawdown condition. The pore pressure changes occurring during sudden drawdown may be estimated by charts and empirical formulas proposed by Bishop (1954), Duncan et al. (1987), and Janbu (1973), see also Chapter 6.

3.6 INFLUENCE OF GEOLOGICAL STRUCTURES ON GROUNDWATER FLOWS

Minor geological structures in slopes can have a major effect on the flow patterns that develop naturally and on the effectiveness of artificial drainage systems. Impervious barriers near the outer extremities of slopes can retard natural drainage, raise the general saturation level, or cause water to build up within the slopes. When this occurs, the stability of slopes can be jeopardized unless artificial drainage systems are in place.

Figure 3.21 illustrates idealized cross sections of geological conditions that are conducive to the buildup of groundwater within slopes. In Figure 3.21*a*, the diagram on the left shows a shallow cover of impervious shale cutting off the natural drainage of a bed of more pervious sandstone, and the diagram on the right describes an impervious rock formation with pervious joints with no natural outlets. In both diagrams, plane *AB* is the inside face of an outer cover of impervious material. Without artificial drainage, water pressures can build up to dangerously high levels on plane *AB* (Figure 3.21*b*),

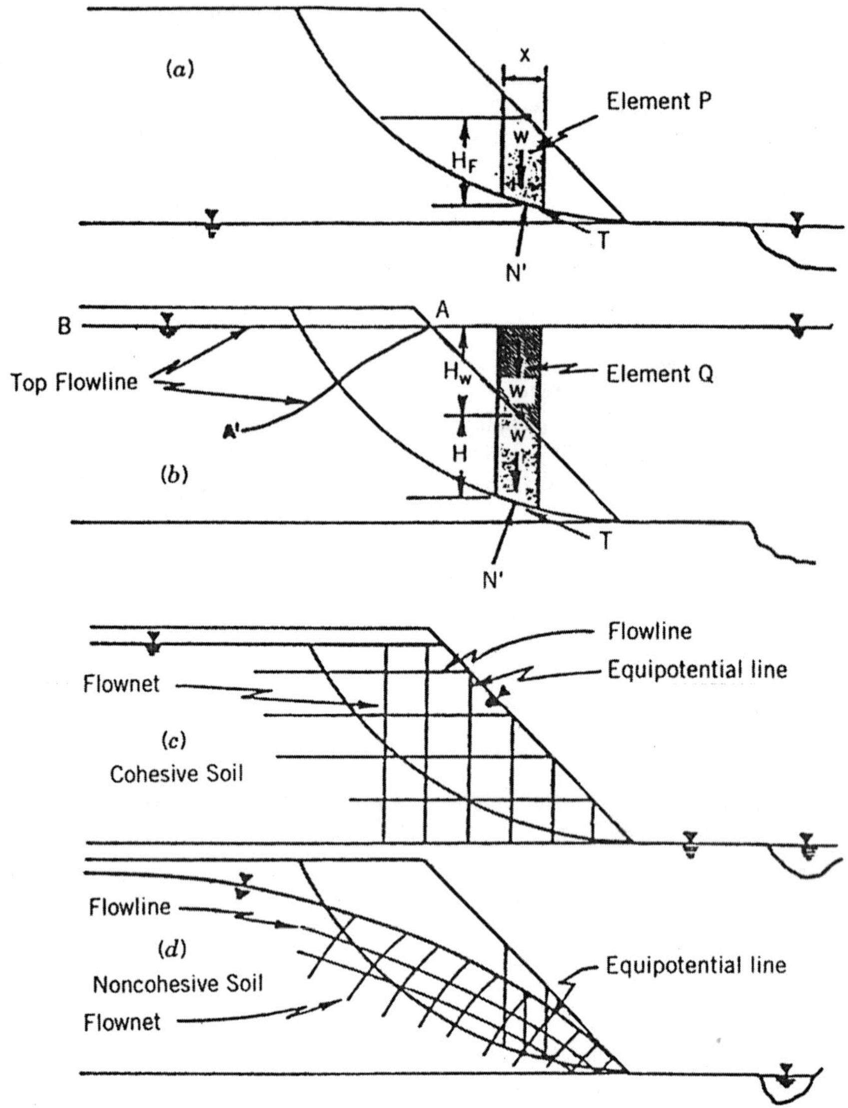

Figure 3.20 Development of a rapid drawdown condition. (*a*) Before a rise in the water level. (*b*) High water level. (*c*) Rapid drawdown in cohesive soils. (*d*) Rapid drawdown in noncohesive soils. (Hopkins et al., 1975.)

causing collapse of the slope if the pressures are not relieved. The hydrostatic pressures may be lowered significantly, as shown in Figure 3.21*c*.

Shallow slope failures may sometimes occur when water in a pervious layer is perched on an impermeable layer, for example, sand overlying a clay zone. This happens often after periods of intense rainfall when the perched water table rises

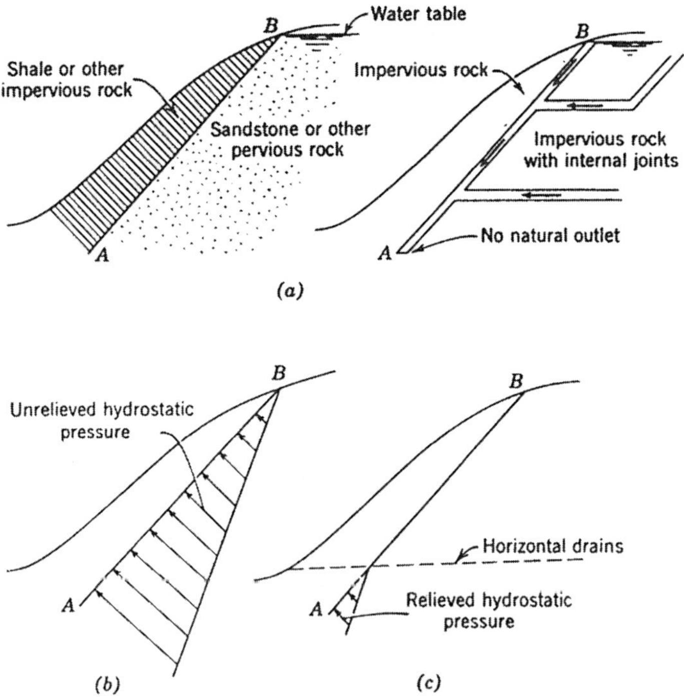

Figure 3.21 Influence of geological structures on flow patterns. (*a*) Internal details conducive to large hydrostatic pressures. (*b*) Unrelieved hydrostatic pressures. (*c*) Pressures relieved by drainage. (Cedergren, 1977.)

to the ground surface. The surface slough-off experienced in residual soils or rocks during and after rainfalls may sometimes be a result of this phenomenon.

3.7 PORE PRESSURES

As mentioned in Section 3.1, subsurface water is divided into zones of positive and negative pore pressures. The dividing line is the groundwater table where the pressure is equal to atmospheric pressure (Table 3.1). Below the groundwater table, the soil is fully saturated, and the pore pressure is above atmospheric pressure and positive in value—positive pore pressure. Above the groundwater table where the soil is unsaturated, the pore pressure is below atmospheric pressure and hence is negative in value—negative pore pressure (soil suction). Any changes in these pore pressures would alter the soil shear strength and therefore have a tremendous effect on the slope stability.

The groundwater table is generally determined from the level of water in an open standpipe or by piezometers (Section 3.7.3). The changes in pore pressures in the two zones affect slope stability in different ways. A schematic diagram of changes in pore pressure in the two zones as a result of rainfall is shown in Figure 3.22.

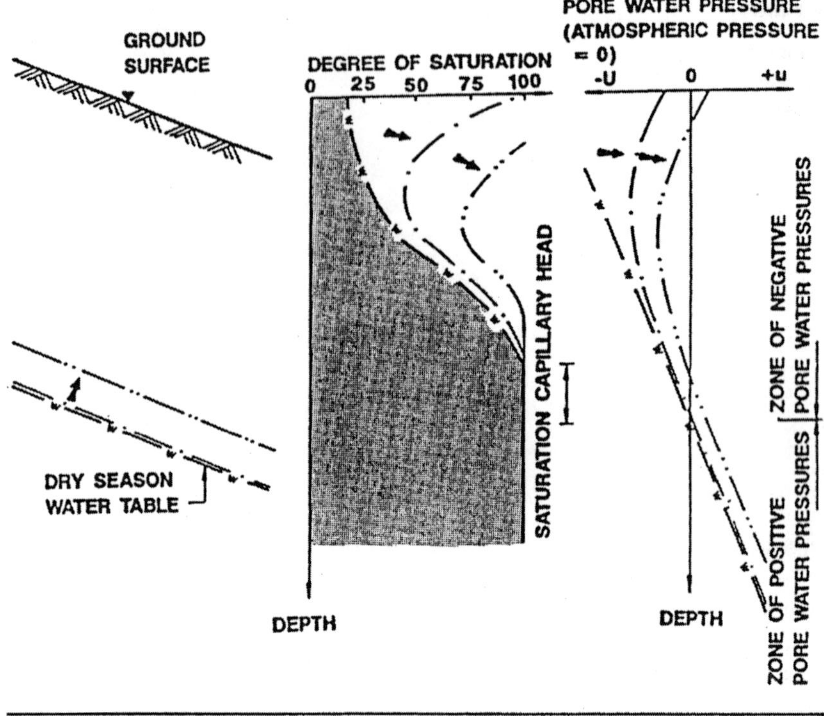

LEGEND:

— w — CONDITIONS AT TIME t_0 (BEFORE THE RAINFALL)

— · — CONDITIONS AT TIME t_1 (IMMEDIATELY AFTER THE START OF RAINFALL

— ·· — CONDITIONS AT TIME t_2 (AFTER A LONG PERIOD OF RAINFALL)

➞ RISE IN WATER TABLE OR INCREASING SATURATION AND PORE WATER PRESSURE DUE TO RAINFALL

Figure 3.22 Typical changes in water table; degree of saturation (S) and pore water pressure (u) are due to rainfall. (Geotechnical Control Office, 1984.)

The response of the groundwater regime to rainfall or snowmelt varies widely from slope to slope, ranging from no response to large and immediate responses. For large projects with slopes whose failure would result in direct social and economic impact, installation of open standpipes and/or piezometers and subsequent regular monitoring, especially during and after heavy rainfall, are imperative for an accurate assessment of the maximum possible rises in pore pressures at a particular location.

3.7.1 Positive Pore Pressures

At the phreatic surface, the pore water is at atmospheric pressure, while below the phreatic surface the pore pressure will be above atmospheric (note that atmospheric pressure is usually taken as the zero pressure datum, and so a positive pore pressure zone exists below the water table). If there is no flow, the pore pressures are hydrostatic,

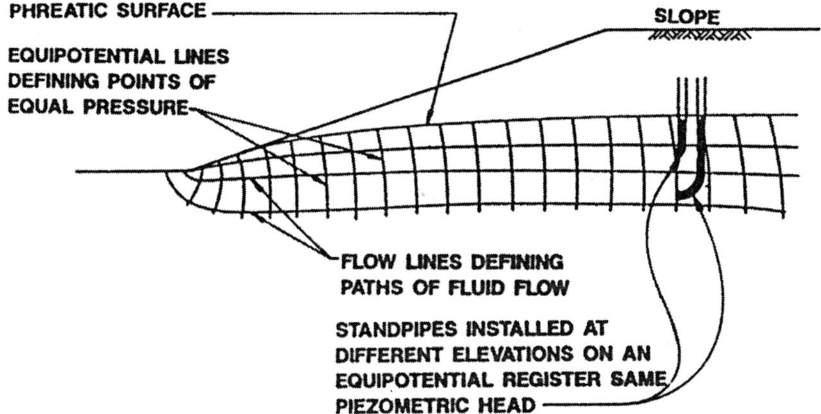

Figure 3.23 Pore pressures in an equipotential line.

and the water level measured in a piezometer within the saturation zone will coincide with the water table. Pore pressures are no longer hydrostatic if there is any flow (Figure 3.23). In this instance, the pore pressure from any point within the soil mass is computed, by means of a flow net, from the difference in head between the point and the free water surface. Calculations of pore pressures in slopes are discussed in Section 3.10.1.

By lowering effective stress, positive pore pressure reduces the available shear strength within the soil mass, thereby decreasing the slope stability. Increases in positive pore pressure can be rapid after a period of heavy rainfall. That is a major reason why many slope failures occur after heavy rainfall. The rate of increase, however, depends on many factors such as the rate of rainfall, the nature of the ground surface, the catchment area, and the soil permeability.

Observation of seepage out of slope faces gives additional information on the position of phreatic surface, and its rise due to rainfall. Care must be taken to distinguish between seepage of groundwater and runoff; otherwise erroneous conclusions can be drawn as to the response of the phreatic surface to rainfall. Normally, this can be accomplished by observing groundwater levels in open standpipes and/or piezometers for a sufficient period of time after rainfall or snowmelt periods. The water levels will eventually reach an equilibrium condition that can be used in slope stability analysis.

Pore pressures below the groundwater table can be assessed using analytical, numerical, analogue, and graphical methods. When the permeability is isostatic, flow nets in the form of flow lines and equipotentials can be sketched on true-to-scale drawings (Figure 3.24*a*). Under anisotropic conditions, transformed sections cna be used (Figure 3.24*b*). Flow-net sketching is explained in many text books, such as Cedergren (1977), Harr (1962), and Lambe and Whitman (1969).

The technique of constructing flow nets for steady-state flow conditions is applicable to two-dimensional homogeneous soil conditions, which are not the conditions usually found in the field. However, flow-net construction is considered to be a useful

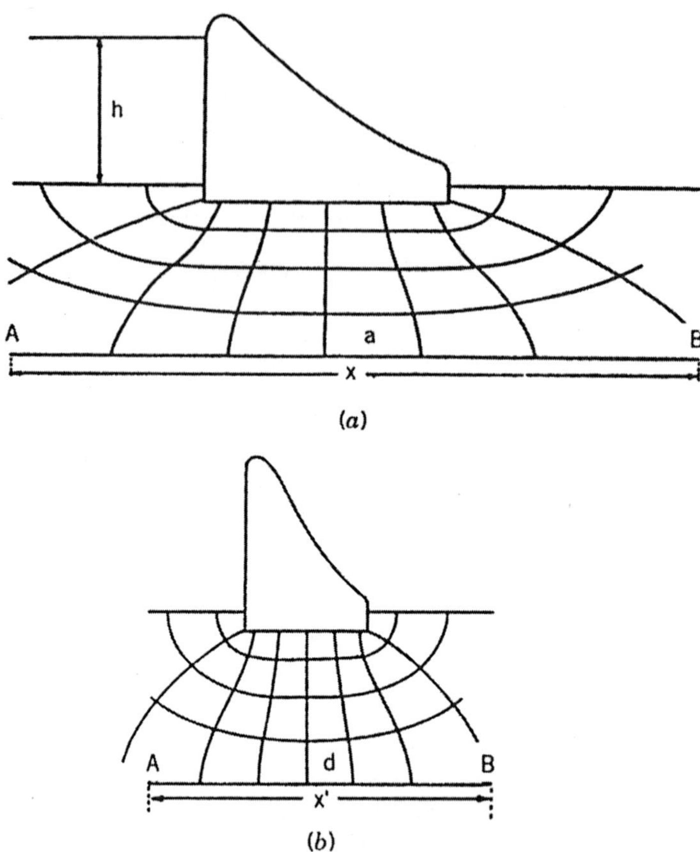

DEVELOP THE TRANSFORMATION FACTOR a.

IF x' = ax, $\frac{\partial h}{x'} = \frac{\partial h}{x}\frac{\partial x}{x'} + \frac{\partial h}{y}\frac{\partial y}{x'}$

BUT $\frac{\partial y}{x'} = 0$

Figure 3.24 (*a*) Actual flow nets for $k_x > k_y$, sketched true to scale. (*b*) Sections of flow net transformed due to soil anisotropy.

first step in the assessment of pore pressure distribution. The flow-net method is applicable only in the positive pore pressure zone and ignores the effect of infiltration into the water table from above.

Various physical analogue methods are available for determining flow nets and pore pressure distributions in slopes (Todd, 1980). These analogue methods are often site-specific and cannot be justified for general use. Numerical techniques, using finite differences or finite elements, provide powerful methods for obtaining pore water distributions in slopes, and they are the only means by which transient flow

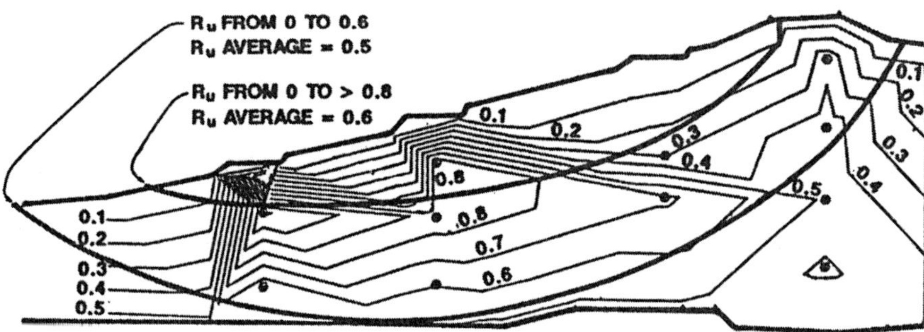

Figure 3.25 Contours of R_u from field measurements. (From Lambe and Silva-Tulla, 1992, reproduced by permission of ASCE.)

situations can be fully modeled. If adequate computer facilities and standard programs are available, these can provide useful design information.

Lambe and Whitman (1969) describe the use of a flow net to determine the pore pressure along a potential failure surface. To obtain the average pore pressure along the potential failure surface, it is necessary to sum the values of pore pressures along the surface and divide the sum by the length of the surface. Figure 3.25 shows contours of R_u obtained from measurement at piezometers in the field (Lambe and Silva-Tulla, 1992). R_u is the pore pressure ratio equal to the ratio of pore pressure to overburden pressure, $u/\gamma_t z$. In practice, the engineer often assumes a constant R_u for the cross section under study and then performs stability analyses to locate the most critical slip surface. Figure 3.26 shows the flow patterns implied by constant R_u values.

A significant difference exists between average pore pressure (the average value of pore pressure for a given stage and geometry) and constant pore pressure for a given section. As pointed out by Lambe and Silva-Tulla (1992), average pore pressure can be correctly used to make a rational stability analysis. However, a search for the section with the minimum factor of safety assuming a constant R_u will always yield meaningless results. A constant value of R_u only exists for a slope with a straight line seepage and a phreatic surface at the ground surface (Lambe and Silva-Tulla, 1992).

3.7.2 Negative Pore Pressures

The negative pore pressure zone is located above the phreatic surface. In this zone, the pore water is continuous or semicontinuous and the pore water pressure is below atmospheric. The magnitude of the negative pore pressure (sometimes called soil suction) is controlled by surface tension at the air–water boundaries within the pores and is governed by grain size. In general, the finer the soil particles, the larger the saturation capillary head, and hence the higher the negative pore pressure. High values of negative pore pressures have been measured in the laboratory in undisturbed samples of residual soils (Wong, 1970).

Negative pore pressures increase the effective stresses within a soil mass and improve the stability of a slope. Ho and Fredlund (1982a) suggested the increase in

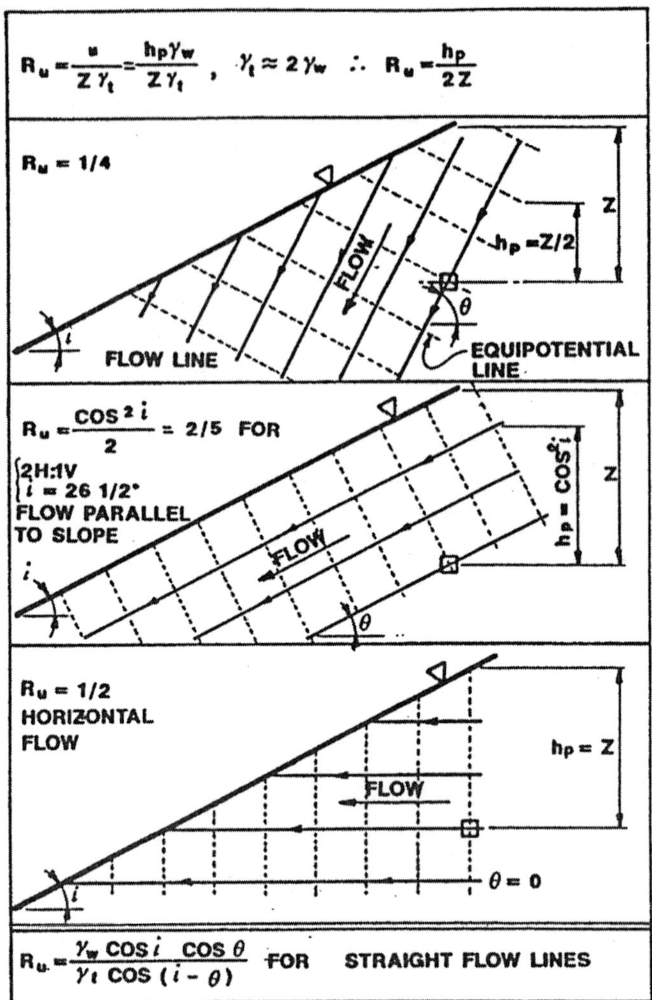

∴ FOR θ = 90°, Ru = 0, i.e. FLOW IS VERTICALLY DOWNWARD FROM THE SLOPE FACE

Figure 3.26 Flow as a function of R_u (for isotropic soil and water table at slope face). (From Lambe and Silva-Tulla, 1992, reproduced by permission of ASCE.)

shear strength due to negative pore pressure (soil suction) may be expressed as

$$C = c' + (u_a - u_w)\tan\phi_b \qquad \text{(Eq. 3-6)}$$

where C = total cohesion of the soil
c' = effective cohesion
$(u_a - u_w)$ = matrix suction
ϕ_b = the slope of the plot of matrix suction when $\sigma - u_a$ is held constant

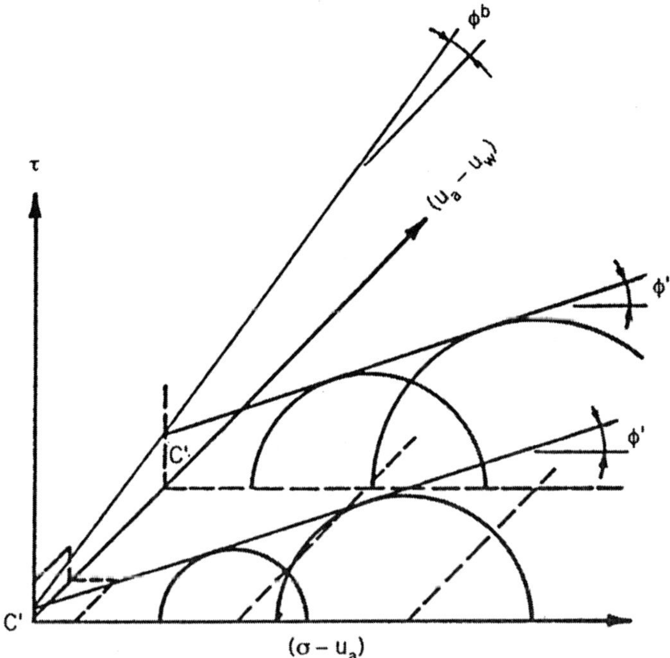

Figure 3.27 Three-dimensional failure surface using stress variables $(\sigma - u_a)$ and $(u_a - u_w)$. (From Ho and Fredlund, 1982*a*, reproduced by permission of ASCE.)

In other words, a metrix suction $(u_a - u_w)$ increases the shear strength by $(u_a - u_w)$ tan ϕ_b.

The increase in soil strength can be represented by a three-dimensional failure surface using stress variables $(\sigma - u_a)$ and $(u_a - u_w)$, as shown in Figure 3.27. These negative pore pressures reduce in magnitude when the degree of saturation increases and become zero when the soils are fully saturated, for example, during and after a heavy rainfall. The major problem in evaluating stability in unsaturated soils is associated with the assessment of the reduction in negative pore pressure and possible positive pore pressure increase as a function of rainfall history. As such, they are usually not considered in the design of slopes even though they are often quoted as the factor enhancing the stability of a steep slope.

3.7.3 Measurement of Pore Pressures

Positive Pore Pressures Positive pore pressures are usually estimated from groundwater conditions that may be specified by one of the following methods.

Phreatic Surface This surface, or line in two dimensions, is defined by the free groundwater level. This surface may be delineated in the field by using open standpipes or piezometers (Figure 3.28*a* and *b*). The most commonly available form of modeling

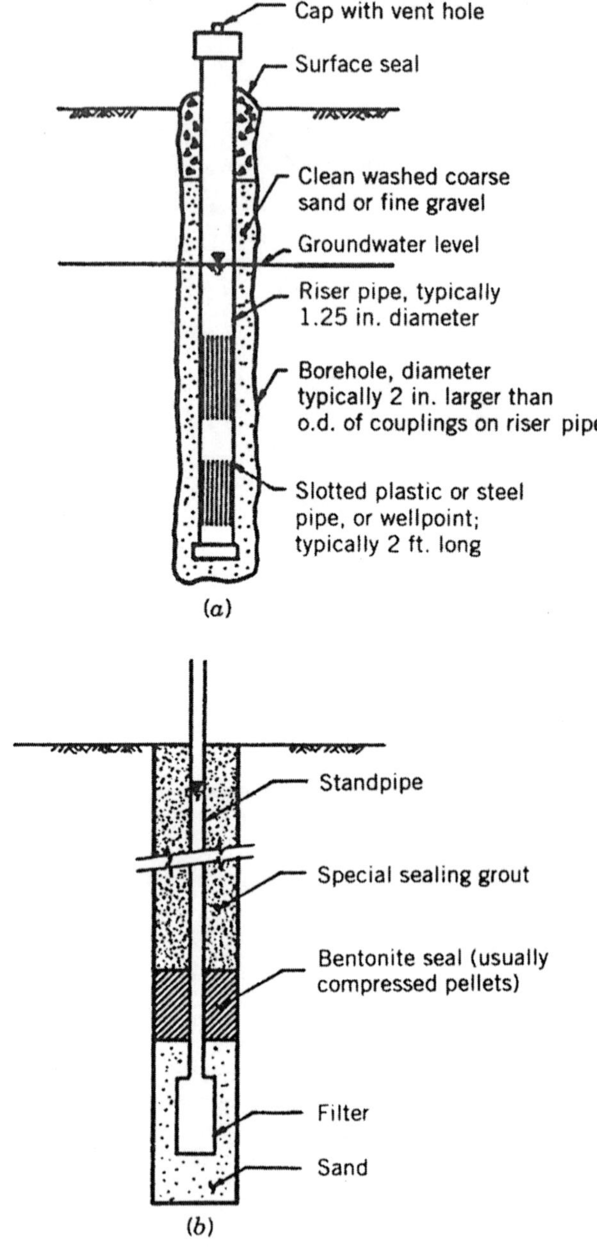

Figure 3.28 (*a*) Schematic of observation well. (*b*) Schematic of open standpipe piezometer installed in a borehole.

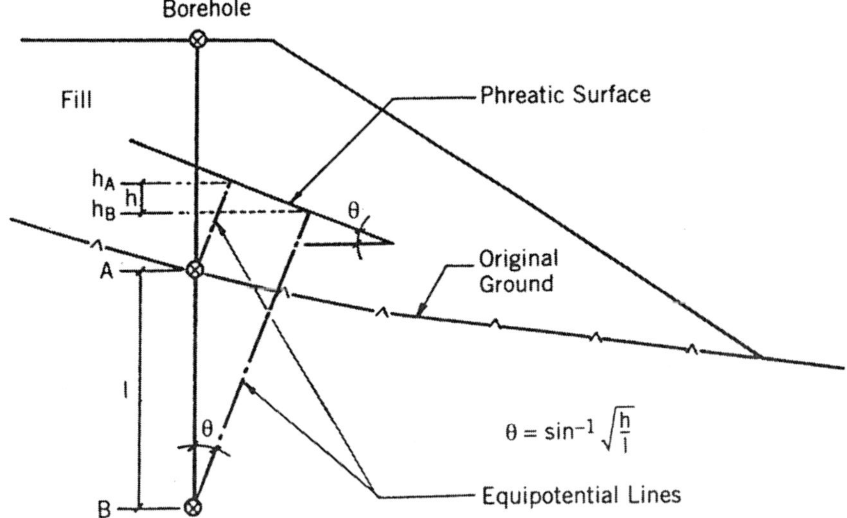

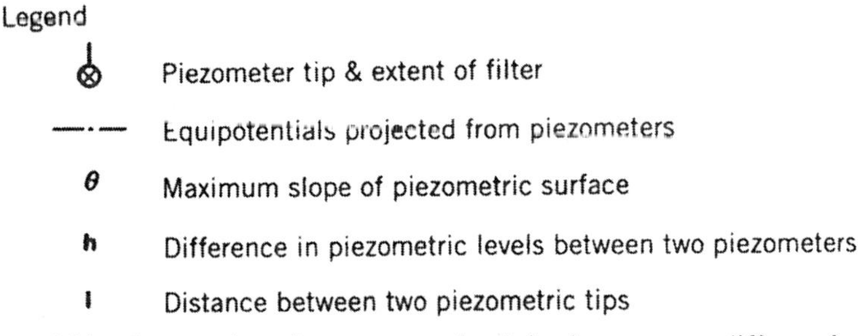

Figure 3.29 Computation of pore pressure heads in piezometers at different depths.

for the groundwater table when using slope stability computer programs is the phreatic surface. The observed water levels in many standpipes and piezometers, installed at different depths, can be used to assess the phreatic surface (Figure 3.29) and whether the groundwater is in a flowing or drawdown condition (Figure 3.30*a* and *b*).

The phreatic surface, which is curved in reality, is usually assumed to be straight for analyzing slope stability. This assumption will provide higher or lower estimates of pore water pressure for a curved (convex) phreatic surface. For the steeply sloping phreatic surface, a convex-shaped phreatic surface, as shown in Figure 3.31, generates an overestimate of the pore water pressures, whereas a concave-shaped surface may lead to an underestimation. This overestimation is entirely due to the assumption of

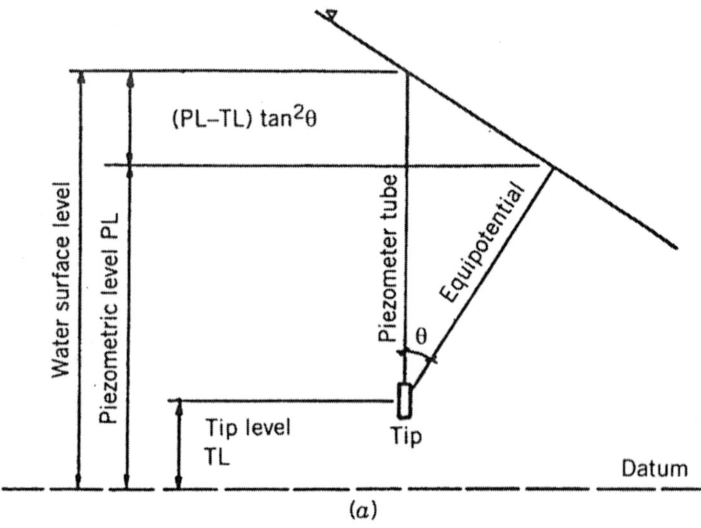

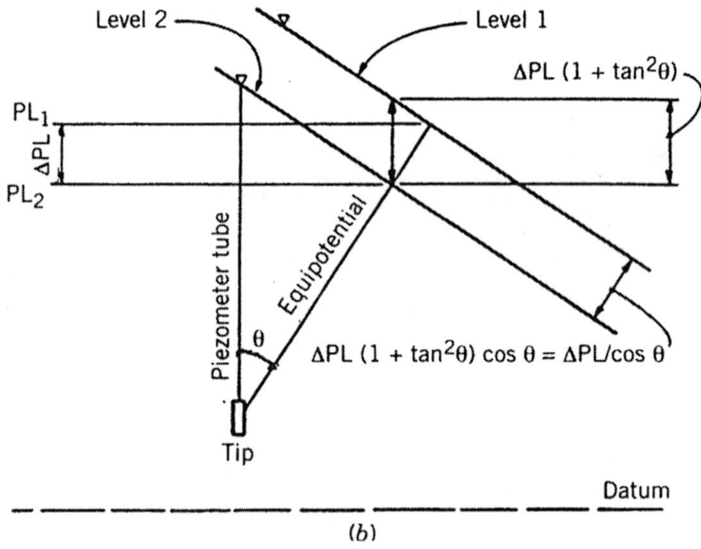

Figure 3.30 Water surface and drawdown relations for a sloping aquifer. (*a*) Steady water level. (*b*) Drawdown water level, ΔPL.

straight equipotential lines intersecting the projected phreatic surface line, *CD*. If the actual phreatic surface (line *AB*) is steeply curved, the equipotential lines must also curve as shown. In this example, a typical solution would use the greater pore water pressure head h_1 rather than the true head, h_2. However, this overestimation is small and would not significantly affect the safety factor of the slope.

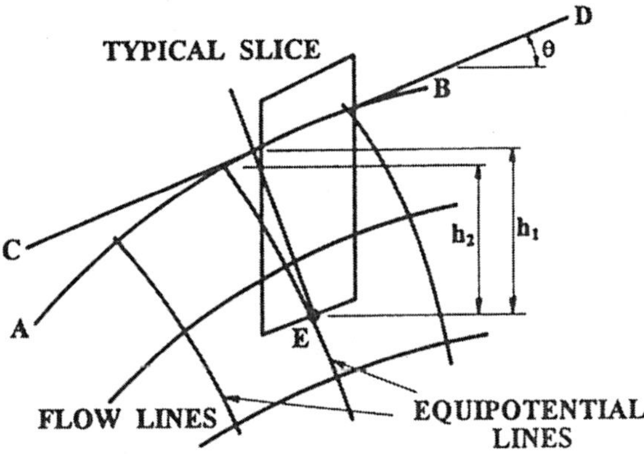

AB - ACTUAL PHREATIC SURFACE

CD - ASSUMED INCLINATION OF PHREATIC SURFACE WITHIN SLICE

Figure 3.31 Phreatic surface and curved equipotential lines.

Piezometric Data This method includes (1) specification of pore pressures at discrete points within the slope and (2) use of an interpolation scheme to estimate the required pore water pressures at any location. The piezometric pressures may be determined from:

Field piezometers
Manually prepared flow net
Numerical solution using finite differences or finite elements

Although this approach is only available in a few slope stability programs, it is the best method for describing the pore water pressure distribution (Chugh, 1981b).

Pore Water Pressure Ratio This is a popular and simple method for normalizing pore water pressures measured in a slope according to the definition

$$r_u = u/\sigma_v \qquad \text{(Eq. 3-7)}$$

where u = the pore pressure
σ_v = the vertical subsurface soil stress at depth z

Effectively, the r_u value is the ratio between the pore pressure and the total vertical stress at the same depth.

This factor is easily implemented, but the major difficulty is associated with the assignment of the parameter to different parts of the slope. Often, the slope will require an extensive subdivision into many regions with different r_u values. This method, if used correctly, will permit a search for the most critical surface. However, it is usually reserved for estimating the factor of safety value from slope stability charts or for assessing the stability of a single surface.

Piezometric Surface This surface is defined for the analysis of a unique, single failure surface. This approach is often used for the back-analysis of failed slopes. Because the combination of a piezometric and failure surface is unique, a search for the critical surface is not possible. The user should note that a piezometric surface is *not* the same as a phreatic surface, as the calculated pore water pressures will be different for the two cases (Section 3.10.1).

Constant Pore Water Pressure This method is used to specify a constant pore water pressure in any particular soil layer. This may be used to examine the stability of fills placed on soft soils during construction, where excess pore water pressures are generated according to consolidation theory.

Negative Pore Pressures Measurement of negative pore pressure is not as common as the measurement of positive pore pressure. Instruments used to measure negative pore pressure in the range of 0 to 11.6 pounds per square inch (80 kilopascals) are called tensiometers. At suctions greater than −11.6 pounds per square inch (−80 kilopascals), water inside the tensiometer cavitates and is lost through the ceramic tip. The reliability of a tensiometer depends on a good contact between the soil and the ceramic tip, and a good seal between the tensiometer tube and the soil.

3.8 WATER LEVELS FOR DESIGN

3.8.1 General

The extent to which infiltration from rainfall reduces the stability of slopes is dependent on a number of factors, such as the original position of groundwater table, the intensity and duration of the rainfall, the antecedent rainfall within the groundwater catchment, the recharge potential of the area, the geology, the degree of saturation, and the topography.

Design of slopes is commonly based on the groundwater conditions that would result from a return period of either 10, 20, or 50 years, which is determined by local authorities. The water level due to a specified-year return period can be determined in two ways (Geotechnical Control Office, 1984). The first involves the analysis of piezometric data taken before, during, and after rainfall. Various methods are available for determination of water levels from piezometric records. These include the statistical correlation of groundwater response with rainfall, groundwater modeling of the aquifer system, and the extrapolation of observed piezometric responses. The second involves the solution of an equation that describes the formation of a wetting band zone of 100 percent saturation (Section 3.8.2), which is dependent on the porosity, permeability, initial and final degrees of saturation of the soil forming the slope, and the percentage of the anticipated return period rainfall that will infiltrate into the ground directly above or behind the slope.

Different aquifer systems have different responses to rainfall depending on their storage characteristics. Some aquifers may display rapid response to intense rainfall, while others may react with a gradual rise in water level during the wet season. Because of this variability in aquifer response, it is recommended that the groundwater conditions to be used in stability analysis be based on water levels measured in the field by piezometers or open standpipes. For sites where there are no piezometer data available, the wetting band approach (Section 3.8.2) may be used to give a rough estimate of groundwater levels.

3.8.2 Wetting Band Approach

The wetting band approach (Lumb, 1962, 1975) assumes that the wetting band descends vertically under the influence of gravity, even after the cessation of rain, until either the main water table or a zone of lower permeability (e.g., a clay stratum) is reached. Under the latter condition, a perched water table will form above the zone of lower permeability, and pore pressure will become positive.

When the descending wetting band reaches the main water table, the surface of the main water table will rise with an increase in pore pressure. The rise in the main water or the thickness of the perched water table will be approximately equal to the thickness of the wetting band. The wetting band thickness is inversely related to the difference between the initial and final degree of saturation of the soil mass. Thicker wetting bands are therefore more likely to occur after a series of heavy rainfall events than after dry spells.

The relationship between rainfall on unprotected slopes, infiltration, and the depth of the wetting front can be approximated by the following equation (Lumb, 1975):

$$h = \frac{kt}{n(S_f - S_0)} \tag{Eq. 3-8}$$

where h = depth of wetting front (thickness of the wetting band after time t)
k = coefficient of permeability
t = duration of rainfall
n = porosity
S_f = final degree of saturation
S_0 = initial degree of saturation.

Figure 3.32 gives an example of variation in the value of $(S_f - S_0)$ and the degree of saturation. The value k for the surface may be determined by carrying out infiltration tests within a sample tube driven into the soil surface (Figure 3.33). During the test, water is fed into the exposed surface at a carefully controlled rate so that minimum ponding occurs on the surface. The total volume of water infiltrating

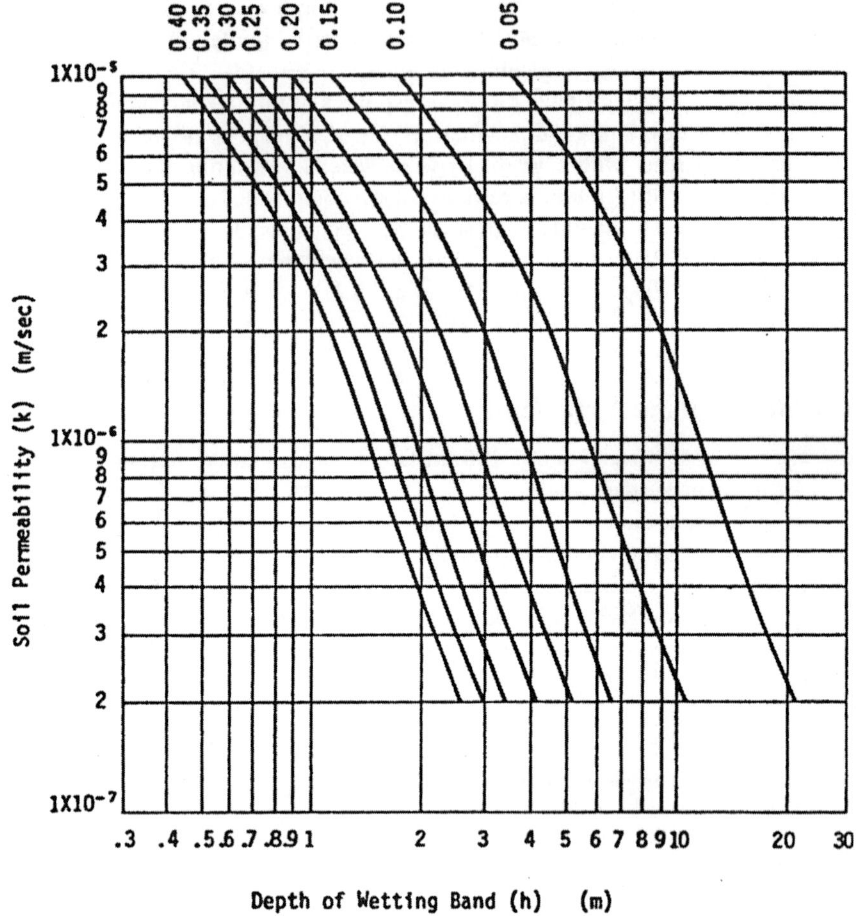

Figure 3.32 The effect of permeability and degree of saturation on wetting band thickness due to 10 year storm. Note: (1) this chart has been prepared assuming runoff is 50 percent and porosity, n, is 40 percent; (2) curves have been plotted for the various degrees of saturation $(S_f - S_0)$ shown. (Geotechnical Control Office, 1984.)

is noted at various time intervals and is plotted against time. Figure 3.33 shows a field infiltrometer, while Figure 3.34 presents typical results from the infiltration test.

The assumptions made in Equation 3-8 result in an extremely simplified model for infiltration. Throttling on infiltration often arises when there exists a very thin band of impermeable layer in the soil profile, or the surface permeability is lower than that of underlying material. Similar analytical difficulties are introduced as a result of the degree of saturation changing throughout the soil profile (Geotechnical Control Office, 1979).

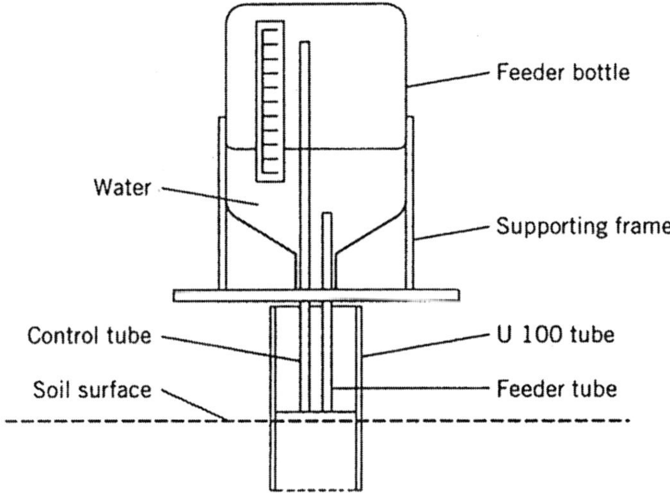

Figure 3.33 Diagram of field infiltrometer. (Lam, 1974.)

The wetting band approach is only applicable to situations where the rise in groundwater level is due to rainfall infiltration. Sloping ground, downslope flow, and the differences in aquifer responses are not accounted for in this simplified method. If the intensity of rainfall is at least sufficient to cause infiltration at the limiting rate, the thickness of the wetting band will be dependent upon the duration of the rainstorm, that is, the longer the rainstorm, the thicker the wetting band.

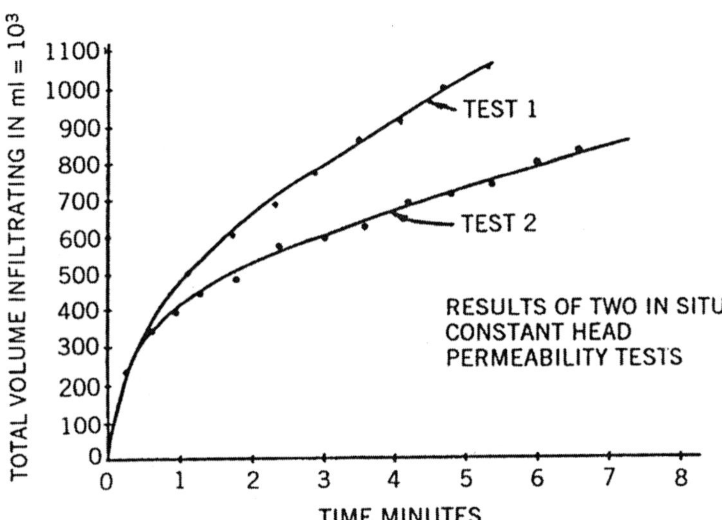

Figure 3.34 Typical results from field infiltration. (Geotechnical Control Office, 1979.)

3.9 FIELD IDENTIFICATION AND INTERPRETATION OF GROUNDWATER CONDITIONS

3.9.1 Field Identification of Groundwater Conditions

It is essential to learn as much as possible about the groundwater conditions during the field investigation. Of primary importance is determination of these points (Prellwitz, 1990):

(1) If groundwater exists, is it hydrostatic or flowing?

(2) If it is flowing, is it in a confined or unconfined aquifer?

(3) What are the upper and lower limits and slope of the aquifer?

(4) What are the aquifer characteristics (soil type and permeability, rock discontinuities)?

(5) What is the proximity of the aquifer to the existing or potential failure surface for stability analysis?

(6) Most important, what is the highest phreatic surface for an unconfined aquifer and/or piezometric surface for a confined aquifer to use in the stability analysis? If it is an existing failure, how high was the groundwater at the time of failure?

All of the aforesaid aspects are related to the engineering geology and the hydrogeomorphology of the slopes under study. In practice, subsurface conditions are very complex and so are the groundwater flows. Any groundwater data should be interpreted by an experienced geotechnical engineer or an engineering geologist. To develop a groundwater model to be used for slope stability analysis, the investigator has to form a hypothesis early in the investigation and gather the data to see if it can be verified. The investigator must also be ready to form a new hypothesis, if necessary, as additional data are obtained. The investigation should not be concluded until the investigator is confident that a workable understanding of the groundwater flows has been developed.

Other data pertinent to groundwater movement may be noted and documented during investigation. Evidence such as a rust-stained zone might indicate that an area has been subjected to repeated drying and wetting, which results in oxidation in an aerobic environment. Another indicator might be the presence of a fully saturated gray slick zone, which results from weathering in a reducing and anaerobic environment. Such a zone is frequently present at a confining layer and can help to determine the lower limits of an unconfined aquifer if a distinct change in soil type is not noted.

3.9.2 Interpretation of Groundwater Conditions

Piezometric Readings As said earlier, groundwater is generally monitored by means of open standpipes and/or piezometers. The drill holes should be cased with at least 1-inch PVC plastic pipe and slotted to allow groundwater to enter only in the anticipated aquifer zone.

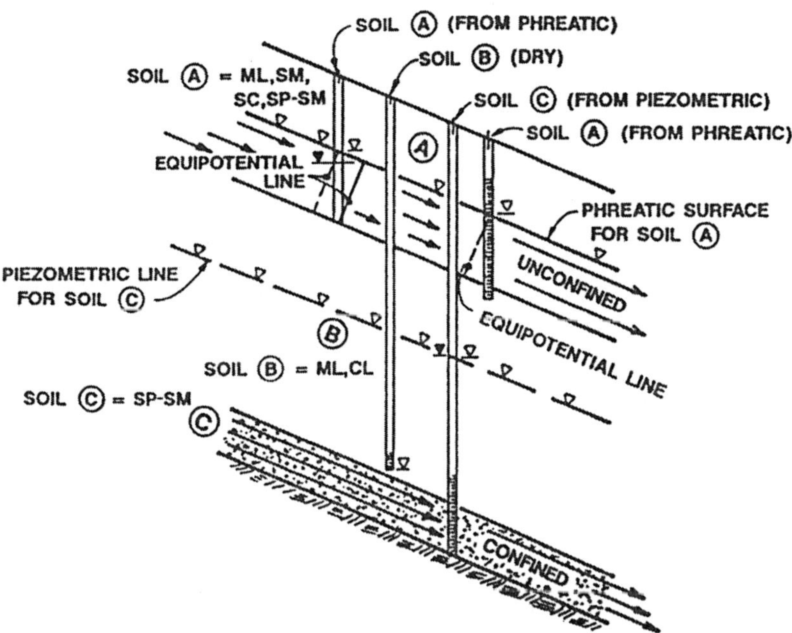

Figure 3.35 Groundwater flows in confined and unconfined aquifers. (Prellwitz, 1990.)

Once the piezometers are in place, questions often arise as to how to determine whether the aquifer is confined or unconfined and how to find the limits and slope of the aquifer. As shown in Figure 3.35, if an unconfined aquifer is encountered during drilling, free water should appear in the casing at all depths below the phreatic surface as long as drilling continues within the aquifer. Free water in the casing should be referenced to the depth of highest confinement. If this is the bottom of the casing and the water is flowing, the elevation to which the free water rises should be less as the casing is advanced. This is because equipotential lines from further down slope are lower as depth increases. The water in the casing will rise to the vertical height of the equipotential line intercepted at the bottom. The bottom of the unconfined aquifer should be recognized as the confining surface of lower permeability is encountered.

If a confined aquifer is encountered, it should be below a confining layer of lower permeability, and groundwater will probably enter the casing only after this confining layer is penetrated, and will rise to some height above the bottom of the casing. A confined aquifer may also exist in clayey soils without a change in soil type; for example, it may be only a zone of open fissures due to prior slope failure.

Borehole Drilling Groundwater levels are normally taken in borehole drilling. These readings present a lot of valuable information regarding the groundwater flows within the soil mass. Unfortunately, their significance is often overlooked by borehole inspectors. The following examples (Examples 3.1 and 3.2) present a field study of groundwater conditions interpreted from the groundwater readings taken from boreholes, each installed with either an open standpipe and/or a piezometer at selected depths (Prellwitz, 1990).

Example 3.1 (See Figure 3.36*a*) Complete the subsurface profile using the data provided in Figure 3.36*a*. Determine if there are confined or unconfined aquifers present, sketch the appropriate phreatic surface or piezometric line for each, and label the aquifer soil units (Prellwitz, 1990).

***Solution to Example* 3.1** (see Figure 3.36*b*) Subsurface profile was completed using the data provided in Figure 3.36*a*. One unconfined aquifer and one confined aquifer were determined. Their corresponding phreatic surface and piezometric line were plotted with soil units labeled (Prellwitz, 1990).

Example 3.2 (See Figure 3.37*a*) Construct phreatic and piezometric surface(s) for computer analysis. The annulus between the casing and the drill hole has been effectively sealed for each observation well from the top of the perforations to the ground surface. For each soil unit, indicate the appropriate water surface to use in pore pressure calculations and whether the surface should be treated as a phreatic surface or a piezometric line. Note that the piezometric rise in an observation well is referenced to the point of highest effective confinement (highest perforation, seal, upper or lower confining surface, etc.) (Prellwitz, 1990).

***Solution to Example* 3.2** (see Figure 3.37*b*) Phreatic and piezometric surfaces were constructed based on the data provided in Figure 3.37*a*. Assuming isotropic conditions, soil units A and B would have the same phreatic surface W1 given the same unconfined aquifer. Soil unit C is a confining layer with groundwater measured close to the bottom of the observation wells. Hence, it can be assumed that there is no flow or pressure within this confining layer. Soil unit D is a confined aquifer subject to flowing water under piezometric head, and hence the water surface should be treated as a piezometric line. The table summarizes the results of the observation well data in Figure 3.37a (Prellwitz, 1990).

3.10 GROUNDWATER IN SLOPE STABILITY ANALYSIS

3.10.1 Developing a Groundwater Model from the Field Data

As said earlier, groundwater is one of the major factors in slope stability analysis. The model required must provide data for pore pressure calculation on the anticipated critical failure profile. This can be done by constructing the phreatic surface for an

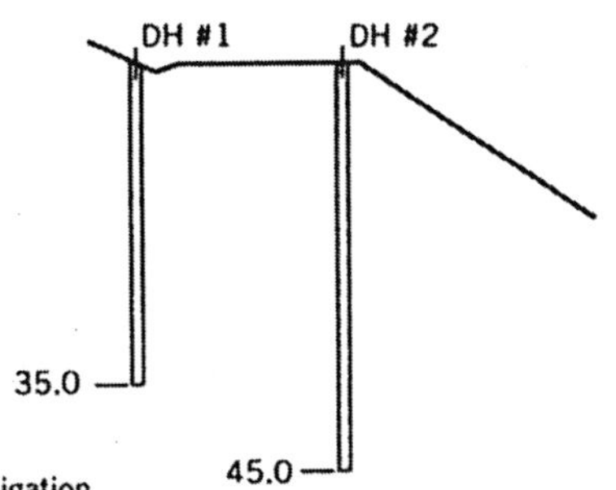

DH#1 Investigation

Casing Depth	Soil or Rock Unit	Moist	Water Level
5.0	1=SM	Moist	None
10.0	1	Saturated	7.2
15.0	1	Saturated	8.3
20.0	1	Saturated	9.2
20.9	1/2 Contact	(SM/CL–ML)	
25.0	2=CL–ML	Moist	None
27.1	2/3 Contact	(CL–ML/SP–SM)	
26.0	3=SP–SM	Saturated	4.9
29.1	3/4 Contact	(SP–SM/CL)	
30.0	4=CL	Moist	None
35.0	Bottom CL	(Weathered shale)	

DH#1 Piezometer Monitoring

Perf.		Sealed		Water Level
From	To	From	To	
5.0	22.0	0.0	5.0	6.7
27.0	30.0	0.0	5.0	4.9
		20.0	27.0	

DH#2 Investigation

Casing Depth	Soil or Rock Unit	Moist	Water Level
5.0	5=SM fill	Moist	None
10.0	5/1 Contact	(SM/SM)	
15.0	1=SM	Moist	None
20.0	1	Saturated	17.5
25.0	1	Saturated	18.4
30.0	1	Saturated	19.3
30.9	1/2 Contact	(SM/CL–ML)	
35.0	2=CL–ML	Moist	None
37.0	2/3 Contact	(CL–ML/SP–SM)	
38.0	3=SP–SM	Saturated	12.9
39.0	3/4 Contact	(SP–SM/CL)	
40.0	4=CL	Moist	None
45.0	Bottom CL	(Weathered shale)	

DH#1 Piezometer Monitoring

Perf.		Sealed		Water Level
From	To	From	To	
5.0	22.0	0.0	5.0	16.9
36.0	40.0	0.0	5.0	12.9
		30.0	36.0	

(*a*)

Figure 3.36 (*a*) Slope profile for Example 3.1. (*b*) Plotted subsurface soil/rock profile for Example 3.1.

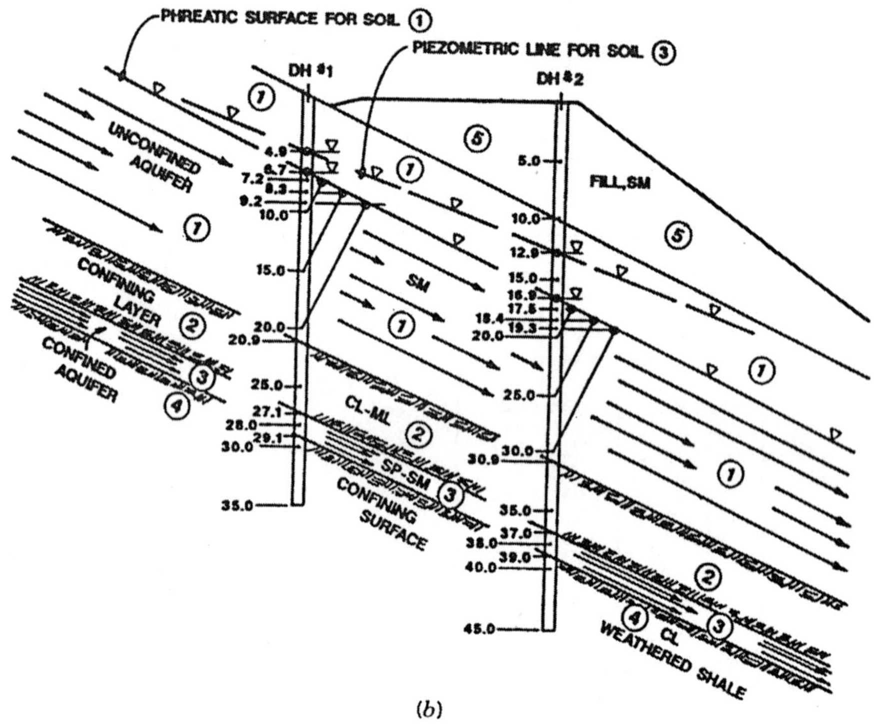

(*b*)

Figure 3.36 (*Continued*)

unconfined aquifer and the piezometric line for a confined aquifer. The pore pressure from any point is computed from the difference in head between that point and the water surface (or the phreatic surface), h_w. Figure 3.38 shows a comparison between phreatic and piezometric pore pressure calculations.

Whether a phreatic or piezometric assumption should be used depends upon the problem. This has been illustrated in the previous problems (Examples 3.1 and 3.2). It also varies with the stability analysis methods and depends on each individual computer program (Chapter 6). The infinite slope equation and most computer programs follow the phreatic method. A few computer programs (such as XSTABL) have an option allowing the user to choose one or the other. It is essential for the engineer to know whether the phreatic or piezometric assumption should be used, as it can have a significant effect on the results of the stability analysis.

3.10.2 Groundwater Effects on Slope Stability

Reduction in Shear Strength Saturation of a soil will decrease the frictional shear strength. This is due to the buoyant reduction in normal force required for frictional shear strength by the pore pressure (the effective stress principle). Saturation of soil may also destroy capillarity and "apparent cohesion" on the cohesive

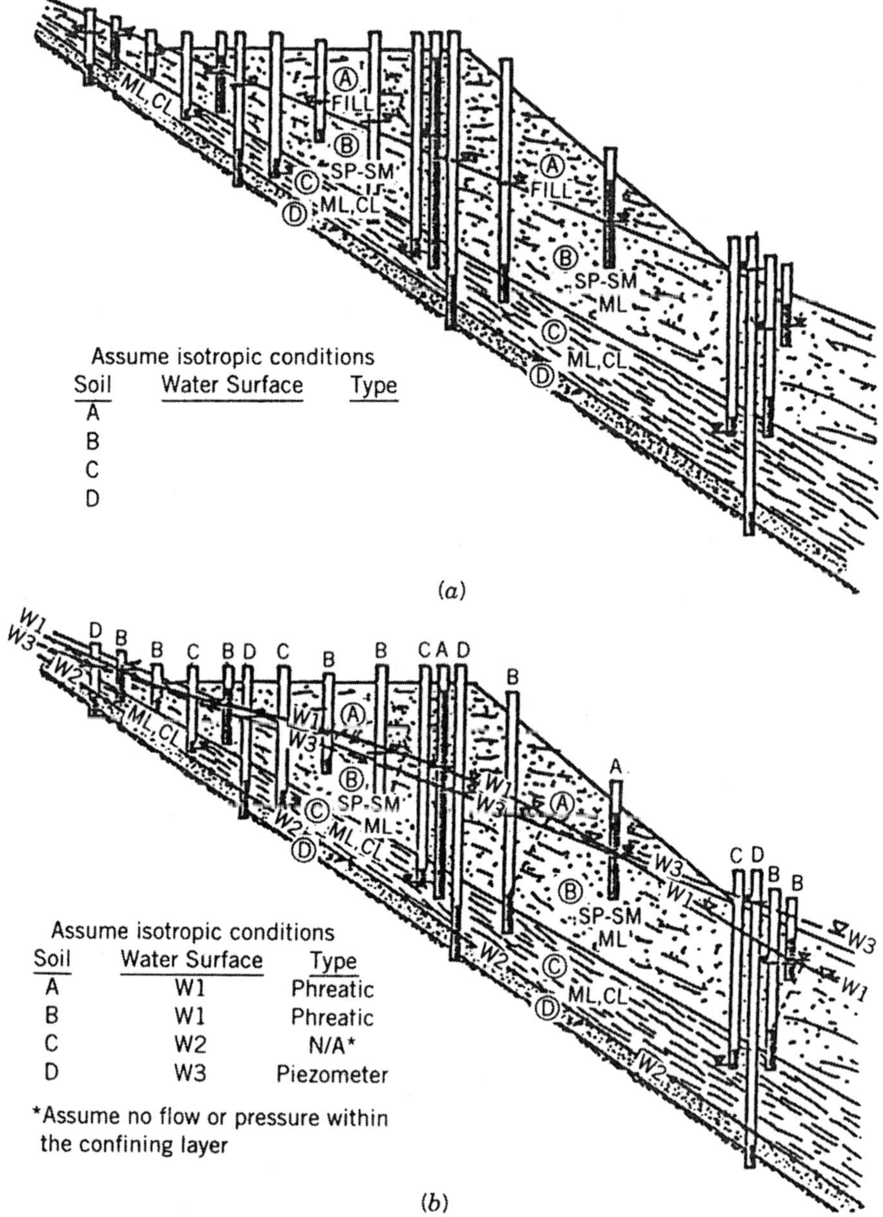

Figure 3.37 Groundwater data from observation wells and piezometers for Example 3.2. (*b*) Constructed phreatic and piezometric surfaces for computer analysis for Example 3.2.

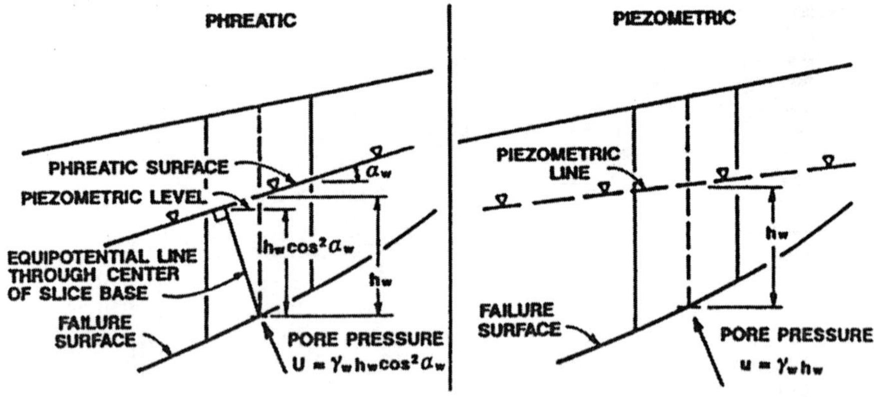

Figure 3.38 Comparison between phreatic and piezometric pore pressure calculations.

component of the soil, or may reduce the dry strength of a cohesive soil. Figure 3.39 illustrates the reduction in frictional shear strength because of the presence of groundwater.

Reduction in Frictional Strength

With Groundwater Slice weight, $W = [\gamma_t(d - d_w) + \gamma_{sat} d_w]$

$$b = [120 \times (8.3 - 5.8) + 135 \times 5.8]5;$$

that is, $W = 5,415$ pounds per foot

Pore water force, $U = \gamma_w d_w \cos^2 \alpha_w L$

$$= 62.4 \times 5.8 \times \cos^2(17°) \times 5.5;$$

that is, $U = 1{,}820$ pounds per foot

Frictional strength, $\tau_f = N \tan\phi = (5,415 \times \cos 26° - 1,820) \tan 35°$; that is,

$$\tau_f = \underline{\underline{2{,}134 \text{ pounds per foot}}}$$

Without Groundwater

$$W = \gamma_r db = 120 \times 8.3 \times 5.0 = 4,980 \text{ pounds per foot}$$

$$U = 0 \text{ (since } d_w = 0)$$

$$\tau_f = N \tan\phi = (4,980 \times \cos 26° - 0) \tan 35°$$

$$\tau_f = \underline{\underline{3{,}134 \text{ pounds per foot}}}$$

Notation

W = slice weight in plf

U = pore water force in plf

N = normal force perpendicular to base of slice in plf

τ_f = frictional force at base of slip surface in plf

γ_t = moist unit weight in pcf

γ_{sat} = saturated unit weight in pcf

γ_w = unit weight of water in pcf

α_w = inclination angle of phreatic line

θ = inclination of the base of slip surface

ϕ = angle of internal friction

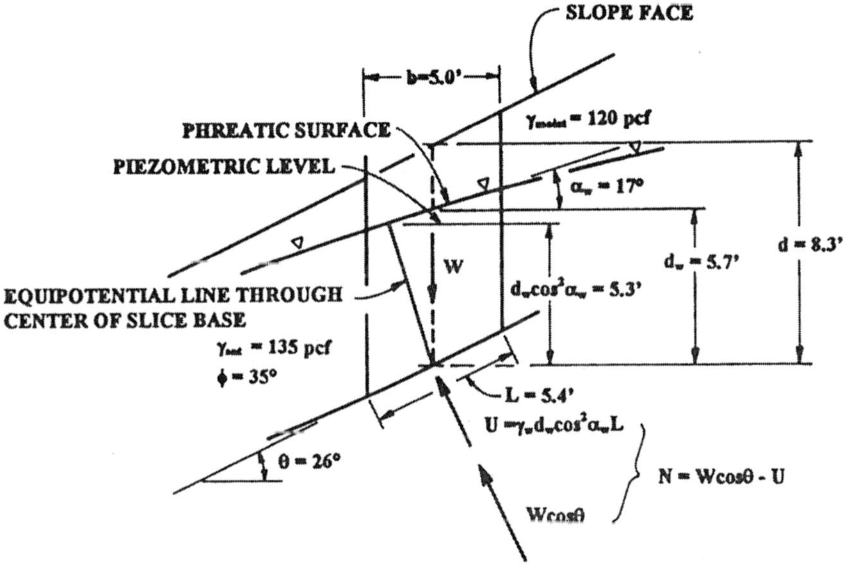

Figure 3.39 Reduction in frictional shear strength due to groundwater (Prellwitz, 1990).

For this slice, groundwater reduces frictional strength by

$$\left[\frac{(3{,}134 - 2{,}134)}{3{,}134}\right] \times 100 \text{ percent} = \underline{\underline{32 \text{ percent}}}$$

Effects of Seepage Direction Many slopes become saturated during periods of intense rainfall or snowmelt, with the water table rising to the ground surface, and water flowing essentially parallel to the direction of the slope. Under this condition, soil element *abcd* in the infinite slope has the submerged weight W_0 and the seepage force F acting as shown in Figure 3.40. A flow net is drawn in Figure 3.40 with flow lines parallel to the slope and equipotentials perpendicular to the slope. Using the hydraulic gradient method, the seepage force F can be determined from the flow net. This seepage force F acts as a driving force in the soil mass, and hence can greatly lower the stability of the slope.

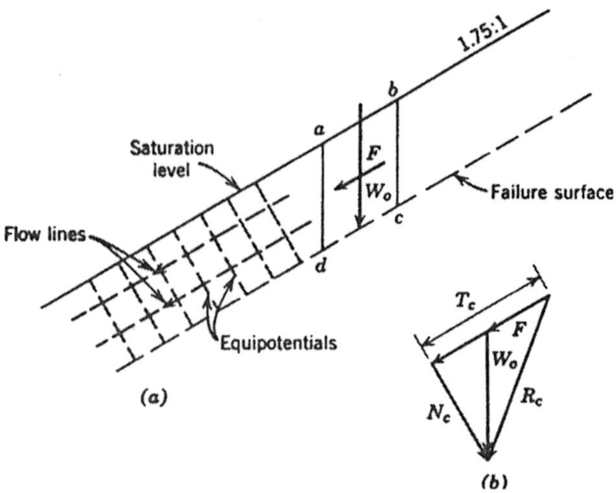

Figure 3.40 Stress conditions in infinite slope with seepage parallel to the slope. (*a*) Diagram of slope. (*b*) Force polygon. (Cedergren, 1977.)

Slopes can be fully saturated, but at the same time free of excess pore pressures and damaging seepage forces. This is the case when the slope is underlain by a highly pervious gravel layer (Figure 3.41), where the flow net consists of vertical flow lines and horizontal equipotentials. Under this seepage condition, the energy of the free water in the soil is consumed harmlessly as the water flows in the vertical direction to the gravel with no pore pressures (compare with Figure 3.26 for which $\theta = 90°$, $R_u = 0$). This serves as the principle of horizontal drainage systems to force seepage into vertical patterns and thus to improve the stability of slopes.

3.10.3 Groundwater in Rock

Groundwater affects the stability of a rock slope in the same way as it affects the stability of a soil slope, that is, it develops hydrostatic pressures in the rock discontinuities. Just like the soil slope, groundwater reduces effective stress and the frictional strength of rock along those discontinuities. How the groundwater should be treated in analysis depends on how intact and porous the rock mass is. For example, a rock mass that is so weathered that water moves through like an unconfined aquifer should be analyzed as a soil under this condition. Likewise, if a porous sandstone is sandwiched by two shales of lower permeability, it would behave like a confined aquifer in soil. For intact rocks that are bounded by discontinuities, two-dimensional plane failure or three-dimensional wedge analysis is usually performed. Groundwater in these analyses is assumed to exert hydrostatic pressures along the discontinuities.

Effects of groundwater on rock discontinuities could be detrimental and should therefore be dealt with judiciously in analyzing rock slope stability. Tension cracks on

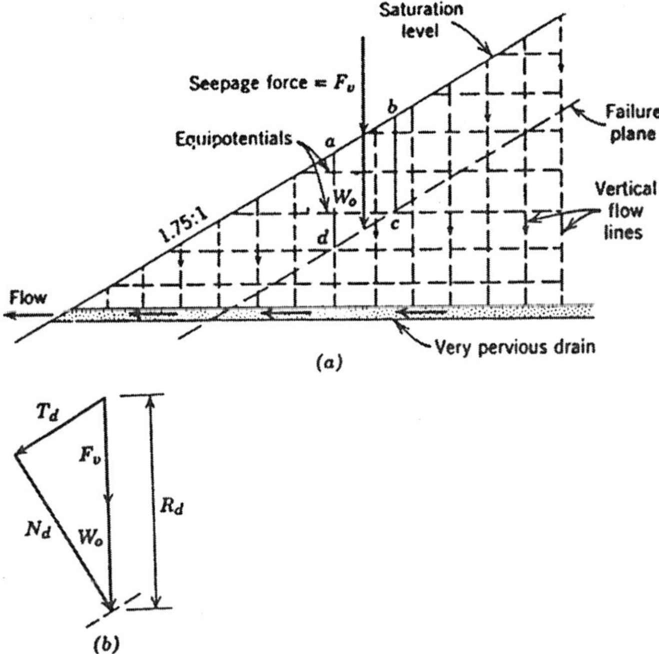

Figure 3.41 Stress conditions in infinite slope with vertical seepage. (*a*) Cross section. (*b*) Force polygon. (Cedergren, 1977.)

top of rock slopes could trap water, which eventually develops hydrostatic pressures in the tension cracks and along rock discontinuities. Figure 3.42 illustrates six variations of how hydrostatic groundwater can exist and how it can be analyzed in practice.

3.11 MONITORING OF GROUNDWATER PRESSURES

3.11.1 Piezometers and Observation Wells

Groundwater pressure in a slope is typically monitored with open standpipe, pneumatic, or vibrating wire piezometers. The choice of piezometer type depends on the predicted pore pressures, access for reading, service life, and response time required. Technical considerations of piezometers should be reliability, simplicity, long-term stability, accuracy, time lag, remote reading capability, and influence of installation technique.

All the methods for monitoring groundwater pressures described in this section require some flow of water into or out of the measuring device before the recorded pressure can reach equilibrium with the actual groundwater pressure. During excavation or drilling of a borehole, a large volume of water may flow before the water level reaches its equilibrium. The rate at which water flows through the soil depends on the

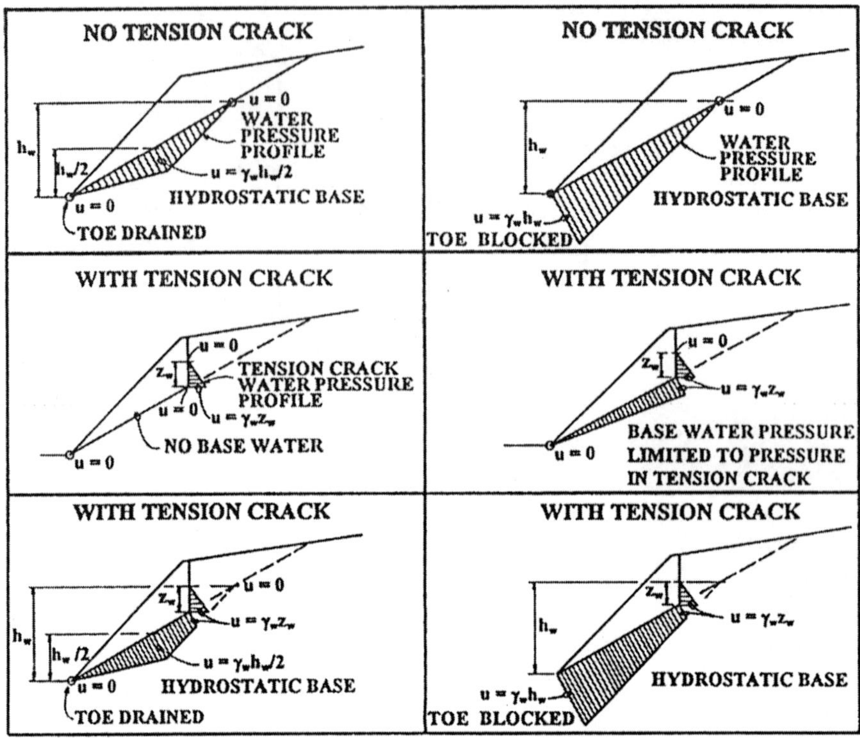

Figure 3.42 Six variations of hydrostatic pressures that exist along rock discontinuities. (Prellwitz, 1990.)

permeability. The time required for a measuring device to indicate the true groundwater pressure is known as the response time, and it depends on the permeability of the soil as well as the porous element, the quantity of water required to operate the device. The selection of a suitable method for measuring the groundwater pressure will largely be determined by the response time.

Observations in Boreholes and Excavations The crudest method to determine the groundwater level is by observation in an open borehole or excavation. The method requires a long response time unless the ground is very permeable. Hence, observations should be made at regular time intervals, until it is established that the water has reached equilibrium. Misleading readings can be obtained if rain and surface runoff are allowed to enter the open hole. Readings taken in a borehole shortly after completion of drilling should be treated with caution, as the groundwater may not have reached its equilibrium.

The reliability of water level observations in boreholes or excavations can be improved by the installation of an observation well, as described in the following section.

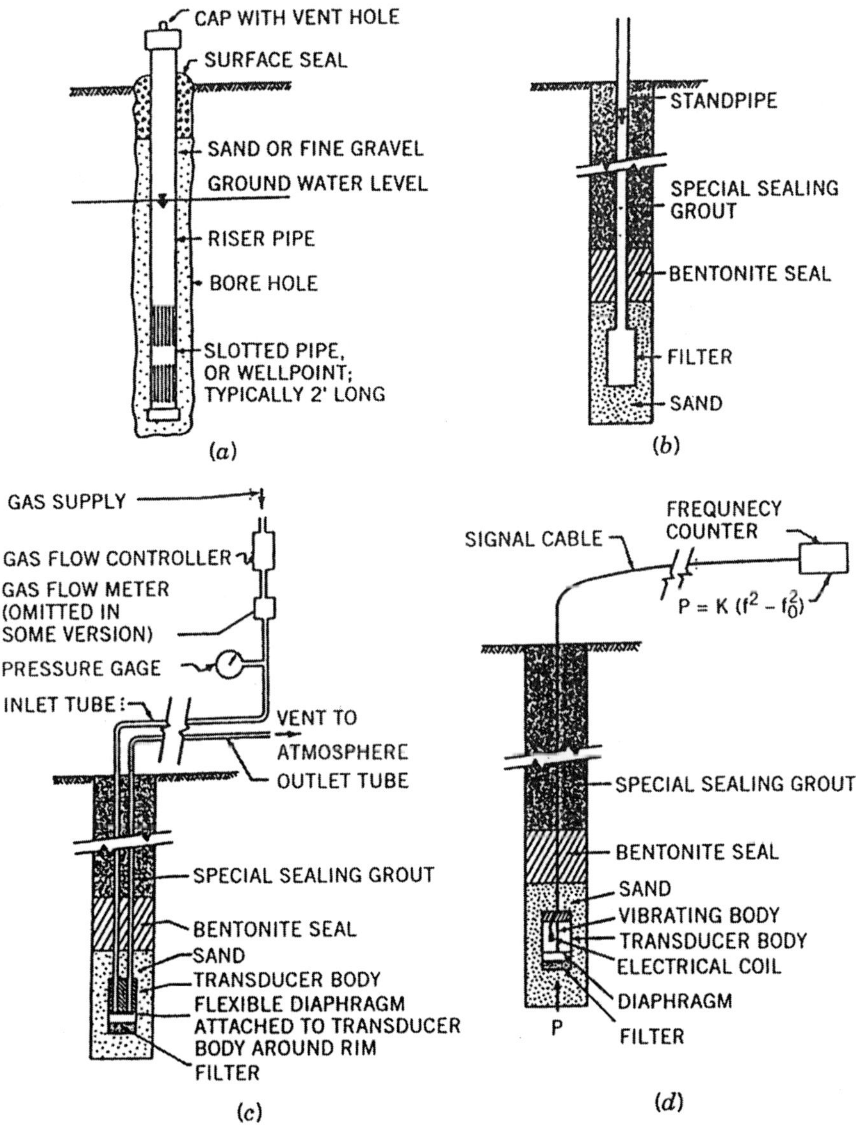

Figure 3.43 Different types of piezometers. (*a*) Observation well (not recommended). (*b*) Open standpipe. (*c*) Pneumatic. (*d*) Vibrating wire.

Observation Well An observation well consists of a perforated section of pipe attached to a riser pipe installed in a sand-filled borehole; see Figure 3.43*a*. The surface seal is needed to prevent surface runoff from entering the borehole. Venting of the pipe cap is required to allow the free flow of water. The water surface elevation is determined by using an electrical dipmeter or a tape and weight.

If the observation well is placed in a homogeneous soil formation with only one groundwater table present, it should adequately provide a source of valid water level information. However, in multilayered soil strata with multiple groundwater inflow sources, observation wells create an undesirable vertical connection between strata, and hence, the reading so taken is not representative or reliable. The water level within the observation wells is likely to correspond to the head in the most permeable zone and will be misleading. If one aquifer is contaminated, installing observation wells may lead to contamination of other aquifers.

Although readings taken in an observation well are more controlled than in an open borehole, standpipe response time is still slow, and if zones of different permeabilities have been penetrated, flow between zones may occur and readings may be misleading, as discussed earlier. These drawbacks can largely be overcome by the installation of standpipe piezometers or other piezometers, as discussed in the following sections.

Open Standpipe Piezometer An open standpipe piezometer (or the open-hydraulic piezometer) is a pipe with a filter. The porous filter element requires sealing so that the instrument responds only to groundwater pressures around the filter element and not to groundwater pressures at other elevations. Piezometers can be installed in fill, sealed in boreholes, or pushed or driven into place. The components are identical in principle to those of the observation well with the addition of seals. The water surface in the standpipe stabilizes at the piezometric elevation. Care must be taken to prevent rainwater runoff from entering open standpipes. Appropriate covers and vents must be used to prevent obstructions.

Figure 3.43*b* shows a sketch of an open standpipe piezometer. The original version developed by Casagrande consisted of a cylindrical porous ceramic tube connected with a rubber bushing to saran plastic tubing (Figure 3.44*a*). High-density porous hydrophilic polyethylene attached to PVC or ABS plastic pipe or polyethylene tubing are the new materials used today. Standpipes are typically sized up 3 inches in diameter.

Flush-coupled Schedule 80 PVC or ABS pipe is a good choice for the standpipe. The coupling should be cemented or threaded with the taper self-sealing threads. In either case the joints must be watertight. Conventional tapered or square pipe threads should be avoided because they cannot sustain a watertight seal.

Open standpipe piezometers are generally considered to be the most reliable. Some of the advantages and limitations of an open standpipe piezometers are as follows:

Advantages

- Simple and reliable
- Long successful performance record
- Self deairing if inside diameter of standpipe is adequate
- Integrity of seal can be checked
- Portable readout devices can be used
- Can be converted to diaphragm piezometer
- Can be used for sampling groundwater
- Can be used to measure permeability

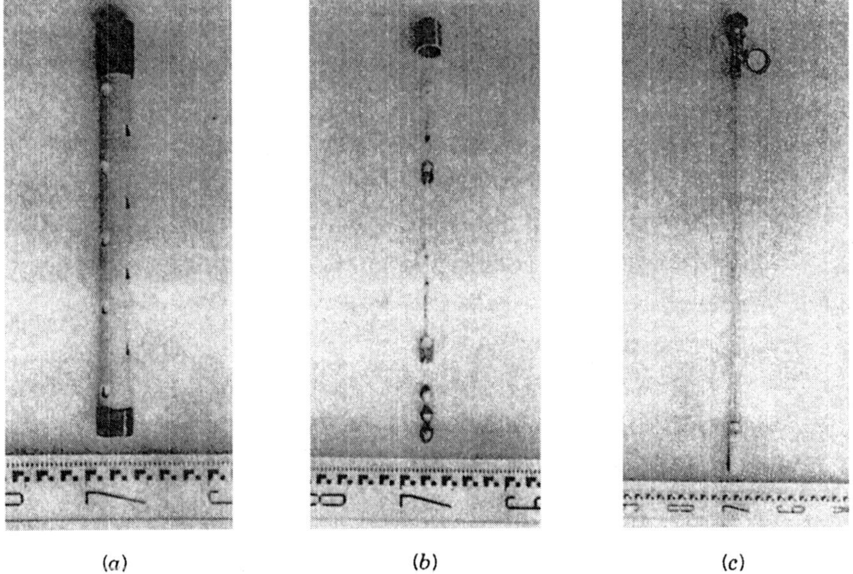

(a) (b) (c)

Figure 3.44 Groundwater pressure measuring equipment. (*a*) Casagrande piezo-meter tip. (*b*) String of piezometer buckets. (British Patent No. 1538487.) (*c*) Tensiometer (jet fill).

Limitations

- Slow response time and long time lag
- Subject to damage by construction equipment and soil compression
- Stick up through fill interrupts construction and interferes with compaction
- Porous filter can plug with repeated water inflow and outflow

Pneumatic Piezometer Pneumatic piezometers consist of a porous tip capped with a flexible diaphragm that controls a pneumatic valve. The valve operates when the pressure in the pneumatic system equals the pressure in the piezometer cavity. The piezometer tip is installed in a sand pocket in the same way as an open standpipe piezometer. Figure 3.43c shows a sketch of the pneumatic piezometer. Advantages and limitations of pneumatic piezometers are as follows:

Advantages

- Only small volume changes required
- Simplicity of operation
- Portable readout devices can be used
- Level of sensor is independent of the recorder
- Long-term stability
- No practical limit in length of leads

- Easy calibration
- Equipment is relatively inexpensive
- No freezing problems
- Installation is simpler than for hydraulic
- Tubes may be purged
- Minimum interference to construction
- Can be used to measure permeability

Limitations

- Deairing of the stone is usually absent
- Operation can be slow
- Head loss can occur with long lead-lengths with some transducers
- Dirt in the leads may foul the valve
- Subatmospheric pressures cannot be measured with some types

Vibrating Wire Piezometer Vibrating wire piezometers consist of a metallic diaphragm separating the pore water from the measuring system. A tensioned wire is attached to the midpoint of the diaphragm. Deflection of the diaphragm causes changes in wire tension, which are measured, recorded, and converted into pressure. Figure 3.43*d* shows a sketch of the vibrating wire piezometer.

Utilizing a vibrating wire transducer introduces potential errors, such zero drift or corrosion. A vibrating wire piezometer should have a dried and hermetically sealed cavity around the wire to minimize corrosion problems. The vibrating wire piezometer should be equipped with an in-place check feature. This feature allows a zero reading and in some cases the calibration to be checked at any time during the life of the piezometer. Some of the advantages and limitations of vibrating wire piezometers are as follows.

Advantages

- Easy to read
- Short time lag
- Minimum interference to construction
- Lead wire effects minimal
- Can be used to measure negative pore water pressures
- No freezing problems

Limitations

- Special manufacturing techniques required to minimize zero drift
- Need for lightning protection should be evaluated

Tensiometer Suction is negative pore water pressure. Measurements of in situ suctions are made using a portable tensiometer (Figure 3.44*c*). These devices are capable of directly measuring pore suction ranging from zero to minus one atmosphere. The instrument has a high-air-entry stone at one end of a metal tube filled with water. A vacuum gage is placed at the other end of the metal tube. When the porous tip is inserted into a small-diameter hole and comes in contact with the soil, there is a tendency for water to be drawn out of the tube and into the soil. The potential for water to be drawn out of the instrument is a measure of the matric suction. The reliability of a tensiometer depends on a good contact between the soil and the ceramic tip and a good seal between the tensiometer tube and the soil. For measurement of soil suction beyond the range of tensiometers, psychrometers may be used. However, psychrometer accuracy is doubtful.

3.11.2 Installation of Piezometers

The typical method of installation of a piezometer in a borehole is shown in Figure 3.45 (Geotechnical Control Office, 1987). The tip should be placed within a sand pocket in the specific zone for which pore pressures are to be measured, referred to as the response zone. The length of the response zone is normally greater than four hole diameters, but preferably not less than 16 inches. Washed sand with particle sizes in the range of 0.2 to 1.2 millimeters is recommended for the response zone in most soils derived from in situ rock weathering. For coarser transported soils (e. g., alluvial and marine sands and gravels), filters should be specifically designed to match the surrounding material using the filter criteria.

Bentonite is normally used to provide a seal above the sand or filter pocket, and if the piezometer is not installed near the base of the borehole, a bentonite seal should be placed beneath the sand or filter pocket. The length of bentonite seals is typically 1.5 to 2 feet, although longer seals may be preferable. Sometimes, bentonite balls, formed from powdered bentonite and water, are used to form the seals. Alteratively, compressed bentonite pellets can be used, but sufficient time should be allowed for the swelling action of the pellets to occur before the grout is placed on top of the seal.

The remaining sections of the borehole, both above the upper seal and beneath the lower seal (if applicable), should be filled with cement-bentonite grout of the same or lower permeability than the surrounding soil. A tremie pipe should be used to place the grout. The volume of grout used should be compared with the volume of the hole to be grouted.

After installation, a well-drained lockable surface box should be provided for every piezometer installation. A response test should also be conducted on each piezometer where possible to check the adequacy of the installation. This can be done by means of falling head tests. Similar response tests should be carried out at intervals during the life of the piezometer. In soft cohesive soils, care should be taken to ensure that the head in response tests does not cause hydraulic fracture in the soil.

The success of groundwater pressure measurements is predicated on the care taken during installation of the piezometers and standpipes. The porous element should be fully saturated and filled with deaired water before installation. Poor sealing of the

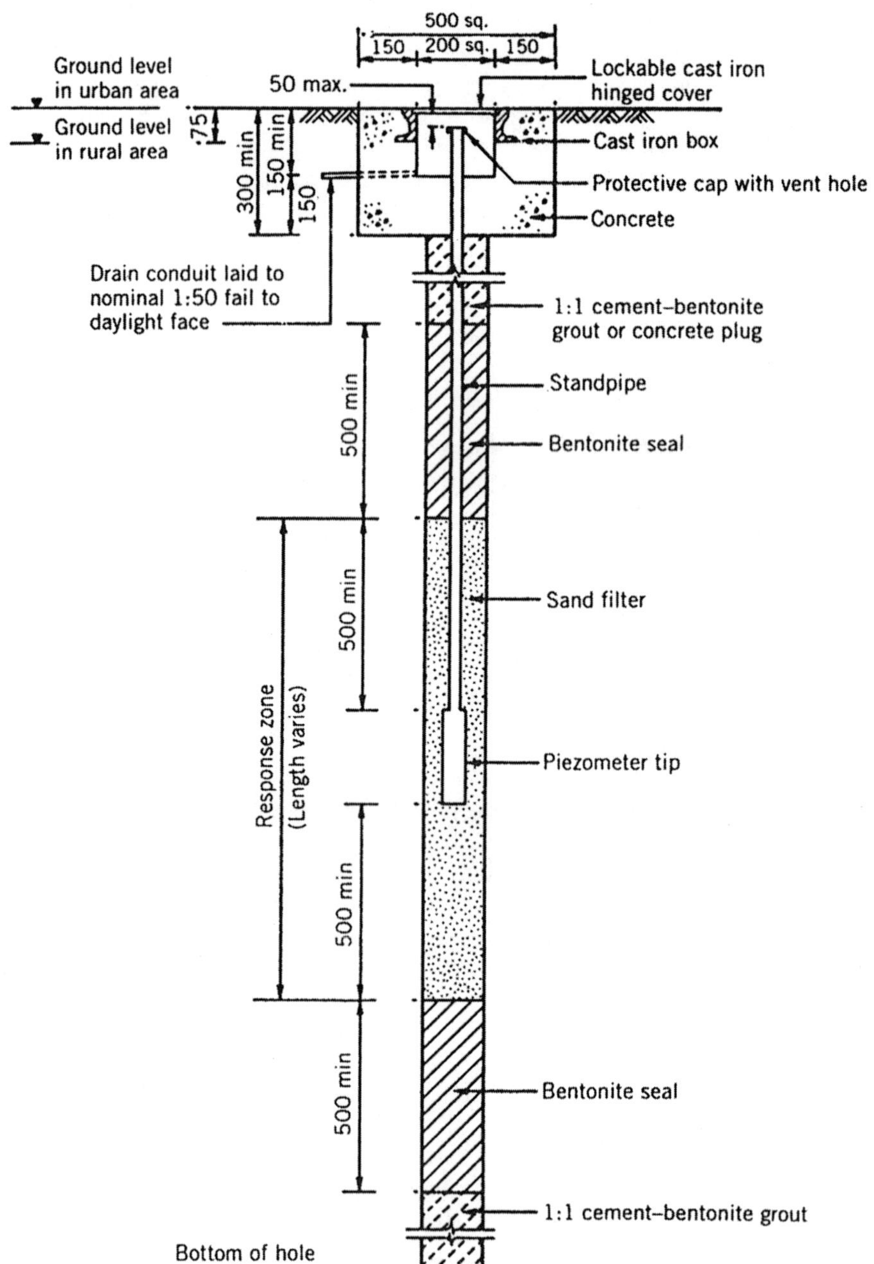

Figure 3.45 Typical installation details of a piezometer in a borehole. Note: (1) scale is diagrammatic; (2) all dimensions are in millimeters.

piezometer will permit water migration from one level to the other, thus rendering the readings meaningless. The installation of more than one piezometer in a single borehole is not generally recommended.

3.11.3 Fluctuating Groundwater Levels

In addition to varying response to rainfall, groundwater levels are usually subject to seasonal changes, in response to tidal variations, or sometimes to abstraction from neighboring wells, or to other causes. Groundwater is usually measured by means of a battery-operated electrical dipmeter, which relies on the conductivity of the groundwater to complete the circuit.

As discussed in Section 3.5.1, the observation of peak groundwater response in piezometers or standpipes or transient water levels can be made by using a string of piezometer "buckets" (Figure 3.18 and Figure 3.44*b*). Care must be taken not to drop the buckets into the borehole (thus rendering the piezometer useless), and not to tangle the bucket string and thus reduce the spacing between the buckets.

An alternative method for recording transient water table in a piezometer or standpipe is the automatic bubbling recorder, or the "bubbler" system. In this system, a small-diameter air line is installed down to the piezometer tip with a small air flow sufficient to produce several bubbles per minute. The air pressure required to release bubbles can be equated to the water pressure produced by the height of water in the standpipe.

3.12 OTHER INSTRUMENTS—RAINFALL GAGES

As mentioned earlier, intense rainfall over short or long periods of time can contribute to slope instability. For large highway project or small isolated slope problems, installing a rain gage may be useful, particularly in a post-failure analysis. Most cities have rainfall gage information. This information should be collected, plotted and included as part of the project records. Documenting the amount of rainfall in the past and present, whether heavier or lighter than normal, is useful in handling claims by the contractor.

REFERENCES

Bishop, A. W., 1954. "The Use of the Slip Circle in the Stability Analysis of Slopes," *Geotechnique*, Vol. 5.

Blyth, F. G. H., and M. H. de Freitas, 1984. *A Geology for Engineers*, 7th ed. New York: Elsevier.

Brand, E. W., 1981. "Some Thoughts on Rain-Induced Slope Failure," *Proceedings of the 10th International Conference on Soil Mechanics and Foundation Engineering*, Stockholm, Vol. 3, pp. 373–376.

Brand, E. W., 1982. "Analysis and Design in Residual Soils," *Proceedings of the ASCE Geotechnical Engineering Division Speciality Conference on Engineering and Construction in Tropical and Residual Soils*, Honolulu, Hawaii, January 11–15.

Brand, E. W., 1989. "Correlation Between Rainfall and Landslides," *Proceedings of the 12th International Conference on Soil Mechanics and Foundation Engineering*, Vol. 5.

Briggs, R. P., J. S. Pomeroy, and W. E. Davies, 1975. *Landsliding in Allegheny County, Pennsylvania*, U. S. Geological Survey, Circular 728.

Bryan, K., 1919. "Classification of Springs," *Journal of Geology*, Vol. 27, pp. 522–561.

Cedergren, H. R., 1977. *Seepage, Drainage & Flownets*, 2nd ed. New York: Wiley.

Chugh, A. K., 1981a. "Multiplicity of Numerical Solutions for Slope Stability Problems," *International Journal for Numerical and Analytical Methods in Geomechanics*, Vol. 5, pp. 313–322.

Chugh, A. K., 1981b. "Pore Water Pressure in Natural Slopes," *International Journal for Numerical and Analytical Methods in Geomechanics*, Vol. 5, pp. 449–454.

Crozier, M. J., and R. J. Eyles, 1980. "Assessing the Probability of Rapid Mass Movement," *Proceedings of the 3rd Australian–New Zealand Conference on Geomorphology*, Vol. 17, pp. 78–101.

Duncan, J. M., A. L. Buchignani, and D. W. Marius, 1987. *An Engineering Manual for Slope Stability Studies*, Department of Civil Engineering, Virginia Polytechnic Institute and State University, March.

Dunn, I. S., L. R. Anderson, and F. W. Kiefer, 1980. *Fundamentals of Geotechnical Analysis*. New York: Wiley.

Eyles, R. J., 1979. "Slip-Triggering Rainfalls in Wellington City, New Zealand," *New Zealand Journal of Science*, Vol. 22, pp. 117–121.

Eyles, R. J., M. J. Crozier, and R. H. Wheeler, 1978. "Landslips in Wellington City," *New Zealand Geographer*, Vol. 34, No. 2, pp. 58–74.

Geotechnical Control Office, 1979. *Geotechnical Manual for Slopes*, 1st ed. Hong Kong: Civil Engineering Services Departments, November.

Geotechnical Control Office, 1984. *Geotechnical Manual for Slopes*, 2nd ed. Hong Kong: Civil Engineering Services Department, May.

Geotechnical Control Office, 1987. *Guide to Site Investigation*. Hong Kong: Civil Engineering Services Department, September.

Harr, M. E., 1962. *Groundwater and Seepage*. New York: McGraw-Hill.

Ho, D. Y. F., and D. G. Fredlund, 1982a. "Increase in Strength Due to Suction for Two Hong Kong Soils," *Proceedings of the ASCE Specialty Conference on Engineering and Construction in Tropical and Residual Soils*, Honolulu, Hawaii, pp. 263–295.

Ho, D. Y. F., and D. G. Fredlund, 1982b, "A Multi-Stage Triaxial Test for Unsaturated Soils," *Geotechnical Testing Journal*, Vol. 5, pp. 18–25.

Ho, D. Y. F., and D. G. Fredlund, 1982c. "Strain Rates for Unsaturated Soil Shear Strength Testing," *Proceedings of the 7th Southeast Asian Geotechnical Conference*, Hong Kong, pp. 787–803.

Hopkins, T. C., D. L. Allen, and R. C. Deen, 1975, *Effects of Warter on Slope Stability*. Lexington, KY: Division of Research, Federal Highway Administration, Department of Transportation, October.

Institution of Civil Engineers, 1976. *Manual of Applied Geology for Engineers*. London: Institution of Civil Engineers.

Janbu, N., 1973. *Slope Stability Computations in Embankment-Dam Engineering*, R. C. Hirschfeld and S. J. Poulos, Eds. New York: Wiley, pp. 47–86.

Krohn, J. P., and J. E. Slosson, 1976. "Landslide Potential in the United States," *California Geology*, Vol. 29, No. 10, October, pp. 224–231.

Lam, K. C., 1974. "Some Aspects of Fluvial Erosion in Three Small Catchments, New Territories, Hong Kong," M. Phil. Thesis., University of Hong Kong.

Lambe, T. W., and F. Silva-Tulla, 1992. "Stability Analysis of an Earth Slope," *Proceedings: Stability and performance of Slopes and Embankments—II*, ASCE Specialty Conference, University of California, Berkeley, California, June, pp. 27–67.

Lambe, T. W., and R. V. Whitman, 1969. *Soil Mechanics*. New York: Wiley.

Leet, L. D., S. Judson, and M. Kauffman, 1978. *Physical Geology*, 5th ed. Englewood Cliffs, New Jersey: Prentice Hall.

Lumb, P., 1962. "Effect of Rainstorms on Slope Stability," *Proceedings of the Symposium on Hong Kong Soils*, Hong Kong, pp. 73–87.

Lumb, P., 1964. "Multi-Stage Triaxial Tests on Undisturbed Soils," *Civil Engineering Public Works Review*, May.

Lumb, P., 1975. "Slope Failures in Hong Kong," *Quarterly Journal of Engineering Geology*, Vol. 8, pp. 31–65.

Meinzer, O. E., 1923. "Outline of Groundwater Hydrology with Definitions," *U. S. Geological Survey Water Supply Paper 494*, 71 pp.

Nilson, T. H., and B. L. Turner, 1975. *Influence of Rainfall and Ancient Landslide Deposits on Recent Landslides (1950–71) in Urban Areas of Contra Costa County, California*. U. S. Geological Survey, Bulletin 1388, 18 pp.

Nilson, T. H., F. A. Taylor, and E. E. Brabb, 1976. *Recent Landslides in Alameda County, California (1940–71); An Estimate of Economic Losses and Correlation with Slope, Rainfall, and Ancient Landslide Deposits*. U. S. Geological Survey Bulletin 1398, 20 pp.

Prellwitz, R. W., 1990. "Groundwater Investigation and Model Development Techniques," Presented at the *Oregon Department of Transportation, 1990 Geotechnical Workshop*, Newport, Oregon, January 31.

Prellwitz, R. W., and R. E. Babbitt, 1984. "Long-Term Groundwater Monitoring in Mountainous Terrain," *Transportation Research Record 965 on Soil Reinforcement and Moisture Effects on Slope Stability*. Washington, DC: TRB.

Radbruch-Hall, D. H., R. B. Colton, W. E. Davies, I. Luccitta, B. A. Skipp, and D. J. Varnes, 1982. *Landslide Overview Map of the Conterminous United States*, U.S. Geological Survey Professional Paper 1183, 25 pp.

Reid, M. E., H. P. Nielsen, and S. L. Dreiss, 1988. "Hydrologic Factors Triggering a Shallow Hilltop," *Bulletin of the Association of Engineering Geologists*, Vol. XXV, No. 3, pp. 349–361.

Slosson, J. E., and J. P. Krohn, 1979. *AEG Building Code Review, Mudflow/Debris Flow Damage, February 1978 Storm—Los Angeles Area, California Geology*, January, pp. 8–11.

Slosson, J. E., and J. P. Krohn, 1982. "Southern California Landslides of 1978 and 1980," in *Storms, Floods, and Debris Flows in Southern California and Arizona 1978 and 1980, Proceedings of a Symposium*, September 17–18, National Academy Press, Washington, D.C., pp. 219–319.

Todd, D. K., 1980. *Groundwater Hydrology*, 2nd ed. New York: Wiley.

Wong, H. Y., 1978. *Soil Strength Parameter Determination*. Hong Kong: Hong Kong Institute of Engineers, February 15, pp. 33–39.

Wong, K. K., 1970. "Pore Water Suction in Hong Kong Soil by Psychrometric Technique," M.Sc. Thesis. University of Hong Kong. Unpublished (available from the University Library).

CHAPTER 4

GEOLOGIC SITE EXPLORATION

4.1 INTRODUCTION

The two principal components of geologic site exploration associated with slopes are surface studies and subsurface investigations. Useful information can be gathered from surface studies and from an examination of the construction records and performance of existing structures in the vicinity of the site. The surface studies should form the first phase of a geologic site exploration, and the subsurface work should be planned only after assessing the results.

Surface studies can be separated into two main stages: (1) desk studies and (2) field studies. Desk studies should be carried out before detailed field studies. The engineer should visit the site during the initial phase of the investigation to get familiar with the site conditions. The planning of the field studies should be based on observations of the site conditions and findings of the desk studies, with emphasis being focused on the potential problem areas.

The flow chart presented in Figure 4.1 shows a series of operations used for planning a geologic site exploration for the design of slopes. This flow chart is intended to be for general guideline purposes, as the sequence of the operations could be altered one way or another according to the nature of the project.

General requirements for site exploration for construction of slopes are given in Table 4.1. This table relates the height of the slope, the slope angle, and the risk category of the slope to be constructed. The explorations needed to fulfill the requirements set out in Table 4.1 are given in Tables 4.2 and 4.3. These tables are intended for general guidance for planning any site exploration of slopes.

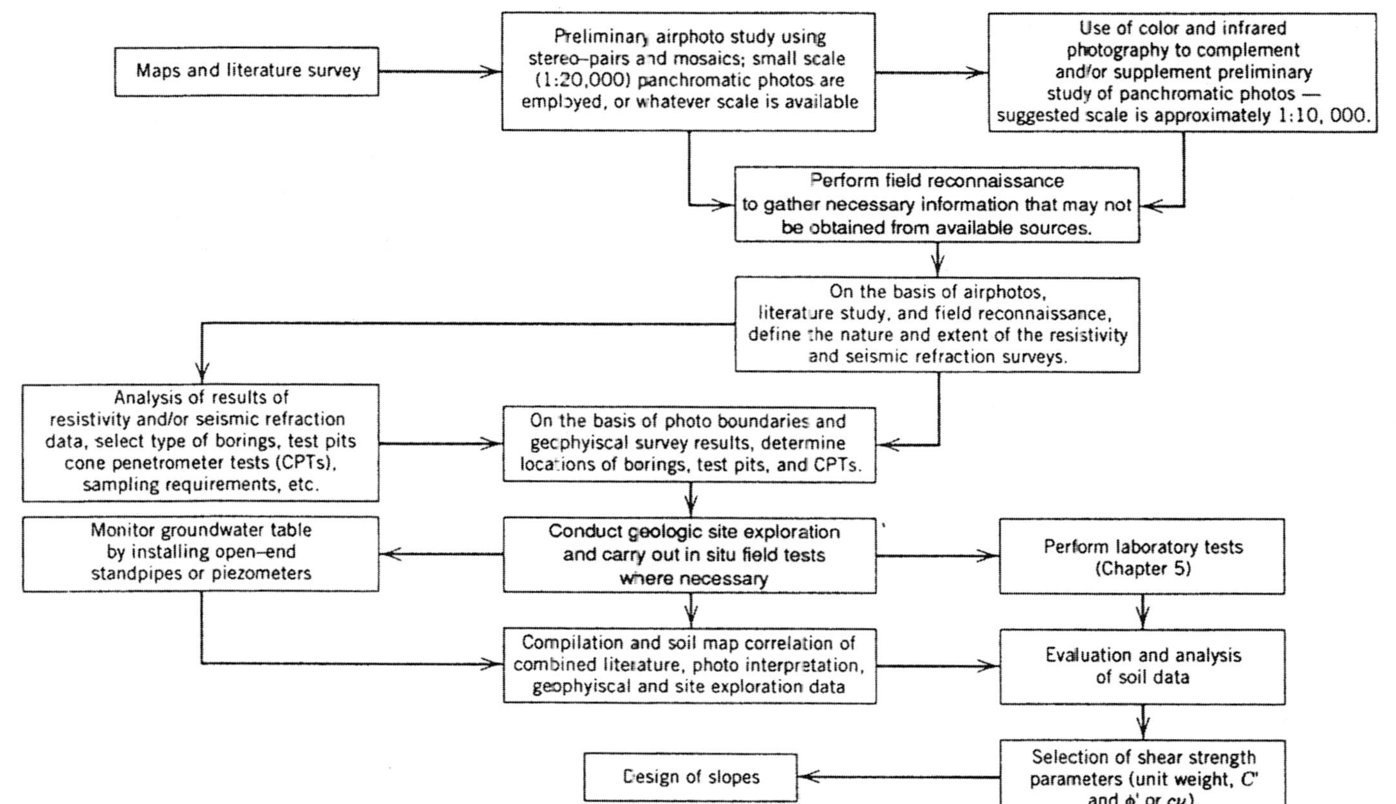

Figure 4.1 Flow chart showing the series of operations used for planning site geologic exploration for the design of slope. (Modified from Mintzer, 1962.)

TABLE 4.1 Guidance on Site Investigation

Risk Category		Formed Slope Classification					Angle of Natural Hillside in the Vicinity of the Site		
			Soil						
Category	a. Loss of life b. Economic Loss	Features	Fill	Cut	Rock	Retaining Wall	0° to 20°	20° to 40° Description of Site Investigation	Greater than 40°
Negligible	a. None expected (no occupied premises)	Height	$\ngtr 25'$	$\ngtr 15'$	$<25'$	$<10'$	Assessment of surrounding geology and topography for indication of stability; visual examination of soil and rock forming the site or to be used for the embankment	As for 0° to 20°; more detailed geology and topography survey; for the steeper slopes information on soil and rock joint strength parameters; survey of hydrological features affecting the site	As for 20° to 40°; area outside confines of site to be examined for instability of soil, rock, and boulders above the site
	b. Minimal structural damage; loss of access on minor roads	Angle	$\ngtr 50°$	$\ngtr 30°$	—	—			
							Specialist advice—Requirement (A)	Specialist advice—Requirement (B)	Specialist advice—Requirement (C)
Low	a. Few (only small occupancy premises threatened)	Height	$\ngtr 50'$	$\ngtr 35'$	$>25'$	$<20'$	Geology and topography survey of site and surrounding area; soil and rock joint strength parameters for foundations and cut slopes; for embankments steeper than 1 on 3, recompacted strength parameters of fill; for cuts, information on groundwater level	As for 0° to 20°; survey of hydrological features affecting the site	As for 20° to 40°; extend outside limits of site to permit analyses of slopes above and below the site
	b. Appreciable structural damage; loss of access on sole access roads	Angle	$\ngtr 60°$	$\ngtr 30°$	—	—			

							Specialist advice—Requirement (B)	Specialist advice—Requirement (B)	Specialist advice—Requirement (C)
High	a. More than a few b. Excessive structural damage to residential and industrial structures; loss of acess on regional trunk routes	Height Angle	> 50′ >60°	>35′ >30°	>50′ —	>20′ —	Detailed geology and topography survey of site and surrounding area; soil and rock joint strength parameters for foundations and cut slopes; recompacted strength parameters for fill; for cuts, information on groundwater level	As for 0° to 20°; survey of hydrological features affecting the site; extend investigation locally outside limits of the site to permit analyses of slopes above and below the site	As for 20° to 40°; extend investigation more widely outside limits of site to permit analyses of stability of slopes above and below the site
							Specialist advice—Requirement (B)	Specialist advice—Requirement (C)	Specialist advice—Requirement (C)

Source: Geotechnical Control Office (1984).

Notes: (1) This table is intended to provide guidance only. Each situation must be assessed on its merits to decide whether or not the recommended investigation procedures are necessary or if peculiar conditions require even more detailed examination.

(2) While this table gives an indication of the requirements for a site investigation under certain general conditions, Table 4.2 gives more precise information on how the above requirements can be met.

(3) For slopes on which there are unstable boulders, the services of an experienced geotechnical engineer or engineering geologist will always be necessary.

(4) Risk category should be assessed with reference to both present use and development potential of the area.

(5) Formed slope classification to be based upon either slope height or angle, whichever gives the highest risk category.

(6) Requirements for specialist advice:

(A) Services of an experienced geotechnical engineer or engineering geologist not necessary.

(B) Services of an experienced geotechnical engineer or engineering geologist to depend on location relative to developed or developable land.

(C) Services of experienced geotechnical engineer or engineering geologist essential.

TABLE 4.2 Content of Site Exploration

Risk Category	Angle of Natural Hillside in the Vicinity of the Site: 0° to 20°	20° to 40°	Greater than 40°
Negligible	B1 D E1 G1	B1 C1 D E1 F1 G1	A B1 C1 D E1 F1 G1
			C2 E2
		G3	G3
Low	A B1 C1 D E1 F1 G1	A B1 C1 D E1 F1 G1	A B1 C1 D E1 F1 G1
	C2 E2 G2	B2 C2 E2 F2 G2	B2 C2 E2 F2 G2
	G3	G3	E3 G3
High	A B1 C1 D E1 F1 G1	A B1 C1 D E1 F1 G1	A B1 C1 D E1 F1 G1
	C2 E2 G2	B2 C2 E2 F2 G2	B2 C2 E2 F2 G2
	E3 G3	E3 G3	E3 G3

Source: Geotechnical Control Office (1984).

A Examination of terrestrial photographs, aerial photos, and geological maps.
B Survey of 1. topographical, geological and surface drainage features.
2. hydrological features.
C Geological mapping of 1. surface features.
2. structures.
D Investigation holes, such as trial pits, boreholes, or drillholes, as appropriate.
E Sampling 1. quality class 4
2. quality class 3
3. quality class 1 or 2 } see Table 4.3.
F Field measurements of 1. groundwater level.
2. permeability.
G Laboratory tests 1. classification tests.
2. density tests for fill materials.
3. Strength tests for soils and rock joints.

Notes: (1) This table is intended to provide guidance only.
(2) Vane testing may be appropriate in marine silts or other fine-grained soils. Installation of instruments for long-term monitoring of displacements and pore pressures should be considered during the site investigation stage.
(3) Chemical tests will be required if aggressive soil/water is suspected in the vicinity of steel or concrete.

4.2 DESK STUDY

4.2.1 Available Existing Data

Maps and Plans The maps and plans discussed in this section are of topographic nature. Other maps and plans include geologic maps and soil survey maps and are discussed in subsequent sections.

Topographic maps and plans can be used to identify geomorphological forms and drainage patterns. This information can give an indication of the materials to be found on the site. In addition, topographic maps provide information on the accessibility of the site and the terrain, both of which may determine the types of equipment to be used for exploration work. Maps do not have the detail of aerial photographs, but can enable a trained observer to surmise relevant information about the geology of a site

TABLE 4.3 Sample Quality Classes

Quality Class	Purpose	Soil Properties Obtainable	Typical Sampling Procedure
1	Laboratory data on undisturbed soils	Total strength parameters Effective strength parameters Compressibility Density and porosity Water content Fabric Remolded properties	Piston thin-walled sampler with water balance Air-foam flush triple-tube core barrel Block samples
2	Laboratory data on undisturbed insensitive soils	Density and porosity Water content Fabric Remolded properties	Pressed or driven thin- or thick-walled sampler with water balance Water flush triple-tube core-barrel
3	Fabric examination and laboratory data	Water content Fabric Remolded properties	Pressed or driven thin- or thick-walled samplers Water balance in highly permeable soils SPT liner samples
4	Laboratory data on remolded soils, sequence of strata	Remolded properties	Bulk and jar samples
5	Approximate sequence of strata only	None	Washings

Source: Geotechnical Control Office (1984).

on the basis of landforms and drainage patterns shown. The amount of information that can be derived from such maps depends on the areas involved and the amount of detail shown. General characteristics of the soil and rock are commonly revealed by topography. Drainage patterns of soil deposits often give an indication of particle size and the likelihood of inundation.

The major source of topographic maps is the United States geological Survey (USGS). The USGS publishes a series of quadrangle maps, known as the National Topographic Map Series, which covers the United States and its territories and possessions. Each map covers a quadrangle area bounded by lines of latitude and longitude. Maps covering areas of 7.5′ of latitude by 7.5′ of longitude and plotted to scales of 1:24,000 and 1:31,680. A complete list of all USGS maps is presented in the U.S. Geological Survey and the supplements thereto, which are published monthly.

Topographic maps are also produced by the Army Map Service and the United States Coast and Geodetic Survey (USC&GS). Other sources of topographic information include: the U.S. Army Corps of Engineers, which publishes topographic maps and charts of some rivers and adjacent shores plus the Great Lakes and their connecting waterways; the U.S. Forest Service, which publishes forest reserve maps; and the Hydrographic Office of the Department of the Navy, which publishes nautical and aeronautical charts.

Geological Maps Geological maps can be used to obtain information on materials and geological structures that affect the site. Geological maps are extremely useful as part of the site exploration, but they are often based on isolated exposures and boreholes so that much of their detail is conjecture rather than fact. This should be kept in mind by the engineer.

Geological maps include (1) bedrock geology maps, (2) structural geology maps, (3) surficial geology maps, (4) tectonic maps, (5) earthquake data maps, and (6) other useful maps such as the glacial map of the United States and the loessial soils of the United States.

The major source of geologic maps and information comes from the USGS, which has published books, maps, and charts in various forms since 1879. "Indexes to Geologic Mapping in the United States" is the most useful series available; it comprises a map of each state on which are shown the areas for which geologic maps have been published. The maps distributed by the USGS include a geologic map of the United States at a scale of 1:2,500,000, and other series of maps like the Geologic Quadrangle Maps of the United States at a scale of 1:24,000, Folios of the Geologic Atlas of the United States, and the Mineral Resources Maps and Charts.

Geologic information is also available from state and local governmental agencies, the Association of Engineering Geologists, the Geological Society of America, the Geotechnical Engineering Division of the American Society of Civil Engineers, and local universities.

Soil Survey Maps The soil surveys conducted by various governmental agencies are also a useful source of information for the engineer planning a subsurface exploration program. These surveys normally consist of the mapping of surface and near-surface soils over a large expanse of land, and are of two types: agricultural and engineering. Since both types usually cover an entire county, the information contained in them is generalized. However, this information, published in the form of maps and text, is particularly useful for highways projects.

Agricultural soil surveys conducted by the Soil Conservation Service (SCS) of the U.S. Department of Agriculture (USDA) are presented in the form of reports that contain a description of the areal extent, physiography, relief, drainage patterns, climate, and vegetation as well as the soil deposits of the area covered. The soil survey maps are usually plotted as overlays on aerial photographs at relatively large scales. They are prepared on a county basis and illustrate the soil cover to a depth of about 7 feet. The shallow depth depicted limits their usefulness in many engineering studies. In some states, engineering supplements to the agricultural survey reports have been prepared by local authorities. These supplements provide data on the drainage characteristics of the materials and anticipated engineering problems. County soil survey reports prepared by USDA usually show soil characteristics from depths of 3 to 15 feet.

Very few state engineering soil surveys are available. An example is the Engineering Soil Survey of New Jersey (Rogers, 1950). The survey reports comprise a general volume and an individual volume for each county in the state. The general volume describes the climate, physiography, geology, and soils of the state, the mapping and soil testing techniques used; and the symbolic notation used for the identification of the various soil types. Each county volume comprises a text, which includes general data on the physiography, surface drainage, and geology of the area, and a soils map, which delineates the areal extent of the various materials by means of symbolic notation according to AASHTO Designation M 145–49.

State agencies (State Engineers and Water Resource Departments) may have on file drilling logs for water wells in the area. Information from these sources can be obtained at relatively low cost and serves as a useful guide in planning the extent of an exploration program.

Landslide Records Alger and Brabb (1985) have compiled a bibliography of maps and reports for landslides throughout the United States—6,500 references are listed on a state by state basis. Many state highway departments, geologic surveys, and university departments have gathered records of landslides that occurred in their states. Within California, USGS (Taylor and Brabb, 1986) has compiled a map of the state showing where landslide inventory and susceptibility maps have been prepared. This reference also lists more than 750 landslide publications for the state of California.

Each landslide record may consist of the following items: (1) location of the landslide, (2) date and time of occurrence, (3) geometry of the slope before and after the landslide (which is accompanied by a photograph), (4) material of the slope, (5) possible cause that triggered the landslide, and (6) rainfall data. All these items may be presented on a landslide card, as shown in Figure 4.2. Locations of the landslides are usually summarized in a state or county map for future reference. These records are essential for the engineers planning exploration programs as well as decision making regarding slope stability at the site. Maps prepared by Saul (1973) have individual landslides numbered, and a table is then provided that lists the geologic units involved, along with their relative ages, probable causes, evidence of movement, and other data regarding types of landslides involved.

Many cities, counties, utility districts, flood control agencies, or other local agencies may keep records of landslides within their jurisdiction. These records are occasionally not available for public review.

Details of particular landslides can sometimes be obtained from local residents. The qualitative description of such incidents may be reasonably accurate; however, the details of timing are often less reliable.

Seismic Records Seismic records that are used to assess earthquake hazard are often of a historical nature. Examples range from the three great earthquakes that

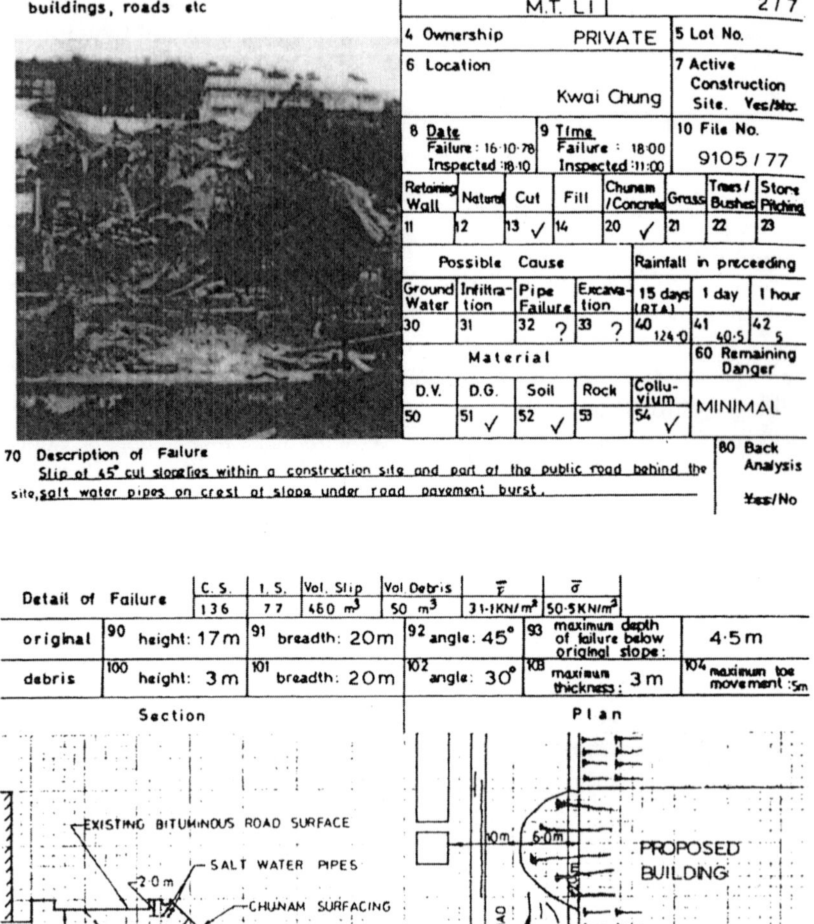

1 Photograph / Sketch of failure relative to buildings, roads etc	2 Inspected by M.T. LI	3 Failure Number 2/7
	4 Ownership PRIVATE	5 Lot No.
	6 Location Kwai Chung	7 Active Construction Site. Yes/No
	8 Date Failure: 16·10·78 Inspected: 18·10	9 Time Failure: 18:00 Inspected: 11:00
	10 File No. 9105/77	

Retaining Wall	Natural	Cut	Fill	Chunam / Concrete	Grass	Trees / Bushes	Stone Pitching
11	12	13 ✓	14	20 ✓	21	22	23

Possible Cause				Rainfall in preceding		
Ground Water	Infiltration	Pipe Failure	Excavation	15 days (RTA)	1 day	1 hour
30	31	32 ?	33 ?	40 124·0	41 40·5	42 5

Material					60 Remaining Danger
D.V.	D.G.	Soil	Rock	Colluvium	MINIMAL
50	51 ✓	52 ✓	53	54 ✓	

70 Description of Failure
Slip of 45° cut slope lies within a construction site and part of the public road behind the site, salt water pipes on crest of slope under road pavement burst.

80 Back Analysis Yes/No

Detail of Failure	C.S.	I.S.	Vol. Slip	Vol. Debris	$\bar{\tau}$	$\bar{\sigma}$
	136	77	460 m³	50 m³	31·1 KN/m²	50·5 KN/m²

original	90 height: 17 m	91 breadth: 20 m	92 angle: 45°	93 maximum depth of failure below original slope:	4·5 m
debris	100 height: 3 m	101 breadth: 20 m	102 angle: 30°	103 maximum thickness: 3 m	104 maximum toe movement: 5 m

Figure 4.2 Landslide card. (From Malone and Shelton, 1982, reproduced by permission of ASCE.)

occurred in the 1811–1812 near New Madrid, Missouri, to the Loma Prieta Earthquake that occurred in Santa Cruz, California, in 1989, and Northridge Earthquake near Northridge, California, in 1994. During the past 60 years, a second source of data has emerged that can pinpoint the location of earthquake epicenters and determine their magnitudes by means of sensitive seismographs located around the United States. This more complete set of data gives a much better picture of seismic activity. Figure 4.3 shows a plot of U.S. earthquakes of intensity V and above on the modified Mercalli scale through the last six decades.

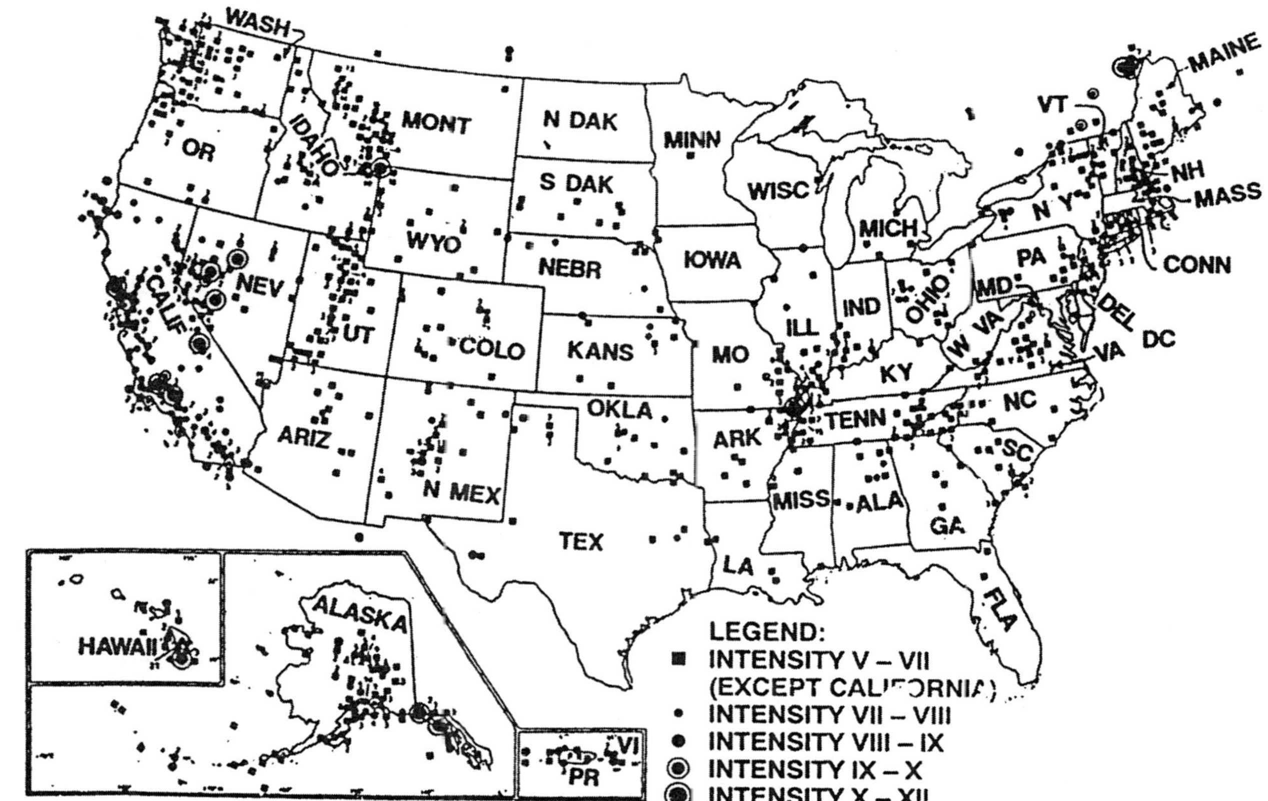

Figure 4.3 Map of the United States show ng earthquake incidences of intensity V and above. (Hunt, 1984a, adapted from National Oceanic and Atmospheric Administration, 1970.)

Records of slope failures as a result of earthquakes are documented by the USGS and some state highway departments. With these records, USGS geologists and geotechnical engineers and the state highway departments have developed techniques for mapping areas most likely to be hazardous because of earthquake-triggered earth movements, and such maps are a valuable tool for regional planning.

Published Literature Valuable information on the geology of a site may sometimes be found from published articles in engineering and geologic journals or university publications. The majority of states have geological surveys or equivalent agencies responsible for the gathering and the dissemination of geologic information. The data may take the form of geologic maps, geologic reports, and records of exploration.

Numerous articles are published by geologic organizations, and these publications are referenced in two periodicals: Bibliography of North America Geology, published by USGS, and Bibliography and Index of Geology Exclusive of North America, by the Geological Society of America. The Association of Engineering Geologists and Geologic Society of America publish geologic maps, as well as monthly journals and special volumes that discuss in detail specific geologic topics on locales.

4.2.2 Previous Geologic Explorations

Geotechnical information about a site may be found in records of previous site development. These include information on site formations, site investigations, well borings, foundations used, and previous stability considerations for slopes. These records are generally held by governmental agencies, consultant architects, and engineers from both public and private developments. Records for old developments may be scant or nonexistent.

With previous site investigation data, subsurface profiles and models, such as the cardboard model shown in Figure 4.4, can be simply constructed to elucidate the site geology. This information is very valuable in planning geologic explorations.

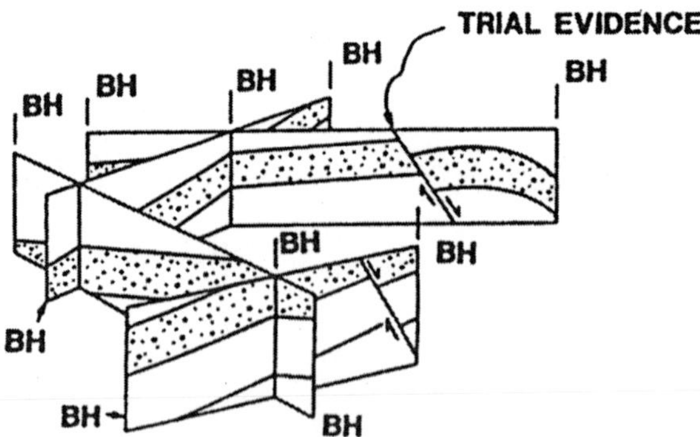

Figure 4.4 A simple three-dimensional model made from cards. (Blyth and de Freitas, 1984.)

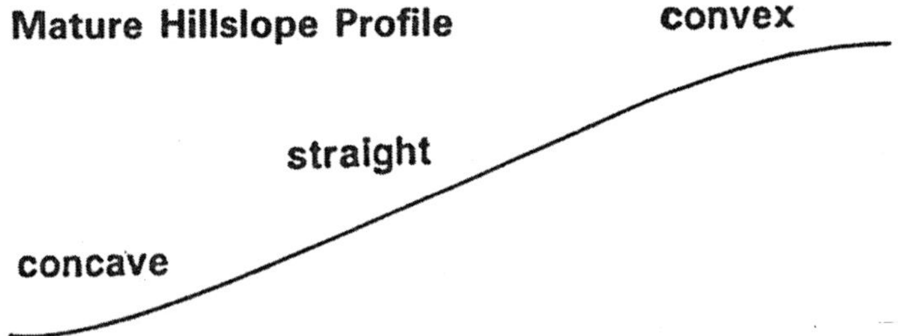

Figure 4.5 Ideal mature hillslope profile, presented by William Morris Davis in 1907. (From Rogers, 1989, reproduced by permission.)

Occasionally, the extent of existing information is such that no additional borings are required.

4.2.3 Identification of Landslide-Prone Terrains through Topographic Expressions

Slope forms and profiles are largely controlled by lithologic and hydrologic requirements, such as slope angle, roughness, types of soil/rock, and quantity of runoff, and by mass movement processes such as landslides. Experienced geologists or geotechnical engineers can distinguish whether a slope is or is not experiencing landslippage by comparing two basic slope forms such as those representing equilibrium conditions (Figure 4.5) and those representing out-of-equilibrium conditions (Figure 4.6). Figure 4.5 is a famous concave-straight-convex mature hilltop profile formulated by William Morris Davis, a geomorphologist in early 1900s. In reality, slope form is more complicated than the model presented in Figure 4.5. All slopes tend to undergo changes due to tectonic uplifts and erosions. For landslide-prone slopes, these tendencies can sometimes be discerned from the topographic maps. However, such

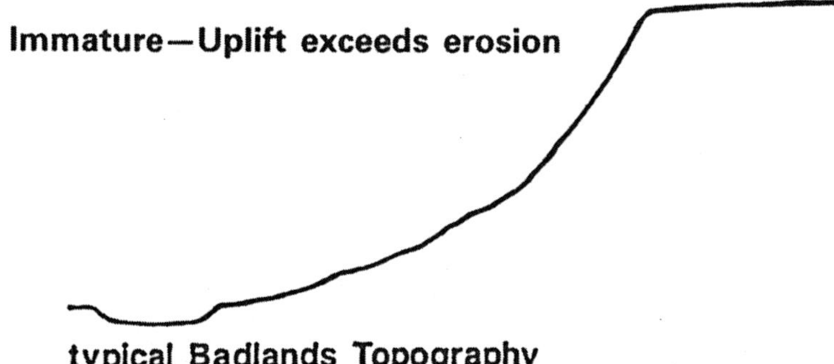

Figure 4.6 Slopes under disequilibrium due to erosion at extremely high rates or sudden uplift. Note the concave slope section dominates the profile with a short straight section and with no recognizable convex crown. (From Rogers, 1989, reproduced by permission.)

discernment depends on the following factors:

(1) Scale and contour interval of the topographic map
(2) Recency of landsliding
(3) Scale and type of slippage

Naturally, topographic expression of landslides is a function of slope steepness, height, and the type of slippage being experienced. In order to discern topographic distortion ascribable to landslide, the map scale must be comparable to the scale of landsliding being evaluated (Rogers, 1989). In other words, the disturbance must be large enough to be noticed on the map. One example of this is a shallow landslide confined to the soil regolith, which may not cause sufficient disturbance to be noticed on a large-scale map like a U.S. Geologic Survey 7.5′ or 15′ quadrangle. On the other hand, large landslides may only be discernible on larger scale maps covering sizable areas (Figure 4.7*a* and *b*). In addition, Figures 4.8 and 4.9 illustrate topographic examples of various styles and scales of landsliding as normally seen on different scales of topographic maps.

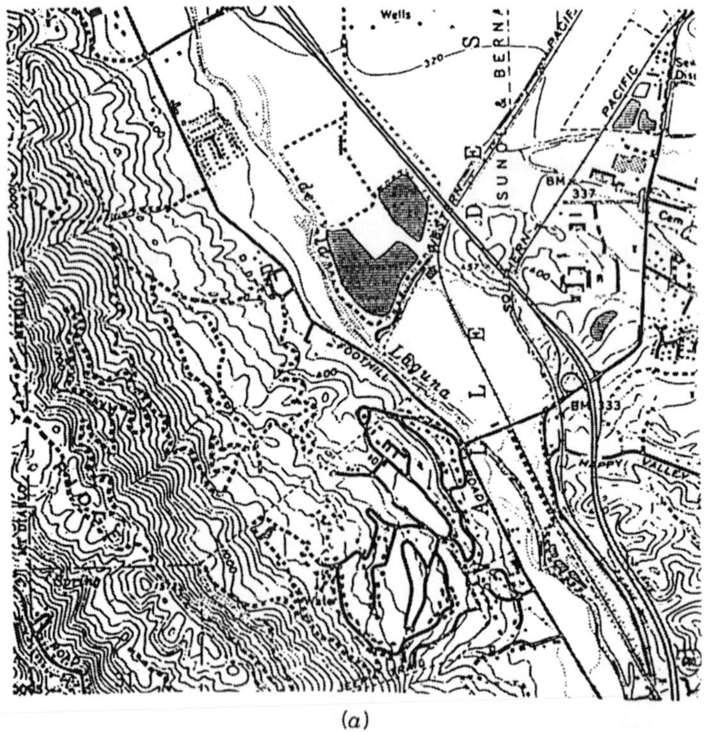

(*a*)

Figure 4.7 (*a*) Portion of the USGS Dublin 7.5′ 1:24,000 scale map with contour interval of 40 feet. At such a gross scale, only the largest of landslide features may be discernible.

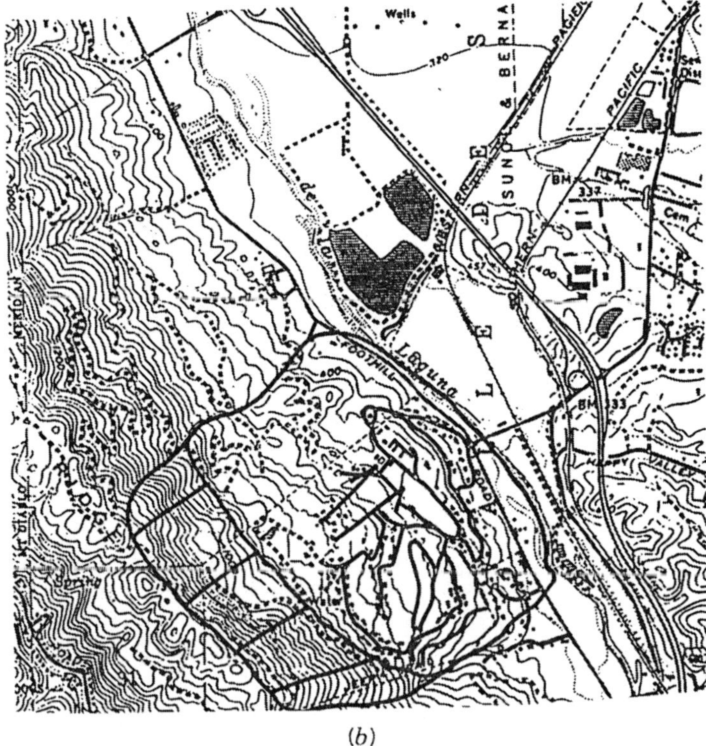

(*b*)

Figure 4.7 (*b*) Overlay showing position of a large Pleistocene-age landslide complex emanating from the Sunol Ridge at left. The crenulated contours with numerous wide benches, hummocky surfaces, and knobs are not typical of an alluvial fan. (From Rogers, 1989, reproduced by permission.)

4.2.4 Air Photos

Recent and past air photos present complete depictions, as well as three-dimensional models, of the site covered. When properly interpreted, they reveal not only the topography, but also considerable information concerning soil, geology, and other natural and engineered features. Advantages of using air photos in the planning stage of a project are:

- *Bird's Eye View of an Area* An overall perspective is obtained of a large area, which gives a three-dimensional view of the area under a pocket or mirror stereoscope.
- *Better Understanding of Drainage and Topography* Important relationships between drainage topography and other natural and engineered elements that are difficult to correlate properly on the ground become obvious in air photos.
- *Better Trace of Surface and Near Surface Runoff* Surface and near-surface drainage channels can be traced.

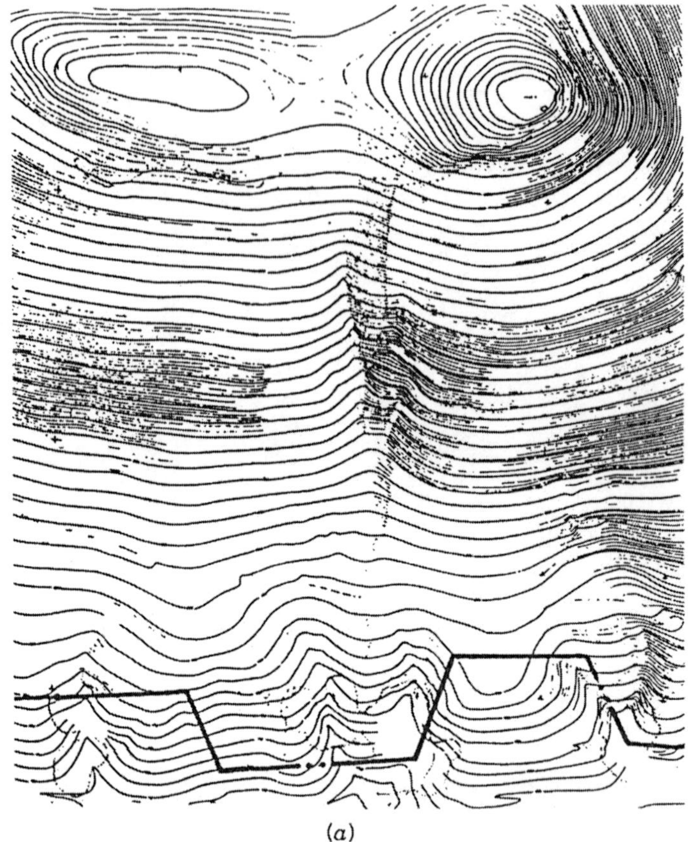

(a)

Figure 4.8 (*a*) Air-flown orthophoto topographic map at a scale of 1:1,200 (1 inch = 100 feet) with a contour interval of 2 feet.

- *Better Understanding of Geologic Formation* Soil and rock formations can be seen and evaluated in their "undisturbed" state.
- *Delineation of Old and Existing Slides* Boundaries of old or existing slides can be readily delineated on air photos. The amount of movement is easily determined from the offset of linear features such as roads, highways, railways, tracks, and so on, as long as they continue into undisturbed areas (Figure 4.10).
- *Easy Identification of Features* Continuity or repetitions of features are emphasized.
- *Effective Planning of Exploration Programs* Routes for field surface and subsurface explorations can be effectively planned.
- *Unblocked View of Areas Covered By a Moderate Vegetation* A moderate vegetative cover does not blanket details to the photo-interpreter as it does to the ground observer.

(*b*)

Figure 4.8 (*b*) Debris flow source area, flow chutes, and accumulation fan, seen from top to bottom. Such complexes are easy to discern by noting the asymmetric drainage patterns as presented. (From Rogers, 1989, reproduced by permission.)

Despite the advantages, aerial photography has its limitations, which include:

- *Personal Experience* The usefulness of air photos increases as the individual interpreter's experience and specific knowledge of the area under study increase. An inexperienced interpreter should be particularly careful in a new, complex area in which he or she has little background knowledge. When this situation arises, individuals with local knowledge should be consulted.
- *Scale* The scale of ordinary aerial photography (1:15,000 to 1:30,000) is adequate for the study of most terrain and slide problems. However, in geologically complex areas or in areas where landslides are rather small, a scale of 1:5,000 to 1:10,000 is more desirable.
- *Terrain Development* The usefulness in tracing terrain development such as landslide histories is greatly handicapped by a developed or built up area.

(a)

Figure 4.9 (*a*) Tight crenulations of topographic contours are often indicative of perturbations associated with active landsliding. The scale of contour interval must be small with respect to the scale of the pertubations. For example, a 10-foot contour interval is not likely to define a 5-foot gully network.

- *Ground Investigation* The use of air photos cannot and should not entirely replace ground investigation. Through careful planning with air photos, however, the surface and subsurface exploration necessary for a site study can be profitably reduced to a minimum.

Air photos can be panchromatic (black and white), color, or infrared. Black and white photographs are most commonly used. They generally are low in cost and have versatility for the maximum number of uses. In addition, they are the easiest to use in the field for delineation of observed ground features. However, the use of color photographs is preferred. This is because the color photos are particularly valuable for assessing differences in moisture and drainage conditions, and for identifying soil or rock materials and the type of vegetative cover. The delineation of landslide-prone terrain is also a special feature easily predictable on color photography. Color infrared photos have special usage where vegetation studies and water conditions are significant in assaying soil types, and are useful for specialized information, such as extent of wet, soft subsoil areas.

GEOLOGIC SITE EXPLORATION

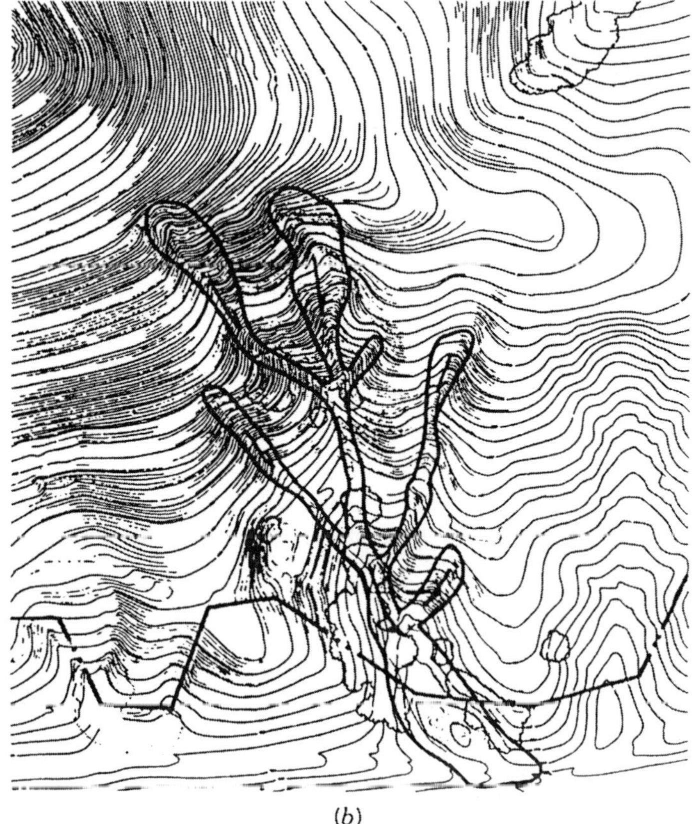

(*b*)

Figure 4.9 (*b*) This figure shows coalescing debris flow complex as seen on a 1:1,200 scale map. (From Rogers, 1989, reproduced by permission.)

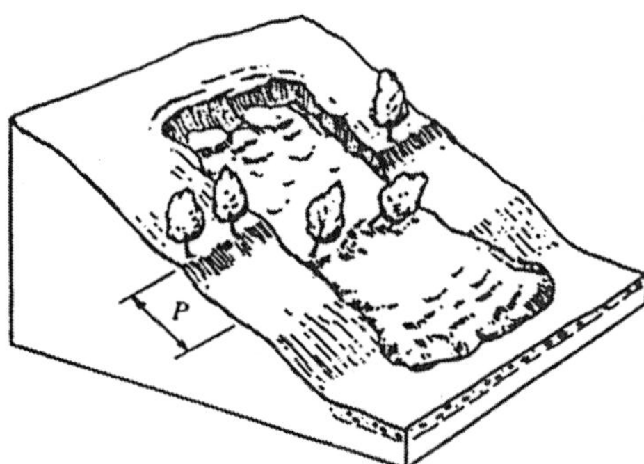

Figure 4.10 The amount of movement (P) that can be determined from the displacement of linear features. (From Zaruba and Mencl, 1982, reproduced by permission of Elsevier Science.)

More sophisticated remote-sensing methods, including satellite imagery, infrared imagery, and radar, are now available for engineering use. These methods can be used to determine large landslides. The satellite imagery can identify regional physiography, geologic structure, most landforms, land-use practices, and the distribution of vegetation. These features, in conjunction with the tonal patterns present on the imagery, provide clues to the types of surface materials present, the surface moisture conditions, and the possible presence of buried valleys. The infrared imagery when combined with aerial photography, can provide information on the surface and near-surface moisture and drainage conditions, the presence or absence of loose material cover, rocks susceptible to sliding, and changes in soil surface temperature.

Availability and access to existing photography are excellent. In the United States, most air photos have been taken for federal and state agencies. The National Cartographic Information Center (NCIC) of USGS publishes a catalog, "Aerial Photography Summary Record System Catalog," of existing, in-progress, and planned aerial photography of the United States. Other government agencies that also publish status maps of aerial photography include the Agricultural Stabilization and Conservation Service (ASCS), the USDA, and the National Archives in Washington, DC. Several state highway and transportation departments similarly have review index maps that indicate the type and extent of photographic coverage available. Some state agencies, for example, the Iowa Geological Survey, publish lists of all the aerial coverage available for their states from all sources.

The NCIC address, and those of other U.S. agencies that have large holdings of aerial photography, are given below.

National Cartographic Information Center
U.S. Geological Survey
507 National Center
Reston, Virginia 22092

Aerial Photography Field Office
Agricultural Stabilization and Conservation Service
U.S. Department of Agriculture
2222 West, 2300 South
Post Office Box 30010
Salt Lake City, Utah 84125

Forest Service
U.S. Department of Agriculture
Washington, D.C. 20250
Headquarters, Defense Mapping Agency
U.S. Naval Observatory
Building 56
Washington, D.C. 20305
Coastal Mapping Division, C-3415

National Ocean Survey
National Oceanic and Atmospheric Administration
6001 Executive Boulevard
Rockville, Maryland 20852

EROS Data Center
U.S. Geological Survey
Sioux Falls, South Dakota 57198

National Archives
Eighth Street and Pennsylvania Avenue, NW
Washington, D.C. 20408

Maps and Surveys Branch
Tennessee Valley Authority
210 Haney Building
Chattanooga, Tennessee 37401

Air Photo Interpretation The use of air photos for slope stability analysis is important for the following reasons:

(1) All areas can be thoroughly searched for signs of instability—a time-consuming and tiring procedure when done in the field.
(2) Many features that can be seen on the photographs are obscured to the ground observer by vegetation, topography, and cultural objects.
(3) Aerial relationships and large features are not always depicted on topographical maps and cannot be observed in the field because an overview is necessary for their detection.

As discussed earlier, black and white air photos are most commonly used. A much better sensing medium would be colored infrared imagery, as it provides a better signature, enhances vegetation changes, and allows seepage to be clearly seen. As the imagery is not that common, the ensuing discussion on recognition of unstable slopes and landslides is mostly confined to black and white air photos.

Recognition of Unstable Slope Forms

(1) Cliff and Rock Outcrops:
  (a) Cliffs are generally recognizable by their steepness and lack of vegetation.
  (b) Unvegetated rock is light in tone but turns grey as it becomes covered in moss or lichen.
  (c) Rock outcrops or areas with only a thin cover of topsoil appear hummocky.

Beware of vegetation hiding outcrops or boulders. If the vegetation is high and only a few boulders can be seen, you can be sure that many more will be beneath the vegetation.

(2) Slopes covered with large quantities of loose soil or rock. Under this category come boulder fields, colluvium, and fans.
   (a) A boulder-strewn slope is recognizable by speckles of light or grey tone (the boulders).
   (b) Colluvium is associated with hilly and mountainous terrain, below cliffs or rock outcrops, usually in valleys, and forming fans when lower ground is reached; it appears much smoother in texture than the mountainous terrain above, and is usually speckled with boulders.
   (c) Fans, recognizable by their fan shape, are usually associated with streams and gullies and tend to be situated at points where slope angle suddenly decreases.

(3) Slope being oversteepened, either by cutting at the base for construction reasons, or by rivers, streams, and gullies, can also be unstable. The latter process can be recognized by:
   (a) Deep incision of the stream in its upper reaches.
   (b) Unusually large amounts of debris in the stream beds.
   (c) Light tones at the head of the stream where exposed rock can be seen.

(4) Slopes with comparatively higher moisture content can be seen by:
   (a) Abrupt changes to a darker tone.
   (b) Abrupt changes in vegetation, usually to a more luxuriant cover.

Recognition of Slope Failure Forms Six forms of slope failure or movement indicative of possible future slope failure are recognizable here: falls, slides, slumps, flows, creep, and gullying.

Differences in mode of movement occur because of changes in material and moisture content, falls generally being associated with rock and little moisture, flows being associated with small single size material with high moisture content. When looking in areas with certain geological and geomorphological characteristics, it is therefore important to be looking for the appropriate signs of instability.

(1) Falls are free falls of debris from the source to a lower level accomplished suddenly. They can be recognized by:
   (a) Light-toned scars on the cliff or outcrop where the face has been newly exposed.
   (b) Boulders at the base of the toe of the cliff or outcrop.
   (c) A hummocky appearance at the base of the cliff.

(2) Slides are mass movements along shear zones where the strength of the slope has failed. The criteria for recognizing these failures are:
   (a) The form is linear in plan.
   (b) The head of the slide is in the form of a concave crescent soil or rock scar, light in tone for recent slides but enhanced by shadow in old slides.
   (c) The slides are sharply outlined by scarps.
   (d) The surface of the slide is hummocky.

(e) There is *no* accumulation of debris at the toe.

(3) Slumps are mass movements along a plane of failure lubricated by excess moisture. The imagery characteristics for slumps are similar to that of slides on black and white photographs except that there is an accumulation of debris at the toe.

(4) Flows also have imagery characteristics very similar to those of slides and slumps. The debris from a flow generally travels a greater distance than slides or slumps, stopping when the flow becomes obstructed or the angle of slope decreases sufficiently. There is usually a toe of accumulated material.

(5) Creep, as its name implies, is a slow form of mass movement that, although it itself does not endanger life, is indicative of possible future failure. Imagery characteristics are:

(a) Slopes showing effect of creep are often scoured by erosion.

(b) Low-angled slopes may have a mottled pattern indicative of small depressions formed by the creep.

(c) Downslope leaning of trees, fences, and posts.

(6) Gullying also indicates an unstable slope. Gullies are usually:

(a) Parallel to each other and perpendicular to the slope with very few or no tributaries.

(b) Relatively short, deeply incised, and arrowhead shaped.

(c) Light in tone where rock or subsoil has been freshly exposed.

Mapping of Air Photos Most of the air photos are printed on tropical paper (the mat finish ones). Grease pencil and blunt lead pencil can be used for drawing on the photographs and can be erased with a soft eraser. On the glossy photographs, only grease pencils can be used. Sharp pencils, colored crayons, or biros should not be used on any air photos.

The procedure of using air photos for slope analysis can be summarized as follows:

Step 1 Look at 1:25,000 photographs for all years.

Step 2 Map in unstable slope forms on the latest photograph, using the identification hints discussed above.

Step 3 Search all areas for signs of failure and mark on the photographs.

Step 4 Look at larger scale photographs (where available) for all years for further detail.

Step 5 Transfer details to appropriate maps.

4.3 FIELD STUDY

4.3.1 Site Reconnaissance

The purpose of site reconnaissance is to observe, recognize, and record those surficial features that may affect the stability of slopes along a proposed highway alignment.

Site reconnaissance is used to supplement and check the accuracy of the information gathered from desk studies and air photos, and to complete the preliminary study of a proposed alignment. The features that are parts of site reconnaissance—supplies and equipment, field mapping, and site observations—are discussed below.

Supplies and Equipment Supplies and equipment for site reconnaissance include a camera, a field log book, a pocket scale, a pen, a pocket penetrometer, a torvane, a geological hammer, a geological compass, a slope level, and a 100- to 200-foot-long measuring tape. Photographs should be taken of existing natural features and structures that may be of some relevance to the exploration of the site. The unconfined compression strength of the in situ clayey materials can be measured by a pocket penetrometer. The shear strength of a clayey soil can be determined using a torvane. The degree of hardness of in situ decomposed rock can be determined roughly by a geological hammer or more accurately by a Schmidt hammer. The inclinations of existing slopes can be measured by a slope level. Dip angles and dip directions of joints are determined by means of a geological compass.

Field Mapping Field mapping is perhaps the most important phase of the field investigation. It is often based on the topographic map so that field notes can be recorded directly on a field copy of the map. Sometimes, in a remote area, the topographic map will not be available. In this case, the geologist or engineer should return to the field after the initial map is prepared to locate the key features. The topographic map can be modified as needed, based on the field mapping, as features need to be added, removed, or relocated.

Key features to be observed during field mapping are features of bedrock outcrops within the study area as well as in surrounding areas; for example, bedrock structures, lithology, unit thickness, or other geological data should be recorded. In the case of landslide areas, important features to be recorded are (1) head scarps, lateral scarps, secondary scarps, slickenside orientations, (2) freshness of scarp features as indicated by the steepness, vegetative cover, crack width, and so on, and (3) existing conditions of the toe area—whether it is still moving on top of an undisturbed ground surface, or the slide is still intact. A field guide to locate various features in relation to landslide types is presented in Table 4.4. The recency of movement should also be evaluated. Table 4.5 presents a good comparison between features associated with active and inactive landslides (Crozier, 1984).

During field mapping, numerous photographs should be taken of the study or slide area as soon as possible. This is because features may change as continued movement occurs which may obscure key data. These photographs should be fully documented with dates and weather conditions so that comparisons can be made between the conditions in the same location on different dates. It is strongly recommended that cameras with a date imprint on the negatives be used to avoid confusion at a later time over dates of photographs.

Site Observations The purposes of site observations are to confirm the basic geology and drainage patterns of the region and the site, to verify the three-dimensional concept of the terrain developed from the literature review, and to identify the presence of faults, such as scarps and straight valleys, and geologic anomalies such as soft clayey seams and slickensided planes. Furthermore, examination of regional patterns of topography and drainage can provide valuable information about probable subsurface structures of the site. This information will be used to develop a subsequent site exploration program.

The amount of time spent on each slope investigation is determined by its topography, geologic environment, and the amount of anticipated excavation or construction. The air photo investigation should limit the areas in which detailed site observation is necessary. Stable areas, such as upland surfaces, can be subjected to a brief examination, while the potentially unstable areas should be investigated in full detail.

Experience has shown that the most effective method of site observation consists of two basic steps.

(1) A rapid reconnoitering of the entire site with special attention to the position of each slope with respect to streams, drainages, and so on, and the larger overall relationships of slopes to each other. This step can be accomplished in a vehicle or on foot. Combined with air photo interpretation, rapid reconnaissance should restrict the remaining observations to the potentially unstable areas and slopes.

(2) A detailed investigation of potentially unstable slopes accompanied by walking over each slope and observing those features that give an indication of its stability. The data gathered in these two steps serves as a basis for planning the subsurface exploration of the area.

During site visits, there may be features that are relevant to the future planning and design of slopes. The following paragraphs describe common features that may be observed during site reconnaissance.

Unnaturally regular topographic forms often indicate the presence of an engineered slope. Such features may include truncated valleys and ridges, and planar slopes. Slumping and settlement are often indications of the presence of fills on the site. Old developments on steep hillsides could have been formed by excavating on one side of the site and filling on the other (Figure 4.11). The overall effect on the topography may not be sufficiently great to allow these features to be identified as engineered slopes by reference to the topography alone.

Vegetation, particularly on old slopes, can make the visual identification of those which are engineered difficult, if not impossible. However, certain types of vegetation often characterize slopes with landslide deposits. Plants such as cattails and alder often thrive in a wet environment and may therefore be indicators of spring activity, water ponds, or general poor drainage. Other plants thrive in loosened soils, which may have been disturbed by landslide movements.

Tree trunk orientation can often be helpful in assessing the type and recency of movement. Moving ground may push a tree and cause it to lean downhill. If the deflection is not recent, the trunk may curve back in a more upright position several feet above the ground. In rotational slides, trees that drop in scarp areas or are on the

TABLE 4.4 Field Guide to Landslide Features

Type of Motion	Crown	Main Scarp	Flanks	Toe
Fall or topple	Has cracks behind scarp	Is nearly vertical, fresh, active, and spalling on surface	Are often nearly vertical	Is irregular
Slide Rotational Slump	Has numerous cracks that are mostly curved concave toward slide	Is steep, bare, concave toward slide, and commonly high; may show striae and furrows on surface running from crown to head; may be vertical in upper part	Have striae with strong vertical component near head and strong horizontal component near foot; have scarp height that decreases toward foot; may be higher than original ground surface between foot and toe; have en echelon cracks that outline in early stages	Is often a zone of earth flow of lobate form in which material is rolled over and buried; has trees that lie flat or at various angles and are mixed into toe material
Translational block	Has cracks, most of which are nearly vertical and tend to follow contour of slope	Is nearly vertical in upper part and nearly plane and gently to steeply inclined in lower part	Have low scarps with vertical cracks that usually diverge downhill	Plow or overrides ground surface

Dry flow	Has no cracks	Is funnel shaped at angle of repose	Have continuous curve into main scarp	Composed of tongues; may override low ridges in valley
Wet flow Debris avalanche Debris flow	Has few cracks	Typically has serrated or V-shaped upper part; is long and narrow, bare, and commonly striated	Are steep and irregular in upper part; may have levees built up in lower parts	Spreads laterally in lobes; if dry, may have a steep front about 3 feet high
Earth flow	May have a few cracks	Is concave toward slide; in some types is nearly circular and slide issues through narrow orifice	Are curved; have steep sides	Is spreading and lobate; consists of material rolled over and buried; has trees that lie flat or at various angles and mixed into toe material
Sand or silt flow	Has few cracks	Is steep and concave toward slide; may have variety of shapes in outline: nearly straight, gentle arc, circular, or bottle shaped	Commonly diverge in direction of movement	Is spreading and lobate

Source: This table was extracted from Rib and Liang (1978), published in Transportation Research Board Special Report 176—Landslides Analysis and Control. Edited by R. L. Schuster and R. J. Krizek (1978).

TABLE 4.5 Features Indicating Active and Inactive Landslides

Active	Inactive
Scarp, terraces, and crevices with sharp edges	Scarps, terraces, and crevices with rounded edges
Crevices and depressions without secondary infilling	Crevices and depressions infilled with secondary deposits
Secondary mass movement on scarp faces	No secondary mass movement on scarp faces
Surface-of-rupture and marginal shear planes show fresh slickensides and striations	Surface-of-rupture and marginal shear planes show old or no slickensides and striations
Fresh fractured surfaces on blocks	Weathering on fractured surfaces of blocks
Disarranged drainage system; many ponds and undrained depressions	Integrated drainage system
Pressure ridges in contact with slide margin	Marginal fissures and abandoned levees
No soil development on exposed surface-of-rupture	Soil development on exposed surface-of-rupture
Presence of fast-growing vegetation spp.	Presence of slow-growing vegetation spp.
Distinct vegetation differences "on" and "off" slide	No distinction between vegetation "on" and "off" slide
Tilted trees with no new vertical growth	Tilted trees with new vertical growth above inclined trunk
No new supportive, secondary tissue on trunks	New supportive, secondary tissue on trunks

Source: Crozier (1984).

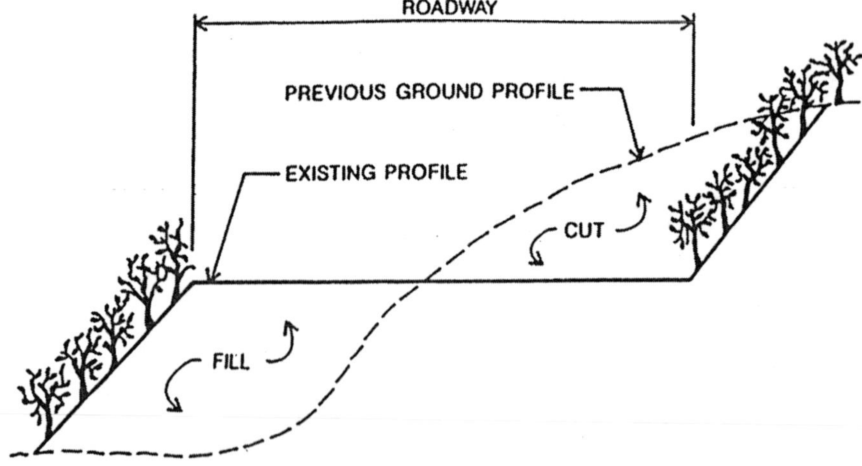

Figure 4.11 Cut and fill obscured by vegetation.

upper portion of a large toe bulge may lean in an uphill direction (Zaruba and Mencl, 1982)

One striking hydrologic phenomenon within a landslide is the many new ponds that are created on the landslide surface. An example of this is the Manti Landslide in Utah (Flemming et al., 1988; Williams, 1988), where numerous ponds of varying dimensions surfaced within a span of about 4 years. Post-slide ponds tend to have distinctly different shapes (more irregular), whereas the pre-slide ponds are usually subround to round. Measurements of pond migration on the landslide are normally made directly on time-sequential aerial photographs. The measurements are made either (1) from an object of fixed position on stable terrains outside the landslide, in line with the direction of slide movement, or (2) a baseline of fixed position drawn between two such fixed objects and about perpendicular to the slide direction.

Sites that have experienced previous landslides may be identified by means of aerial photograph interpretation. However, many of the fine details and more subtle evidences of slope movement cannot be identified at the small scales of aerial photographs and maps, and can be detected only by site observation and survey.

For any site visit, it is of paramount importance to identify previous ground movement. Quite often, the first sign of ground movement is recorded by settlement of a roadway (Figure 4.12), or, depending on the location of the roadway within the moving mass, a bulge in the pavement and broken paved ditches (Figure 4.13). Other signs indicative of ground movements are opening of tension cracks on the roadways (Figure 4.14), minor failure in an embankment (Figure 4.15), material falling onto the roadway from an upper slope (Figure 4.16), or the progressive failure of the region below a fill, which may lead to a larger landslide that could endanger the roadway. Table 4.4 lists some features that can aid in the recognition of common slope movements.

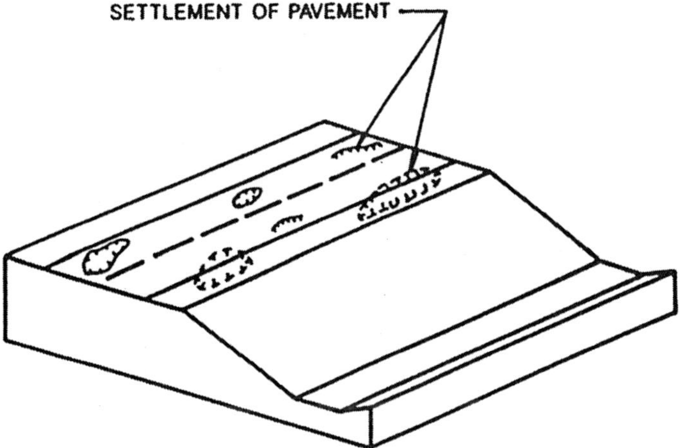

Figure 4.12 Settlement of roadway. (Federal Highway Administration, 1988.)

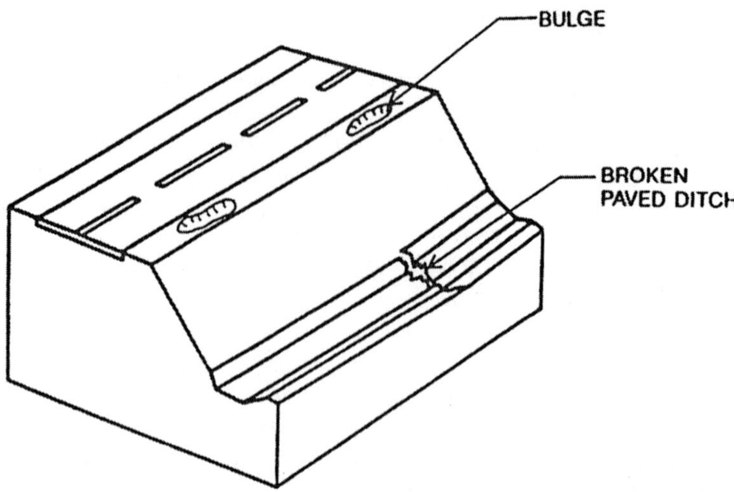

Figure 4.13 Bulge of pavement and broken paved ditch. (Federal Highway Administration, 1988.)

The observation tips presented below are intended to aid the engineer in identifying features that may indicate past and future ground movements that may eventually lead to slope instability. These features, however, should be reviewed judiciously by experienced geotechnical engineers and/or engineering geologists. Once they are identified, these observations could result in a reason for concern or a false sense of impending slope failure and subsequent implementation of costly slope remedial work that may not be required.

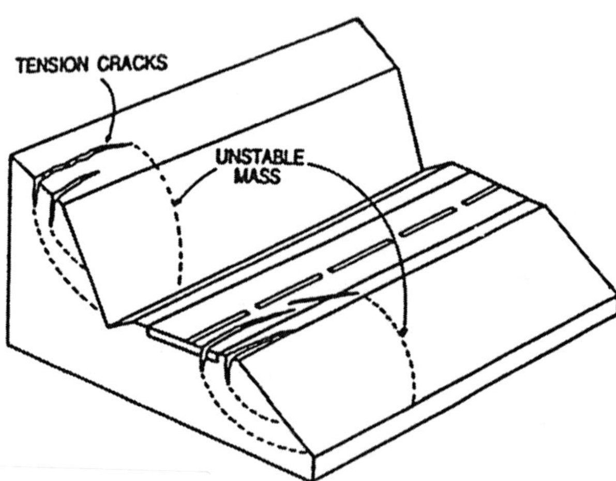

Figure 4.14 Development of tension cracks at top of roadway or cut slope. (Federal Highway Administration, 1988.)

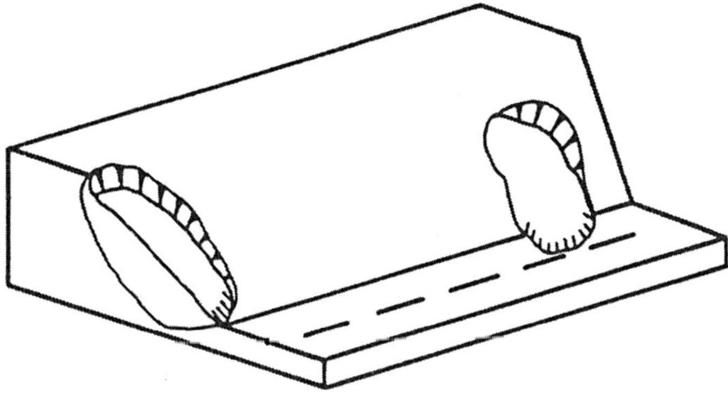

Figure 4.15 Minor failures of embankment. (Federal Highway Administration, 1988.)

Observation Tips

(1) *Look for Ground Movements* Signs of ground movements are evidenced by formation of tension cracks (Figure 4.14), hummocky surfaces of slopes (Figure 4.17), breakage of pipes or power lines (Figure 4.18), tilting of trees (Figure 4.19), spalling or other signs of distress in highway structures, such as guardrails (Figure 4.20), cracking of drainage channels on slope berms (Figure 4.13), closure of expansion joints in bridge plates or rigid pavements, and loss of alignment of building foundations.

(2) *Identify Patterns of Surface Cracks* Surface cracks are not necessarily normal to the direction of ground movement (Figure 4.21), as is commonly assumed by some. For example, cracks near the crown are indeed normal to the direction of horizontal movement, but the cracks along its flank are nearly parallel to it. Small echelon cracks commonly develop in the surface soil before other signs of rupture take place. Cracks parallel to slopes are indicative of a block slide (Figure 4.22), whereas cracks in a horseshoe pattern in plan indicate a slump slide (Figure 4.21)

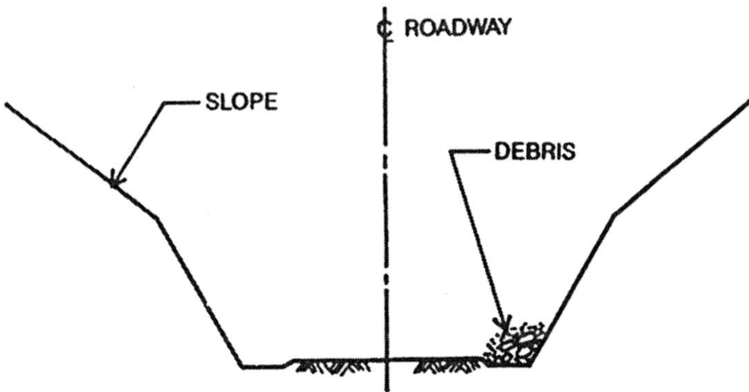

Figure 4.16 Falling debris from an upper slope near roadway.

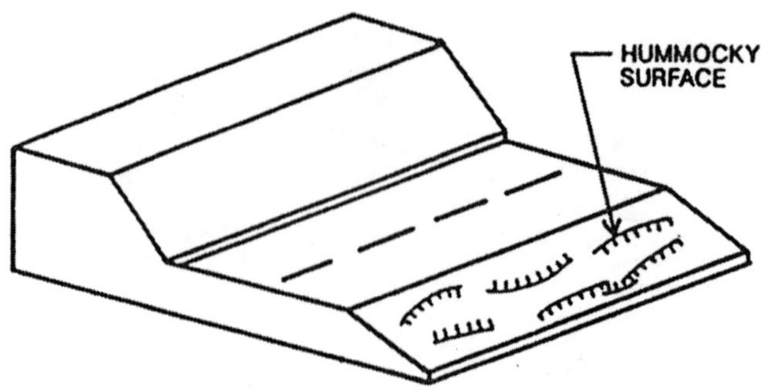

Figure 4.17 Hummocky surfaces of slope face.

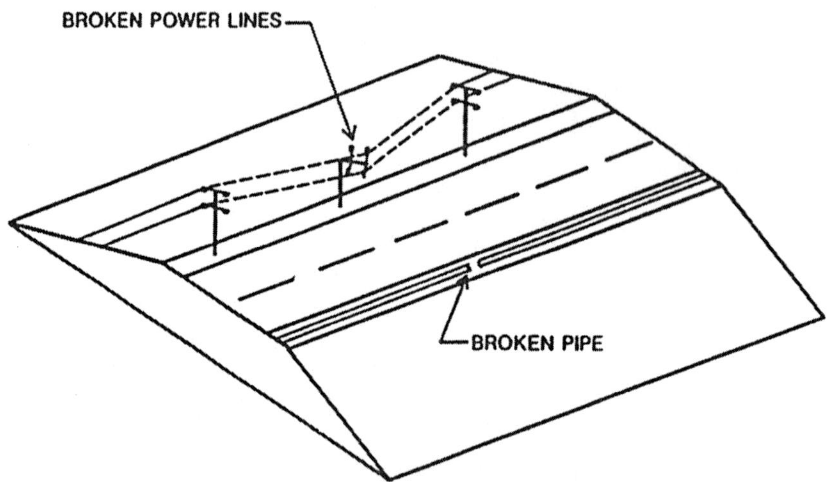

Figure 4.18 Breakage of pipes or power lines.

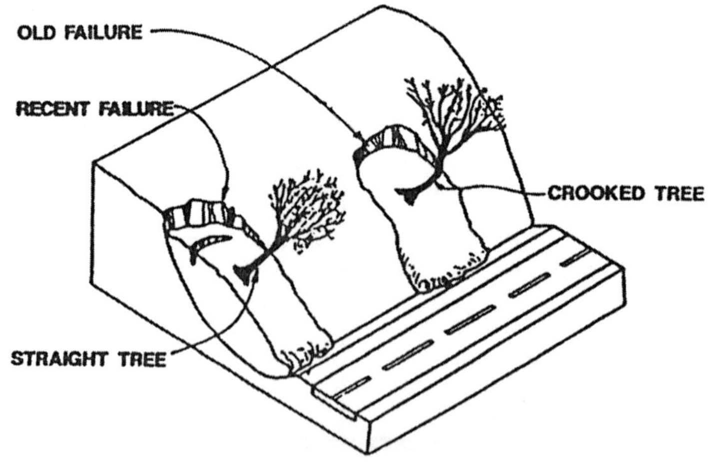

Figure 4.19 Tilted or curved trees. (Federal Highway Administration, 1988.)

UNSTABLE MASS
DIP IN GUARDRAIL

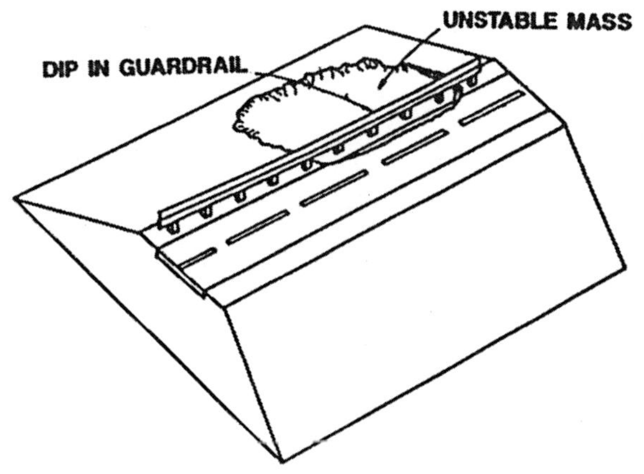

Figure 4.20 Dip in guardrail. (Federal Highway Administration, 1988.)

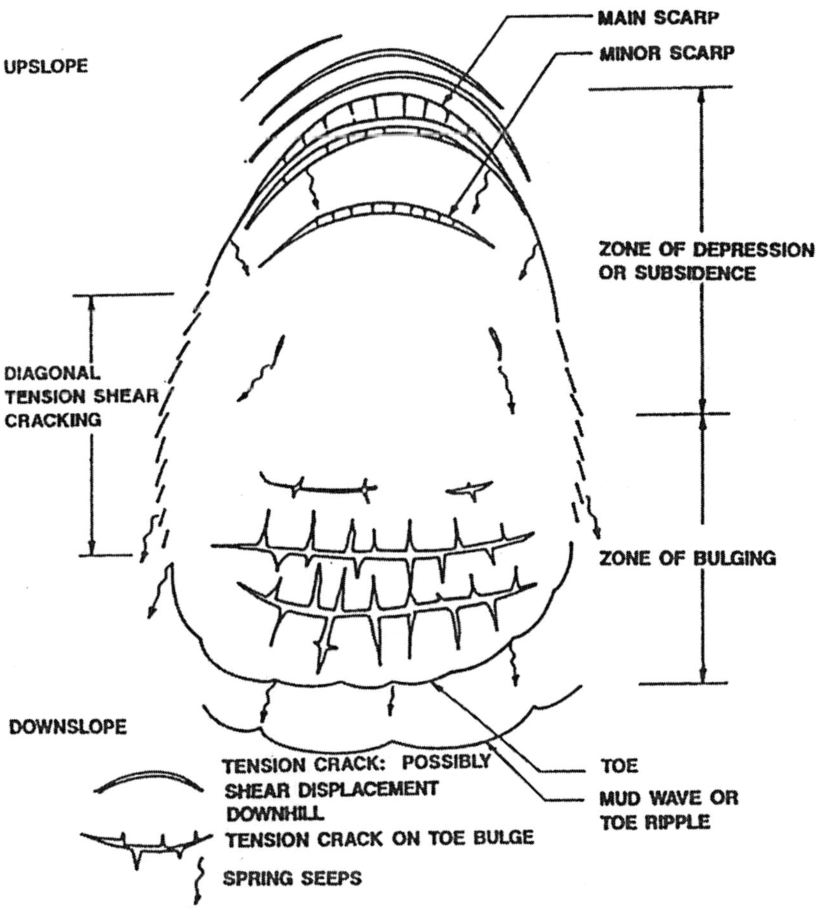

MAIN SCARP
MINOR SCARP
UPSLOPE
ZONE OF DEPRESSION OR SUBSIDENCE
DIAGONAL TENSION SHEAR CRACKING
ZONE OF BULGING
DOWNSLOPE
TENSION CRACK: POSSIBLY SHEAR DISPLACEMENT DOWNHILL
TENSION CRACK ON TOE BULGE
SPRING SEEPS
TOE
MUD WAVE OR TOE RIPPLE

Figure 4.21 Cracks, bulges, scarps, and springs. (Schuster and Krizek, 1978.)

Figure 4.22 Block slide in cohesive materials near Portage, Montana. (Schuster and Krizek, 1978.)

(3) *Look for Troublesome Hydrogeologic or Soil Formations* If the formation has alternate weak and competent soil layers, slides may occur along the weak layers (Figure 4.23). Other areas have soils that are subject to liquefaction, for example, the saturated loose sandy soils in California. Some soil areas are subject to erosion, for example, river banks, toes of hills, toes of embankments, and steep hillsides. Erosion removes support from the toe of engineered and natural structures, and landforms.

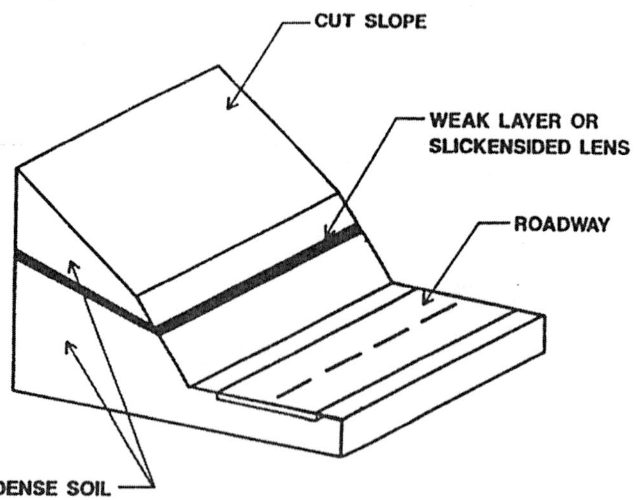

Figure 4.23 Sliding along weak layer or slickensided lens. (Federal Highway Administration, 1988.)

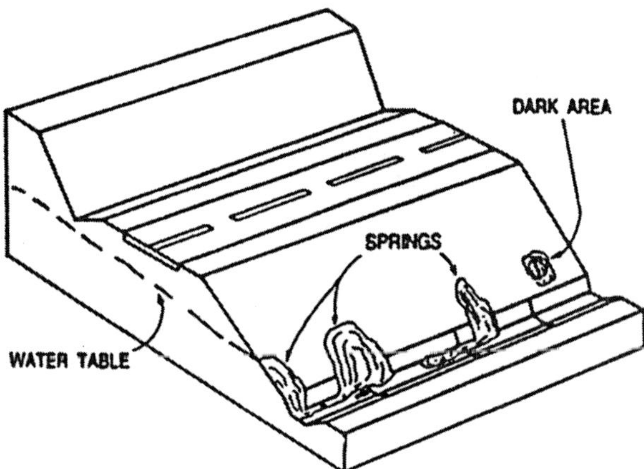

Figure 4.24 Naturally occurring springs on highway slopes. (Federal Highway Administration, 1988.)

Naturally occurring springs located at toes or crests of slopes may soften the soil, causing it to lose strength and allowing the slopes to fail (Figure 4.24). Often locations of springs can be found in densely vegetated areas (Figure 4.25). River banks, natural escarpments, quarries, and highway and railway cuts may reveal, through the presence of seeps or springs, information on groundwater flow in the area.

Fills most likely to be unstable are (1) those constructed in stream valleys where the depth of highly weathered material is the greatest, and (2) those constructed on hillside areas where the potential sliding surface is inclined.

(4) *Determine Existing Drainage Patterns* Site drainage is one of the most important factors involving slope instability. Subsurface water may saturate and weaken the soils of embankments, foundations, and natural soils. The result often leads to

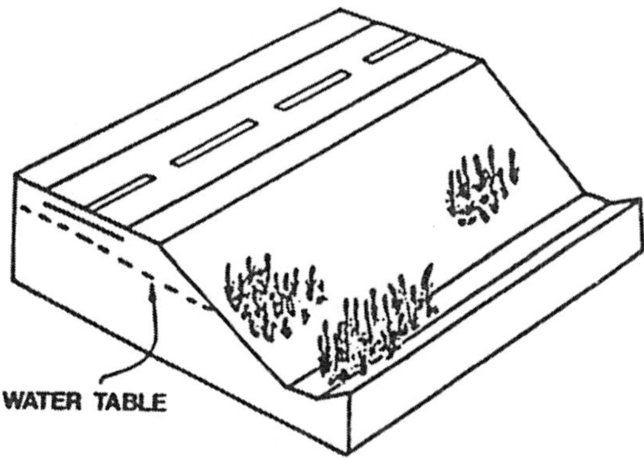

Figure 4.25 Cattails or willow trees warn of subsurface seepage. (Federal Highway Administration, 1988.)

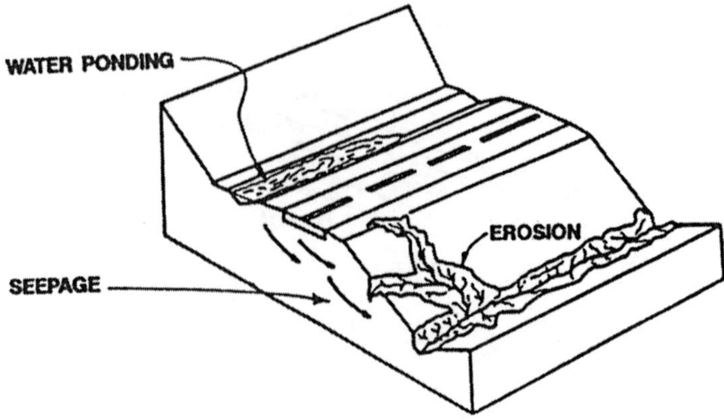

Figure 4.26 Poorly drained highway locations. (Federal Highway Administration, 1988.)

a landslide. Surface water, if not properly drained away from the slope, also may saturate the soil (Figures 4.26 and 4.27). Therefore, it is necessary to look for any drainage flow that may have a potentially adverse effect on slope stability.

During the field reconnaissance, all stream courses, channels, nullahs, ditches, catchpits, and culverts should be mapped. The details, sizes, and conditions should be plotted on the geotechnical site plan. This information will prove useful when assessing surface drainage characteristics of the existing site, and how these existing surface drainage measures will have to be modified or improved to accommodate the future slope stability.

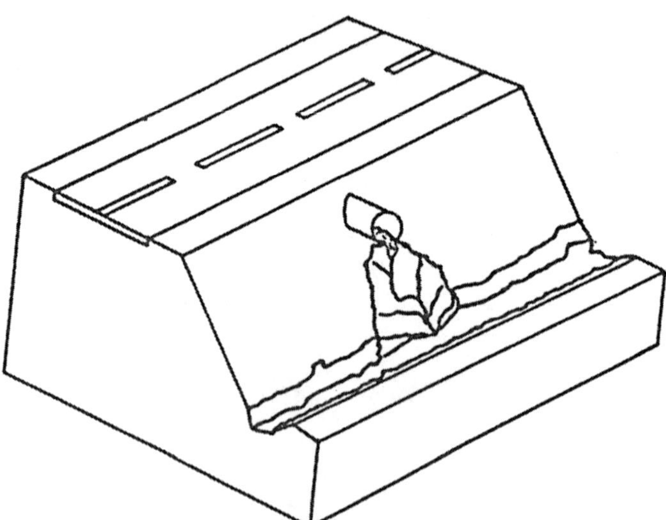

Figure 4.27 Slope erosion, caused by discharge from drainage structure. (Federal Highway Administration, 1988.)

Slope instability along an existing roadway may sometimes be attributed to the inadequate maintenance of existing drainage features. Therefore, all the existing drainage features should be checked for inadequacy and leakage.

(5) *Always Take Note of Natural or Engineered Earth Structures (i.e., Natural, Cut, or Fill Slopes and Retaining Structures) in the Vicinity of Site* These structures often give clues as to (1) the most likely and practical way of designing, constructing, and remediating slopes, (2) the potential problems that may occur after construction, and (3) the types of remedial measures to be undertaken should the slope experience instability.

(6) *Use Common Sense to Explain Features Associated with Ground Movements and to Determine the Causes of Ground Movements* Ground movements occur if the ground experiences "something" that undermines its equilibrium. This "something" could be natural causes, such as weathering, intense rainfall, and existence of soft layers, or human causes, such as undercutting toes of slopes, overstressing the ground, and so on.

All observations should be recorded in writing and photographs so that they can be reviewed at a later time in the office. It is stressed that *all* observations should be recorded. An observation that seems insignificant at the time often serves as the key to the solution of a difficult design, construction, or remediation problem later on. A small hole on a slope that is thought to be an animal burrow may turn out to be an exit tunnel, for instance. Another example is daylighting relict joints in residual soil slopes, which may be an adverse factor that will trigger slope instability.

4.4 EXPLORATION METHODS

4.4.1 Introduction

Field exploration methods inherently involve excavation techniques designed to provide data concerning the geologic conditions of the site. These techniques can be divided into two groups, based on the relative size of excavation: borings and large excavations.

Boring methods utilize the drilling of a hole from the ground surface to determine various geologic conditions, the most notable being the soil or rock stratigraphy and groundwater elevation. During boring operations, soil samples are taken for on-site field identification and testing as well as future laboratory testing for such properties as Atterberg limits, moisture content, internal angle of friction, cohesion, and shear strength (Chapter 5). During drilling of the boring, close attention should be paid to the drilling operation by a geologist or geotechnical engineer, who also logs and identifies soil samples as they are recovered. Samples are taken at régular intervals (typically every 5 feet) or continuously, if there is a need for more detailed data. Continuous sampling is discussed later in this section. After borings are completed, they can be utilized for field permeability and other types of tests in the underlying material or converted into observation wells for careful groundwater monitoring.

Numerous methods have been developed to advance borings to the required depths including: auger drilling, wash boring, percussion drilling, rotary wash drilling, and hammer drilling. Of these methods, auger drilling and rotary wash drilling are the most common and are discussed later in this section.

Larger excavations are utilized when it is desirable to observe the soil stratigraphy in situ on a larger scale. These excavations are large enough for an engineer or geologist to enter for close examination. In addition, large samples can be recovered for more detailed testing, such as plate load tests, direct shear tests, and so on. Larger excavations can be classified into three main groups: test pits, test trenches, and large boreholes. Large boreholes and test pits are discussed later in this section.

4.4.2 Auger Drilling

Auger drilling is one of the most common methods of advancing a boring. A power drill rig provides rotational movement and downward pressure to advance the auger. Drill rigs can be (1) tripod-mounted for areas with space constraints and limited platform, (2) skid-mounted for rugged terrains, (3) barge-mounted for offshore drilling, and (4) truck-mounted for access to sites on land. A typical truck-mounted drill rig is shown in Figure 4.28. Augers are fitted with a cutting head at the tip and an adaptor at the opposite end. The adaptor permits additional augers to be attached, thus advancing the hole to its required depth. This type of auger system is called a continuous-flight auger. There are two different types of continuous-flight augers: solid stem and hollow stem.

Figure 4.28 Truck-mounted drill rig.

Solid-stem augers are the fastest method for drilling a borehole, particularly when sampling is done at longer intervals. Because of the nature of the solid stem, the augers have to be removed whenever sampling is desired. This can be a lengthy undertaking if the hole has been drilled relatively deep. Solid-stem augers are not effective in granular (sandy) soils or soils below the water table because removal of the augers causes caving of the hole. Hence, solid-stem augers are used for shallow borings and borings above the water table. Solid-stem augers generally range from 2.5 to 4.5 inches in diameter.

Hollow-stem augers provide access for sampling through the stem of the auger. This facilitates sampling without removing the augers from the hole. In addition, the augers serve as a casing to prevent caving of the borehole whether above or below the water table. Below the water table, water must be added to the hollow-stem to counteract the hydrostatic forces on the outside of the augers. Figure 4.29 shows a schematic diagram of the hollow-stem auger. When sampling is desired, the center

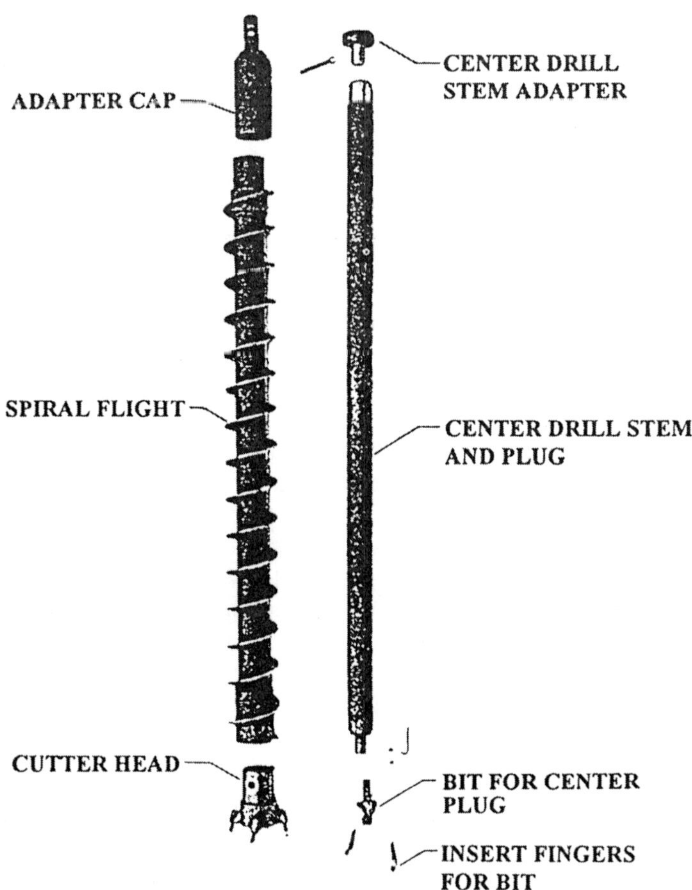

Figure 4.29 Schematic diagram of a hollow-stem auger.

rod is inserted into the stem of the auger. The center rod, like the augers themselves, comes in sections, is put together with a sampler at the end, and is placed at the proper depth. When sampling is complete, the rods are pulled back through the auger, and the sampler is removed from the bottom of the rods.

Care must be taken when using augers in sandy soils, especially below the groundwater table. Sand at the tip of the bottom auger will be disturbed within one diameter, due to the cutting action of the carbide tooth bit. If this disturbed soil is tested using the standard penetration test (SPT), blow count results will not be accurate. The water level in the hollow-stem augers should be above the groundwater table to negate the effects of excess hydrostatic pressure and prevent a "quick" condition from developing at the bottom of the hole. If this should occur, the sample zone will be severely disturbed and the augers should be advanced to the next sample depth. Drilling mud may be required to keep the hole stable from that point on.

Sizes of hollow-stem augers typically range from 6.25 inches OD (2.5 inches ID) to 9 inches OD (3.25 inches ID). A typical auger drilling operation is shown in Figure 4.30.

4.4.3 Rotary Wash Drilling

Another common method of advancing a boring is rotary wash drilling. This method utilizes a series of drill rods attached to a cutting bit that rotates with a downward pressure to shear away material and advance the hole. Drilling fluid is applied under pressure to cool the cutting bit and remove cuttings from the bottom of the hole. The fluid is constantly circulated through the hole using a wash tub, which also collects the cuttings. Drilling fluid, commonly referred to as drilling mud, is usually a slurry of water and bentonite, but other drilling fluids are used depending upon the nature of the underlying soil. The drilling fluid also helps to keep the hole open during the drilling and sampling operations. Casing also is driven or drilled into place, usually just above the next sample depth, to keep the hole open.

When sampling is desired, the rods are withdrawn from the hole and the cutting bit is replaced with a sampler. One distinct advantage of rotary wash drilling is its ability to handle virtually any type of soil. Also, SPT blow counts are typically not affected by hydrostatic pressures because water is always circulating throughout the hole. However, for this to be true, every time the drill rods are removed, drilling fluid must be added to the hole to replace the loss of volume left by the drill steel. One minor disadvantage of this method is that it requires heavier truck-mounted equipment and a larger setup area for the wash tub. This constraint may make it difficult to employ the rotary wash system in tight areas. A typical rotary wash setup is shown in Figure 4.31.

Wire-Line Drilling For drilling deep boreholes, say over 100 feet deep, wire-line drilling is commonly used. It is a rotary-type drilling method in which the core device is an integral part of the drill rod string, which serves as a casing. Core samples are obtained by removing the inner barrel assembly from the core barrel portion of the drill rod. The inner barrel is released by a retriever lowered by a wire-line through the drilling rod.

Figure 4.30 Typical auger boring operation.

Odex Drilling Odex drilling is the tradename for a drilling technique developed by Atlas Copco and Sandvik. Sometimes used to drill through caving soils, boulders, fractured rocks, or weathered rocks overlying rock formations, it is also useful in drilling through landslide debris, which is usually a mixture of boulders, cobbles, gravels, sand, silt, and clay.

The method uses a drill bit with an eccentric reamer to drill a hole larger than the outside diameter of the casing. The drilling method uses a top drive hammer for 3-inch diameter holes and a down-the-hole hammer for 4.5-inch holes. Figure 4.32 shows the drill strings for both the top drive and the down-the-hole (rotary drive) hammers. Odex drilling can keep the casing right behind the bit and facilitates driving the casing. It makes use of a percussion hammer using air to blow out the cuttings, and is a very noisy operation. Large loss of ground could be experienced when Odex drilling

Figure 4.31 Typical rotary wash setup.

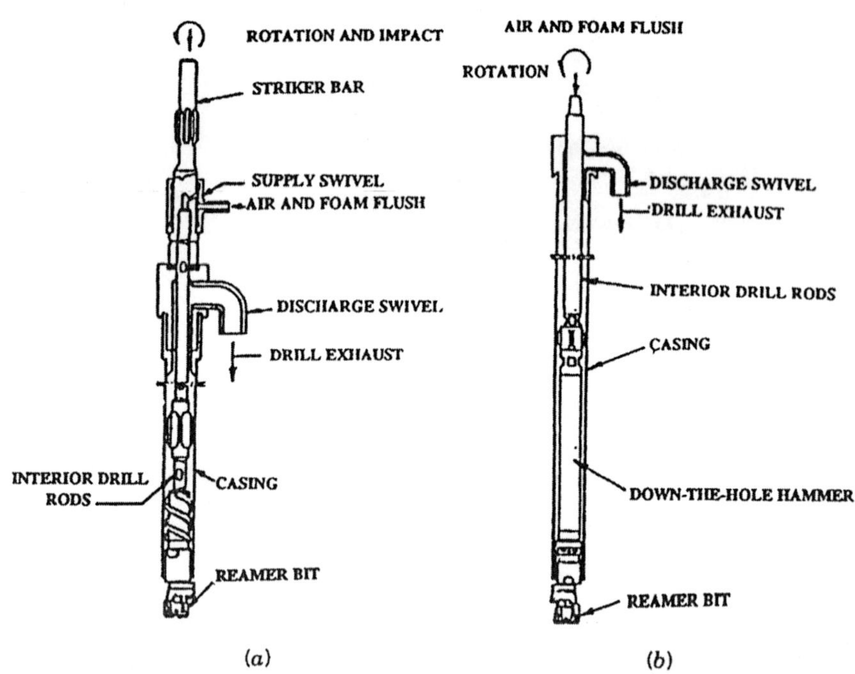

Figure 4.32 Odex drill string arrangements. (*a*) Top drive Odex drilling. (*b*) Rotary drive Odex drilling.

is used to drive casing through a sand and/or gravel layer below the groundwater table.

4.4.4 Limitations of Auger and Rotary Wash Drilling

The drawback of flight auger borings and small-diameter cored holes is that they generally do not provide detailed information that can be obtained through downhole geologic inspection of large-diameter bucket auger borings (see Section 4.4.6). Flight auger borings and small-diameter core holes may not recover critical, but discrete, clay lenses, landslide shear planes, and fault zones. If such key discontinuities are missed in the subsurface investigation, the cost effectiveness of small-diameter borings is thrown into question, and some critical data needed for design considerations will be missing. In general, landslide slip planes are soft and easily washed away, thereby eluding recovery and recognition. Furthermore, it is basically impossible to obtain in-place moisture content determination in core soil samples recovered from rotary wash drilling (due to water injection). Some core holes may be successful, but that success relies heavily upon the experience of the drilling crew and the experience and ability of the inspector. When logging both auger and rotary wash drilling, only half of the subsurface data is obtained, and the engineer does not know where the cuttings are coming from unless continuous sampling is being used.

4.4.5 Sampling in the Ground

Sampling The main purposes of sampling are: (1) to establish the subsurface geological profile in detail and (2) to supply both disturbed and undisturbed materials for laboratory testing.

There are basically three techniques for obtaining soil samples, namely: (1) taking disturbed samples from the drill tools or from excavating equipment in the course of boring or excavation; (2) drive sampling, in which a tube or split tube sampler having a sharp cutting edge at its lower end is forced into the ground, either by static thrust or by dynamic impact; and (3) taking block samples specially cut by hand from a test pit or shaft. The selection of these techniques depends on the quality of the sample that is required and the character of the ground, particularly with regard to the extent to which disturbance occurs before, during, or after sampling.

It should be borne in mind that the overall behavior of the ground is often dictated by planes or zones of weakness that may aggravate slope stability (e.g., slickensided surfaces, relict joints, kaolin seams, gouge zones, discontinuities, etc.). Samples obtained by methods 2 and 3 above will often be sufficiently intact to enable the ground structure within the sample to be examined. The quality of such samples can vary considerably, depending on the technique and the ground conditions, and most will exhibit some degree of disturbance. These intact samples are usually taken in a vertical direction, but specially orientated samples may be required to investigate particular features.

Disturbed samples are often taken at 5-foot intervals in borings. The quality of these samples depends on the technique used for sinking the borehole and on whether the ground is dry or wet. When disturbed samples are taken from below water in a borehole, there is a danger that the samples obtained may not be truly representative of the ground. This is particularly the case with granular soils containing fines, which tend to be washed out of the tool.

Continuous Sampling In cases when extremely detailed subsurface data is a necessity, continuous sampling may be employed. Instead of the typical sampling interval of 5 feet, samples are taken throughout the entire depth of the hole. Hence the boring is advanced through sampling and not through drilling. When sampling is completed at a given depth, the sample is removed and a new sampler is inserted. This method provides the most detailed information possible during exploration, since information is obtained continuously. For example, any changes in soil strata at all depths can be logged with a very high level of accuracy. However, the cost of this method is generally higher than exploration methods mentioned previously because it takes a longer time to complete a borehole. In some states, it is common practice to take continuous undisturbed samples, and it may be the most economic choice in those locations. When the need for detailed information is paramount, or outweighs the higher cost, this method is chosen.

Handling and Labeling of Samples Samples retrieved from boreholes should be treated with great care. It must be stressed that the usefulness of the results of the laboratory tests depends on the quality of the samples at the time they are retrieved. It is therefore important to establish a satisfactory procedure for handling and labeling the samples, and also for their storage and transport, so that they remain as close to the in situ conditions as possible, and can readily be identified and drawn from sample stores when required.

All samples should be labeled immediately after being taken from a borehole. If they are to be preserved at their natural moisture content, they will at the same time have to be sealed in an airtight container or coated in wax. The label should show all necessary information about the sample, and an additional copy should be kept separately from the sample; this latter is normally recorded on the daily field report.

4.4.6 Large Boreholes

Large boreholes provide an opportunity to examine subsurface conditions in situ. Boreholes with diameters up to 96 inches can be excavated, with diameters of 30 inches or greater suitable for an engineer or a geologist to enter the hole and to study the soil stratigraphy. Holes are excavated using large-diameter solid-stem augers or bucket augers. When the solid-stem auger is used, the auger is advanced to the height of the flight or until the flight is filled with soil. The auger is then withdrawn from the hole, and the soil is removed from the auger by reverse rotation of the drill. This process is repeated until the hole is advanced to the required depth. Larger diameter solid-flight augers can be used on any type of soil or rock if the appropriate cutting teeth are used.

Bucket Auger Boreholes Bucket augers are another alternative for large-diameter boreholes. These augers cut the soil in the same manner as solid-flight augers, but the cuttings are collected in a bucket rather than on the flight of the auger. When the bucket is filled with material, it is removed, emptied, and placed in the hole for another pass. Bucket augers are not suitable for excavations in rock or excavations in sandy soil below the water table.

When the boring is completed, an engineer or a geologist can be lowered to examine and log the subsurface conditions, and to take bulk samples if desired. Bucket auger drill holes used for downhole geologic inspection are typically 24 to 30 inches in diameter (Figure 4.33). This would allow the engineer or geologist to comfortably fit into the hole with all the equipment that is required for comfort as well as safety. In downhole inspection, the engineer or geologist is equipped with safety harness, sampling tools, rock hammer, and Brunton compass, and then lowered down the borehole in an aluminium cage or on a swinglike stand (Figure 4.34). The hole must be kept free of hazardous gases and well ventilated if necessary before anyone enters it. A continuous air supply is normally provided when hydrogen sulfide gas is encountered at intolerable level. Hazards associated with personnel entering these boreholes should not be overlooked. They should be addressed in accordance with the State and Federal Safety Standards.

Bucket auger borings have been used for collecting valuable geologic structural and lithologic details, especially in landslide investigations. Downhole logging permits accurate measurements of bedding, fracture, and shear orientations, as well as slikenside bearings on landslide surfaces. Bulk samples of shear surfaces can be obtained in sufficient quantity to permit laboratory tests such as residual shear strength testing on remolded samples. However, downhole logging is prohibited below caving of saturated and fractured materials or groundwater table.

4.4.7 Test Pits

Test pits provide an alternative for soils to be examined in their natural conditions and locations. They can be excavated by hand to limited depths. They often are excavated by construction equipment, such as a backhoe, and trenches can be up to 30 feet deep. Hand-excavated test pits are a viable alternative when accessibility limitations to a site prevent the use of a backhoe. For deeper test pits and locations where there are no access constraints, the backhoe is the preferred equipment for excavation. Sometimes, test pits are excavated below the groundwater table, in which case the groundwater has to be pumped out of the excavation. Once excavation is completed, a geologist or an engineer enters the test pit to examine and log the soil stratigraphy and take samples, if required. The pit must be stable or shored for safe inspection to take place. Although the information obtained from a test pit is limited to the upper layers of a soil deposit, test pits are a fast, economical method to examine subsurface conditions. Figure 4.35 is an example of a test pit log. A previous landslide failure plane can be examined using this method.

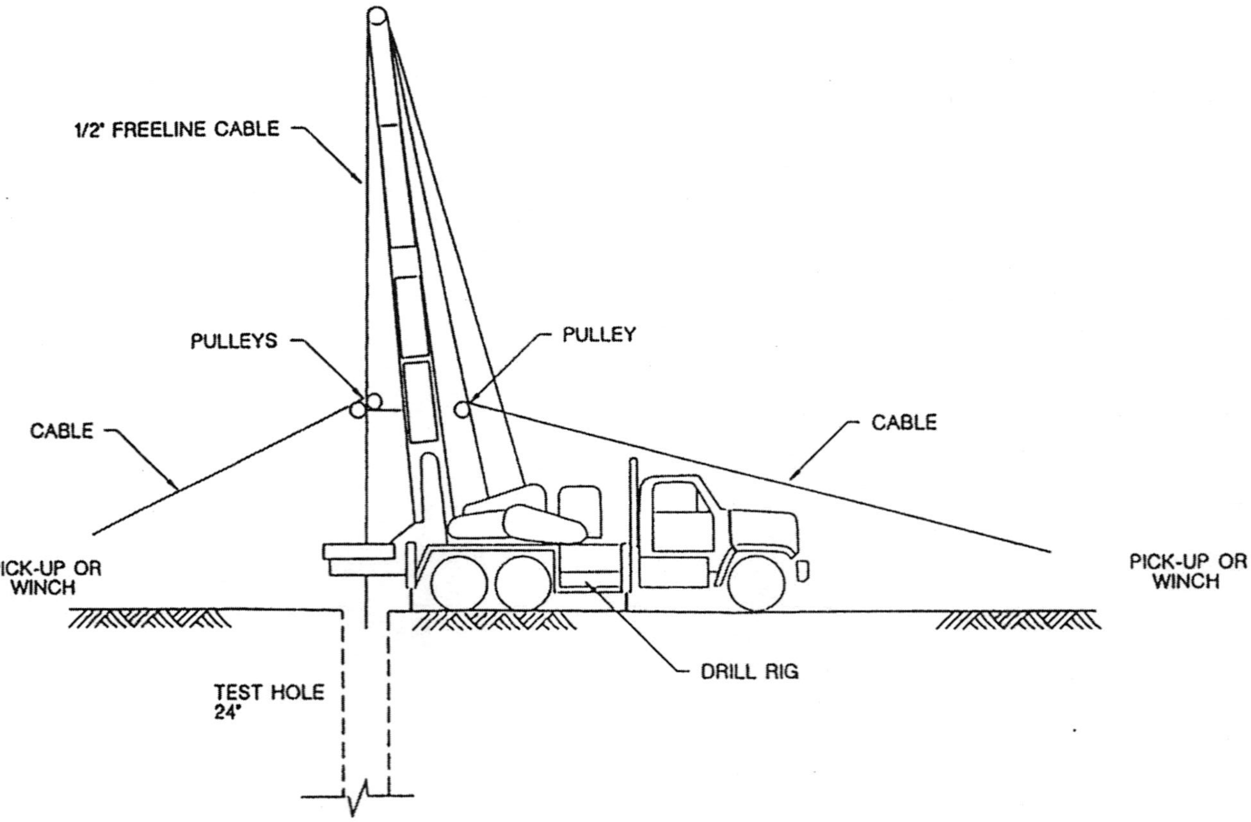

Figure 4.33 Schematic of bucket auger hole setup. (Scullin, 1994.)

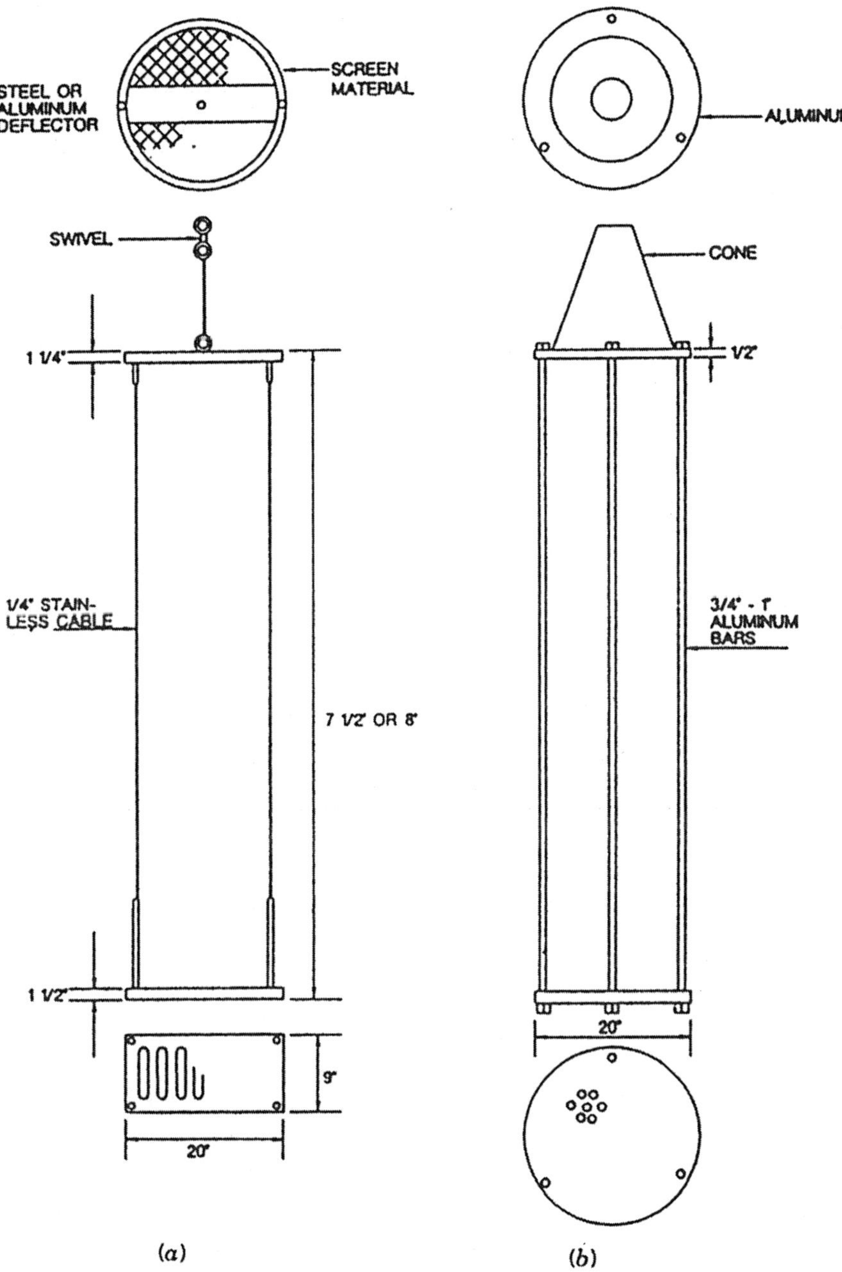

Figure 4.34 (*a*) Schematic of cable-type swing cage. (*b*) Schematic of aluminium strut cage. (Scullin, 1994.)

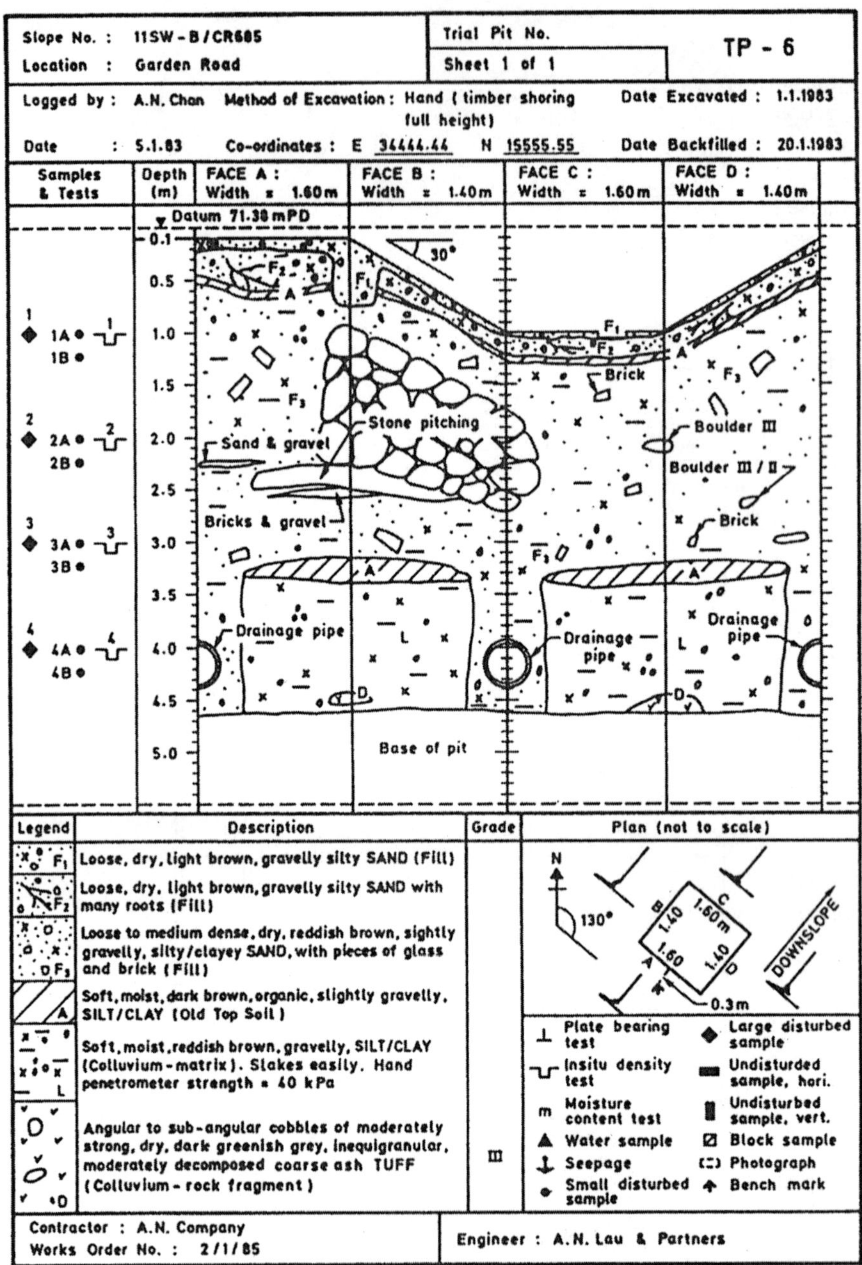

Figure 4.35 Test pit log. (Geotechnical Control Office, 1987.)

4.5 TESTING METHODS

Testing methods used in the slope stability analysis consist of (1) in situ testing, (2) geophysical testing, and (3) laboratory testing. These methods generally fall into three principal categories, namely (1) tests for determining soil strengths, (2) tests for identifying soil structure and fabric, and (3) tests for identifying soil mineralogy. Because of its important role in slope stability analysis, the laboratory testing is not discussed in this section, but is separately addressed in detail in Chapter 5.

4.5.1 In Situ Testing

In situ tests carried out during the exploration stage can be used to assess strength, deformation properties, and permeability of soils and rocks. Details of many of the tests considered here are given by the United States Bureau of Reclamation and the American Society for Testing and Materials (1982). Types of in situ testing fall into three categories: (1) borehole tests, (2) large-scale pit tests, and (3) geophysical tests. These tests are discussed in the following sections.

Borehole Tests Although in situ borehole tests suffer from the limited volume of material tested, they do allow the soil to be tested without the disturbance produced by removing a sample from the ground, transporting it to the laboratory, and preparing it for testing (Chapter 5). In situ tests are performed in the same boreholes that are drilled for identifying the soil strata and for securing small-diameter samples. Common tests performed in situ are (1) the SPT, (2) Dutch cone test, (3) pocket penetrometer test, (4) torvane test, and (5) vane shear test. Other borehole tests include (1) pressuremeter test, (2) borehole shear test, (3) dilatometer test, and (4) permeability test.

Standard Penetration Test (SPT) This test is most commonly used to give an approximate relative measure of the density of soils. The procedure is described in ASTM D1586-84 and AASHTO T-206. A split-tube sampler (Figure 4.36) with $1\frac{3}{8}$-inch internal diameter, 2-inch outer diameter, and 27 inches long is driven 18 inches in three 6-inch increments by dropping a 140-pound hammer a distance of

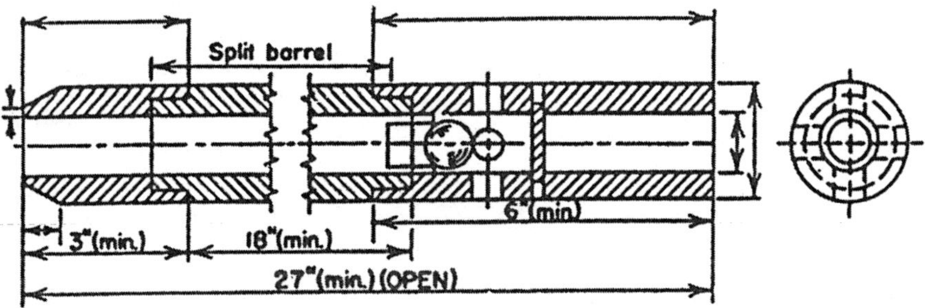

Figure 4.36 Standard penetrometer sampler.

30 inches. The sum of the blows for the second and third increments is the standard penetration resistance (N), which is expressed in blows per foot.

The test and its interpretation have been reviewed by Fletcher (1965) and Nixon (1982). The SPT results can be significantly affected by the testing techniques. Therefore, the following points should be noted while carrying out the test and interpreting the results.

(1) The borehole casing should not be ahead of the borehole, and water balance should be maintained if carrying out the test below the water table.

(2) Large-diameter rods ($1\frac{9}{16}$ inches or equivalent) or smaller rods with rod supports should be used to reduce energy dissipation.

(3) The accuracy of a cathead winch is dependent on the skill of the operator.

(4) Higher blowcounts will be obtained in a gravel- and cobble-strewn formation because added blows are required to drive the spoon when gravel or cobbles plug the end of the split-spoon. In such cases, it is advisable to select the lowest number of blows recorded in the formation to evaluate its density.

(5) Deep excavations can have a relaxing effect on the blow count of the test. Stresses from the weight of the overburden, as well as the soil density, affect the blows on the sample spoon. For example, removal of 15 feet or more will relieve the pressure noticeably. Therefore, the test results should be corrected to account for lower overburden pressure.

(6) The test can be misleading in cases where fine sand or inorganic silts are encountered using rotary wash drilling methods above the water table. The use of water in the drilling operation may soften or loosen the formation so that the blows on the spoon are deceptively low when compared to the actual density of the formation.

(7) In view of the present state of knowledge, SPT results in weathered rock should be used only to give a crude indication of relative strength. The empirical relationships developed for transported soils between blow count values and foundation design indices, relative density, and shear strength are not valid for weathered rocks. Corestones, for example, can give misleadingly high values that are unrepresentative of the soil mass.

The results of the SPT are generally correlated empirically with soil properties measured either by laboratory tests or by field tests of the same material. Table 4.6 presents a correlation between relative density for sand and a correlation of penetration resistance with consistency for clay. Curves for determining friction angle, ϕ, versus effective overburden pressures are given in Figure 4.37.

Cone Penetration Test (CPT) The cone penetration test has been used in the United States since the mid-1970s. Although many cone penetration tests have been developed, the Dutch cone test has become the most popular. The cone penetrometer consists of a 60° cone with a projected area of 10 square centimeters (Figure 4.38). The cone is pushed into the soil at a rate of 10 to 20 millimeters per second (2 to

TABLE 4.6 SPT Blow Count versus Relative Density of Sand and Consistency of Clay

Relative Density of Sand		Strength of Clay	
Penetration Resistance N (blows/ft)	Density	Penetration Resistance N (blows/ft)	Consistency
0–4	Very loose	<2	Very soft
5–10	Loose	2–4	Soft
11–24	Medium	4–8	Medium
25–50	Dense	8–15	Stiff
>50	Very Dense	15–30	Very stiff
		>30	Hard

Source: Terzaghi and Peck (1967), modified by AASHTO.

4 feet per minute) by hydraulic pressure applied to a drill rod extending from the cone to the ground surface. The penetration resistance, q_c, is found by dividing the measured force by the 1,000 square millimeter cone area. Readings are usually taken at 20-centimeter intervals, although a continuous reading of both cone resistance and friction can be given with a continuously advanced electrical cone penetrometer. The procedure is described in ASTM D3441-86.

The cone resistance can be used to estimate bearing capacity and density, but the results are badly affected if the penetrometer impinges on particles larger than the cone. Therefore, the equipment may not be suitable for a weathered rock or formation strewn with cobbles or gravel, but it is highly suitable for soft clays and marine sediments.

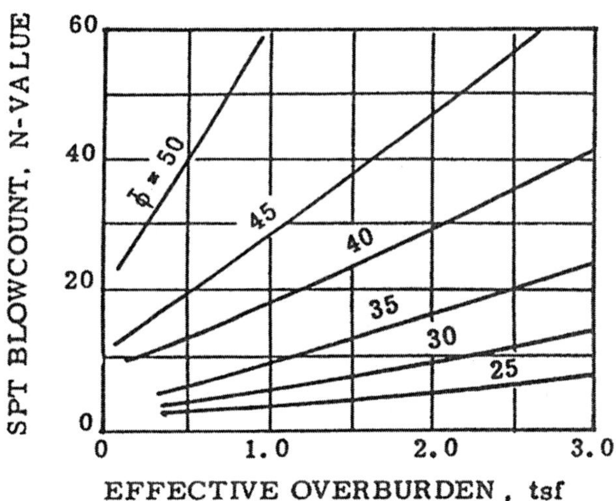

Figure 4.37 SPT blow counts versus effective friction angle for sand. (From Schmertmann, 1975, reproduced by permission of ASCE.)

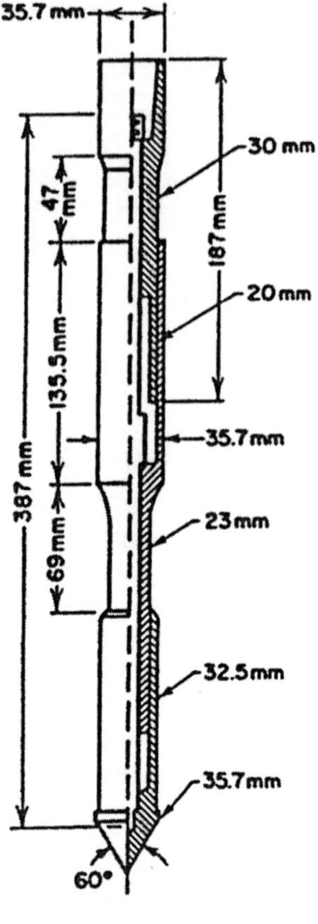

Figure 4.38 Mechanical friction cone penetrometer.

Curves for determining ϕ are given in Figure 4.39. These curves are designed for cohesionless sands but are also applicable to fine-grained soils with low cohesion values.

The general application of cone penetrometer tests (CPTs) in slope stability problems is to determine undrained shear strengths of cohesive soils. Undrained shear strength, s_u, is related to cone resistance, q_c, by the following relationship:

$$s_u = \frac{q_c - \sum \gamma_n z_n}{N_c} \qquad \text{(Eq. 4-1)}$$

where s_u = undrained shear strength
q_c = cone resistance = R_p/A
R_p = point resistance of cone
A = projected area of cone

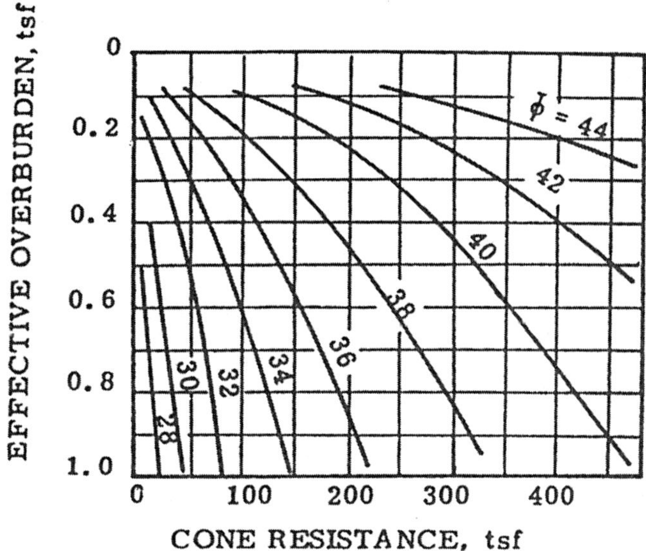

Figure 4.39 Cone resistance versus effective friction angle for sand. (Trofimendkov, 1974)

γ_n = total unit weight of layer n
N_c = bearing capacity factor or cone factor
z_n = depth of layer n

The bearing capacity factor, N_c, is not a constant, but varies with the rate of cone penetration, types of the cone tips, the stress–strain properties, and sensitivity of the clays. Table 4.7 illustrates this point. Schmertmann (1978) suggested the use of $N_c = 10$ for cylindrical tips and $N_c = 16$ for Delft mechanical tips. To ensure the reliability of s_u, it is necessary to correlate it against actual s_u back-calculated from failures. However, given the scarcity of back-calculated case histories in which s_u data are available, correlations will likely have to be made from common laboratory testing methods, such as those from triaxial tests, direct shear tests, unconfined compression tests, or vane tests.

Discussion of interpretation of CPT results is given by FHWA Reports numbers TS-77-207 and TS-78-209 (Federal Highway Administration, 1977; Schmertmann, 1978).

Pocket Penetrometer Tests Unconfined compressive strength of a clay can be determined by means of a pocket penetrometer in the field. The pocket penetrometer is manually pushed down into the clayey soils to a predetermined distance (usually about 1/2 inch) and the pressure in tons per square foot required to reach the preset distance is measured. This test gives a very crude value of the compressive strength. Hence, its value should be compared with others for a reliable estimate of the unconfined compressive strength.

TABLE 4.7 Summary of N_c

Test Site	Depth (m)	Range, τ_f(t/m^2)	Plasticity, I_p(%)	Sensitivity	Cone Factor, N_k
Sundland	4–9	2–2.5	22–28	10–15	17–18
Drammen	9–14	2–5	~10	~2	20
	14–22	2.5–4	10	3–4	15.5
Dansvigs gate	5–10	2–3	20–25	6–9	14–15
Drammen	11–30	2–4	10–11	2–4	14–16
Børresens gate	5.5–12	3–2	~15	15–25	16–20
Drammen	12–30	1.3–2.5	~5	50–160	20–24
Onsøy	1–9	1.2–1.4	20–30	5–10	16–18
	10–20	1.8–4.8	35–40	4–7	13–18
Skå-Edeby	1–4	0.6–1.2	45–80	6–10	8–9
	4–12	0.8–2.0	30–50	10–15	10–12
Gøteborg	3–10	1.5–2.5	50–60	15–24	13.5–14.5
	10–21	2.5–4.2	50–55	13–19	13–14
	21–30	4.5–5.5	~40	13–17	13–14

Source: Lunne and Eide (1976) and Schmertmann (1978).

Vane Shear Test A vane shear test is performed by pushing a vane that is fastened to an extension rod into the undisturbed bottom of a borehole to a depth of about two feet (Figure 4.40). Vertical blades at the end of the rod produce a vertical cylindrical surface of shear when rotated. The torque required to initiate continual rotation is a measure of the peak undrained strength of the soil, and the torque required to maintain rotation after several revolutions measures the residual or disturbed strength. To minimize end effects, the length of the vane should be at least twice its width. The blades should be sufficiently thin such that there is a minimum soil disturbance due to displacement and at the same time thick enough to withstand bending under load. A reference test using only the torque rod without the vane is required so that the torque necessary to overcome rod friction can be subtracted from the total torque measured when the soil is tested. Standard test procedures are described in ASTM D2573-78.

It must be borne in mind that the test cannot be expected to give "exact" values for the undrained shear strength. It is necessary to perform the test in the same way every time. This means that careful installation procedures, the same rate of strain of 0.1 degree per second, and the same delay before the test is started, preferably not more than 5 minutes, should be used (Flaate, 1966). Erratic results are obtained if the soil contains gravel or any large particles. Hence, the test is limited to measuring the undrained strength of clays and marine sediments.

Caution should be exercised when one interprets the peak strength; in some cases, the strength measured in a vane test has been found to be 30 percent greater than that measured by other methods (Sowers and Royster, 1978).

The undrained shear strength, or the cohesion for total stress analysis with $\phi = 0$, is equal to one-half of the unconfined compressive strength. The relationship between the undrained shear strength and the SPT blow count has been studied by a number

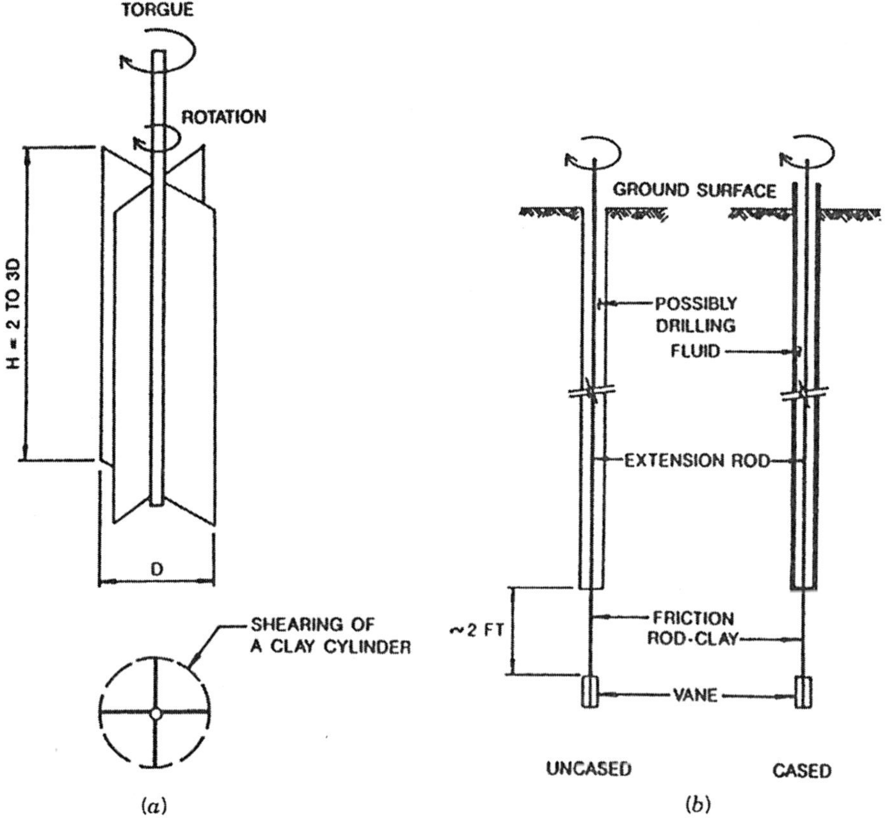

Figure 4.40 Field vane test. (*a*) Principle of the vane test. (*b*) Installation procedures. (Flaate, 1966.)

of researchers and practicing engineers in the past few decades. For example, if the clay is normally consolidated with $N <$ depth (in feet)/5, Schmertmann (1975) suggests:

$$s_u \geq N/15 \qquad \text{(Eq. 4-2)}$$

where s_u is the undrained shear strength in tons per square foot and N is the SPT blow count per foot of penetration.

A crude estimate for the undrained shear strength of cohesive soils versus SPT blow count is also shown in Figure 4.41. These correlations are, however, not meaningful for medium to soft clays where effects of disturbance are excessive (Naval Facility Engineering Command, 1982).

Strength values measured using field vane shear tests should be corrected for the effects of anisotropy and strain rate using Bjerrum's correction factor, 1, which is shown in Figure 4.42, that is $(S_u)_{\text{field}} = (S_u)_{\text{vane}} \times \mu$.

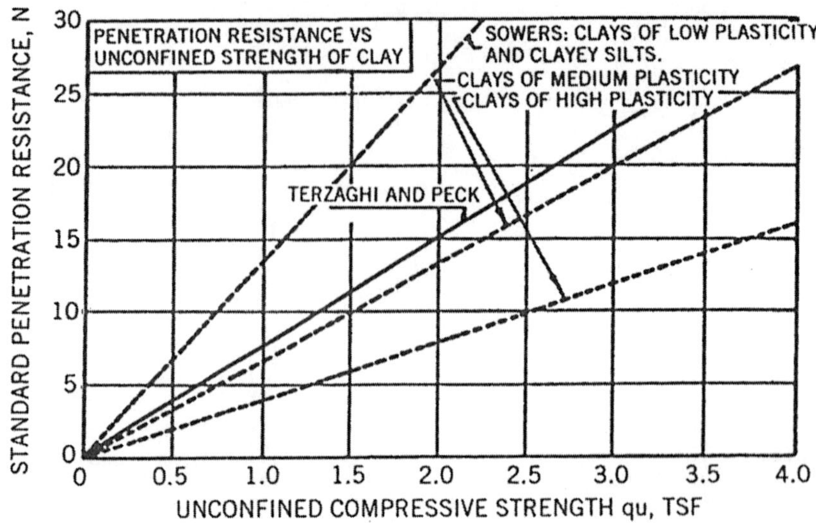

Figure 4.41 Relationship between SPT *N*-values and unconfined compressive strength. (Naval Facility Engineering Command, 1982.)

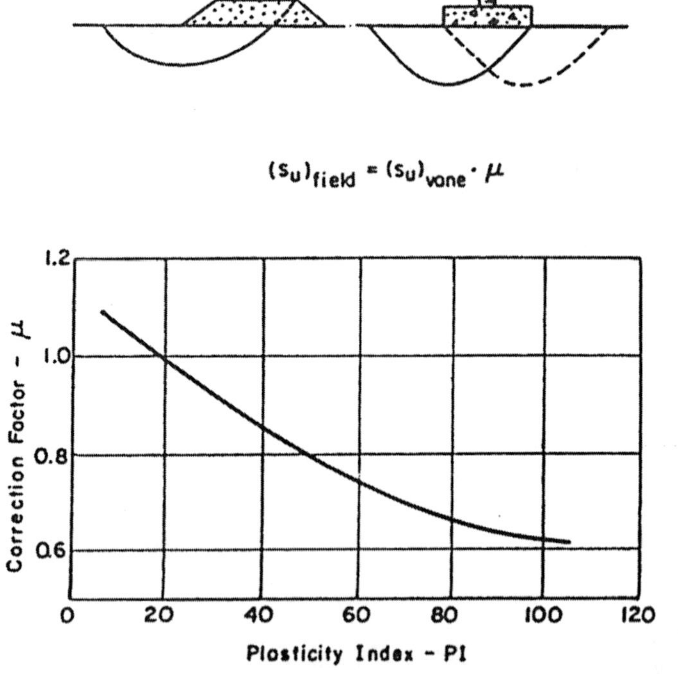

Figure 4.42 Correction factor for vane strength. (From Bjerrum, 1972, reproduced by permission of ASCE.)

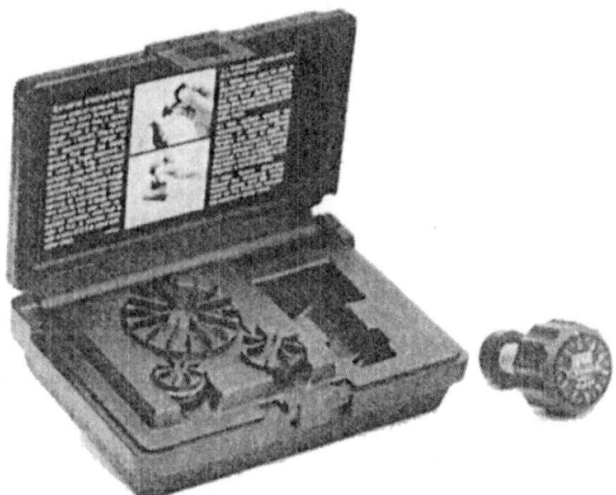

Figure 4.43 The torvane. (Slope Indicator Co.)

Torvane Tests The torvane (Figure 4.43), which was designed by the Slope Indicator Company, is a soil testing instrument for the rapid determination of shear strength of cohesive soils either in the field or laboratory. Applications for evaluation of shear strength include sides and bottom of test pits, end of Shelby tube samples, or chunk samples from test pits. Hence, it is meant to be used for shallow inspection purposes.

The torvane consists of a disc (vane) with blades on the lower surface that is pressed into the soil to be tested. A torque is applied to the disc when the upper knob is rotated with finger pressure. The torque is resisted by shear stresses in the clay or soil across the lower face and around the circumferential area of the blades.

Just like the pocket penetrometer, the torvane permits a rapid determination of a large number of strength values with different orientation of failure planes. It does not specifically indicate exact shear strength characteristics, but rather identifies strength variations with depth and zones of weakness in the subsoil. The strengths obtained from the torvane tests must be calibrated against a torvane shear strength correlation before they can be compared with other field and laboratory tests, because of the depth effect. The tests give a crude approximation of the undrained shear strength of the soil mass and should be compared with others such as the vane shear test and the borehole shear test for reasonable estimate of the undrained shear strength.

Pressuremeter The pressuremeter (Figure 4.44) consists of a probe that, when placed in a borehole, can be inflated. The volume changes of the probe (the expansion of which is limited to that in the radial plane) can be measured by means of a surface volume meter to which the probe is connected. A pressure versus volume change graph can be plotted and converted into a stress–strain curve. From the test results, a limit pressure, which reflects the ultimate bearing capacity, is determined. A deformation modulus may also be determined, from which an estimation of settlement can be made.

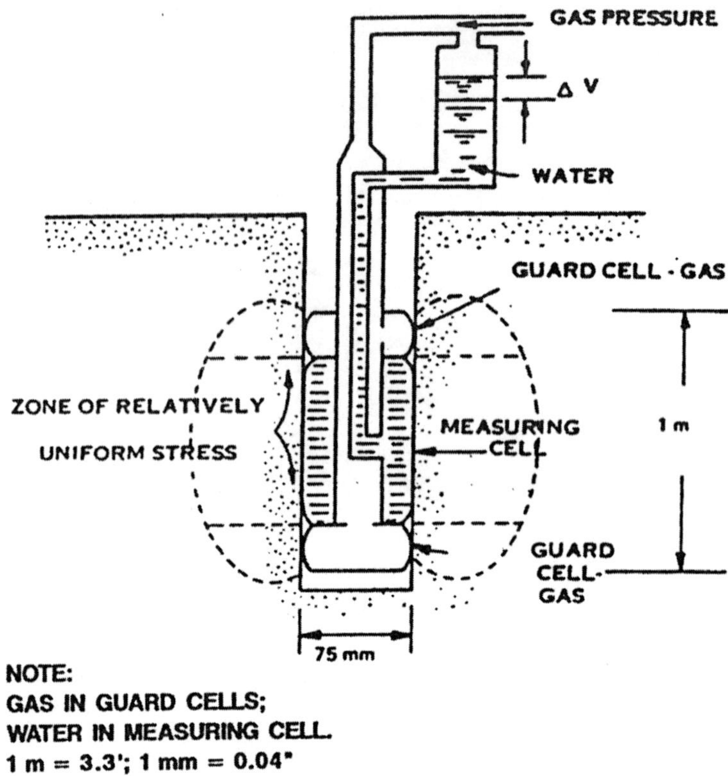

Figure 4.44 Pressuremeter for borehole dilation test (not to scale). (Schuster and Krizek, 1978.)

Tests are normally carried out at 5-foot intervals in the boreholes. If the seating area for the pressuremeter is oversized, or if the walls of the hole are not smooth, interpretation of results becomes difficult. Such conditions can be sensed by the volume needed to inflate the membrane.

The self-boring pressuremeter is an improvement over Menard's pressuremeter in that the disturbance to the in situ soil caused by predrilling of the borehole is greatly reduced. The undrained shear strength of a saturated clay is evaluated from pressure-radial expansion data for the expandable test cell. Details of the theory and test procedures for pressuremeters are discussed by Ladd et al. (1977) and Schmertmann (1975).

The pressuremeter test can determine the lateral stress, the undrained stress–strain properties of cohesive soils, and the peak angle of internal friction of cohesionless soils under drained conditions. Although the test offers the best opportunity for measuring in situ horizontal stress and stress history, it is an elaborate device. Discussion of interpretation of results is given by the Federal Highway Administration (1980), Jamiolkowski et al. (1985), and Wroth (1984).

It must be borne in mind that the interpretation of this test is largely empirical and certainly open to question in variable materials. Typically, the length of the device limits the stressed zone to a length of about 2 feet and diameter of about 1 foot. The use of the test results without confirmation by other means would be unwise.

Borehole Shear Test The borehole shear test was first developed by Kansas Highway Commission, as described by Handy and Fox (1967), for the purpose of in situ strength measurements at any depth of a borehole. It was subsequently modified by Wineland (1975). The test equipment consists of a serrated cylinder that has been split lengthwise. It is lowered into a 76-millimeter diameter borehole and the cylinder is expanded into the opposite sides of the borehole by applying pressure from the surface through a pipe system (Figure 4.45). The test proceeds by applying a known pressure and then pulling the cylinder vertically until slip occurs. The cylinder pressure is incremented and the test repeated several times until sufficient data are obtained to plot a Mohr–Coulomb's rupture envelope. The worm-gear pulling force is read on the hydraulic gages to obtain the shear stress.

The borehole shear test is unique among in situ test methods in giving direct evaluation of soil cohesion, c, and friction angle, ϕ, at a particular depth. This test is particularly useful in active slide areas, where samples for laboratory testing are difficult to obtain. It can yield large amounts of c and ϕ data in a short time, and thus is adaptable for probability risk analysis based on strength variability.

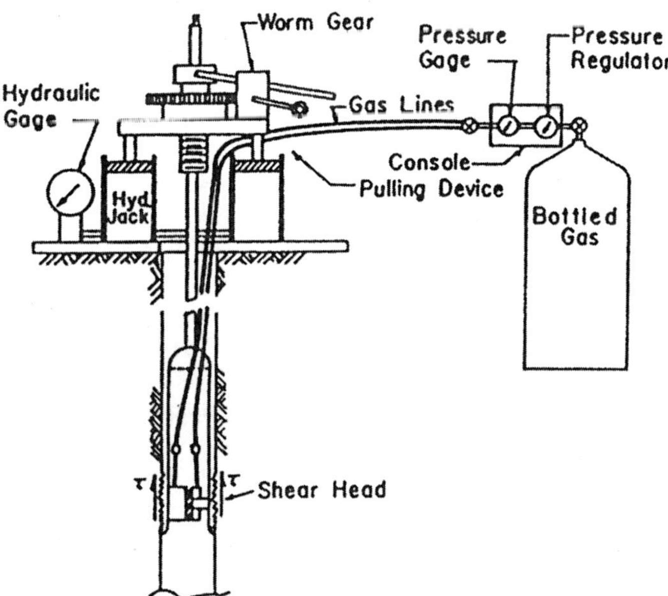

Figure 4.45 Borehole shear test. (From Wineland, 1975, reproduced by permission of ASCE.)

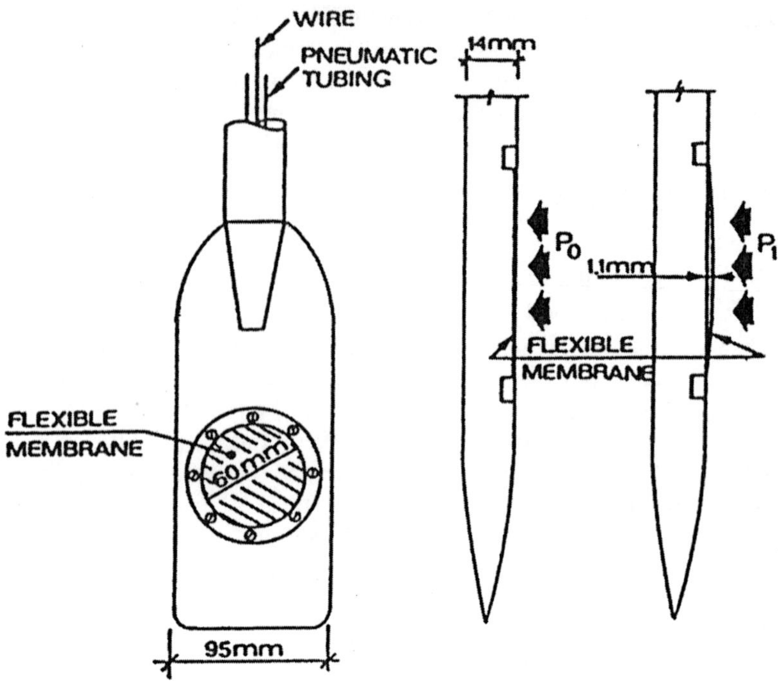

Figure 4.46 Dilatometer device setup.

Dilatometer The dilatometer consists of a stainless steel blade 14 millimeters thick, 95 millimeters wide, and 150 millimeters long, a sharp cutting edge, and a 60-millimeter diameter stainless steel membrane centered on and flushed with one side of the blade (Figure 4.46). It is attached to a string of drill rods, and pressed or driven into the ground in 15- to 30-centimeter increments. The force or blows required to cause penetration provide information similar to CPTs and SPTs. From empirical correlations, quantitative estimates of the horizontal earth pressure coefficient, K_0, the overconsolidation ratio, OCR, and the tangent one-dimensional compressibility modulus can be obtained (Schmertmann, 1983).

Field Permeability Tests Field permeability tests are particularly valuable in the case of materials such as sands or gravels, undisturbed samples of which are difficult to obtain for testing in the laboratory. The coefficient of permeability, k, can be calculated from the results of rising-, falling-, or constant-head tests carried out in boreholes or standpipe piezometers. The test procedures can affect the natural permeability of the material being tested. In tests involving a flow of water into the soil, fines from the water in the holes can be washed into the soil, reducing the permeability. To minimize this effect, only clean water should be used. Less commonly in tests where water flows into the borehole, fines can be removed from the soil, and if the water level is reduced too far below the water table, piping can occur and the derived values of k will be too high.

There are no rigid requirements as to the number and spacing of permeability tests required. This depends on the complexity of the subsoils and the nature of the project.

Large-Scale Pit Tests One of the major limitations of laboratory tests (Chapter 5) is their inability to integrate variations in the soil, particularly in zones with weak and hard spots. This can be overcome by large-scale in-place tests performed in test pits or trenches excavated to the strata of interest or zones of slickendsiding. Such tests include (1) plate load test and (2) large-scale direct shear test. These tests permit a large volume of soil to be evaluated under the conditions present within the total mass without the problems of sample disturbance and exposure inherent in small-scale sampling and laboratory testing.

Plate Load Test For a plate load test, a pit is excavated to the surface of the stratum in question, and a rigid square or circular plate is placed on the ground. The plate is loaded incrementally so that at least 10 successively greater loads are applied before the plate shears the soil beneath it. The detailed procedure is described in many standard soil mechanics textbooks (e.g., Sowers and Sowers, 1970).

The results of such a test can be interpreted in terms of soil bearing capacity to give shear strength of the soil along a curvilinear surface. However, there are many different interpretations of such tests, all yielding different values for the shear strength parameters, Therefore, the test has limited value in determining the strength of the soil involved in the stability of a slope. It is more useful for estimating the bearing capacity than the shear strength of soil.

The results of a plate load test can be grossly affected by the presence of boulders immediately below the test area. The accuracy of the application of the results to the prediction of the behavior of full-size structures is dependent upon the size of the plate used for the test. Usually a 1-foot diameter plate is adequate, applied to the plate by jacking against a reaction beam or heavy piece of equipment. The test cannot be made on steep slopes because accommodation of the equipment is difficult if not impossible.

Large-Scale Direct Shear Test A large-scale direct shear test can be performed in a pit or a trench at the level of the questionable weak stratum. The pit or trench should be large enough to allow engineers and technicians to work around the sides without disturbing the soil to be tested in the center. All the soil within the pit or the trench is excavated, leaving a block of sample above the bottom of the pit or the trench. The size of the block is determined by the geologist's or the geotechnical engineer's evaluation of the variation of the soil strength. It should be large enough to be as representative of the stratum as a whole. A double box is placed around the block to be tested. If there is a definite plane of weakness, the sides of the box should be perpendicular to that plane, and the plane should be between the top and bottom halves of the box. Such a setup is shown in Figure 4.47. Good contact between the box and the soil must be secured. This can be accomplished either by trimming the soil carefully or by pouring plaster to fill the space between the box and the soil.

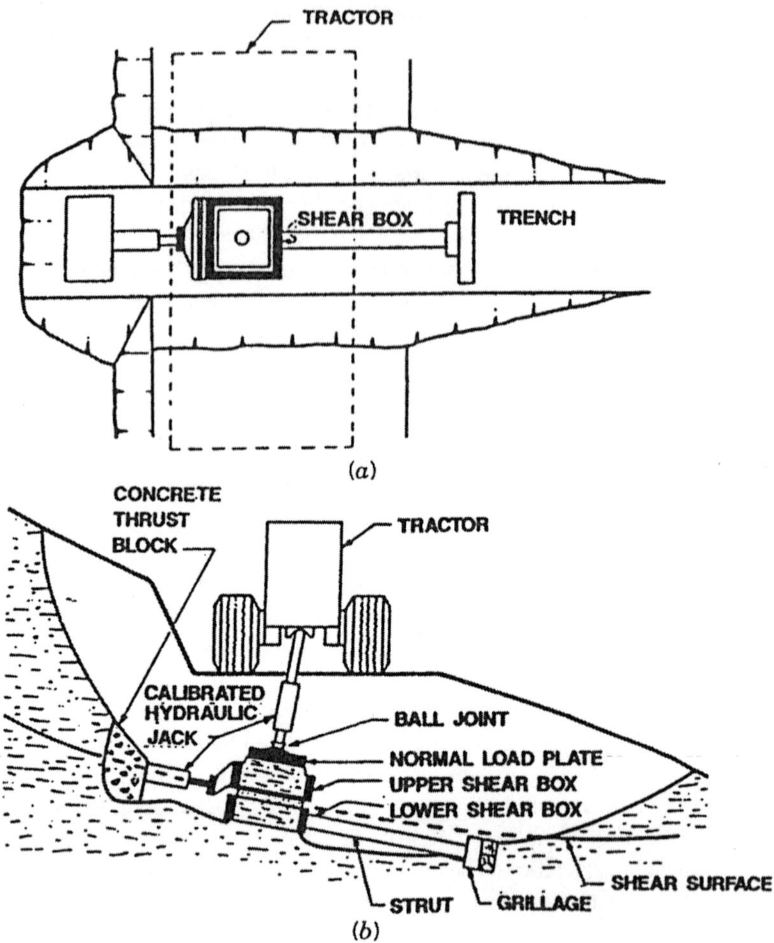

Figure 4.47 Direct shear test of strength along failure surface. (*a*) Plan view. (*b*) Cross section. Note: trench shore for safety and support of tractor. (Schuster and Krizek, 1978.)

A normal load is placed on the block by means of a plate. The load is applied by jacking against heavy machinery above the pit or trench. The bottom half of the box is anchored securely in place by packed soil, struts, and so on, in the bottom of the pit or trench. The upper half is then jacked sideways by a calibrated system so that the amount of lateral movement and the load causing the movement can be measured. The same surface can be tested at several different normal loads if the test for each vertical load increment is stopped soon after peak strength is reached or significant movement develops.

One of the drawbacks of this test is the difficulty in including the effects of changing water pressure. Experience by Sowers and Royster (1978) showed that meaningful shear test data can be obtained from the large-scale shear test if done during the wet season. Alternatively, correlating the large-scale shear test results with those of smaller laboratory tests on similar soils may make it possible to extend the large-scale shear test data to include the effects of changing water pressure.

4.5.2 Geophysical Testing

The primary use of geophysics is to extend borehole data and to eliminate unnecessary drilling. Geophysical exploration is of greater value when started from known but limited borehole information or soil profiles. Geophysical tests in soil/rock exploration are usually low in cost. Specialist competency, knowledge of the geology, and experience, along with calibration-type field data concerning the subsurface strata, are necessary for obtaining reliable results from geophysical survey methods. For typical highway projects where slopes under construction are usually very long in alignment, it is cheaper to use geophysical testing in conjunction with boreholes than closely spaced borings.

Geophysical testing consists of (1) the seismic reflection and refraction tests, (2) the earth-resistivity test and electromagnetic test, (3) radar survey, (4) gravity survey, (5) magnetic survey, (6) radiometric survey, and (7) borehole logging. Table 4.8 lists their brief descriptions and engineering applications. Of these tests, the earth-resistivity test and the seismic refraction test are the most commonly used, and are discussed in this section. Other geophysical techniques are mainly of value in mineral exploration and reconnaissance for oil and gas, and have seldom found engineering application in highway projects. Hence they are not discussed in further detail.

The amount of geophysical production per day is controlled by accessibility, landforms, and the nature of the engineering data desired. Accessibility is controlled primarily by weather, relief, and vegetation, and can present problems in all terrain types. Production rates in resistivity and seismic surveys are reduced in areas with densely vegetated slopes and few roads because it is difficult to move from station to station.

Earth-Resistivity Test The earth-resistivity test is based on the fact that different materials offer different resistance to the passage of an electric current. Thus it is possible to infer the stratification and lateral extent of subsurface deposits through the determination of different resistances to electric current. Several methods involving different electrode arrangements have been developed for making field resistivity measurements. The most common one is the Wenner arrangement (Figure 4.48). The procedure of the field resistivity measurements using the Wenner arrangement is described in ASTM G57-84.

In resistivity work, the factors controlling production are terrain and type of profile desired. Two and one-half miles of profile can be obtained per day by continuous profiling at constant spacing over flat terrain. This would be reduced to $1\frac{1}{2}$ miles in upland terrain. Depth profiling at continuous spacing of 400 feet would yield $\frac{3}{4}$ of a mile of profile per day on fairly flat terrain. In very difficult terrain the average continuous section at 400-foot spacing would be approximately $\frac{1}{2}$ mile.

TABLE 4.8 Geophysical Methods

Method	Field Operations	Quantities Measured	Computed Results	Applications
Seismic tests	Reflection and refraction surveys using: *on land*—several trucks with seismic energy sources, detectors, and recording equipment; *at sea*—one or two ships. Data-processing equipment in central office. Two-man refraction team using a sledgehammer energy source	Time for seismic waves to return to surface after reflection or refraction by subsurface formations	Depths of reflecting or refracting formations, speed of seismic waves, seismic contour maps	Exploration for oil and gas, regional geological studies Superficial deposit surveys, site investigation for engineering projects, boundaries, material types and elastic moduli
Electrical and electromagnetic tests	Ground self-potential and resistivity surveys, ground and airborne electromagnetic surveys, induced polarization surveys	Natural potentials, potential drop between electrodes, induced electromagnetic fields	Anomaly maps and profiles, position of ore-bodies, depths to rock layers	Exploration for minerals; site investigations
Radar survey	Ground survey 2-man. Portable microwave source	Induced reflections from surfaces in ground	Depths to reflecting surfaces	Shallow engineering projects: frozen ground at depth

Gravity survey	*Land surveys* using gravity meters; *marine surveys* using gravity or submersible meters	Variations in strength of Earth's gravity field	Bougner anomaly and residual gravity maps; depths to rocks of contrasting density	Reconnaissance for oil and gas; detailed geological studies
Magnetic survey	Airborne and marine magnetic surveys, using magnetometers; Ground magnetic surveys	Variations in strength of Earth's magnetic field	Aero- or marine magnetic maps or profiles; depth to magnetic minerals	Reconnaissance for oil and gas, search for mineral deposits; geological studies at sites
Radiometric survey	Ground and air surveys using scintillation counters and gamma-ray spectrometers; geiger counter ground surveys	Natural radioactivity levels in rocks and minerals; induced radioactivity	Iso-rad maps, radiometric anomalies, location of mineral deposits	Exploration for metals used in atomic energy plant
Borehole logging	Seismic, gravity, magnetic, electrical, and radiometric measurements using special equipment lowered into borehole	Speed of seismic waves, vertical variations in gravity and magnetic fields; apparent resistivities, self-potentials	Continuous velocity logs; resistivity and thickness of beds; density; gas and oil, and K, Th, U content. Salinity of water	Discovery of oil, gas, and water supplies; regional geological studies by borehole correlation. Applicable to site investigations

Source: Blyth and de Freitas (1984).

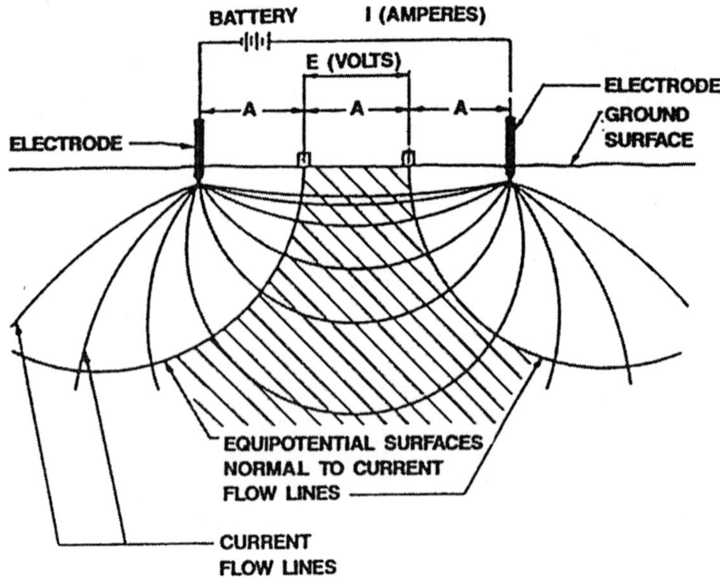

Figure 4.48 Diagram showing theory of earth-resistivity methods. (Winterkorn and Fang, 1975.)

Seismic Refraction Test The seismic refraction technique yields the compression wave velocity of the soil and rock. If the density is known, the dynamic modulus of elasticity in compression can be calculated. The seismic refraction technique can be used to investigate subsurface conditions from ground surface to depths of approximately several hundred feet. Figure 4.49 shows a schematic diagram of the test.

Data from the film records are often plotted in the form of time–distance graphs, as shown in the upper part of Figure 4.50. Point A on this figure is the source of the seismic impulse. Points D_1 through D_{12} represent the locations of the detectors,

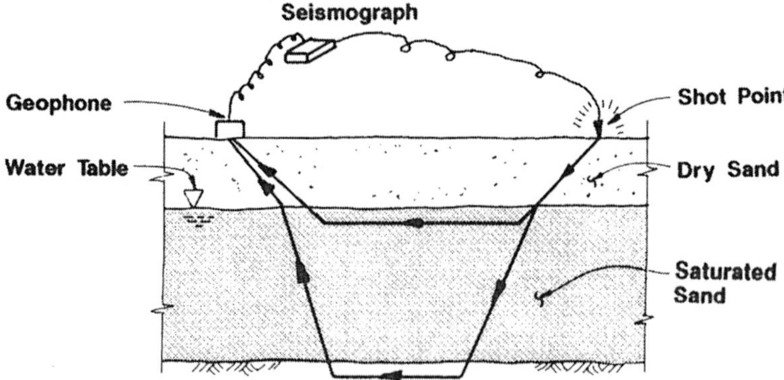

Figure 4.49 Schematic diagram showing theory of seismic refraction method.

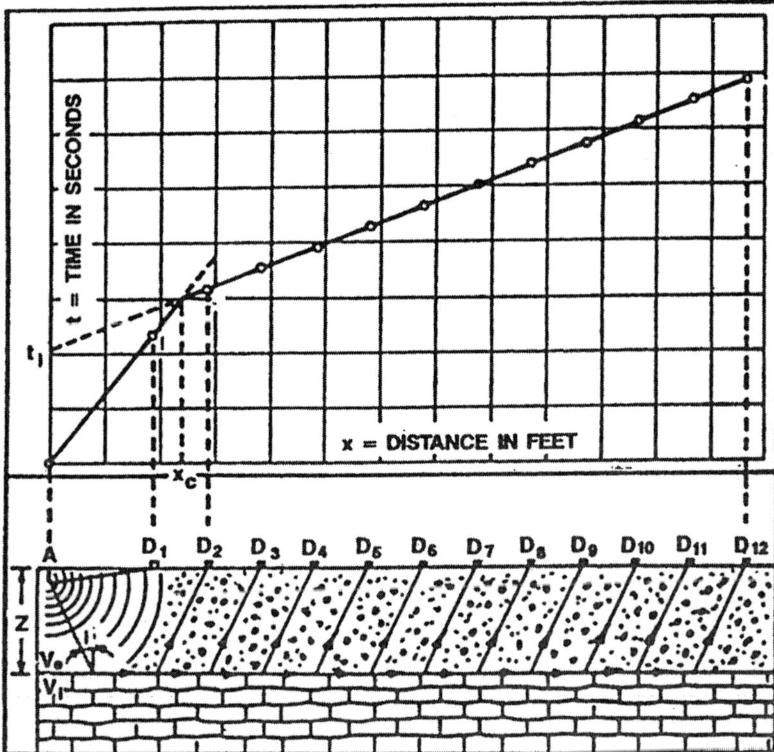

Figure 4.50 Schematic representation of refraction of seismic energy at a horizontal interface and the resultant time–distance graph. (Winterkorn and Fang, 1975.)

or geophones, whose spacing is dependent on the amount of detail required and the depth to the strata being investigated. In general, the distance between D_1 and D_{12} is three to four times the depth to be investigated. The geophones are connected by cable to recording devices, which may be truck-mounted or may be portable units placed at the ground surface. A high-speed camera is used to record the time at which the seismic impulse is generated and the time of arrival of the wave front at each geophone. A continuous profile may be obtained by moving the geophones along a line while generating a new impulse from the same source point each time the geophones are moved. For each shot, the time of initiation of the wave and the first arrival times are recorded.

The slope of the plot in Figure 4.50 represents the velocity of the seismic wave as it passes through the various subsurface materials present. Detailed investigation procedures for seismic refraction studies, formulas used for depth determinations, and a table of the range of seismic wave velocities for various materials are presented by Jakosky (1950).

In seismic refraction work, with highly portable seismic equipment, continuous profiles covering $\frac{1}{2}$ mile per day could be obtained in upland terrain. This might be

TABLE 4.9 Typical Seismic Velocities of Earth Materials

Material	Velocity (ft/sec)
Dry silt, sand, loose gravel loam, loose rock, talus, and moist fine-grained topsoil	600–2,500
Compact till, indurated clays, gravel below water table, compact clayey gravel, cemented sand, and sand-clay	2,500–7,500
Rock, weathered, fractured, or partly decomposed	2,000–10,000
Shale, sound	2,500–11,000
Sandstone, sound	5,000–14,000
Limetone, chalk, sound	6,000–20,000
Igneous rock, sound	12,000–20,000
Metamorphic rock, sound	10,000–16,000

Source: Peck et al. (1974).

extended to $\frac{3}{4}$ mile per day in flat terrain. A production of 8 to 12 stations is estimated if a discontinuous seismic survey is desired.

Application of Earth-Resistivity Tests and Seismic Refraction Tests in Landslide Investigation Combined earth-resistivity and seismic refraction surveys have been successfully used to locate the depth of the rupture surfaces in landslides (Cummings and Clark, 1988; Palmer and Weisgarber, 1988). The basis of the interpretations is the very different physical properties of the sliding materials compared to the underlying undisturbed sediments or bedrock. Distinguishing these materials visually is difficult because they usually look similar. The electrical resistivity surveys detect the increased water content in the slide debris, whereas the seismic refraction surveys detect the decreased stiffness or rigidity of the sliding mass relative to the underlying undisturbed sediments or bedrock (Cummings and Clark, 1988). Tables 4.9 and 4.10 give the typical seismic velocities and resistivity values for earth materials.

In the Stumpy Basin Landslide (Palmer and Weisgarber, 1988), the use of a seismic refraction and resistivity survey was successful in defining the shear surface. In

TABLE 4.10 Typical Resistivity Values of Earth Materials

Material	Resistivity (ohm-cm)
Clay and saturated silt	0–10,000
Sandy clay and wet silty sand	10,000–25,000
Clayey sand and wet silty sand	25,000–50,000
Sand	50,000–150,000
Gravel	150,000–500,000
Weathered rock	100,000–200,000
Sound rock	150,000–4,000,000

Source: Peck et al. (1974).

general, the undisturbed sediments have higher velocities and resistivity values than those of underlying sliding materials. The drops in velocity and resistivity may be a function of one or more of the following three factors (Palmer and Weisgarber, 1988):

(1) The sediments that undergo expansion upon shearing would have a significant decrease in velocity because of the expansion, which in turn affects elastic properties and density. Increased water content due to expansion and increased porosity upon shearing would lower resistivity value of the sheared materials.

(2) The presence of shear planes in the upper mobile zone may act to decrease the average velocity of this layer. Increased water content caused by groundwater barriers as a result of shear zones would have lower resistivity value for the sliding soil mass.

(3) Alteration by leaching and groundwater through weathering may decrease the velocity and resistivity in the sliding zone.

To verify the validity of the interpretations of the electric resistivity and seismic refraction data, a drill hole, for example, a bucket auger hole, is normally drilled to permit a geologist or geotechnical engineer to go down the borehole and to identify the basal rupture surface.

4.5.3 Downhole Geophysics Logging

Downhole geophysics logging has been used for many years in oil field investigations. However, it has become necessary to utilize this technology to improve the three-dimensional perspective in the boring locations, especially in the investigation of landslides to supplement continuous core information. Table 4.11 presents various downhole geophysics log types and their usages. Some of these can be used only in cased or uncased holes, while others can be used in both cased and uncased holes.

Of these downhole geophysics logging, gamma ray, dipmeter, and neutron loggings have become extremely helpful in evaluating landslide subsurface conditions at depth. The conventional small-diameter core holes are prepared for geophysical wire-line logging by reaming each boring with a $7\frac{1}{4}$-inch tri-cone rock bit, since dipmeter tools require a minimum 6-inch diameter hole to operate. The dipmeter requires an open hole, whereas the gamma ray and neutron logs can be recorded in either a cased and an uncased boring.

Gamma Ray Logging The gamma ray tool provides a measurement of the natural radiation within formation materials and is especially useful in lithology identification and correlation. Radioactive elements tend to concentrate in marine shales and clays, whereas sands and basalts generally exhibit low radiation levels. On this basis, lithology identification is enhanced, especially in sections of the boring where little or no recovery of cores are made.

TABLE 4.11 Downhole Geophysics Log Types versus Usages

Log Type or Device	Uses: Correlation	Lithology	Bed Thickness	Clay and Shale Content	Total Porosity	Effective Porosity	Bulk Density	Water Table	Chemical and Physical Properties	Moisture	Infiltration	Direction, Velocity Groundwater	Dispersion, Dilution	Source and Movement of Water in Well	Cementing	Well Construction	Casing Corrosion	Casing Leaks or Plugging	Hole Diameter	Depth Control	Dip and Strike of Formations	Fish Location	Hole Inclination
Spontaneous potential	○	○	○	○					○														
Short normal	○	○	○	○																			
Long normal	○	○		○		○		○															
Long lateral		○		○		○																	
Induction log	○	○	○	○																			
Microlog		○	○	○																			
Caliper log																•	•		○ •				
Gamma ray log	○ •	○ •	○ •	○ •																			
Neutron log	○ •	○ •	○ •	○ •	○ •			○ •		○ •	○ •												

Spaced neutron log	○ ●	○ ●	○	○ ●	○ ●	○				○ ●	○ ●												
Density log		○ ●	●	○ ●			○ ●								●	●							
Sonic log	○ ●	○ ●			○										●								
Temperature log (differential)											●	○ ●	●	○ ●	●			●					
Dipmeter																					○		○
Sidewall cores	○	○		○	○	○	○		○														
Flowmeter log												○ ●		●									
Fluid conductivity									●				●										
Downhole camera	○	○	○					○ ●						●		●	●	●		●		○ ●	
Collar locator																				●		○	
Directional survey																							○

Dipmeter Logging The Dips of planar features could be easily measured in core samples; the strike or dip direction of these features, however, could not be measured without orientation of the core samples. The dipmeter log can provide this missing component as well as attitudes in zones of unrecovered core. Since dipmeter tools require a minimum 6-inch diameter hole to operate, the conventional small diameter core holes are prepared for geophysical wire-line logging by reaming each boring with a $7\frac{1}{4}$-inch tri-cone rock bit.

Using six equally spaced micro-resistivity electrodes mounted on a six-arm hydraulically actuated tool, the wireline dipmeter tool records resistivity data from the borehole materials. These data are then downloaded to a computer to compute three-point problems from correlations in the resistivity data at selected depth intervals. Although only three points are needed, the additional three points are to provide a check of data quality for each interval in which an altitude is computed. The resultant plot (Figure 4.51) gives accurate attitudes from which inferences can be made with regard to bedding structures, landslide surfaces, discontinuities, and faults.

Neutron Logs Neutron logs make use of an americium neutron source to bombard the formation and measure neutrons that have been thermalized by collisions with the formation. The logs take into account the neutron count rate, which is related to total water content surrounding the tool for a constant borehole size. The logs register higher counts for lower water content because lower water content capture fewer log energy neutrons and results in a higher neutron count rate at the detector.

Total water content in a saturated formation is controlled by the clay content because clay minerals contain a significant volume of bound water. In view of the inverse count versus water content relationship mentioned above, this means that lower neutron count rates are associated with higher clay content. Sheared zones or zones subject to chemical alteration and weathering often carry a lot of water and/or conduits of water passage. Hence, they can be easily detected by means of neutron logs.

4.5.4 Mineralogy Tests

All soils are made up of minerals. Because each mineral has its own physical and chemical properties, it is often important to know the mineralogy and petrology, as they may have a bearing on the treatment method to be chosen, particularly for landslide cases. For example, fine-grained materials that contain a large proportion of sodium- or potassium-bearing clay minerals may be stable, whereas if the sodium or potassium is replaced by calcium ions, as by the percolation of groundwater through a marine clay, the same material may become extremely sensitive.

The most useful methods used to analyze soil mineralogy are (1) X-ray diffraction, (2) differential thermal analysis, and (3) petrographic analysis, which are discussed as follows.

(1) *X-Ray Diffraction* This test is made on clay mineral soils when the fine grain size or lack of identification features makes usual methods of identification impossible. The diffraction pattern of the unknown mineral is compared with the patterns of known

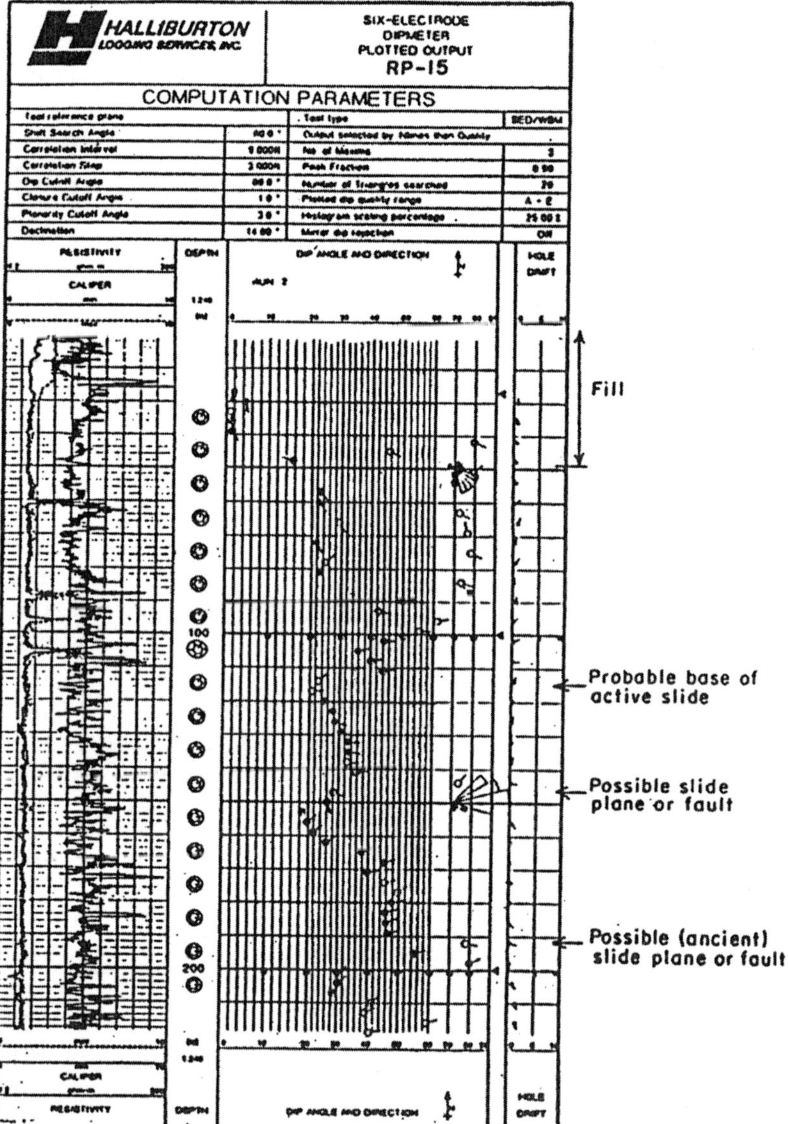

Figure 4.51 Dipmeter and resistivity logs from a borehole. "Tadpoles" indicate dip angle (distance from left margin) and direction (tail). Resistivity and caliper logs are displayed in left margin. (Crowther et al., 1991.)

minerals until a near-perfect match, and thus identification, is made. An example of using X-ray diffraction to compare clay minerals of a striated sheared sample and that taken a few millimeters from the sliding surface (both within the marl layer) is shown in Figure 4.52*a* and *b*. The striated sliding sample in Figure 4.52*b* shows a

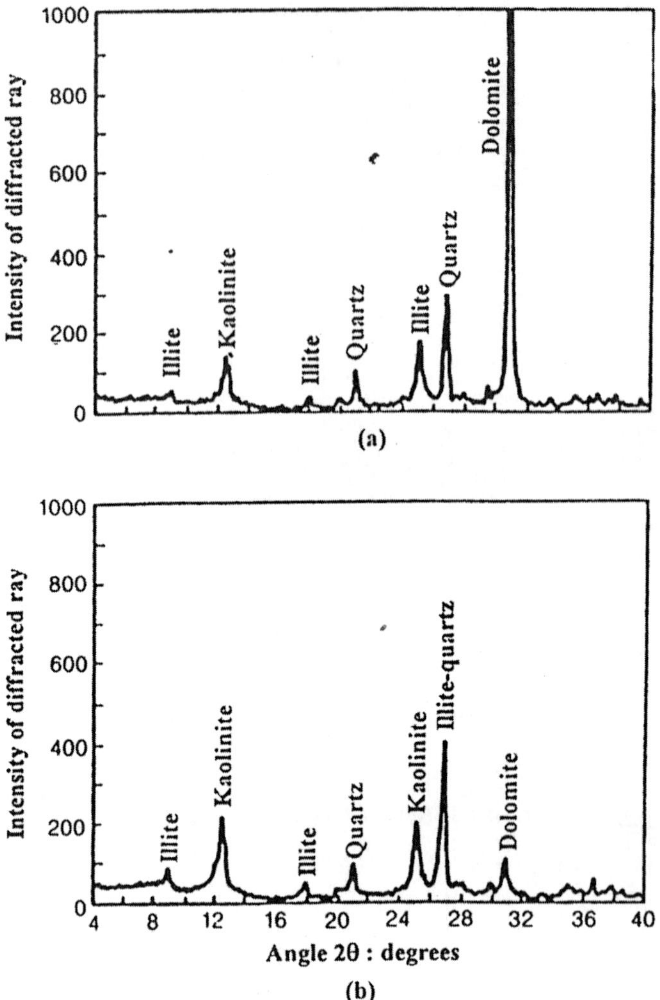

Figure 4.52 Diffractograms of two neighboring samples within the marl layer. (*a*) Intact marl in brown and beige. (*b*) Striated sliding surface in gray and green. θ = angle of diffracted ray. (From Alonso and LLoret, 1993, reproduced by permission of the Institution of Civil Engineering.)

depletion of dolomite as compared to that of the main body of the layer. A possible mechanism leading to this change is the removal of dolomite through dissolution and transportation by percolating acidic water through the fractured limestone overlying the more impervious marl layer (Alonso and LLoret, 1993).

(2) *Differential Thermal Analysis* This is test devised for mineral identification utilizing the thermal properties of minerals when heated at a uniform rate from room temperature to temperature near or at 1,0000°C. Exothermic and endothermic

reactions in a given mineral take place at typical temperatures and with typical minerals, thus permitting identification of the sample.

(3) *Petrographic Analysis* This comprises identification tests utilizing the petrographic microscope, when the mineral grains are sufficiently large, and employs the use of thin sections and polished sections. Here the intrinsic physical properties are determined optically, and identification of the mineral grains and their relationship to each other is made.

4.5.5 Radiocarbon Dating

Radiocarbon dating is sometimes used to determine (1) the time when the preexisting valley began to fill and (2) the rate at which landslide debris accumulated. Carbon sufficient for dating is, however, limited to wood fragments, which may not often be obtained from borehole samples. These data, along with the surface morphology, do provide an indication of the origin, source, and age of the material or landslide debris.

The formation of ponds attributable to landslide movements (Chapter 2) leads to a possibility of estimating the time of past landslide activities by means of dating pond sediments. A variety of methods is available for dating lakes and ponds (Winter and Wright, 1977). These methods generally consist of analyzing minerals, biological matter, and chemicals from cores of the bed sediments. Adam (1975) established the date of pond formation and the dates of subsequent movement of a landslide in California by analyzing pollen and radiocarbon dates from the sediments of a landslide pond.

Apart from scientific interest, dating of past landslide activities would have important practical application. Knowledge of the approximate frequency of previous landslide activities in an area could be useful to planners and land developers, foresters, farmers, and other involved in land use.

4.6 EXPLORATION PROGRAM DESIGN

4.6.1 Locations and Number of Boreholes

The objective of an exploratory program is to establish in reasonable detail the stratigraphy, together with a basic knowledge of the engineering properties of the overburden soil and rock formations that will have an effect upon slope performance. Hence, the number and spacing of the boreholes should be such as to properly achieve this objective at a minimum cost. In other words, there is no definite requirement as to the number and spacing of boreholes so long as the objective is achieved.

Building codes often stipulate the number of borings required for structures, and this number is mostly based on area per borehole. However, there is no definite stipulation required for slopes. In a strict sense, rigid rules for the number and spacing of boreholes cannot and should not be established for slope investigations.

For highway projects, which have large longitudinal extent compared to their width, boreholes generally fall into two categories. The first consists of shallow boreholes along the alignment. These boreholes are to verify delineations of large areas of similar materials along the alignment as suggested by air photos and geologic mapping. The second category includes boreholes for major structures such as bridges, retaining walls, embankments, and cut slopes. These are generally deeper and more closely spaced. Boreholes for retaining walls, embankments and cut slopes are located to give detailed longitudinal and transverse subsurface profiles. The exact spacing will depend on the anticipated variability of the soil profile. However, for typical highway projects, general spacings for borings are 200 to 400 feet, with at least one boring in each landform. For cases where an area is suspected of slope instability, boreholes are usually laid out in a grid pattern that includes representative positions up and down the slope and along the length of the slope.

Emphasis should be placed on locating boreholes to develop typical geologic cross sections; for example, in the case of a river valley, the boreholes should be located on the crest and toe of the valley, and on lines perpendicular to the valley axis. The spacing of the borings on the cross section lines should be much closer than the spacing along the valley line (Figure 4.53), since geologic variations will occur in shorter distances across the valley than up and down the valley. Shallow rock is sometimes encountered

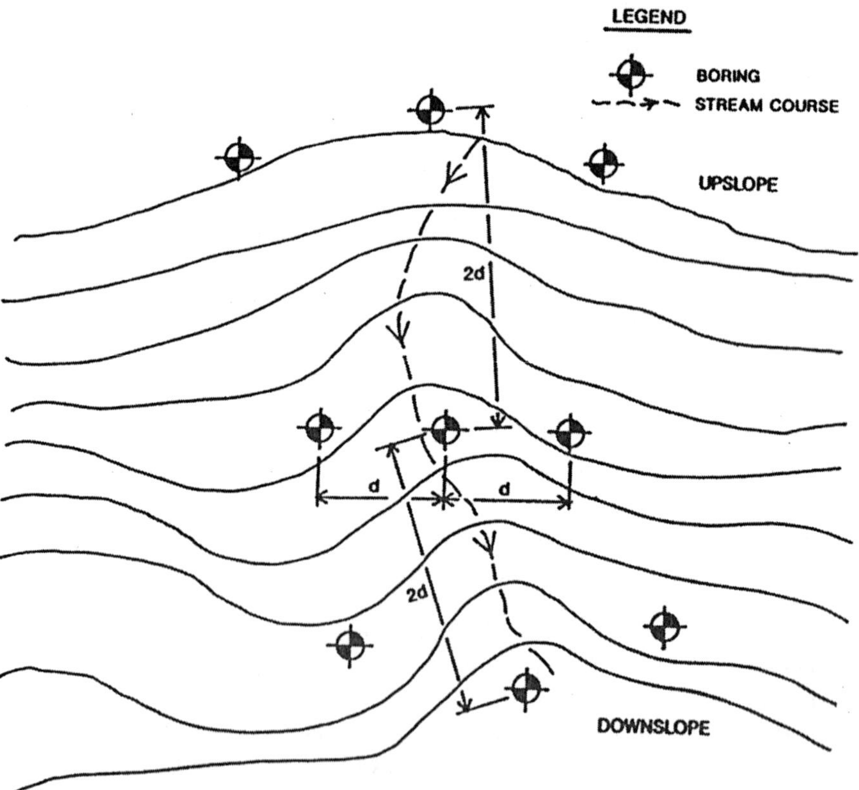

Figure 4.53 Example of planning boring locations at a river valley.

along the stream course as the overburden soils are often eroded by the running water. However, deeper overburden soils would be expected on both sides and at the toe of the valley.

In the case of a proposed cut slope extending from soil into rock, the level of bedrock along the face of the cutting is important. Consideration must be given to obtaining the subsurface profile by additional drilling and geophysical means. Oriented boreholes may be made to determine the strike/dip of joints of the rock from which the slope is cut.

Boreholes in landslide areas should be located both within and outside the slides. Within the slide, the disturbed conditions above and the undisturbed conditions below the slide plane can be studied. Outside the slide, the undisturbed conditions within the same depth as the slide materials inside the slide can be studied, because they may reflect the conditions that existed within the slide mass before failure. The borings should be located to maximize the ability to draw strategic cross sections. The number of borings needed in landslide areas would vary depending on the extent of the slide areas, the risk category of the slopes under study, and the amount of subsurface data available. There is no clear guideline as to the most optimal number of boreholes required in any slide area. However, general rules of thumb are to locate boreholes along the centerline of the slide and to have a borehole at the foot or bulge zone of the slide area. The reasons of placing a borehole at the foot or bulge zone of the slide area are twofold: one is that the area usually experiences continuing movement once a slide has occured; the other is that the area is critical in the design of resisting structures to mitigate the movement (Sowers and Royster, 1978).

4.6.2 Depth of Boreholes

There is no rigid rule in determining the depths to which boreholes are carried. However, a major factor that controls the depth is the nature of the anticipated subsurface conditions based on geology, soil surveys, previous explorations, and the configuration of the highway.

For typical highway projects, highway embankment borings are generally extended to a depth equal to twice the embankment height, whereas cut borings are extended at least 15 feet beyond the anticipated depth of cut (Figure 4.54). However, where deep cuts are to be made or where embankments underlain by weak subsurface soil layers in soft clays or clays are to be constructed, the depth will depend primarily on the existing and proposed topography and the nature of the subsoil on a case-by-case basis. Therefore, no basic guidelines can be given, except that boreholes should be deep enough to provide information on materials that could potentially cause problems with respect to stability. In other words, the depth of the borehole should be made through the problem soils into more competent soils and bedrock if within the zone of interest (Figure 4.54).

For boreholes in landslide areas, the depth of boreholes should not be less than the depth of the slide, which may be estimated in advance based on the slide geometry so data collection can be maximized at critical locations. The depth of boreholes should be beyond the sliding zone and is often dictated by the competent material below the

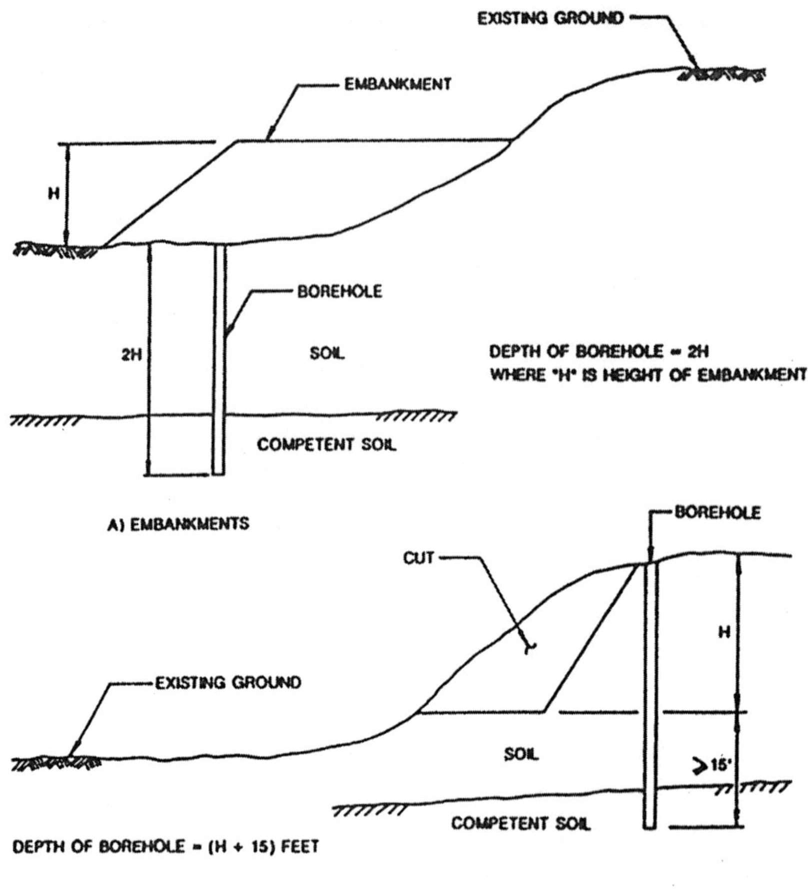

Figure 4.54 Recommended depth of borings. (*a*) For embankments. (*b*) For cut slopes. Note: in all situations borings should be made through problem soils into competent soils.

sliding zone. Any restraining structures to mitigate movements of the slide will need to be sunk into this competent material for lateral support.

REFERENCES

Adam, D. P., 1975. "A Late Holocene Pollen Record from Pearson's Pond, Weeks Creek Landslide, San Francisco Peninsula, California," *U.S. Geological Survey Journal of Research*, Vol. 3, No. 6, pp. 721–731.

Alger, C. S., and E. E. Brabb, 1985. "Bibliography of United States Landslide Maps and Reports," *U.S. Geological Survey*, Open File Report 85–585.

Alonso, E. E., and A. LLoret, 1993. "The Landslide of Cortes de Pallas, Spain," *Geotechnique*, Vol. 43, No. 4, pp. 507–521.

American Society for Testing and Materials, 1982. *Annual Book of ASTM Standards*. Philadelphia.

Bjerrum, L., 1972. "Embankments on Soft Ground," *Proceedings, Specialty Conference on Performance of Earth and Earth-Supported Structures*, Purdue University, Lafayette, Indiana, ASCE, June 11–14.

Blyth, F. G. H., and M. H. de Freitas, 1984. *A Geology for Engineers*, 7th ed. London: Elsevier.

Brand, E. W., 1982. "Landslides in Hong Kong 1978–1980," *Proceedings of the ASCE Speciality Conference on Engineering and Construction in Tropical and Residual Soils*, Honolulu, ASCE.

Crowther, D. D., M. B. Phipps, T. L. Slosson, and J. E. Slooson, 1991. "Use of Multiple Working Hypotheses and Multiple Geologic/Geophysical Technologies to Analyze a Complex Landslide," *Proceeding's of the 27th Symposium on Engineering Geology and Geotechnical Engineering*, Utah State University, Logan, Utah, pp. 20-1–20-15.

Crozier, M. J., 1984. *Field Assessment of Slope Instability*, D. Brunsden and D. B. Prior, Eds. New York: Wiley.

Cummings, D., and B. R. Clark, 1988. "Use of Seismic Refraction and Electrical Resistivity Surveys in Landslide Investigations," *Bulletin of the Association of Engineering Geologists*, Vol. XXV, No. 4, pp. 459–464.

Federal Highway Administration, 1992. *U.S. Department of Transportation, Soils and Foundations Workshop Manual*, R. S. Cheney and R. G. Chassie, Eds., Washington, DC, November.

Federal Highway Administration, 1977. *Cone Penetration Test—Performance and Design Guidelines*, TS-77-207, Washington, DC.

Federal Highway Administration, 1980. *Evaluation of Self-Boring Pressuremeter Tests in Boston Blue Clay*. Interim Report, RD-80-052.

Federal Highway Administration, 1988. *U.S. Dept. of Transportation, Highway Slope Maintenance and Slide Restoration Workshop*, T. C. Hopkins, D. L. Allen, R. C. Deen, and C. G. Grayson, Eds. Report No. FHWA-RT-88-040, Washington, DC, December.

Flaate, K., 1966. "Factors Influencing the Results of Vane Tests," *Canadian Geotechnical Journal*, Vol. 3, No. 1, pp. 18–31.

Flemming, R. W., R. B. Johnson, and R. L. Schuster, 1988. "The Restriction of the Manti Landslide, Utah", U.S. Geological Survey Professional Paper 1311, pp. 1–22.

Fletcher, G. A., 1965. "Standard Penetration Test: Its Uses and Abuses," *Journal of the Soil Mechanics and Foundations Division*, ASCE, Vol. 91, No. SM4, July, pp. 67–75.

Geotechnical Control Office, 1984. *Geotechnical Manual for Slopes*, 2nd ed. Hong Kong: Civil Engineering Services Department, May.

Geotechnical Control Office, 1987. Hong Kong: Civil Engineering Services Department.

Handy, R. L., and N. S. Fox, 1967. "A Soil Borehole Direct Shear Test Device," *Highways Research News*, No. 27, pp. 42–51.

Hunt, R. E., 1984a. *Geotechnical Engineering Investigation Manual*. New York: McGraw Hill.

Hunt, R. E., 1984b. *Geotechnical Engineering Techniques and Practices*. New York: McGraw Hill.

Jakosky, J. J., 1950. *Exploration Geophysics*. Newport Beach, California: Trija Publishing.

Jamiolkowski, M., C. C. Ladd, J. T. Germaine, and R. Lancellotta, 1985. "New Developments in Field and Laboratory Testing of Soils," *Proceedings of 11th International Conference on Soil Mechanics and Foundation Engineering*, San Francisco, California.

Ladd, C. C., R. Foott, K. Ishihara, F. Schlosser, and H. G. Poulos, 1977. "Stress Deformation and Strength Characteristics," State-of-the-Art Report, *Proceedings of the 9th International Conference of Soil Mechanics and Foundation Engineering*, Tokyo, Vol. 2, pp. 421–494.

Lunne, T., and O. Eide, 1976. "Correlations Between Cone Resistance and Shear Strength in Some Scandinavian Soft to Medium Stiff Clays," *Canadian Geotechnical Journal*. Vol. 13, No. 4, pp. 430–441.

Mintzer, O. W., 1962. "Terrain Investigation Techniques for Highway Engineers," Annual Report No. 196-1, Engineering Experiment Station, Ohio State University, Columbas, Ohio, September.

National Oceanic and Atmospheric Administration, 1970. *Environmental Data Service*, revised ed. Washington, D. C.: U.S. Department of Commerce.

Naval Facility Engineering Command, 1971. *Soil Mechanics, Foundations, and Earth Structures*, Design Manual DM-7. Alexandria, Virginia: Department of the Navy, March.

Naval Facility Engineering Command, 1982. *Soil Mechanics, Foundations and Earth Structures and Deep Stabilization, and Special Geotechnical Construction*. Design Manual DM 7-1, 7-2 and 7-3. Alexandria, Virginia: Department of the Navy, May.

Nixon, I. K., 1982. "Standard Penetration Test," State of the Art Report, *Proceedings of the Second European Symposium on Penetration Testing*, Amsterdam, Vol. 1. pp. 3–24.

Palmer, D. F., and S. L. Weisgarber, 1988. "Geophysical Survey of the Stumpy Basin Landslide, Ohio," *Bulletin of the Association of Engineering Geologists*, Vol. XXV, No. 3, pp. 363–370.

Peck, R. B., W. E. Hanson, and T. H. Thornburn, 1974. *Foundation Engineering*. New York: Wiley.

Rib, H. T., and T. Liang, 1978. "Recognition and Identification," in *Landslides: Analysis and Control*, Special Report 176, R. Schuster and R. J. Krizek, Eds. Washington, DC: Transportation Research Board, National Academy of Sciences.

Rogers, F. C., 1950. "Engineering Soil Survey of New Jersey," Report No. 1, Engineering Research Bulletin No. 15. College of Engineering, Rutgers University, Ann Arbor, MI: Edwards Brothers.

Rogers, D., 1989. "Topographic Expression of Landslide-Prone Terrain," Reprint from the *9th National Short Course on Slope Stability and Landslides*, San Diego, California.

Saul, R. B., 1973. "Geology and Slope Stability of the S. W. 1/4 Walnut Creek Quadrangle, Contra Costa County California," California Division of Mines and Geology, Map Sheet 16.

Schmertmann, J. H., 1975. "Measurement of the In Situ Shear Strength," *Proceedings: Specialty Conference on In Situ Measurement of Soil Properties*, ASCE, Raleigh, North Carolina, State-of-the-Art Report, Vol. 2, pp. 57–138.

Schmertmann, J. H., 1978. *Guidelines for CPT Performance and Design*, FHWA-TS-78-209. Washington, DC: Federal Highway Administration.

Schmertmann, J. H., 1983. "Revised Procedure for Calculating K_0 and OCR from DMT's with $I_D > 1.2$ and which Incorporates the Penetration Force Measurements to Permit Calculating the Plain Strain Friction Angle," DMT Workshop 16–18 March, Gainsville, Florida.

Schuster, R. L., and R. J. Krizek, Eds., 1978. *Landslides: Analysis and Control*, Special Report 176. Washington, DC: Transportation Research Board, National Academy of Sciences.

Scullin, C. M., 1994. "Subsurface Exploration Using Bucket Auger Borings and Down-Hole Geologic Inspection," *Bulletin of the Association of Engineering Geologists*, Vol. XXXI, No. 1, pp. 91–105.

Sowers, G. B., and G. F. Sowers, 1970. *Introductory Soil Mechanics and Foundations*, New York: Macmillan.

Sowers, G. F., and D. L. Royster, 1978. "Field Investigation," in *Landslides: Analysis and Control*, Special Report 176, R. L. Schuster and R. J. Krizek, Eds. Washington, DC: Transportation Research Board, National Academy of Sciences, Chapter 4.

Taylor, F., and E. E. Brabb, 1986. "Map Showing the Status of Landslide Inventory and Susceptibility," *U.S. Geological Survey*, Open File Report 86–100.

Terzaghi, K., and R. B. Peck, 1967. *Soil Mechanics in Engineering Practice*, New York: Wiley.

Trofimenknov, J. G., 1974. "Penetration Testing in the USSR," *European Symposium on Penetration Testing*, Stockholm, Sweden, Vol. 1, pp. 147–154.

Transportation Research Board, 1975. *Treatment of Soft Foundations for Highway Embankments*, Report No. 29, Washington, DC.

Wineland, J. D., 1975. "Borehole Shear Device," *Proceedings of Special Conference In Situ Measurement of Soil Properties*, ASCE, Vol. 1, Raleigh, North Carolina, pp. 511–522.

Winter, T. C., and H. E. Wright, Jr., 1977. "Paleohydrologic Phenomena Recorded by Lake Sediments: EOS," *Transaction of the American Geophysical Union*, Vol. 58, No. 3, pp. 188–196.

Williams, G. P., 1988. "Stream-Channel Changes and Pond Formation at the 1974–76 Manti Landslide, Utah," U. S. Geological Survey Professional Paper 1311, pp. 47–69

Winterkorn, H. F., and, H. Y. Fang, 1975. *Foundation Engineering Handbook*. Van Nostrand Reinhold. New York.

Wroth, C. P., 1984. "The Interpretation of In Situ Soil Tests," *Geotechnique*, Vol. 34, No. 4, pp. 449–489.

Zaruba, Q., and V. Mencl, 1982. *Landslides and Their Control*. New York: Elsevier Scientific.

CHAPTER 5

LABORATORY TESTING AND INTERPRETATION

5.1 INTRODUCTION

The limit equilibrium methods used for evaluating the stability of slopes require an accurate and reliable estimate of the in situ shear strength of the slope materials. However, the shear strength parameters are strongly influenced by many complex conditions, including the in situ state of stress, drainage, overconsolidation ratio, loading rates, and soil composition. The objective of this chapter is to provide the reader with a suitable understanding of shear strength concepts such that appropriate laboratory tests can be used to determine the appropriate in situ strength for use with the slope stability analysis methods discussed in Chapter 6.

Shear strength of soils has been discussed in great detail over the past 60 years, but it was the 1960 "Boulder Conference" that first set the tone for this important parameter. Many papers summarizing the current state of knowledge were presented at this meeting. Of the many papers presented, the following three papers are worthy of further study: (1) "The Relevance of the Triaxial Test to the Solution of Stability Problems" by Bishop and Bjerrum (1960), which was a further elaboration of previous work by Bishop and Henkel (1957), (2) "The Strength of Compacted Cohesive Soils," by Seed et al. (1960), and (3) "Comparison of Shear Strength Characteristics of Normally Consolidated Clays" by Bjerrum and Simons (1960). It is important that geotechnical engineers consult these papers and also read the considerable discussions that followed. The deliberations from the 1960 Boulder Conference were further extended later with papers by Skempton (1964) on the long-term stability of clay slopes and the sixth Rankine lecture by Bishop (1966) on the strength of soils.

Some of these deliberations were carried to the 1967 Oslo Conference, with more papers on the shear strength of clays by Bishop (1967), Bjerrum and Kenney (1967), Kenney (1967a, 1967b), and Skempton and Petley (1967) concerning progressive failure and stiff, fissured clays. In subsequent years, the 1969 World Conference provided us with an important state-of-the art paper by Skempton and Hutchinson (1969) titled, "Stability of Natural Slopes and Embankment Foundations," which was later followed by "First-Time Slides in Overconsolidated Clays" by Skempton (1970). The paper, "Analysis and Design Relating to Embankments," by Johnson (1974), also provides an excellent summary of shear strength concerns as well as the analytical concepts.

In more recent years, our knowledge has been further extended by the state-of-the art summaries by Saada and Townsend (1981) on shear strength, by Jamiolkowski et al. (1985) concerning recent developments, and the twenty-second Terzaghi lecture presented by Ladd (1991). If the reader wishes to concentrate more on the laboratory testing phase, Volume 3 of the "Soil Laboratory Testing" series by Head (1998) concerning effective stress testing should be seriously considered as a prime reference. We recommend that the serious geotechnical engineer review the many papers discussed above as a supplement to the material presented in this chapter.

After presenting the concept of *effective stress*, the first part of this section provides a brief introduction to the use of Mohr's circle, the Mohr–Coulomb failure criteria, and stress paths. This is followed by a discussion of the factors that affect the shear strength of soils and the laboratory test procedures commonly used to determine the strength parameters for slope stability analysis. Additionally, in view of the complex interdependent nature of soils, other related laboratory tests are also reviewed within the overall scope of slope stability analysis.

5.2 EFFECTIVE STRESS CONCEPTS

A saturated soil mass consists of two distinct phases: the soil skeleton and the water-filled pores between the soil particles. Any stresses imposed on such a soil will be sustained by the soil skeleton and the pore water. Typically the skeleton can transmit normal and shear stresses at the interparticle points of contact and the pore water can only exert a hydrostatic pressure that is equal in all directions. The stresses sustained by the soil skeleton are known as *effective stresses* and the hydrostatic stress from the water in the voids is known as *pore water pressure.*

It is the effective stress that controls the behavior of soil rather than the total stress or pore water pressure. Thus if the soil particles are to be packed into a denser arrangement, it is the effective stress that must be increased rather than the total stress. If the total stress were to be increased, this increase would be sustained by an identical increase in the pore water pressure, leaving the effective stresses unchanged.

This correlation of effective stress with soil behavior, especially compressibility and strength, is known as the *principle of effective stress.* The effective stress, σ', acting on any plane within the soil mass is defined by:

$$\sigma' = \sigma - u \quad \text{(Eq. 5-1)}$$

where σ = *total stress* acting on the plane
u = pore water pressure

The total stress is equal to the total force per unit area acting perpendicular to the plane, and the pore water pressure may be determined from the groundwater conditions, as discussed later. It should be noted that the effective stress cannot be calculated directly. It is always calculated indirectly with information about the total stress and the pore water pressure.

Most natural, saturated soils derive their strength from the friction at the interparticle contacts. Since the shear stresses at the particle contact are frictional, the strength is directly controlled by the effective stresses. This shear strength of a soil is implemented via the Mohr–Coulomb failure envelope, which may be determined using a variety of laboratory and field tests.

5.3 MOHR CIRCLE

In a two-dimensional soil mass, the stresses at a point may be represented by the conceptual, infinitely small, element A, shown in Figure 5.1. If these three stresses, σ_x, σ_y, and τ_{xy}, are plotted in the shear-normal stress (i.e., $\tau - \sigma$) space, with the same scales used for each axis, the circle drawn through these two points is known as the *Mohr circle of stress.* All combinations of σ_x, σ_y, and τ_{xy}, for any plane passing through this point A, must lie on the Mohr circle.

In order to reduce the number of variables describing the state of stress on a defined plane from three, the *principal stresses,* σ_1 and σ_3, may also be used to define the

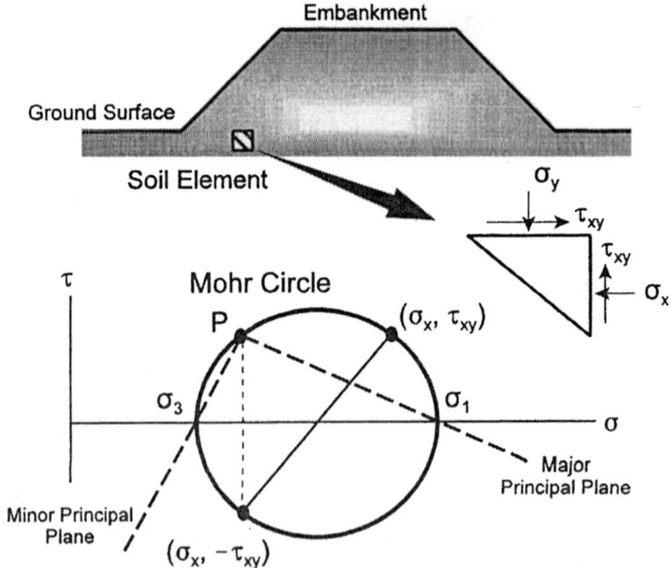

Figure 5.1 Mohr circle showing state of stress at element under an embankment.

stress regime at this point. The principal stresses are the *major* (maximum) and *minor* (minimum) σ values defined by the Mohr circle. The planes on which these orthogonal principal stresses act are similarly defined as the major and minor principal planes. As the maximum and minimum points of the Mohr circle on the σ-axis correspond to zero shear stress, only normal stresses (σ_1 and σ_3) will act on the principal planes, which will be oriented in the directions shown in Figure 5.1.

The magnitude of the principal stresses and the inclination of the principal planes can be determined graphically or by using equations published in most textbooks on soil mechanics. This concept is useful within the framework of shear strength, as the state of stress in the ground can be related to the available strength via a failure criterion.

5.4 MOHR–COULOMB FAILURE CRITERION

A failure theory is required to relate the available strength of a soil as a function of measurable properties and the imposed stress conditions. The *Mohr–Coulomb* failure criterion is commonly used to describe the strength of soils. Its main hypothesis is based on the premise that a combination of normal and shear stresses creates a more critical limiting state than would be found if only the major principal stress or maximum shear stress were to be considered individually. This is illustrated by the Mohr–Coulomb failure envelope, as shown in Figure 5.2.

In this figure, Mohr circle *A* plots well below the Mohr–Coulomb envelope, which indicates a "safe" state-of-stress. In contrast, Mohr circle *B* is *tangential* to the

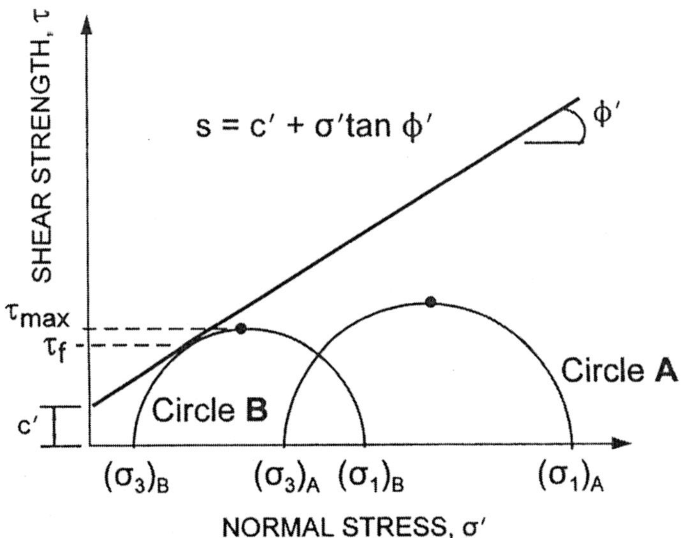

Figure 5.2 Mohr–Coulomb failure envelope for shear strength of soils.

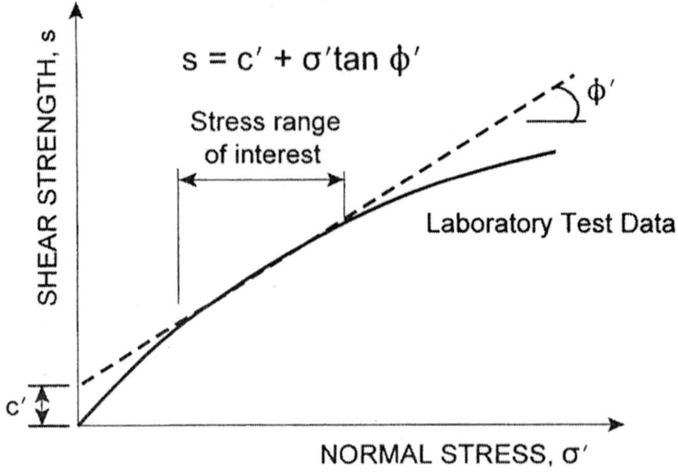

Figure 5.3 Typical nonlinear Mohr–Coulomb envelope of shear strength.

Mohr–Coulomb failure envelope, suggesting that a critical stress combination, σ_n and τ_f, has been reached and the state of stress corresponds to failure. So even though the stress combination, σ_n and τ_{max}, for circle A is obviously greater than that of circle B, it is circle B that is on the verge of failure. Finally, it should be noted that states of stress represented by Mohr circles that extend beyond the Mohr–Coulomb envelope cannot exist.

This criterion may be represented by a curve, as shown in Figure 5.3, and can be expressed also as a power function of the form

$$s = A(\sigma')^b \qquad \text{(Eq. 5-2)}$$

where A and b are constants determined as part of the curve-fitting procedure following laboratory testing. The power function is a popular approach but there are several other forms of expressing such a nonlinear envelope as well (e.g., Hoek and Bray, 1977). If one chooses not to define the Mohr–Coulomb envelope as a curve, a straight line may be used to approximate the envelope within a selected stress range, as shown in Figure 5.3.

For the straight-line case, the Mohr–Coulomb failure envelope may be described by

$$s = c' + (\sigma - u)\tan\phi' \quad \text{or} \quad s = c' + \sigma' \tan\phi' \qquad \text{(Eq. 5-3)}$$

where c = intercept on the strength axis (often called *cohesion*)

ϕ' = angle of internal friction related to the slope of the Mohr–Coulomb line shown in Figure 5.3

5.4.1 Mohr–Coulomb Failure Envelope—Unsaturated Soils

The Mohr–Coulomb failure criterion can also be extended to include partially saturated soils. In contrast to a fully saturated soil, applied stresses in an *unsaturated* soil will be sustained by the soil skeleton, pore water *and* air in the voids. The additional influence of the pore air pressures can readily be included within the effective stress formulation using the equation proposed by Fredlund et al. (1978):

$$s = c' + (\sigma - u_a)\tan\phi' + (u_a - u_w)\tan\phi^b \qquad \text{(Eq. 5-4)}$$

where u_a = pore air pressure
u_w = pore water pressure
ϕ^b = slope of the plot of matrix suction versus the shear stress line if the term $(\sigma - u_w)$ is held constant, as depicted in Figures 5.4 and 5.5

From these figures, one can readily see that as the matric suction term $(u_a - u_w)$ increases, due to an increase in suction, there is a commensurate increase in the apparent cohesion. This increase in apparent cohesion will be a function of the ϕ^b angle, as determined from laboratory tests. As a soil approaches saturation, $u_a \approx u_w$ and Equation 5.4 simplifies to the more usual expression given by Equation 5.3 for saturated soils.

The shear strength of unsaturated soils can be readily accommodated within conventional slope analyses by using the *total cohesion* method. With this approach a

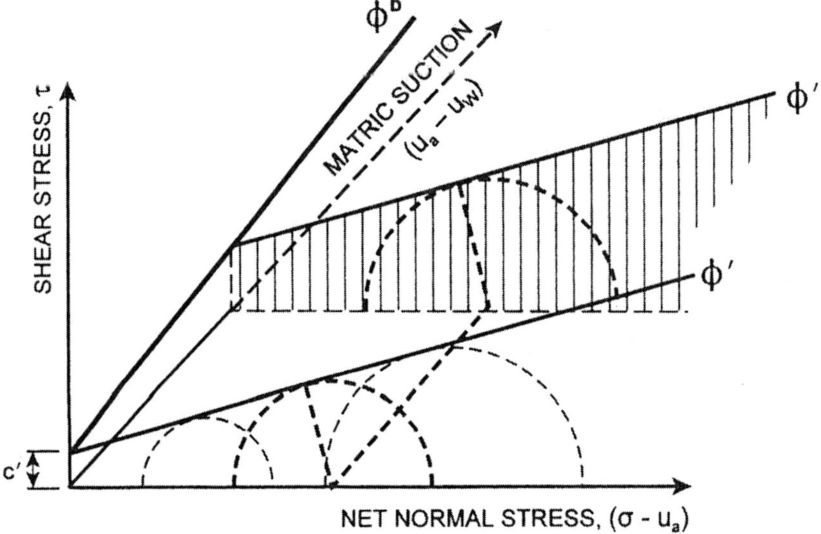

Figure 5.4 Extended Mohr–Coulomb failure envelope for unsaturated soils. (From Fredlund and Rahardjo, 1993, with permission.)

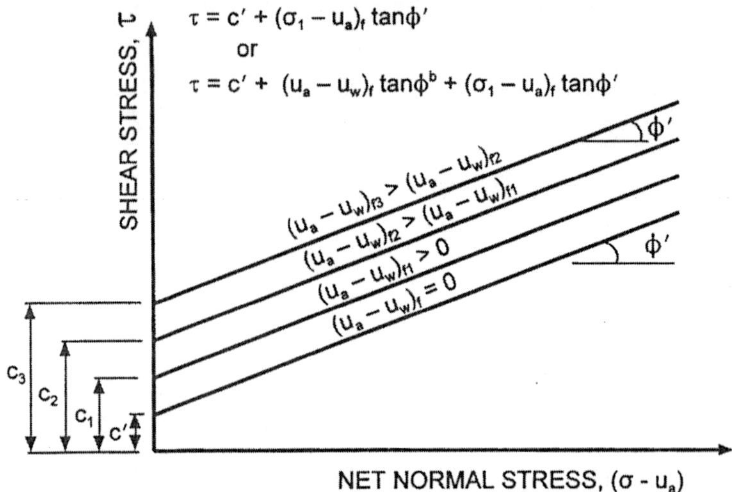

Figure 5.5 Horizontal projection of contour lines of the failure envelope onto the τ versus $(\sigma - u_a)$ plane. (From Fredlund and Rahardjo, 1993, with permission.)

modified value of c' is used to represent the effect of negative pore pressures within the slope. For such cases, the shear strength of the unsaturated soil may be written as:

$$
\begin{aligned}
s &= c^* + (\sigma - u_a)\tan\phi' \\
&= [c' + (u_a - u_w)\tan\phi^b] + (\sigma - u_a)
\end{aligned}
\qquad \text{(Eq. 5-5)}
$$

where the c^* term represents the total (modified) cohesion value to be used to describe the Mohr–Coulomb envelope for the unsaturated soil. Typical values of c', ϕ', and ϕ^b for a series of soils are given in Table 5-1.

The above concepts can be used to explain successfully the behavior of many compacted embankments as the water content in the material either increases or decreases from the initial, as-constructed value. Krahn et al. (1989) discuss the interesting case of a railroad embankment at the Notch Hill site in central British Columbia, Canada, where the side slopes began to fail several years after construction. The embankment material, Tappen-Notch Hill silt (see Table 5.1), was tested in the saturated and unsaturated conditions. In situ field suction measurements in test pits were also made with a Quikdraw tensiometer, which is capable of measuring suctions up to about 85 kilopascals. The field suction measurements generally ranged from about 13 to 30 kilopascals, down to depths of about 4 meters. Tensiometer measurements of the original material, compacted close to its optimum moisture content, indicated matric suction values of about 50 kilopascals. At this level, the factor of safety for the $1\frac{1}{2}$ H–1V slope would have exceeded 3.0, and for zero suction, would have dropped to about unity.

So in effect, the delayed *shallow* failures were a result of the dissipation of the suction within the unsaturated embankment. With time, Krahn et al. (1989) speculate

TABLE 5.1 Typical Data from Unsaturated Tests

Soil Type	c' (kPa)	ϕ' (degrees)	ϕ^b (degrees)	Test Procedure
Compacted shale: $w = 18.6\%$	15.8	24.8	18.1	Constant water content triaxial
Boulder clay: $w = 11.6\%$	9.6	27.3	21.7	Constant water content triaxial
Dhanauri clay: $w = 22.2\%$, $\gamma_d = 15.5$ kN/m^3	37.3	28.5	16.2	Consolidated drained triaxial
Dhanauri clay: $w = 22.2\%$, $\gamma_d = 14.5$ kN/m^3	20.3	29.0	12.6	Constant drained triaxial
Dhanauri clay: $w = 22.2\%$, $\gamma_d = 15.5$ kN/m^3	15.5	28.5	22.6	Consolidated water content triaxial
Dhanauri clay: $w = 22.2\%$, $\gamma_d = 14.5$ kN/m^3	11.3	29.0[a]	16.5	Constant water content triaxial
Madrid grey clay: $w = 29.0\%$	23.7	22.5	16.1	Consolidated drained direct shear
Undisturbed decomposed granite: Hong Kong	28.9	33.4	15.3	Consolidated drained multistage triaxial
Undisturbed decomposed granite: Hong Kong	7.4	35.3	13.8	Consolidated drained multistage triaxial
Tappen-Notch Hill silt: $w = 21.5\%$, $\gamma_d = 15.6$ kN/m^3	0.0	35.0	16.0	Consolidated drained multistage triaxial
Compacted glacial till: $w = 12.2\%$, $\gamma_d = 17.8$ kN/m^3	10.0	25.3	7.0–25.5	Consolidated drained multistage triaxial

Source: Fredlund and Rahardjo (1993).

[a] Average value.

that the suction would diminish near the surface due to moisture infiltration and general moisture migration due to thermal gradients.

5.4.2 Mohr–Coulomb Envelope in *p–q* Space

The Mohr–Coulomb envelope may also be plotted in a slightly different configuration within a p–q space, defined by:

$$p = \frac{1}{2}(\sigma_1 + \sigma_3) \qquad \text{or} \qquad p' = \frac{1}{2}(\sigma_1' + \sigma_3')$$
$$q = \frac{1}{2}(\sigma_1 - \sigma_3) \tag{Eq. 5-6}$$

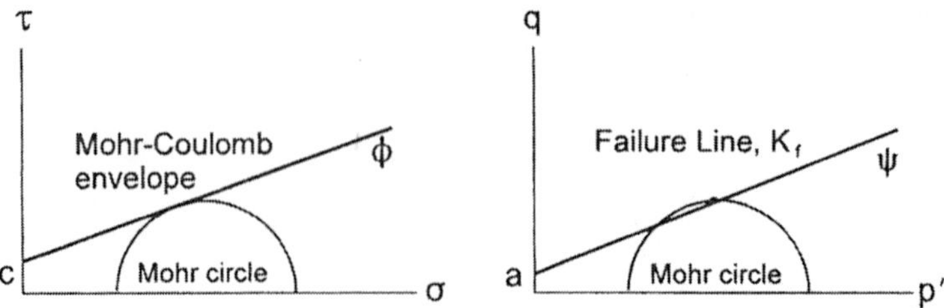

Figure 5.6 Relationship between the $p-q$ and $\sigma-\tau$ space. (Note: $a = c\cos\psi$ and $\tan\psi = \sin\phi$.)

In this configuration, the equivalent Mohr–Coulomb failure envelope is known as the *failure line*, K_f, as shown in Figure 5.6. The main advantage of the p–q space (or p'–q for effective stresses) is that a Mohr circle can be represented by a point corresponding to the p, q coordinates defined by Equation 5-6, above. Then failure is defined by cases where the (p, q) stresses plot on or outside the *safe* region. The main advantage of the p–q configuration is that one does not have to fit the critical Mohr circle tangentially to a failure envelope.

The parameters used to define the K_f line can be readily derived from the conventional Mohr–Coulomb parameters, c and ϕ, using the following:

$$q = a + p\tan\psi \qquad \text{or} \qquad q = a + p'\tan\psi \qquad \text{(Eq. 5-7)}$$

and where

$$a = c\cos\phi \qquad \text{and} \qquad \tan\psi = \sin\phi \qquad \text{(Eq. 5-8)}$$

5.5 EFFECTIVE/TOTAL STRESS ANALYSIS

The shear strength of the soil along the failure surface is a function of the effective stress. However, as discussed earlier, the effective stress can only be calculated indirectly if the pore water pressures are known. These pore water pressures will have to be estimated for both the short- and long-term conditions. Excess pore water pressures in granular soils tend to dissipate rapidly during construction and thus only effective stress analyses should be considered. However, in fine-grained soils the dissipation of excess pore water pressures becomes a consolidation phenomenon.

The short- or long-term conditions are based on the ability of the soil in the slope to reach equilibrium conditions with respect to volume changes that are a reflection of stress changes affecting the slope. Such changes may result in the volume expansion of the soil in the case of an excavation or contraction for soils underlying a newly constructed embankment. The rate of loading or unloading will also influence the selection of pertinent strength parameters of the soil. If the loading rate is slow with respect to the rate of consolidation, the short-term stability of the slope is unlikely to be a critical condition.

If the pore water pressures are known, the analysis may be performed using effective stress principles with drained strength parameters. Although the effective stress approach is realistic, a total stress analysis may have to be employed if the pore water pressures are unknown or difficult to determine. For most practical problems, the total stress analysis is used for short-term stability problems and the effective stress approach is used to assess the long-term stability with the knowledge that any excess pore water pressures generated during the loading have probably dissipated completely. The strength parameters $\phi_u = 0$ and c_u (undrained strength) are used for total stress analysis assuming that the soil behavior is exclusively "cohesive." For effective stress analyses, c' and ϕ', along with the pore water pressure, u, will be required for determining the factor of safety.

The examples shown in Figures 5.7 and 5.8 illustrate the reasons for selecting a total or effective stress analysis for an embankment design and an excavation. Figure 5.7 shows the predicted variation of pore water pressure, at point P in a fine-grained soil, with an increasing height of embankment construction. During the construction phase, the pore water pressure increase is difficult to predict at all locations along the failure surface. As these pressures cannot be predicted, a total stress analysis using unconsolidated, undrained (UU) values of shear strength are recommended to assess the stability of the slope during and immediately after construction. Data from a consolidated undrained (CU) test may also be used to assign the undrained strengths. Any excess pore water pressures generated by the construction will eventually dissipate, and then the embankment loads will be carried by the soil skeleton. For this case, the factor of safety will be a critical minimum value at the end of construction, when the excess pore pressures are likely to be at their highest. The continuous reduction of pore water pressure leads to an increase in effective stress, strength, and the factor of safety. This construction example illustrates the short-term stability problem where the factor of safety is a minimum at the end of construction.

If pore water pressures are monitored during the construction phase of an embankment, an effective stress analysis may be performed using drained strength parameters. This approach of using measured pore water pressures for an effective stress analysis during construction is rapidly gaining popularity as it allows a more accurate and reliable assessment of the stability of the embankment and supporting soils.

Figure 5.8 illustrates similar effects for an excavation where the loads on the soil in the slope are reduced, with a subsequent generation of *negative* excess pore water pressures, that is, suction. As the excess pore pressures are negative, the effective stresses are temporarily increased and the soil displays an apparent increase in shear strength. However, such negative pore water pressures will dissipate with time and lead to a subsequent reduction in effective stress, strength, and the factor of safety. Since the minimum factor of safety occurs at the end of the consolidation phase when the negative excess pore water pressures have dissipated completely, this example reflects the long-term case.

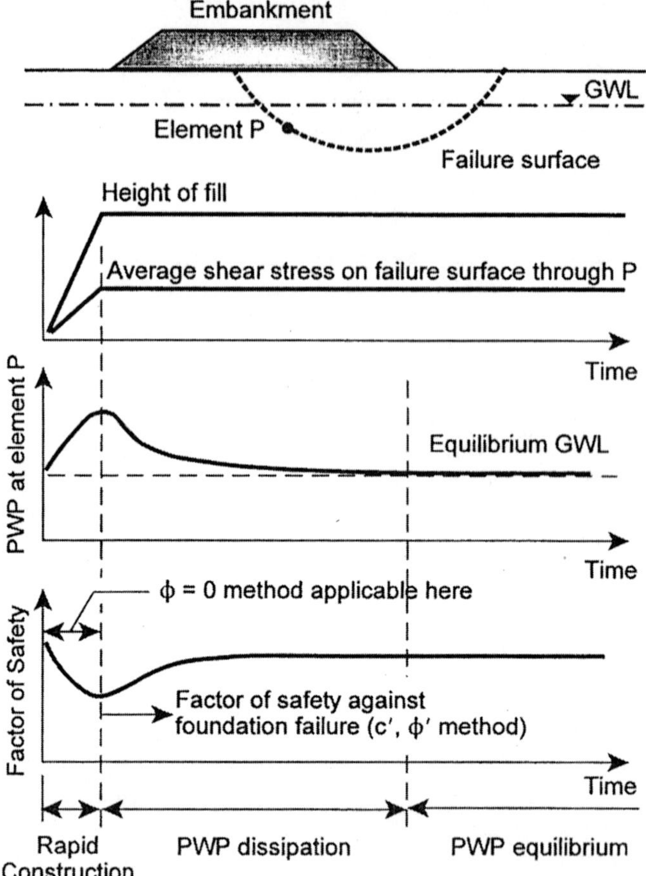

Figure 5.7 Changes in pore water pressure and factor of safety of an embankment on soft clay during and after construction. (After Bishop and Bjerrum, 1960, reproduced by permission of ASCE.)

5.5.1 Factors of Safety

Point *A* in Figure 5.9 shows a Mohr–Coulomb representation of the state of stress acting at a point located along a potential failure surface. At this stress level, the element will mobilize some of the strength available for equilibrium. The current state of practice considers that the effective stress, factor of safety (FOS), will be given by

$$\text{FOS}_{\text{eff}} = \frac{s_{\text{eff}}}{\tau_m} \qquad \text{(Eq. 5-9)}$$

where τ_m = mobilized stress
s_{eff} = maximum strength

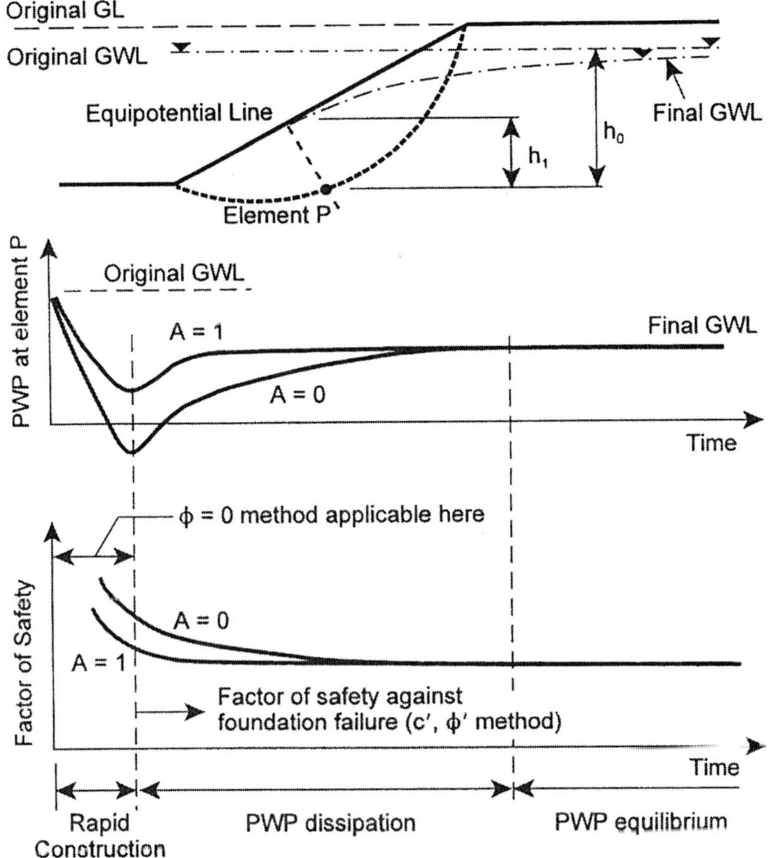

Figure 5.8 Changes in pore water pressure and factor of safety of an excavation in soft clay during and after construction. (After Bishop and Bjerrum, 1960, reproduced by permission of ASCE.)

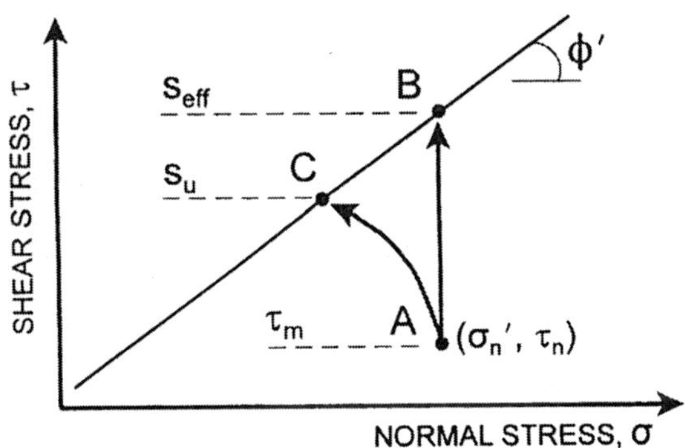

Figure 5.9 Effective and undrained factors of safety.

However, if s_{eff} represents the available strength, it is based on the presumption that excess pore water pressures will *not* be generated (or failure occurs very, very slowly) as the material passes from the mobilized state to the failure state, as shown by line *A–B*. Also, with this stress path, the effective normal stress remains unchanged as the element goes from the stress state at *A* to *B*.

In reality, if the material does fail, it will follow the stress direction *A–C*, reaching its limiting strength at point *C*. For this stress path, the undrained FOS will be given by

$$\text{FOS}_{\text{undr}} = \frac{s_u}{\tau_m} \tag{Eq. 5-10}$$

where s_u is the undrained strength for a sample consolidated to the stresses corresponding to point *A*. As point *C* will always be below point *B* (effective stress) for normally consolidated clays, the undrained FOS is always less than the drained, or effective, FOS. For staged construction where the field pore water pressures are being monitored, a marginal FOS based on an effective stress analysis is often used to control the construction. Ladd (1991) considers the use of such a marginal drained FOS as *unsafe*, because undrained failure is more likely. If the pore pressure parameters are available, the theoretical undrained FOS may be determined from the more readily computed drained FOS value (Moore, 1970).

5.6 STRESS PATHS

Terzaghi (1943) states:

> We test the clay in the laboratory under conditions of pressure and drainage similar to those under which the shear failure is likely to occur in the field and we introduce the values c and ϕ thus obtained into our equations. It is obvious that the success of this procedure depends chiefly on the degree to which the experimenter has succeeded in imitating the field conditions.

In view of Terzaghi's concerns, Lambe (1964, 1967) proposed the use of the *stress path method* to provide a rational approach to study field and laboratory behavior. It was argued that as the stability and deformation characteristics are strongly influenced by the stress history and pore water pressures, the engineer should attempt to simulate, *as closely as possible*, the expected behavior in the field during the laboratory testing for the determination of strength and deformaiton parameters.

The stress path is used to show successive states of stress in a p–q (or for effective stresses, p–q) stress space. The stress point on a Mohr circle corresponding to maximum shear stress, defined by p or p' and q calculated using Equation 5-6, are connected as the typical element progresses from one state of stress to another, as shown in Figure 5.10. For clarity, the Mohr circles shown in Figure 5.10*b* are usually omitted from the stress path diagram.

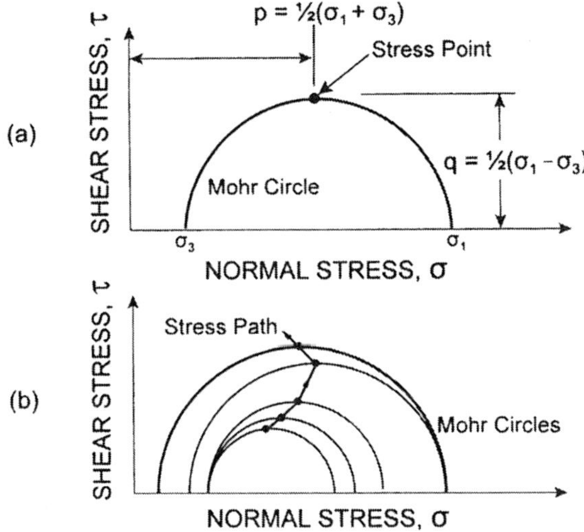

Figure 5.10 The location of (*a*) stress point and (*b*) the stress path.

The stress path affecting a typical soil element during the construction of an embankment is shown in Figure 5.11. From this figure, we can see that the following three stress paths can be plotted:

Effective Stress Path, or ESP This path represents the coordinates (p', q), and is intended to show the *true* behavior of the selected soil element. The strength and stability will be directly related to the ESP and its closeness to the failure line, K_f.

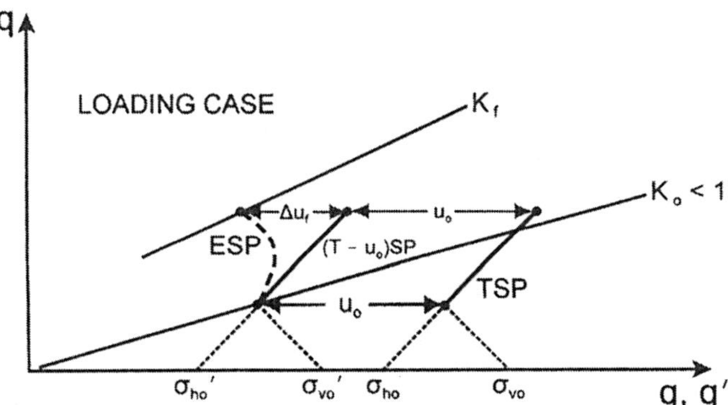

Figure 5.11 Typical stress path under an embankment loaded to failure in a normally consolidated clay.

Total Stress minus Static Pore Pressure, or $(T - u_0)SP$ This path illustrates the total stress path followed by the soil element, with allowance made for the initial pore water pressure (u_0) due to the *static* groundwater level. If the groundwater levels do not change, the difference between the ESP and $(T - u_0)$SP will be the excess pore water pressures, Δu, generated as the soil strains.

Total Stress Path, or TSP This path is represented by a plot of the (p, q) coordinates, and is intended to show the effect of total stresses only.

From these stress paths one can readily view the behavior of typical soil elements. For example, one can see that the $(T - u_0)$SP is always inclined at 45° to the horizontal axis and follows the increasing vertical load applied by the embankment construction. If the excess pore water pressure (Δu) can be estimated from field piezometers or calculations, the ESP can be plotted on the same p, p'–q space. From the plot of the ESP, one can readily see the possibility of instabilities as the ESP approaches the critical failure line, K_f.

Typically, the ESP and $(T - u_0)$SP are the ones used to follow stresses in the field and laboratory samples. Figure 5.12 presents a simple case of an isotropically consolidated sample in a *drained* triaxial test affected by different stress conditions. For this drained case, the ESP and the $(T - u_0)$SP locations are identical, as the rate of loading is slow enough to ensure complete dissipation of excess pore water pressures.

Figure 5.13 illustrates the typical drained stress paths that can be simulated in a triaxial test performed on anisotropically consolidated samples such that $\sigma_{1c}/\sigma_{3c} = K_0$. Then if additional stresses are applied to simulate the conditions displayed in the figure, one can expect the $(T - u_0)$ stress paths to follow the 45° lines shown.

In an undrained triaxial test, the $(T - u_0)$SP will be the same as in a drained test, but as excess pore water pressures are generated, the ESP will follow a different line,

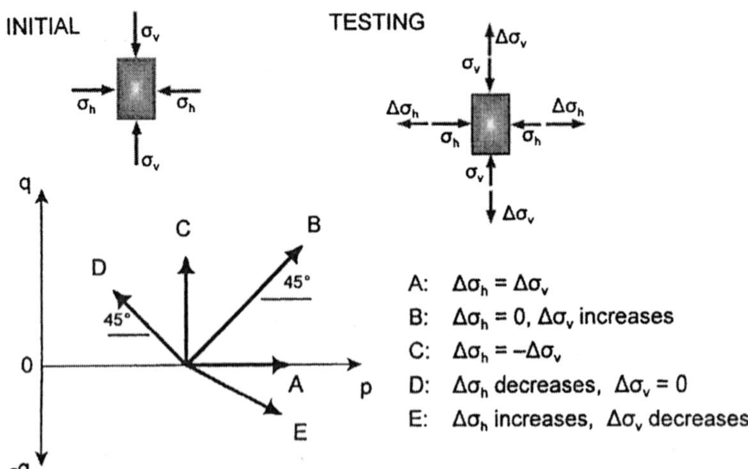

Figure 5.12 Drained stress paths in a triaxial test on isotropically consolidated sample.

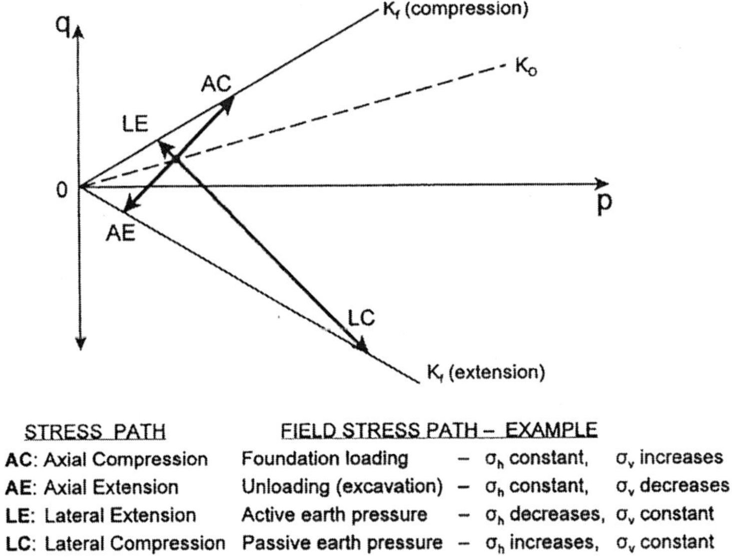

STRESS PATH	FIELD STRESS PATH – EXAMPLE		
AC: Axial Compression	Foundation loading	– σ_h constant,	σ_v increases
AE: Axial Extension	Unloading (excavation)	– σ_h constant,	σ_v decreases
LE: Lateral Extension	Active earth pressure	– σ_h decreases,	σ_v constant
LC: Lateral Compression	Passive earth pressure	– σ_h increases,	σ_v constant

Figure 5.13 Drained stress paths in a triaxial test conducted on an anisotropically consolidated sample.

as shown in Figure 5.14. For normally consolidated soils, the ESP will terminate at the K_f line to the left of the $(T - u_0)$SP because of the positive excess pore water pressures, as shown in Figure 5.14*a*. In contrast, for heavily overconsolidated soils, negative excess pore water pressures are generated, and so the ESP veers off to the right of the $(T - u_0)$SP near the failure line, as shown in Figure 5.14*b*.

5.6.1 Typical Field Stress Paths

An understanding of the field stress paths is necessary to identify the critical elements that are likely to affect the shear strength. Once the potential changes in the field have been reviewed, appropriate laboratory tests can be assigned to determine the in situ strength and deformation characteristics required for design and analysis. Within the slope stability framework, the two main types of design loading conditions are: (1) embankment loading and (2) excavations.

Figure 5.15*a* shows the stress path for an embankment placed on a soft, normally consolidated clay foundation. An element located directly below the centerline will exist at the stress state, corresponding to point *A* in the figure. As the height of the embankment increases, the $(T - u_0)$SP will follow the case simulated by axial compression and follow the path AB. As the loading is relatively rapid, and the fine-grained clay is unlikely to drain, excess pore water pressures, Δu, are developed in the foundation soils. Thus the ESP will curve to the left, along AC, towards the failure line, K_f. If the embankment loading were to continue, the ESP would intersect the K_f line and result in failure at point D. The corresponding point reached by the

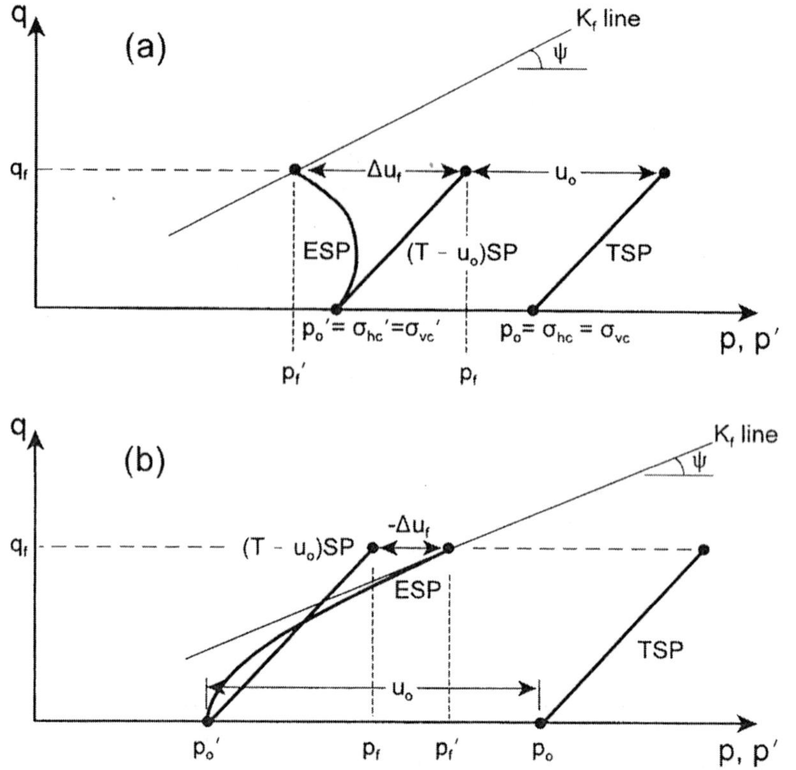

Figure 5.14 Stress paths for the hydrostatically consolidated axial compression tests on (*a*) normally consolidated clays (*b*) overconsolidated clays.

$(T - u_0)$SP would intersect an imaginary total stress failure line (if one existed!) at E, which would correspond to the undrained strength of the foundation soils. If the embankment construction is halted prior to failure, the excess pore water pressures will slowly dissipate and the ESP will move from point C toward B, effectively increasing the factor of safety. This is the main reason why the end-of-construction condition is the most critical condition for normally consolidated soils.

If the same soil layer were excavated, the stress paths shown in Figure 5.15*b* would result as the decrease in vertical load would lead to a $(T - u_0)$SP that follows the axial extension path from A to C in a triaxial test, as shown in Figure 5.15*b*. The stress relief in the foundation soils leads to the generation of *negative* excess pore water pressures, and so the ESP trends to the right of the $(T - u_0)$SP. However, in this case the negative pore water pressures will dissipate with time and the ESP point *B* will move to the left along the path BD. So with time, the ESP will approach closer and closer to the failure line, and failure will result independent of any changes in total stress. This is the main reason why steep excavations can be expected to remain stable for short periods. If long-term stability is required, as in the case of a highway cut,

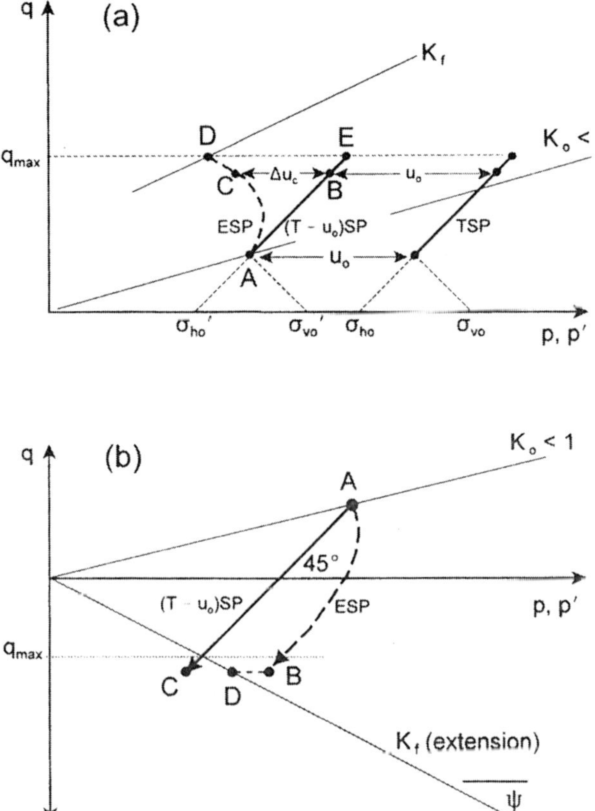

Figure 5.15 (*a*) Stress path for foundation loading of normally consolidated clay. (*b*) Stress path for an excavation in normally consolidated clay.

the engineer must ensure that the $(T - u_0)$SP does not intersect the effective stress failure line. For the case shown in Figure 5.15*b*, the maximum undrained strength should be limited to q_{max}, which corresponds to the intersection of the $(T - u_0)$SP with the failure line as shown.

5.7 SHEAR STRENGTH OF SOILS

The accurate determination of representative soil shear strengths of the materials of a slope are essential to meaningful slope stability analyses. Although it is possible in some circumstances for satisfactory strength measurements to be made in situ, laboratory measurements of strength are by far the most common for fine-grained soils that can be sampled reliably. However, the values of shear strength determined from laboratory tests are dependent upon many factors, particularly the type of soil, quality of the test samples, the size of test samples, and the testing methods. These tests will typically determine the stress–strain curves for the expected soil conditions, thus allowing the engineer to select appropriate strain compatible strength values.

5.7.1 Shear Strength of Granular Soils

Granular soils are classified as materials that are comprised predominantly of sands and gravels and do not display any *cohesive* behavior under unconfined conditions. The shear strength of saturated *cohesionless* soils is derived exclusively from interparticle friction and is essentially a function of the initial voids ratio and confining stress. Thus high strengths are anticipated for granular soils with low void ratios (i.e., at high unit weights) for typical confining pressures. During typical loading and unloading of saturated granular soils, the higher permeability allows rapid movement of water in the pores, effectively preventing the buildup of excess pore water pressures. However, if the rate of loading is very high, for example during earthquakes, it is possible that the volume changes will not occur quickly enough to prevent the buildup of excess pore water pressures.

If a granular material is unsaturated, a small amount of *apparent cohesion*, under unconfined conditions, may exist due to negative pore water pressures (i.e., suction) within the soil particles. However, such cohesion is of a temporary nature and should not be relied upon for design of slopes.

Undisturbed samples of granular soils below the groundwater level are very difficult to obtain and thus laboratory strength tests are rarely performed to determine their shear strength. Typically, the strength is assigned on the basis of in situ tests such as the standard penetration test (SPT) or the cone penetration test (CPT) described in Chapter 4.

5.7.2 Shear Strength of Fine-Grained Soils

Soils that consist predominantly of fine-grained clayey particles may have considerable cohesive strength under unconfined conditions. This cohesive behavior is usually due to inherent negative pore water pressures within the soil mass that lead to increased effective stresses, which in turn simulate the effect of a confined sample.

The influence of the difference between the permeability of granular and fine-grained soils is the main feature that needs to be understood for slope stability analysis. If there is a change in the effective stresses due to construction, volume changes will occur in all soils. However, as the finer grained soils have a much lower permeability than granular soils, the tendency for water to move in and out of the pores is severely restricted in the clayey soils. Thus it will take a much longer time for such volume changes to occur in clayey soils in comparison to the granular soils with their higher permeabilities. If these volume changes cannot occur, pore water pressures increase and the effective stresses (and shear strength) are reduced accordingly. As the increase in pore water pressures due to such conditions cannot be calculated reliably, the analysis may be performed using an *undrained* Mohr–Coulomb failure envelope with $\phi = 0$.

Stiff, Fissured Clays The presence of fissures in an overconsolidated clay will also affect the in situ shear strength parameters. Figure 5.16 shows a photograph of a 300-millimeter diameter stiff clay sample that failed along preferential fissures in a triaxial test. It should be apparent that if a smaller sample of this material had been tested, the strength of the intact material would have been much higher. In the field, the distribution of the fissures, and their continuity, will control the real shear strength for slope stability. So we must select a large enough sample for testing, such that it accurately reflects the field conditions, and includes the fissures and joints for a more realistic assessment of the strength. Because it excludes the fissures, a small sample will overestimate the strength by a factor of up to ten (Johnson, 1974).

After reviewing five slope failures in stiff, fissured clays, Skempton and Petley (1967) determined the shear strength along principal slip surfaces and fissures using a triaxial test. On the basis of these careful tests, they concluded that for *first time* slides, the maximum average (bulk) shear strength of the fissured material closely approximates the shear strength of the same material in its normally consolidated state. In such cases, the fissures introduce an inherent "weakness," which effectively reduces the cohesion intercept, c', to zero, with only a very small change of less than 1.5° in the ϕ' value. Once slip does occur, Skempton and Petley (1967) found that the strength along the failure surface drops to the residual level with small movements of less than 5 millimeters. This effectively suggests that residual strengths should be used for the repair of landslides in stiff clays.

Residual Strength This drop to a lower *residual* shear strength is attributed to the platelike shape of clay minerals. These minerals have a tendency to align themselves in a direction parallel to a shear plane created at large strains. With such a particle arrangement, the shear strength along this *realigned* zone may be substantially less than the strength of the adjacent undisturbed material. The strength of this realigned materials is known as the *residual strength.*

5.7.3 Stress–Strain Characteristics of Soils

Typical strength testing of soils reveals two distinct forms of stress–strain curves, as shown in Figure 5.17. This figure presents a plot of the applied shear force as a function of shear strain for soils exhibiting *brittle* and *nonbrittle* behavior. Typically, dense granular soils and heavily overconsolidated soils exhibit brittle behavior, which is indicated by a distinct *peak* strength at low strains followed by a gradual decrease with increasing strains to a noticeably lower *residual* strength. The change from maximum to minimum strength is attributed to dilatant behavior initially, which is then followed by contraction as the material approaches the critical state.

Sensitive soils will also display a brittle stress–strain curve similar to Figure 5.17. Such soils have a metastable, brittle structure that tends to break down with increasing strain, generating positive excess pore water pressures. The large strain structure of these soils is similar to what would be seen if the soil had been remolded completely. For such soils, the strength will have to be selected with great care as any single localized failure can quickly propagate to a much larger instability due to *progressive*

Figure 5.16 Failure of a 600 × 300 mm diameter sample of "weathered" blue London Clay in an undrained triaxial compression test. (From Bishop, 1967.)

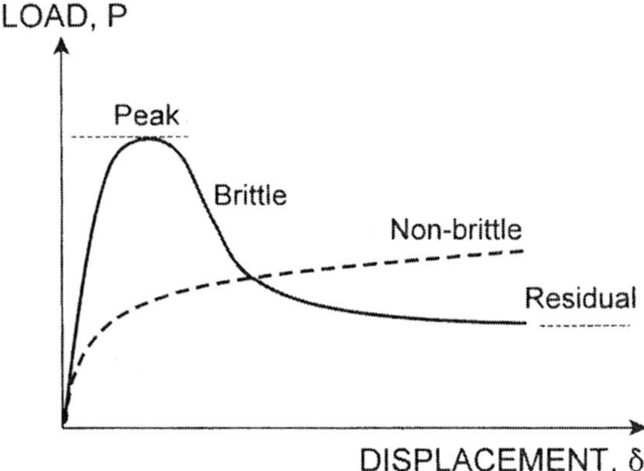

Figure 5.17 Typical stress–strain curves for soils.

failure. Loose granular soils and normally consolidated clays typically display a nonbrittle behavior and there is no marked reduction in strength with increasing strains, as shown in Figure 5.17.

For the analysis of a slope consisting of materials with different stress–strain curves, for compatibility one must select values of strength that correspond to a common level of strain. In other words, a mixture of peak values of strength at very low strains and strengths at 10 or 20 percent strain for the more ductile materials is likely to generate misleading factors of safety.

5.7.4 Discrepancies between Field and Laboratory Strengths

There are at least six ways in which the sample strength measured in the laboratory can differ from the field or in situ strength (Skempton and Hutchinson, 1969). These include: (1) sampling, (2) sample orientation, (3) sample size, (4) rate of shearing, (5) softening upon removal of load by excavation, and (6) progressive failure. All of these possibilities are discussed in the following sections.

In addition to the factors mentioned above, the shear strength of a given soil is also dependent upon the degree of saturation, which may vary with time in the field. Because of the difficulties encountered in assessing test data from unsaturated samples, it is recommended that laboratory test samples be saturated prior to shearing in order to measure the *minimum* shear strengths. Unsaturated samples should only be tested when it is possible to simulate in the laboratory the exact field saturation (i.e. matric suction) and loading conditions relevant to the design. Alternatively, if the matric suction can be measured in situ by means of tensiometers, the an estimate of the unsaturated strength can be made using ϕ' and ϕ^b.

Sample orientation can be important where the soil stratum contains discontinuities or fissures, as often found in residual soils. Where failure in the field could occur along discontinuities or relict joints, account of this fact must be taken when deciding on the orientation of the laboratory samples for testing.

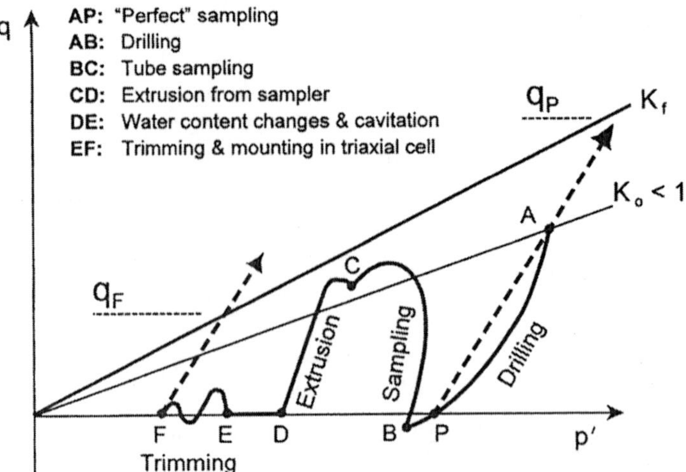

Figure 5.18 Stress path during soil sampling and sample preparation of a normally consolidated clay. (After Ladd and Lambe, 1963, ASM, reprinted with permission.)

Stability predictions based on laboratory shear strengths may have serious limitations for both slightly and heavily overconsolidated clays of high plasticity. This is attributed to the difficulty of obtaining representative samples, measurement of reliable pore pressures, the impact of fissuring, and the gradual degradation of strength with time for overconsolidated clay shales.

Sampling Technique Block or large diameter samples usually provide the most reliable laboratory shear strength results. However, the cost of procuring these types of samples is relatively high, and thus slope analyses are usually based on samples obtained using thin-walled piston samplers (recommended) or Shelby tubes. The impact of obtaining a sample from the ground is effectively illustrated in Figure 5.18. If an ideal sample could be recovered, the state of stress of the extruded sample should correspond to point A. However, due to the drilling, sampling, and other laboratory phases, the sample is ready for testing at a stress state corresponding to point F.

These changes are likely to affect the sample's water content, voids ratio, and structure, which will lead to a poor estimation of in situ shear strength. Table 5.2 provides a summary of the various features that are likely to *disturb* the soil samples and thus affect their properties, as measured in laboratory tests. In general, sampling and disturbance will tend to reduce the measured strength of soil. In soft clays, even the most perfect sampling technique will lead to some reduction in undrained strength due to the changes in total stresses inevitably associated with sampling from the ground. The effect of sample disturbance is most severe in soft sensitive soils and appears to become more significant as depth of the sample increases. Stiff intact clays are perhaps the easiest materials to sample, unless they contain cobbles or gravels.

In preparing a sampling program, it is essential that the best possible samples be procured for laboratory testing. It is recommended that the extent of sample

TABLE 5.2 Sources of Sample Disturbance in Cohesive Soils

Condition	Item	Remarks
Stress relief	Change in stresses due to drilling hole	Excessive reduction in σ_v due to light drilling mud causes excessive deformations in extension Overpressure causes excessive deformations in compression
	Eventual removal of in situ shear stress	Resultant shear strain should usually be small
	Eventual reduction (removal) of confining stress	Loss of negative u (soil suction) due to presence of coarser grained materials Expansion of gas (bubbles and/or dissolved gas)
Sampling technique	Sample geometry: Diameter/length Area ratio Clearance ratio Accessories, i.e., piston, coring tube, inner foil, etc.	These variables affect: Recovery ratio Adhesion along sample walls Thickness of remolded zone along interior wall
	Method of advancing sampler	Continuous pushing better than hammering
	Method of extraction	To reduce suction effect at bottom of sample, use vacuum breaker
Handling procedures	Transportation	Avoid shocks, changes in temperature, etc.
	Storage	Best to store at in situ temperature to minimize bacteria growth, etc. Avoid chemical reactions with sampling tube Opportunity for water migration increases with storage time
	Extrusion, trimming, etc.	Minimize further straining (i.e., do it carefully!)

Source: Jamiolkowski et al. (1985).

disturbance be investigated using procedures outlined by Ladd et al. (1977). If at all possible, corrections should be considered to reflect the loss of strength associated with sampling disturbance on a project-by-project basis.

Sample Orientation—Anisotropy Most natural soil deposits exhibit anisotropic soil behavior. For slope stability analyses, this anisotropy affects the shear strength of the soil, as used for the calculation of the factor of safety. In a limit equilibrium analysis, (1) the original consolidation stresses and (2) the orientation of the failure plane with respect to the principal in situ stress are found to be the most significant

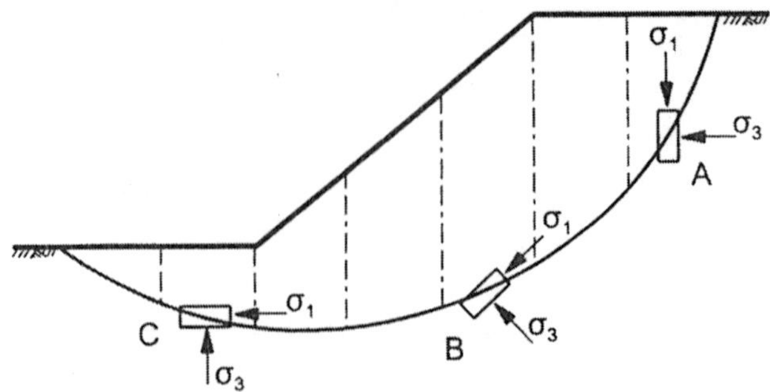

Figure 5.19 Directional strength anisotropy.

parameters. The overall concept of this anisotropy is presented in Figure 5.19. In this figure, three soil elements (A, B, and C) are shown along with the orientation of the principal stresses at failure and the failure plane itself. In general, the shear strength varies as a function of the angle between the principal planes and the failure plane. For San Francisco Bay mud (Duncan and Seed, 1966), the strength along the failure plane for element C was about 75 percent of the tested undrained strength for element A. The drained strength, as estimated from effective stresses, was tested as $\phi = 38^\circ$ and 35° for elements A and C, respectively.

Interestingly, for the same San Francisco Bay mud, the isotropically consolidated samples had an angle of internal friction (ϕ') of 34.5°. Most triaxial tests are performed on isotropically consolidated samples and the orientation of the failure plane would most likely correspond to element A in Figure 5.19. Thus the effective strength for the San Francisco Bay mud can be expected to range between 34.5° and 38°, depending on whether the sample was consolidated isotropically or anisotropically.

The vast majority of conventional strength tests are made (1) on triaxial compression specimens with a vertical axis, (2) on shear box specimens with a horizontal shear plane, or (3) by means of in situ vane tests measuring the undrained shear, which is controlled essentially by the strength along vertical planes. Because of anisotropy, the strength along a slip surface in the ground may vary considerably from the laboratory strength measured by conventional testing, due solely to differences in orientation.

As pointed out by Lowe (1967), Lowe and Karafiath (1960b), and Lee and Morrison (1970), anisotropically consolidated specimens conform better to probable field conditions and it is claimed that they yield higher shear strengths. It may be practicable on larger projects to perform anisotropically consolidated tests (Johnson, 1974). Possible effects of anisotropic consolidation can be examined using several approximate and preliminary methods even if only isotropic consolidation is used. The procedure developed by Taylor, described in detail by Lowe and Karafiath (1960b), or the method proposed by Skempton and Bishop (1954) may be used to compute anisotropically consolidated strengths from results of isotropically consolidated tests.

For most slope analyses, the shear strength values from *isotropically* consolidated tests are used to model the slope materials. This effectively underestimates the strength of soils consolidated at stress ratios, $K \geq 1.0$, and thus leads to a conservative estimate of the factor of safety.

Sample Size Ideally, samples should be sufficiently large to contain a representative selection of all the particles and all the discontinuities in the soil. This is particularly true for fissured clays, for which the sample size can play an important role. For 38 × 76-millimeter triaxial samples, a wide scatter is usually found among the results due principally to fissures, which may or may not be present in the test specimen. In this case, samples of at least 102-millimeter diameter should be tested and an average strength should be selected on the basis of a considerable number of tests.

Over the range of effective normal pressures typically encountered in landslide problems (say 25 to 125 kilopascals, there is little difference between the strengths in the large shear box and those obtained from 38-millimeter triaxial specimens. The latter are, however, considerably less than the strength of intact clay (Skempton and Hutchinson, 1969).

Rate of Shearing Conventional triaxial tests are routinely run at relatively rapid strain rates as compared to the field loading conditions. Studies by Duncan and Buchignani (1973) and Skempton and Hutchinson (1969) indicated that lower strengths would be obtained for specimens sheared at lower shearing rates in the laboratory. For instance, Duncan and Buchignani (1973) stated that, for San Francisco Bay mud, the shearing resistance is only about 70 percent of the value measured in conventional triaxial tests for loads maintained a week or longer. Other data reported by Skempton and Hutchinson (1969) and Kulhawy and Mayne (1990) are in agreement.

For triaxial tests in compression, Ladd and Foott (1974) indicated that for each log cycle increase in strain rate, there is a 10 percent increase in the undrained strength. In other words, if a sample is tested at a strain rate of 1.0 percent per hour and has an undrained strength of 100 kilopascals, then the same sample tested at a strain rate of 10.0 percent per hour will now have an undrained strength of 110 kilopascals, that is, a 10 percent increase.

The accelerated rate of shearing in the laboratory thus tends to *overestimate* the in situ shear strength that may be mobilized during movement of the slope. This overestimation is usually evident if back-calculated strength values from observed failures are compared with high quality laboratory test data. However, for actual design, this overestimation will be partially offset by an underestimation of strength due to sampling disturbance, as well as the many uncertainties that are accounted for by the final design factor of safety.

Softening The stress relief associated with removal of load by excavation will initiate a process of softening as water is attracted by the negative excess pore pressures. Because of the low permeability of clays, the final state of equilibrium under the reduced effective stresses (long-term condition) may not be attained until many years after excavation. The lateral expansion followed by load reduction may cause

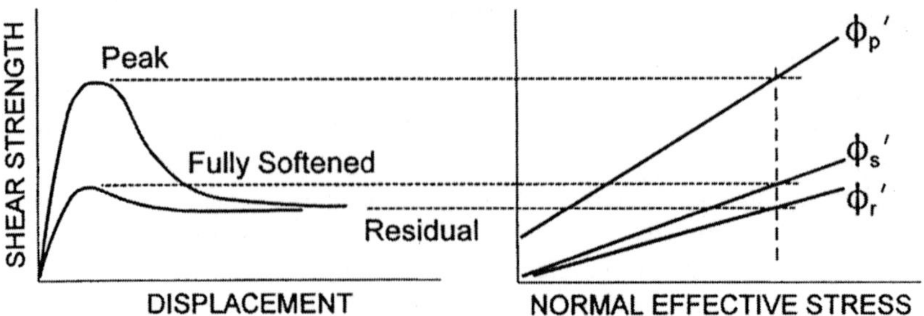

Figure 5.20 Idealized stress-displacement response of stiff fissured clay. (After Skempton, 1970, with permission.)

some opening of fissures and an increase in mass permeability and coefficient of consolidation, c_v. Thus negative excess pore pressures usually dissipate much faster than positive excess pore pressures. Softening may proceed from the face of the open fissures under zero effective stress, thereby causing a reduction in average strength, which in turn allows more deformation to occur. Other fissures then open up and the process continues, and the strength of the clay mass could fall to a much lower value until overall failure occurs.

The fully softened condition corresponds to the strength of the same material, but for normally consolidated conditions. So at low strains, there will be a small peak, which then drops to the residual strength at larger strains. If the softened clay has not been sheared previously (by failure, for example), its effective angle of friction, ϕ' , will remain essentially constant, but its effective cohesion, c', will tend to zero (Skempton and Hutchinson, 1969). For stiff fissured clays, the effect of internal softening will result in strengths that are lower than those measured in conventional drained tests. If the softened clay is sheared past its peak, ϕ' will also decrease and the discrepancy may be even larger (Figure 5.20).

Progressive Failure As mentioned in an earlier section, the magnitude of the mobilized strength along a failure surface is far from uniform along its entire length. If at some time, the shear stress exceeds the available strength in small zone along the failure surface, the excess loading will have to be transferred to adjacent zones. However, if the soils exhibit *brittle* behavior, the stress transfer is likely to lead to failure in adjacent elements as well. Thus failure, having been initiated at a single point, generally progresses until the entire mass fails.

Therefore, the peak strength, in a first-time slide, must be reached at some point before being receted at other points. For brittle clays, that is, those in which the maximum deviator stress, $(\sigma_1 - \sigma_3)$, is reached at 6 percent axial strain or less (Department of Army, 1970), the strength at these points must decrease as further movements take place. The mobilized strength will continue to reduce until overall failure occurs. The more brittle the clay, the greater the difference is likely to be between the mobilized strength and the average peak strength along the slip surface. The brittleness is usually

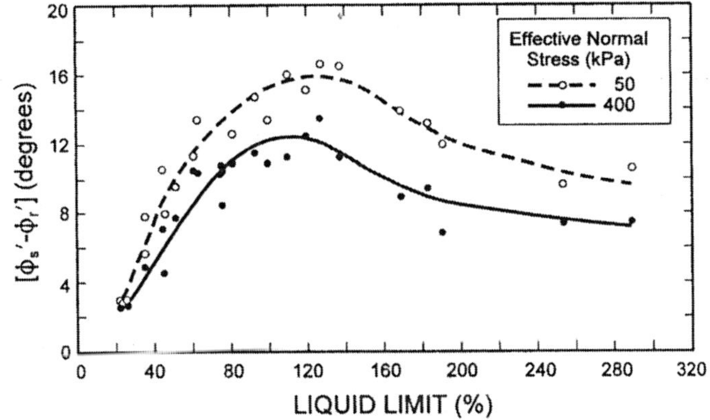

Figure 5.21 Difference between secant fully softened (ϕ_s) and residual friction (ϕ_r) angles. (From Stark and Eid, 1997, reprinted by permission of ASCE.)

quantified by a *brittleness* index (Bishop, 1967), I_b:

$$I_b = \frac{s_p - s_r}{s_p} \qquad \text{(Eq. 5-11)}$$

where s_p = peak strength defined earlier in Figure 5.17
s_r = residual strength defined earlier in Figure 5.17

The possibility of progressive failure will be directly related to the value of this index, with higher I_b values being indicative of potential problems.

Stark and Eid (1997) have also studied the peak and residual shear strength behavior of soils involved with landslides. The results of their studies are summarized in Figure 5.21, which shows the measured change in the ϕ values based on peak (low strain) and residual (large strain) shear strengths as a function of liquid limit for two confining pressures. On the basis of these results, the difference between residual and peak ϕ values can be as high as 12 to 16°.

Once a progressive failure is initiated, it may proceed slowly or rapidly. There are numerous records of cuts and natural slopes remaining stable or undergoing very slow creep movements for many years before the final period of accelerated movement and failure.

5.7.5 Strength Testing

Strength testing requires the careful selection and preparation of a representative soil samples and the appropriate testing procedures and equipment. The *triaxial* compression (TC) test is one of the most common laboratory tests used for the determination of shear strength for slope analysis. Depending on the test conditions, it can be used to determine either the total and/or effective strengths.

Occasionally, if drainage conditions are not critical, the *direct shear* (DS) test may be selected for its operational simplicity. Due to the thin specimen used, drained conditions can be expected to exist for most materials except for highly plastic clays. The DS test can be completed in about a fifth to a tenth of the time required to run a drained TC test. Therefore, the direct shear strength results are usually reported in terms of effective stress. Furthermore, it is more convenient to subject a sample of clay to large strains and to cycles of strain in the direct shear machine than in the triaxial machine if ultimate strengths (i.e., residual strengths) are required for the analysis. In addition to multiple reverse DS tests, the ring shear (Bishop et al., 1971) and Bromhead ring shear (Bromhead, 1979) tests can be used to determine the residual strengths corresponding to large displacements or strains.

Several *direct simple shear* (DSS) testing devices have been developed, but the one described by Bjerrum and Landva (1966) is one of the most commonly used for testing undisturbed samples. The test can be performed under undrained, consolidated undrained, and drained conditions. Although the general state of stress within the sample is indeterminate, the average normal and shear stresses acting on a horizontal plane can be evaluated with sufficient accuracy for practical applications (Ladd and Edgers, 1972; Prevost and Hoeg, 1976).

If economic constraints prevent the use of triaxial tests, unconfined compression tests may be performed on good quality samples to determine the undrained strength. However, it should be noted that in view of the many changes experienced by the soil specimen during the sampling and preparation stage (see Figure 5.18), the laboratory strength will often give a *poor* estimate of the in situ strength! The general requirement for developing a testing program is based on the type of loading conditions expected and whether the subsoil behavior follows *drained* or *undrained* behavior. Figure 5.22 presents a simplified approach for selecting appropriate tests to determine the Mohr–Coulomb parameters, c and ϕ.

For unsaturated soils, special tests are used to determine the shear strengths. These include: (1) modified DS test, (2) undrained (U) test, (3) constant water content (CW) test, (4) controlled suction (CS) test, and (5) dead load test (Brand, 1981). These unsaturated tests are much more complicated than the conventional triaxial tests where saturated specimens are tested because of the additional variable, the *pore air pressure*, u_a, which needs to be monitored and controlled during the test.

For the special laboratory requirements for testing unsaturated soils, the reader should review the state-of-the-art texts by Bishop and Henkel (1957) and Fredlund and Rahardjo (1993). Other information about testing unsaturated soils can also be found in Fredlund (1981), Gan et al. (1988), and Ho and Fredlund (1982a,b, and c).

Studies by Gan et al. (1988) and Shen (1985) revealed that, for most practical applications, the direct shear apparatus modified to monitor suction may be the ideal apparatus in view of its simplicity and short testing time. Similar tests can also be run in a triaxial test, but the suction control at the base of the specimen and soil suction monitoring at the top of the specimen cannot be carried out simultaneously as the top must be vented to atmospheric pressure during the test.

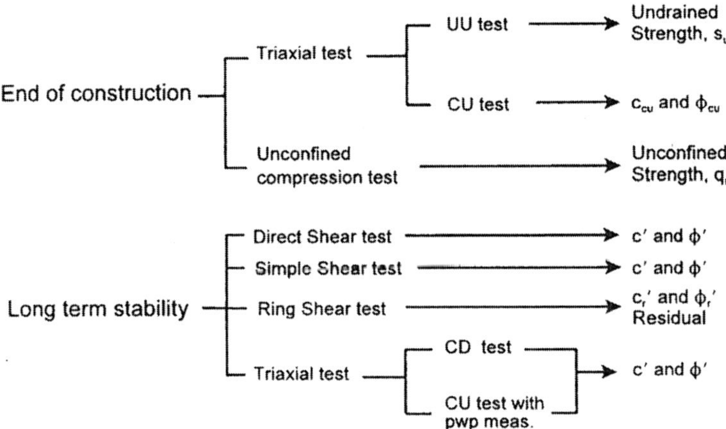

Figure 5.22 Selection of laboratory testing procedures.

5.7.6 Selection and Preparation of Test Samples

As mentioned earlier, it is important that the tests be conducted on soil samples prepared from undisturbed samples that are as representative as possible of the in situ materials. For slope analysis, it is recommended that the shear strength be determined by testing samples obtained from *at least* one-, two-, and three-quarter depths along the potential zone of failure. For evaluating failures, samples should be obtained from the actual failure plane.

Where possible, block or large samples should be used and obtained from the correct depth relevant to slope design. Usually, however, samples are obtained from boreholes using Shelby tubes, piston samplers, and so on. These samples should be checked for disturbance before they are used for laboratory testing. One of the most cost-effective ways to check tube samples is through the use of X-rays.

Radiography A radiograph is produced by passing X-rays through a sample and then recording the transmission intensity on film. The intensity of the image is a function of the density of the materials in the sample, with low-density particles in the soil, producing darker images. Based on considerable use of radiography at the Massachusetts Institute of Technology (MIT), Jamiolkowski et al. (1985) indicate that radiography can show:

- Variations in soil types, especially granular, versus cohesive materials
- Macrofabric features resulting from bedding planes, varves, fissures, shear planes, and so on
- Presence of "intrusions" such as sand lenses, stones, shells, calcareous nodules, peaty materials, drilling mud, and so on

- Voids and cracks due to gas pockets
- Variations in the degree of sample disturbance, ranging from barely detectable curvature adjacent to the sample edges to gross disturbance evidenced by completely contorted appearance and large voids and cracks (most often occurring at the ends of the tube)

Radiography provides a nondestructive method for examining the sample tube so that the least disturbed portion may be selected for testing. The overall cost of radiography is relatively small, averaging to about $60 per tube in areas where such services are available. The advantages of such a radiographic image are considerable, and so we can expect such services to become more widespread in the future.

Selection of test samples should not be based on boring logs alone, but requires personal inspection of the samples and close teamwork with the laboratory personnel and the design engineers. This cooperation must be continued throughout the testing program since, as quantitative data become available, changes in the initial allocation of samples or the securing of additional samples may be necessary.

Laboratory samples for shear testing should be large enough to minimize boundary effects. The minimum sample dimension should be at least six times the size of the largest particle contained in the sample, and the ratio of height to diameter should be between 2 and 3. Triaxial test samples should generally be at least 35 millimeters in diameter. Samples for the direct shear tests are commonly 60×60 millimeters or 100×100 millimeters square and 20 millimeters thick.

It is essential that preparation and handling of the test specimens should be done in such a way as to preserve the natural structure and water content of the material. Test specimens should be prepared in a humidity room. During the preparation of specimens, it is necessary to record a complete description of the material and to judge whether the material is truly undisturbed. Any indication of disturbance of boring samples (strata deformed at periphery or distortions concentric with axis of sample) must be noted and its influence assessed during test interpretation.

Soils with gravel or cobble-sized particles often precludes the preparation of small-scale test samples that are truly representative of the in situ conditions. Hence, test samples must be made from the fine-grained matrix material alone. This usually results in the measurement of minimum strength of the overall material. It has been shown (Holtz and Gibbs, 1956; Holtz, 1960; Henscher et al., 1984, Siddiqi et al., 1987) that the mass strength increases with the proportion of large-size particles (Figure 5.23). On the other hand, the preparation of a test sample of weak material often proves to be difficult, and the resulting test samples tend to represent the stronger portions of the undisturbed samples. Hence, design engineers should be aware of this and select judiciously the most appropriate shear strength parameters for design of soil slopes.

5.7.7 Laboratory Test Conditions

Almost all practical design stability analyses make use of results from triaxial compression tests. This practice is acceptable where conventional safety factors are used

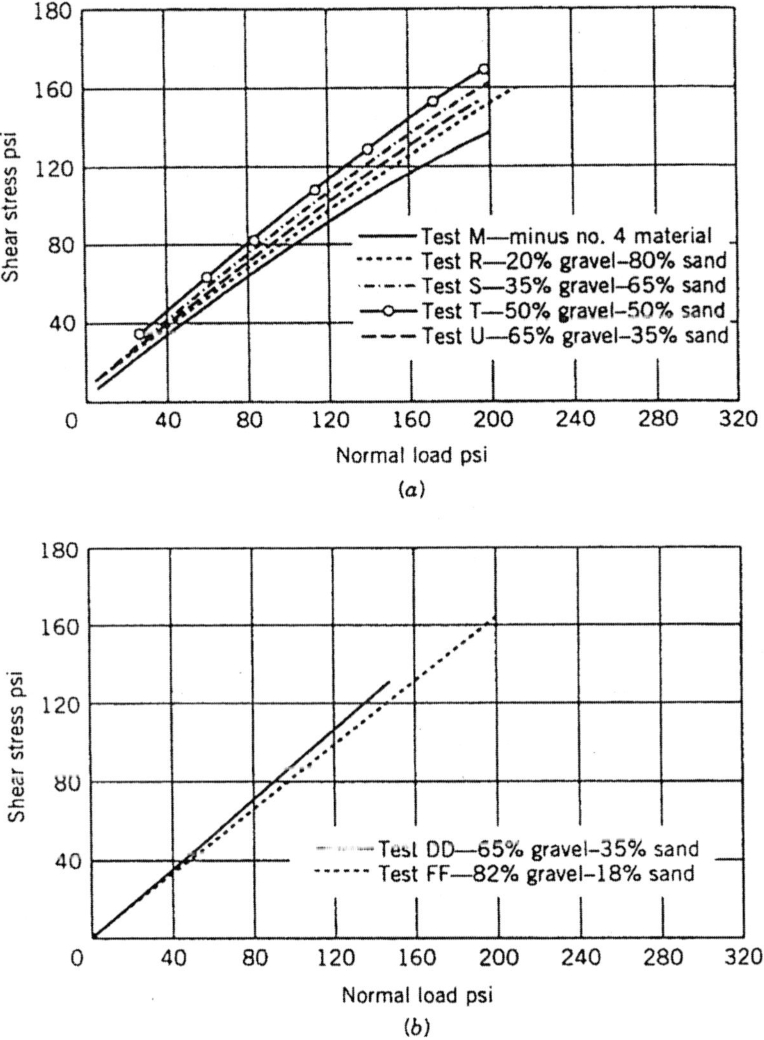

Figure 5.23 Increase in mass strength due to large size particles. (From Holtz and Gibbs, 1956, reproduced by permission of ASCE.)

for design. However, this may not be the case when low safety factors are involved, or if a refined analysis, one that corresponds as closely as possible to field conditions, is desired.

For any embankments, cuts, and natural slopes, the soil may fail by compression, simple shear, and/or extension, as shown in Figure 5.24. Compression and extension tests can be performed using triaxial or plane strain equipment. Differences between the results of triaxial compression or extension tests and direct simple shear tests are illustrated in Figure 5.25 (Bjerrum, 1972; Ladd et al., 1972; Johnson, 1974). If

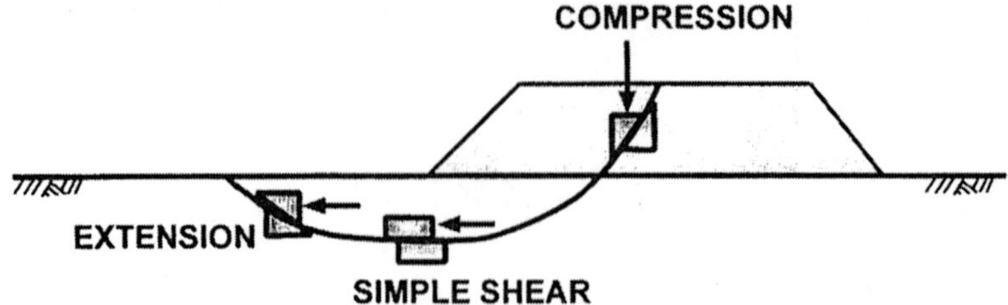

Figure 5.24 Laboratory test conditions relative to field conditions.

average strength along a potential slip surface is assumed to be an average of triaxial compression, direct simple shear, and triaxial extension tests, the conventional use of triaxial compression tests for the entire slip surface may *overestimate* the average shear strengths by 20 to 30 percent.

Because of the two-dimensional nature of typical failure conditions in highway slopes, the triaxial test conditions do not simulate plane strain conditions. Plane strain tests can only be performed with specialized equipment, which is not always available for use on routine projects. Typically, plain strain tests generally give about 5 to 8 percent larger strengths than triaxial compression tests, and possibly as much as 10 to 5 percent more for triaxial *extension* tests (Jamiolkowski et al., 1985).

As discussed above, uncertainties exist when determining the shear strength of soils in laboratories. Unless these effects are evaluated, it is evident that the adopted design safety factor must compensate for tangible, and possibly substantial, uncertainties.

5.7.8 The SHANSEP Method

The *S*tress *H*istory and *N*ormalized *S*oil *E*ngineering *P*roperties (SHANSEP) approach was developed at MIT, as described by Ladd and Foott (1974) and more recently in

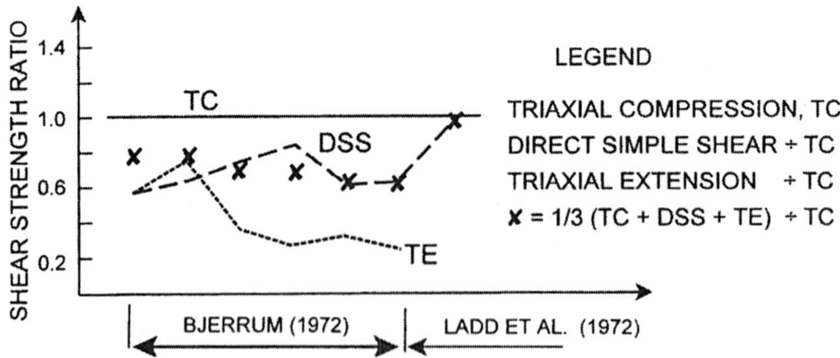

Figure 5.25 Comparison of triaxial compression and extension tests and direct simple shear tests for normally consolidated clays. (From Johnson, 1974, reproduced by permission of ASCE.)

much greater scope by Ladd (1991). The method was developed for soft clays and it attempts to normalize the undrained shear strength, s_u, with respect to the in situ *vertical* effective stress, σ'_{vo}. This approach is based on the fact that soft clays with the same overconsolidated ratio (OCR), but different consolidation stresses, display similar strength and stress–strain characteristics. The normalization procedure thus provides a dimensionless constant, linking the existing effective vertical stress to the undrained shear strength.

The method uses a series of consolidation tests and CK_0U triaxial (axial compression and axial extension) and direct simple shear tests to develop a relationship between the OCR and undrained strength normalized with respect to the vertical effective stress, σ'_{vo}. Ideally, plane strain tests are recommended, but a small correction permits use of the more routine triaxial testing equipment. The results of the laboratory testing are presented in the form:

$$\frac{s_u}{\sigma'_{vo}} = S \cdot (\text{OCR})^m \qquad \text{(Eq. 5-12)}$$

where the constant S corresponds to the normalized strength for normally consolidated soils with an OCR $\approx$ 1.0. The constant m, known as the strength increase exponent, is determined from the relationship between normalized strength and OCR, as shown in Figure 5.26, which is then replotted onto the doubly normalized curve presented in Figure 5.27. One can see that the curves for the six clays all lie in a relatively narrow band. With such data, the undrained strength can then be determined according to the vertical effective stress, σ'_{vo}, or possibly related directly to depth.

The SHANSEP procedure should only be used for reasonably homogenous clay deposits that can be suitably characterized by a normalized strength. The methods is not appropriate for sensitive, normally consolidated soils or cemented clays. The mechanical overconsolidation of these soils will break their structure and the measured undrained strengths will be greatly underestimated by the CK_0U tests. In the same context, the methods is also not recommended for *true*, normally consolidated soils. In conclusion, the SHANSEP method provides a great deal of data about the stress history and the undrained strength profile according to the imposed field loading conditions. However, the CK_0U testing is relatively expensive and the quality of the estimated undrained strength will still depend on uncertain preconsolidation stress, σ'_{vm}, determined from the conventional $e \log \sigma'_v$ data from consolidation tests.

A modified procedure that uses isotropic (CIU) tests, rather than the more expensive CK_0U tests, may also be used to establish the SHANSEP relationship. With this approach, isotropic consolidation is used to bring the sample's water content as close as possible to the same water content as would exist in the anisotropically consolidated (i.e., K_0-condition) sample. The sample is then sheared at this water content and its strength related to the isotropic consolidation stress. For more information about this procedure, the reader should consult the original paper by Coatsworth (1985) for the theoretical aspects and Head (1998) for a complete description of the required laboratory procedures.

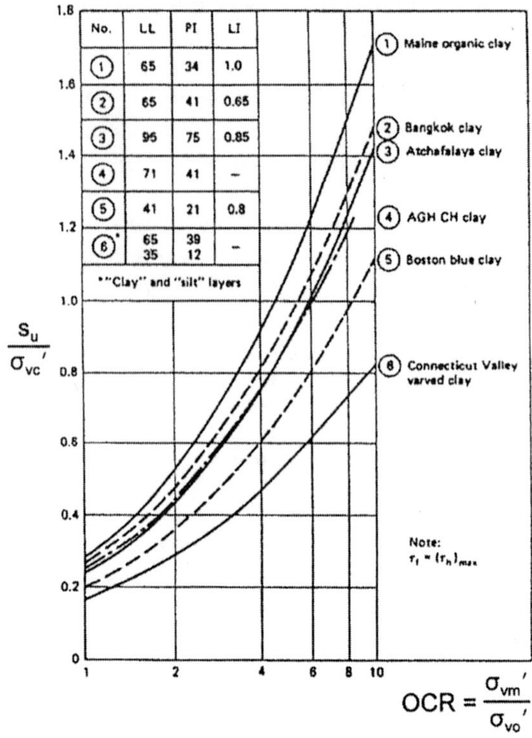

Figure 5.26 Undrained strength ratio versus OCR from direct simple shear tests on six clays. (After Ladd et al., 1977, with permission.)

5.7.9 Triaxial Tests

Common laboratory tests for determining the shear strength of soils include the triaxial test, the direct shear test, and the unconfined compression test. Of these, the triaxial test is one of the most commonly used methods for determining the shear strength of soil (Saada and Townsend, 1981). Figure 5.28 shows a photograph of a triaxial frame with a triaxial chamber. This commercial equipment includes a data acquisition system and can also be set up to automatically follow a stress path defined by the user.

The triaxial test can be used to determine either the total or effective peak strength parameters in accordance with ASTM D2850-87 and AASHTO T234-85 (for UU tests) or ASTM D4767-88 (for CU tests). This type of test permits control of the applied principal stresses and the drainage conditions. Accurate measurements of pore water pressure can be made on saturated samples. Hence, it can be used for determining either the total strength or the effective strength. Table 5.3 presents the types of tests to be run for different soil types and construction.

As indicated in Table 5.3, effective strength parameters for use in slope stability analysis can be determined from a consolidated drained (CD) test or consolidated undrained (CU) test with pore pressure measurements. The CU test, without any pore

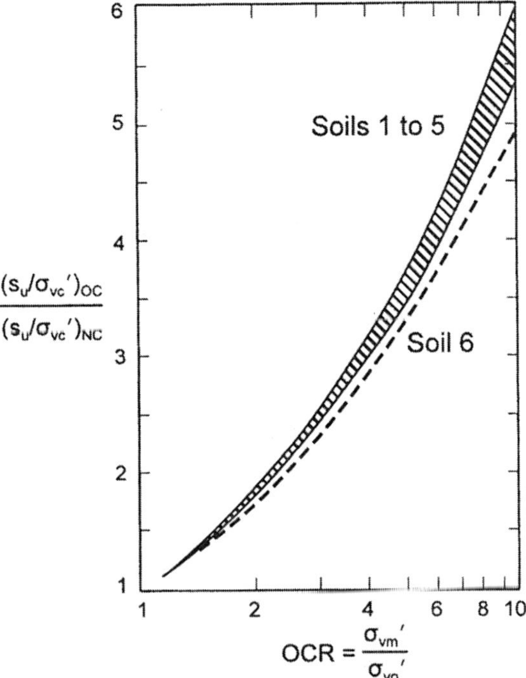

Figure 5.27 Relative increase in undrained strength ratio with OCR based on six clays presented in previous figure. (After Ladd et al., 1977, with permission.)

water pressure measurements, is routinely used to determine the undrained (total stress) strengths. For undrained strengths, the quicker (and cheaper) unconsolidated undrained (UU) test may also be considered if good quality samples are available.

The soil type, defined as fine- or coarse-grained in Table 5.3, is based on the material's ability to dissipate excess pore water pressures. On this basis, Duncan (1996) recommends that soils with a hydraulic conductivity, k, greater than 10^{-6} meter per second be treated as coarse-grained, and soils with k less than 10^{-9} meter per second as fine-grained. If one considers consolidation parameters, then a similar demarcation can be made by using the time factor, T_v(where $T_v = c_v t / H_d^2$). If the calculated T_v is greater than 3.0, the material may be considered coarse, and for T_v less than 0.01, fine-grained. Of course there is a considerable range of hydraulic conductivity, that is, 10^{-9} meter per second $< k < 10^{-6}$ meter per second where the drainage conditions are uncertain and so the more conservative undrained assumption may have to be presumed unless previous, local experience indicates otherwise.

The triaxial test is performed on a cylindrical sample in a triaxial chamber, as shown in Figure 5.29. The specimen is first trimmed and then covered with a rubber membrane before it is placed in the triaxial chamber. Water is introduced into the chamber and a predetermined confining pressure is applied. For isotropically consolidated tests, an all-around confining pressure of 70 to 75 percent of the vertical

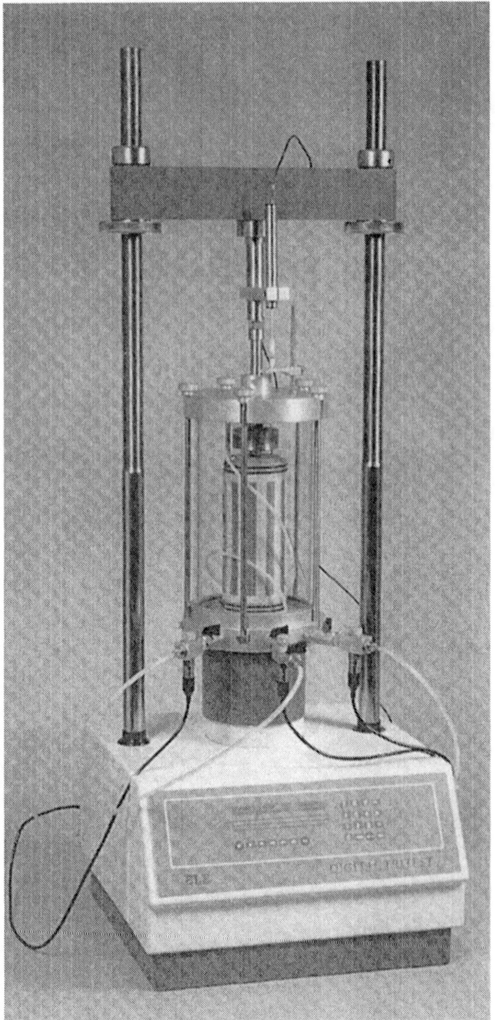

Figure 5.28 A triaxial loading frame capable of stress path testing. (Courtesy ELE Engineering, Inc.)

effective stress in the field is recommended to bring the sample close to its approximate in situ state. After consolidation (if specified), a vertical axial stress is then applied, and the deformation and loading dials are read and recorded until the specimen fails, as indicated by a decrease in reading of the loading dial. If the specimen does not fail, the test is usually continued to a strain of about 15 to 20 percent.

There are three types of triaxial tests: (1) the consolidated undrained (CU) test, in which the sample is first consolidated to a predetermined pressure, σ_{1c} and σ_{3c}, and no drainage is permitted during shearing of the sample; (2) the consolidated drained (CD) tests, in which the sample is first consolidated to a predetermined consolidation

TABLE 5.3 Selection of Strength Tests

Soil Type	Type of Construction	Type of Testing and Strength[a]
Fine-grained soils with:	Short-term (end of construction)	UU or CU triaxial tests for undrained strengths (c_u) at appropriate in situ stress levels
$k < 10^{-9}$ m/s or $T_v < 0.01$	Staged construction	CU triaxial test for undrained strengths at appropriate stress levels—undrained strength, c_u
	Long-term	CU triaxial tests with pore pressure measurements for c' and ϕ', or CD triaxial test for effective strength parameters, c' and ϕ'
Coarse-grained soils with:	Short and long term	Strength parameter (ϕ') obtained from in situ field tests, direct shear, or triaxial testing
$k > 10^{-6}$ m/s or $T_v > 3.0$	Rapid loading: e.g., earthquakes, blasts, or drawdown conditions	CU triaxial tests for undrained strengths (c_u) at appropriate in situ stress levels

[a] *Notes:* CD = consolidated drained
CU = consolidated undrained
UU = unconsolidated undrained.

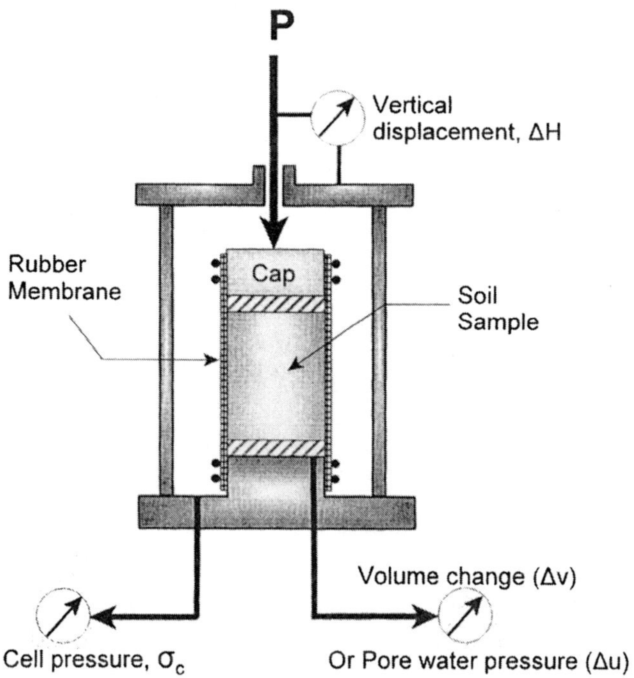

Figure 5.29 Schematic diagram of triaxial testing apparatus.

pressure, σ_{1c} and σ_{3c}, and then drainage is permitted during shearing; and (3) the unconsolidated undrained (UU) tests, in which the sample in the pressure chamber is subjected to a confining pressure without allowing the sample to consolidate (drain). No drainage is permitted during application of the axial load for the undrained tests.

The sample may be consolidated using one of the following:

Isotropic (CIU, CID) This test uses an all around confining pressure, $\sigma_{1c} = \sigma_{3c}$.

Anisotropic (CAU, CAD) For this test, the consolidation stresses, $\sigma_{1c} \neq \sigma_{3c}$. This stress condition should be achieved by first consolidating the sample isotropically (i.e., $K = 1.0$) to the σ_{3c} level, followed by a *very* slow increase in the vertical stress up to σ_{1c}.

At-Rest (K_0) consolidation (CK_0U, CK_0D) Here, the stress ratio, $K_0 = \sigma_{1c}/\sigma_{3c}$, is maintained for zero lateral strains, and K_0 is the at-rest coefficient of lateral earth pressure. The computer controlled equipment for this type of specialized consolidation may not be available for routine triaxial tests.

Soil samples that are unsaturated in the field can either be fully saturated prior to testing, which will give lower bound values for shear strength parameters, or tested unsaturated, which will model the strength of the soil at the degree of saturation at which the test is conducted. Shear strength will, however, be overestimated if the degree of saturation in the field exceeds the saturation level in the laboratory sample. The method of testing must therefore be taken into account when using laboratory derived strength parameters for analysis and design.

Multistage CU tests may be used to generate multiple Mohr circles from a single sample, unlike a single stage test, which can only provide *one* Mohr circle. Multistage CU tests are commonly used in residual soils to eliminate inherent variability between specimens. It is, however, important that the multistage specimen not be deformed excessively during the earliest stages; otherwise the measured deviator stress may be somewhat equivocal and lead to unreasonable strength parameters (Lumb, 1964; Wong, 1978). As a general rule, it is best to carry out only two stages of testing at confining pressures corresponding to the range of pressure relevant to design problems (generally less than 100 kilopascals for shallow failure in residual soils), making sure that the first stage is stopped before any appreciable strain-softening takes place.

Multistage CD tests are usually not recommended for soft to medium stiff clays because of the likelihood of overstraining specimens. Even for CU tests, significant errors in the assessment of shear strength will result if the strain to failure for each test stage is excessive (e.g., 15 percent). To obtain reliable strength data, it is essential not to overstrain the test specimen at each stage. For multistage tests on low plastic silts (micaceous) and clayey silts (low to medium plasticity), unpublished studies by Oregon Department of Transportation (ODOT) indicate that test results have been consistent and reliable if strain is limited to 5 percent per stage in three-stage triaxial testing. Also, organic silts have often required strains of up to 10 percent and, therefore, the tests should be limited to two stages, and occasionally to only a single stage.

TABLE 5.4 Estimated Times to Failure in Triaxial Tests

Type of Test	No Side Drains	With Side Drains
Undrained, CU	$0.51 \times t_{100}$	$1.8 \times t_{100}$
Drained, CD	$8.5 \times t_{100}$	$14 \times t_{100}$

Source: Bishop and Henuel (1962).

Filter paper side drains, often used to accelerate consolidation, should not be routinely used in triaxial tests because they may lead to errors in strength measurement. Corrections for the effect of membrane penetration should be applied according to established standards of practice.

The strain rate for drained tests with pore pressure monitoring should be such that the pore water pressure fluctuation is negligible, and in any case, it should not be greater than 5 percent of the effective confining pressure. If the consolidation stage of the test is monitored and the time for about 100 percent consolidation, t_{100}, found, then the time to failure may be estimated from Table 5.4 (Bishop and Henkel, 1962). For an unconsolidated undrained tests, it is desirable to keep the strain rate from exceeding 2 percent per minute, which will usually lead to failure in about 10 to 20 minutes.

The mode of failure of a triaxial sample should be observed and, if the sample fails on a distinct plane, the angle of inclination of the plane of failure should be measured. Theoretically, all samples should fail on planes at $45° + \frac{1}{2}\phi'$ to the minor principal stress, which will be horizontal in a triaxial cell. However, samples containing discontinuities may fail on other planes and, under these circumstances, an analysis of the shear strength parameters must consider failure along preexisting discontinuities.

For undrained tests, failure can be defined either as the maximum deviator stress, $(\sigma_1' - \sigma_3')_f$, or as the maximum obliquity ratio, σ_1'/σ_3'. For fully drained tests, these two criteria coincide. This discrepancy between the two definitions may be relevant for very refined analysis, but for most highway analyses, the strengths from either approach are acceptable for design.

Figure 5.30 shows the Mohr circles at failure and the Mohr–Coulomb envelope, based on the $(\sigma_1' - \sigma_3')_f$ criterion, for consolidation stresses that spanned the preconsolidation stress (σ_{vm}'). So for consolidation stresses below σ_{vm}', negative excess pore water pressures are generated (circles A and B) during shear as the sample behaves as an overconsolidated soil. For circles C and D, where the consolidation stress is higher than σ_{vm}', positive pore water pressures are generated during the test. According to Hirschfield (1963), the "break" in the total stress envelope (point z) occurs at about twice the σ_{vm}' for typical clay soils.

Back-Pressure Saturation Effects It is common practice to use back-pressure to achieve saturation in triaxial compression tests. This practice is believed to give more conservative results than can be expected in the field, but this may not always be the case. It has been found (Johnson, 1974) that if an undisturbed specimen in a triaxial compression test is saturated using high back-pressures, it will lead to a greater deviator stress, $(\sigma_1 - \sigma_3)_f$, than would occur with a lower back-pressure at

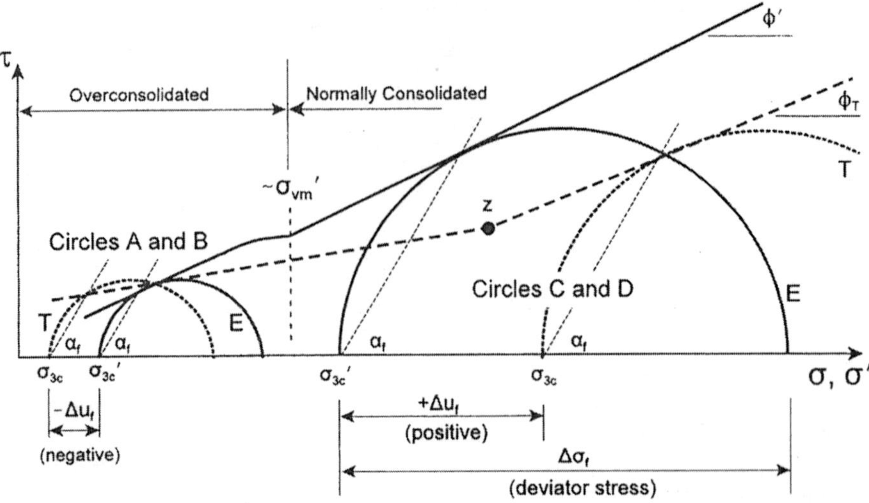

Figure 5.30 Mohr failure envelopes over a range of stresses spanning the preconsolidation stress, σ'_{vm}.

the same confining pressure. This effect is illustrated in Figure 5.31, which shows the results from four triaxial tests that used back-pressures of zero to 150 pounds per square inch (1,037 kilopascals). To avoid such excessively high back-pressures, it is beneficial if a high degree of saturation (>90 percent) is achieved by first percolating deaired water under a small hydraulic gradient through the specimen until air stops bubbling from it.

Back-pressure is also used to prevent pore water cavitation at low stresses in dilatant materials, such as compacted soils and overconsolidated in situ materials tested at stresses less than their preconsolidation stress. In such soils, negative pore pressures are generated during shear, and thus there is a need to ensure that the original pore

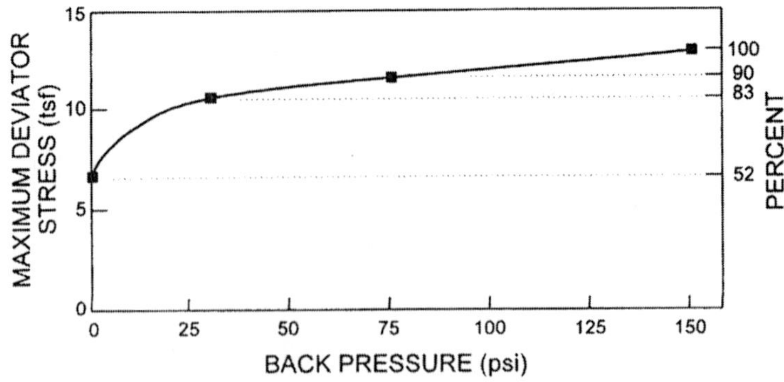

Figure 5.31 Effect of back pressure on deviator stress. (From Johnson, 1974, reproduced by permission of ASCE.)

pressure is of a large enough magnitude to prevent the development of possible cavitation. The negative, excess pore water pressure will result in an undrained strength that is greater than the effective shear strength. The availability of this apparent additional undrained strength for end-of-construction stability evaluation is uncertain, and thus should not be used for design (Lowe, 1967).

Other Corrections The results obtained from a triaxial test are strongly affected by the loading end-plates—their smoothness, size, and effective permeability—and the flexible membrane used to constrain the sample. The use of lubricated end-plates is one way to avoid these problems. As for the membrane, corrections can be made to account for membrane stretching and penetration effects (see Head, 1998).

5.7.10 Direct Shear Test

The direct shear test is one of the oldest and simplest tests for determining the shear strength of a soil sample. Even though the stress conditions within the sample are unknown, it can be used to reliably determine the residual shear strength (Saada and Townsend, 1981). The direct shear test should be performed in accordance with the procedures set forth in ASTM D3080-90 and AASHTO T 236-90. Figure 5.32 shows a schematic diagram of the direct shear box and Figure 5.33 shows a photograph of a commercial direct shear testing device. The device shown includes a data acquisition system and can determine residual shear strengths on 102 × 102 millimeter samples.

The sample is placed between two porous stones to facilitate drainage and the normal load is applied to the sample by placing weights in a hanger system, or by a hydraulic piston. The shear force is supplied by the piston driven by an electric motor. The horizontal displacement is measured by a horizontal dial and the shear force by

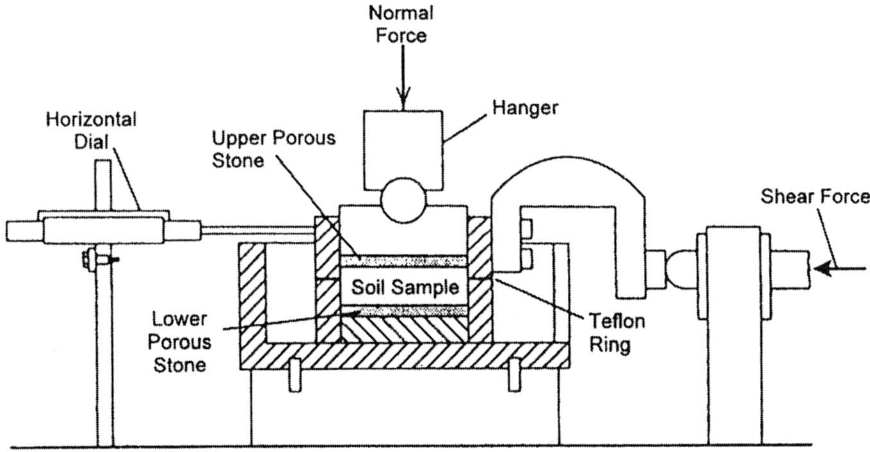

Figure 5.32 Schematic diagram of a direct shear test box assembly.

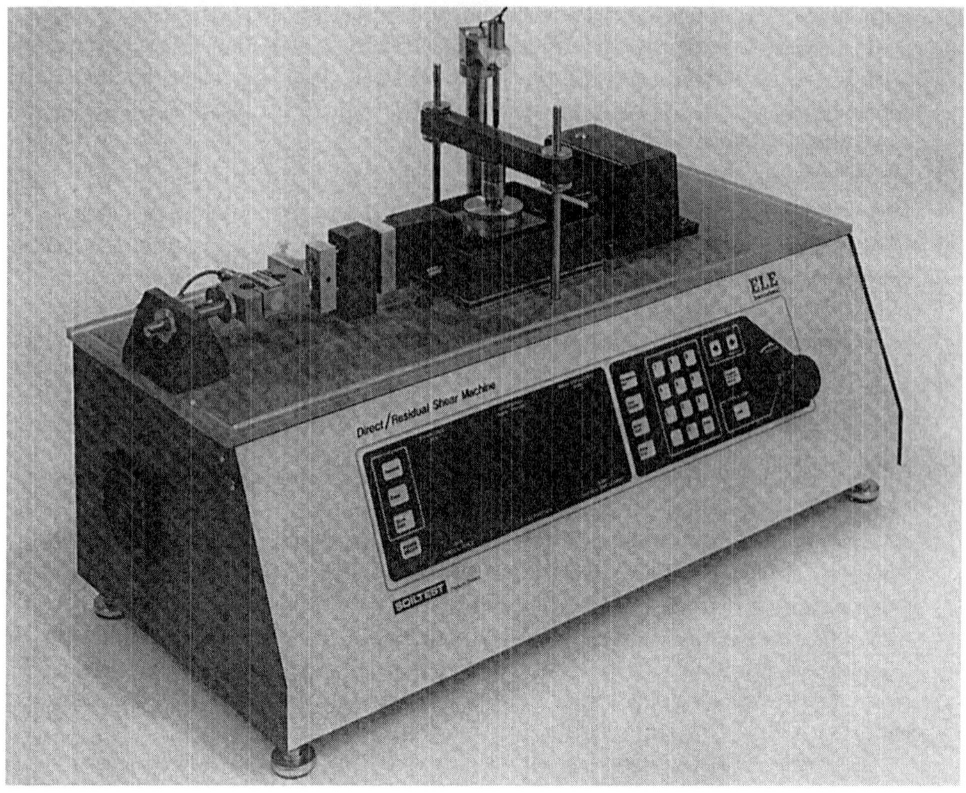

Figure 5.33 A direct shear testing device. (Courtesy ELE Engineering, Inc.)

a proving ring and load dial, which are not shown in the figure. The shear strength can be measured on any predetermined plane in a soil mass by trimming specimens at the correct appropriate orientation.

Specimens for direct shear tests cannot be brought to full saturation, but a high degree of saturation can be achieved by immersing specimens in water for a sufficiently long period of time prior to testing. This soaking process probably represents most closely the conditions to which the material is subjected in the field under steady infiltration.

The rate of horizontal displacement for the shear test should be slow enough to ensure that drained conditions prevail. A maximum rate of 0.075 millimeters per minute is considered appropriate for drained tests on 20-millimeter thick soil specimens. Accurate measurements should be made throughout the test of the shear force, the relative displacement of the two halves of the box, and the vertical movement of the specimen in the box.

The results of the direct shear test are often plotted as shear stress versus normal stress, from which the effective cohesion and the effective angle can be obtained. Moreover, variations of shear stress and horizontal displacement may sometimes be plotted as well.

As mentioned earlier, ultimate (residual) shear strength of clay shale material can be measured by means of a repeated direct shear test under a drained condition (Townsend and Gilbert, 1976). In this test, a soil specimen acted on by a normal stress is repeatedly sheared by reversal of the direction of shear until a minimum, that is, residual, shear stress is determined. The concept of the test is that the soil sample is acted upon by normal and shear stresses until, after large shear deformation, disaggregation and progressive increase in parallel orientation of soil particles in the direction of shearing occur, and a surface or a thin zone of remolded material is formed. Thus a minimum drained shearing resistance is offered.

Since the soil is sheared forward and then backwards until a minimum shear resistance is measured, the sample does not experience continuous shear in one direction, as would be the case for field conditions. However, the reversal approach allows use of conventional direct shear test equipment, without unduly affecting the results.

The shear surface can be formed by cutting a plane surface through an intact specimen with a fine wire (recommended), or sometimes by shearing a completely remolded specimen. Any irregularities of the shear surface would introduce an added resistance, which would not be a measure of the shear strength of the material.

Ring Shear Test A ring shear test is used to shear a soil sample in a single, continuous direction to determine the residual strength. In a ring shear test, an annular sample is sheared horizontally into two discs, or rings, and the decrease in frictional resistance that occurs, as one disc is rotated relative to the other, leads to the residual strength. Procedures for conducting a ring test can be found in Bishop et al. (1971) and Bromhead (1979). A schematic diagram of the Bromhead ring shear device is shown in Figure 5.34. In this figure, note the two proving rings used to monitor the torque applied to the ring-shaped sample. Figure 5.35 shows a photograph of a Bromhead ring shear device, as used at Loughborough University, U.K.

The main advantage of the ring shear test is that it continuously shears the sample in one direction for any magnitude of displacement, which can reach totals of several meters (Stark and Contreras, 1998). This continuous shearing action is preferable to using the shear reversal approach used in a direct shear test device. Other advantages of the test include a constant cross-sectional area of the sample during shear and minimal laboratory supervision during shear with an automatic data acquisition system.

A number of ring shear devices have been developed (e.g., La Gatta, 1970; Bishop et al., 1971; Bromhead, 1979). Comparative studies by Hawkins and Privet (1985) and Townsend and Gilbert (1976) have shown that similar values of residual strength are found using different types of equipment. Furthermore, the specimen preparation procedures used for the determination of residual strength do not affect the final results (La Gatta, 1970; Bishop et al., 1971; Bromhead, 1979).

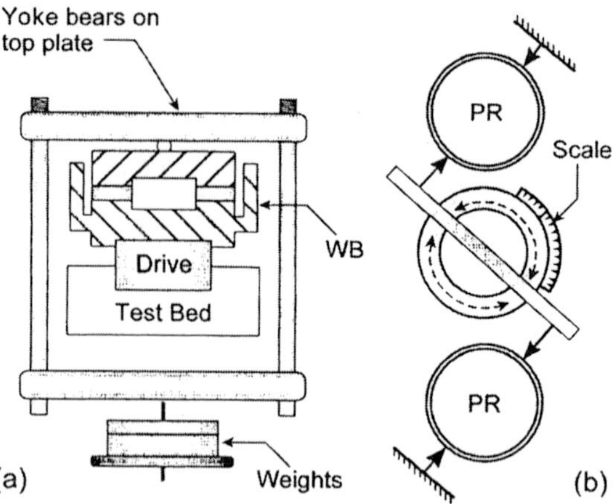

Figure 5.34 Features of a ring shear apparatus. (a) Section. Sample turret confines sample between top and bottom plates (WB = waterbath) (b) Plan. Shear load measured by Proving Ring (PR) and displacement by scale. (From Bishop et al., 1971, reproduced by permission of the Institution of Civil Engineers.)

Figure 5.35 A Bromhead ring shear device for determining residual strength.

More recently, Stark and Contreras (1996, 1998) presented details about a modified Bromhead device to perform consolidated *undrained* shear tests where a constant volume is maintained by a closed-loop system. In this device, the constant volume is maintained by automatically varying the vertical load (normal stress) such that there are no measurable vertical deformations.

5.7.11 Direct Simple Shear (DSS) Test

The direct simple shear test is conducted on a cylindrical sample, which may be encased in a wire-reinforced rubber membrane (Norwegian Geotechnical Institute development) or retained inside a series of thin rings (Swedish Geotechnical Institute development). A schematic of the disk-shaped sample used in a DSS device is shown in Figure 5.36 and a photograph of the Geonor device used at MIT is shown in Figure 5.37. The flexible membrane, or the rings, allow shear deformation to be distributed fairly uniformly through the sample. In the test, the sample is consolidated anisotropically under a vertical stress, and deformed by application of a shear stress. The test can be performed under undrained (UU), consolidated undrained (CU), and drained (CD) conditions. The procedures for conducting the DSS test are described by Bjerrum and Landva (1966). For undrained shear tests, the vertical normal pressure must be varied to maintain the required conditions of constant volume.

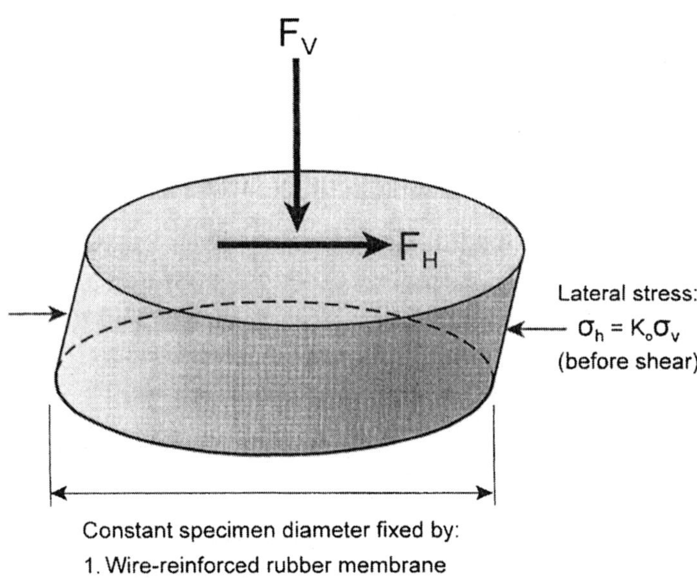

Figure 5.36 Schematic diagram of a direct simple shear (DSS) sample.

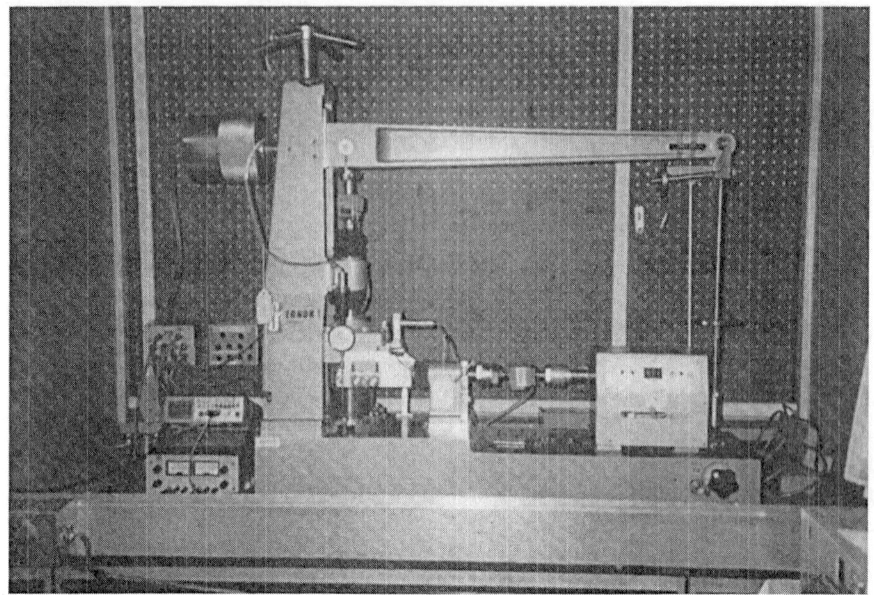

Figure 5.37 A simple shear device. (Courtesy of MIT.)

The Mohr circles associated with the DSS test are presented in Figure 5.38. In such a test, pure shear stresses are applied at the top and bottom of the sample during the deformation. Unlike a direct shear test, the failure plane will not be horizontal in a DSS test. This "freedom" to choose the inclination of a failure plane is believed to be a more realistic condition during slope movements.

The DSS test has gained considerable popularity over the last 20 years as a possible replacement for the direct shear test. The main advantage offered by the DSS device is that the stresses within the sample are likely to be fairly uniform, which can be expected to provide more realistic values of shear strength. However, serious reservations have been expressed about the stress distribution within the sample by Ladd (1981), Poulos (1981), La Rochelle (1981), and Saada and Townsend (1981).

5.7.12 Unsaturated Tests

In soils that are unlikely to become saturated, it might be possible to use higher shear strengths determined using unsaturated tests. These tests are not usually performed and so there is a general lack of familiarity as regards equipment, procedures, and results. Interested readers should consult the text by Fredlund and Rahardjo (1993) before proceeding with the tests that are briefly described below.

Direct Shear Test The shear box can be modified to carry out strength tests on unsaturated soils with control or measurement of suction. This procedure has the

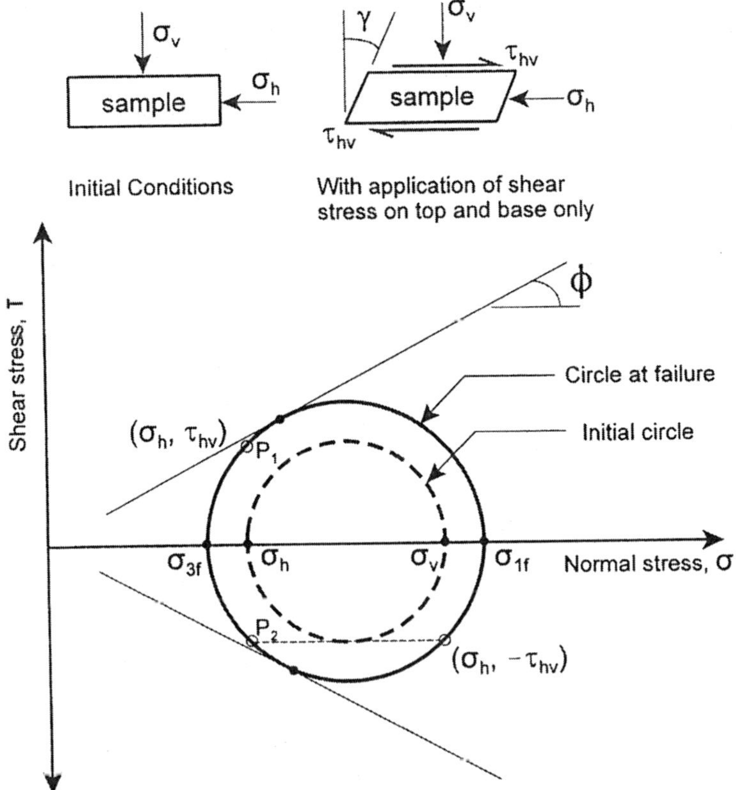

Figure 5.38 Mohr circles for a direct simple shear test.

advantage of requiring less time than the triaxial test due to the small thickness of the specimen. A schematic drawing for the modified direct shear device proposed by Gan et al. (1988) is presented in Figure 5.39.

Undrained Test In the undrained test, neither pore water nor pore air is allowed to drain in or out of the test specimen during shear. Pore water pressure is monitored through a porous ceramic base plate and pore air pressure is monitored through a layer of glass fiber cloth at the top of the specimen (Bishop and Bjerrum, 1960). This test has its limitations in practice because of the difficulty in measuring pore air pressure accurately and because of the problem of diffusion of pore air through the rubber membrane.

Controlled Suction Test The controlled suction test is a drained test with respect to both air and water of the specimen. Both water and air are allowed to enter or exit from the specimen in order to maintain constant water and air pressures. The cell, air, and water pressures are controlled separately, making various

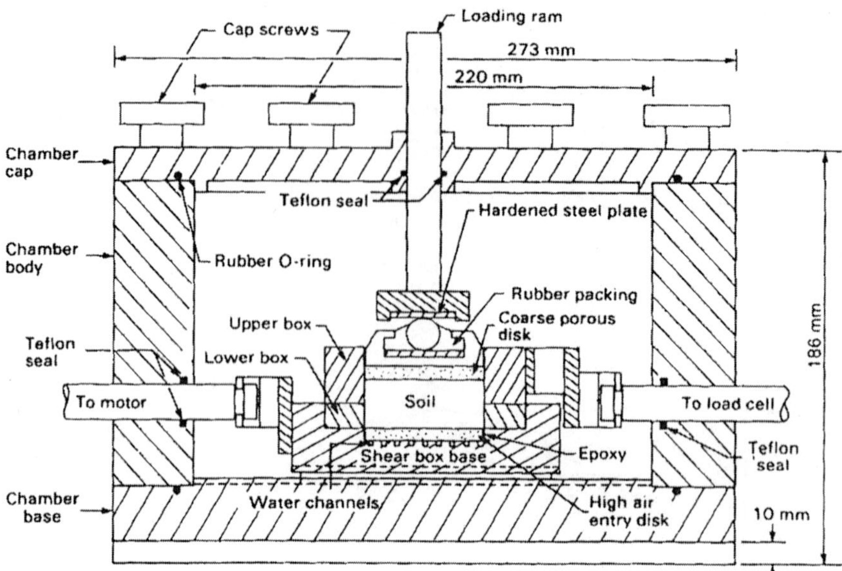

Figure 5.39 Schematic of a modified direct shear device for testing unsaturated soils. (From Gan et al., 1988, with permission.)

combinations of $(\sigma_n - u_a)$ and $(u_a - u_w)$ possible. The arrangement of apparatus is generally similar to that described in Fredlund (1981) and Ho and Fredlund (1982a and b).

Constant Water Content Test In the constant water content test, a specimen is drained with respect to air but undrained with respect to water, that is, the specimen is sheared at a known constant water content with no drainage of pore water while the pore air pressure is vented to a constant pressure (normally atmospheric). The arrangement is essentially the same as the controlled suction test except that instead of being connected to a mercury pot, the base of the specimen is connected to a pore pressure transducer where pore pressure is measured. The high air entry ceramic at the base of the specimen is still necessary.

Dead Load Test The dead load test is a stress controlled test designed to simulate the stress path of a rain-induced slope failure in the field (Brand, 1981). A test specimen is first anisotropically consolidated in a cell fitted with a dead load hanger to the expected anisotropic stresses in the field. Gradual reduction of suction within the specimen, by percolation, induces failure. The test was first conducted in Hong Kong in the early 1980s. Procedures for the dead load test can be found in Howat and Shen (1981) and Shen (1985).

5.8 PORE PRESSURE PARAMETERS

5.8.1 Skempton's Parameters

Effective stress analysis requires knowledge of pore pressures in the field. These pore pressures can be estimated if the changes in stress within the soil can be estimated reliably. For this estimation, the pore pressure parameters A and B, proposed by Skempton (1954), may be used to calculate the excess pore pressures:

$$\Delta u = B[\Delta\sigma_3 + A(\Delta\sigma_1 - \Delta\sigma_3)] \qquad \text{(Eq. 5-13)}$$

where Δu = excess pore pressure
A = pore pressure parameter A
B = pore pressure parameter B
$\Delta\sigma_1$ = change in major principal stress
$\Delta\sigma_3$ = change in minor principal stress

Pore pressure parameters A and B must be determined from laboratory tests or selected from past experience. For saturated soils, B approaches unity but its value decreases drastically with a decrease in the degree of saturation. Values of the parameter A are strain dependent (see, e.g., Figure 5.40) and usually reach a maximum at failure. Normally consolidated soils tend to generate positive excess pore water pressures during shear and will have a positive value of A. In contrast, heavily consolidated soils can be expected to generate *negative* excess pore water pressures and are thus characterized by a negative A parameter. Typical values of the A parameter at failure (A_f) are given in Table 5.5.

The value of A_f is greatly influenced by: (1) the level to which the soil has previously strained, (2) the initial stress in the soil, (3) the stress history of the soil, and (4) the type of total stress path to which the soil is subjected or the type of stress change, for example, load or unload (Lambe and Whitman, 1969).

5.8.2 Henkel's Parameters

An alternative approach for calculating the excess pore pressures, Δu, has been presented by Henkel (1960). The effects of the intermediate principal stresses, not directly considered by Skempton's method, are included with this procedure in the following form:

$$\Delta u = B\frac{\Delta\sigma_1 + \Delta\sigma_2 + \Delta\sigma_3}{3} + \frac{a}{3}\sqrt{(\Delta\sigma_1 - \Delta\sigma_2)^2 + (\Delta\sigma_2 - \Delta\sigma_3)^2 + (\Delta\sigma_3 - \Delta\sigma_1)^2} \qquad \text{(Eq. 5-14)}$$

and B and a are known as Henkel's coefficients. The B parameter is the same as the one used for Skempton's procedure and will depend on the degree of saturation. The

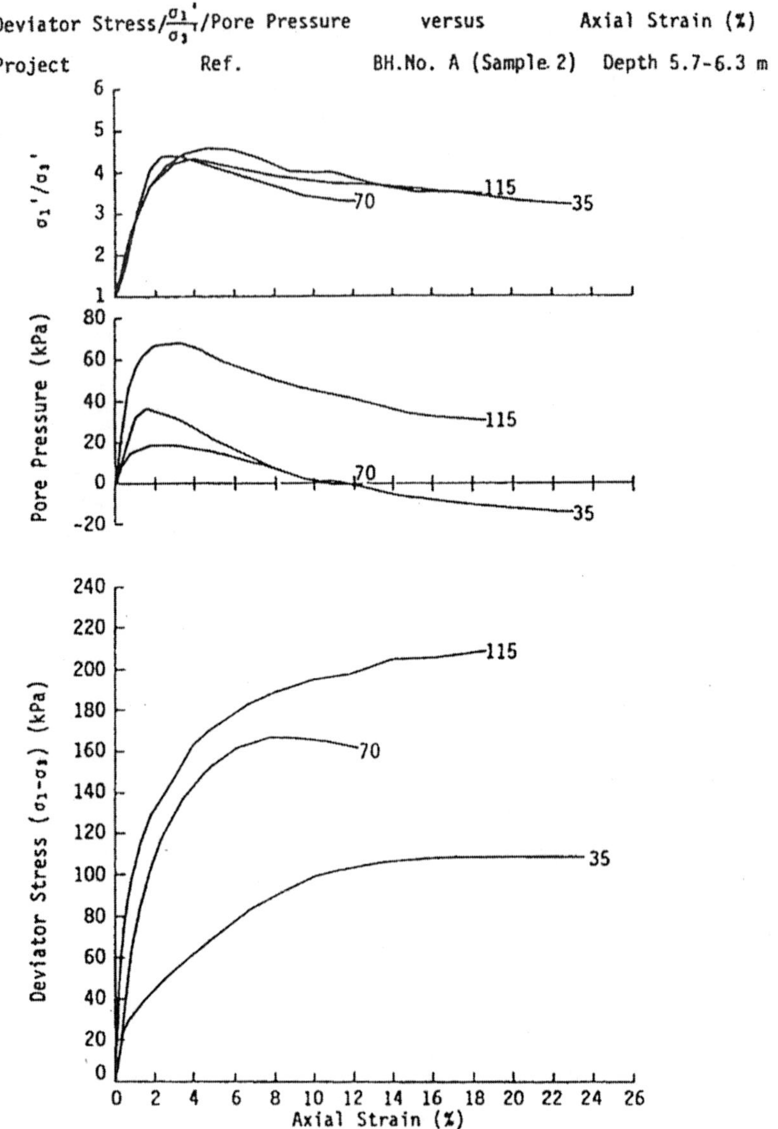

Figure 5.40 Typical triaxial test data. (Geotechnical Control Office, 1984.)

a parameter can be related to Skempton's *A* parameter using:

$$A_{AC/LC} = \frac{1}{3} + a\frac{\sqrt{2}}{3}$$

$$A_{AE/LE} = \frac{2}{3} + a\frac{\sqrt{2}}{3} \quad \text{(Eq. 5-15)}$$

TABLE 5.5 Values of A_f for Typical Soils

Type of Clay	Skempton's A Parameter at Failure (A_f)
Highly sensitive clays	0.75–1.5
Normally consolidated clays	0.5–1.0
Compacted sandy clays	0.25–0.75
Lightly overconsolidated clays	0.0–0.5
Heavily overconsolidated clays	−0.5–0.0
Compacted clay-gravels	−0.25–+0.25

Source: After Skempton (1954), reproduced by permission of ASCE.

where the A parameters would have been determined from the appropriate *AC, LC, AE*, or *LE* triaxial tests. However, it should be noted that the same limitations and reservations expressed for Skempton's parameters apply to Henkel's parameters.

5.9 INTERPRETATIONS OF STRENGTH TESTS

5.9.1 Triaxial Tests

The results of the triaxial tests can be plotted either as Mohr circles (Figure 5.1) or as a series of points representing the maximum shear stresses on the Mohr circles having the coordinates p, p', and q. The p'–q plot is preferred because it is clearer and simpler to construct and can also be used to plot stress paths (Lambe and Whitman, 1969). Graphical examples of typical test data collected during a triaxial test are shown in Figures 5.40 and 5.41. For this example, three CU triaxial tests, with pore water pressure measurements, were performed at three confining pressures. In this case the shear strength of the sample was reported as $c' = 3.1$ kilopascals and $\phi' = 36.6°$.

Strength envelopes determined from triaxial tests will often be nonlinear (Figure 5.3), and they will sometimes exhibit an apparent breakpoint in the region of a definite "critical" pressure. This is because the stress–strain behavior of the material is dependent upon the confining pressure under which it is sheared. Specimens that are tested at low confining pressures in the triaxial test tend to dilate during shear. At high confining pressures, specimens tend to compress. These different stress–strain behaviors are indicated clearly by the different shapes of their respective stress paths. In some soils, the critical pressure can be considered analogous to the maximum past pressure.

In the interpretation of triaxial test data, especially in the low stress range, the following sources of error should be kept in mind:

- Test specimens tend to bulge like a barrel at high strains, which leads to an overestimation of the shear strength.
- The saturation process prior to shear can lead to specimen disturbance in the form of unintended volume change. Dense (strong) materials tend to swell during

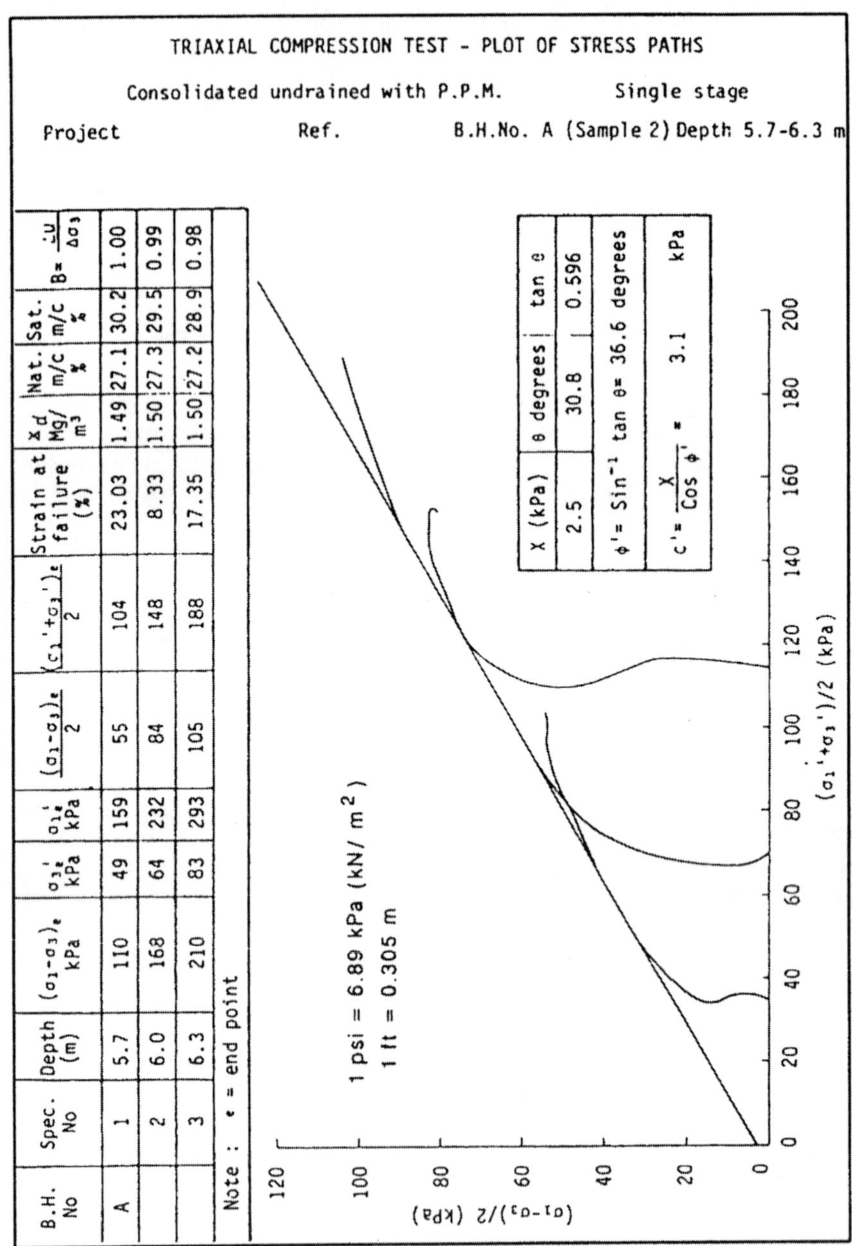

Figure 5.41 Typical stress path from a triaxial test. (Geotechnical Control Office, 1984.)

saturation, which frequently results in loss of strength. Loose (weak) materials may occasionally compress, giving misleadingly high shear strengths.

For undrained tests, failure can be defined either as the maximum deviator stress or as the maximum obliquity. For fully drained tests, these two criteria coincide. Research by Bureau of Reclamation on the shear characteristics of soils led to the selection of the maximum obliquity as the failure criterion for evaluating triaxial tests (Holtz, 1947).

It is important to remember that when a nonlinear strength envelope is simulated by a linear envelope, the linear portion of the envelope must span the appropriate design stress range.

Consolidated Undrained Test Data The practice of plotting a failure envelope that is tangential to a series of total stress failure circles is *not* correct as it implies that total stresses control strength (Figure 5.42). In reality, one must determine the total stresses acting on the single, unique failure plane that must be inclined at an angle of $(45 + \frac{1}{2}\phi')$ from the horizontal, where ϕ' is the angle of internal friction. These total stresses on the failure plane, as shown in Figure 5.42, are the ones that should be used to develop the undrained failure envelope. From Figure 5.42, one can see that the incorrect envelope is slightly unconservative. The undrained failure envelope is defined by the parameters, c_{cu} and ϕ_{cu}.

The undrained strength parameters, c_{cu} and ϕ_{cu}, should not be used directly in slope analysis. Instead, these parameters are used to determine the undrained shear strength, s_u, which will be a function of the in situ stresses.

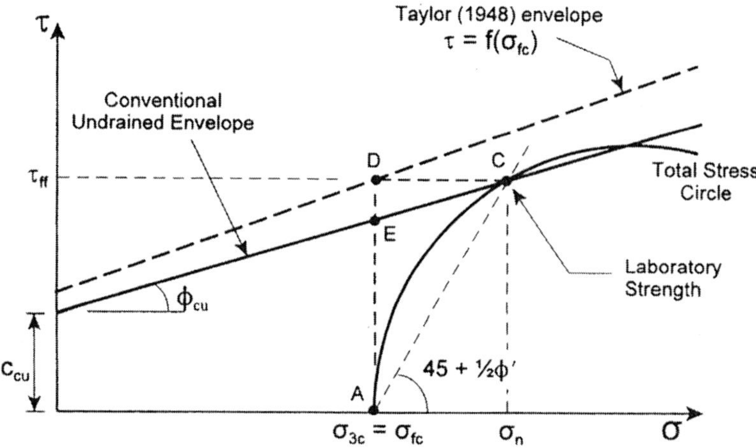

Figure 5.42 Consolidated-undrained design envelope, isotropic consolidation. (After Lowe, 1967, reproduced by permission of ASCE.)

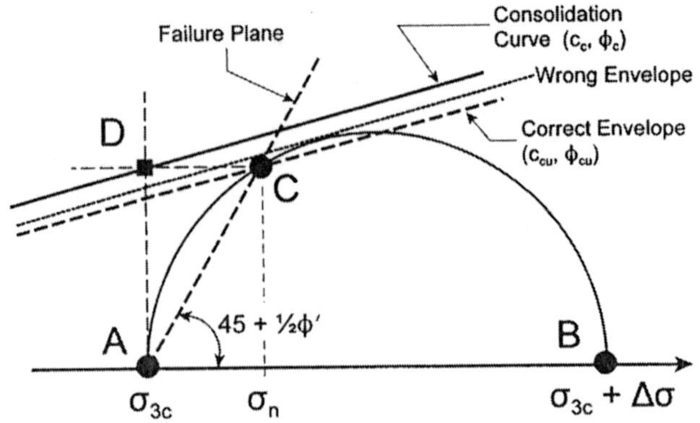

Figure 5.43 Correct approach for the consolidated undrained envelope.

In a CU test, the sample is initially consolidated to an isotropic consolidation stress, σ_{3c}, and then sheared under undrained conditions. The procedure to find the shear strength, referring to Figure 5.43, is: (1) draw the Mohr circle AB; (2) draw the failure plane, which is inclined at an angle of $(45 + \frac{1}{2}\phi')$ from point A, which has the coordinates $(\sigma_{3c}, 0)$; (3) extend this failure plane until it intersects the Mohr circle at C; and (4) the shear strength will then correspond to the ordinate value of AD. Please note that for this approach to work, ϕ' must be known. If pore pressures are not monitored during the CU test, an estimate of ϕ' will be necessary.

A relationship between c_{cu}, ϕ_{cu}, ϕ', and σ_{3c} can then be developed in the form

$$s_u = [c_{cu} + \sigma_{3c} \tan \phi_{cu}] \frac{\tan\left(45 + \frac{1}{2}\phi'\right)}{\tan\left(45 + \frac{1}{2}\phi'\right) - \tan \phi_{cu}} \qquad \text{(Eq. 5-16)}$$

which gives the undrained strength as a function of c_{cu}, ϕ_{cu}, ϕ', and σ'_{3c}. From Figure 5.43, we can also note that if the c_{cu} and ϕ_{cu} values had been determined from the tangential envelope approach, the *real* undrained strength would have been overestimated.

Alternatively, we can just plot the undrained strength as a function of the consolidation stress and determine the parameters of this straight line, c_c and ψ_c, as shown in Figure 5.43, such that

$$s_u = c_c + \sigma_{3c} \tan \psi_c \qquad \text{(Eq. 5-17)}$$

Before we can use Equations 5-16 or 5-17 to determine the undrained shear strength, the in situ consolidation stresses, σ_{3c}, must be evaluated. For practical convenience, the isotropic consolidation stress used in the laboratory is presumed to be equivalent to the mean, in situ stress. For a typical element at any depth, the horizontal effective stresses will equal $K_0\sigma'_v$, where K_0 is the coefficient of at-rest lateral earth pressure and σ'_v is the vertical effective stress. Then the mean of the two horizontal and one vertical stress components will be given by:

$$\sigma_{3c} = \tfrac{1}{3}[\sigma'_v + K_0\sigma'_v + K_0\sigma'_v]$$
$$= \tfrac{1}{3}\sigma'_v[1 + 2K_0] \qquad \text{(Eq. 5-18)}$$

So for a typical value $K_0 = 0.5$, the equivalent isotropic consolidation stress will be equal to $0.67\sigma'_v$. To illustrate the procedure for selecting undrained strengths from CU data, consider the following example.

Example 5.1 At a site where the groundwater level (GWL) is at a depth of 5 meters, an undisturbed sample was obtained from a depth of 10 meters. Two CU tests, with pore water pressure measurements, were performed on this sample with the following results:

Test 1: $\sigma_{3c} = 100$ kilopascals, $\Delta\sigma_f = 108.6$ kilopascals, ... So strength $s_u = 47.0$ kilopascals.

Test 2: $\sigma_{3c} = 200$ kilopascals, $\Delta\sigma_f = 164.6$ kilopascals, ... So strength $s_u = 71.3$ kilopascals.

Given: $\gamma_m = 16$ kilonewtons per cubic meter, $\gamma_{\text{sat}} = 18$ kilonewtons per cubic meter, and $\phi' = 30°$, determine the undrained strength at a depth of 10 meters.

The Mohr circles for the two CU tests are shown in Figure 5.44. The shear strength for each test was calculated using the details provided in Figure 5.43. The intersection of the failure surface line with the Mohr circles indicated shear strengths of 47.0 and 71.3 kilopascals, as shown in Figure 5.44, which also shows the total stress envelope as it passes through the intersection points on the circles. The consolidation envelope links the initial consolidation stress to the shear strength, and is also shown in Figure 5.44 for this data. The following undrained strength parameters can be determined

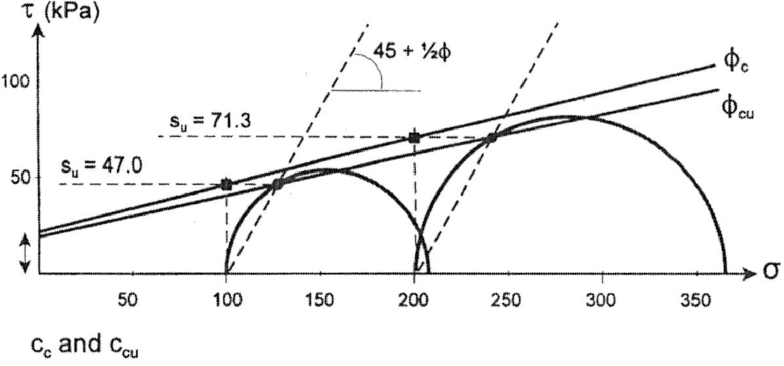

Figure 5.44 Example CU test data for determining the total stress parameters.

from Figure 5.44 for $\sigma_{3c} > 50$ kilopascals:

CU envelope:	$\phi_{cu} = 12^\circ$	$c_{cu} = 20.0$ kPa
Consolidation stress:	$\phi_c = 13.6^\circ$	$c_c = 22.8$ kPa
Tangential (wrong):	$\phi_{cu} = 12.6^\circ$	$c_{cu} = 21.2$ kPa

At 10 meters, the vertical stresses are:

$$\sigma_v = 170 \text{ kPa} \qquad u_0 = 50 \text{ kPa} \qquad \therefore \sigma_v' = 120 \text{ kPa}$$

For an at-rest coefficient of $K_0 = 0.5$, the average consolidation stress (in the ground), using Equation 5-18, can be estimated as:

$$\begin{aligned} \sigma_{3c} &= \tfrac{1}{3}\sigma_v' [1 + 2K_0] \\ &= 0.67 \times 120 = 80 \text{ kPa} \end{aligned}$$

So with the knowledge that $\sigma_{3c} = 80$ kilopascals, the undrained strength may be estimated using Equation 5-17:

$$s_u = 22.8 + 80 \tan(13.6^\circ) = \mathbf{42.2\ kPa}$$

So for slope stability analysis, an undrained shear strength of 42.6 kilopascals may be assigned at a depth of 10 meters. Similar calculations can be repeated for a series of soil layers, each with an undrained strength corresponding to the s_u value corresponding to its middepth. If we had used the data from the incorrect tangential fit, then Equations 5-17 would have predicted an undrained strength of 44.9 kilopascals, an overestimate of about 5 percent.

5.9.2 Direct Shear Tests

The results of the direct shear test are often plotted with shear stress versus normal stress, from which the effective cohesion, c', and the effective angle of friction, ϕ', can be obtained, as shown in Figure 5.45, for defining the Mohr–Coulomb envelope. This relationship is often nonlinear. Sometimes, variations of shear stress and horizontal displacement may also be plotted.

In calculating the shear and normal stresses, corrections should be made for the change in area of the shear plane throughout the test. One drawback of the direct shear test is the rapid changes in moisture content as shearing progresses. The time to failure in the drained direct shear test must be long enough to achieve essentially complete dissipation of excess pore pressure at failure.

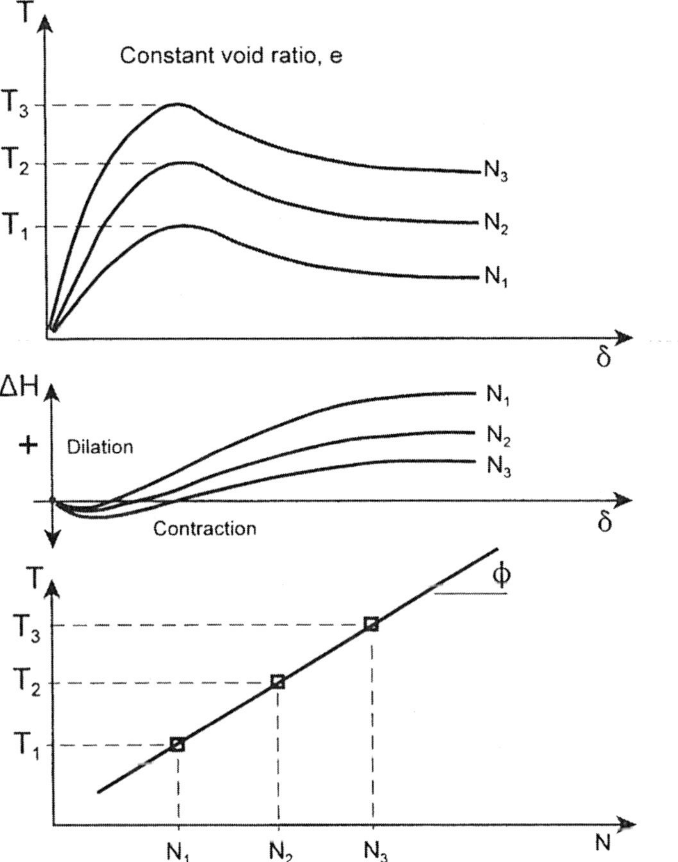

Figure 5.45 Typical direct shear test results performed at three different normal loads. N_1, N_2, and N_3.

When testing undisturbed firm or stiff clays, particularly at low normal loads, it may not be possible to transfer the required shear force to the specimen by means of the standard porous stone. The use of dentated porous stones, wire cloth, or abrasive grit between the stone and the specimen may be necessary to effect the transfer of shear stress (Department of Army, 1970).

Measuring residual strengths in the laboratory using multireversal shear box tests has sometimes given results different from those obtained from ring shear tests (Bishop et al., 1971). Data obtained by the conventional reversal shear box test, the ring shear test, or the Bromhead ring test for determination of ultimate (residual) strength indicate that the residual failure envelopes are often curved and that the curvature is most pronounced below about 200 kilopascals effective normal stress and in soils with high clay fraction (Hawkins and Privett, 1985). Therefore, residual strength tests should be performed in a stress range relevant to field conditions. For a shallow slide

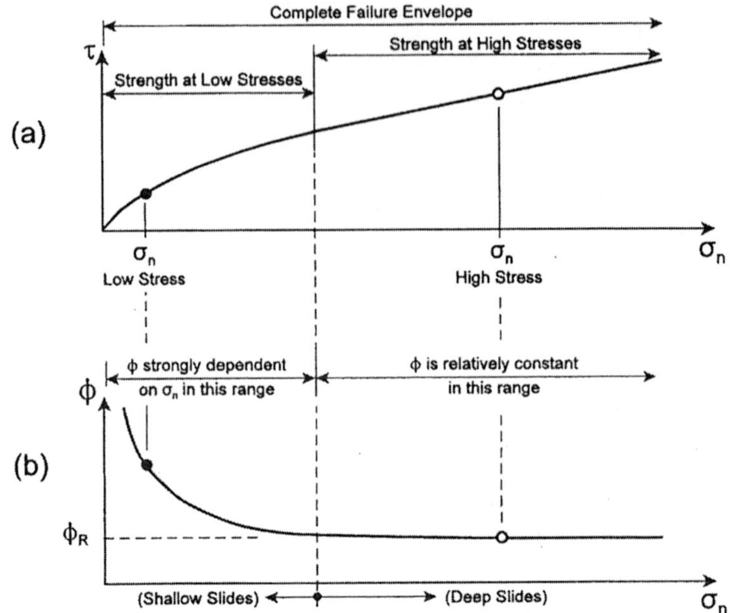

Figure 5.46 (a) Curved Mohr–Coulomb failure envelope, (b) ϕ values derived from the curved envelope. Note the higher ϕ values at lower normal stresses in comparison to the almost constant ϕ value at higher stresses.

of 3 to 10 feet, the residual friction angle may be higher than for a deep slide (Figure 5.46).

Results of direct shear tests and ring shear tests on cohesive soils reveal that the drained residual strength is related to the type of clay minerals and the quantity of clay-size particles. The liquid limit is used as an indicator of clay mineralogy, and the clay-size fraction indicates quantity of particles smaller than 0.002 millimeter. Therefore, increasing the liquid limit and clay-size fraction would decrease the drained residual strength.

Practical correlations between residual friction angle, gradation, clay fraction, clay mineralogy, and the index properties of cohesive soils are widely reported in the literature. Presented in Table 5.6 is a comparison of proposed and existing empirical correlations for drained residual friction angle. These empirical correlations are very useful in determining the drained residual friction angle given the basic soil properties such as index properties, clay fraction, and clay mineralogy. Based on many years of research, laboratory and field testing (Bishop et al. 1971; Collotta et al., 1989; Filz et al., 1992; Lambe, 1985; Lupini et al., 1981; Mesri and Cepeda-Diaz, 1986; Skempton, 1964, 1985; Voight, 1973), the following conclusions can be drawn:

- Residual strength behavior changes significantly as the clay content of cohesive soil increases.

TABLE 5.6 Comparisons of Existing Empirical Correlations for Drained Residual Friction Angle

Solid Index Properties	Reference
Liquid limit and clay fraction	Stark and Eid (1994)
Clay fraction	Collotta et al. (1989) Lupini et al. (1981) Skempton (1964, 1985)
Plasticity index	Kanji (1974) Lambe (1985) Mitchell (1993) Voight (1973)
Liquid limit	Bishop et al. (1971) Mesri and Cepeda-Diaz (1986) Mitchell (1993)

- The proportion of platy particles to spherical particles in the soil and the coefficient of interparticle friction of the platy particles are factors that control the residual shear strength of the cohesive soils.
- Correlations between residual strength and soil index properties and/or gradation cannot be general and should be used judiciously.

Stark and Eid (1994) suggested that the drained residual strength failure envelope is nonlinear for cohesive soils with a clay fraction greater than 50 percent, and a liquid limit between 60 and 220 percent. They also proposed that the drained residual friction angle is a function of liquid limit, clay fraction, and effective normal stress, as shown in Figure 5.47. To model the effective stress-dependent behavior of the

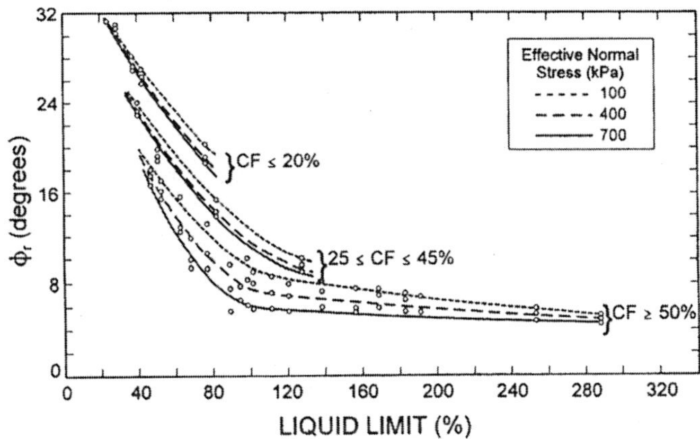

Figure 5.47 Empirical relationship between liquid limit and residual friction angle. (From Stark and Eid, 1994, reproduced by permission of ASCE.)

residual strength, Stark and Eid recommended that the nonlinear failure envelope or a secant residual friction angle corresponding to the average effective normal stress on the slip surface be used in a stability analysis.

5.9.3 Unsaturated Tests

Constant water and controlled suction tests are adequate for testing unsaturated soils. It is preferable to test unsaturated soils at their in situ moisture contents and in situ stress conditions. Dead load tests may be valuable for obtaining threshold suction (i.e., minimum suctions required for stability) and for studying the mechanism of failures that are initiated by infiltration.

Fredlund's stress state variables may be used to assign unsaturated strengths, which may be used directly for slope analysis using the *total cohesion* approach. To use this approach it is important to obtain the "correct" saturated ϕ' value at low confining pressures and the correct ϕ^b value from appropriate unsaturated tests. It must be noted that it is difficult to obtain the "corrected" saturated ϕ' value. Laboratory evidence (Howat and Shen, 1981) has shown that the conventional practice of back-saturation can destroy the fabric of an originally unsaturated soil specimen to such an extent that both dense and loose specimens can become medium dense. Typical values of ϕ^b for seven different soils were given earlier in Table 5.1, which also includes values of conventional c' and ϕ' Mohr–Coulomb parameters.

If it can be shown that suction is maintained in a slope after heavy rainfall, then it may be included in conventional stability analyses as an increase in apparent strength. If the variation of suction as a function of moisture content is high, as in the case of the volcanic and granitic residual soils in Hong Kong, then a small change in moisture content will result in a large change in suction. For such cases, it may be prudent to disregard the effects of suction for the slope stability analyses.

5.9.4 Selection of Design Shear Strengths

The in situ strength of the soil that will be used to evaluate the factor of safety (FOS) for a slope must selected carefully with full consideration given to the many complex facets that may have affected the laboratory determination of shear strength. Table 5.7 attempts to present a quantitative appraisal of the many features that may lead to an overestimation (unconservative) or an underestimation (conservative) of the in situ shear strength near a potential failure zone in a slope.

Johnson (1974) states:

> Evaluations of this type have little meaning unless they are done for a specific site and conditions, but even then required data are usually not available to permit reliable conclusions.

Thus it is desirable that the engineer consider each of the factors listed in Table 5.7 and assign a quantitative factor of confidence or uncertainty to each factor, as it may have influenced the reported laboratory test data.

TABLE 5.7 Factors Influencing Design Shear Strengths

Factor	Influence (percent)	Remarks
Sample disturbance of foundation materials (for relatively good, undisturbed samples)	−(5 to 20)	Remolding may increase strength of slickensided specimens. Disturbance is greatest for deep borings and soft soils
Effect of fissures in clays, especially highly overconsolidated clays and clay shales—effects not reflected in small samples	+(25 to 1,000)	Generally a factor for highly overconsolidated soils only
Rough caps and bases in laboratory tests	+5	
Triaxial compression tests instead of compression, simple shear, and extension tests	+(20 to 30)	Especially important for foundation soils
Triaxial instead of plane strain tests	−(5 to 8)	
Back-pressure saturation	Depends on field conditions	May cause grossly excessive strengths in CU tests at low confining stresses; conservative at high confining stresses
Conventional plotting of CU test data as total stress envelopes	−(15 to 20)	Effect may be eliminated by plotting data according to Taylor's method
Isotropic, instead of anisotropic, consolidation in CU triaxial compression tests		Values shown assume test envelopes for isotropic consolidation interpreted as τ_f versus of σ'_{fc}; i.e., as used by designers in stability analysis
(a) $A_f > \frac{1}{4}$	−(0 to 30)	
(b) $A_f < \frac{1}{4}$	+(0 to 20)	
Anisotropic material behavior—use of vertical instead of suitably inclined test samples	+(10 to 40)	

(*continued*)

TABLE 5.7 *(Continued)*

Factor	Influence (percent)	Remarks
Conventional rates of shear in the laboratory	+(5 to 200)	Effect depends on rate of testing, soil type, rate of consolidation in field, etc.
Progressive failure	+(0 to 20)	Depends on soil; mainly a factor for foundation soils. May be more serious than shown for some soils

Source: Johnson (1974), reproduced by permission of ASCE.

After evaluating the quality of the laboratory data, it is strongly recommended that the engineer compare the appraised shear strength values with available correlations and local experience. For granular soils, there are several correlations between field measurements such as SPT and CPT and drained strength. These have been discussed in the previous chapter on field exploration.

For fine-grained soils such as clays, the engineer may use the correlation between the plasticity index (PI) and the *peak* drained angle of internal friction shown in Figure 5.48. Alternatively, if presheared soils are being evaluated, Figure 5.47 and 5.49 may be used to correlate the PI with the residual angle of internal friction. Also, it should be noted that the cohesion component, c_r, will be negligible at large displacements where residual strength controls.

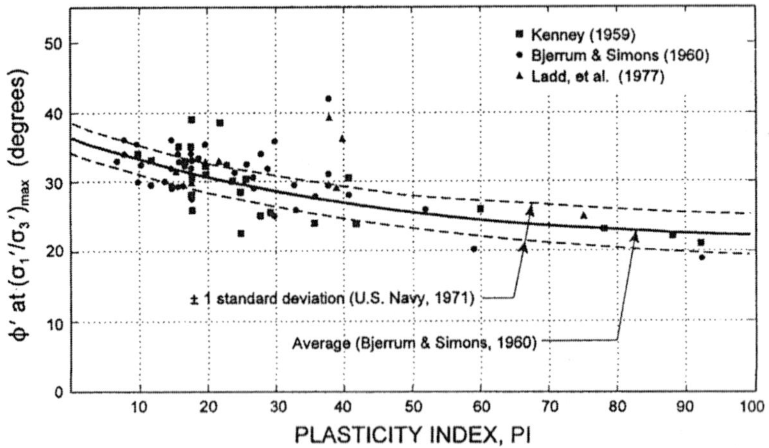

Figure 5.48 Correlation between plasticity index (PI) and angle of internal friction, ϕ'.

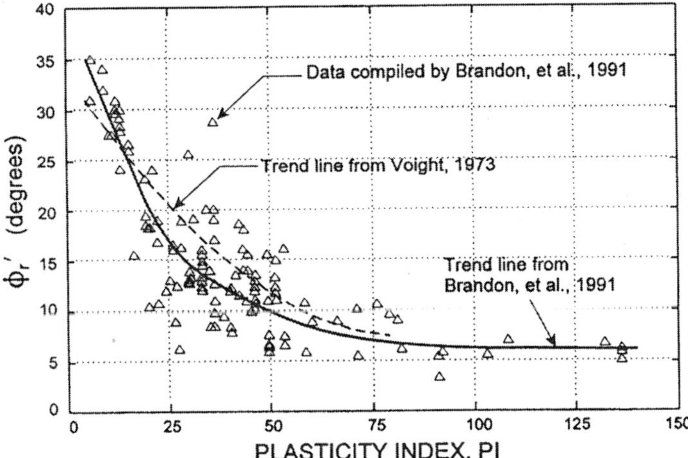

Figure 5.49 Correlation between plasticity index (PI) and the residual angle of friction, ϕ'_r. (From Filz et al., 1992.)

The undrained strength of a fine-grained soil will depend on the type of loading (i.e., compression or extension) and the overconsolidated ratio (OCR). For an OCR ≥ 4.0, the undrained strength in compression will be greater than the drained strength in the short-term, but this higher strength should not be used if the slope is expected to maintain long-term stability. Also, overconsolidated soils, will exhibit a cohesion value, c, below their preconsolidation stress. For normally consolidated soils, the undrained strength in compression will generally be about 50 percent of the drained strength, or $\phi_{cu} = \frac{1}{2}\phi'$.

As our database of undrained strengths grows, it has been found that for normally consolidated clays, the c_u/σ'_{vm} (or c_u/σ'_{vo}) is nearly a constant value equal to 0.23 ± 0.04 (Jamiolkowski et al., 1985; Mesri 1975). Typical values of the constants S and m, used by the SHANSEP approach, are given in Table 5.8.

If the engineer is reviewing laboratory test data for unsaturated soils, Table 5.1 may be used to assess the quality of the reported magnitude of the ϕ^b parameter. Admittedly, this table only includes data from seven different soils from eleven studies, but this database is likely to grow as more and more tests are performed for slopes that can be expected to remain unsaturated during their life.

5.10 OTHER PROPERTIES

In this section, several supplementary laboratory tests are reviewed on the basis of their indirect influence within the overall framework of slope stability and practical geotechnical engineering.

TABLE 5.8 Typical SHANSEP Parameters

Soil Type	Strength Ratio, S	Strength Exponent, m
Sensitive marine clays, $I_P < 30\%$, $I_L > 1$	0.20	1.0
Homogenous CL and CH sedimentary clays of low to moderate sensitivity, $I_P = 20$–80%	0.22	0.8
Northeastern U.S. varved clays	0.16 (DSS mode)	0.75
Sedimentary deposits of silts and organic soils (Atterberg limits plot below *A* line, *but excluding peats*) and clays with shells	0.25	0.8

Source: Ladd (1991), reproduced by permission of ASCE.

5.10.1 Consolidation Tests

Consolidation tests, performed according to ASTM D2435-90, are used to determine the compressibility (C_s and C_c), preconsolidation pressures (σ'_{vm}), and coefficient of consolidation (c_v) for fine-grained soils. Soils that contain a mixture of sands and gravels can be tested in a large oedometer or in a triaxial test. The tests are usually carried out on undisturbed samples but recompacted samples of fill may be used also.

Data from consolidation tests are used to develop the stress history of the subsoils and possibly to assign undrained shear strengths according to the SHANSEP procedure. For analysis of embankment stability over soft ground, the rate of consolidation, c_v, is used to estimate the rate of strength gain. The coefficient of consolidation, c_v, normally decreases with increasing consolidation; that is, with increasing effective stress. Thus it is important that the engineer select an average value for the anticipated stress range. It is suggested that the c_v be selected for a stress range that corresponds to approximately three-fourths of the difference between the initial and the final effective stresses (Transportation Research Board, 1975).

5.10.2 Permeability Tests

Laboratory permeability of a soil sample can be measured by constant-head or falling-head tests, according to ASTM D2434-68 and ASTM D5084-90, respectively. ASTM D2434-68 concerns a test for determining the coefficient of permeability of granular soils using a constant head permeameter, whereas ASTM D5084-90 discusses the procedure for determining the coefficient of permeability of clayey to silty soils using a flexible wall permeameter by either a constant-head or a falling-head method.

Both test methods apply to one-dimensional, laminar flow of water within saturated soil specimens. It is assumed that Darcy's law is valid and that the coefficient of permeability is essentially unaffected by hydraulic gradient. ASTM D5084-90 can provide a means for determining coefficient of permeability at a controlled level of effective stress.

The correlation between results obtained from these methods and the coefficients of permeabilities of in-place field materials has not been fully investigated. Experience has shown that flow patterns in small test specimens are not necessarily the same as those seen on larger scales and that coefficients of permeability measured on small test specimens are not necessarily the same as larger-scale values. Therefore, the results should be applied to field situations with caution. The results obtained from field tests, if correctly conducted, will generally be closer to the true permeability of the in situ material than the results of the laboratory tests.

Knowledge of the ability of a geologic formation to transmit water is essential in the planning of drainage systems to enhance slope stability or to correct landslide activities. This can be accomplished by carrying out laboratory and in-place permeability tests. The in-place permeability tests are discussed in Chapter 4.

5.10.3 Compaction Tests

The relationship between dry unit weight and moisture content for a soil used for fill may be determined by the standard compaction test as detailed in ASTM D698-78 and ASTM D1557-78. The two methods differ according to the size and drop-height of a rammer, and the number of compacted layers used in the tests. ASTM D698-78 specifies a 24.5-newton (5.5-pound) rammer with a 300-millimeter (12-inch) drop for compacting three layers of soil in a 102- or 154-millimeter mold, while ASTM D1557-78 recommends a 44.6-newton (10-pound) rammer with an 457-millimeter (18-inch) drop for compacting five layers of soil in a 102- or 154-millimeter mold.

Both test methods are applicable to soils with at least 70 percent or more passing the 19-millimeter ($\frac{3}{4}$-inch) sieve by weight. One of the hazards of these tests lies in the fact that they do not correctly determine the dry unit weight of soils that contain large proportion of oversized particles such as gravels and cobbles. A correction for oversized particles should be applied to account for gravels and cobbles. Another problem is that the standard compaction tests use the impact method, which simulates the action of a sheepsfoot roller. There is no field equivalent of static compaction that consists of pressing the soil into a mold with a given uniform pressure over the entire surface.

Some soils are susceptible to crushing during a compaction test; the extent to which this occurs can be checked by carrying out particle size distribution analyses on a sample before and after the test.

Proper compaction of fill is the most critical aspect in ensuring the stability of fill embankments. The properties of compacted fill to a significant extent depend on the method of compaction, the compaction effort, and the moisture content at which they are compacted. Different soil types have different dry unit weights versus optimum moisture content relationships (Figure 5.50). In general, compacted granular soils will have higher dry unit weights, in the range of 18 to 21 kilonewtons per cubic meter (115 to 135 pounds per cubic foot), than those of clayey to silty soils, which are in the range of 13.5 to 18 kilonewtons per cubic meter (85 to 115 pounds per cubic foot).

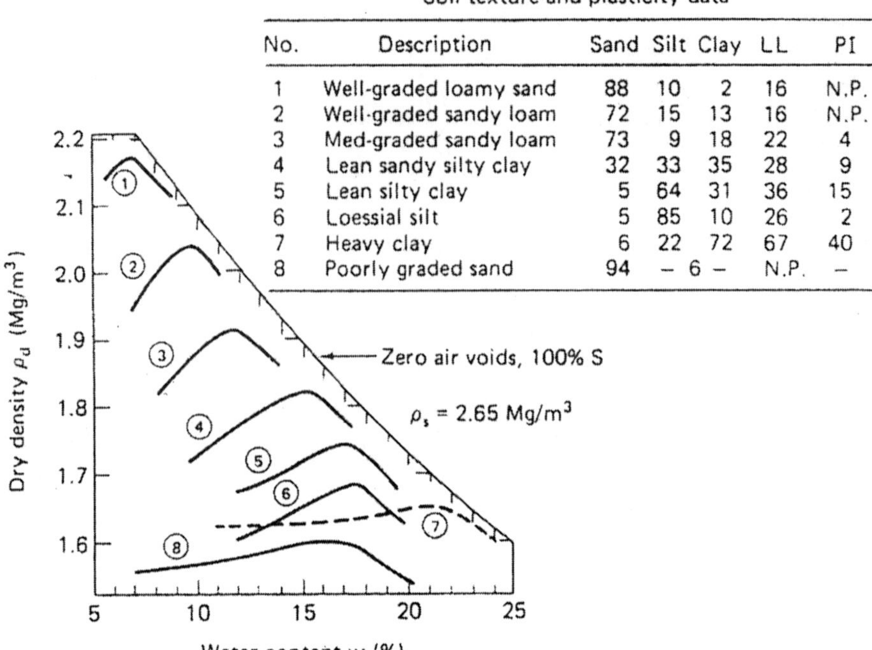

Soil texture and plasticity data

No.	Description	Sand	Silt	Clay	LL	PI
1	Well-graded loamy sand	88	10	2	16	N.P.
2	Well-graded sandy loam	72	15	13	16	N.P.
3	Med-graded sandy loam	73	9	18	22	4
4	Lean sandy silty clay	32	33	35	28	9
5	Lean silty clay	5	64	31	36	15
6	Loessial silt	5	85	10	26	2
7	Heavy clay	6	22	72	67	40
8	Poorly graded sand	94	– 6 –		N.P.	–

Figure 5.50 Water content–dry density relationship for eight soils compacted according to the standard Proctor method. (From Johnson and Sallberg, 1960.)

The corresponding optimum moisture contents for the granular and silty to clayey soils are generally in the order of 5 to 15 percent and 20 to 35 percent, respectively.

5.10.4 Classification Tests

Common Test Methods Common classification tests include the determination of moisture content (w), Atterberg limits (LL and PL), specific gravity, and particle size distribution. Samples chosen for the laboratory testing should be as representative as possible of the materials encountered in the subsurface investigation. They should be located within the presumed slip circle of the slope under consideration. Moisture content and specific gravity are used in the calculation of other soil properties, such as compressibility, dry unit weight, and degree of saturation. Liquid and plastic limit test results may give a quick estimate of the undrained shear strength of a soil sample through the use of empirical formulas. Particle size analyses are required for the design of filters, liners, and compaction control.

Moisture Content The standard method is described in ASTM D2216-90. The soil sample is first weighed using a balance and is then dried in an oven at 110°C.

Soils containing halloysitic clays, gypsum, or calcite, dehydrate or lose water from crystallization if dried at the standard temperature of 105° to 110°C. If the presence of a significant amount of these minerals is suspected, the effect on the determination of moisture content can be assessed by drying at various temperatures. Similarly, organic soils may decompose by oven drying at 110°C. The resulting water content of these organic soils may be determined at other temperatures.

In a humid environment, oven-dried samples reabsorb water very easily. It is therefore important that they be cooled in a desiccator before weighing.

Atterberg Limits Atterberg limits tests are normally made on the fraction of soil sample that passes the U.S. No. 40 sieve (425 microns). The tests are done in accordance with ASTM D4318-84.

For the determination of liquid limit (LL), about 150 to 200 grams of soil passing No. 40 sieve are dried at room temperature or in an oven at a temperature no higher than 60°C. The material is then placed level in a special cup and a groove is formed in the soil. The number of drops required to close the groove is recorded. The liquid limit is the water content (w_L) at which two halves of a soil cake prepared in a standardized manner in the cup of a standardized device flow together for a distance of 13 millimeters along the bottom of the groove when the cup is dropped 25 times from a distance of 10 millimeters at the rate of 2 drops per second onto a hard rubber pad.

For the determination of plastic limit (PL), about 20 grams of soil is selected from the material prepared for the liquid limit test. This portion of soil is rolled between the palm or fingers and a ground glass plate with just sufficient pressure to roll the mass into a thread of uniform diameter throughout its length until the diameter reaches 3 ± 0.5 millimeters. The plastic limit is the lowest moisture content (w_P) expressed as a percentage of the oven-dry weight of a soil at which the soil can be rolled into threads of 3-millimeter diameter without the threads breaking into pieces.

The LL and PL quantitatively describe the effect of varying water content on the consistency of fine-grained soils. With increasing water content, fine-grained soils pass from the solid to semisolid to plastic and then to a liquid state.

Soils containing significant proportions of halloysitic clays must be tested without previous drying and rewetting because the results obtained from a dried sample differ from those obtained from the sample in its natural condition. Testing of samples without drying is preferable for all soils, and a test report must state if the sample was dried.

Particle Size Analyses Particle size analyses can be determined by means of the procedures laid down in ASTM D422-63. Soil particles larger than approximately 0.075 millimeter (coarse fraction retained on the No. 200 sieve) are generally separated into particle-size ranges using a series of sieves. The grain size distribution of the soil is then determined by measuring the dry weight of material retained on each sieve. The size of particles less than 0.075 millimeter (fine fraction) is generally determined by a sedimentation process, using a hydrometer to secure the necessary data.

5.10.5 Interpretations of Classification Tests

Moisture Content Moisture content determined on soils obtained from borings is termed the natural moisture content (w_n). It varies with seasonal groundwater level fluctuations. When located under the groundwater table, the soil is fully saturated and the pore pressure is equal to the product of the depth below the groundwater table and the unit weight of water (Chapter 3). Cohesion of clayey soil is in part governed by moisture content. The natural moisture content may be used as an aid in predicting settlements and more importantly strength relative to soil slope stability analyses. A plot of w_n versus depth may detect the height of capillary rise, perched water, or the water table location if all pertinent factors are taken into account. In the same type of soil and below the groundwater table, changes in water content would be indicative of unit weight, grain size, and strength changes, which are the parameters required for soil slope stability analyses.

Moisture contents of clays and silty sands are generally in the range of 20 to 40 percent and 6 to 19 percent, respectively. Some residual soils, containing halloysite and allophane, often have much higher moisture content (in the order of 60 to 150 percent) as compared to common clays.

As mentioned above, cohesionless granular soils have lower moisture contents than those of cohesive clayey soils. In general, soil moisture contents do not have any detrimental effect on slope stability. But for clayey soils, the degree of saturation may affect slope stability. The presence of negative pore pressure (or soil suction) in unsaturated or partially saturated clayey soils would enhance the slope stability. As indicated in Section 5.4.1, the matrix suction $(u_a - u_w)$ increases the shear strength by $(u_a - u_w)$ tan ϕ^b where ϕ^b is the angle of internal friction with respect to matrix suction.

Atterberg Limits The plastic limit (PL) and liquid limit (LL) determined from Atterberg limit tests quantitatively describe the effect of varying water content on the consistency of fine-grained soils. The difference between the liquid limit and the plastic limit is called the plasticity index, PI (PI $= w_L - w_P$). The ratio of $(w_n - w_P)$ over PI is called the liquidity index, LI, where w_n is the natural water content. Low values of the LI suggest soils close to the plastic limit, and high values of LI are indicative of sensitive, metastable soils.

The plasticity index represents the range of water content in which the soil remains plastic. A plastic soil has a large value of PI. In general, the higher the PI, the greater the amount of clay particles present, and the more plastic the soil. The more plastic a soil, (1) the more compressible it will be; (2) the higher its shrinkage-swell potential is, and (3) the less permeable it will be.

A plasticity chart (Figure 5.51) proposed by Casagrande (1948) is routinely used in general practice for laboratory classification of fine-grained soils. The plasticity index and the liquid limit of the soil can be represented by a straight line (A line) from which soil classifications can be made.

It must be noted that all of the limits and indices, with the exception of the shrinkage limit, are determined on soils thoroughly worked into a uniform soil-water mixture. The limits therefore give no indication of particle fabric or of residual bonds between

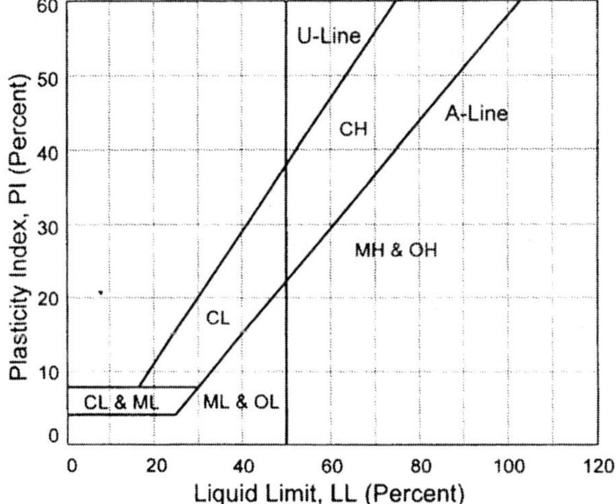

Figure 5.51 Plasticity chart. (ASTM D2487)

particles that may have been developed in the natural soil but that have been destroyed in preparing the specimen for the determinations of the limits. This is particularly true for residual soils like lateritic/saprolitic soils and andisols where they lie under the A line. Hence, the engineering characteristics of the residual soils are likely to be different from these of sedimentary soils.

A soil with natural water content, w_n, near the plastic limit (PL) is probably overconsolidated. Some stress had to be applied in the past to squeeze water out of the soil since its original deposition or formation. Similarly, clay may also be assumed to be overconsolidated if the liquidity index (LI) is less than 0.7 (Cheney and Chassie, 1982).

Overconsolidated clays generally show stress–strain relations that suggest general strain softening and are often of low permeability such that the time required to develop the fully drained condition may be many years. For the overconsolidated clays, the undrained strength is greater than the drained strength as the excess pore pressure immediately after construction is negative and will dissipate with time. As water content increases, strength decreases. Under field conditions, the long-term or drained conditions for overconsolidated clays are more critical than the short-term or undrained conditions.

There is no doubt that the values of LL and PL are associated with the mineralogical compositions of the clay fraction. The latter is studied by means of chemical, X-ray, and differential thermal analyses. Since the clay fraction may be present in a larger or smaller proportion, LL and PL are also affected by the amounts of clay particles. Therefore, values for the LL and PL cannot be directly correlated with the mineralogical composition without considering the proportion of the clay particles. This leads to the evaluation of the "activity" of clays, which is equal to the ratio of the plasticity index and the percentage by weight of particles smaller than 0.002 millimeter.

Knowledge of the plasticity of soils is particularly important from a slope stability standpoint because it is related to consistency and hence the strength of the soil in question, which governs the stability of a slope. This is particularly important in clayey soils because the shear deformation moduli as well as the shear strength change considerably with water content. Thus, for example, an increase of about 1 percent in the water content of a stiff plastic Neogene clay (with an illitic clay content) produces a decrease in shear strength of about 15 percent (Zaruba and Mencl, 1982).

Such an increase in the water content may follow a volume increase, as often occurs along slip surfaces. Therefore, it is essential to examine the water content data wherever clay is identified as this may enable the designer to locate the position of the slip surface. If samples cannot be obtained, downhole geophysical logging, as described in Chapter 4, may be used to assess water content variations qualitatively.

5.10.6 Shrink/Swell Potential

Some clayey soils exhibit a swelling potential upon absorbing water. Such soils may be identified by their high plasticity index (PI), fine particle characteristics, and mineralogy, such as montmorillonite, and are often called expansive soils. Other soils would expand upon the hydration of anhydrite to gypsum, or when they convert to ice as in permafrost.

It has been found (Holtz and Gibbs, 1954) that the colloid content of a soil and the Atterberg limit tests provide satisfactory indicators of the expansive characteristics of clays when considered together. Soils with high plasticity indices and colloid contents would have higher swell potential (see Table 5.9). On the other hand, a low shrinkage limit would show that a soil could begin volume change at a low moisture content, resulting in a higher swell potential.

The expansion of the soil volume is associated with the generation of swelling pressure, which may be determined using an oedometer as described in ASTM D4546-85. Similarly, some clayey soils have a shrinkage potential that, upon dehydrating, would shrink or reduce in volume. This is caused by capillary tension. For shrinkage to be a possibility, the soil in question has to be located above the normal water table. Additionally, at the time of construction, the soils must be at a water content appreciably higher than the shrinkage limit, which is defined as the water content at the point that shrinkage ceases, and the soil is no longer saturated.

Shrinkage can be measured by drying the soil and computing the relation between saturated water content and volume. In general, the lower the shrinkage

TABLE 5.9 Volume Change Potential

Volume Change	Shrinkage Limit	Plasticity Index (PI)
Probably low	12 or more	0–15
Probably moderate	1–12	15–30
Probably high	0–10	30 or more

Source: From Holtz and Gibbs (1954), reproduced by permission of ASCE.

limit, and the higher the plasticity index (PI), the greater the potential for shrinking (Table 5.9).

Shrink/swell is important from a slope stability analysis standpoint. This is because cohesive soils containing montmorillonitic minerals swell, and tensile cracks could develop subsequently. Water percolating into these cracks not only exerts an additional driving force into the slope, but also decreases the overall shear strength of the clayey soil. In addition, soils with a high percentage of montmorillonitic minerals quickly lose their cohesive component of strength upon saturation.

5.10.7 Slake Durability

The purpose of a slake durability test is to determine the effects of alternate drying and wetting on the durability of the soil/rock materials in the slope. Many shales and pyroclastic rocks and some argillaceous sandstones, conglomerates, and limestones deteriorate rapidly when subjected to periodic drying and saturation. The drying out of rock valley walls through a long period of exposure, particularly in arid and semiarid areas, followed by saturation, may result in sufficient deterioration to endanger the stability of the valley walls.

As for rocklike, clayey soils, these will disintegrate into a soft wet mass when a period of drying beyond the shrinkage limit is followed by a wet period with possible inundation by water. Slaking is considered as the principal weathering mechanism of stiff to very stiff rocklike clayey soils. The slaking process can penetrate into the rocklike soils depending on the presence of water channels and exposure to the atmosphere.

Samples suspected of slaking should be tested to determine their slake durability index (SDI), as in accordance with ASTM D4644-87 (Figure 5.52). Soils and rocks with high slake durability indexes (80 to 100 percent) would undergo little to no alteration when exposed to air or water (the SDI value is based on the second cycle

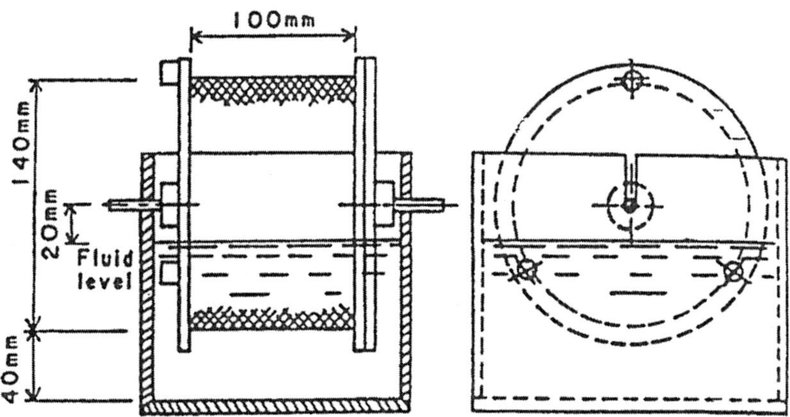

Figure 5.52 Slake durability testing device. (From ASTM D4644-87, copyright ASTM, reprinted with permission.)

slake durability). Materials with low slake durability indexes (0 to 40 percent) are likely to soften upon drying and wetting, and hence protective measures and support should be undertaken during excavation. Rocks containing weaker bentonite marker and bentonitic shale beds would also have lower slake durability. Thus protection of the rocks with a thin shotcrete layer may be advisable, and local stabilization and strengthening of bentonite and shale beds may be necessary. Soils with an SDI lower than 10 percent are of very low slake durability.

Based on the above, there is no doubt that slaking does affect slope stability, especially in clayey shale. According to Perry and Andrews (1982), slope instability due to slaking is attributed to inherent grain size, which results in the destruction of original rock structure and production of a sediment mass consisting of fine-grained particles. Disintegration of rocklike clayey soils and rocks by this slaking mode can take a few days to several years or longer.

Among all the rocks, shales, and pyroclastic rocks (like tuff) are the ones that undergo rapid deterioration upon drying and wetting. Among different types of shales, compaction shales are more susceptible to deterioration than cemented shales. Low-grade clay shales of the compaction type, like the Pierre shale of South Dakota, undergo complete disintegration after several cycles of drying and wetting. However, high-grade, coarser-textured compaction shales in Pennsylvania, West Virginia, and Ohio may require many cycles of drying and wetting before they undergo any changes (Burwell and Moneymaker, 1950).

The problems of the pyroclastic rocks (the ones that were formed from the fragmentary materials blown from the vents of volcanoes) are similar to those of shales. They are subject to rapid disintegration upon exposure to alternate drying and wetting, and, therefore, protective measures must be resorted to during slope construction to prevent serious spalling and raveling of the slope. Examples of this are slopes composed of tuff in Oregon, which suffer serious disintegration upon exposure and are of questionable stability.

5.10.8 Collapsibility

Soils that are susceptible to large decreases in bulk volume when they become saturated are called collapsing soils. Gibbs and Bara (1967) used a criterion to determine whether soils are susceptible to collapsibility; soils are more prone to collapse when the void space in the soil is greater than that corresponding to the liquid limit, which is indicated as Case I in Figure 5.53. Case III, on the contrary, indicates that soils with void space of less than the amount for the liquid limit would always be plastic even when saturated and would not be subject to collapse unless loaded.

Collapse may be triggered by water alone or by saturation and loading acting together. Soils with collapsible grain structures may be residual, water deposited, or eolian. In most cases, the deposits are characterized by loose structures or bulky shaped grains, often in the silt to fine sand size range. In dry climates, these soils are partially saturated and have cemented bonds at the contacts, which may break upon saturation. Hence, when their strengths are evaluated as in the case for slope stability analyses, consideration should be given to the water content under operating conditions.

In residual soils, collapsible grain structures form as a result of leaching of soluble and colloidal material. Water and wind deposited collapsible soils are usually found in arid and semiarid regions. Collapsible soils are often encountered in the midwestern and western United States.

5.10.9 Dispersivity

Dispersive soils are usually high in adsorbed sodium and disperse or deflocculate easily and rapidly in water of low salt content. The higher the percentage of sodium cation, the higher the susceptibility to dispersion. Highly dispersive clays frequently have the same Atterberg limits, gradation, and compaction characteristics as nondispersive clays. These clays plot above the A line on the plasticity chart and are generally of low to medium plasticity (LL = 30 to 50 percent, CL-CH classifications). Such clays generally have high shrink-swell potential, low resistance to erosion, and have low permeability in an intact state.

Dispersive clay soils are identified by running (1) the pinhole test, (2) the crumb test, and (3) the Soil Conservation Service dispersion test. Of these three tests, the pinhole test is the most reliable test (Sherard et al., 1976a and b). The test is described

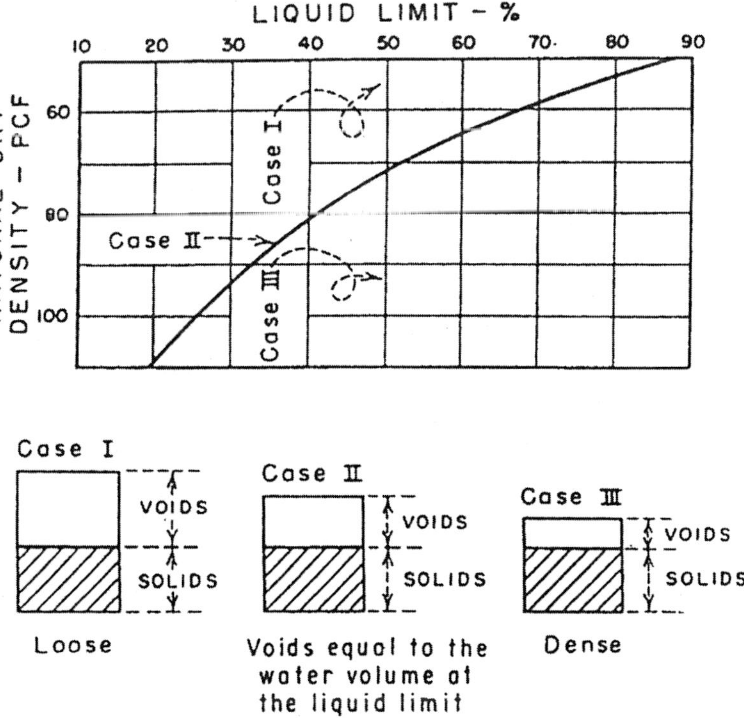

Figure 5.53 Criterion for evaluating potential for soil collapsibility. (From Gibbs and Bara, 1967, reproduced by permission of ASCE.)

in detail in ASTM D4647-87. Test specimens can be either disturbed or undisturbed soils.

This test has, however, several limitations:

- It is not applicable to soils with less than 12 percent fraction finer than 0.005 millimeter and with a plasticity index less than or equal to 4 (Ryker, 1977; Sherard et al., 1976b). This is because such soils generally have low resistance to erosion regardless of dispersive characteristics.
- The natural moisture content of the soil sample must be preserved to achieve consistent laboratory and field results (ASTM D4647-87).
- The test may not identify some dispersive clays in which the pore water contains less than 0.4 milligram per liter of total soluble salts that are more than 80 percent sodium salts (ASTM D4647-87).
- It may not be applicable to undisturbed samples of highly sensitive clays (Dascal et al., 1977).
- It uses distilled water at a pH of 5.5 to 7.0 as the eroding fluid. The use of water with various ionic concentrations and combinations will give different results (Arulanandan et al., 1975; Statton and Mitchell, 1977).

Dispersive soils are troublesome in terms of slope stability. The underlying soil mass of the slope often suffers from internal erosions when subject to localized water seepage. The internal erosions are not easily detected at the beginning until failures occur. Failures often take long period of time and happen in a progressive manner. Other problems involving dispersive soils include rainfall erosion on slopes, and erosion of channels (both lined and unlined) constructed in dispersive soils.

5.10.10 Chemical Tests

Sulfate Content Soils may contain certain amounts of naturally occurring sulfate. The amount would be significantly increased if they are subjected to polluting groundwater from industrial effluent or other contaminants. Sulfate tests are therefore necessary as sulfate can attack cement in either cement-stabilized soils or concrete. Procedures for determining total sulfate content of soils can be found in local test methods (e.g., California Test Method 417).

Total sulfate contents of more than 0.2 percent by weight in soil and 300 parts per million in groundwater are potentially aggressive. It is important to note that sulfate content is subject to seasonal and other variations, and condition of sampling.

Water used for circulation during rotary wash drilling may cause alterations in the chemical properties of groundwater. Sulfate tests on samples of groundwater taken when rotary wash drilling was used may, therefore, give results that are not representative of the ground conditions. Hollow stem augers should be used for drilling if the chemical properties of the groundwater are critical.

Sulfates are common in clays and peat. The sulfate contents for most of the other types of soils are small. Where present, it is aggressive to concrete/metallic water-carrying conduits and pipes. Special attention should be paid to potential corrosion of such conduits and pipe when they are located on the crest of a slope, as leakage under those circumstances may trigger slope instability.

Hydrogen Sulfide and Other Sulfides Mineral sulfides are contained in the soil; in addition, iron sulfide can be mineralized from organic sulfur compounds (remains of vegetation or animals)—for example, in clayey soils of tidal mud flats, sulfides can be mineralized from sewage. Hydrogen sulfide may be oxidized into free sulfuric acid by the action of oxygen and water on the one hand and by certain soil bacteria on the other (note: the presence of hydrogen sulfide frequently characterizes a soil as anaerobic). The sulfuric acid in turn could expedite the weathering process of the native soil/rock, causing a decrease in shear strength.

Acidity The acidity of soil and groundwater affects the rate of corrosion of metals and the deterioration of concrete. The standard electrometric method of determining acidity using a pH meter is described in ASTM G51-77. Generally, the acidity of most soils is negligible. Hence this test is usually carried out only in soils that are suspected of being contaminated.

Most soils have a pH value of 5 to 9. However, this factor in itself is of little value as a guide for the degree of aggressiveness if, for instance, the aeration factor is left out of consideration. The concentration of hydrogen ions indicates only the already dissociated portion and is not a good indication of the total active quantity of acid present in the soil. In soils having a pH value of <5 to 5, the impact of steel pipes or structures increases as the pH value decreases. Galvanized steel will not be affected in alkaline soils ($pH > 7$).

Carbonate Calcium carbonate acts as a cementing agent between individual mineral particles in most soil and rock. This would enhance the binding of clay minerals and increase the apparent cohesion of the material. Determination of carbonate can be made in accordance with ASTM D4373-90.

As for the case of corrosion, the action of calcium carbonate in the soil is strongly neutralizing. It renders the acids present in the soil largely harmless and lowers or removes the risk of corrosion. It also stimulates the formation of protective layers on metal pipes. Many soils or soils containing marl, due to their calcium carbonate content, are usually weakly aggressive or nonaggressive.

Specific Electrical Resistance The specific electrical resistance (opposite to the electrical conductivity) of the soil is one of the most important factors in determining soil aggressiveness. As the specific surface increases, the electrical conductivity of a water-bearing soil will increase, and the specific resistance will decrease (Table 5.10).

Dry soils are practically nonconducting. Electrical conductivity occurs only upon the access of water by way of dissociation of the electrolytes present. The conductivity then increases as the water content increases up to the point of saturation, thereafter remaining fairly constant. The exactly opposite progression applies to the specific resistance.

TABLE 5.10 Specific Electrical Resistance (Ω-cm)

Specific Resistance	Appraisal
<1,000	Very strongly aggressive
1,000–2,000	Strongly aggressive
2,300–5,000	Moderately aggressive
5,000–10,000	Weakly aggressive
>10,000	Nonaggressive

The electrical resistivity is also influenced by the salinity of the soil and may also be altered by the salinity of the groundwater. Groundwater salinity lies generally between 150 and 500 milligrams per liter, corresponding to a change in specific resistance of between 4,660 and 1,440 ohm-centimeters. The resistivity of potable water lies between about 2,000 and 2,500 ohm-centimeters, and that of sea water near 20 ohm-centimeters. Laboratory measurements of electrical resistivity, as a rule, produce lower results than in practice, especially where the soil contains a high proportion of sand.

Redox Potential The redox potential (reduction-oxidation potential) is the potential of a platinum electrode in the soil referenced to the value of the standard hydrogen electrode. It is also an indication of the aerobic or anaerobic states of soils. In the case of anaerobic soils, the redox potential is 0 millivolt, so that a risk of sulfate reduction exists if, besides the sulfate-reducing bacteria, sulfates and organic substances are present to promote bacteria growth. A low redox potential may develop in the presence of organic sewage or liquid manure. The relationship between redox potential and aggressiveness of the soil is shown in Table 5.11.

5.10.11 X-Ray Diffraction Analysis

Occasionally, we need to determine the composition of the soil material at the failure plane. If the clay minerals are the main concern, an X-ray diffraction analysis can identify these minerals and also give an indication of their overall percent presence. For this test, a small representative sample (say 20 grams) is dried, and then ground in a porcelain mortar and pestle so that the entire sample passes through a 115-mesh (124-micron) sieve. For a complete investigation, the sample is analyzed untreated and then after following several pretreatments, which consist of using different manganese- or potassium-based solvents and heating the sample. For more

TABLE 5.11 Redox Potential and Aggressiveness of Soils

Redox Potential (mV)	Appraisal
−100–+100	Strongly aggressive
+100–+200	Moderately aggressive
+200–+400	Weakly aggressive
+400–+500	Nonaggressive

information about X-ray diffraction and about the pretreatments, the reader should consult Moore and Reynolds (1989) and Whittig and Allardice (1986).

Interpretation The results of the X-ray diffraction are presented as a series of charts that plot the intensity as a function of the double angle for each pretreatment, as shown in Figure 5.54. For major peaks, Bragg's law may be used to calculate the basal spacings, which are often reported in angstroms. For example, in the case of montmorillonite minerals, the air-dried sample has a basal spacing of 12 to 15 angstroms. After pretreatment with ethylene glycol or glycerol, they expand to a basal value of 17 to 18 angstroms. If the sample is then oven-dried, the interlayer water is removed, and the spacing drops to about 10 angstroms. Similar changes occur for material prepared with magnesium or potassium solvents and at different temperatures. For example, the uppermost curve for the landslide material in Figure 5.54 corresponds to the X-ray diffraction result for sample pretreated with potassium and then heated to 550°C. In this case, the basal spacing of 10.1 angstroms indicated the presence of montmorillonite.

A quantitative estimate of the amount of different minerals present in a soil mass can also be made by comparing the relative peaks of the X-ray diffraction chart. Mitchell (1993) considers these estimates as only semiquantitative, unless corrections are applied for the influence of the differences in mass absorption coefficients of the various minerals present. Also, particle orientations, sample size, surface texture of the sample, and many other factors can affect the intensity of the diffracted rays. Essentially, X-ray diffraction can identify the mineral according to its basal spacing, but the area under the peak, or its intensity, can only provide an estimate of the overall percent composition.

By studying the clay mineralogy of landslide materials, it may be possible to determine the origins of the materials and also whether the material is heavily weathered through chemical weathering and alteration as a result of faulting.

5.11 QUALITY CONTROL/QUALITY ASSURANCE

Quality control/quality assurance of the shear strength parameters, from both in situ and laboratory tests, should be maintained in an attempt to minimize any human errors in identifying the soil type, and in obtaining and evaluating the shear strength parameters.

Tests for soil classification as well as for shear strength parameters should always be carried out in accordance with the standards laid down by ASTM or AASHTO, and should be conducted by competent personnel under the close supervision of a geotechnical engineer. Only tests done by standard procedures can be rationally evaluated and used confidently in slope stability analysis.

Laboratory soil samples are obtained from the ground by sampling from boreholes and sealing and transporting these samples to the laboratory. The degree of disturbance affecting the samples varies depending on the types of soils, sampling methods, and the skill of the driller. At best, some disturbances will occur from the

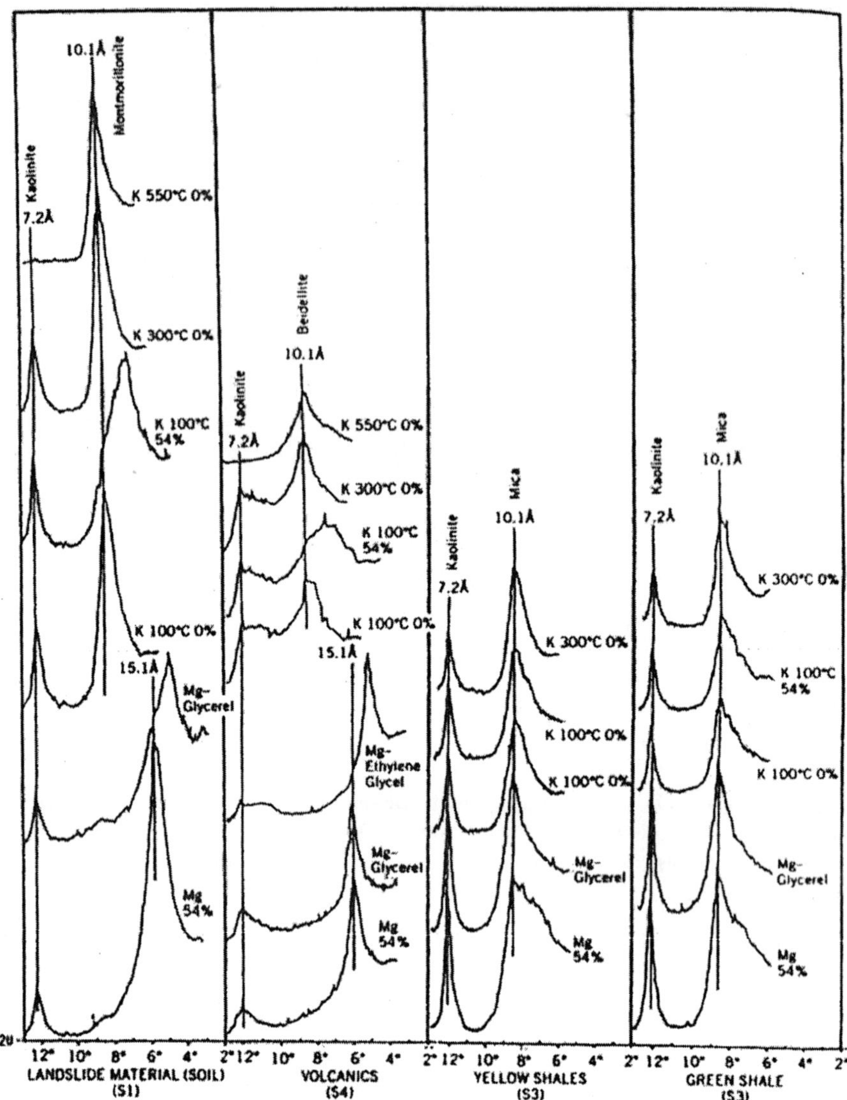

Figure 5.54 X-ray diffraction curves for the (001) planes of four soil samples S_1, S_2, S_3, and S_4. (After Tavitian, 1993. Reprinted from: A Anagnostopoulos., F. Schlosser, N. Kalteziotis, and R. Frank, Eds., *Geotechnical engineering of hard soils—soft rocks/La geotechnique des sols indureés—roches tendre*—Proceedings of an international symposium under the auspices of the International Society for Soil Mechanics and Foundation Engineering (ISSMFE), Athens, Greece, September 20–23, 1993. 1993–94. c. 2000 pp., Hfl. 350/US$195.00. Please order from A. A. Balkema, Old Post Road, Brookfield, Vermont 05036 (telephone: 802-276-3162; telefax: 802-276-3837).

removal of in situ stresses during sampling and laboratory preparation for testing. The disturbance generally tends to reduce the shear strength obtained from unconfined or unconsolidated tests, and to increase the strength obtained from consolidated tests. Hence, the shear strengths obtained from laboratory testing should be, where possible, compared to those measured in situ from the field.

It is essential to evaluate the quality of soil samples. This should be done by an experienced geotechnical engineer. High sample and test qualities are important because of their influence on subsequent stability analyses. Poor-quality samples, when tested in the laboratory, yield values for the preconsolidation stress and for the coefficient of consolidation that are too low. These effects may lead the designer to conclude erroneously that foundation treatment is required.

Because of the abundance of field and laboratory test data, it is important to rank the quality of specific tests on specific samples. Furthermore, it is essential to plot these data versus depth to determine the pattern of strength variation for each test type with depth and to assess the reliability of the data, that is, a CU test result that is lower than the UU test result at the same depth should be suspect. The general trend of in situ shear strength results should be to increase with depth in the same soil deposit. Clays that have been overconsolidated may only exhibit this increase at greater depths as the amount of preconsolidation increases shear strength.

The evaluation of shear strength parameters is no trivial task. These parameters are variable entities that must be evaluated carefully by experienced geotechnical engineers. It is essential to use common sense, enhanced by experience, to select the most appropriate shear strength parameters for evaluating the stability of a slope.

REFERENCES

Al-Khafaji, A. W., and O. B. Andersland, 1992. *Geotechnical Engineering and Soil Testing.* Philadelphia, PA: Saunders.

Arulanandan, K., P. Loganathan, and R. B. Krone, 1975. "Pore and Eroding Fluid Influences on Surface Erosion of Soil," *Journal of the Geotechnical Engineering Division*, ASCE, Vol. 100, No. GT-6, pp. 51–66.

Bishop, A. W., 1966. "The Strength of Soils as Engineering Materials," Sixth Rankine Lecture, *Geotechnique*, Vol. 16, No. 2, pp. 91–128.

Bishop, A. W., 1967. "Progressive Failure—With Special Reference to the Mechanism Causing It," Panel Discussion, *Proceedings of the Geotechnical Conference*, Oslo, Norway, Norwegian Geotechnical Institute, Vol. 2, pp. 142–150.

Bishop, A. W., and L. Bjerrum, 1960. "The Relevance of the Triaxial Test to the Solution of Stability Problems," *Research Conference on Shear Strength of Cohesive Soils*, Boulder, Colorado, ASCE, pp. 437–501, June.

Bishop, A. W., and D. J. Henkel, 1957. *The Measurement of Soil Properties in a Triaxial Test*, 1st ed. Edward Arnold. London.

Bishop, A. W., and D. J. Henkel, 1962. *The Measurement of Soil Properties in a Triaxial Test*, 2nd ed. Edward Arnold, 228 pp., London.

Bishop, A. W., G. E. Green, V. K. Garga, A. Andersen, and J. D. Brown, 1971. "A New Ring Shear Apparatus and Its Application to the Measurement of Residual Strength," *Geotechnique*, Vol. 21, No. 2, pp. 273–328.

Bjerrum, L., 1972. "Embankments on Soft Ground," *Proceedings, Speciality Conference on Performance of Earth and Earth-Supported Structures*, Purdue University, Lafayette, Indiana, ASCE, pp. 1–54, June.

Bjerrum, L., and T. C. Kenney, 1967. "Effect of Structure on the Shear Behavior of Normally Consolidated Quick Clays," *Proceedings of the Geotechnical Conference*, Oslo, Norway, Norwegian Geotechnical Institute, Vol. 2, pp. 19–27.

Bjerrum, L., and A. Landva, 1966. "Direct Simple Shear Tests on a Norwegian Quick Clay," *Geotechnique*, Vol. 26, No. 1, pp. 1–20.

Bjerrum, L., and N. E. Simons, 1960. "Comparison of Shear Strength Characteristics of Normally Consolidated Clays," *Research Conference of Shear Strength of Cohesive Soils*, Boulder, Colorado, ASCE, pp. 711–727, June.

Brand, E. W., 1981. "Some Thoughts on Rain-Induced Slope Failure," *Proceedings of the Tenth International Conference on Soil Mechanics and Foundation Engineering*, Stockholm, Sweden, Vol. 3, pp. 373–376.

Brandon, T. L., G. M. Filz, and J. M. Duncan, 1991. "Review of Landslide Investigation, Phase I—Part B, Olmsted Locks and Dam," A Report to the Louisville Distinct of the U.S. Army Corps of Engineers, June.

Bromhead, E. N., 1979. "A Simple Ring Shear Apparatus," *Ground Engineering*, Vol. 12, pp. 40–44.

Brown, G., 1961. *The X-Ray Identification and Crystal Structures of Clay Minerals*. London: Mineral Society (Clay Mineral Group).

Brown, G., and R. Farrow, 1956. "Vapor Glycerolation," *Clay Mineralogy Bulletin*, Vol. 63, pp. 43–45.

Burwell, E. B., and B. C. Moneymaker, 1950. "Geology in Dam Construction, Application of Geology to Engineering Practice," *Geological Society of America, Berkey Volume*, Harvard Soil Mechanics Series, S. Paige, Ed. Cambridge, Massachusetts: Harvard University, pp. 11–43.

Casagrande, A., 1948. "Classification and Identification of Soils," *Transactions, ASCE*, Vol. 113, pp. 901–992.

Cheney, R. S., and R. G. Chassie, 1982. *Soils and Foundations Workshop Manual*. Washington, D.C.: Federal Highway Administration, November, 340 pages.

Coatsworth, A. M., 1985. "A Rational Approach to Consolidated Undrained Triaxial Testing," *Proceedings of the 20th Regional Meeting, Engineering Group*, Guildford, U.K., Geological Society, Vol. 1, September 1984.

Collotta, T., R. Cantoni, V. Pavesi, E. Ruben, and P. C. Moretti, 1989. "A Correlation Between Residual Friction Angle, Gradation, and the Index Properties of Cohesive Soils," *Geotechnique*, Vol. 39, No. 2, pp. 343–346.

Dascal, O., G. Pouloit, and J. Hurtubise, 1977. "Erodibility Tests on a Sensitive Cemented Marine Clay," *Symposium on Dispersive Clays, Related Piping, and Erosion in Geotechnical Projects*, ASTM STP 623, Philadelphia, Pennsylvania: ASTM, pp. 74–93.

Department of Army, 1970. *Laboratory Soils Testing*, Engineer Manual, EM 1110-2-1906, Washington, DC.

Duncan, J. M., 1996. "State of the Art: Limit Equilibrium and Finite-Element Analysis of Slopes," *Journal of the Geotechnical Engineering Division*, ASCE, Vol. 122, No. GT-7, pp. 577–596.

Duncan, J. M., and A. L. Buchignani, 1973. "Failure of Underwater Slope in San Francisco Bay," *Journal of Soil Mechanics and Foundation Division*, ASCE, Vol. 99, No. SM9, pp. 687–703.

Duncan, J. M., and H. B. Seed, 1966. "Strength Variation Along Failure Surfaces in Clay," *Journal of Soil Mechanics and Foundation Division*, ASCE, Vol. 95, No. SM6, pp. 81–104.

Duncan, J. M., and T. D. Stark, 1992. "Soil Strengths from Back Analysis of Slope Failures," *Proceedings of the ASCE Specialty Conference on Stability and Performance of Slopes and Embankments II*, Berkeley, California, Geotechnical Special Publication 31, ASCE, pp. 890–904.

Duncan, J. M., A. L. Buchignani, and D. W. Marius, 1987. *An Engineering Manual for Slope Stability Studies*, Department of Civil Engineering, Virginia Polytechnic Institute and State University, Blacksburg, Virginia, March.

Filz, G. M., T. L. Brandon, and J. M. Duncan, 1992. "Back Analysis of the Olmsted Landslide Using Anisotropic Strengths," *Transportation Research*, Record 1343, TRB, National Research Council, pp. 72–78.

Fredlund, D. G., 1981. "The Shear Strength of Unsaturated Soil and Its Relationship to Slope Stability Problems in Hong Kong," *Hong Kong Engineer*, Vol. 9, No. 4, pp. 37–45.

Fredlund, D. G., and H. Rahardjo, 1993. *Soil Mechanics for Unsaturated Soils*. New York: Wiley.

Fredlund, D. G., N. R. Morgenstern, and A. Widger, 1978. "Shear Strength of Unsaturated Soils," *Canadian Geotechnical Journal*, Vol. 15, No. 3, pp. 313–321.

Gan, K. M., D. G. Fredlund, and H. Rahardjo, 1988. "Determination of the Shear Strength Parameters of an Unsaturated Soil Using the Direct Shear Test," *Canadian Geotechnical Journal*, Vol. 25, No. 3, pp. 500–510.

Geotechnical Control Office, 1984. *Geotechnical Manual for Slopes*, Civil Engineering Services Department, Hong Kong, May.

Gibbs, H. J., and J. P. Bara, 1967. "Stability Problems of Collapsing Soils," *Journal of the Soil Mechanics and Foundation Division*, ASCE, Vol. 93, No. 4, July, pp. 577–594.

Hawkins, A. B., and K. D. Privett, 1985. "Measurement and Use of Residual Shear Strength of Cohesive Soils," *Ground Engineering*, Vol. 18, No. 8, pp. 22–29.

Head, K. H., 1998. *Manual of Soil Laboratory Testing—Effective Stress Tests*, 2nd ed., Vol. 3. New York: Wiley, 442 pp.

Henkel, D. J., 1960. "The Shear Strength of Saturated Remolded Clays," *Proceedings of the Research Conference on Shear Strength of Cohesive Soils*, Boulder, Colorado, ASCE, pp. 553–554, June.

Henscher, S. R., J. B. Massey, and E. W. Brand, 1984. "Application of Back Analysis to Some Hong Kong Landslides," *Proceedings of the 4th International Symposium on Landslides*, Vol. 1, Toronto.

Hirschfeld, R. C., 1963. "Stress-Deformation and Strength Characteristics of Soils," Harvard University (unpublished), Cambridge, Massachusetts, 87 pp.

Ho, D. Y. F., and D. G. Fredlund, 1982a. "Increase in Strength Due to Suction for Two Hong

Kong Soils," *Proceedings of the ASCE Specialty Conference on Engineering and Construction in Tropical and Residual Soils*, Honolulu, Hawaii, pp. 263–295.

Ho, D. Y. F., and D. G. Fredlund, 1982b. "A Multi-Stage Triaxial Test for Unsaturated Soils," *Geotechnical Testing Journal*, Vol. 5, pp. 18–25.

Ho, D. Y. F., and D. G. Fredlund 1982c. "Strain Rates for Unsaturated Soil Shear Strength Testing," *Proceedings of the 7th Southeast Asian Geotechnical Conference*, Hong Kong, pp. 787–803.

Hoek, E., and J. W. Bray, 1977. *Rock Slope Engineering*. London: Institute of Mining and Metallurgical Engineering.

Holtz, W. G., 1947. "The Use of the Maximum Principal Stress Ratio as the Failure Criterion in Evaluating Triaxial Shear Tests on Earth Materials," *5th Annual Meeting of the American Society for Testing Materials*, Vol. 47, pp. 1067–1076, June.

Holtz, W. G., 1960. Discussion to session "Testing Equipment, Techniques, and Errors," *Proceedings of the Research Conference on Shear Strength of Cohesive Soils*, Boulder, Colorado, ASCE, pp. 997–1002, June.

Holtz, W. G., and H. J. Gibbs, 1954. "Engineering Properties of Expansive Clays," Award-Winning ASCE Papers in Geotechnical Engineering, 1950–1959, *Journal of the Soil Mechanics and Foundation Division*, ASCE, pp. 256–264.

Holtz, W. G., and H. J. Gibbs, 1956. "Triaxial Shear Tests on Previous Gravelly Soils," *Journal of the Soil Mechanics and Foundation Division*, ASCE, Vol. 82, No. SM1, pp. 1–22.

Holtz, R. D., and W. D. Kovacs, 1981. *An Introduction to Geotechnical Engineering*. Englewood Cliffs, New Jersey: Prentice-Hall.

Howat, M. D., and J. M. Shen, 1981. *Suction Shear Strength and Suction Volume Change in Anisotropically Confined Granitic Fill Materials*, Geotechnical Control Office, Materials Division Report No. 27, Hong Kong.

Jackson, M. L., 1956. *Soil Chemical Analysis—Advanced Courses*. Madison, Wisconsin: published by the author.

Jamiolkowski, M., C. C. Ladd, J. T. Germaine, and R. Lancellotta, 1985. "New Developments in Field and Laboratory Testing of Soils," *Proceedings of 11th International Conference in Soil Mechanics and Foundation Engineering*, San Francisco, California.

Johnson, J. J., 1974. "Analysis and Design Relating to Embankments," *Proceedings of the Conference on Analysis and Design in Geotechnical Engineering*, University of Texas, Austin, Texas, ASCE, pp. 1–48, June 9–12.

Johnson, A. W., and J. R. Sallberg, 1960. "Factors That Influence Field Compaction of Soils," *Bulletin 272*, Highway Research Board, 206 pp.

Kanji, M. A., 1974. "The Relationship Between Drained Friction Angles and Atterberg Limits of Natural Soils," *Geotechnique*, Vol. 24, No. 4, pp. 671–674.

Kenney, T. C., 1967a. "The Influence of Mineral Composition on the Residual Strength of Natural Soils," *Proceedings of the Geotechnical Conference*, Vol. 1, Oslo, Norway, Norwegian Geotechnical Institute, pp. 123–129.

Kenney, T. C., 1967b. "The Strength Along Structural Discontinuities in Stiff Clays," *Proceedings of the Geotechnical Conference*, Vol. 2, Oslo, Norway, Norwegian Geotechnical Institute, pp. 29–46.

Kenney, T. C., 1959. Discussion of "Geotechnical Properties of Glacial Lake Clays," by T. H. Wu, *Journal of the Soil Mechanics and Foundations Division*, ASCE, Vol. 85, No. SM3, pp. 67–69.

Kenney, T. C., 1977. "Residual Strengths of Mineral Mixtures," *Proceedings of the 9th International Conference of Soil Mechanics*, Vol. 1, Tokyo, pp. 155–160.

Krahn, J., D. G. Fredlund, and M. J. Klassen, 1989. "Effect of Soil Suction on Slope Stability at Notch Hill," *Canadian Geotechnical Journal*, Vol. 26, No. 2, pp. 269–278.

Kulhawy, F. H., and Mayne, P. W., 1990. *Manual on Estimating Soil Properties for Foundation Design*, Electric Power Research Institute, Palo Alto, California, EL-6800, Project 1493-6, Final Report, August.

Ladd, C. C., 1972. "Test Embankment on Sensitive Clay," *Proceedings of the Speciality Conference on Performance of Earth and Earth-Supported Structures*, Purdue University, Lafayette, Indiana, ASCE, pp. 101–128, June.

Ladd, C. C., 1981. "Discussion on Laboratory Shear Devices," *Laboratory Shear Strength of Soil*, ASTM Special Technical Publication 740, R. N. Young and F. C. Townsend, Eds. Philadelphia, Pennsylvania: ASTM, pp. 643–652.

Ladd, C. C., 1991. "Stability Evaluation During Staged Construction," *Journal of the Geotechnical Engineering Division*, ASCE, Vol. 117, No. GT4, pp. 540–614.

Ladd, C. C., and L. Edgers, 1972. "Consolidation Undrained Direct Simple Shear Tests on Saturated Clays," Department of Civil Engineering, Massachusetts Institute of Technology, Cambridge, Massachusetts, *Research Report R72–82*, No. 284.

Ladd, C. C., and R. Foott, 1974. "New Design Procedure for Stability of Soft Clays," *Journal of the Geotechnical Engineering Division*, ASCE, Vol. 100, No. GT7, pp. 763–786.

Ladd, C. C., and T. W. Lambe, 1963. "The Strength of Undisturbed Clays Determined from Undrained Tests," *Laboratory Shear Testing of Soils*, ASTM Special Technical Publication No. 361, pp. 342–371.

Ladd, C. C., D. H. Connell, L. Edgers, and C. E. Williams, 1972. "Engineering Properties of Soft Foundation Clays at Two South Louisiana Levee Sites," Research Report R72-26, Soils Publication 304 to U. S. Army Corps of Engineers, December.

Ladd, C. C., R. Foott, K. Ishihara, F. Schlosser, and H. G. Poulos, 1977. "Stress Deformation and Strength Characteristics," State-of-the-Art Report, *Proceedings of the 9th International Conference of Soil Mechanics and Foundation Engineering*, Vol. 2, Tokyo, pp. 421–494.

La Gatta, D. P., 1970. "Residual Strength of Clays and Clay Shales by Rotational Shear Tests," *Harvard Soil Mechanics Series*, No. 86. Cambridge, Massachusetts: Harvard University.

Lambe, T. W., 1964. "Methods of Estimating Settlement," *Journal of Soil Mechanics and Foundation Division*, ASCE, Vol. 90, No. SM5, pp. 43–67.

Lambe, T. W., 1967. "Stress Path Method," *Journal of Soil Mechanics and Foundations Division*, ASCE, Vol. 93, No. SM6, pp. 309–331.

Lambe, T. W., 1985. "Amuay Landslides," *Proceedings, 11th International Conference on Soil Mechanics and Foundation Engineering*. Rotterdam, The Netherlands: A. A. Balkema, pp. 137–158.

Lambe, T. W., and R. V. Whitman, 1969. *Soil Mechanics*. New York: Wiley.

La Rochelle, P., 1981. "Limitation of Direct Simple Shear Test Devices," *Laboratory Shear Strength of Soil*, ASTM Special Technical Publication 740, R. N. Young and F. C. Townsend, Eds. Philadelphia, Pennsylvania: ASTM, pp. 653–658.

Lee, K. L., and R. A. Morrison, 1970. "Strength of Anisotropically Consolidated Compacted Clay," *Journal of Soil Mechanics and Foundation Division*, ASCE, Vol. 96, No. 5M6, pp. 2025–2043.

Lowe, J., 1967. "Stability Analysis of Embankments," *Journal of Soil Mechanics and Foundation Division*, ASCE, Vol. 93, No. 5M4, July, pp. 1–34.

Lowe, J., and T. C. Johnson, 1960. "Use of Back Pressure to Increase Degree of Saturation of Triaxial Test Specimens," *Proceedings of the Research Conference on Shear Strength of Cohesive Soils*, Boulder, Colorado, ASCE, pp. 819–836, June.

Lowe, J., and L. Karafiath, 1960a. "Stability of Earth Dams Upon Drawdown," *Proceedings of the 1st Pan American Conference on Soil Mechanics and Foundation Engineering*, Mexico City, pp. 537–552.

Lowe, J., and L. Karafiath, 1960b. "Effect of Anisotropic Consolidation on the Undrained Shear Strength of Compacted Clays," *Proceedings of the Research Conference on Shear Strength of Cohesive Soils*, Boulder, Colorado, ASCE, pp. 837–858, June.

Lumb, P., 1964. "Multi-Stage Triaxial Tests on Undisturbed Soils," *Civil Engineering Public Works Review*, May.

Lupini, J. F., A. E. Skinner, and P. R. Vaughan, 1981. "The Drained Residual Strength of Cohesive Soils," *Geotechnique*, Vol. 31, No. 2, pp. 181–213.

MacEwan, D. M. C., 1944. "Identification of the Montomorillonite Group of Minerals by X-Rays," *Nature*, Vol. 154, pp. 577–578.

Mesri, G., 1975. Discussion of "New Design Procedures for Stability of Soft Clays," by C. C. Ladd and R. Foott, *Journal of the Geotechnical Engineering Division*, ASCE, Vol. 101, No. GT4, pp. 409–412.

Mesri, G., and M. E. M. Abdel-Ghaffar, 1993. "Cohesion Intercept in Effective Stress-Stability Analysis," *Journal of the Geotechnical Engineering Division*, ASCE, Vol. 119, No. GT-8, pp. 1229–1249.

Mesri, G., and A. F. Cepeda-Diaz, 1986. "Residual Shear Strength of Clays and Shales," *Geotechnique*, Vol. 36, No. 2, pp. 269–274.

Mitchell, J. K., 1993. *Fundamentals of Soil Behavior*, 2nd ed. New York: Wiley.

Moore, P. J., 1970. "The Factor of Safety against Undrained Failure of a Slope," *Soils and Foundations*, Japan, Vol. 10, No. 3, pp. 81–91.

Moore, D. M., and R. C. Reynolds, Jr., 1989. *X-Ray Diffraction and the Identification and Analysis of Clay Minerals*. New York: Oxford University Press, 332 pp.

Perry, E. F., and D. E. Andrews, 1982. "Slaking Modes of Geologic Materials and Their Impact on Embankment Stabilization," *Transportation Research Record 873*, Washington, DC.

Poulos, S. J., 1981. "Discussion of Soil Testing Practices," *Laboratory Shear Strength of Soil*, ASTM Special Technical Publication 740, R. N. Young and F. C. Townsend, Eds. Philadelphia, Pennsylvania: ASTM, pp. 659–666.

Prevost, J. H., and K. Hoeg, 1976. "Reanalysis of Simple Shear Soil Testing," *Canadian Geotechnical Journal*, Vol. 13, No. 4, pp. 418–429.

Ross, C. S. 1960. "Montmorillonite Group Minerals," *Proceedings of the 9th National Conference on Clay and Clay Minerals*, A. Swineford, Ed., London: Pergamon, pp. 225–229.

Ross, C. S., and M. M. Mortland, 1966. "A Soil Beidellite," Soil Science Society of America, Vol. 30, pp. 337–343.

Ryker, N. L., 1977. "Encountering Dispersive Clays on S.C.S. Projects in Oklahoma," *Symposium on Dispersive Clays, Related Piping and Erosion in Geotechnical Projects*, ASTM STP 623, pp. 370–389.

Saada, A. S., and F. C. Townsend, 1981. "State of the Art: Laboratory Strength Testing of Soils," *Laboratory Shear Strength of Soil*, ASTM Special Technical Publication 740, R. N. Young and F. C. Townsend, Eds. Philadelphia, Pennsylvania: ASTM, pp. 7–77.

Seed, H. B., J. K. Mitchell, and C. K. Chan, 1960. "The Strength of Compacted Cohesive Soils," *Proceedings of the Research Conference on Shear Strength of Cohesive Soils*, Boulder, Colorado, ASCE, pp. 877–964, June.

Shen, J. M., 1985. "GCO Research into Unsaturated Shear Strength 1978–1982," Special Project Division, *Report No. RR 1/85*, Geotechnical Control Office, Hong Kong, February.

Sherard, J. L., R. S. Decker, and N. L. Ryker, 1972. "Piping in Earth Dams of Dispersive Clay," *Proceedings of the Specialty Conference on Performance of Earth and Earth-Supported Structures*, Purdue University, Lafayette, Indiana, ASCE, pp. 589–626, June 11–14.

Sherard, J. L., L. P. Dunnigan, and R. S. Decker, 1976a. "Identification and Nature of Dispersive Soils," *Journal of the Geotechnical Engineering Division*, ASCE, Vol. 102, No. GT4, pp. 287–301.

Sherard, J. L., L. P. Dunnigan, R. S. Decker, and E. F. Steele, 1976b. "Pinhole Test for Identifying Dispersive Soils," *Journal of the Geotechnical Engineering Division*, ASCE, Vol. 102, No. GT1, January, pp. 69–85.

Siddiqi, F. H., R. B. Seed, C. K. Chan, H. B. Seed, and R. M. Pyke, 1987. "Strength Evaluation of Coarse-Grained Soils," EERC, University of California, Berkeley, California, *Report UCB/EERC–87/22*, December.

Skempton, A. W., 1954. "The Pore Pressure Coefficients A and B," *Geotechnique*, Vol. 4, No. 4, pp. 143–147.

Skempton, A. W., 1964. "Long-Term Stability of Clay Slopes," *Geotechnique*, Vol. 14, No. 2, pp. 77–101

Skempton, A. W., 1970. "First-Time Slides in Overconsolidated Clays," *Geotechnique*, Vol. 20, No. 3, pp. 320–324.

Skempton, A. W., 1977. "Slope Stability of Cuttings in Brown London Clay," *Proceedings of the 9th International Conference on Soil Mechanics and Foundation Engineering*, Vol. 3, Tokyo, pp. 261–270.

Skempton, A. W., 1985. "Residual Strength of Clays in Landslides, Folded Strata, and the Laboratory," *Geotechnique*, Vol. 35, No. 1, pp. 3–18.

Skempton, A. W., and A. W. Bishop, 1954. "Soils" in *Building Materials, Their Elasticity and Inelasticity*, M. Reiner, Ed. Amsterdam: North-Holland, Chapter X.

Skempton, A. W., and J. N. J. Hutchinson, 1969. "Stability of Natural Slopes and Embankment Foundations," *Proceeding of the 7th International Conference on Soil Mechanics and Foundation Engineering*, State-of-the-Art Volume, Mexico City, pp. 291–340.

Skempton, A. W., and D. J. Petley, 1967. "The Strength along Structural Discontinuities in Stiff Clays," *Proceedings of the Geotechnical Conference*, Vol. 2, Oslo, Norway, Norwegian Geotechnical Institute, pp. 29–46.

Stark, T. D., and I. A. Contreras, 1996. "Constant Volume Ring Shear Apparatus," *Geotechnical Testing Journal*, ASTM, Vol. 19, No. 1, pp. 3–11.

Stark, T. D., and I. A. Contreras, 1998. "Fourth Avenue Landslide during 1964 Alaskan Earthquake," *Journal of the Geotechnical Engineering Division*, ASCE, Vol. 124, No. GT-2, February, pp. 99–109.

Stark, T. D., and H. T. Eid, 1994. "Drained Residual Strength of Cohesive Soils," *Journal of the Geotechnical Engineering Division*, ASCE, Vol. 120, No. GT-5, May, pp. 856–871.

Stark, T. D., and H. T. Eid, 1997. "Slope Stability Analysis in Stiff Fissured Clays," *Journal of the Geotechnical Engineering Division*, ASCE, Vol. 123, No. GT-4, April, pp. 335–343.

Statton, C. T., and J. K. Mitchell, 1977. "Influence of Eroding Solution Composition on Dispersive Behavior of a Compacted Shale Sample," *Symposium on Dispersive Clays, Related Piping and Erosion in Geotechnical Projects*, ASTM, 5TP623, pp. 398–407.

Tavitian, C. J., 1993. "Clay Mineralogy of the Aaqoura Earthflow on Mount Lebanon, Lebanon," *Geotechnical Engineering of Hard Soils—Soft Rocks*, A. Anagnostopoulos, F. Schlosser, N. Kalteziotis, and Frank, Eds. Rotterdam, The Netherlands: Balkema, pp. 1183–1189.

Taylor, D. W., 1948. *Fundamentals of Soil Mechanics.* New York: Wiley.

Terzaghi, K., 1943. *Theoretical Soil Mechanics.* New York: Wiley.

Terzaghi, K., R. B. Peck, and G. Mesri, 1996. *Soil Mechanics in Engineering Practice*, 3rd ed. New York: Wiley.

Townsend, F. C., and P. A. Gilbert, 1976. "Effects of Specimen Type on the Residual Strength of Clays and Clay Shales," *Soil Specimen Preparation for Laboratory Testing*, STP 599. Philadelphia, Pennsylvania: ASTM, pp. 43–65.

Transportation Research Board, 1975. *Treatment of Soft Foundations for Highway Embankments*, Report No. 29, Washington, DC.

U.S. Navy, 1971. *Soil Mechanics. Foundations, and Earth Structures*, Design Manual 7. Alexandria, Virginia: Naval Facility Engineering Command, March.

Voight, B., 1973. "Correlation Between Atterberg Plasticity Limits and Residual Shear Strength of Natural Soils," *Geotechnique*, Vol. 23, No. 2, pp. 265–267.

Walker, G. F., 1958. "Reactions of Expanding-Lattice Clays with Glycerol and Ethylene Glycol," *Clay Mineralogy Bulletin*, Vol. 3, pp. 302–313.

Weir, A. H., and R. Green-Kelly, 1962. "Beidellite," *American Mineralogy*, Vol. 47, pp. 137–146.

Whittig, L. D., and W. R. Allardice, 1968. "X-ray diffraction techniques," *Methods of Soil Analysis*, Chapter 12, Agronomy No. 9, Part 1, 2nd ed., American Society of Agronomy, Madison, Wisconsin, pp. 331–362.

Wong, H. Y., 1978. *Soil Strength Parameter Determination.* Hong Kong: Hong Kong Institute of Engineers, February 15, pp. 33–39.

Zaruba, Q., and V. Mencl, 1982. *Landslides and Their Control*, New York: Elsevier.

CHAPTER 6

SLOPE STABILITY CONCEPTS

6.1 INTRODUCTION

Once the slope geometry and subsoil conditions have been determined, the stability of a slope may be assessed using either published chart solutions or a computer analysis. Most of the computer programs used for slope stability analysis are based on the limiting equilibrium approach for a two-dimensional model, although some also allow three-dimensional analysis.

Other, more complex programs that use the finite element or boundary element methods are also available, and allow the engineer to perform refined, two- or three-dimensional slope evaluations. However, such analyses require a relatively complete model of the subsoils and their constitutive parameters determined by an extensive program of laboratory tests. Concerns about the laboratory testing, lack of familiarity with the methodology, and the requirements for extensive computing for each analysis have generally restricted the use of the finite element approach to earth embankment dams only.

This chapter reviews the mechanics of the limit equilibrium approach, discussing the classical closed-form solutions as well as the popular method of slices. The derivation of several procedures that use the method of slices is presented, along with useful design charts for single material slopes. A final section discusses the evaluation of slopes using probabilistic methods, which are gaining popularity in engineering practice.

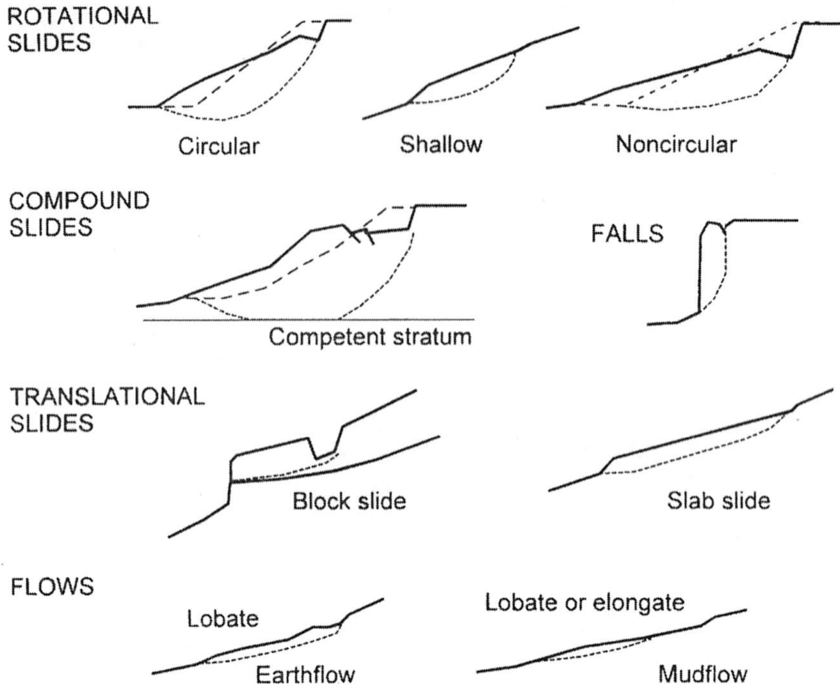

Figure 6.1 Types of mass movements in clay slopes. (Adapted from Skempton and Hutchinson, 1969, with permission.)

6.2 MODES OF FAILURE

Terzaghi and Peck (1967) state, "Slides may occur in almost every conceivable manner, slowly or suddenly, and with or without any apparent provocation." These slope failures are usually due either to a sudden or gradual loss of strength by the soil or to a change in geometric conditions, for example, the steepening of an existing slope. Figure 6.1 illustrates the typical slides that can be expected to occur in soil slopes. These usually take the form of either: (1) translational, (2) plane or wedge surface, (3) circular, or (4) noncircular, or *a combination of these types*. The aspect ratio used to differentiate between the translational and rotational surfaces is shown in Figure 6.2. With this definition a "grey" area, where $0.1 < D/L < 0.15$, has been left to account for the case of a combined rotational and translational failure.

Table 6.1 summarizes the geologic conditions that influence the shape and development of the potential failure surface. The planar failure surfaces are usually expected in slopes where a soil layer, or relict jointing, with a relatively low strength strongly influences the shape of the failure surface. The translational type of failure occurs in shallow soils overlying relatively stronger materials and circular failure surfaces usually occur in slopes consisting of homogeneous materials.

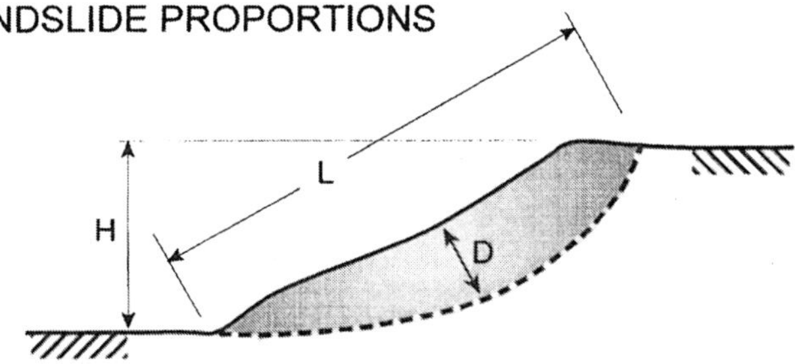

Figure 6.2 Aspect ratio of failure mass.

As most soils are generally heterogeneous, noncircular surfaces, consisting of a combination of planar and curved sections, are most likely. Often, retrogressive failures consisting of multiple curved surfaces can occur in layered soils, as shown in Figure 6.3. Such failures are typical where the first slip tends to oversteepen the slope, which then leads to additional failures. The main items required to evaluate the stability of a slope are: (1) shear strength of the soils, (2) slope geometry, (3) pore pressures or seepage forces, and (4) loading and environmental conditions.

The shear strength aspects have been discussed in great detail in the last chapter and are not repeated here except within the context of the factor of safety. Typically,

TABLE 6.1 Geologic Factors Controlling Shape of Potential Failure Surface

Geologic Conditions	Potential Failure Surface
Cohesionless soils;	Translational with small depth/length ratio
Residual or colluvial soils over shallow rock	
Stiff fissured clays and marine shales within the upper, highly weathered zone	
Sliding block	Single planar surface
Interbedded dipping rock or soil	
Faulted or slickensided material	
Intact stiff to hard cohesive soil on steep slopes	
Sliding blocks in rocky masses	Multiple planar surfaces
Weathered interbedded sedimentary rocks	
Clay shales and stiff fissured clays	
Stratified soils	
Side-hill fills over colluvium	
Thick residual and colluvial soil layers	Circular or cylindrical shape
Soft marine clays and shales	
Soft to firm cohesive soils	

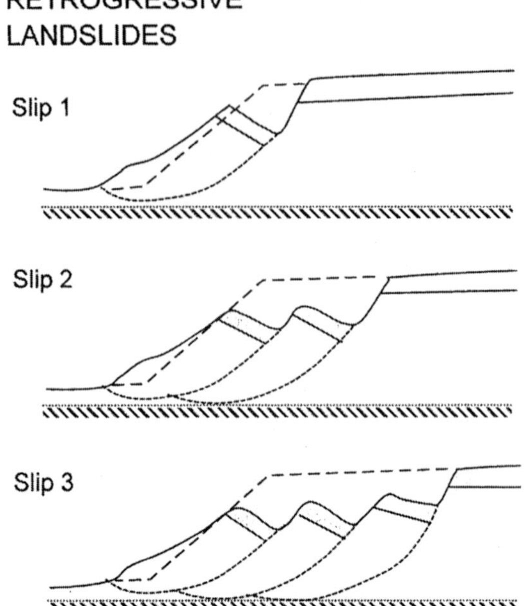

Figure 6.3 Typical retrogressive slide. (Adapted from Skempton and Hutchinson, 1969, with permission.)

the engineer will have developed a subsoil model of the slope and foundation soils through a site exploration study and subsequently selected appropriate shear strengths for the soils. Ideally, the shear strengths should be provided as undrained strenght, s_u, or the more typical Mohr–Coulomb parameters, c and ϕ. The slope geometry may be known for existing, natural slopes or may be a design parameter for embankments and cut slopes. A major contributor to many slope failures is the change in effective stress caused by pore water pressures. These tend to alter the shear strength of the soil along the shear zone and may even cause destabilizing forces to develop if surface runoff infiltrates cracks at the head of the failure surface. Later sections discuss the methods used to account for pore water pressure effects within the analytical procedures.

6.3 FACTOR OF SAFETY CONCEPTS

An understanding of the role of the factor of safety (FOS) is vital in the rational design of slopes. One well-recognized function of the FOS is to account for uncertainty and thus to guard against ignorance about the reliability of the items that enter into the analysis, such as strength parameters, pore pressure distribution, and stratigraphy. In general, the lower the quality of the site investigation, the higher the desired FOS should be, particularly if the designer has only limited experience with the materials in question. Another role of the FOS is that it constitutes an empirical tool whereby

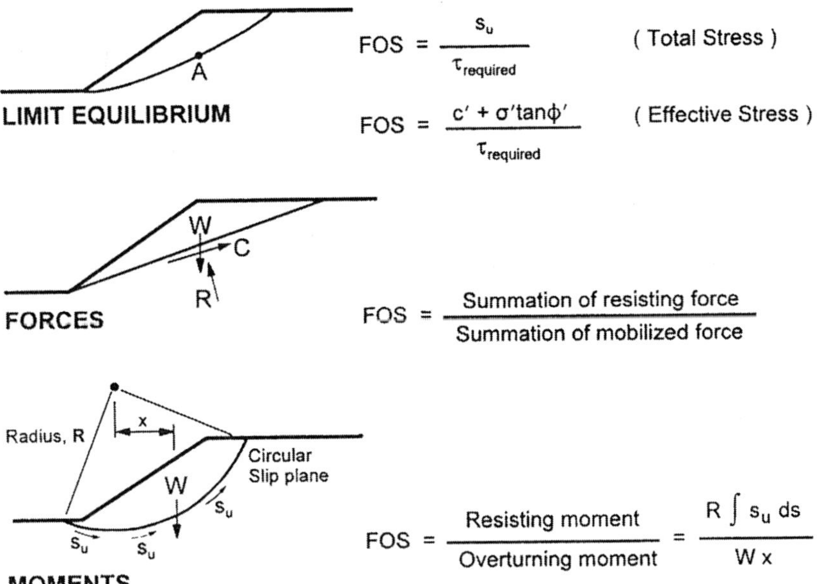

Figure 6.4 Various definitions of factor of safety (FOS).

deformation stability performances are limited to tolerable amounts within economic restraints. In this way, the choice of the FOS is greatly influenced by the accumulated experience with a particular soil mass. Since the degree of risk that can be taken is also greatly influenced by experience, the actual magnitude of the FOS used in design will vary with material type and performance requirements.

In most limit equilibrium analyses, the shear strength required along a potential failure surface *just* to maintain stability is calculated and then compared to the magnitude of available shear strength. In this case the FOS is assumed to be constant for the entire failure surface. For example, at point A in the upper slope shown in Figure 6.4, this average FOS will be given by the ratio of available to required shear strength. Thus a constant proportion of available strength is mobilized at *every* point on the failure surface to resist potential sliding.

If τ_{req} is the required shear strength, then

$$\tau_{req} = \frac{s_u}{F} \quad \text{for total stresses}$$
$$\tau_{req} = \frac{c'}{F_c} + \frac{\sigma' \tan\phi}{F_\phi} \quad \text{for effective stresses} \qquad \text{(Eq. 6-1)}$$

where s_u = the total stress strength
c' and ϕ' = effective stress strength parameters
F = the factor of safety for total stresses
F_c and F_ϕ = the factors of safety for effective stresses.

The adoption of F_c and F_ϕ allows different proportions of the cohesive (c') and frictional (ϕ') components of strength to be mobilized along the failure surface. However, most limit equilibrium methods assume $F_c = F_\phi$, implying that the same proportion of the c' and ϕ' components are mobilized at the same time along the shear failure surface.

Another definition of FOS often considered is the ratio of total resisting forces to total disturbing (or driving) forces for planar failure surfaces or the ratio of total resisting to disturbing moments, as in the case for circular slip surfaces. However, one must realize that these different values of the FOS obtained using the three methods, that is, mobilized strength, ratio of forces, or ratio of moments, will not give identical values for c–ϕ soils.

For typical slope designs, the required FOS (nonseismic) are usually in the 1.25 to 1.5 range. Higher factors may be required if there is a high risk of loss of life or uncertainty regarding the pertinent design parameters. Likewise, a lower FOS may be used if the engineer is confident of the accuracy of input data and if the construction is being monitored closely.

6.4 PORE WATER PRESSURES

If an effective stress analysis is to be performed, pore water pressures will have to be estimated at relevant locations in the slope. These pore pressures are usually estimated from groundwater conditions that may be specified by one of the following methods:

(1) *Phreatic Surface* This surface, or line in two dimensions, is defined by the free groundwater level. This surface may be delineated, in the field, by using open standpipes as monitoring wells. This is the most commonly programmed method available.

(2) *Piezometric Data* Specification of pore pressures at discrete points, within the slope, and use of an interpolation scheme to estimate the required pore water pressures at any location. The piezometric pressures may be determined from:

(a) Field piezometers,

(b) A manually prepared flow net, or

(c) A numerical solution using finite differences or finite elements.

Although this approach is only available in a few slope stability programs, it is the best method for describing the pore water pressure distribution (Chugh, 1981b).

(3) *Pore Water Pressure Ratio* This is a popular and simple method for normalizing pore water pressures measured in a slope according to the definition

$$r_u = \frac{u}{\sigma_v} \qquad \text{(Eq. 6-2)}$$

where u is the pore pressure and σ_v is the *total* vertical subsurface soil stress at depth z. Effectively, the r_u value is the ratio between the pore pressure and the total vertical stress at the same depth. This factor is easily implemented, but the major difficulty is associated with the assignment of the parameter to different parts of the slope. Often, the slope will require an extensive subdivision into many regions with different r_u values. This method, if used correctly, will permit a search for the most critical surface. However, it is usually reserved for estimating the FOS value from slope stability charts or for assessing the stability of a single surface, as presented in an example problem later.

(4) *Piezometric Surface* This surface is defined for the analysis of a unique, single failure surface. This approach is often used for the back-analysis of failed slopes. Because the combination of a piezometric and a failure surface is unique, a search for the critical surface is not possible. The user should note that a piezometric surface is *not* the same as a phreatic surface, as the calculated pore water pressures will be different for the two cases.

(5) *Constant Pore Water Pressure* This approach may be used if the engineer wishes to specify a constant pore water pressure in any particular soil layer. This may be used to examine the stability of fills placed on soft soils during construction where excess pore water pressures are generated according to the consolidation theory.

6.4.1 Phreatic Surface

If a phreatic surface is defined, the pore water pressures are calculated for the steady-state seepage conditions according to the sketch shown in Figure 6.5. This concept is based on the assumption that *all* equipotential lines are straight and perpendicular to the segment of the phreatic surface passing through a slice-element in the slope. Thus if the inclination of the phreatic surface segment is θ, and the vertical distance between the base of the slice and the phreatic surface is h_w, the pore pressure is given by

$$u = \gamma_w \times (h \cos^2 \theta) \qquad \text{(Eq. 6-3)}$$

where γ_w is the unit weight of water. This is the easiest approach for defining the phreatic surface with minimal data.

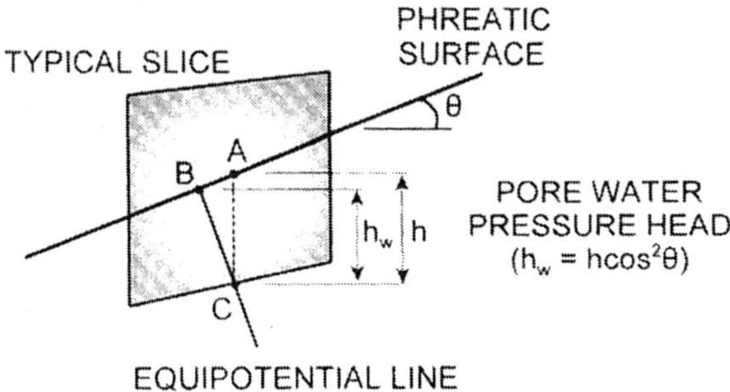

Figure 6.5 Calculation of pore water pressure head from phreatic surface.

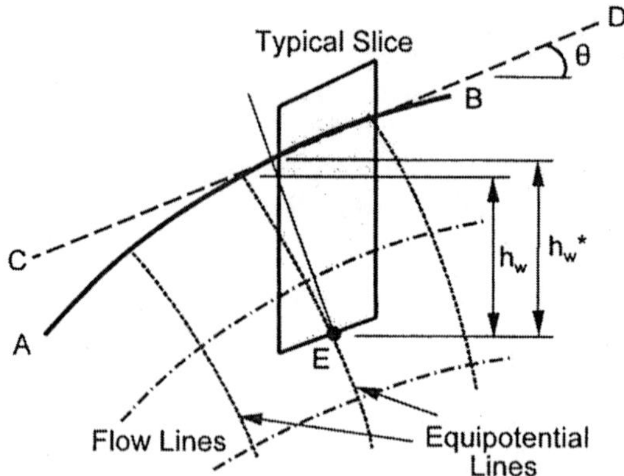

Figure 6.6 Phreatic surface and curved equipotential lines.

This is a reasonable assumption for a sloping *straight-line* phreatic surface, but will provide *higher* or *lower* estimates of pore water pressure for a curved (convex) phreatic surface. For the steeply sloping phreatic surface, a convex-shaped phreatic surface (as shown in Figure 6.6) generates an overestimate of the pore water pressures, whereas a concave-shaped surface may lead to an underestimation. This feature is shown in Figure 6.6. This overestimation is entirely due to the assumption of straight equipotential lines intersecting the projected phreatic surface line, *CD*.

If the actual phreatic surface, line *AB*, is steeply curved, the equipotential lines must also curve, as shown. In this example, a typical solution would use the *greater* pore water pressure head, h_w, rather than the true head, h_w^*. However, this overestimation is small and will only affect a few slices within the sliding mass. Also, this conservatism is not expected to significantly affect the FOS. In cases where the user is concerned about this problem, a pore water pressure grid may be used to simulate the pore water pressure distribution more accurately.

6.4.2 Piezometric Surface

A piezometric surface may be specified to correspond to a single, *unique* failure surface in the slope. This feature is typically used to back-analyze slope failures where the pore water pressures may have been determined from in situ measurements. Figure 6.7 shows the approach taken to calculate the pore water pressure for a piezometric surface, where the vertical distance (elevational head) is taken to represent the pressure.

There are some computer programs that oversimplify the analysis by misinterpreting a phreatic surface as a piezometric surface. With this erroneous assumption, the overestimated pore pressure head is incorrectly taken as the vertical distance between the phreatic surface and base of slice, as shown in Figure 6.7, rather than the correct approach of Figure 6.5.

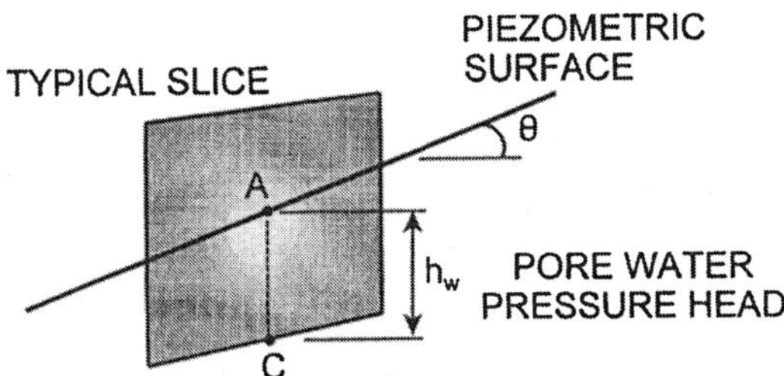

Figure 6.7 Calculations of pore water pressure head for specified piezometric surface.

6.4.3 Example

An example of an effective stress analysis of a slope affected by seepage is presented to show the advantages and disadvantages of using the various procedures for specifying pore water pressures. The example slope is presented in Figure 6.8 and consists of an earth embankment ($\gamma = 125$ pounds per cubic foot, $\phi' = 32°$, and $c' = 90$ pounds per square foot) founded on firm stratum (Lambe and Whitman, 1969).

Analysis of Critical Surface The most critical surface located in a search analysis is shown in Figure 6.9, along with the phreatic and corresponding *piezometric* surfaces. The reader should note that the piezometric surface, determined from the flow net presented earlier in Figure 6.8, suggests that pore water pressures exist *only between points A and B on the failure surface*.

A summary of the FOS values for the different methods used to specify the pore water pressures is presented in Table 6.2. The approach adopted for calculating the r_u value for the slope is presented in Figure 6.10. The failure surface is represented by

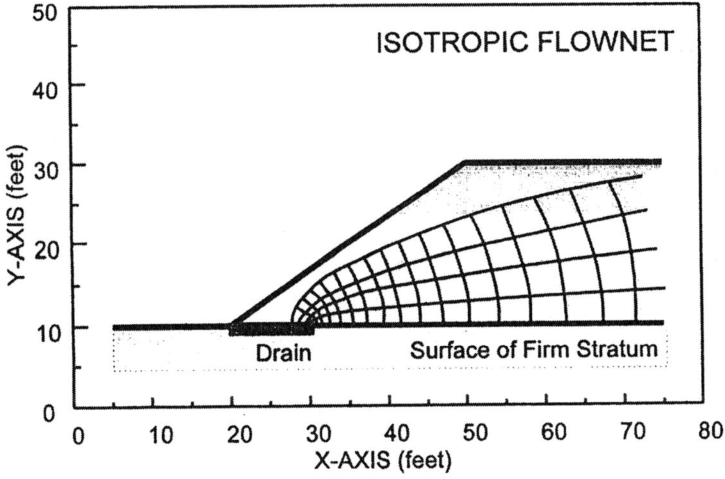

Figure 6.8 Example embankment with flownet.

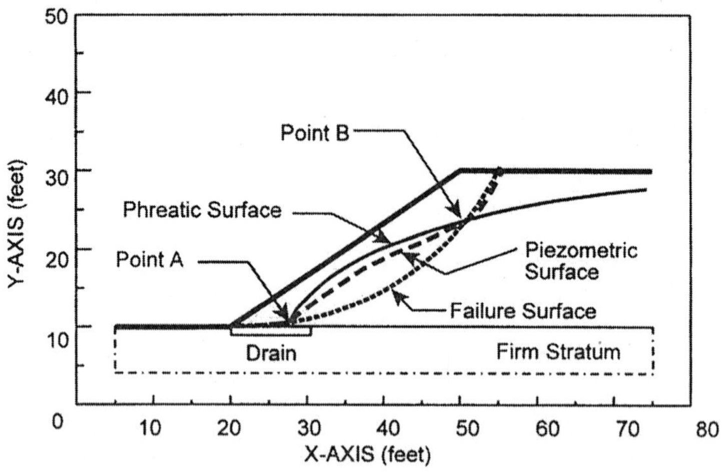

Figure 6.9 Comparison of piezometric and phreatic surfaces applicable for the analysis of the critical surface.

the arc *AFEDC* and the piezometric surface by the line *FGD*. For this example, the pore pressure ratio will be a function of the ratio between the shaded area *FGDEF* and the failed mass, *ABCDEFA*, and is given by

$$r_u = \frac{\text{Area } FGDEF}{\text{Area } ABCDEFA} \times \frac{\gamma_w}{\gamma_s} \qquad \text{(Eq. 6-4)}$$

where γ_s = unit weight of soil
γ_w = unit weight of water

As expected, the computed FOS value is almost identical to the FOS for the piezometric surface.

Finally, if the slope analysis had *incorrectly* used the phreatic surface to simulate the piezometric surface, an FOS value for the selected circular surface would have been calculated as a conservative value of 1.233. This reduction is directly attributable to the overestimation of the pore pressures along about 85 percent of the failure surface.

TABLE 6.2 Summary of FOS Values for Different Pore Water Pressure Specifications

Type	FOS
Phreatic surface	1.300
Pore pressure grid	1.299
Piezometric surface	1.291
Pore pressure ratio, $r_u = 0.157$	1.300
Incorrect piezometric surface	1.233

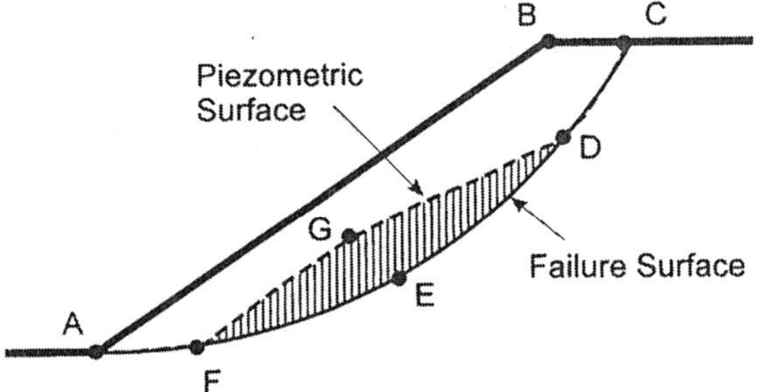

Figure 6.10 Use of piezometric surface to determine pore pressure coefficient, r_u.

This example illustrated the use of four methods that may be used to assign pore water pressures for slope stability analysis. Of these, three are suitable for problems where a program is used to determine the critical surface with the lowest FOS using an automatic search procedure. The analysis of the circular surface revealed that the FOSs were consistent for each of the methods employed and varied within a very small range of 1.291 to 1.301. This range is small enough to have been influenced by minor uncertainties concerning the manual development of a flow net and the determination of the piezometric surface for the circular failure surface. The small FOS range strongly suggests that a phreatic surface may be used to effectively model the pore water pressures within a slope. However, the engineer should be discouraged from using computer programs to search for a critical surface using a piezometric surface to simulate the variation in pore pressures. In such cases, the pore water pressure will be overestimated and an unrealistically low FOS value will be reported by the program.

6.4.4 Negative Pore Pressures

There may be cases where an engineer wishes to use *negative* pore pressures to take advantage of the apparent cohesive strength available due to suction within the soil in the slope. As recommended in the previous chapter, the influence of suction should be included by increasing the *total cohesion* according to the measured values of matric suction within the slope. In some cases, actual negative pore pressures have been used in slope analysis to increase the shear strength of the soil. This method is not recommended as it only affects the frictional component via the $(\sigma - u)\tan\phi$ term and may not generate reliable values of strength.

6.5 BLOCK ANALYSIS

A block analysis may be used to estimate the FOS against sliding in situations where the shearing strength of an embankment fill is greater than that of the foundation soils, as shown in Figure 6.11. In this case, it is necessary to investigate the stability along

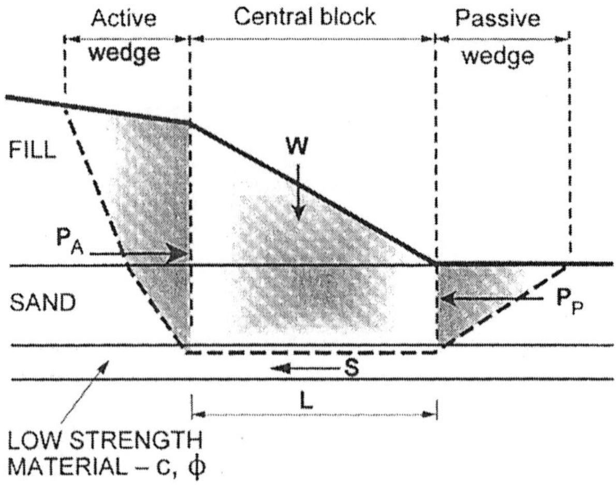

Figure 6.11 Sliding block analysis.

a surface of failure passing through the foundation of the embankment in addition to the usual studies for possible failure within the embankment itself.

A predominantly planar failure surface develops if the weak foundation layer is relatively thin. Stability may be analyzed by means of a sliding block shearing through the weak foundation layer. The block analysis is fairly simple and straightforward, and can be quickly performed by hand calculation.

For the analysis, the potential sliding block is divided into three parts (Figure 6.11): (1) an *active* wedge at the head of the slide, (2) a *central* block, and (3) a *passive* wedge at the toe. The FOS, defined by 1 would be computed by summing forces horizontally to give:

$$F = \frac{\text{Horizontal resistance forces}}{\text{Horizontal driving forces}}$$

$$= \frac{P_p + c'_m L + (W\cos\alpha - U)\tan\phi'_m}{P_a} \qquad \text{(Eq. 6-5)}$$

where P_a = active force (driving)
P_p = passive force (resisting)
L = resisting force due to cohesion of clay
α = inclination of the base of the central block
c'_m and ϕ'_m = strength parameters of the soil at the base of the central block, with effective weight $(W - U)$

The active and passive lateral earth pressures used in the block analysis are calculated using

$$\sigma_{A/P} = K_{A/P}\sigma'_v \mp 2c_m\sqrt{K_{A/P}} \qquad \text{(Eq. 6-6)}$$

where K_A = active and passive earth pressure coefficient
K_P = passive earth pressure coefficients
σ'_v = vertical effective stress
c_m = mobilized cohesion parameter

The earth pressure coefficients may be estimated using the Rankine expression:

$$K_A = \frac{1 - \sin\phi_m}{1 + \sin\phi_m} \quad \text{and} \quad K_P = \frac{1 + \sin\phi_m}{1 - \sin\phi_m} \qquad \text{(Eq. 6-7)}$$

The above expressions are suitable for cases where the backslope is horizontal. For other cases, the engineer should consult a text on soil mechanics (retaining wall analysis) for appropriate *K* values for nonhorizontal backslopes.

For cases where the active or passive wedge passes through more than one soil type with different soil strengths or unit weights, appropriate coefficients, K_a or K_p, should be selected for calculating the forces from each soil layer. The total forces, that is, active and passive, may then be calculated by adding the contribution of each layer.

Several trial locations of the active and passive wedges must be checked to determine the minimum FOS. When the weak layer has considerable thickness, the failure plane must be assumed at different depths to find its critical location, which gives the lowest FOS or the highest required strength. Note that wedge-type failures are assumed at the head and toe of the slide, similar to what occurs against vertical planes, which are treated as "imaginary" retaining walls. The active and passive forces are computed in the same way as for retaining wall problems.

6.5.1 Example

An example slope, shown in Figure 6.12, will be analyzed to illustrate the proposed procedure. With this approach, the *central block* is assumed to be a rigid body, which is affected by lateral earth and hydrostatic forces acting on its boundaries. The active forces on the left boundary, marked P_A in Figure 6.12, tend to destabilize the block, and the passive force on the right-hand side provides resistance, along with any strength, *S*, developed along the base of the sliding block. Such an analysis follows an iterative procedure where the mobilized *c* and ϕ values are adjusted until a common FOS is calculated for the slope. If an initial factor of safety, $F_\phi = F_c = 1.3$, is assumed, then the following lateral forces may be calculated:

(1) Active force, P_{A1}:

$$\tan\phi_m = (\tan\phi) \div F_\phi = 19.73^\circ$$

$$K_A \approx (1 - \sin\phi_m) \div (1 + \sin\phi_m) = 0.495$$

$$P_{A1} = 0.5\gamma H^2 K_A = 0.5 \times 120 \times (10)^2 \times 0.49 = 2{,}971 \text{ lb}$$

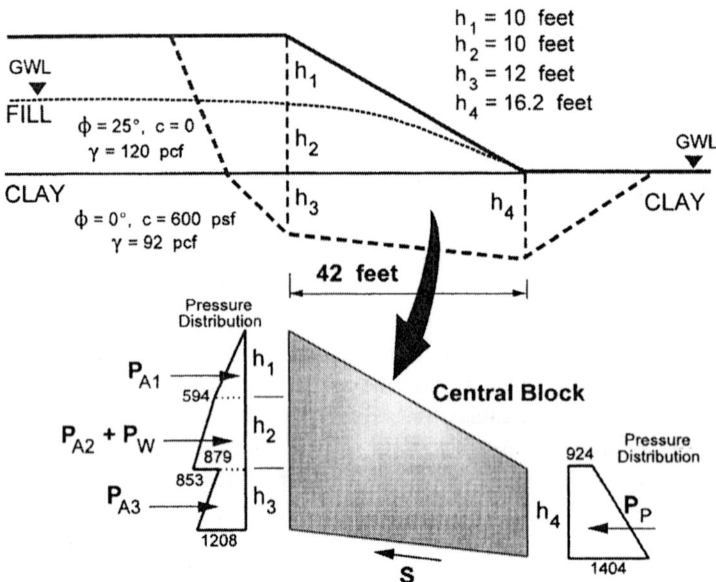

Figure 6.12 Block analysis—example problem.

(2) Active force, P_{A2}:

$$K_A \approx (1 - \sin\phi) \div (1 + \sin\phi) = 0.495$$

$$P_{A2} = 7{,}368 \text{ lb}$$

$$\text{Water force, } P_w = 0.5 \times 62.4 \times (10)^2 = 3{,}120 \text{ lb}$$

(3) Active force, P_{A3}:

$$K_A \approx (1 - \sin\phi) \div (1 + \sin\phi) = 1.0$$

$$c_m = \frac{600}{1.3} = 462 \text{ psf}$$

$$P_{A3} = 0.5 \times (853 + 1{,}208) \times 12 = 12{,}366 \text{ lb}$$

(4) Passive force, P_P:

$$K_P \approx (1 + \sin\phi) \div (1 - \sin\phi) = 1.0$$

$$P_P = \tfrac{1}{2}(924 + 1{,}404) \times 16.2 = 18{,}857 \text{ lb}$$

(5) Horizontal component of sliding resistance along base:

$$c_m L = 460 \times 42 = 19{,}385 \text{ lb}$$

Note: If the soil under the central block had $\phi > 0$, then the effective weight of the block would be used to increase the sliding resistance via the $(W \cos\alpha - U)\tan\phi_m$ term.

(6) Please note that the water forces in the clay layer will not be used as they are equal and opposite for the active and passive sides of the central block.

(7) Factor of safety:

$$\begin{aligned} F &= \frac{\text{Horizontal resistance}}{\text{Driving forces}} \\ &= (18{,}857 + 19{,}385) \div (2{,}971 + 7{,}368 + 3{,}120 + 12{,}366) \\ &= 1.48 \end{aligned}$$

Now as we had originally assumed an FOS $= F_\phi = F_c = 1.3$, the above exercise should be repeated with another estimate of F_ϕ until the newly computed value is in close agreement (within ± 5 percent of assumed value is perfectly acceptable). If the exercise is repeated with $F_\phi = 1.37$, the reader will find that the newly computed FOS value will be 1.37.

6.6 INFINITE SLOPE ANALYSIS

A slope that extends for a relatively long distance and has a consistent subsoil profile may be analyzed as an infinite slope. The failure plane for this case is parallel to the surface of the slope and the limit equilibrium method can be applied readily.

6.6.1 Infinite Slopes in Dry Sand

A typical slice for a slope in *dry sand* is shown in Figure 6.13 along with its free body diagram. The weight of the slice (with a unit dimension into the page) is given by

$$W = \gamma b h(1) \qquad \text{(Eq. 6-8)}$$

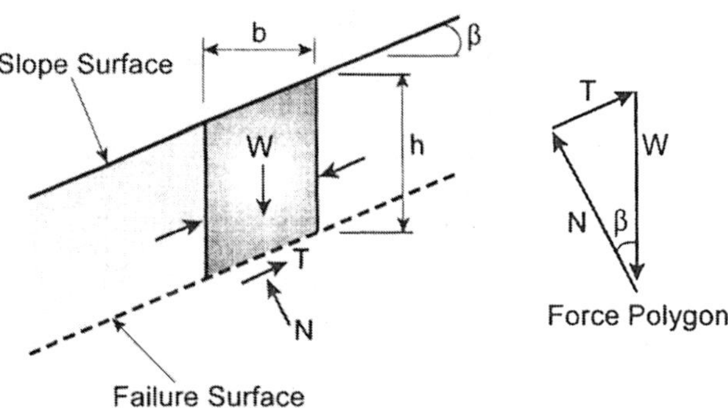

Figure 6.13 Infinite slope failure in dry sand.

Next the normal, N, and driving, T, forces are determined:

$$N = W\cos\beta \qquad \text{and} \qquad T = W\sin\beta \tag{Eq. 6-9}$$

The available frictional strength along the failure plane will depend on ϕ and is given by

$$S = N\tan\phi \tag{Eq. 6-10}$$

Then if we consider the FOS as the ratio of available strength to strength required to maintain stability (limit equilibrium), the FOS will be given by

$$F = \frac{N\tan\phi}{W\sin\beta} = \frac{\tan\phi}{\tan\beta} \tag{Eq. 6-11}$$

The FOS is independent of the slope height and depth, z, and depends only on the angle of internal friction, ϕ, and the angle of the slope, β. Also, at $F = 1$, the maximum slope angle will be limited to the angle of internal friction, ϕ.

6.6.2 Infinite Slope in c–ϕ Soil with Seepage

If a saturated slope, in cohesive c–ϕ soil, has seepage parallel to the slope surface as shown in Figure 6.14, the same limit equilibrium concepts may be applied to determine the FOS, which will now depend on the effective normal force, N'. From Figure 6.14, the pore water force acting on the base of the typical slice will be given by

$$U = (\gamma_w h\cos^2\beta)\frac{b}{\cos\beta} = \gamma_w bh\cos\beta. \tag{Eq. 6-12}$$

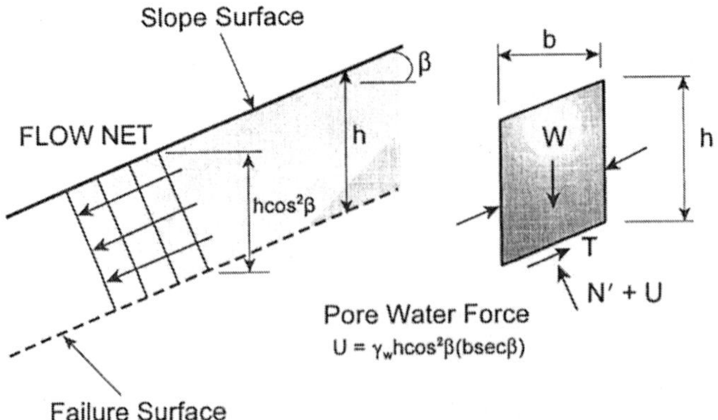

Figure 6.14 Infinite slope failure in c–ϕ soil with parallel seepage.

The available frictional strength along the failure plane will depend on ϕ' and the effective normal force and is given by

$$S = c'b \sec\beta + (N - U)\tan\phi' \quad \text{(Eq. 6-13)}$$

So the FOS for this case will be

$$F = \frac{cb \sec\beta + (N - U)\tan\phi'}{W \sin\beta} \quad \text{(Eq. 6-14)}$$

If we substitute $W = \gamma_{sat}bh$ into the above expression and rearrange, the FOS will be given by

$$F = \frac{c' + h(\gamma_{sat} - \gamma_w)\cos^2(\beta)\tan\phi'}{\gamma_{sat}\, h \sin\beta\cos\beta} \quad \text{(Eq. 6-15)}$$

where $\gamma' = (\gamma_{sat} - \gamma_w)$. For a $c' = 0$ soil, the above expression may be simplified to give

$$F = \frac{\gamma'}{\gamma_{sat}} \times \frac{\tan\phi'}{\tan\beta} \quad \text{(Eq. 6-16)}$$

From Equation 6.16 one can see that for a *granular* material, the FOS is still independent of the slope height and depth, h, but is reduced by the factor γ'/γ_{sat}. For typical soils, this reduction will be about 50 percent in comparison to dry slopes.

The above analysis can be generalized if the seepage line is assumed to be located at a height of $(m \times h)$ above the failure surface. In this case, the FOS will be given by

$$F = \frac{c' + h\cos^2\beta[(1 - m)\gamma_m + m\gamma']\tan\phi'}{h\sin\beta\cos\beta[(1 - m)\gamma_m + m\gamma_{sat}]} \quad \text{(Eq. 6-17)}$$

and γ_{sat} and γ_m are the saturated and moist unit weights of the soil below and above the seepage line. The above equation may be readily reformulated to determine the critical depth of the failure surface for any seepage condition and a c'–ϕ' soil.

6.7 PLANAR SURFACE ANALYSIS

Planar failure surfaces usually occur in slopes with a thin layer of soil that has relatively low strength in comparison to the overlying materials. Also, this is the preferred mode of failure for jointed materials that may dip toward proposed excavations.

A planar failure surface can be readily analyzed with a closed-form solution that depends on the slope geometry and the shear strength parameters of the soil along the failure plane. For the slope shown in Figure 6.15, three forces—weight, W, mobilized shear strength, S_m, and the normal reaction, N—need to be determined in order to evaluate the stability.

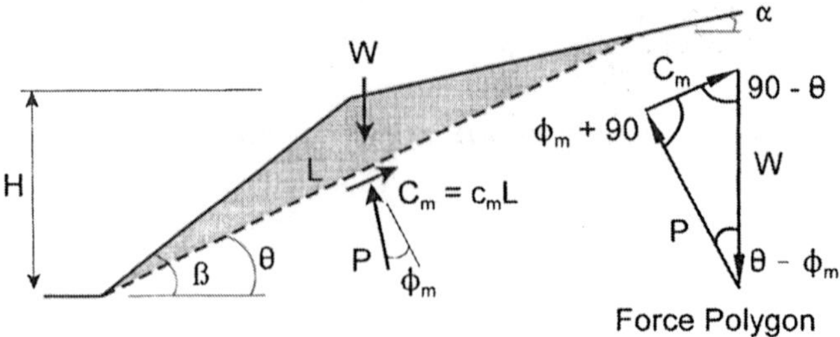

Figure 6.15 Planar (block) failure surface.

The weight of the wedge may be determined from the geometry using

$$\begin{aligned} L &= \frac{H}{\sin\beta} \times \frac{\sin(\beta-\alpha)}{\sin(\theta-\alpha)} \\ W &= \frac{1}{2}\gamma H^2 \left[\frac{\sin(\beta-\theta)}{\sin^2\beta} \cdot \frac{\sin(\beta-\alpha)}{\sin(\theta-\alpha)}\right] \end{aligned} \qquad \text{(Eq. 6-18)}$$

The angle, α, in the above equation is the inclination of the backslope, with respect to the horizontal. The normal force, N, and the mobilized strength, S_m, will be given by

$$N = W\cos\theta \qquad \text{and} \qquad S_m = W\sin\theta \qquad \text{(Eq. 6-19)}$$

and if the FOSs with respect to cohesion, F_c, and friction, F_ϕ, are used such that the mobilized shear strength contributions are given by

$$c_m = \frac{c}{F_c} \qquad \text{and} \qquad \tan\phi_m = \frac{\tan\phi}{F_\varphi} \qquad \text{(Eq. 6-20)}$$

then by equating the mobilized strength calculated using the Mohr–Coulomb criterion and Equation 6-20, the following relationship may be developed:

$$\begin{aligned} W\sin\theta &= c_m L + W\cos\theta\tan\phi_m \\ c_m &= \frac{W}{L}\left[\sin\theta - \cos\theta\tan\phi_m\right] \\ &= \frac{\gamma H^2}{2L}\frac{\sin(\beta-\theta)}{\sin^2\beta}\left[\sin\theta - \cos\theta\tan\phi_m\right] \\ &= \frac{1}{2}\gamma H\left[\frac{\sin(\beta-\theta)[\sin\theta-\cos\theta\tan\phi_m]}{\sin\beta}\right] \end{aligned} \qquad \text{(Eq. 6-21)}$$

Interestingly, the influence of the inclination of the backslope, α, is eliminated in the above equation, and so it does not affect the calculations directly. This equation allows the user to calculate the magnitude of the "cohesive" resistance required to satisfy equilibrium. However, this would only provide the solution for a failure surface inclined at an angle θ. To determine the critical slope, θ_{crit}, Equation 6.21 must provide the *maximum* value of the mobilized cohesive resistance. With the assumption that γ, β, and H are constant in Equation 6-21, the first derivative with respect to the failure plane angle, θ, is given by

$$\begin{aligned}
&\frac{\partial}{\partial\theta}[\sin(\beta-\theta)(\sin\theta-\cos\theta\tan\phi_m)] \\
&\quad = -\cos(\beta-\theta)[\sin\theta - \cos\theta\tan\phi_m] + \sin(\beta-\theta)[\cos\theta+\sin\theta\tan\phi_m] \\
&\quad = -[\sin\theta\cos(\beta-\theta) - \cos\theta\sin(\beta-\theta)] \\
&\qquad + \tan\phi_m[\cos\theta\cos(\beta-\theta) + \sin\theta\sin(\beta-\theta)] \\
&\quad = -\sin(2\theta-\beta) + \tan\phi_m[\cos(2\theta-\beta)] = 0
\end{aligned} \qquad \text{(Eq. 6-22)}$$

From the above equation, one can calculate

$$\theta_{crit} = \frac{\beta+\phi_m}{2} \qquad \text{(Eq. 6-23)}$$

which allows the critical value of c_m to be calculated using

$$c_m = \frac{1}{4}\gamma H\left[\frac{1-\cos(\beta-\phi_m)}{\sin\beta\cos\phi_m}\right] \qquad \text{(Eq. 6-24)}$$

The above equation may also be manipulated to determine the critical height of a slope by substituting $c_m = c$ and $\phi_m = \phi$ (i.e., FOS $= 1$) to give

$$H_{crit} = \frac{4c}{\gamma}\left[\frac{\sin\beta\cos\phi}{1-\cos(\beta-\phi)}\right] \qquad \text{(Eq. 6-25)}$$

For the case of a $\phi = 0$ analysis and a vertical slope (i.e., $\beta = 90^\circ$), the above equation gives a critical height of $4c/\gamma$.

For a typical analysis, the procedure requires a trial-and-error solution for a c–ϕ soil such that the factors of safety with respect to the cohesion and friction are equal. This is typically accomplished using the following steps:

(1) Assume an FOS against frictional resistance, F_ϕ.
(2) Compute the ϕ_m value.
(3) Calculate the mobilized cohesive value, c_m, using Equation 6-21.
(4) Calculate the FOS, $F_c = c/c_m$.
(5) Repeat steps 1 to 4 until $F_\phi = F_c$.

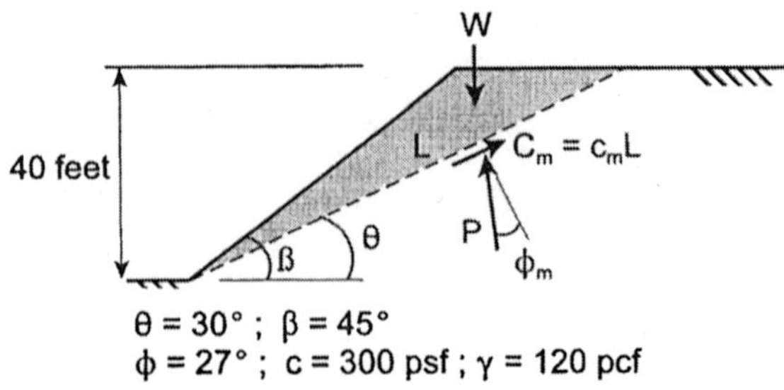

Figure 6.16 Planar wedge analysis—an example.

6.7.1 Planar Surface Example

The slope shown in Figure 6.16 will be analyzed to illustrate the procedures that may be used to determine the FOS for a potential planar failure surface. The following results are noted according to the items listed above:

(1) Assume $F_\phi = 1.3$.

(2) The mobilized friction angle, $\phi_m = 21.40$.

(3) Using Equation 6-21, $c_m = 141.05$ pounds per square foot.

(4) So $F_c = 300 \div 141.05 = 2.127$.

(5) As F_c is not equal to the assumed F_ϕ, choose a new value of F_ϕ and repeat steps 2 to 4.

If the above exercise is repeated for a few times, the final FOS will be found to be 1.566. An iterative technique can be readily set up in an electronic spreadsheet, as shown in Table 6.3. Once programmed, it offers an elegant approach to solving the problem for different geometric configurations and soil parameters. In fact if one varies the slope of the failure surface, that is, θ, a critical plane corresponding

TABLE 6.3 Use of Electronic Spreadsheet for Planar Wedge Solution—First Iteration

Planar Wedge Analysis

Height, H	= 40	**FOS (ϕ)**	**= 1.3**
Slope angle	= 45	FOS (c)	= 2.1269
Backslope angle	= 0	Weight, W	= 70276.88
Failure surface angle	= 30	Length, L	= 80.00
Friction angle	= 27	ϕ (mob)	= 21.4
Cohesion	= 300	c (mob)	= 141.05
Unit weight	= 120		

A trial value of **FOS (ϕ) = 1.3** leads to a back-calculated FOS (c) = 2.1269. The iteration must proceed with such trial-and-error values until FOS (ϕ) = FOS (c).

to the lowest FOS could be determined as well. For this case, θ_{crit} was inclined at approximately 31.5° and had an FOS = 1.556.

6.8 CIRCULAR SURFACE ANALYSIS

Circular failure surfaces are found to be the most critical in slopes consisting of homogeneous materials. There are two methods, the $\phi_u = 0$ and *friction circle*, that may be used to calculate the FOS for a slope.

6.8.1 Circular Arc(ϕ_u = 0) Method

The simplest circular analysis is based on the assumption that a rigid, cylindrical block will fail by rotation about its center and that the shear strength along the failure surface is defined by the undrained strength. As the undrained strength is used, the angle of internal friction, ϕ, is assumed to be zero (hence the $\phi_u = 0$ method). The FOS for such a slope (Figure 6.17) may be analyzed by taking the ratio of the resisting and overturning moments about the center of the circular surface.

If the overturning and resisting moments are given by Wx and $c_u LR$, respectively, the FOS for the slope may be given by

$$F = \frac{c_u LR}{Wx} \qquad \text{(Eq. 6-26)}$$

where c_u = undrained shear strength
R = radius of circular surface
W = weight of sliding mass
x = horizontal distance between circle center, O, and the center of the sliding mass.

If the undrained shear strength varies along the failure surface, the $c_u L$ term must be modified and treated as a variable in the above formulation.

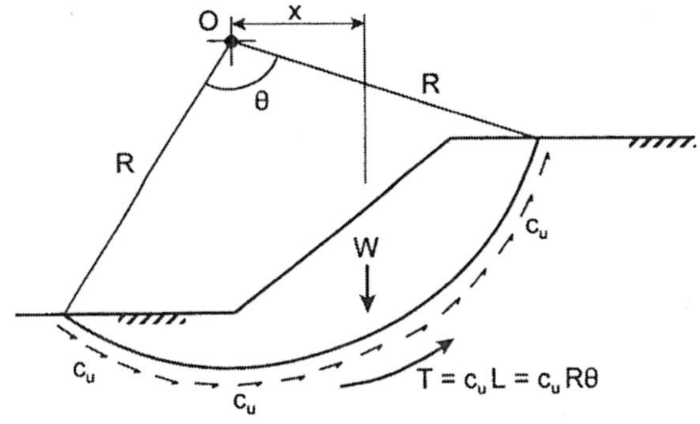

Figure 6.17 Circular failure surface in a $\phi_u = 0$ soil.

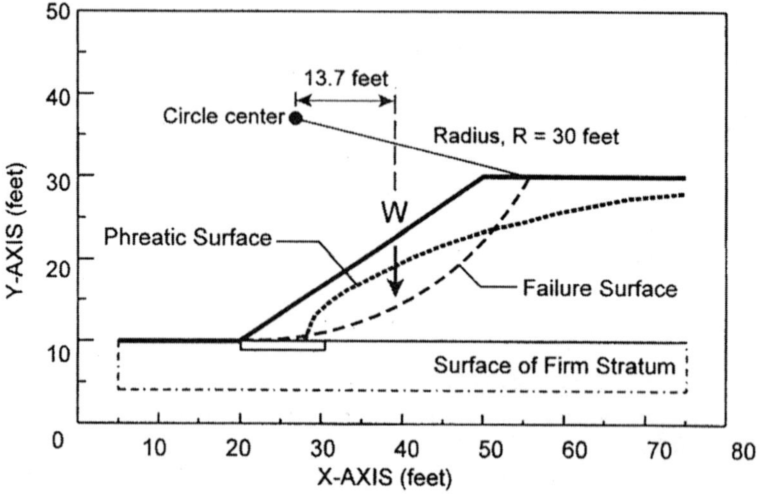

Figure 6.18 Example problem—$\phi_u = 0$ method.

6.8.2 $\phi_u = 0$ Example

For this example, the circular failure surface shown in Figure 6.18 will be analyzed for the case where the soil parameters are: $c = 1{,}000$ pounds per square foot, $\phi_u = 0$, and $\gamma = 125$ pounds per cubic foot. The phreatic surface, shown in the figure, was neglected for this example. The failure surface configuration has:

(1) Radius, $R = 30$ feet.
(2) Arc length, $L_{\text{arc}} = 42.3$ feet.
(3) Weight of the slide material, $W = 26.5$ kips.
(4) Location of W, $x = 13.7$ feet.

With the above data, the FOS for the $\phi_u = 0$ method can be readily calculated as

$$\begin{aligned} F &= (1{,}000 \times 42.3 \times 30) \div (26{,}500 \times 13.7) \\ &= 3.495 \end{aligned}$$

6.8.3 Friction Circle Method

This method is useful for homogeneous soils with $\phi_u > 0$, such that the shear strength depends on the normal stress. In other words, it may be used when both cohesive and frictional components for shear strength have to be considered in the calculations. The method is equally suitable for total or effective stress types of analysis in homogeneous soils.

The method attempts to satisfy the requirements of complete equilibrium by assuming that the direction of the resultant of the normal and frictional component of

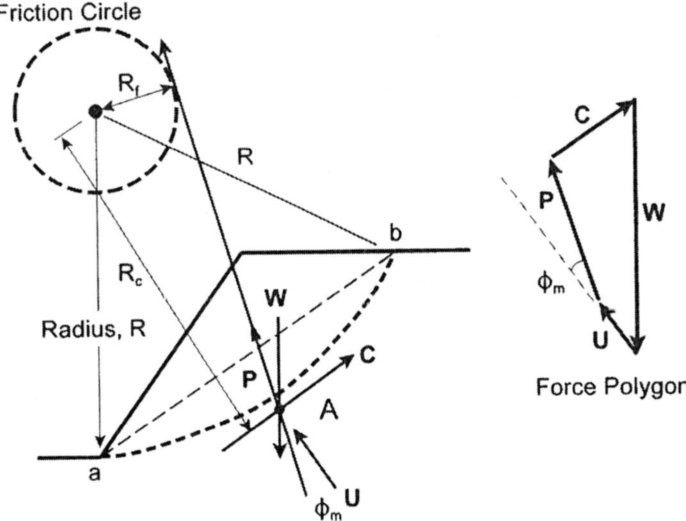

Figure 6.19 Friction circle procedure.

strength mobilized along the failure surface. This direction corresponds to a line that forms a tangent to the *friction circle*, with a radius, $R_f = R \sin \phi_m$. This is equivalent to assuming that the resultant of all normal stresses acting on the failure surface is concentrated at one point. This assumption is *guaranteed* to give a lower bound FOS value (Lambe and Whitman, 1969). The cohesive shear stresses along the base of the failure surface, line *ab* in Figure 6.19, will have a resultant, C_m, that acts parallel to the direction of the chord "*ab*". Its location may be found by taking moments of the distribution and the resultant, C_m, about the circle center. This line of action of resultant C_m can be located using

$$R_c = \frac{L_{\text{arc}}}{L_{\text{chord}}} \times R \qquad \text{(Eq. 6-27)}$$

where R = radius of failure circle
R_c = perpendicular distance from the circle center to force, C_m
L_{arc} = length of the circular arc defining the failure surface
L_{chord} = length of the chord defining the failure surface

The actual point of application, A, is located at the intersection of the *effective* weight force, which is the resultant of the weight and any pore water forces. The resultant of the normal and frictional (shear) force, P, will then be inclined parallel to a line formed by a point of tangency to the friction circle and point A. As the direction of C_m is known, the force polygon can be closed to obtain the value of the mobilized cohesive force. Again, the final FOS is computed with the assumption $F_\phi = F_c =$ FOS.

The solution procedure is usually followed graphically, although there are a few numerical schemes that have been established to generate the answer directly. For general ease of use, the following modified procedure is recommended:

(1) Calculate weight of slide, W.
(2) Calculate magnitude and direction of the resultant pore water force, U (may need to discretize slide into slices).
(3) Calculate perpendicular distance to the line of action of C_m.
(4) Find effective weight resultant, W', from forces W and U, and its intersection with the line of action of C_m at A.
(5) Assume a value of F_ϕ.
(6) Calculate the mobilized friction angle, $\phi_m = \tan^{-1}(\tan\phi / F_\phi)$.
(7) Draw the friction circle, with radius $R_f = R \sin\phi_m$.
(8) Draw the force polygon with W' appropriately inclined, and passing through point A.
(9) Draw the direction of P, tangential to the friction circle.
(10) Draw direction of C_m, according to the inclination of the chord linking the endpoints of the circular failure surface.
(11) The closed polygon will then provide the value of C_m.
(12) Using this value of C_m, calculate $F_c = cL_{\text{chord}}/C_m$.
(13) Repeat steps 5 to 12 until $F_c \approx F_\phi$.

6.8.4 Friction Circle Example

The example shown earlier in Figure 6.18 (Lambe and Whitman, 1969), with the phreatic surface included, will be analyzed to illustrate the proposed approach for calculating the FOS using the friction circle method. The soil parameters for the example homogeneous, single soil slope are: $\phi' = 32°$, $c' = 90$ pounds per square foot, and $\gamma = 125$ pounds per cubic foot. The following values were calculated for one cycle of the computation:

(1) $W = 26.5$ kips.
(2) $U = 4.9$ kips, inclined at 29.4° (from vertical).
(3) $R_c = R \times L_{\text{arc}}/L_{\text{chord}} = 30 \times (41.3 \div 38) = 32.9$ feet
(4) See Figure 6.20 for resultant W' and location of A.
(5) Assume $F_\phi = 1.34$.
(6) Then $\phi_m = 25$ degrees.
(7) Radius for friction circle, $R_f = 12.6$ feet.
(8)–(11) See Figure 6.20 for final force polygon.
(12) $F_c = (38 \times 90) \div 3{,}150 = 1.09$.
(13) As $F_c \neq F_\phi$, steps 5 to 12 need to be repeated.

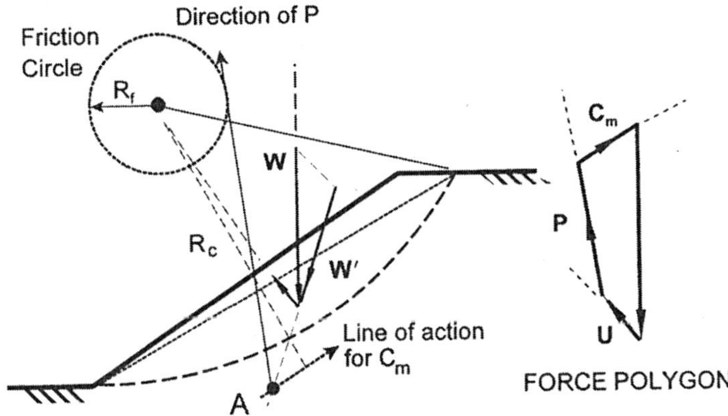

Figure 6.20 Example solution for friction circle method.

If the calculation cycle is repeated with improving values of F_ϕ, a final FOS of 1.27 is calculated for this slope.

6.9 METHOD OF SLICES

The methods discussed earlier do not depend on the distribution of the effective normal stresses along the failure surface. However, if the mobilized strength for a c–ϕ soil is to be calculated, the distribution of the effective normal stresses along the failure surface must be known. This condition is usually analyzed by discretizing the mass of the failure slope into smaller slices and treating each individual slice as a unique sliding block. The *method of slices* is used by most computer programs, as it can readily accommodate complex slope geometries, variable soil conditions, and the influence of external boundary loads. This section presents the formulation of several methods of analysis, including the powerful general limit equilibrium (GLE) method.

All limit equilibrium methods for slope stability analysis divide a slide-mass into n smaller slices, as shown in Figure 6.21. Each slice is affected by a general system of forces, as shown in Figure 6.22. The thrust line indicated in the figure connects the points of application of the interslice forces, Z_i. The location of this thrust line may be assumed, as in the rigorous Janbu method (Janbu, 1954a, 1954b, 1973), or its location may be determined using a rigorous method of analysis that satisfies complete equilibrium. The popular simplified methods of analysis neglect the location of the interslice force because complete equilibrium is not satisfied for the failure mass.

For this system, there are $(6n - 2)$ unknowns, as listed in Table 6.4. Also, since only four equations can be written for the limit equilibrium of the system, the solution is statically indeterminate. However, a solution is possible providing the number of unknowns can be reduced by making some simplifying assumptions. One of the common assumptions is that the normal force on the base of the slice acts at the midpoint,

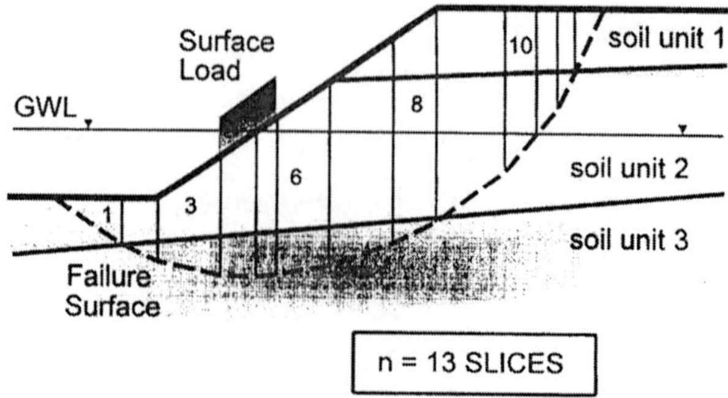

Figure 6.21 Division of potential sliding mass into slices.

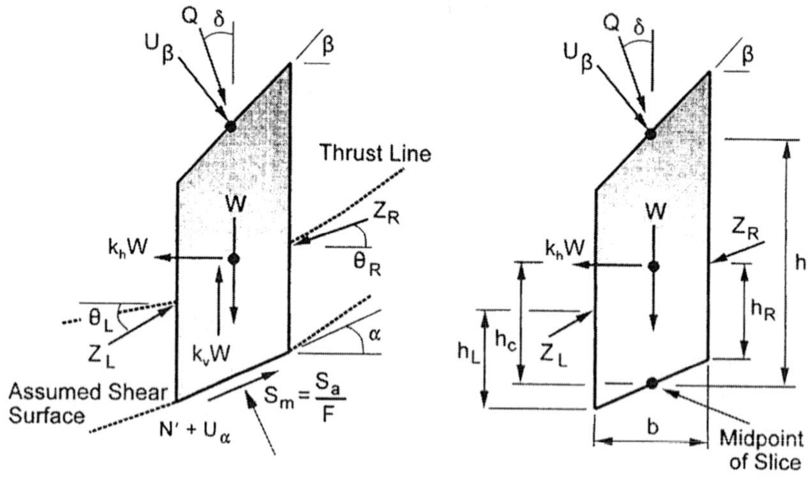

F = factor of safety
S_a = available strength
= $C + N' \tan\phi$
S_m = mobilized strength
U_α = pore water force
U_β = surface water force
W = weight of slice
N' = effective normal force
Q = external surcharge
k_v = vertical seismic coefficient
k_h = horiz. seismic coefficient

Z_L = left interslice force
Z_R = right interslice force
θ_L = left interslice force angle
θ_R = right interslice force angle
h_L = height to force Z_L
h_R = height to force Z_R
α = inclination of slice base
β = inclination of slice top
δ = inclination of surcharge
b = width of slice
h = average height of slice
h_c = height to centroid of slice

Figure 6.22 Forces acting on a typical slice.

TABLE 6.4 Equations and Unknowns Associated with the Method of Slices

Equations	Condition
n	Moment equilibrium for each slice
$2n$	Force equilibrium in two directions (for each slice)
n	Mohr–Coulomb relationship between shear strength and normal effective stress
$4n$	**Total number of equations**
Unknowns	Variable
1	Factor of safety
n	Normal force at base of each slice, N'
n	Location of normal force, N'
n	Shear force at base of each slice, S_m
$n-1$	Interslice force, Z
$n-1$	Inclination of interslice force, θ
$n-1$	Location of interslice force (line of thrust)
$6n-2$	**Total number of unknowns**

thus reducing the number of unknowns to $(5n - 2)$. This then requires an additional $(n - 2)$ assumptions to make the problem determinate. It is these assumptions that generally categorizes the available methods of analysis (Sharma and Lovell, 1983).

Table 6.5 lists the common methods of analysis and the conditions of static equilibrium that are satisfied in determining the FOS. The assumptions made by each of these methods, to render the problem determinate, are also summarized below.

TABLE 6.5 Static Equilibrium Conditions Satisfied by Limit Equilibrium Methods

Method	Force Equilibrium x	Force Equilibrium y	Moment Equilibrium
Ordinary method of slices (OMS)	No	No	Yes
Bishop's simplified	Yes	No	Yes
Janbu's simplified	Yes	Yes	No
Lowe and Karafiath's	Yes	Yes	No
Corps of Engineers	Yes	Yes	No
Spencer's	Yes	Yes	Yes
Bishop's rigorous	Yes	Yes	Yes
Janbu's generalized	Yes	Yes	No
Sarma's	Yes	Yes	Yes
Morgenstern–Price	Yes	Yes	Yes

Ordinary Method of Slices (OMS) This method (Fellenius, 1927, 1936) neglects all interslice forces and fails to satisfy force equilibrium for the slide mass as well as for individual slices. However, this is one of the simplest procedures based on the method of slices.

Bishop's Simplified Method Bishop (1955) assumes that all interslice shear forces are zero, reducing the number of unknowns by $(n-1)$. This leaves $(4n-1)$ unknowns, leaving the solution overdetermined as horizontal force equilibrium will not be satisfied for one slice.

Janbu's Simplified Method Janbu (1954a, 1954b, 1973) also assumes zero interslice shear forces, reducing the number of unknowns to $(4n-1)$. This leads to an overdetermined solution that will not completely satisfy moment equilibrium conditions. However, Janbu presented a correction factor, f_0, to account for this inadequacy.

Lowe and Karafiath's Method Lowe and Karafiath (1960) assume that the interslice forces are inclined at an angle equal to the average of the ground surface and slice base angles, that is, $\theta = \frac{1}{2}(\alpha + \beta)$, where θ is the assumed inclination of the interslice force on the right-hand side of the typical slice shown in Figure 6.22. This simplification leaves $(4n-1)$ unknowns and fails to satisfy moment equilibrium.

Corps of Engineers Method The Corps of Engineers (1970) approach considers the inclination of the interslice force as either (1) parallel to ground surface (i.e., $\theta = \beta$), or (2) equal to the average slope angle between the left and right endpoints of the failure surface. The approach is similar to the one proposed by Lowe and Karafiath (1960) and presents an overdetermined system where moment equilibrium is not satisfied for all slices.

Spencer's Method Spencer (1967, 1973) proposes a method that rigorously satisfies static equilibrium by assuming that the resultant interslice force has a constant, but unknown, inclination. These $(n-1)$ assumptions again reduce the number of unknowns to $(4n-1)$, but the unknown inclination is an additional component that subsequently increases the number of unknowns to match the required $4n$ equations.

Bishop's Rigorous Method Bishop (1955) assumes $(n-1)$ interslice shear forces to calculate an FOS. Since this assumption leaves $(4n-1)$ unknowns, moment equilibrium cannot be directly satisfied for all slices. However, Bishop introduces an additional unknown by suggesting that there exists a unique distribution of the interslice resultant force, out of a possible infinite number, that will rigorously satisfy the equilibrium equations.

Janbu's Generalized Method Janbu (1954a, 1954b, 1973) assumes a location of the thrust line, thereby reducing the number of unknowns to $(4n-1)$. Sarma (1979) points out that the position of the normal stress on the last (uppermost) slice is not used and hence moment equilibrium is not satisfied for this last slice. However, similar to the rigorous Bishop method, Janbu's generalized method also suggests that the actual location of the thrust line is an additional unknown, and thus equilibrium can be satisfied rigorously if the assumption selects the correct thrust line.

Sarma's Method Sarma (1973) uses the method of slices to calculate the magnitude of a horizontal *seismic coefficient* needed to bring the failure mass into a state of limiting equilibrium. This allows the procedure to develop a relationship between the seismic coefficient and the *presumed* FOS. The static FOS will then correspond to the case of a zero seismic coefficient. Sarma uses an interslice force distribution function (similar to Morgenstern and Price, 1965) and the value of the seismic coefficient can be calculated directly for the presumed FOS. All equilibrium conditions are satisfied by this method. However, it should be noted that the critical surface corresponding to the statis FOS (for a zero seismic coefficient) will often be different than the surface determined using the more conventional approach where the FOS is treated as an unknown.

Morgenstern–Price Method Morgenstern and Price (1965) proposed a method that is similar to Spencer's method, except that the inclination of the interslice resultant force is assumed to vary according to a "portion" of an arbitrary function. This additional "portion" of a selected function introduces an additional unknown, leaving $4n$ unknowns and $4n$ equations.

A general limit equilibrium (GLE) formulation (Chugh, 1986; Fredlund et al., 1981) can be developed to encompass most of the assumptions used by the various methods and may be used to analyze circular and noncircular failure surfaces. In view of this universal applicability, the GLE formulation has become one of the most popular methods as its generalization offers the ability to model a discrete version of the Morgenstern and Price (1965) procedure via the function used to describe the distribution of the interslice force angles. The method can be used to satisfy either force and moment equilibrium, or if required, just the force equilibrium conditions. With this approach, Spencer's (1973) method can be implemented directly via the use of a constant interslice force function.

The GLE procedure relies on the selection of an appropriate function that describes the variation of the interslice force angles to satisfy complete equilibrium. In contrast, Janbu's general procedure of slices (GPS) assumes the location of the thrust line and then proceeds to calculate the interslice force angles required to satisfy complete equilibrium. Intuitively, it is perhaps easier to assume the location of the thrust line, but in reality, the computations are very sensitive and can often lead to numerical problems, such as the failure to converge to a successful solution of the FOS.

The main difficulty in using the GLE procedure is related to the requirement that the user verify the reliability and "reasonableness" of the calculated FOS. This additional complexity prevents the general use of the GLE method for automatic search procedures that attempt to identify the critical failure surface. However, single failure surfaces can be analyzed and the detailed solution examined for reasonableness. For judging the viability of the reported FOS values, a substantially higher level of familiarity and understanding of the underlying assumptions, algorithms, and objectives is now required for practical problems.

Conversely, the simplified Bishop (1955) and Janbu (1954a, 1954b, 1973) methods are popular because an FOS value can be quickly calculated for most surfaces. However, these methods do not satisfy complete force and moment equilibrium, and thus different values of the FOS will be calculated in comparison with the methods that satisfy conditions of complete equilibrium. For circular failure surfaces, the FOS computed by Bishop's method is usually greater than the value from Janbu's formulation. Bishop's FOS value is also generally within 5 percent of the FOS value that may be calculated using a more rigorous approach, such as the GLE method. So for the analysis of circular failure surfaces, the simplified Bishop's method is strongly recommended for analysis. However, Janbu's method is more flexible as the formulation may be applied to evaluate the FOS for circular and noncircular surfaces.

In order to calculate an FOS for a slope, it is important that the geotechnical engineer be familiar with the formulation used by the limit equilibrium methods. The complexity of these procedures ranges from the simple ordinary method, which is suitable for hand calculations, to the rigorous methods, such as Spencer's method, which really do require the use of a computer. The complete equations for several popular limit equilibrium methods, as well as GLE procedure, are presented next.

6.9.1 Ordinary Method of Slices (OMS)

The OMS was one of the earliest analytical methods that used the method of slices to estimate the stability of a slope. The method assumes that the resultant of the interslice forces for all slices is inclined at an angle that is *parallel* to the base of the slice. Please note that this simplifying assumption fails to satisfy interslice equilibrium where adjacent slices have different base inclinations. This is the main shortcoming of this method and leads to the calculation of inconsistent effective stresses at the base of the slices.

If the slice forces are resolved in a direction perpendicular to the base of the slice shown in Figure 6.22, then

$$\sum F_\alpha = N' + U_\alpha + k_h W \sin\alpha - W(1 - k_v)\cos\alpha - U_\beta \cos(\beta - \alpha) - Q\cos(\delta - \alpha) = 0 \quad \text{(Eq. 6-28)}$$

The above equation may be arranged for N' as

$$N' = -U_\alpha - k_h W \sin\alpha + W(1 - k_v)\cos\alpha + U_\beta \cos(\beta - \alpha) + Q\cos(\delta - \alpha) \quad \text{(Eq. 6-29)}$$

With the FOS against shear failure defined as F, and assumed to be the same for all slices, the Mohr–Coulomb mobilized shear strength, S_m, along the base of each slice is given by

$$S_m = \frac{C + N' \tan\phi}{F} \qquad \text{(Eq. 6-30)}$$

where C and $N' \tan\phi$ are the cohesive and frictional shear strength components of the soil. The overall moment equilibrium of the forces about the center of the circular failure surface for each slice is given by

$$\begin{aligned}\sum M_0 = &\sum_{i-1}^{n}[W(1-k_v) + U_\beta \cos\beta + Q\cos\delta]R\sin\alpha \\ &- \sum_{i=1}^{n}[U_\beta \sin\beta + Q\sin\delta](R\cos\alpha - h) \\ &- \sum_{i=1}^{n}[S_m]R + \sum_{i-1}^{n}[k_h W(R\cos\alpha - h_c)] \\ = 0 \end{aligned} \qquad \text{(Eq. 6-31)}$$

where R = radius of the circular failure surface
h = average height of the slice
h_c = vertical height between center of the base slice and the centroid of the slice

The influence of the internal interslice forces has been excluded from this expression as their net resultant moment will be zero. The above equation may be simplified by dividing throughout by the radius to get

$$\begin{aligned}\frac{\sum M_0}{R} = &\sum_{1}^{n}[W(1-k_v) + U_\beta\cos\beta + Q\cos\delta]\sin\alpha \\ &- \sum_{1}^{n}[S_m] - \sum_{1}^{n}[U_\beta \sin\beta + Q\sin\delta]\left(\cos\alpha - \frac{h}{R}\right) \\ &+ \sum_{1}^{n}\left[k_h W\left(\cos\alpha - \frac{h_c}{R}\right)\right]\end{aligned} \qquad \text{(Eq. 6-32)}$$

If the FOS is assumed to be the same for all slices, substitute Equation 6-30 into Equation 6-32 to give

$$F = \frac{\sum_{i=1}^{n}(C + N'\tan\phi)}{\sum_{i=1}^{n} A_1 - \sum_{i=1}^{n} A_2 + \sum_{i=1}^{n} A_3} \qquad \text{(Eq. 6-33)}$$

where

$$\begin{aligned} A_1 &= (W(1-k_v)+U_\beta \cos\beta + Q\cos\delta)\sin\alpha \\ A_2 &= (U_\beta \sin\beta + Q\sin\delta)\left(\cos\alpha - \frac{h}{R}\right) \\ A_3 &= k_h W\left(\cos\alpha - \frac{h_c}{R}\right) \end{aligned} \quad \text{(Eq. 6-34)}$$

and N' is given by Equation 6-29. This is the formulation that is often used to compute the FOS according to the assumptions of the OMS.

6.9.2 Simplified Janbu Method

The simplified (or modified) Janbu method uses the method of slices to determine the stability of the slide mass. It is based on the forces shown, in Figure 6.22, for the free-body diagram of a typical slice. The simplified procedure assumes that there are no interslice shear forces. The geometry of each slice is described by its height, h, measured along its centerline, its width, b, and by the inclinations of its base and top, α and β respectively.

Janbu's method satisfies vertical force equilibrium for each slice, as well as overall horizontal force equilibrium for the entire slide mass (i.e., all slices). Vertical force equilibrium for each slice is given by

$$\begin{aligned} \sum F_v &= (N'+U_\alpha)\cos\alpha + S_m \sin\alpha - W(1-k_v) - U_\beta\cos\beta - Q\cos\delta \\ &= 0 \end{aligned} \quad \text{(Eq. 6-35)}$$

The above equation may be arranged for N' as

$$N' = \frac{-U_\alpha\cos\alpha - S_m\sin\alpha + W(1-k_v) + U_\beta\cos\beta + Q\cos\delta}{\cos\alpha} \quad \text{(Eq. 6-36)}$$

If the FOS against shear failure is defined as F, and is assumed to be the same for all slices, the Mohr–Coulomb mobilized shear strength, S_m, along the base of each slice is given by

$$S_m = \frac{C + N'\tan\phi}{F} \quad \text{(Eq. 6-37)}$$

where C and $N'\tan\phi$ are the cohesive and frictional shear strength components of the soil. By substituting Equation 6-37 into 6-36, the effective normal force acting at the base of the slice can be determined as

$$N' = \frac{1}{m_\alpha}\left[W(1-k_v) - \frac{C\sin\alpha}{F} - U_\alpha\cos\alpha + U_\beta\cos\beta + Q\cos\delta\right] \quad \text{(Eq. 6-38)}$$

where

$$m_\alpha = \cos\alpha \left[1 + \frac{\tan\alpha \tan\phi}{F} \right] \qquad \text{(Eq. 6-39)}$$

Next, the overall horizontal force equilibrium is evaluated for all slices of the slide mass. In this case, for an individual slice, i,

$$[F_H]_i = (N' + U_\alpha)\sin\alpha + Wk_h - U_\beta \sin\beta - Q\sin\delta - S_m\cos\alpha \qquad \text{(Eq. 6-40)}$$

Then, after substituting for S_m from Equation 6-37 and rearranging, overall horizontal force equilibrium for the slide mass is given by

$$\begin{aligned} \sum_{i=1}^{n}[F_H]_i &= \sum_{i=1}^{n}[(N' + U_\alpha)\sin\alpha + Wk_h - U_\beta\sin\beta] \\ &\quad - \sum_{i=1}^{n}\left[Q\sin\delta + \frac{C + N'\tan\phi}{F}\cos\alpha \right] \\ &= 0 \end{aligned} \qquad \text{(Eq. 6-41)}$$

By rearranging the above equation, the following expression may be obtained:

$$\begin{aligned} &\sum_{i=1}^{n}[(N' + U_\alpha)\sin\alpha + Wk_h - U_\beta\sin\beta - Q\sin\delta] \\ &\quad = \sum_{i=1}^{n}\left[\frac{1}{F}(C + N'\tan\phi)\cos\alpha \right] \end{aligned} \qquad \text{(Eq. 6-42)}$$

Then if each slice has the same FOS, F,

$$F = \frac{\sum_{i=1}^{n}[C + N'\tan\phi]\cos\alpha}{\sum_{i=1}^{n} A_4 + \sum_{i=1}^{n} N'\sin\alpha} \qquad \text{(Eq. 6-43)}$$

where N' is given by Equation 6-38, and the term A_4 is given by

$$A_4 = U_\alpha \sin\alpha + Wk_h - U_\beta\sin\beta - Q\sin\delta \qquad \text{(Eq. 6-44)}$$

Equation 6-43 essentially represents a ratio of the *available* shear strength and the *driving* shear force along the failure surface. This format allows the state of the effective stress to be determined and appropriate corrections implemented if N' is calculated to be less than zero, as discussed in a later section.

The reported Janbu FOS value is calculated by multiplying the calculated F value by a modification factor, f_0,

$$\text{FOS}_{\text{Janbu}} = f_0 \cdot F_{\text{calculated}}$$

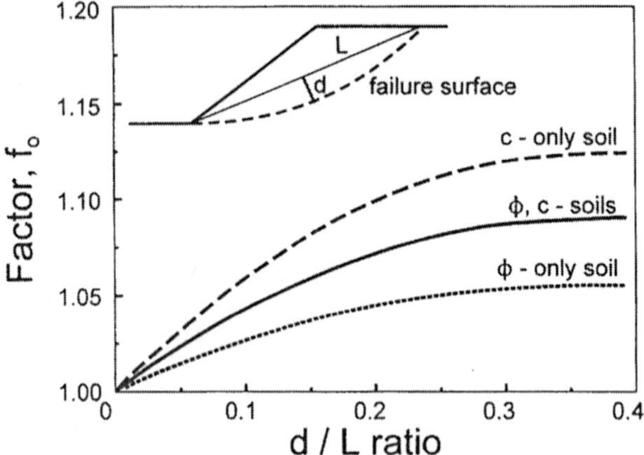

Figure 6.23 Janbu's correction factor for the simplified method.

This modification factor is a function of the slide geometry and the strength parameters of the soil. Figure 6.23 illustrates the variation of the f_0 value as a function of the slope geometry (i.e., d and L) and the type of soil.

These curves were presented by Janbu in an attempt to compensate for the assumption of negligible interslice shear forces ($Z \sin \alpha$) in his formulation for the simplified method. Janbu then performed calculations using his simplified and rigorous (i.e., satisfying complete equilibrium) methods for the same slopes with homogeneous soil conditions. The subsequent comparison between the simplified and rigorous FOS values was used to develop the correction curves shown in Figure 6.23.

There is no consensus concerning the selection of the appropriate f_0 value for a surface intersecting different soil types consisting of c only, ϕ only, or both c and ϕ soils. In cases where such a mixed variety of soils is present, the c–ϕ curve is generally used to correct the calculated FOS value.

For convenience, this modification factor can also be calculated according to the formula

$$f_0 = 1 + b_1 \left[\frac{d}{L} - 1.4 \left(\frac{d}{L} \right)^2 \right] \qquad \text{(Eq. 6-45)}$$

where b_1 varies according to the soil type: c only soils: $b_1 = 0.69$
ϕ only soils: $b_1 = 0.31$
c and ϕ soils: $b_1 = 0.50$

The appropriate b_1 value is selected for use in Equation 6-45 according to the type (i.e., c only, ϕ only, or both c and ϕ) of soil encountered *along* the analyzed failure surface. If a mixed-soil type is encountered, use the c and ϕ soil relationship described by the above expression.

6.9.3 Simplified Bishop Method

The simplified Bishop method also uses the method of slices to discretize the soil mass for determining the FOS. This method satisfies vertical force equilibrium for each slice and overall moment equilibrium about the center of the *circular* trial surface. The simplified Bishop method also assumes zero interslice shear forces. Using the notation shown in Figure 6.22, the overall moment equilibrium of the forces acting on each slice is given by

$$\sum M_0 = \sum_{i=1}^{n}[W(1-k_v)+U_\beta \cos\beta + Q\cos\delta]R\sin\alpha$$
$$-\sum_{i-1}^{n}[U_\beta \sin\beta + Q\sin\delta](R\cos\alpha - h)$$
$$-\sum_{i-1}^{n}[S_m]R+\sum_{i=1}^{n}[k_h W(R\cos\alpha - h_c)] = 0 \qquad \text{(Eq. 6-46)}$$

where R = radius of the circular failure surface
h = average height of the slice
h_c = vertical height between center of base slice and centroid of the slice

The above equation may be simplified by dividing throughout by the radius to get

$$\frac{\sum M_0}{R} = \sum_{1}^{n}[(W(1-k_v)+U_\beta\cos\beta + Q\cos\delta)]\sin\alpha$$
$$-\sum_{1}^{n}[S_m]-\sum_{1}^{n}[U_\beta\sin\beta + Q\sin\delta]\left(\cos\alpha - \frac{h}{R}\right)$$
$$+\sum_{1}^{n}\left[k_h W\left(\cos\alpha - \frac{h_c}{R}\right)\right] \qquad \text{(Eq. 6-47)}$$

Please note that the effective normal and pore pressure forces, acting on the base of the slice, do not affect the moment equilibruim expression since they are directed through the center of the circle. *Thus Bishop's method should not be used to compute an FOS for noncircular surfaces.*

If the FOS is assumed to be the same for all slices, substitute the Mohr–Coulomb criterion from Equation 6-37 into Equation 6-47 to give

$$F = \frac{\sum_{i=1}^{n}(C+N'\tan\phi)}{\sum_{i=1}^{n}A_5 - \sum_{i=1}^{n}A_6 + \sum_{i=1}^{n}A_7} \qquad \text{(Eq. 6-48)}$$

where

$$A_5 = (W(1-k_v) + U_\beta \cos\beta + Q\cos\delta)\sin\alpha$$

$$A_6 = (U_\beta \sin\beta + Q\sin\delta)\left(\cos\alpha - \frac{h}{R}\right) \quad \text{(Eq. 6-49)}$$

$$A_7 = k_h W\left(\cos\alpha - \frac{h_c}{R}\right)$$

Next, forces are summed in the vertical direction for each slice to determine the effective normal force in the same manner as used for Janbu's method:

$$N' = \frac{1}{m_\alpha}\left[W(1-k_v) - \frac{C\sin\alpha}{F} - U_\alpha\cos\alpha + U_\beta\cos\beta + Q\cos\delta\right] \quad \text{(Eq. 6-50)}$$

where m_α is again given by

$$m_\alpha = \cos\alpha\left[1 + \frac{\tan\alpha\tan\phi}{F}\right] \quad \text{(Eq. 6-51)}$$

Equations 6-48 through 6-51 are the expressions that are used to calculate the FOS for circular surfaces according to the simplified Bishop method.

6.9.4 Generalized Limit Equilibrium (GLE) Method

The GLE method is an extension of Spencer's (1973) procedure, which has been generalized by Chugh (1986). The GLE method adopts a function, $\theta_i = \lambda \cdot f(x_i)$ to assign the interslice force angle on the right-hand side of slice i, as shown in Figure 6.22. The function, $f(x_i)$, ranges between 0 and 1 and essentially represents the shape of the distribution used to describe the variation of the interslice foce angles, as shown in Figure 6.24. The adoption of this function satisfies $(n-1)$ assumptions regarding the interslice force angles and the λ value is an additional unknown, which is introduced such that there are the requisite $(n-2)$ unknowns, as discussed earlier. The selected interslice force angle function, $f(x)$, can be set as a constant (i.e., $f(x) = 1.0$) to emulate Spencer's procedure, or any other shape for a discrete version of a Morgenstern–Price solution.

The adopted formulation uses a discrete form of the continuous function, $f(x)$, to calculate the function at each interslice boundary, using the angles labeled θ_L and θ_R for the left and right vertical sides of the slice, as shown in Figure 6.22. Thus for a typical interslice boundary, $\theta_R = \lambda f(x)$, where x is the x coordinate of the right side of the selected slice. This distribution is usually implemented with a function that is normalized with respect to the lateral (horizontal) extent of the failure surface. As the interslice force angle for the left side of the first slice (at the toe) and the right side of the last slice (at the crest) is assumed to be zero, this lateral extent is assumed to range *between* the first and last interslice boundary.

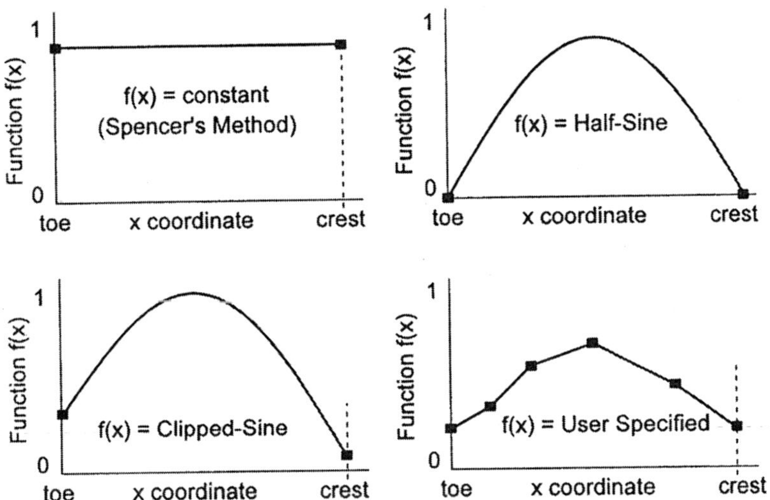

Figure 6.24 Examples of functions used to describe the variation of interslice force angles.

Force Equilibrium The GLE method assumes that the interslice resultant forces, Z_L and Z_R, are inclined at θ_L and θ_R on the left and right side of each slice, as shown in Figure 6.22. These interslice forces are *total* forces as the hydrostatic component along the interslice boundaries is not considered separately. Interslice hydrostatic forces can be considered in an analysis, but are difficult to implement for layered soils and multiple water surfaces. If force equilibrium is considered in a direction parallel to the base of each slice,

$$S_m + Z_L \cos(\alpha - \theta_L) - Z_R \cos(\alpha - \theta_R) - W(1 - k_v) \sin\alpha \\ - W k_h \cos\alpha - U_\beta \sin(\alpha - \beta) - Q \sin(\alpha - \delta) = 0 \qquad \text{(Eq. 6-52)}$$

and if the Mohr–Coulomb strength criterion is adopted such that the mobilized strength becomes

$$S_m = \frac{S_a}{F} = \frac{C}{F} + N' \frac{\tan\phi}{F} = C_m + N' \tan\phi_m \qquad \text{(Eq. 6-53)}$$

then by substituting Equation 6-53 into 6-52, the following expression is derived:

$$\begin{aligned} N' \tan\phi_m &= Z_R \cos(\alpha - \theta_R) - Z_L \cos(\alpha - \theta_L) \\ &\quad + W[(1 - k_v) \sin\alpha + k_h \cos\alpha] - C_m \\ &\quad + U_\beta \sin(\alpha - \beta) + Q \sin(\alpha - \delta) \end{aligned} \qquad \text{(Eq. 6-54)}$$

Next force equilibrium is formulated in a direction normal to the base of the slice:

$$N' + Z_R \sin(\alpha - \theta_R) - Z_L \sin(\alpha - \theta_L) - W(1 - k_v)\cos\alpha$$
$$+ Wk_h \sin\alpha + U_\alpha - U_\beta \cos(\alpha - \beta) - Q\cos(\alpha - \delta) = 0 \quad \text{(Eq. 6-55)}$$

By substituting Equation 6-55 into 6-54, the following force equilibrium equation may be formulated:

$$\begin{aligned} Z_R = {} & A_8 Z_L[\cos(\alpha - \theta_L) + \sin(\alpha - \theta_L)\tan\phi_m] \\ & + A_8\,[W\cos\alpha(1 - k_v)(\tan\phi_m - \tan\alpha) + C_m \\ & - U_\alpha \tan\phi_m - Wk_h(1 + \tan\phi_m \tan\alpha)\cos\alpha \\ & + U_\beta[\cos(\alpha - \beta)\tan\phi_m - \sin(\alpha - \beta)] \\ & + Q[\cos(\alpha - \delta)\tan\phi_m - \sin(\alpha - \delta)]] \end{aligned} \quad \text{(Eq. 6-56)}$$

where the factor A_8 is given by

$$A_8 = \frac{1}{\cos(\alpha - \theta_R)[1 + \tan\phi_m \tan(\alpha - \theta_R)]} \quad \text{(Eq. 6-57)}$$

Moment Equilibrium The conditions for moment equilibrium are satisfied by taking moments of all slice forces about the midpoint of the base of the slice, as shown in Figure 6.22, generating the following expression:

$$Z_L \cos\theta_L \left[h_L - \frac{b}{2}\tan\alpha\right] + Z_L \frac{b}{2}\sin\theta_L - Z_R \cos\theta_R \left[h_R + \frac{b}{2}\tan\alpha\right]$$
$$+ Z_R \frac{b}{2}\sin\theta_R - Wk_h h_c + U_\beta h \sin\beta + Qh\sin\delta = 0 \quad \text{(Eq. 6-58)}$$

Next the above expression is simplified to determine the location of the interslice force, h_R, on the right-hand side of each slice using

$$\begin{aligned} h_R = {} & \frac{Z_L}{Z_R \cos\theta_R}\left[h_L \cos\theta_L - \frac{b}{2}(\cos\theta_L \tan\alpha + \sin\theta_L)\right] \\ & + \frac{1}{Z_R \cos\theta_R}[h(U_\beta \sin\beta + Q\sin\delta) - h_c K_h W] \\ & + \frac{b}{2}[\tan\theta_R - \tan\alpha] \end{aligned} \quad \text{(Eq. 6-59)}$$

The GLE procedure uses Equations 6-56 and 6-59 iteratively to satisfy complete moment and force equilibrium for all slices. Once the FOS has been determined, the *total* normal, vertical, and shear stresses at the base of each slice should be calculated using

$$\sigma_n = \frac{1}{b \sec\alpha}\{Z_L \sin(\alpha - \theta_L) - Z_R \sin(\alpha - \theta_R) + U_\beta \cos(\alpha - \beta)$$
$$- U_\alpha + W\,[(1 - k_v)\cos\alpha - k_h \sin\alpha] + Q\cos(\alpha - \delta)\} \quad \text{(Eq. 6-60)}$$

$$\sigma_v = \frac{W + Q\cos\delta + U_\beta \cos\beta}{b \sec\alpha} \quad \text{(Eq. 6-61)}$$

$$\tau_{\text{base}} = c_m + \sigma_n' \tan\phi_m \quad \text{(Eq. 6-62)}$$

Solution Procedure The GLE solution is computed using the following steps:

(1) Assume an interslice force angle distribution with θ_L for the first slice and θ_R for the last slice set to zero.

(2) Determine the FOS, F, that allows Equations 6-56 and 6-59 to satisfy force equilibrium such that Z_R for the last slice (at the crest) is equal to the boundary force. This force will be equal to the hydrostatic water force in a water-filled crack at the crest of the slope. If there is no water-filled crack, this boundary force will be zero.

(3) Retain the calculated interslice forces, Z_L and Z_R, that were part of the solution for the FOS.

(4) Using the interslice forces from step 3, use Equation 6-59 to calculate the magnitude of the interslice force angles, θ_R, that satisfy moment equilibrium such that h_R for the last slice is zero or equal to the location of the horizontal hydrostatic force in a water-filled crack. These calculations are performed sequentially for each slice, starting with the knowledge that θ_L and h_L for the first slice (at the toe) will be zero.

(5) Repeat steps 2 to 4, until the calculated FOSs and the interslice force angles are within a tolerable limit.

(6) Calculate the total normal, vertical, and shear stresses at the base of each slice, using Equations 6-60 through 6-62, to allow the user to evaluate the reasonableness of the reported FOS.

6.9.5 Janbu's Generalized Procedure of Slices (GPS)

The GLE formulation may also be used to calculate the FOS for Janbu's GPS. The GPS approach satisfies complete force equilibrium for all slices and moment equilibrium for all but the last slice. Force equilibrium is satisfied by implementing Equations 6-54 and 6-56 for the slide mass and moment equilibrium is satisfied for individual slices sequentially, beginning at the toe and ending at the last slice near the crest of the slope. As the location of the thrust line is assumed for this case, moment equilibrium can be satisfied by adjusting the interslice force angles.

Using Equation 6-58, derived for the GLE procedure, moment equilibrium about the center of the slice base is satisfied if

$$Z_R \cos\theta_R \left[h_R + \frac{b}{2}\tan\alpha\right] - Z_R \frac{b}{2}\sin\theta_R = Z_L \cos\theta_L \left[h_L - \frac{b}{2}\tan\alpha\right]$$
$$+ Z_L \frac{b}{2}\sin\theta_L - Wk_h h_c + U_\beta h \sin\beta + Qh\sin\delta \qquad \text{(Eq. 6-63)}$$

In the above equation, θ_R is the only unknown if θ_L for the first slice is assumed to be zero. The interslice forces, Z_L and Z_R, calculated for force equilibrium are considered to be approximately valid. In order to solve this equation for θ_R, Equation 6-63 can be rewritten in the format:

$$A\sin(\psi - \theta_R) = B \qquad \text{(Eq. 6-64)}$$

where the expansion of the above expression gives:

$$A\sin\psi\cos\theta_R - A\cos\psi\sin\theta_R = \left[h_R + \frac{b}{2}\tan\alpha\right]\cos\theta_R - \frac{b}{2}\sin\theta_R \qquad \text{(Eq. 6-65)}$$

and the right side is equivalent to

$$B = \frac{Z_L}{Z_R}\cos\theta_L\left[h_L - \frac{b}{2}\tan\alpha\right] + \frac{b}{2}\frac{Z_L}{Z_R}\sin\theta_L$$
$$- \frac{1}{Z_R}[Wk_h h_c + U_\beta h\sin\beta + Qh\sin\delta] \qquad \text{(Eq. 6-66)}$$

In seeking a solution for θ_R, the above expression may be used to determine ψ initially, and finally θ_R, using the following:

$$A\sin\psi = h_R + \frac{b}{2}\tan\alpha$$
$$A\cos\psi = \frac{b}{2} \qquad \text{(Eq. 6-67)}$$

The above expressions can be combined to give

$$\tan\psi = \frac{2h_R}{b} + \tan\alpha$$
$$A^2 = \left(\frac{b}{2}\right)^2 + \left(h_R + \frac{b}{2}\tan\alpha\right)^2 \qquad \text{(Eq. 6-68)}$$

and as

$$\sin(\psi - \theta_R) = \frac{B}{A} \qquad \text{(Eq. 6-69)}$$

Therefore

$$\theta_R = \psi - \sin^{-1}\left(\frac{B}{A}\right)$$

$$= \tan^{-1}\left(\frac{2h_R}{b} + \tan\alpha\right) - \sin^{-1}\left(\frac{B}{A}\right) \qquad \text{(Eq. 6-70)}$$

This value of θ_R represents the angle that the previously calculated interslice force Z_R must maintain to satisfy moment equilibrium for the general slice. The interslice force angles calculated using Equation 6-70 lag the interslice force values calculated using Equation 6-56 by one step. However, with each new iteration for the FOS, force equilibrium conditions are satisfied, and new interslice forces, and their lines of action, are calculated using Equation 6-70. The iterations are terminated if the difference between the old and the newly calculated FOS value is within a tolerable range. As the interslice force angle on the right side of the last slice is assumed to be zero, and is not calculated using Equation 6-70, moment equilibrium is not implicitly satisfied for this one last slice. Thus although Janbu's GPS does not completely satisfy the conditions of force and moment equilibrium, it does provide the user with a solution procedure that is based on an assumed thrust line rather than a functional description of the distribution of the interslice force angles used by the GLE procedure.

Solution Procedure The FOS according to Janbu's GPS may be calculated using the following steps:

(1) Assume a reasonable interslice force angle for all interslice boundaries.
(2) Using Equation 6-56, calculate the FOS by satisfying the boundary conditions for the right side of the last slice (same as the GLE).
(3) Determine the interslice forces using Equation 6-56.
(4) With the specified location of the thrust line, calculate the interslice force angles required for moment equilibrium (Equation 6-70).
(5) Using the recently calculated interslice force angles, repeat steps 2 to 4 until the change in the FOS value is less than 0.005.

Janbu (1973) recommended that the thrust line be placed generally at about one-third of the interslice height above the failure surface. As a possible modification, it was also suggested that the thrust line may be established at slightly higher positions within the passive zone, near the toe, and at slightly lower locations within the active zone near the crest of the slope. This assumed location of the thrust line is implemented with a function that defines the location of the thrust line with respect to the normalized coordinates along the shear surface.

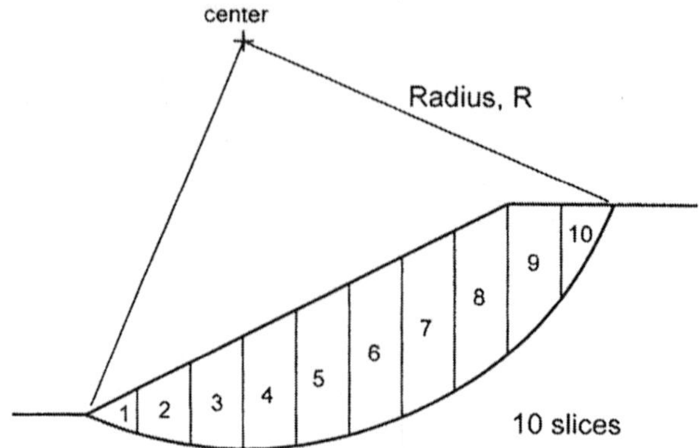

Figure 6.25 Example slope for the method of slices.

6.9.6 Method of Slices—An Example

The FOSs for the 20-meter high 2H–1V slope and the failure surface shown in Figure 6.25 have been calculated using the three simplified methods presented in this section. The center of the circular failure surface is at (35.1, 55), which corresponds to coordinates at the toe of (20, 20), and a radius of 38.1 meters.

Simplified Methods The sliding mass was divided into 10 slices, each 5 meters wide. The weight of each slice was estimated on the basis of *average* height and a unit weight, $\gamma = 16$ kilonewtons per cubic meter. The shear strength of the soil is assumed to be $c = 20$ kilopascals $\phi = 20°$. The pertinent data for these ten slices are given in Table 6.6.

The slice data may be used, along with the concluding equations for each method presented, to calculate the FOS for the selected circular failure surface. The calculations for the OMS are relatively simple and the results are shown in Table 6.7.

TABLE 6.6 Slice Data for Example Problem

Slice	Width (m)	Height (m)	Weight (kN)	c (kPa)	ϕ (deg)	α (deg)	β (deg)
1	5.0	2.0	160.0	20.0	20.0	−19.2	26.57
2	5.0	6.0	480.0	20.0	20.0	−11.4	26.57
3	5.0	9.0	720.0	20.0	20.0	−3.8	26.57
4	5.0	11.5	920.0	20.0	20.0	3.8	26.57
5	5.0	13.5	1,080.0	20.0	20.0	11.4	26.57
6	5.0	14.5	1,160.0	20.0	20.0	19.2	26.57
7	5.0	15.0	1,200.0	20.0	20.0	27.4	26.57
8	5.0	14.5	1,160.0	20.0	20.0	36.2	26.57
9	5.0	12.0	960.0	20.0	20.0	46.2	0.0
10	5.0	5.0	400.0	20.0	20.0	58.6	0.0

TABLE 6.7 Solution of Example Problem Using Ordinary Method of Slices

Slice	Base Length (m)	N' Eq. 6-29 (kN)	$(C' + N' \tan\phi)$ From Eq. 6-33 (kN)	$A - 1$ Eq. 6-34 (kN)
1	5.29	151.1	160.89	−52.6
2	5.10	470.5	273.27	−94.9
3	5.01	718.4	361.70	−47.7
4	5.01	918.0	434.34	61.0
5	5.10	1,058.7	487.35	213.5
6	5.29	1,095.5	504.61	381.5
7	5.63	1,065.4	500.40	552.2
8	6.20	936.1	464.62	685.1
9	7.22	664.5	386.32	692.9
10	9.60	208.4	267.79	341.4
			3,841.28	2,732.4
		Ordinary Method of Slices, FOS = 1.406		

Table 6.8 shows the final iterated solution determined using the simplified Janbu formulation. The failure mass has a d/L ratio of 0.2, which corresponds to a f_0 correction factor of 1.07 from Figure 6.23, or Equation 6-45. This slope has a corrected Janbu FOS of 1.469. These iterative calculations are time consuming and can easily take about an hour each. However, using a suitably programmed electronic spreadsheet, the solutions may be determined very rapidly.

TABLE 6.8 Solution of Example Problem Using Janbu's Method

Slice	N' Eq. 6-38 (kN)	$(C' + N' \tan\theta)\cos\alpha$ from Eq. 6-43 (kN)	$N' \sin\alpha$ (kN)
1	216.2	174.3	−71.1
2	533.1	290.2	−105.4
3	739.5	368.5	−49.0
4	901.3	427.3	59.7
5	1,031.6	468.1	203.9
6	1,099.9	478.1	361.7
7	1,150.9	471.9	529.7
8	1,148.6	437.3	678.4
9	1,000.6	352.1	722.2
10	375.6	171.2	320.6
		3,639.1	2,650.7

Calculated FOS = 1.373
d/L ratio = 0.2
Correction factor, f_0 = 1.07 (Eq. 6-45)
Corrected Janbu FOS = 1.469

TABLE 6.9 Solution of Example Problem Using Bishop's Method

Slice	N' Eq. 6-50 (kN)	$(C' + N' \tan \phi)$ From Eq. 6-48 (kN)	$A - 5$ Eq. 6-50 (kN)
1	210.6	182.5	−52.6
2	528.1	294.2	−94.9
3	737.4	368.6	−47.7
4	903.6	429.1	61.0
5	1,039.0	480.2	213.5
6	1,113.0	511.0	381.5
7	1,170.7	538.7	552.2
8	1,175.8	551.9	685.1
9	1,034.9	521.2	692.9
10	406.9	340.0	341.4
		4,217.5	2,732.4
		Bishop's Method FOS = 1.544	

Table 6.9 presents the results for the calculation of the FOS according to Bishop's method. This solution presented is the final answer as an iterative process is required to solve for the FOS.

Rigorous Factors of Safety The GLE method may also be used to assess the FOS for the example slope. As this method requires an iterative procedure that varies F and θ until force and moment equilibrium conditions are satisfied for all slices, the slope stability computer program XSTABL (I. S. Designs, Inc., 1994) was used to perform these calculations. Using a constant and a half-sine function to describe the variation of the angle of the interslice forces, XSTABL reported the following:

Function	FOS
Constant (Spencer)	1.589
Half-sine	1.588

Realistically, the shape and location of the slip surface within a slope is fundamentally controlled by the effective stresses within the soil mass (Simons and Menzies, 1974). So if effective stresses are calculated along a selected surface, and at the interslice boundaries, using the limit equilibrium method, these are unlikely to be compatible with the physical model. This failure surface–effective stress incompatibility is one of the major concerns in evaluating the reasonableness of the FOS computed using a rigorous method that satisfy complete equilibrium. Thus the engineer has to pay special attention to the results with a view to ensuring that some of the physical assumptions are not grossly violated by the analysis. If an unrealistic FOS value is computed, it is usually because an *unrealistic surface was used to model the potential failure surface.*

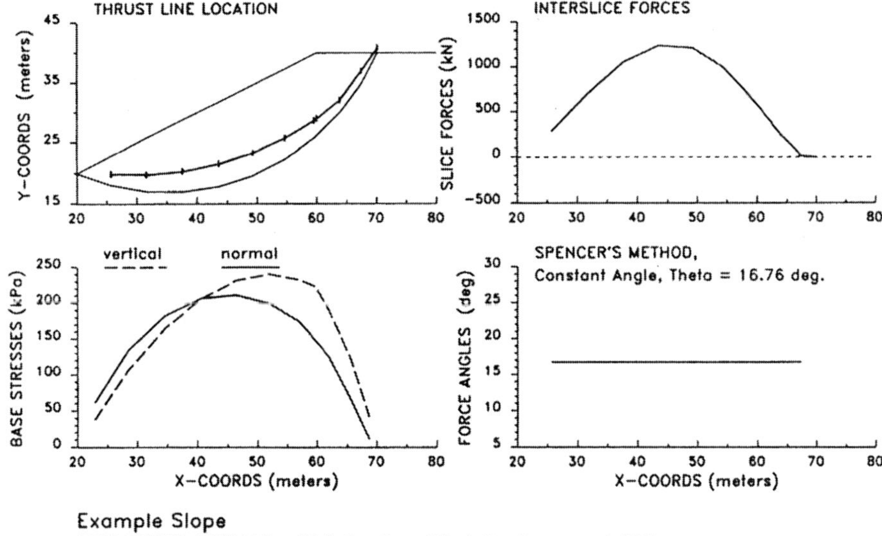

Figure 6.26 Summary plot from XSTABL for Spencer's method.

In view of the very sensitive nature of the GLE method, it is essential that the engineer review the distribution of the stresses, and forces, along the shear surface as well as the location of the thrust line. For the example problem, XSTABL presents the user with the plots shown in Figures 6.26 and 6.27. These plots need to be examined to assess the *reasonableness* of the reported FOS values.

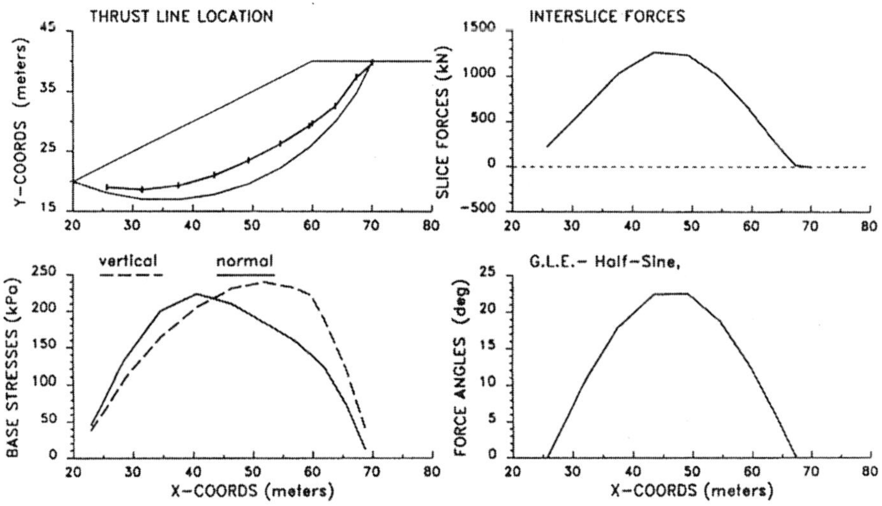

Figure 6.27 Summary plot from XSTABL for an assumed half-sine function.

For example, the thrust line for Spencer's method extends above the ground surface near the crest of the failure mass. This location is considered unreasonable and is probably due to an incompatibilty between the shear surface and the limit equilibrium assumptions. However, the thrust line for the half-sine assumption shown in Figure 6.27 is located entirely within the slide mass, and thus represents a *better* solution. Interestingly, the calculated FOS values are almost identical. This suggests that the selected interslice force angle function does not significantly affect the magnitude of the FOS, only the thrust line location and the interslice forces.

From such data, the user should also look for *smooth* changes in the distribution of the interslice forces and stresses along the failure surface. The normal stresses along the shear surface will typically be greater than the vertical stresses in the lower (toe) part of the slope.

Janbu's GPS method may also be used to determine the FOS for the example slope. For this computation, the user must select an *appropriate* location for the thrust line. The calculations for Janbu's GPS method are very sensitive to this location, and the user may have to experiment with several lines before the procedure converges to a satisfactory FOS solution. Ideally, the thrust line should be located between $0.2h$ and $0.4h$, where h is the height of the side of the slices forming the failure mass. A reasonable solution for the example problem was obtained with a thrust line that linearly varied from $0.4h$ at $x/L = 0.0$ (toe), to $0.25h$ at $x/L = 0.80$, and finally terminated at $0.2h$ at $x/L = 1.0$ (crest).

With the above thrust line, XSTABL used 11 slices and calculated an FOS of 1.591. The interslice forces and angles, and the base stresses are shown in Figure 6.28. The interslice force distribution is smooth, but the base stresses and the interslice force

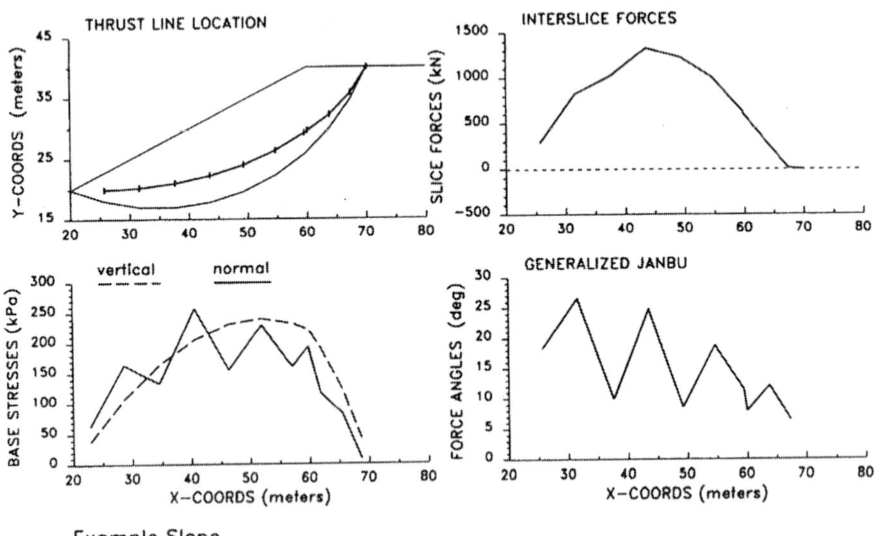

Figure 6.28 Summary plot from XSTABL for Janbu's GPS.

TABLE 6.10 Summary of Calculated Factors of Safety for Example Slope

Calculation Procedure	Manual Calculation	Computer Program XSTABL
Ordinary method	1.406	n/a
Simplified Janbu	1.469	1.532
Simplified Bishop	1.544	1.592
Spencer's method	n/a	1.589
GLE, half-sine function	n/a	1.588
Janbu's GPS	n/a	1.591

angles oscillate considerably. So although a solution was generated and it appears reasonable, the slice-to-slice fluctuations are erratic.

Summary The FOS values computed manually from the data presented in Table 6.6 and the results reported by XSTABL for the same slope are summarized in Table 6.10. The small difference, less than 3 percent, is probably due to the greater accuracy with which the slice data were calculated by the computer program.

6.9.7 Control of Negative Effective Stresses

The limit equilibrium methods that have been developed can sometimes lead to numerical difficulties that are manifested by *negative* normal effective stresses calculated along the failure surface using

$$\sigma' = \frac{N'}{b \sec \alpha}$$

$$= \frac{\cos \alpha}{b m_\alpha}\left[W(1-k_v) - \frac{C \sin \alpha}{F} - U_\alpha \cos \alpha + U_\beta \cos \beta + Q \cos \delta\right] \quad \text{(Eq. 6-71)}$$

where

$$m_\alpha = \cos \alpha \left[1 + \frac{\tan \alpha \tan \phi}{F}\right] \quad \text{(Eq. 6-72)}$$

Negative effective stresses are usually encountered for cases that involve: (1) *high* pore water pressures, (2) a combination of thin slices with a low self-weight and a high "*c* value," and (3) *steep* slice-base angles. Most of these problems are associated with the indeterminacy of the limit equilibrium analysis and the failure to adequately satisfy the conditions for complete static equilibrium.

If a modification of the excessive pore water pressures fails to eliminate the computed negative effective stress, a combination of a small W and a relatively large C component in Equation 6-71 is the problem. In this case the user may decide to follow one of the following suggestions:

Option 1 Proceed with analysis, but enforce the condition that the Mohr–Coulomb shear strength, $\tau = c' + \sigma' \tan \phi$, *always* be greater than or equal to zero.

Option 2 Proceed with analysis without any restrictive conditions concerning the computed shear strength.

Option 3 Conclude that the proposed failure surface is unrealistic because a reasonable FOS cannot be computed using the limit equilibrium method.

In selecting Option 1, the assumption that all slices have the same FOS such that the shear strength, S_m, is given by

$$S_m = \frac{C + N' \tan \phi}{F} \geq 0 \qquad \text{(Eq. 6-73)}$$

will be violated if S_m is arbitrarily adjusted to ensure that it is not less than zero. This restriction implies that for the cases where S_m would have been less than zero, the FOS for the slice is implicitly increased to infinity to ensure zero shear strength. This assumption implicitly contradicts the FOS value used in Equation 6-71 to calculate N', which is subsequently used in Equation 6-73. Although this approach is offered as an option for practical convenience, it is important that users recognize the implication of these assumptions. The alternative, Option 2, does not violate the limit equilibrium assumption of a constant FOS for all slices. Thus it will always generate the lowest factors of safety as the strength mobilized along the failure surface will be a minimum because S_m may be calculated to be less than zero for some slices.

Please note that the FOS values resulting from the selection of Options 1 or 2 will be identical if values of $S_m < 0$ are not computed during the analysis. For other cases where $S_m < 0$ is computed, the FOS values will be different depending on whether the user selects Option 1 or 2. Interestingly, it is also possible that if the ϕ angle for the offending layer is increased, the FOS may be *reduced* due to the mobilization of *negative* frictional strength.

6.9.8 Comparison of Limit Equilibrium Methods

The simplified Bishop and Janbu methods for slope stability analysis have been used extensively since their presentation in the 1950s. Although Bishop's method fails to satisfy horizontal force equilibrium and Janbu's method does not satisfy moment equilibrium, an FOS can be readily calculated for most slopes. However, these FOS values may generally differ by up to ± 15 percent upon comparison with results calculated using procedures that satisfy complete force and moment equilibrium, for example Spencer's method or the Morgenstern–Price method.

The limit equilibrium formulation leads to a statically indeterminate solution and one cannot directly compare these results with a closed-form *correct* solution. Although a direct comparison between the different methods is not always possible, the FOS value determined using Bishop's simplified method for circular surfaces can be

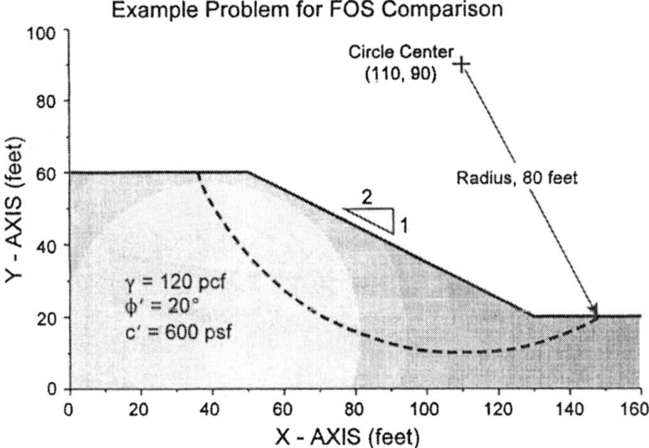

Figure 6.29 Examples slope used for comparison of limit equilibrium methods.

expected to differ by less than 5 percent with respect to the more rigorous Spencer or Morgenstern–Price solutions. The simplified Janbu method, used for noncircular failure surfaces, generally *underestimates* the FOS by as much as 30 percent, with respect to the more rigorous methods. In cases where the failure surface is presumed to be blocky (rather than smooth), considerably greater differences in the FOS are also possible. In contrast, the simplified Janbu method may also *overestimate* the FOS value by as much as 5 percent for some slopes and unusual failure surface configurations.

A good example comparing the different limit equilibrium methods has been presented by Fredlund and Krahn (1977). The failure surface in the slope shown in Figure 6.29 was analyzed and the results are presented in Figure 6.30 as a function of lamda (λ), which can be defined as the ratio of the normal and shear forces acting along the vertical slice boundaries. One can see from this figure that the Spencer and Morgenstern–Price solutions are in good agreement with the simplified Bishop result, whereas the simplified and rigorous Janbu FOS values appear to be slightly lower. The two curves labeled F_m and F_f represent the loci of the points corresponding to the FOS and λ values in satisfying static moment or force equilibrium, respectively. The intersection of these two curves provides a unique combination of FOS and λ that satisfies *complete* static equilibrium within the context of the implied assumptions. For a more complete discussion, and comparison, of the slope analysis methods, the reader should refer to the paper by Fredlund and Krahn (1977).

The methods that satisfy complete equilibrium are more complex and subsequently require a *much* greater level of understanding for successfully assessing the stability of slopes. The user should note that the numerical difficulties experienced with the simplified Bishop (or Janbu) method usually lead to much more serious problems in the Spencer or Morgenstern–Price procedures. These numerical problems often intensify and may lead to unreasonable FOS values and an unrealistic thrust line. Such numerical difficulties generally limit the use of these more complex methods

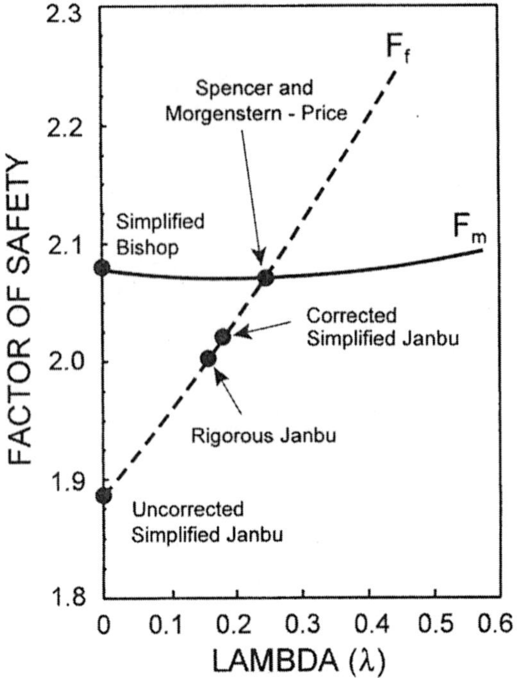

Figure 6.30 Comparison of FOS values calculated using different limit equilibrium methods. The Morgenstern-Price solution uses a uniform distribution of λ. (After Fredlund and Krahn, 1977, with permission.)

to the analysis of single surfaces only as these methods do not appear suitable for automatic surface generation and analysis.

6.10 SELECTION AND USE OF LIMIT EQUILIBRIUM METHODS

6.10.1 Essential First Four Steps

Prior to selecting a method for slope stability analysis, it is essential that the following four steps be carried out:

(1) Attempt to visualize the probable shape of slip surface or slip surfaces. Special attention must be given to the existence of major discontinuities, existing slip surfaces, stratification, nonhomogeneity, tension cracks, and open joints. In homogeneous soil slopes without discontinuities, assume a slip surface of circular shape unless local experience dictates otherwise. In embankments, give consideration to method of construction, zones of different materials, and nature of foundations in order to visualize the probable shape of the slip surfaces.

(2) Distinguish clearly between first-time slides and possible renewed movements along existing slip surfaces. Rely only on the residual strength along parts of assumed slip surfaces that correspond to existing, or prior, shear zones.

(3) Make decisions on relative FOS values with respect to cohesion and friction. Whenever possible, compare strength parameters from back-analyses of case records with those from laboratory and field tests. Examine the reliability of data concerning strength parameters and pore water pressures. Consider the possibility of artesian pressures and perched water tables by examining significant geological details. Give consideration to seepage, submergence, and drawdown conditions when appropriate.

(4) Make decisions concerning the use of effective or total stress types of analysis. In particular, consider the type of materials, whether analysis is for short-term or long-term conditions, whether reliable estimates of pore pressures can be made in advance, and whether pore pressures are to be monitored in the field.

6.10.2 Selection of Analysis Method

The geometry and relative locations of different soil types within the earth mass and the shapes of the slip surfaces in general dictate the method selected for determining the stability of a slope. In many instances, especially for a major project, several methods should be used to assess the stability.

(1) For long uniform slopes where the failure surface is parallel to the ground surface, simple infinite slope equations given in Section 6.9.2 are fairly accurate.

(2) For shallow, long planar failure surfaces that are not parallel to the ground surface, the simplified Janbu approach will give reliable results.

(3) For planar failure surfaces, a block analysis can be used to determine the FOS and critical failure surface location. Higher accuracy may be obtained by means of a GLE solution.

(4) For surfaces that can be approximated by arcs of circles, preliminary studies are facilitated by the use of stability charts. For greater accuracy, the simplified Bishop method provides reasonable results.

(5) For slip surfaces of arbitrary shape, a generalized procedure of slices can be used. Preliminary studies may be made using Janbu's simplified procedure (without interslice forces) and correction factors. For more accurate studies, Janbu's GPS method, Spencer's procedure, the Morgenstern–Price method, or Sarma's method are available for a more rigorous analysis.

Limit equilibrium methods have provided a very useful technique for the analysis of slopes and other geotechnical engineering problems. However, the procedure does have several inherent weaknesses, which include:

(1) Incipient failure is assumed at an overall FOS equal to one, which is highly influenced by many variables associated with geological details, material parameters, pore water pressures, and so on.

(2) An assumption of constant FOS along the entire slip surface is an oversimplification, especially if different soil materials exist along the failure surface.

(3) The stress–strain relationship of the soil is neglected, that is, stress-deformation increments and/or decrements within a slope are not simulated by consideration of static equilibrium by itself.

Most of the limit equilibrium methods have been programmed for use on personal computers. In general, the user should thoroughly understand the slope stability method used in available computer programs in order to recognize potentially erroneous results. Results obtained from slope stability computer programs should always be reviewed carefully and interpreted with judgement based on good geotechnical practice.

6.10.3 Considerations for All Types of Analyses

All slope stability analyses should consider the following points:

- Inclusion of tension cracks and open joints, if any, should not be overlooked in analysis. Where appropriate, assume water in these cracks.
- Sensitivity analyses may be made by varying one parameter at a time, for example, cohesion parameter, friction parameter, groundwater table, and so on. Plot each parameter against the computed FOS to assess its overall influence.
- Where the strength envelope is curved, exercise great care in selecting appropriate c' and ϕ' values. Selected shear strength values must correspond to stress levels appropriate to the problem being analyzed. For shallow slip surfaces, select c' and ϕ' from the portion of the strength envelope in the low normal stress range; for deep slip surfaces, select them in the high normal stress range. If possible, use a computer program that allows use of a nonlinear Mohr–Coulomb envelope.
- Consider the possibility of progressive failure occurring and especially the characteristics of slope disturbance, tension cracks, strain-softening, and nonuniform stress–strain distribution in initiating or accelerating progressive failure.
- Consider possibilities of delayed failure due to decrease in shear strength parameters with time, increase of pore pressures with time, and other factors.

6.11 DESIGN CHARTS

Slope stability charts are useful for preliminary analysis, to compare alternates that can later be examined by more detailed analyses. Chart solutions also provide a rapid means of checking the results of detailed analyses. Engineers are encouraged to use these charts before using a computer program to determine the approximate value of the FOS as it allows some quality control and a check for the subsequent computer-generated solutions.

Another use for slope stability charts is to back-calculate strength values for failed slopes to aid in planning remedial measures. This can be done by assuming an FOS of unity for the conditions at failure and solving for the unknown shear strength. Since soil strength usually involves both cohesion, c', and friction, ϕ', there are no unique values that will give an FOS equal to one. As such, selection of the most reasonable c' and ϕ' depends on local experience and judgement. Since the friction angle is usually within a narrow range for many types of soils and can be obtained by testing with a certain degree of confidence, the cohesion, c', is generally varied in practice while the friction angle is fixed for the back-calculation of slope failures.

The major shortcoming in using design charts is that most of them are for ideal, homogeneous soil conditions, which are not encountered in practice. The charts have been devised using the following general assumptions:

(1) Two-dimensional limit equilibrium analysis
(2) Simple homogeneous slopes
(3) Slip surfaces of circular shapes only

It is imperative for the user to understand the underlying assumptions for the charts before using them for the design of slopes.

Regardless of the above shortcomings, many practicing engineers use these charts for nonhomogeneous and nonuniform slopes with different geometrical configurations. To do this, one must use an average slope inclination, weighted averages of c and ϕ, and values of s_u calculated on the basis of the proportional length of slip surface passing through different relatively homogeneous layers. Such a procedure is extremely useful for preliminary analyses and saves both time and expense. In most cases, the results should be checked by performing detailed analyses using more suitable and accurate methods, for example, the method of slices.

6.11.1 Historical Background

Some of the first slope stability charts were published by Taylor (1937, 1948). These well-known classical charts are strictly applicable only for analysis in terms of a *total stress* approach. Pore pressures or pore pressure ratios, r_u, are not considered in these charts. In deriving the charts, Taylor used the friction circle method. Since then, various charts have been successively presented by Bishop and Morgenstern (1960), Hunter and Schuster (1968), Janbu (1968), Morgenstern (1963), Spencer (1967), Terzaghi and Peck (1967), and others, as summarized in Table 6.11.

6.11.2 Stability Charts

Taylor's Charts Taylor (1948) developed slope stability charts, shown in Figures 6.31 and 6.32, for soils with $\phi = 0$ and $\phi > 0$. As shown in these charts, the slope has an angle β, a height H, and base stratum at a depth of DH below the toe, where D is a depth ratio. The charts can be used to determine the *developed* cohesion, c_d, as shown by the solid curves, and nH, which is the distance from the toe to the failure circle, as indicated by the short dashed curve.

TABLE 6.11 Summary of Slope Stability Charts

Author	Parameters	Slope Inclinations (Degrees)	Analytical Methods	Notes
Taylor (1948)	c_u	0–90	$\phi = 0$	Undrained analysis
	c, ϕ	0–90	Friction circle	Dry slopes only
Bishop and Morgenstern (1960)	c, ϕ, r_u	11–26.5	Bishop	One of the first to include effects of water
Gibson and Morgenstern (1962)	c_u	0–90	$\phi = 0$	Undrained analysis with c_u increasing linearly with depth; zero strength at ground level
Spencer (1967)	c, ϕ, r_u	0–34	Spencer	Toe circles only
Janbu (1968)	c_u c, ϕ, r_u	0–90	$\phi = 0$ Janbu GPS	Extensive series of charts for seepage and tension crack effects
Hunter and Schuster (1968)	c_u	0–90	$\phi = 0$	Undrained analysis with c_u increasing linearly with depth; finite strength at ground level
Chen and Giger (1971)	c, ϕ	20–90	Limit analysis	
O'Connor and Mitchell (1977)	c, ϕ, r_u	11–26	Bishop	Extended Bishop and Morgenstern (1960) to include $N_c = 0.1$
Hoek and Bray (1977)	c, ϕ	0–90	Friction circle	Includes groundwater and tension cracks
	c, ϕ	0–90	Wedge	Three-dimensional analysis of wedge block
Cousins (1978)	c, ϕ, r_u	0–45	Friction circle	Extension of Taylor (1948)
Charles and Soares (1984)	ϕ	26–63	Bishop	Nonlinear Mohr–Coulomb failure envelope, $\tau = A(\sigma')^b$
Barnes (1991)	c, ϕ, r_u	11–63	Bishop	Extension of Bishop and Morgenstern (1960); wider range of slope angles

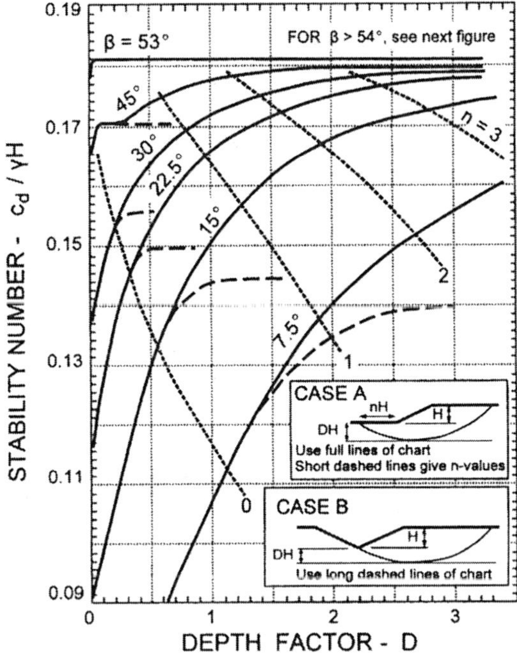

Figure 6.31 Taylor's chart for $\phi = 0$ conditions and for slope angles less than 54°. (Taylor, 1948, with permission.)

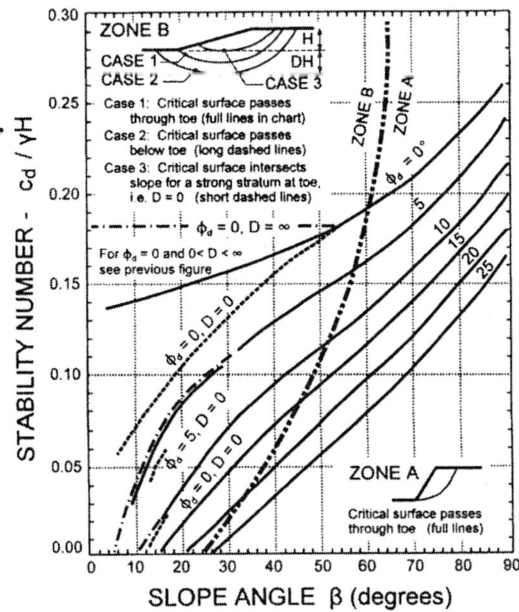

Figure 6.32 Taylor's chart for soils with friction angle (Taylor, 1948, with permission.)

If there are loadings outside the toe that prevent the circle from passing below the toe, the long dashed curve should be used to determine the developed cohesion. Note that the solid and the long dashed curves converge as n approaches zero. The circle represented by the curves on the left of $n = 0$ does not pass below the toe, so the loading outside the toe has no influence on the developed cohesion.

Example 1 Given a slope with height $H = 40$ feet, $DH = 60$ feet, slope angle $\beta = 30°$, cohesion $c = 1{,}200$ pounds per square inch, and total unit weight, $\gamma = 120$ pounds per cubic foot, determine:

(a) The FOS and the distance from the toe to the point where the most critical circle appears on the ground surface.
(b) The FOS if there are heavy loadings outside the toe.

Solution

(a) For $D = 60/40 = 1.5$, and $\beta = 30°$, from the solid curve in Figure 6.31,

$$N = \frac{c_d}{cH} = 0.176$$

and so

$$c_d = 0.176\,(120)\,40 = 844.8 \text{ psf}$$

$$\text{FOS} = \frac{c}{c_d} = \frac{1{,}200}{844.8} = 1.42$$

Form the short dashed lines in Figure 6.31, $n = 2.09$, for the $\beta = 30°$, curve and so the distance between the toe and the failure circle $= nH = 2.09\,(40) =$ 83.6 feet.

(b) The toe of the slope is loaded, so Case B applies and if the dashed curve in Figure 6.31 for $\beta = 30°$ is extended horizontally, then for $D = 1.5$, $N = c_d/cH = 0.1557$, and so

$$c_d = 0.1557(120)40 = 747.4 \text{ psf}$$

The FOS with the toe loaded (Case B) is

$$\text{FOS} = 1{,}200/747.4 = 1.61$$

Spencer's Charts Spencer's (1967) charts, shown in Figure 6.33, are based on solutions computed using the rigorous Spencer's method, which satisfies complete equilibrium. Typically, these charts are used to determine the required slope angle for a specific (preselected) FOS. The charts assume that a firm stratum is at a great depth below the slope, and they presents solutions for three different pore pressure ratios,

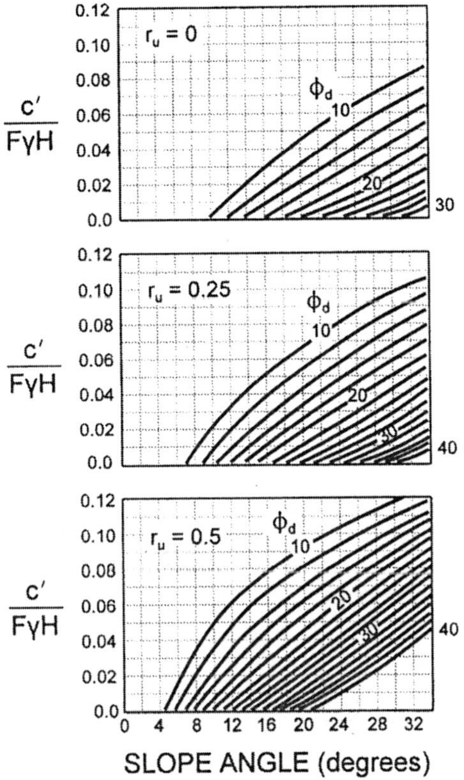

Figure 6.33 Stability chart for different pore pressure ratios. (Spencer, 1967, with permission.)

$r_u = 0, 0.25$, and 0.5. In using the charts, the *developed* friction angle is calculated using

$$\phi_d = \tan^{-1}\left[\frac{\tan\phi}{F}\right] \quad \text{(Eq. 6-74)}$$

where F is the FOS.

Example 2 Given a slope with height $H = 64$ feet, $c' = 200$ pounds per square foot, $\phi' = 30°$, $\gamma = 125$ pounds per cubic foot, and a pore water pressure ratio, $r_u = 0.25$, determine the maximum slope angle β for an FOS of 1.5.

Solution With $c'/(F\gamma H) = 200/(1.5 \times 125 \times 64) = 0.0167$,

$$\phi_d = \tan^{-1}\frac{\tan 30°}{1.5} = 21°$$

and with $r_u = 0.25$, and from Figure 6.33, the maximum slope angle, $\beta = 21^\circ$, for an FOS of 1.5.

Example 3 For the same materials as in Example 2, determine the FOS for a slope with inclination $\beta = 16^\circ$.

Solution

FIRST ITERATION Assume that FOS, $F = 1.6$; then

$$N = \frac{c'}{F\gamma H} = \frac{200}{1.6 \times 125 \times 64} = 0.0156$$

from Figure 6.33, with $r_u = 0.25$, and $\phi_d = 18^\circ$, so that

$$F = \frac{\tan(30)}{\tan(18)} = 1.78$$

As the assumed FOS and the calculated values differ substantially, the iteration is repeated with a newly assumed FOS.

SECOND ITERATION Assume a new FOS, $F = 1.7$; then

$$N = \frac{c'}{F\gamma H} = \frac{200}{1.7 \times 125 \times 64} = 0.0147$$

from Figure 6.33, with $r_u = 0.25$, $\phi_d = 19^\circ$, such that

$$F = \frac{\tan(30)}{\tan(19)} = 1.68$$

The assumed and calculated FOS values are close enough to the presumed value of 1.7. The FOS for the 16° slope is equal to 1.7.

Janbu's Charts In 1968, Janbu published stability charts for slopes in soils with uniform strength throughout the depth of the soil layer assuming $\beta = 0$ and $\phi > 0$ conditions. Those charts (Janbu, 1968), which extended the range of the original 1954 charts, are presented in Figures 6.34 through 6.37. This series of charts accounts for several different conditions and provides factors for surcharge loading at the top of the slope, for submergence, and for tension cracks that can be expected to influence the design of typical slopes. Two examples are used to illustrate the use of these charts and are taken from Duncan and Buchignani (1975) and Duncan et al. (1987).

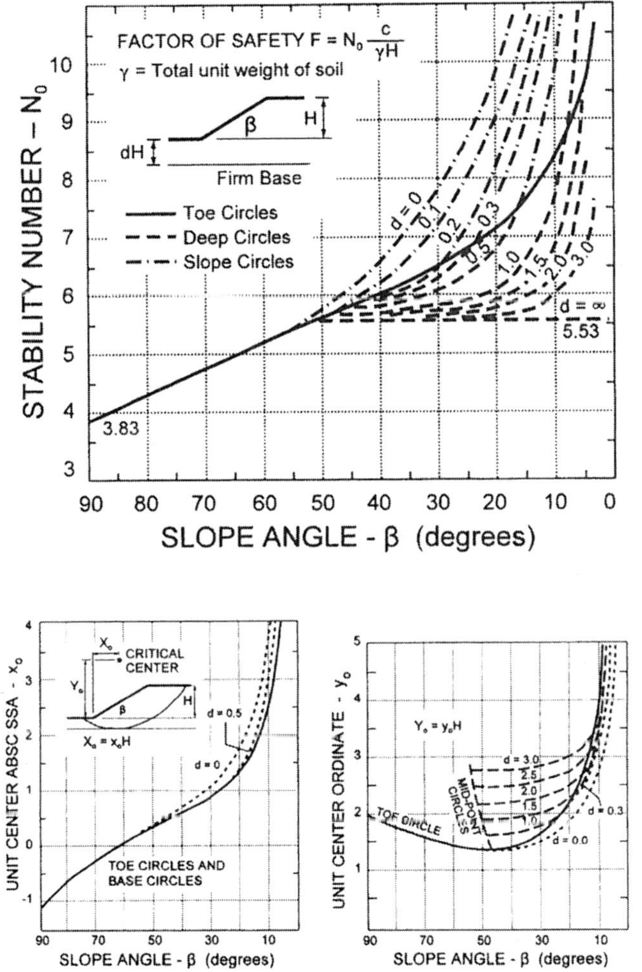

Figure 6.34 Stability charts for $\phi = 0$ soils. (Janbu, 1954, with permission.)

Example 4 ($\phi' = 0$ soil) The stability of the 50° slope with a height of 24 feet shown in Figure 6.38 will be evaluated using Janbu's charts. The subsoils within the slope consist of an upper and lower layer with the following soil parameters:

Soil Layer	Shear Strength (psf)	Unit Weight (pcf)
Upper layer	600	120
Lower layer	400	100

The water level is 8 feet above the toe. Determine the FOS of the circular failure surface that is tangential to a horizontal line through the toe.

REDUCTION FACTORS FOR TENSION CRACKS – μ_t

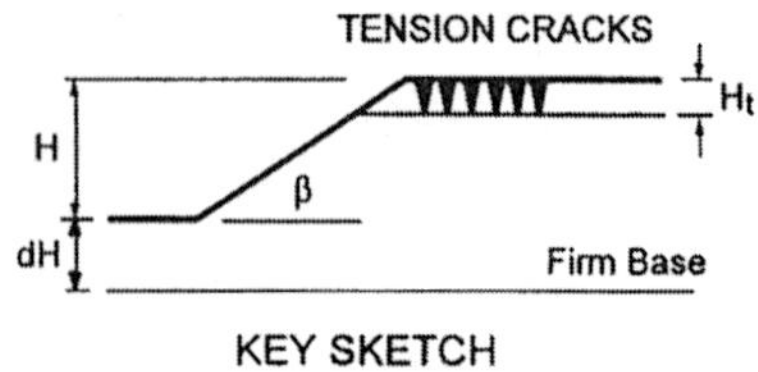

NO HYDROSTATIC PRESSURE IN CRACK

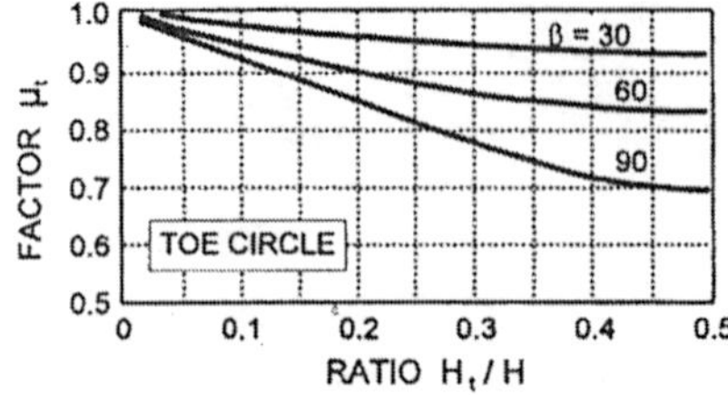

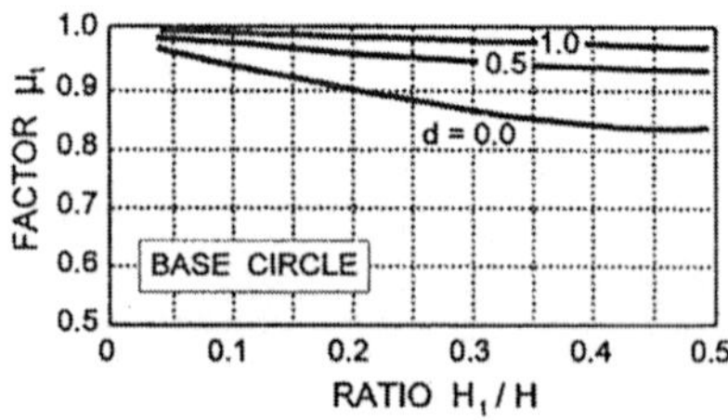

FULL HYDROSTATIC PRESSURE IN CRACK

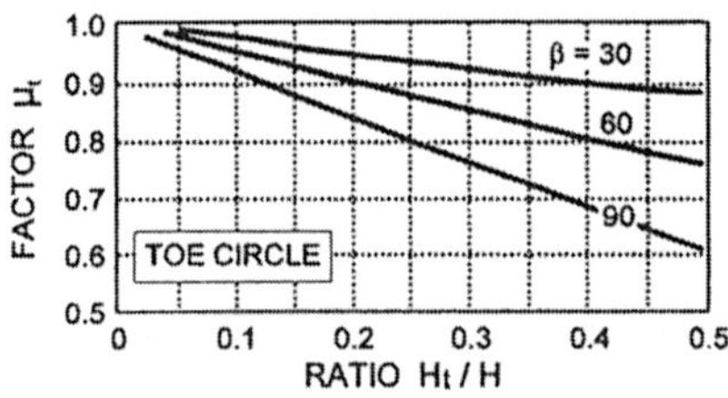

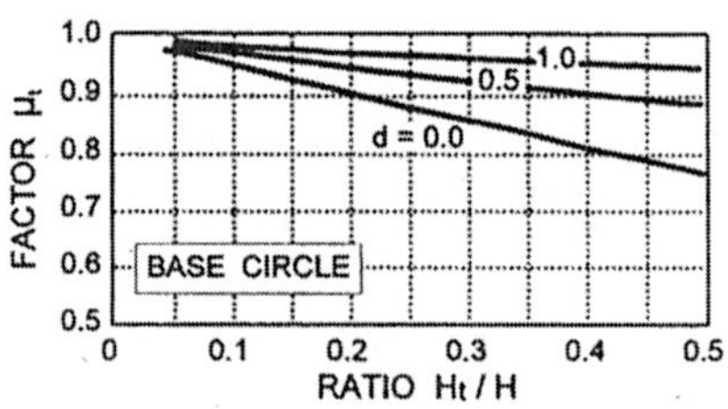

Figure 6.35 Reduction factor for $\phi = 0$ and $\phi > 0$ soils. (Janbu, 1968, with permission.)

Solution For a slip circle tangent to a horizontal line through the toe,

$$d = 0, \qquad \frac{H_w}{H} = \frac{8}{24} = 0.33$$

From Figure 6.34, $x_0 = 0.35$, $y_0 = 1.4$ for the critical circle that intersects near the toe, so

$$X_0 = 24 \times 0.35 = 8.4 \text{ ft}, \qquad \text{and} \qquad Y_0 = 24 \times 1.4 = 33.6 \text{ ft}$$

Next, determine the average shear strength available along the circular failure surface according to the proportion of the failure surface located within each soil layer:

$$C_{\text{ave}} = [(22)(600) + (62)(400)] \div [22 + 62] = 452 \text{ psf}$$

From Figure 6.35, $\mu_w = 0.93$ for $\beta = 50°$ and $H_w/H = 0.33$,

$$\begin{aligned} P_d &= [(110)(24) - (62.4)(8)] \div [(1)0.93(1)] \\ &= (2{,}640 - 499)/0.93 = 2{,}302 \end{aligned}$$

REDUCTION FACTORS FOR SURCHARGE – μ_q

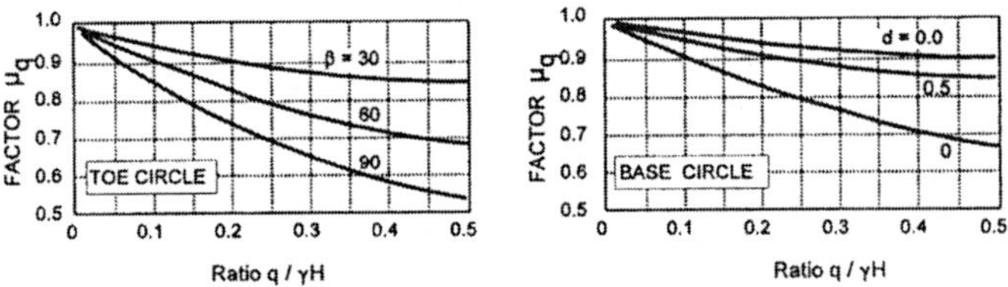

SUBMERGENCE – μ_w AND SEEPAGE – μ'_w

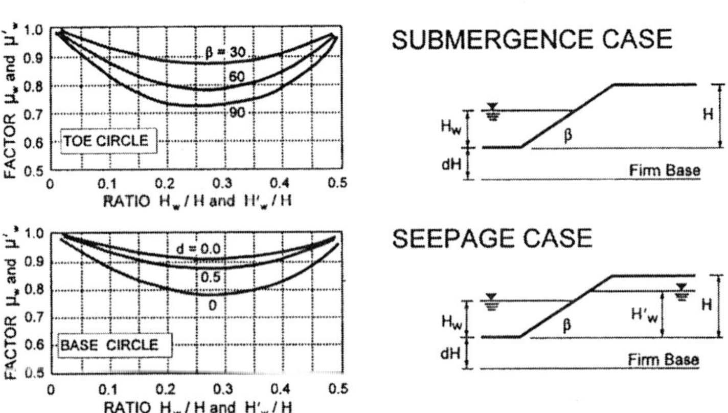

Figure 6.36 Reduction factors for $\phi - 0$ and $\phi > 0$ soils. (Janbu, 1968, with permission.)

Then from Figure 6.34, $N_0 = 5.8$ for $d = 0$ and $\beta = 50°$, and the FOS,

$$F = (5.8 \times 452) \div 2{,}302 = 1.14$$

Example 5 ($\phi' > 0$ soil) Assume a 1.5 H–1V slope of height 40 feet, as shown in Figure 6.39. The slope is composed of two soil layers. The shear strength parameters γ, ϕ', and c' of the upper layer are 115 pounds per cubic foot, 35°, and 100 pounds per square foot, and for the lower layer, they are 115 pounds per cubic foot, 30°, and 150 pounds per square foot. The groundwater table is 30 feet above the toe of slope. The slope is submerged by water to a depth of 10 feet. Determine the FOS of a slip circle passing through the toe of the slope.

Solution For a toe slip circle using effective stress,

$$\mu'_w = 0.95 \text{ for the ratio } \frac{H'_w}{H} = 0.75 \text{ from Figure 6.36}$$

$$\mu_w = 0.96 \text{ for the ratio } \frac{H_w}{H} = 0.25 \text{ from Figure 6.36}$$

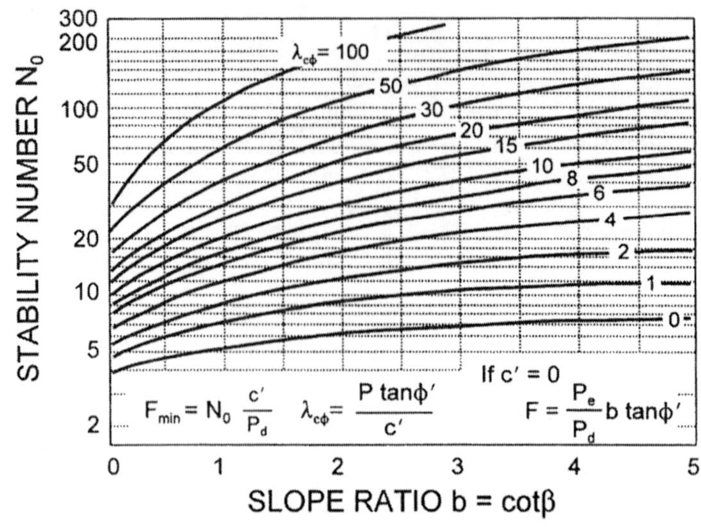

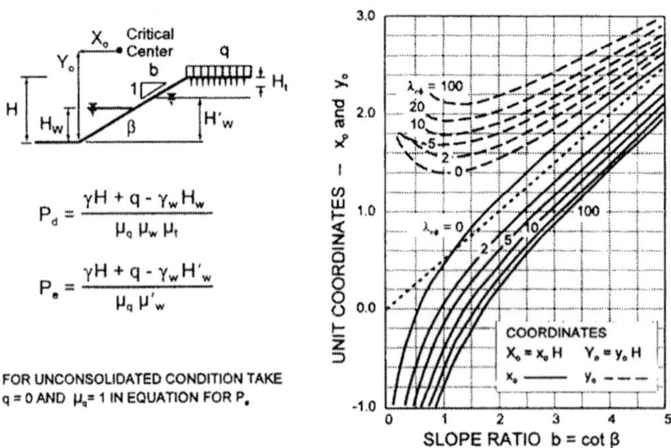

Figure 6.37 Stability charts for $\phi > 0$ soils. (Janbu, 1968, with permission.)

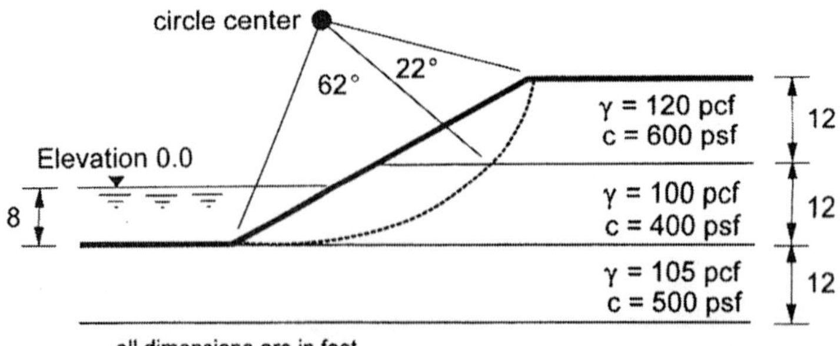

Figure 6.38 Slope used for Example 4.

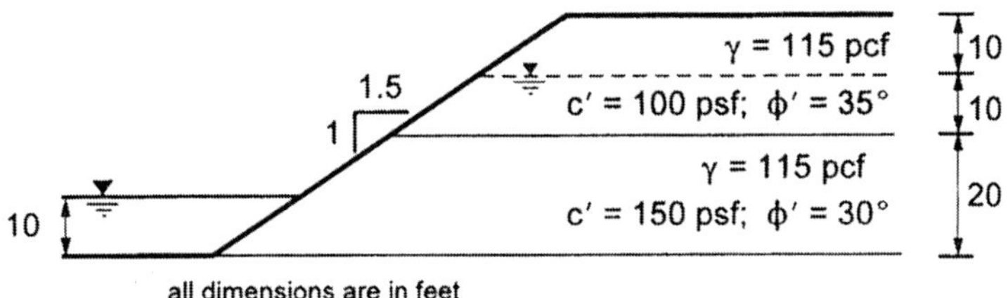

Figure 6.39 Slope used for Example 5.

Then

$$P_d = [115(40) - (62.4)(10)] \div [(1)0.96(1)] = 4,142 \text{ psf}$$
$$P_e = [115(40) - (62.4)(30)] \div [(1)0.95(1)] = 2,872 \text{ psf}$$

Estimate the shear strength as

$$c_{\text{ave}} = 120 \text{ psf} \qquad \text{and} \qquad \tan\phi_{\text{ave}} = 0.64$$

Then

$$\lambda_{c\phi} = 2,872(0.64) \div 120 = 15.3$$

From Figure 6.37, the location of the critical circle is given as

$$x_0 = 0.0 \qquad \text{and} \qquad y_0 = 1.9 \quad \text{for } b = 1.5$$

and so,

$$\lambda_{c\phi} = 15.3, \qquad X_0 = 0 \text{ ft}, \qquad Y_0 = 1.9 \times 40 = 76 \text{ ft}$$

Then based on the portion of the failure surface in each layer,

$$C_{\text{ave}} = [19(100) + (42)150] \div (19 + 42) = 134 \text{ psf}$$
$$\tan\phi_{\text{ave}} = [(19)\tan 35^\circ + (42)\tan 30^\circ] \div (19 + 42) = 0.62$$
$$\lambda_{c\phi} = (2,870)(0.62) \div (134) = 13.3$$

From Figure 6.37, $x_0 = 0.02$ and $y_0 = 1.85$ for $b = 1.5$ and

$$\lambda_{c\phi} = 13.3, \qquad X_0 = 0.02 \times 40 = 0.8 \text{ ft}, \qquad Y_0 = 1.85 \times 40 = 74 \text{ ft}$$

This circle is close enough to the previous iteration, so the values of $\lambda_{c\phi} = 13.3$ and $c_{\text{ave}} = 134$ pounds per square foot are retained as part of the final solution. From Figure 6.37, $N_{cf} = 35$ for $b = 1.5$ and $\lambda_{c\phi} = 13.3$, and the FOS, $F = 35(134) \div (4,141) = 1.13$.

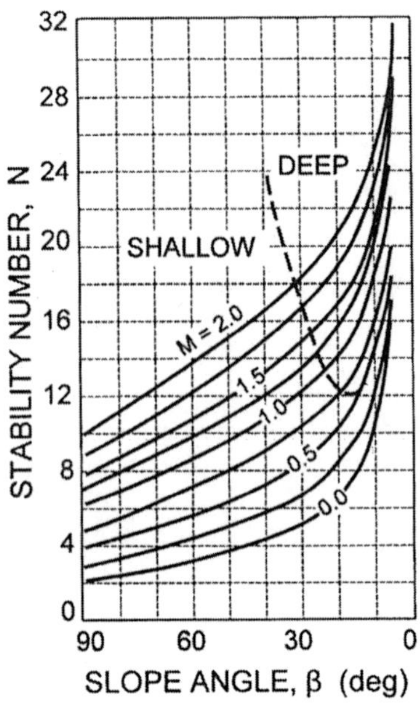

Figure 6.40 Stability charts for $\phi = 0$ and strength increasing with depth. (Hunter and Schuster, 1968, with permission.)

Hunter and Schuster Chart Hunter and Schuster (1968) developed a stability chart (Figure 6.40) that takes into account the case where the shear strength increases linearly with depth, under a $\phi = 0$ condition. The steps required to determine the FOS are:

(1) Calculate the ratio, $M = c_{top} \div (c_{base} - c_{top})$, where c_{top} and c_{base} are the undrained strength values at the top and base of the slope.

(2) Determine the stability number, N, from Figure 6.40 for M factor and slope inclination, β.

(3) Calculate the FOS: $F = (c_{base} - c_{top})N \div (\gamma H)$, where H is the height of slope and the unit weight, γ, may be taken as

$$\gamma = \gamma_{buoyant} \quad \text{for a fully submerged slope}$$

$$\gamma = \gamma_{moist} \quad \text{for a slope with no water on its surface}$$

$$\gamma = \gamma_{average} \quad \text{for a partially submerged slope}$$

Example 6 Given a 45° fully submerged slope of $H = 100$ feet, unit weight, $\gamma = 100$ pounds per cubic foot, and an undrained shear strength that increases linearly from 150 pounds per square foot at the top to 1,150 pounds per square foot at the base of the slope, determine the FOS for this slope.

Solution

(1) M = 150 ÷ (1,150 − 150) = 0.15.
(2) From Figure 6.40, for $M = 0.15$ and $\beta = 45°$, $N = 5.1$.
(3) The FOS, $F = (1{,}150 - 150)\,(5.1) \div (100 \times 100) = 1.36$.

Interestingly, if the average shear strength of this slope is taken as 650 pounds per square foot, and if the buoyant unit weight is used with Taylor's chart, the FOS is about 0.96.

6.12 SEISMIC ANALYSIS

Earthquake ground motions are capable of inducing large destabilizing inertial forces, of a cyclic nature, in slopes and embankments. Also, the shear strength of the soil may be reduced due to the transient loads (i.e., cyclic strains) or due to the generation of excess pore water pressures. The combined effect of the seismic loads and the changes in shear strength will result in an overall decrease in the stability of the affected slope.

Typically, cyclic loads will generate excess pore water pressures in loose, saturated cohesionless materials (gravels, sands, and nonplastic silts), which may liquefy with a considerable loss of preearthquake strength. However, cohesive soils and dry cohesionless materials are not generally affected by cyclic loads to the same extent. If the cohesive soil is not sensitive, in most cases it appears that at least 80 percent of the static shear strength will be retained during and after the cyclic loading (Makdisi and Seed, 1978). For a more complete discussion of the cyclic shear strength of soils, the reader should refer to the previous chapter.

In general, four methods of analysis have been proposed for the evaluation of the stability of slopes during earthquakes (Houston et al., 1987). In increasing order of complexity and expense, these are:

(1) *Pseudostatic Method* The earthquake's inertial forces are simulated by the inclusion of a static horizontal and vertical force in a limit equilibrium analysis.
(2) *Newmark's Displacement Method* This method is based on the concept that the actual slope accelerations may exceed the static yield acceleration at the expense of generating permanent displacements (Newmark, 1965).
(3) *Postearthquake Stability* This is calculated using laboratory undrained strengths determined on representative soil samples that have been subjected to the cyclic loads comparable to the anticipated earthquake (e.g., Castro et al., 1985).
(4) *Dynamic Finite Element Analysis* This is a coupled two- (or three-) dimensional analysis using an appropriate constitutive soil model that will provide details concerning stresses, strains, and permanent displacements (e.g., Finn, 1988; Prevost et al., 1985).

From the above list, the first two methods have become well established in general geotechnical engineering practice, mainly due to their ease of implementation, familiarity, and financial economics. The postearthquake stability method is simple to implement, but requires extensive dynamic laboratory testing to determine the shear strength of the soils along some of the preselected potential failure surfaces in the slope. The finite element analysis is also expensive, as it requires extensive laboratory testing to identify the parameters for the constitutive model and considerable computational resources. Due to their ready implementation, only the first two procedures are discussed here. For more information about the last two methods, please consult the cited references.

6.12.1 Pseudostatic Method

The pseudostatic method offers the simplest approach for evaluating the stability of a slope in an earthquake region. In its implementation, the limit equilibrium method is modified to include horizontal and vertical *static* seismic forces that are used to simulate the potential inertial forces due to ground accelerations in an earthquake. These seismic forces are assumed to be proportional to the weight of the potential sliding mass times horizontal and vertical seismic coefficients, k_h and k_v, expressed in terms of the acceleration of the underlying earth (in units of g), as shown in Figure 6.41. It is recommended that only the most critical surface, as identified by a static analysis, should be reanalyzed using pseudostatic seismic coefficients as it will be the most stressed region within the slope.

Typically, the seismic force is presumed to act in a horizontal direction only, that is, $k_v = 0.0$, inducing an inertial force, $k_h W$, in the slope, where W is the weight of the potential sliding mass. An FOS is then calculated using conventional methods. The greatest difficulty with this procedure involves the selection of an appropriate seismic coefficient and the value of an acceptable FOS.

The magnitude of the seismic coefficient should effectively simulate the nature of the expected earthquake forces, which will depend on: (1) earthquake intensity, for example, peak ground acceleration (PGA), (2) duration of shaking, and (3) frequency content. Of course as a very conservative assumption, one can select a seismic coefficient that is equal to the peak ground acceleration expected at the slope. However, this conservatism will lead to a very uneconomic evaluation and possible numerical

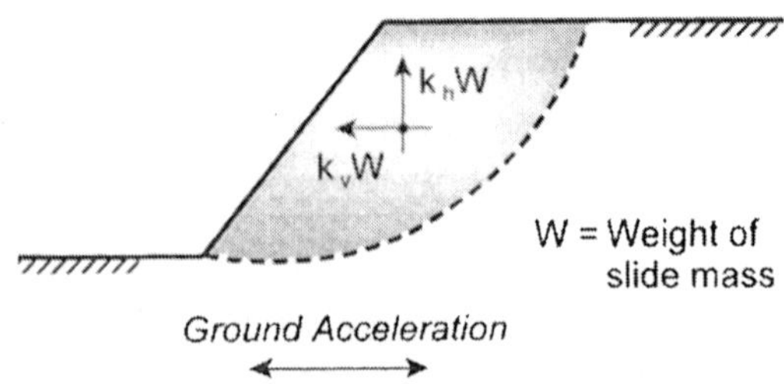

Figure 6.41 Pseudostatic limiting equilibrium analysis for seismic loads.

TABLE 6.12 Typical Seismic Coefficients and FOSs Used in Practice

Seismic Coefficient	Remarks
0.10	Major earthquake, FOS > 1.0 (Corps of Engineers, 1982)
0.15	Great earthquake, FOS > 1.0 (Corps of Engineers, 1982)
0.15–0.25	Japan, FOS > 1.0
0.05–0.15	State of California
0.15	Seed (1979), with FOS > 1.15 and a 20 percent strength reduction
$\frac{1}{3}$–$\frac{1}{2}$ PGA[a]	Marcuson and Franklin (1983), FOS > 1.0
$\frac{1}{2}$ PGA	Hynes-Griffin and Franklin (1984), FOS > 1.0 and a 20 percent strength reduction

[a] PGA = peak ground acceleration, in units of g.

difficulties for $k_h \geq 0.4$. The selection of such coefficients, therefore, must be rationalized if slopes are to be designed economically. Some typical seismic coefficients that have been used for evaluating the seismic stability of slopes are given in Table 6.12.

It is recommended that the results of a pseudostatic analysis be presented in the form shown in Figure 6.42. This plot shows the variation of the FOS as a function of the horizontal seismic coefficient for the critical surface identified for the static condition. In examining Figure 6.42, one can readily compare the FOS corresponding to different seismic coefficients and also determine the critical seismic coefficient (or yield acceleration), k_y, that generates an FOS equal to one.

For the example slope analyzed, the critical coefficient, k_y, is about 0.4. This result may be interpreted such that if the expected seismic coefficient at the site is less than 0.4, the slope can be expected to perform satisfactorily during ground shaking. Conversely, the slope may be overly stressed if the seismic coefficient at the site happens to exceed 0.4.

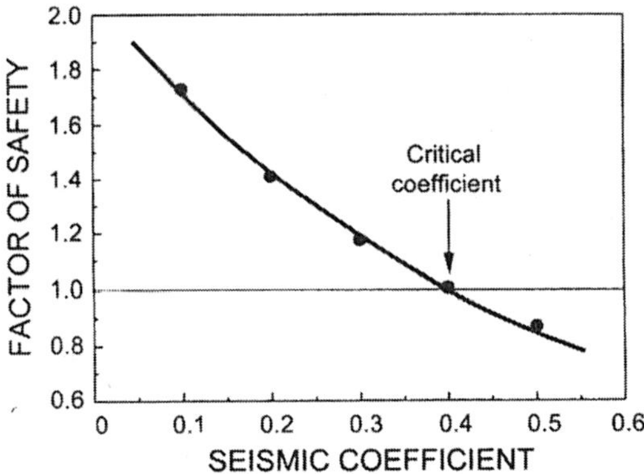

Figure 6.42 Variation of factor of safety with horizontal seismic coefficient, k_h, for example slope.

TABLE 6.13 Guidelines for Pseudostatic Analysis

Condition	Remarks
$k_y >$ PGA	Slope can be expected to survive the design earthquake
$\frac{1}{2}$PGA $< k_y <$ PGA	Minor damage is possible
$k_y < \frac{1}{2}$PGA	Slope may be unstable

If the magnitude of the expected PGA is known for the slope, the calculated value of critical acceleration, k_y, may be used to reach the conclusions suggested in Table 6.13.

6.12.2 Newmark's Displacement Method

The procedure proposed by Newmark (1965) extends the simple pseudostatic approach by directly considering the acceleration time history (accelerogram) of the slide mass within the slope. This accelerogram, selected to represent a realistic model of the ground motions expected at the site, is then compared with the yield acceleration to determine permanent displacements.

Newmark's method assumes: (1) existence of a well-defined slip surface, (2) a rigid, perfectly plastic slide material, (3) negligible loss of shear strength during shaking, and (4) permanent strains occur if the dynamic stress *exceeds* the shear resistance. Also, the slope is only presumed to deform in the downslope direction, thus implying infinite *dynamic* shear resistance in the upslope direction. The procedure requires that the value of a yield acceleration or critical seismic coefficient, k_y, be determined for the potential failure surface using conventional limit equilibrium methods. This value can be readily determined from analyses that generate a graph such as the one presented in Figure 6.42. The main difficulty associated with this method is related to the selection of an appropriate accelerogram that simulates the motions of the slide mass. However, once this has been selected, the permanent displacements are calculated by double integration of the portions of the accelerogram that exceed the yield acceleration for the critical failure surface.

A simple framework for a FORTRAN program capable of performing this double integration is given in Table 6.14. The code is well commented, and can be easily adapted to other programming languages. The yield acceleration and the accelerogram details (number of points, time increment, value of g) form the input data for the program. Positive or negative values of the yield acceleration may be specified with this version. The input format requires each acceleration value to be on a separate line. The input/output functions of this program have been deliberately retained in a simple format so that changes can be made as needed. The main loop cycles through all the acceleration values and computes the velocity (variable, V1) and the cumulative displacement (variable, D1) at the end of each time increment (variable, DT). The logic statements ensure that only the "excess" accelerations are considered in calculating the permanent displacements.

TABLE 6.14 FORTRAN Program for Newmark's (1965) Method

```
C  FORTRAN PROGRAM FOR COMPUTING NEWMARK DISPLACEMENTS
C
C  INPUT DATA:
c
C  YIELD = yield acceleration
C          (in same units as the acceleration values)
C   NPTS = number of acceleration points
C     DT = time increment between acceleration values
C      G = value of g in suitable units,
C          e.g. 9.81 m/s/s; 32.2 ft/s/s; 386 in/s/s etc.
C    ACC = acceleration value, in units of g
C          (if accel. values have units, set g=1.0)
C
C  OUTPUT DATA:
C
C      D1 = computed permanent displacement reported in
C           same units as acceleration values and ''G''
C
      READ(5,*) YIELD,NPTS,DT,G         ! read/write data
      WRITE(6,*) YIELD,NPTS,DT,G

      A0=0.0                            ! initialize data
      V0=0.0
      D0=0.0
      V1=0.0
      YIELD=G*YIELD                     ! convert g-units
      DT2=DT*DT/6.0

      DO K=1,NPTS
        READ(5,*) ACC
        ACC=G*ACC                       ! convert accelerations
        IF(YIELD.LT.0.0) ACC=-ACC       ! flip accelerogram
                                        ! if YIELD negative

        A1=ACC-ABS(YIELD)

        IF(V1.LT.0.0001 .AND. A1.GT.0.0) THEN
          XT=DT*A1/(ABS(A1)+ABS(A0))
          V1=0.5*XT*A1                  ! first exceedance of YIELD
          D1=DT2*A1
        ELSE
          V1=V0+0.5*DT*(A0+A1)          ! compute new velocity
          IF(V1.GT.0.0) THEN
            D1=V0*DT+DT2*(2.0*A0+A1)      ! compute increment
          ELSE                            ! for +ve velocities
            XT=DT*V0/(ABS(V1)+ABS(V0))
            D1=0.5*XT*V0                ! linear approximation
            V1=0.0
          ENDIF
        ENDIF
        D1=D0+D1                        ! compute displacement
        A0=A1                           ! save current values for
        V0=V1                           ! next cycle of LOOP
        D0=D1
      ENDDO
      WRITE(6,*) D1                     ! output displacement
      STOP
      END
```

A more sophisticated FORTRAN program for Newmark's method, DISPLMT, has been published by Houston et al. (1987). This version operates on personal computers under the MSDOS operating system. It includes options for specifying a variable yield acceleration function, including variation as a function of time and displacement. Screen graphics are also provided to allow the user to observe the downslope movements of a hypothetical Newmark sliding block as a function of time. For a simple interpreted BASIC version, please refer to Jibson (1993).

Permanent Displacement The reported permanent displacements represent the motion of the center of gravity of the slide mass. For a planar slip-surface, the direction of this permanent displacement will be parallel to the slip surface. For the typical nonplanar failure surface, the direction of the permanent displacements is not immediately obvious. In such cases, the initial direction of the block's motion may be determined by considering the free-body forces that exist along the boundary of the slide mass. This direction may be calculated by first determining the resultant of all the shear forces and all the normal forces along the failure surface boundary. This essentially amounts to a vectorial summation of the shear and normal forces at the base of all slices, as determined in a limit equilibrium analysis. The permanent displacements are then assumed to act along the direction of the resultant of the cumulative shear and normal forces (Bromhead, 1992).

The calculations for these displacements are simplistic, and rely on the presumption that the failure mass within the slope behaves as a perfectly rigid body during the earthquake shaking. If displacements are essential for design, a numerical procedure, in conjunction with a plastic constitutive model, should be used to investigate the dynamic behavior.

6.12.3 Accelerogram Selection for Newmark's Method

As the slide mass is assumed to be a rigid body, the accelerogram must be selected such that it effectively models the seismic motions of the block. Realistically, the material in the slide block is not rigid and will thus deform, and have different motions within its mass. What is needed here is a procedure for estimating the *average* motions for the sliding block based on a site-specific response analysis of the slope.

A typical response analysis consists of selecting an accelerogram to represent expected motions on bedrock, which should effectively simulate the intensity, duration, and frequency content of the shaking motions. Then by using a numerical model, these bedrock motions are propagated through the overlying soil layers. Results from such an analysis can provide acceleration, stress, and strain time histories within the geometric model of the slope. For Newmark's method, the required site-specific accelerograms within the slide mass can be determined using a two-dimensional finite element analysis program such as FLUSH (Lysmer et al., 1975). Alternatively, a simpler approach is to perform a series of one-dimensional analyses for several soil profiles within the slide mass using a program like SHAKE (Schnabel et al., 1972). Only small differences have been reported between analyses performed using one- and two-dimensional analyses (Schnabel et al., 1972).

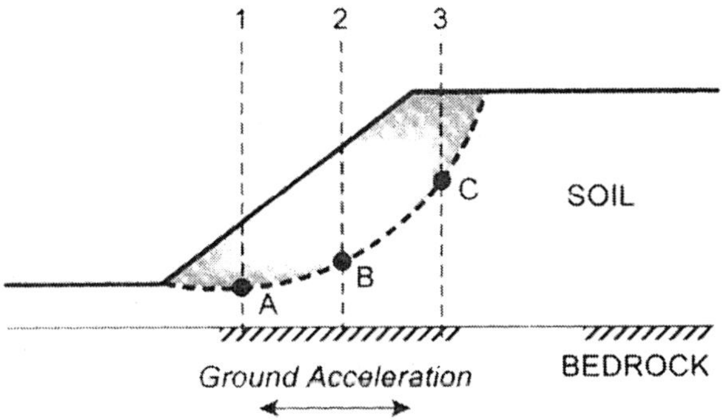

Figure 6.43 Earthquake response analysis of slope to determine motions along failure surface boundary.

Figure 6.43 presents an example of a response analysis for a slope. The bedrock motions, in the form of an accelerogram, are input at the bedrock elevation and the response accelerograms are computed for several points along the failure surface, shown as *A*, *B*, and *C* in Figure 6.43. The computed accelerations along the failure surface may be amplified, or attenuated, in comparison with the input bedrock motions depending on the material properties. The response accelerations should be determined for at least three locations: near the toe, midheight, and near the crest; however, if conditions warrant, additional locations may also be selected for the analysis. For a one-dimensional analysis, several vertical profiles may be selected along the failure surface, as indicated by the dashed lines 1, 2, and 3 in Figure 6.43.

The single accelerogram for use with Newmark's sliding block is then calculated according to a weighted average of the response motions computed for n points at locations designated as $i = 1, 2, \ldots, n$. This average, known as the horizontal equivalent acceleration (HEA) time history, is calculated using

$$\mathrm{HEA}(t) = \frac{\sum_{i=1}^{i=n} m_i a_i(t)}{\sum_{i=1}^{i=n} m_i} \qquad \text{(Eq. 6-75)}$$

where m_i = mass of unit column directly above point i.

$a_i(t)$ = acceleration response at point i

It is also possible to use a previously recorded accelerogram that has been suitably scaled to reflect the anticipated intensity of the motions of the slide block. The scaling factor for this approach is simply a substitute for the response analysis and will depend on the geometric and material properties of the slope. This scale factor is generally reported to be in the 2.0 to 3.5 range (Hynes-Griffin and Franklin, 1984).

6.12.4 Computed Permanent Displacements

In order to avoid the response analysis portion of the proposed analysis, two popular procedures have been proposed that allow the designer to estimate the magnitude of the permanent displacements without directly performing a double integration according to Newmark's method. The first simplification was contributed by Franklin and Chang (1977) in the form of a summary of the cumulative displacements calculated for 354 horizontal acceleration time histories. With these data, Newmark displacements could be estimated without having to perform a double integration. Of course, the engineer will still have to include a scale factor to account for possible amplifications within the slide mass.

The second contribution was provided in the report by Makdisi and Seed (1977, 1978, 1979). This simplified procedure can implicitly include the response analysis effects and provides graphs that allow an estimate of the permanent displacements. Thus with bedrock motions known in the vicinity of the slope, this method allows the engineer to calculate quickly the maximum accelerations and deformation in the slope. These procedures are discussed next.

Franklin and Chang (1977) In 1977, Franklin and Chang double integrated 354 acceleration time histories, of which 348 had been recorded during actual earthquakes, for the permanent displacements according to Newmark's method. Their results are presented in Figure 6.44, which shows the variation between a normalized coefficient and the computed Newmark displacements for the 354 available records. The data are presented in the form of three curves: (1) mean, (2) mean $+\sigma$, and (3) an upper bound envelope to all results. The mean and the mean $+\sigma$ curves were developed using regression methods.

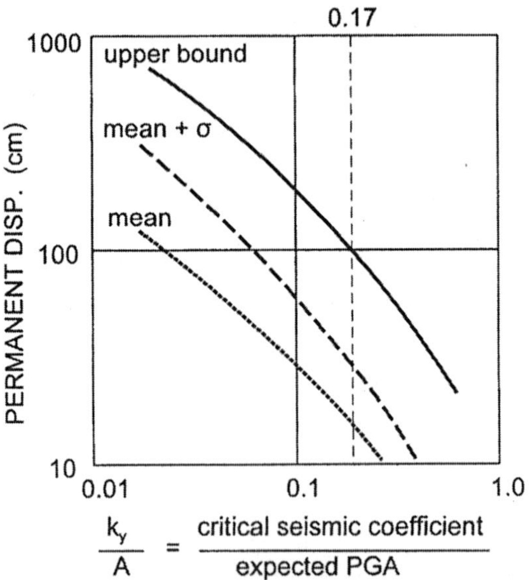

Figure 6.44 Permanent displacement as a function of k_y/A based on 354 horizontal accelerograms. (Hynes-Griffin and Franklin, 1984.)

As an example, the slope that had a yield acceleration of $0.4g$ will be considered for this analysis. If the site can expect a PGA of $0.5g$, the ratio $k_y/\text{PGA} = 0.8$. Then by using the graph presented in Figure 6.44, the upper bound for the maximum amount of deformation may be estimated as approximately 20 centimeters. The mean and the mean $+ \sigma$ values will be considerably less than the 20-centimeter upper bound value. If the slide mass experiences ground motions greater than the $0.5g$ PGA, the slope can be expected to experience greater displacements.

Makdisi and Seed (1977, 1978) This method uses the concept proposed by Seed and Martin (1966) for calculating the dynamic response of an earth embankment based on the embankment properties and input in the form of a response spectra, as shown in Figure 6.45. With this approach, the following three computations are made:

(1) *Yield Acceleration* (k_y) This is calculated for FOS = 1, using traditional limit equilibrium methods.
(2) *Average Acceleration* (k_{max}) Average acceleration within the failure mass is computed.
(3) *Permanent Displacement* (U) This is calculated for the ratio $k_y/k_{\max}$ using Newmark's method.

A simplified method, with strain compatible parameters, is used to evaluate the crest acceleration. Then the relationship shown in Figure 6.46 is used to estimate the average acceleration within the slide mass, as a proportion of crest acceleration. The permanent displacement is then estimated using ground motion accelerograms and Newmark's method.

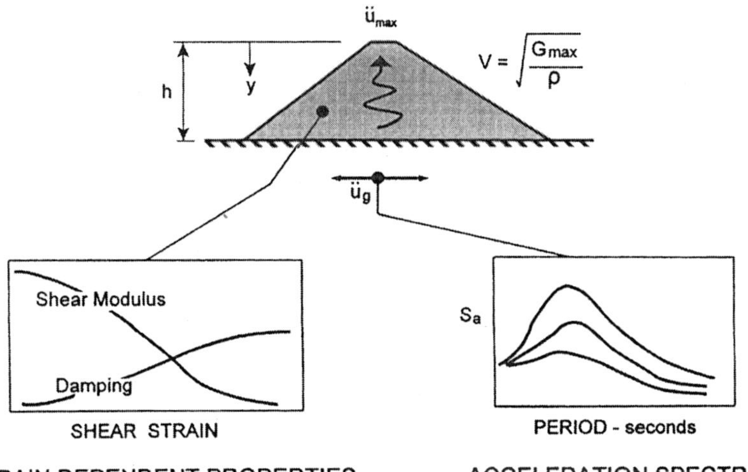

Figure 6.45 Framework for calculating the approximate maximum acceleration at the crest of an embankment.

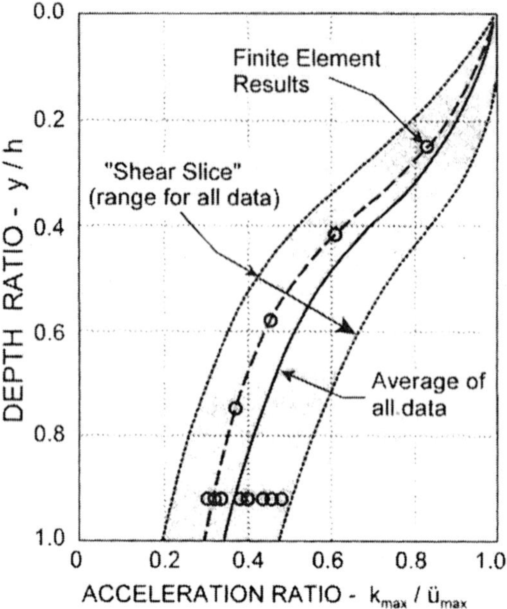

Figure 6.46 Variation of maximum acceleration ratio with depth of sliding mass. The finite element results are shown as open circles. (After Makdisi and Seed, 1978, with permission.)

Information concerning the variation of shear modulus and damping ratio as a function of shear strain for the embankment material is used to estimate the first three modal frequencies. On the basis of these three modes, the crest acceleration is estimated using modal superposition and the elastic acceleration response spectra of the base motions. Briefly, the method involves the following steps:

(1) Estimate the average shear strain level within the embankment.

(2) For the selected shear strain level, determine the value of the reduced shear modulus, G, and the damping ratio, λ. Calculate the equivalent shear wave velocity for this strain level using $V = (G/\rho)^{1/2}$, where ρ = mass density of the embankment material.

(3) Calculate the three modal periods using

$$T_1 = 2.618\frac{H}{V} \qquad T_2 = 1.138\frac{H}{V} \qquad T_3 = 0.726\frac{H}{V} \qquad \text{(Eq. 6-76)}$$

where H = embankment height

V = shear wave velocity determined in step 2 above

(4) Using the base motion accelerogram, compute the spectral accelerations corresponding to the damping ratio, λ, and the three modal periods. These values can be determined directly using a suitable computer program, or they may be estimated from an acceleration response spectrum determined for several different levels of damping ratio. These spectral accelerations, S_{a1}, S_{a2}, and S_{a3} will correspond to the three periods, T_1, T_2, and T_3 and the appropriate damping ratio.

(5) The maximum acceleration occurs at the crest of the embankment and may be estimated using

$$\ddot{u}_{\max} = \left[2.56S_{a1}^2 + 1.12S_{a2}^2 + 0.74S_{a3}^2\right]^{1/2} \qquad \text{(Eq. 6-77)}$$

(6) Finally, calculate the average shear strain using

$$\gamma_{\text{average}} = 0.195\left[\frac{H}{V^2}\right]S_{a1} \qquad \text{(Eq. 6-78)}$$

(7) If the γ_{average} value computed in the last step is not approximately the same as the value assumed in step 1, repeat steps 1 through 6 again, but this time using the newly computed average shear strain value as the "new" estimate.

Once the maximum acceleration and the natural period, T_1, have been determined for a compatible average shear strain, the Newmark displacement can then be estimated using the following remaining steps:

(8) Determine the yield acceleration of the potential failure surface, using reduced static shear strengths if cyclic strength degradation is expected for the embankment materials. Also determine the y/H ratio for the critical failure surface, where y is the maximum depth of the sliding mass and H is the height of the embankment.

(9) Using Figure 6.46 and the y/H ratio, determine the $k_{\max}/\ddot{u}_{\max}$ ratio, and hence the value of $k_{\max}$, using the known maximum acceleration from step 5.

(10) For the given earthquake magnitude and known ratio of $k_y/k_{\max}$, use Figure 6.47 to estimate the range of potential permanent displacement. These ranges are given for three different earthquake magnitudes as one would expect larger events to generate longer durations of ground shaking.

In Figure 6.47, the ratio $k_y/k_{\max}$ will predict three ranges of displacements. The correct range will depend on the size of the event simulated by the chosen response spectra. In preparing Figure 6.47, Makdisi and Seed (1977, 1978) used the

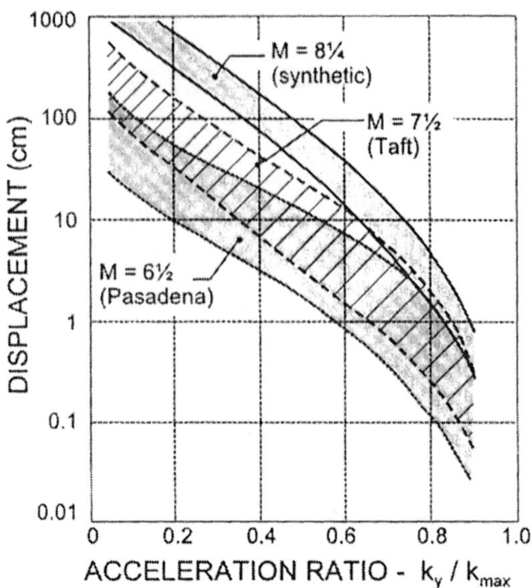

Figure 6.47 Variation of permanent displacement with yield acceleration—summary of all data. (After Makdisi and Seed, 1978, with permission.)

following accelerograms:

(1) Caltech and Lake Hughes records from the 1971 San Fernando ($M = 6.5$) event, with scaled PGA values of $0.2g$ and $0.5g$ (a total of five cases)

(2) Taft record from the 1952 Kern County ($M = 7.5$) event, with scaled PGA values of $0.2g$ and $0.5g$ (a total of three cases)

(3) A synthetic accelerogram, representative of an $M = 8.25$ event, with scaled PGA values of $0.4g$ and $0.75g$ (a total of four cases)

In view of the original data, this method should not be used for cases where the PGA, at the base of the embankment, is less than $0.2g$ for $M = 6.5$ and $M = 7.5$ events and less than $0.4g$ for the $M = 8.25$ event. Although the PGA ranges for the $M = 6.5$ and $M = 7.5$ events are the same, the duration of the Taft record was longer than the Caltech or Lake Hughes record, and thus, generated greater displacements.

Example To illustrate the application of the Makdisi and Seed (1977, 1978) procedure, a 46-meter high embankment will be analyzed to determine the permanent displacements for a peak ground acceleration of $0.2g$. The critical failure surface for this problem had a y/H ratio of 0.85 and a critical yield acceleration of $0.14g$. The average properties of the embankment material are: (1) Maximum shear modulus, $G_{max} = 175{,}000$ kilopascals and unit weight, $\gamma = 20.4$ kilonewtons per cubic meter.

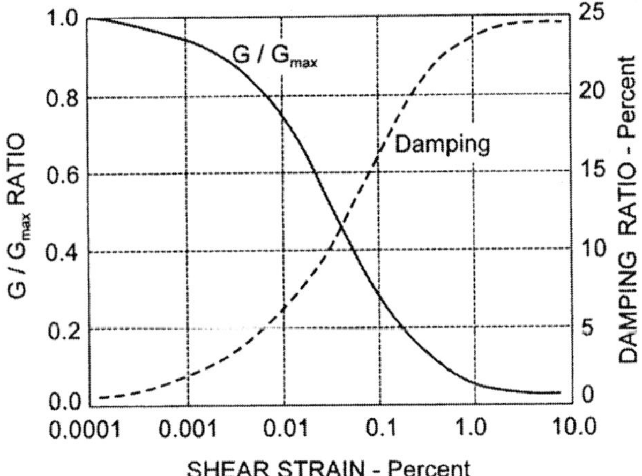

Figure 6.48 Typical strain dependent shear modulus and damping ratio parameters used for response calculations.

The variation of the shear modulus and the damping ratio for this material follows the curves given in Figure 6.48. The ground motions will be simulated using the north-south component of the Taft record from the 1952 Kern County earthquake. The normalized acceleration response spectra for damping ratios between 5 and 20 percent are given in Figure 6.49.

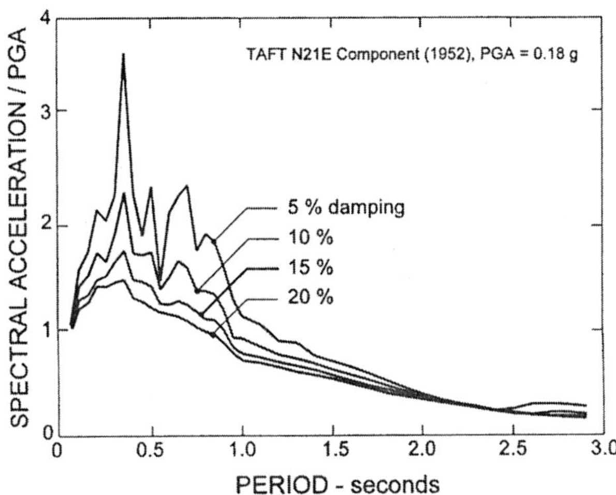

Figure 6.49 Normalized acceleration response spectra—N21E Taft accelerogram recorded during the 1952 Kern County, California, earthquake.

Solution

FIRST ITERATION

Step 1: Assume an average shear strain of 0.06 percent.

Step 2: From Figure 6.48 and $\gamma_{\text{average}} = 0.06$ percent, $G/G_{\max} = 0.4$, that is, $G =$ 70,000 kilopascals. Damping ratio, $\lambda = 13$ percent. Equivalent shear wave velocity at the presumed strain level will be

$$\begin{aligned} V &= \sqrt{(0.4 \times 175{,}000 \times 9.81 \div 20.4)} \\ &= 183.5 \text{ m/s} \end{aligned}$$

Step 3: The three modal periods will be

$$\begin{aligned} T_1 &= 2.618 \times 46 \div 183.5 = 0.656 \text{ seconds} \\ T_2 &= 1.138 \times 46 \div 183.5 = 0.285 \text{ seconds} \\ T_3 &= 0.726 \times 46 \div 183.5 = 0.182 \text{ seconds} \end{aligned}$$

Step 4: Using the normalized response spectra presented in Figure 6.49,

$$\begin{aligned} \text{For } T &= 0.656, \qquad S_{a1} = 0.2 \times 1.30 = 0.26g \\ \text{For } T &= 0.285, \qquad S_{a2} = 0.2 \times 1.60 = 0.32g \\ \text{For } T &= 0.182, \qquad S_{a3} = 0.2 \times 1.45 = 0.29g \end{aligned}$$

Step 5: The maximum acceleration will then be

$$\begin{aligned} \ddot{u}_{\max} &= \sqrt{[2.56 \times (0.26)^2 + 1.12 \times (0.32)^2 + 0.74 \times (0.29)^2]} \\ &= 0.591g \end{aligned}$$

Step 6: The average shear strain is estimated as:

$$\begin{aligned} \gamma_{\text{average}} &= 0.195 \times 46 \times 0.26 \times 9.81 \div (183.5)^2 \\ &= 0.00068 \quad (\text{i.e., } 0.068 \text{ percent}) \end{aligned}$$

Step 7: As the 0.068 percent value differs from the assumed value of 0.06, steps 1 through 6 are repeated with an assumed $\gamma_{\text{average}} = 0.07$ percent.

SECOND ITERATION

Step 1: Assume an average shear strain of 0.07 percent.

Step 2: From Figure 6.48 and $\gamma_{\text{average}} = 0.07$ percent, $G/G_{\max} = 0.36$, that is,

$G = 63{,}000$ kilopascals. Damping ratio, $\lambda = 14$ percent. Equivalent shear wave velocity,

$$V = \sqrt{(0.36 \times 175{,}000 \times 9.81 \div 20.4)}$$
$$= 174.1 \text{ m/s}$$

Step 3: The three modal periods will be

$$T_1 = 2.618 \times 46 \div 174.1 = 0.692 \text{ seconds}$$
$$T_2 = 1.138 \times 46 \div 174.1 = 0.301 \text{ seconds}$$
$$T_3 = 0.726 \times 46 \div 174.1 = 0.192 \text{ seconds}$$

Step 4: Using the normalized response spectra presented in Figure 6.49,

$$\text{For } T = 0.692, \quad S_{a1} = 0.2 \times 1.22 = 0.244g$$
$$\text{For } T = 0.301, \quad S_{a2} = 0.2 \times 1.60 = 0.32g$$
$$\text{For } T = 0.192, \quad S_{a3} = 0.2 \times 1.47 = 0.294g$$

Step 5: The maximum acceleration will then be

$$\ddot{u}_{\max} = \sqrt{[2.56 \times (0.244)^2 + 1.12 \times (0.32)^2 + 0.74 \times (0.294)^2]}$$
$$= 0.575g$$

Step 6: The average shear strain is estimated as

$$\gamma_{\text{average}} = 0.195 \times 46 \times 0.244 \times 9.81 \div (174.1)^2$$
$$= 0.00071 \quad \text{(i.e., 0.071 percent)}$$

Step 7: As the 0.071 percent value is almost the same as the assumed value of 0.07, further iterations are not necessary.

Step 8: The critical failure surface has the y/H ratio of 0.85 and the yield acceleration was reported as $0.14g$.

Step 9: Using the average curve given in Figure 6.46, for $y/H = 0.8$,

$$\frac{k_{\max}}{\ddot{u}_{\max}} \approx 0.4 \quad \text{or} \quad k_{\max} = 0.4 \times 0.575 = 0.23g$$

Step 10: The ratio, $k_y/k_{\max} = 0.14 \div 0.23 = 0.61$. With this ratio and the curves given in Figure 6.47, the permanent displacements may then be estimated as given in

TABLE 6.15 Permanent Displacements for Example Embankment

Earthquake Magnitude	Expected Permanent Displacements (cm)
6.5	1–8
7.5	2–14
8.25	—[a]

[a] PGA = 0.2g is outside the range for a $M = 8.25$ event.

Table 6.15. Note that a value for the $M = 8.25$ range is not offered, as it is less than the range of PGA 0.4 to 0.75 that was considered by the study. Also, the input PGA for this case was $0.2g$, which is the value for the lower range of the 6.5 and 7.5 magnitudes. Hence the lower values, 1 and 2 centimeters, from Table 6.15, should be reported as the permanent displacements for this example embankment.

6.12.5 Tolerable Permanent Displacements

Only a full, nonlinear numerical analysis can provide details about the actual permanent deformations that a slope or embankment may experience during an earthquake. The magnitude of the displacements computed by Newmark's approach are a qualitative reflection of the impact that the seismicity of the site will have on the stability of the slope.

The tolerable levels of displacement that have been used to differentiate between safe and unsafe behavior have been varied. Some examples are:

(1) Hynes-Griffin and Franklin (1984) suggest that up to 100-centimeter displacements may be acceptable for well-constructed earth dams.

(2) Wieczorek et al. (1985) used 5 centimeters as the critical parameter for a landslide hazard map of San Mateo County, California.

(3) Keefer and Wilson (1989) used 10 centimeters for coherent slides in southern California.

(4) Jibson and Keefer (1993) used a 5- to 10-centimeter range for landslides in the Mississippi Valley.

(5) The State of California (1997) finds slopes acceptable if the Newmark displacement is less than 15 centimeters. A slope with a Newmark displacement greater than 30 centimeters is considered *unsafe*. For displacements in the "grey" area between 15 and 30 centimeters, engineering judgement is required for assessment.

A single value for the critical permanent displacement that can be used to evaluate the performance of a slope during an earthquake is difficult to establish at this time.

Jibson (1993) suggests that the engineer link the critical, or acceptable, value of the permanent displacements to the parameters of the problem and the characteristics of the slope materials. Thus slopes made up of ductile, plastic materials may be able to sustain displacements that are much larger than those for slopes consisting of brittle, sensitive materials. In conclusion, the following statement made by Houston et al. (1987), in their discussion of Newmark's method, appears to offer the best suggestion:

> It probably should be viewed as a tool to assist the engineer in deciding whether the probable slope movement is: (1) a fraction of an inch, or (2) a few inches, or (3) a few feet. This level of distinction is usually adequate to enable an engineering or management decision.

6.13 OTHER FACTORS AFFECTING SLOPE STABILITY ANALYSIS

Besides the shear strength parameters and the groundwater table, other factors that would affect the stability analyses of a slope include tension cracks, foundation loads from nearby structures, presence of leaking water or sewer lines, and presence of trees and shrubs on slopes. The following sections discuss the influence of these factors.

6.13.1 Effect of Tension Cracks on Stability Analysis

The existence of tension cracks is likely to increase the tendency of a soil to fail. The length of failure surface along which shear strength can be mobilized is reduced; in addition a crack may be filled with water (e.g., due to rainfall). The additional force due to water pressure in a crack increases the tendency for a slip to occur.

Cracks open up in cohesive soils due to the soil's low tensile strengths; the depth of these cracks can theoretically be found by

$$z_c = \frac{2c}{\gamma} \tan^2\left(45 + \tfrac{1}{2}\phi\right) \qquad \text{(Eq. 6-79)}$$

where z_c = depth of tension crack
c = cohesion
ϕ = angle of friction
γ = unit weight of the soil material.

The question often arises as to whether effective or undrained strength parameters should be applied to the above equation. The depth of a tension crack based on c' and ϕ' is likely to be much less than that based on $c = s_u$ and $\phi = 0$, because c' is often considerably less than s_u. For embankments and undisturbed natural slopes, it would be logical to use the effective stress parameters, c' and ϕ', to calculate z_c, because undrained conditions due to sudden removal of lateral support probably have not occurred. On the other hand, undrained strength parameters for cut slopes may have to be used because tension cracks would be formed in the short term after a cut is made and their depth would be unlikely to decrease thereafter.

The importance of tension cracks and their effect on stability is not always emphasized in soil mechanics. This may be due in part to conclusions reached by some that the effect of tension cracks on the for embankment slopes is negligible (e.g., Spencer, 1967, 1973). Such conclusions must be understood in terms of the small depth of tension cracks predicted by the equation for z_c. In cut slopes, tension cracks could extend to considerable depth and exert a significant influence on the value of FOS as in the case of the slips at Bradwell (Skempton and La Rochelle, 1965).

Quite apart from the effect of tension cracks on a conventional limit equilibrium analysis, there is the importance of progressive failure. Tension cracks may be in some cases a consequence of the initiation of progressive shear failure. As such, the existence of tension cracks deserves careful attention.

6.13.2 Effects of Vegetation

Vegetation on cut and fill slopes, in the form of trees/shrubs or grass, aids erosion control and adds considerably to the landscape amenity. The vegetation on a slope often enhances slope stability. However, this effect is a complex interaction of mechanical and hydrological factors that are difficult to quantify. As yet, there are no firm design rules, but in recent years, there has been rapid growth in local experience with a variety of vegetation types, and this knowledge can be of valuable assistance to the designer.

As a general rule, a single species of vegetation should not be planted in isolation. On typical slopes, vegetation cover should not consist only of grass, but should include trees and shrubs. Consideration should also be given to appropriate vegetation management techniques to assist the natural succession process.

It is widely appreciated that slope vegetation is beneficial in terms of erosion control and its contribution to landscape quality. However, the overall effect of vegetation as a slope stabilizing mechanism cannot be easily categorized as beneficial or adverse, as illustrated in Table 6.16. The results of detailed research studies carried out elsewhere

TABLE 6.16 Effects of Vegetation on Slope Stability

Beneficial Effects
Interception of rainfall by foliage, including evaporative losses
Depletion of soil moisture and increase of soil suction by root uptake and transpiration
Mechanical reinforcement by roots
Restraint by buttressing and soil arching between tree trunks
Arresting the roll of loose boulders by trees
Surcharging the slope with large heavy trees
Adverse Effects
Increased capacity for rainwater infiltration
Superficial root reinforcement may be rapidly eroded in a heavy storm
Surcharging the slope with large heavy trees

provide support for the use of vegetation as a net beneficial mechanism, primarily through the effect of root reinforcement (Gray, 1978; Gray and Leiser, 1982: Gray and Megahan, 1981; Gray and Sotir, 1996, Sotir and Gray, 1989).

Until such time as further data from local studies are available, it is recommended that designs for new highway slopes that include vegetative surface protection should allow for direct infiltration while neglecting the beneficial effects provided by the vegetation. For stability analyses of existing slopes that have an existing tree and shrub cover, it may be appropriate to investigate and quantify the beneficial factors given in Table 6.16. Guidance on suitable methods for assessing these factors is given by Gray (1978) and Gray and Leiser (1982).

Where it can be clearly demonstrated that direct surface infiltration is a critical control of slope stability, and where stability cannot be improved by flattening the slope, it may be worthwhile to consider covering the slope face with rigid surface protection (e.g., shotcrete) in order to reduce infiltration.

6.13.3 Foundation Loads on Slopes

The stability of a slope can be affected by excavation for construction of foundations on or adjacent to the slope. As a general rule, slope stability affected by foundations should be checked if the slope angle is greater than $\frac{1}{2}\phi'$. Where this is so, the foundation can be considered as an equivalent line load or a surcharge imposing horizontal and vertical loads and incorporated into the stability analysis (see Figure 6.50).

The stability of a slope can also be impaired by excavation for the construction of shallow foundations on or adjacent to the slope and the demolition of structures

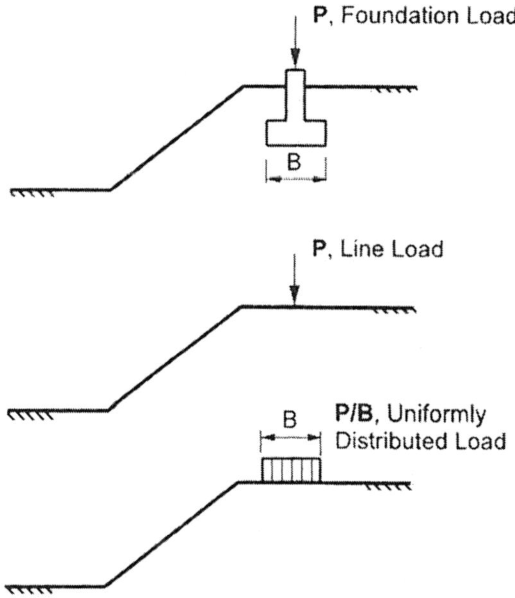

Figure 6.50 Foundation loads on crest of a slope.

supporting the toe of the slope. Both these effects should be considered during the analysis. In order to minimize the short-term instability of a slope, excavations should be as small as possible and should be properly shored.

Lateral loads due to deep foundations may affect slope stability. However, in a slope that has an acceptable FOS against failure, the effects of the lateral loads are usually not considered during design of most foundations (Schmidt, 1977). On the other hand, high lateral loading can be transferred to foundations in situations where there is significant ground movement (i.e., where the slope above or below the foundations soils or where the slope in front of the foundation is excavated). Under these circumstances, lateral loads on deep foundations should be prevented. This can be achieved by either:

- Stabilizing potentially unstable slopes before or during construction of foundations, or
- Providing an annular sleeve around the foundation. Such an annular sleeve is a space of sufficient width that can be air-filled or can contain a suitable compressible material.

6.14 THREE-DIMENSIONAL ANALYSIS

Most slope failure analysis are performed on a two-dimensional model even though the shape of the slope failure in the field is truly three-dimensional. Generally, two-dimensional analysis has been successful as it yields a conservative estimate of the FOS (Duncan, 1996). The FOS is underestimated as the slope failure is assumed to extend to infinity in the third dimension, and the end effects are ignored. However, there is a current trend toward implementing three-dimensional analysis for routine analysis (Koerner and Soong, 1999; Mitchell et al., 1993; Stark and Eid, 1998).

The three-dimensional analytical approach has its roots in the 1970s, with papers by Baligh and Azzouz (1975) and Hovland (1977). Hovland modified the familiar two-dimensional method to the third dimension by substituting *columns* for the more familiar slices (see Figure 6.51). As we jump from two to three dimensions, there are considerably more unknowns in comparison to the equations that can be generated from equilibrium and Mohr–Coulomb conditions. And of course, the subsoil model now has to be presented in three dimensions as well. Considerable work in this area has been completed and evaluated by Duncan (1996), Hungr et al. (1989), and Stark and Eid (1998). On the basis of these studies, the following factors are worth considering:

(1) FOS values from three-dimensional analysis are always greater than those from the two-dimensional analysis.
(2) Back-calculated strengths are, thus, likely to be lower for three-dimensional analysis.
(3) Effects are more pronounced for materials with $c > 0$.
(4) For $c = 0$ analyses, the effects are less pronounced.

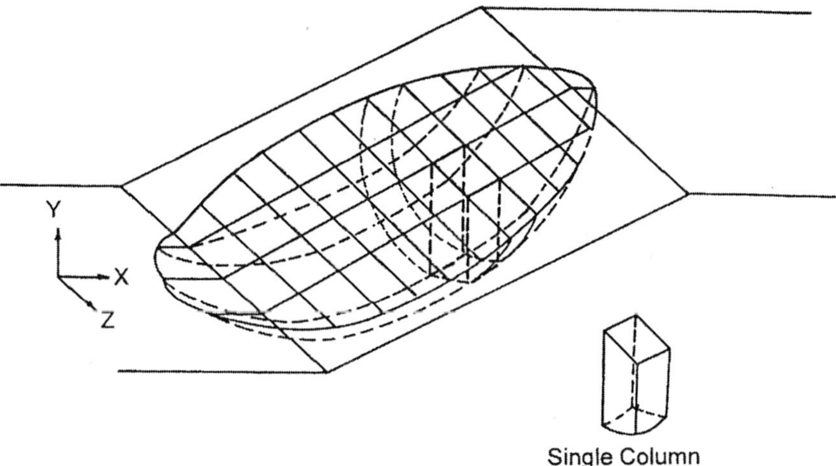

Figure 6.51 Features of three-dimensional slope analysis.

(5) It is very difficult to delineate a three-dimensional failure surface in the field for use in a three-dimensional analysis.

(6) The three-dimensional subsoil model requires considerably greater effort to prepare and input into available computer programs.

Currently, there are three computer programs available for three-dimensional analysis: CLARA, TSLOPE3, and 3D-PCSTABL. For more information and a comparison of these computer programs, please see Stark and Eid (1998). Lam and Fredlund (1993) reported excellent results using a computer program they developed but the authors have not yet released the software for general use.

In conclusion, there are several limitations in the familiar two-dimensional analysis that the three-dimensional analyses overcome. However, one major limitation of implementing three-dimensional analysis is the selection of the failure surface shape with a view to its kinematical behavior. For two-dimensional analysis, the slices are logically expected to slide *downhill*. But when you have all these columns, the inter-column kinematics become very important and, so far, has received very little attention. New state-of-the-art programs are expected to become available in the next decade, and hopefully, these will provide us with the graphical visualization tools so that the kinematics of the columnar elements can be isolated and followed during the analysis. Until these tools become available, the authors are not satisfied with the performance of current computer programs for three-dimensional analysis of slopes.

6.15 ROCK SLOPE STABILITY

The stability of soil slopes may be analyzed readily using two-dimensional limit equilibrium analysis and a representative value of the material's shear strength, which may or may not include the effect of inherent discontinuities. Such a two-dimensional, continuum-based analysis is not appropriate for rock slopes, as the stability will be controlled by the shear strength along the joints as well as the three-dimensional geometric interaction of the jointing and bedding patterns. The predominant modes of rock slope failure that have been observed tend to be a two-dimensional planar wedge, a three-dimensional wedge, or toppling, although deep seated circular, and multiple plane failures have also been reported in the literature.

The classical three-dimensional wedge is the most common type of failure in rock slopes. It occurs as a result of two (or several) intersecting discontinuities that combine to form a wedge, which subsequently slides down a V-shaped notch formed by the discontinuities (Hoek, 1973). If the strike and dip of the rock discontinuities is summarized on a polar stereographic projection, potential failure wedges can be identified for further examination (for further information, please consult: Goodman, 1976, 1980; or Hoek and Bray, 1977).

The stability of such wedges will be controlled by the shear strength along the interfaces between the wedge and intact rock. The magnitude of the available shear strength along joints and interfaces is very difficult to determine due to the inherent variability of the material and the difficulties associated with sampling and laboratory testing. Occasionally, depending on the critical nature of the project, direct-shear field tests may be performed on joints in an effort to determine reliable strength parameters. Factors that directly or indirectly influence the strength include (Bromhead, 1992):

(1) The planarity and smoothness of the joint's surfaces. A smooth planar surface will have a lower strength than an irregular and rough surface.
(2) The inclination of the joint with respect to the slope.
(3) The openness of the discontinuity, which can range from a small fissure to a readily visible joint. In reviewing this data, the engineer should consider the potential effects of stress-relief due to an excavation that may further open the existing (preconstruction) discontinuities.
(4) The extent of the weathering along the surfaces and the possible infill of the joint with weaker material (such as clay, fault gouge, etc.). A cementing infill may potentially increase the strength of the joint, whereas a soft clay infill may reduce the strength of the joint to the same level as the clay material itself. Such infills may also change the seepage pattern, improving or degrading the drainage, which will be manifested by an increase or decrease in pore water pressures within the joints.

Once the geometry of the three-dimensional wedge has been ascertained and the shear strength along the interfaces determined or estimated, its stability can be evaluated rapidly using simple computer programs (e.g. Abramson, 1985; Hendron et al., 1971; Kovari and Fritz, 1975; Watts and West, 1985). It is recommended that a range of strength parameters and several different geometric configurations be considered for the evaluation of the stability of rock slopes.

6.16 THE FINITE ELEMENT METHOD (FEM)

The limit equilibrium method allows engineers to evaluate the stability of slopes quickly. However, these procedures are the same whether the analysis considers (1) slope of newly constructed embankment, (2) slope of a recent excavation, or (3) an existing natural slope. The stresses within these slopes are strongly influenced by K_0, the ratio of lateral to vertical normal effective stresses, but conventional limit equilibrium procedures ignore this important feature (Chowdhury, 1981). In reality, the stress distributions within these three slopes would be different and hence significantly influence their stability.

The finite element method (FEM) bypasses many of the deficiencies that are inherent with the limit equilibrium methods. It was first introduced to geotechnical engineering by Clough and Woodward (1967), but its use has been limited to the analysis of complex earth structures, such as large earth dams, as indicated by the summary presented by Duncan (1996). For typical cases, the FEM can incorporate *incremental construction* for embankments and excavations in an attempt to simulate the stress history of the soil within the slope. However, the quality of the FEM is directly dependent on the ability of the selected constitutive model to realistically simulate the nonlinear behavior of the soil within the slope. For new embankment designs, the data may be collected from laboratory tests. For excavations and natural slopes, the constitutive model can only really be developed on the basis of high quality field tests that are further supported by field observations.

The FEM essentially divides the soil continuum into discrete units, that is, *finite elements* (see Figure 6.52). These elements are interconnected at their nodes and at predefined boundaries of the continuum. The *displacement method* formulation of the FEM is typically used for geotechnical applications and presents results in the form of displacements, stresses, and strains at the nodal points. There are many two- and three-dimensional computer programs available for finite element analysis of slopes and embankments.

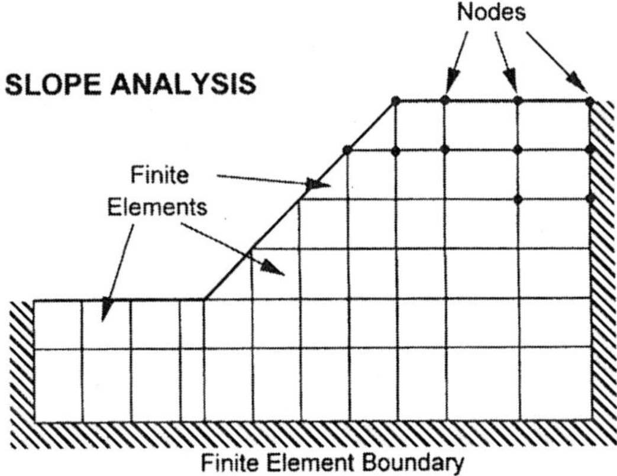

Figure 6.52 Definitions of terms used for finite element method (FEM).

In selecting a suitable program, the user must consider: (1) the implementation of the constitutive model(s), (2) the availability of different types of finite elements (e.g., triangular, quadrilateral, or isoparametric), and (3) the laboratory and field test data required for defining the properties of the soils. Once a program has been selected it may be used to effectively determine slope stresses and deformations that may be used to evaluate the stability of the slope.

Although the FEM presents a very powerful technique for the geotechnical engineer, paradoxically it also introduces complexities that have resulted in its limited use to solve practical problems. Wong (1984) mentions the difficulty associated in developing an FOS against failure. In a conventional limit equilibrium approach, failure may be described as the condition where the driving forces (or moments) exceed the resisting forces (or moments) and is usually manifested by an FOS of less than unity. In the FEM, the soil is modeled as a set of discrete elements and the failure condition will be a progressive phenomenon where not all elements fail simultaneously. So failure can span a wide range that extends from the point where yield occurs first to the final failure state where all elements have effectively failed. Some of the popular failure criteria (Wong, 1984) in use are:

Bulging of Slope Line (Snitbhan and Chen, 1976) This criterion is described by the horizontal displacements of the surface of the slope, and is established by specifying a maximum tolerable limit for these horizontal displacements.

Limit Shear (Duncan and Dunlop, 1969) In this case, the computed FEM stresses along a potential failure surface are used directly to estimate an FOS. This FOS value would correspond to the ratio of available strength along the failure surface compared to stresses calculated using the FEM.

Nonconvergence of the Solution (Zienkiewicz, 1971) This may be indicative of a collapse of the elements under the imposed loading conditions. With this approach, the shear strength parameters are reduced until nonconvergence or numerical instabilities occur. The FOS is then reported as the ratio of the actual, available strength and the lowest strength value that produced a viable solution. This approach has been recently used by Dawson et al. (1999) and Griffiths and Lane (1999).

Depending on which failure criterion is selected, the difference in the failure loads can be significant. With the lack of an obvious failure criterion, the interpretation of FEM results is still a problem and the user must rely on experience and intuition to understand the ability of the numerical model to predict the behavior of the real physical model of the slope. In view of the uncertainty and a lack of familiarity with the FEM, the complex approach is not used for the design and analysis of typical highway slopes and embankments.

6.16.1 Example of FEM Analysis of Slopes

The slope shown in Figure 6.53 was analyzed with a finite element program, and the stresses, σ_x, σ_y, and τ_{xy}, were determined at unique points within the slope and foundation soils. The distribution of these stresses is shown in Figure 6.54. If

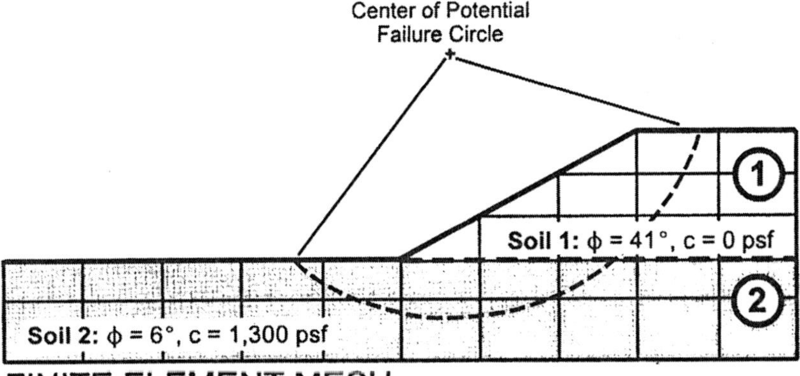

Figure 6.53 Slope used for FEM example.

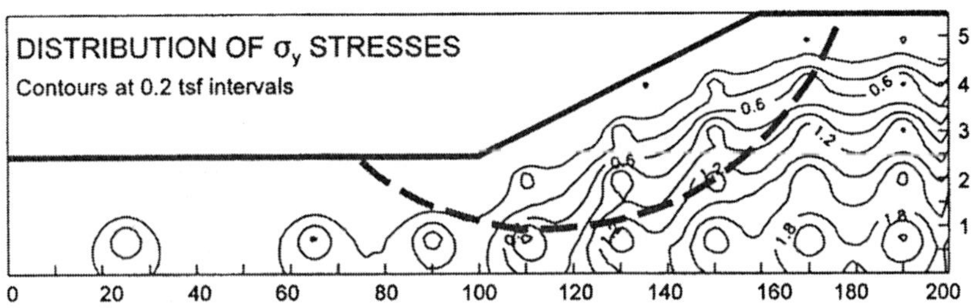

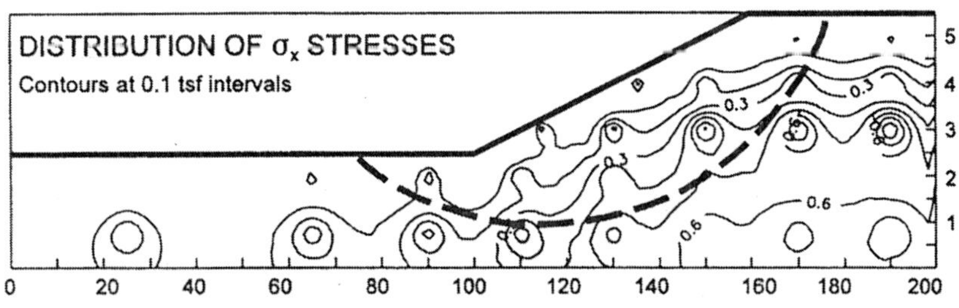

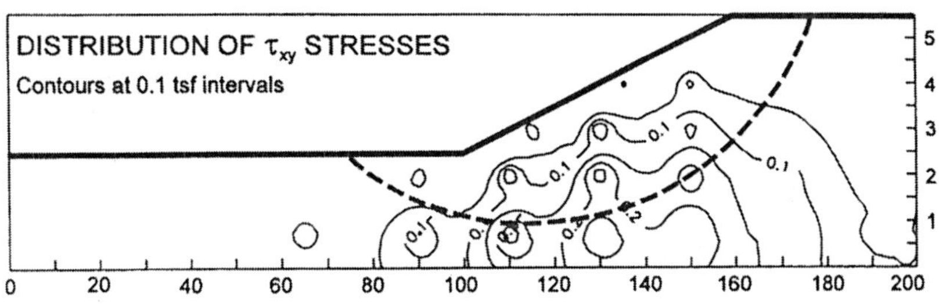

Figure 6.54 Distribution of stresses, σ_x, σ_y, and τ_{xy}.

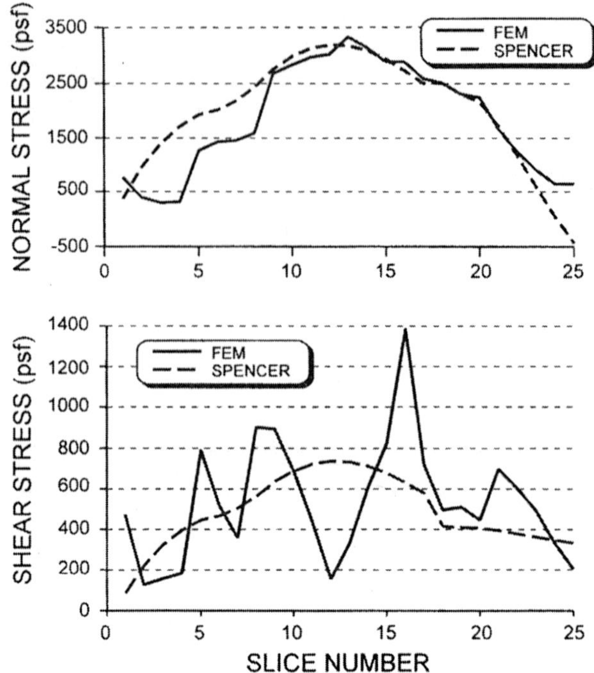

Figure 6.55 Distribution of slice stresses from FEM and Spencer analysis.

a potential failure surface is placed within the stress field, the stresses, σ_n and τ_n, may be effectively determined, and a conventional FOS calculated within the same definition as used for the limit equilibrium methods.

The distribution of the σ_n and τ_n stresses is presented in Figure 6.55 along with the results obtained from a Spencer's solution for the same slope geometry and soil parameters. The limit equilibrium stresses in these figures are different, especially the τ_n values. For this illustrative example, the FE mesh used was very coarse and the interpolated values of τ_n are highly variable. By using these stresses, the average FOS value or the local FOS *for each slice* may be determined. The local and average FOS values for the 25 slices are shown in Figure 6.56. In conclusion, the Spencer FOS and the average FEM values are in very good agreement, but it should be acknowledged that the FEM results provides a more thorough report of the stresses within the slope and also do not generate the negative normal stresses at the base of the uppermost slices that are predicted by the Spencer's solution.

In conclusion, Griffiths and Lane (1999) showed concern that the FEM is not used more often for routine slope stability analysis in practice. According to Duncan (1996), finite element analyses are very costly as they require considerable effort and expense to develop the model parameters, and then to make sense of the results. Although computers have become more prevalent, costs associated with the actual computational times are only about 10 percent of the overall cost. Although the FEM

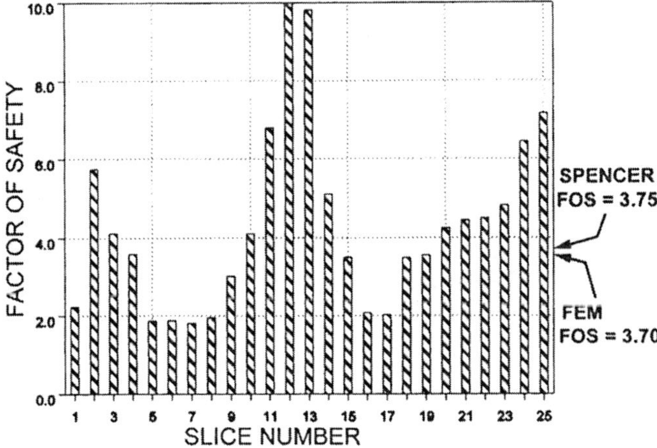

Figure 6.56 Local and Average FOS values from FEM and Spencer analyses.

can provide considerable insight into the behavior of earth structures, it is this overall expense in comparison to limit equilibrium methods that has precluded its use for routine geotechnical problems.

6.17 COMPUTER ANALYSIS

All of the limit equilibrium methods discussed earlier have been implemented within many computer programs that are operational on mainframe and personal computers. These programs can readily discretize the soil mass above the failure surface into slices and employ one of several methods to calculate the FOS.

6.17.1 Available Computer Programs

The many computer programs currently available for analyzing the stability of slopes may be separated into four distinct categories. These categories are:

(1) *STABL Programs* The initial program of this group originated at Purdue University in 1975 (Siegel, 1975). Subsequent versions that are a superset of the original are available as PC_STABL, XSTABL, and GEOSLOPE, for example.

(2) *Progams from the University of Texas* These programs originated with SSTAB1 and have been subsequently published as SSTAB2, UTEXAS, UTEXAS2, and UTEXAS3.

(3) *Progams from the University of California, Berkeley* STABR, STABGM, SLOPE8R, GEOSOFT.

(4) *Other Programs* PC-SLOPE, SLOPE/W, CLARA, GALENA, GSLOPE, TSLOPE.

Program Availability A license to use XSTABL for slope stability analysis may be obtained from: Interactive Software Designs, Inc., 953 N. Cleveland Street, Moscow, Idaho 83843, U.S.A. A license to use UTEXAS3 for slope stability analysis may be obtained from: Shinoak Software, 3406 Shinoak Drive, Austin, Texas 78731, U.S.A. For more information about the SLOPE/W program, please contact: GEO-SLOPE International, # 830, 633 6th Avenue, S.W., Calgary, Alberta, Canada, T2P 2Y5.

6.18 PROBABILISTIC ANALYSIS OF SLOPES

The current state of practice relies on using calculated FOS values in the design process to account for uncertainties associated with soil parameters, site stratigraphy, ignorance, and the potential consequences if the slope fails. Under ideal conditions, a FOS of at least 1 should ensure safe design, but because of all the uncertainties, a higher value of the FOS is desirable for design recommendations. Based on past experience, a FOS in the range of 1.3 to 1.5 is usually recommended as it represents a level of acceptable safety that minimizes failures. Unfortunately, the use of such a single value of the design FOS does not effectively relate to the quality of the site characterization. In other words, if a site is well characterized by a comprehensive geotechnical site investigation, with extensive in situ tests, and high quality laboratory testing, then guidelines should permit a lower FOS for design, because a considerable amount of uncertainty has been eliminated from the analysis. However, this is not the case currently, as the same minimum FOS often is required for all cases covering different levels of uncertainty associated with the site characterization. Similarly, two slopes with the same FOS may have considerably different levels of safety depending on how well the site has been characterized with a view to minimizing uncertainties.

To account for such levels of uncertainty, the probabilistic method should be considered for assessing the performance of the slope (Li and Lumb, 1987; Chowdhury and Xu, 1994). Such an analysis can take into account uncertainties associated with the site stratigraphy, soil parameters, and even the method of analysis. Of course, the final result here will be the probability of failure, p_f, or alternatively, the probability of unsatisfactory performance (Corps of Engineers, 1997).

6.18.1 Sources of Uncertainty

There are many sources of uncertainty in slope stability analysis. These uncertainties include our lack of knowledge and inability to model precisely the following features:

Spatial uncertainties:

1. Site topography
2. Site stratigraphy and variability
3. Geologic origins and characteristics of subsurface materials
4. Groundwater levels

Data uncertainties:

1. In situ soil and/or rock characteristics
2. Engineering properties
3. Soil and/or rock behavior

Additionally, it is impossible to predict with certainty future environmental conditions such as external loading, earthquakes, and precipitation-induced pore water pressures. The amount of uncertainty varies from one analysis to another and from one site to another. Every analysis is different for this reason.

In this section, the influence of spatial uncertainties will not be considered due to its complexity, implementation difficulties, and subsequent interpretation of the results. For more information concerning this subject, the reader should refer to Chowdhury (1984), Lacasse and Nadim (1996, 1997), Li and Lumb (1987), Phoon and Kulhawy (1999b), Sharp et al. (1981), and Vanmarke (1977a and b).

6.18.2 Basic Probability Concepts

Uncertain parameters (i.e., physical attributes or characteristics) that affect the stability of a slope are treated as random variables in the probabilistic formulation. Such a parameter was previously presumed to have a single, unique value for the deterministic analysis, but will now have a range of values as defined by its probability density function (PDF). The important properties of random variables are described below.

Mean Value For a random variable X, with a PDF defined by the function $f(x)$, the mean, μ_X, or expected value, $E[X]$, is defined by

$$E[X] = \mu_X = \int_{\text{all } x} x f(x)\,dx \qquad \text{for continuous functions, or}$$

$$E[X] = \mu_X = \frac{1}{n}\sum x_i \qquad \text{for discrete functions} \qquad \text{(Eq. 6-80)}$$

Variance The variance describes the extent of the range of the random variable about the mean and is calculated using

$$\text{Var}[X] = \int_{\text{all } x} (x - \mu_X) f(x)\,dx \qquad \text{for continuous functions, or}$$

$$\text{Var}[X] = \frac{1}{n-1}\sum (x_i - \mu_X)^2 \qquad \text{for discrete functions} \qquad \text{(Eq. 6-81)}$$

Standard Deviation The standard deviation, σ_x, is related to the variance, Var[X], by

$$\sigma_X = \sqrt{\mathrm{Var}[X]} \qquad \text{(Eq. 6-82)}$$

Coefficient of Variation This is the ratio of the standard deviation and the mean value given by

$$V_X = \frac{\sigma_X}{\mu_X} \times 100\% \qquad \text{(Eq. 6-83)}$$

Correlation If a pair of random variables (X and Y, for example) depend on each other, the variables X and Y are considered to be correlated, and their covariance is defined by

$$\mathrm{Cov}[X, Y] = E[(X - \mu_X)(Y - \mu_Y)] \quad \text{for continuous functions, or}$$

$$\mathrm{Cov}[X, Y] = \frac{1}{n}\sum (x_i - \mu_X)(y_i - \mu_Y) \quad \text{for discrete functions} \qquad \text{(Eq. 6-84)}$$

This covariance is very similar to the variance. If the covariance is normalized by the standard deviation of the X and Y variables, the correlation coefficient, ρ_{XY}, may be described by

$$\rho_{XY} = \frac{\mathrm{Cov}[X, Y]}{\sigma_X \sigma_Y} \qquad \text{(Eq. 6-85)}$$

The correlation coefficient ranges in value from -1 to $+1$ (i.e., $-1 \leq \rho_{XY} \leq +1$). The case $\rho_{XY} = 1$ indicates a perfect positive, linear correlation between the variables X and Y such that high values of X imply high values of Y. The case $\rho_{XY} = -1$ indicates a negative, or inverse, correlation, where high values of Y occur for low values of X. If the two random variables are linearly independent, then $\rho_{XY} = 0$.

Probability Density Function (PDF) The PDF defines the distribution of the random variable and can take many shapes, but the most common ones used in geotechnical applications are the normal and lognormal, although the triangular distribution is also gaining popularity. These distributions, along with their definitive parameters, are shown in Figure 6.57.

The PDF for the normal distribution with a mean, μ, and standard deviation, σ, is defined by

$$f(x; \mu, \sigma) = \frac{1}{\sigma\sqrt{2\pi}} \exp\left[-\frac{1}{2}\left(\frac{x-\mu}{\sigma}\right)^2\right] \qquad \text{(Eq. 6-86)}$$

This distribution is symmetric about the mean, and the random variable can take on values between $-\infty$ and $+\infty$. Although this range may appear to be vast, the

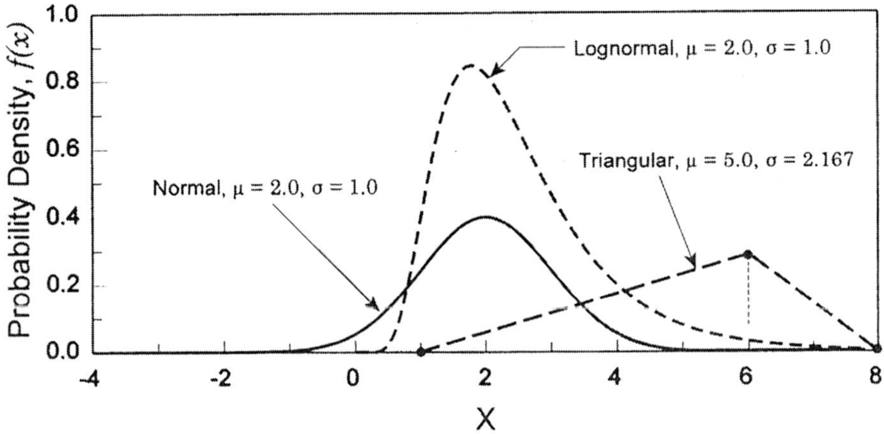

Figure 6.57 Comparison of lognormal, normal, and triangular PDFs.

random variables tend to congregate around the mean such that 68.3 percent will be found within $\pm 1\sigma$ about the mean, 95.5 percent will be found within $\pm 2\sigma$ about the mean, and 99.7 percent will be found within $\pm 3\sigma$ about the mean. So, although theory indicates that the values of a normally distributed random variable may range between $-\infty$ and $+\infty$, 99.7 percent of the values will be positive, provided the mean is positive and is at least greater than $3 \times \sigma$.

For a lognormally distributed random variable, X, with a mean, μ_X, and a standard deviation, σ_X, its natural logarithm, $\ln(X) = Y$, will be normally distributed. The PDF equation for the normal distribution is modified and uses transformed values of the mean, μ_N, and standard deviation, σ_N, given by

$$\sigma_N = \sigma_{\ln X} = \sqrt{\ln\left(1 + V_X^2\right)}$$
$$\mu_N = \mu_{\ln X} = \ln(\mu_X) - \tfrac{1}{2}\sigma_N^2 \qquad \text{(Eq. 6-87)}$$

where $\mu_X = E[X]$ (mean or expected value of the lognormally distributed variable, X)

$\sigma_X = \sqrt{\mathrm{Var}[X]}$ (standard deviation of the lognormallly distributed variable, X)

$V_X = \dfrac{\sigma_X}{\mu_X}$ (coefficient of variation)

These may be used to generate the lognormal distribution according to

$$f(y; \mu_N, \sigma_N) = \frac{1}{\sigma_N x \sqrt{2\pi}} \exp\left[-\frac{1}{2}\left(\frac{\ln(x) - \mu_N}{\sigma_N}\right)^2\right] \qquad \text{(Eq. 6-88)}$$

The lognormal distribution ranges between zero and infinity, and so is particularly suited for parameters that cannot take on negative values.

A triangular distribution (see Figure 6.57) has a fixed range of values, with its apex occurring at the most likely value, which is called the mode. The statistical parameters of a triangular distribution for a random variable, X, which ranges between a and b, with a mode at c, are

$$E[X] = \mu_X = \frac{1}{3}[a + b + c] \qquad \text{(Eq. 6-89)}$$

$$\text{Var}[X] = \sigma_X^2 = \frac{1}{18}[a^2 + b^2 + c^2 - ab - ac - bc] \qquad \text{(Eq. 6-90)}$$

6.18.3 Reliability Index

The component input parameters in a slope stability analysis are modeled as random variables and are used to estimate the PDF of the factor of safety. This PDF is characterized by its mean value, μ_F, and standard deviation, σ_F. Although the PDF of the FOS, F, can take on any shape, it is usually assumed to be either normally or lognormally distributed, as shown in Figure 6.58. The total area under the curve for both the normal and lognormal PDF distributions is always equal to 1.0.

If one considers FOS values less than one to represent failure, the shaded areas of these distributions will represent the probability of failure, p_F. Variables that comply

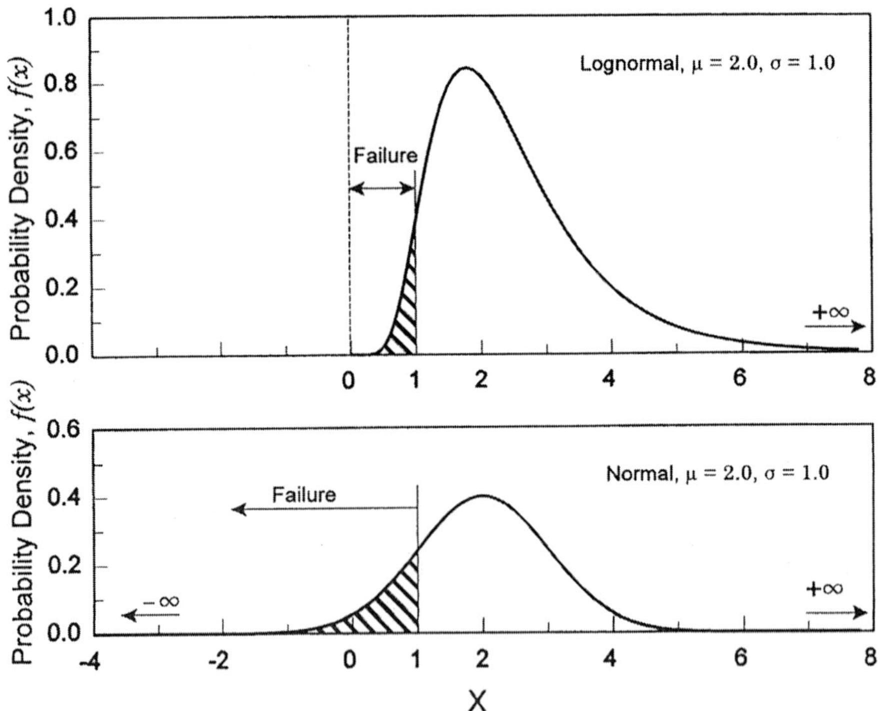

Figure 6.58 Definition of probability of failure for lognormal and normal PDFs.

with a lognormal distribution will always take on values greater than zero, and as this is the case for the factor of safety, this distribution is often considered to be more appropriate. Examples will be presented at the end of this section that illustrate the relative difference between assuming that the PDF is normally or lognormally distributed.

For a lognormal PDF of F, the reliability index, β, is defined by (Wolff, 1996)

$$\beta = \frac{|\ln(F_c) - \mu_N|}{\sigma_N} \qquad \text{(Eq. 6-91)}$$

where F_c = the critical factor of safety corresponding to *unsatisfactory* performance, or 1.0 for the case of failure.

$$\sigma_N = \sigma_{\ln F} = \sqrt{\ln(1 + V_F^2)}$$

$$\mu_N = \mu_{\ln F} = \ln(\mu_F) - \tfrac{1}{2}\sigma_N^2$$

$\mu_F = E[F]$ (mean or expected value of a lognormally distributed FOS)

$\sigma_F = \sqrt{\mathrm{Var}[F]}$ (standard deviation of a lognormally distributed FOS)

$V_F = \dfrac{\sigma_F}{\mu_F}$ (coefficient of variation of FOS)

For the special case of the probability of failure, $F_c = 1.0$, and Equation 6-91 simplifies to

$$\beta = \frac{-\mu_N}{\sigma_N} = \frac{\ln\left[\mu_F/\sqrt{1 + V_F^2}\right]}{\sqrt{\ln(1 + V_F^2)}} \qquad \text{(Eq. 6-92)}$$

The reliability index, β, describes safety as it is a measure of the number of standard deviations separating the best estimate of F, and the prescribed critical value of the FOS, F_c.

If the PDF of F is presumed to be normal, the reliability index is defined by

$$\beta = \frac{|F_c - \mu_F|}{\sigma_F} \qquad \text{(Eq. 6-93)}$$

where F_c = the critical factor of safety corresponding to *unsatisfactory* performance, or 1.0 for the case of failure.

$\mu_F = F[F]$ (mean or expected value of a normally distributed FOS)

$\sigma_F = \sqrt{\mathrm{Var}[F]}$ (standard deviation of a normally distributed FOS)

Figure 6.59 shows the PDFs of an example where calculations resulted in different means and standard deviations for two slopes, Slope A and Slope B, and the PDFs are presumed to be lognormally distributed for F. In this example, Slope A has a mean FOS, $\mu_F = 1.3$, and Slope B, has a higher mean FOS, $\mu_F = 1.5$. In a deterministic

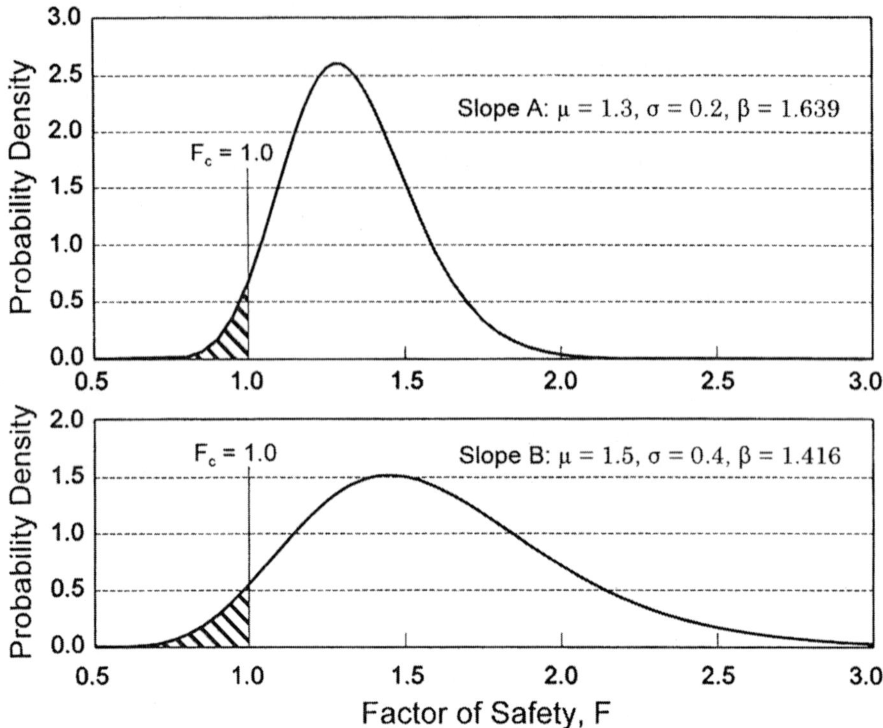

Figure 6.59 Comparison of safety for two slopes, where Slope *A* has a higher reliability index due to its smaller σ_F value.

sense, it would appear that Slope *B* is safer than Slope *A*. However, the FOS PDF for Slope *A* has a much smaller standard deviation in comparison to Slope *B*, which leads to reliability index values of $\beta_A = 1.639$ and $\beta_B = 1.416$ for the critical $F_c = 1.0$ assumption. As Slope *A* has a greater reliability index, Slope *A* can be considered safer than Slope *B* as there is less uncertainty associated with Slope *A*.

The reliability index may be used to calculate the probability of failure by presuming that the FOS has a PDF that corresponds to either a lognormal or normal distribution. The probability of unsatisfactory performance, p_F, may then be estimated using the standardized normal distribution tables provided in most text books on statistics. Table 6.17, based on Corps of Engineers (1997), provides values of the probability of unsatisfactory performance, p_F, for example values of the reliability index, β, along with a subjective description of the level of risk.

The reliability index values for the case shown in Figure 6-59 are based on the presumption that the PDF of the FOS follows a lognormal distribution. Using the computed β values of 1.639 and 1.416, the probability of failure, p_f, is 0.0506 and 0.0784 for Slope *A* and Slope *B*, respectively. Descriptively, on the basis of Table 6.17, Slope *A* would be classified as "unsatisfactory" and Slope *B* would be assigned a "hazardous" rating.

TABLE 6.17 Example Values for β and p_F

Performance Level	Reliability Index, β	Probability, $p_F = P(F < F_c)$
High	5.0	2.871×10^{-7}
Good	4.0	3.169×10^{-5}
Above average	3.0	0.00135
Below average	2.5	0.00621
Poor	2.0	0.02275
Unsatisfactory	1.5	0.06681
Hazardous	1.0	0.15866

Source: Based on Corps of Engineers (1997).

6.18.4 Probabilistic Formulation for Slopes

The main intent of a probabilistic analysis is to estimate the reliability index, β, as discussed in the last section, using the mean and standard deviation of the FOS, F. Alternatively, the PDF of the FOS, F, may be estimated and the probability, p_F, calculated directly. The same FOS equations, presented in Section 6.9, may be used to develop the resultant joint PDF with parameters, μ_F and σ_F. For example, we can use Equation 6-43 (Simplified Janbu method) to describe the FOS response due to variations in the input parameters, namely c, ϕ, W, U, N, etc. Such predictive models are called *performance functions* and can be represented by an equation of the form

$$F = g(X_1, X_2, \ldots, X_n) + e \qquad \text{(Eq 6-94)}$$

where X_i = component input variables, described as random variables
e = model error that accounts for the error between the response of the adopted model and the real world behavior (Christian et al., 1994).

If the reliability index is to be calculated using Equations 6-91 or 6-93, we must calculate the mean and standard deviation of the PDF of the FOS in response to the input random variables. The performance function, with continuous random variables as the input parameters, must be integrated to generate the joint PDF for a closed form solution. Although this may be possible for simple linear performance functions, the usual FOS equation cannot be integrated directly, because it is complicated, requires iterations, and cannot be simplified easily. So, we must resort to deriving approximate solutions using numerical methods. Most numerical methods will not provide us with a PDF of the FOS, but will provide us with the expected value, $E[F]$, and the variance, $\text{Var}[F]$, for use in calculating the reliability index, β.

6.18.5 Probabilistic Analysis of Performance Function

As a direct integration of the adopted performance function for the FOS is not feasible, the expected value, $E[F]$, and the variance, $\text{Var}[F]$, of the FOS may be estimated by using (1) Taylor series method, or (2) Rosenblueth's point estimate method, (3) Monte

Carlo simulation, or (4) Fourier analysis. The results of a Monte Carlo simulation and the Fourier analysis will provide an estimate of the PDF of the FOS as well. Of these four methods, the first three have been used extensively, while the Fourier analysis method has been used primarily for block-slide models to analyze rock slope stability.

Taylor Series Method This method is based on the Taylor series expansion of the performance function about the expected values of the random variables (Hahn and Shapiro, 1967). If only the first-order terms of the expansion are retained, the expected value of the performance function for independent random variables may be approximated by

$$E[F] = \mu_F \approx g(E[X_1], E[X_2], \ldots, E[X_n]) + e \qquad \text{(Eq. 6-95)}$$

where the function $g(X_1, X_2, \ldots)$ is determined using input parameters that correspond to the mean values of the component random variables. For the case where pairs of random variables show some dependency (that is $|\rho_{XY}| > 0$), the expected value may be approximated by

$$E[F] = \mu_F \approx g(E[X_1], E[X_2], \ldots, E[X_n]) + \sum_{i=1}^{k} \sum_{j=1}^{k} \frac{\partial^2 F}{\partial X_i \, \partial X_j} \operatorname{Cov}(X_i, X_j) + e \qquad \text{for } i < j \qquad \text{(Eq. 6-96)}$$

where e represents modeling error. The contribution of the summation term is usually relatively small, and being of second-order can be omitted from the calculations.

The Taylor series may also be used to estimate the variance of the response of the performance function to the input random variables. Again, if we limit the series to first-order terms only, the variance can be approximated by

$$\operatorname{Var}[F] = \sigma_F^2 \approx \sum_{i=1}^{k} \left[\left(\frac{\partial F}{\partial X_i} \right)^2 \operatorname{Var}[X_i] \right] + 2 \sum_{i=1}^{k} \sum_{j=1}^{k} \left[\left(\frac{\partial F}{\partial X_i} \frac{\partial F}{\partial X_j} \right) \operatorname{Cov}[X_i, X_j] \right] + V[e] \qquad \text{for } i < j$$

(Eq. 6-97)

where the partial derivative, $\partial F / \partial X_i$ is evaluated for each random variable, based on the selected performance function. If all random variables are independent, all covariances will be zero, and the second summation in Equation 6-97 reduces to zero. The term $V[e]$ represents the variance of the unspecified modeling error.

With the expected value and the standard deviation known, the reliability index can be readily calculated using Equation 6-91 or 6-93. The main advantage of the Taylor series method is that as the terms are summed for the variance, one can readily see the relative contribution of uncertainty projected by each component random variable. Although this method will give an exact solution for a linear function, the first-order approximations introduce errors for nonlinear performance functions, such as the ones typically used for the FOS. In considering one of these formulations, the partial derivatives cannot be derived directly and numerical methods must be used to estimate these derivatives.

For example, the Corps of Engineers (1997), recommends the use of the finite difference approximation of the derivative such that

$$\frac{\partial F}{\partial X_1} = \frac{g(X_1+\sigma_1, X_2, X_3, \ldots, X_n) - g(X_1-\sigma_1, X_2, X_3, \ldots, X_n)}{2\sigma_1} = \frac{\Delta F_1}{2\sigma_1}$$
$$\frac{\partial F}{\partial X_2} = \frac{g(X_1, X_2+\sigma_2, X_3, \ldots, X_n) - g(X_1, X_2-\sigma_2, X_3, \ldots, X_n)}{2\sigma_2} = \frac{\Delta F_2}{2\sigma_2}$$
$$\vdots$$

(Eq. 6-98)

If each of the above derivatives is squared and then multiplied by the variance corresponding to the random variable, the expression simplifies to

$$\left(\frac{\partial F}{\partial X_1}\right)^2 \text{Var}[X_1] = \left[\frac{\Delta F}{2\sigma_1}\right]^2 \sigma_1^2 = \tfrac{1}{4}[\Delta F_1]^2 \quad \text{(Eq. 6-99)}$$

In the above equation $X_2, X_3, \ldots, X_n$ are the mean values of the component random variables. So for independent random variables, the variance will be given by the simplified expression

$$\text{Var}[F] = \sigma_F^2 \approx \frac{1}{4}\sum_{i=1}^{k} [\Delta F_i]^2 + V[e] \quad \text{(Eq. 6-100)}$$

and for dependent variables, substituting $\text{Cov}[X_i, X_j] = \rho_{ij}\sigma_i\sigma_j$ from Equation 6-85

$$\text{Var}[F] = \sigma_F^2 \approx \frac{1}{4}\sum_{i=1}^{k} [\Delta F_i]^2 + \frac{1}{2}\sum_{i=1}^{k}\sum_{j=1}^{k} \left[\Delta F_i \, \Delta F_j \rho_{ij}\right] + V[e] \qquad \text{for } i < j$$

(Eq. 6-101)

The ΔF differences are obtained over two standard deviations. In using this version of the Taylor series method, the performance function will have to be evaluated once for the expected value, and then $2n$ times again, for the n random variables for a total of $(2n+1)$ evaluations.

Rosenblueth's Point Estimate Method This point estimate procedure was originally proposed by Rosenblueth (1975) where the PDFs of the random variables are simulated by "point" masses located at plus or minus one standard deviation from the mean values. This method is direct, gives reasonably accurate results quickly, and has been discussed in detail by McGuffey et al. (1981), Harr (1987), Wolff (1996), and Corps of Engineers (1997). In this section, details for determining the expected value of the FOS, F, and it variance will be presented for two and three random variables. For more than three variables, the presentation of the equations gets more complex, but the approach is the same.

For the case of two random variables, X and Y, the performance function for the FOS, F, will take the form

$$F = g(X, Y) \quad \text{(Eq. 6-102)}$$

where the uncertainty in the random variables is described by their standard deviation, σ_X and σ_Y, and their mean values are taken as μ_X and μ_Y. For the general case, we can also presume that the variables are dependent, with a correlation coefficient of ρ_{XY}, and that the PDFs are all symmetric.

For two random variables, the first step requires the evaluation of the FOS, F, at four unique combinations of the random variables, as shown below

$$\begin{aligned} F_{++} &= g[(\mu_X + \sigma_X), (\mu_Y + \sigma_Y)] \\ F_{+-} &= g[(\mu_X + \sigma_X), (\mu_Y - \sigma_Y)] \\ F_{-+} &= g[(\mu_X - \sigma_X), (\mu_Y + \sigma_Y)] \\ F_{--} &= g[(\mu_X - \sigma_X), (\mu_Y - \sigma_Y)] \end{aligned} \quad \text{(Eq. 6-103)}$$

In the above set of expressions, the subscripts indicate the values of X and Y that are used to evaluate the function. F_{++}, for example, represents the value of the function $g(X, Y)$, as evaluated for $X = \mu_X + \sigma_X$ and $Y = \mu_Y + \sigma_Y$. Similarly, F_{+-} represents the value of the function $g(X, Y)$, as evaluated for $X = \mu_X + \sigma_X$ and $Y = \mu_Y - \sigma_Y$. Once these expressions are evaluated, the expected FOS will be given by their sum as

$$E[F] = \mu_F = P_{++}F_{++} + P_{+-}F_{+-} + P_{-+}F_{-+} + P_{--}F_{--} \quad \text{(Eq. 6-104)}$$

where

$$\begin{aligned} P_{++} &= P_{--} = \tfrac{1}{4}(1 + \rho_{XY}) \\ P_{+-} &= P_{-+} = \tfrac{1}{4}(1 - \rho_{XY}) \end{aligned} \quad \text{(Eq. 6-105)}$$

The variance of the FOS will be given by

$$\text{Var}[F] = E[F^2] - (E[F])^2 \quad \text{(Eq. 6-106)}$$

where $E[F]$ = expected value of the FOS from Equation 6-104, and

$$E[F^2] = P_{++}F_{++}^2 + P_{+-}F_{+-}^2 + P_{-+}F_{-+}^2 + P_{--}F_{--}^2 \quad \text{(Eq. 6-107)}$$

So for two variables, the FOS values are evaluated for four conditions using the adopted performance function.

The above equations for two variables can be extended to three random variables, X, Y, and Z, which will lead to an expected FOS, $E[F]$, given by the summation of eight products

$$E[F] = \mu_F = P_{+++}F_{+++} + P_{++-}F_{++-} + P_{+-+}F_{+-+} + P_{+--}F_{+--}$$
$$+ P_{---}F_{---} + P_{--+}F_{--+} + P_{-+-}F_{-+-} + P_{-++}F_{-++}$$
(Eq. 6-108)

where

$$\begin{aligned} P_{+++} &= P_{---} = \tfrac{1}{8}(1 + \rho_{XY} + \rho_{YZ} + \rho_{ZX}) \\ P_{++-} &= P_{--+} = \tfrac{1}{8}(1 + \rho_{XY} - \rho_{YZ} - \rho_{ZX}) \\ P_{+-+} &= P_{-+-} = \tfrac{1}{8}(1 - \rho_{XY} - \rho_{YZ} + \rho_{ZX}) \\ P_{+--} &= P_{-++} = \tfrac{1}{8}(1 - \rho_{XY} + \rho_{YZ} - \rho_{ZX}) \end{aligned}$$
(Eq. 6-109)

and the variance is obtained from the expression

$$\text{Var}[F] = E[F^2] - (E[F])^2$$
(Eq. 6-110)

In general, the expected value, $E[F]$, is given by the sum of the product of the P coefficients and the F values, and the variance is computed using Equation 6-110. So, for even more random variables, new P coefficients must be derived to follow the sequence presented for two and three random variables. For each case, the numeric coefficient (in Equation 6-105 or 6-109) is the reciprocal of 2^n, where n is the number of random variables being considered. The signs in front of the correlation coefficients are related to the subscripts of P, with $++$ or $--$ correlation leading to a positive sign.

Calculating the expected value, $E[F]$, and the variance, Var$[F]$, using Rosenblueth's procedure requires 2^n evaluations of the performance function. For only two variables, this is the less than the number required for the Taylor series method. However, for more than two variables, Rosenblueth's method will require more evaluations of the performance function than with the Taylor series method.

Monte Carlo Simulation This method effectively simulates the response of the FOS performance function to randomly selected, discrete values of the component variables. The process is repeated many, many times to obtain an approximate, discrete PDF of the resulting FOS values, F (for example, see Tobutt, 1982, Hammond, et al., 1992, or Chandler, 1996). The component random variables for each calculation are selected from a sample of random values that are based on the selected PDF of the random variables. Although these PDFs can take on any shape, the normal, lognormal, triangular, beta, and uniform distributions are among the most favored for analysis. The Monte Carlo simulation follows a four-step process:

1. For each component random variable being considered, select a random value that conforms to the assigned distribution.
2. Calculate the value of the FOS using the adopted performance function and the input values obtained from step 1.

3. Repeat steps 1 and 2 many times, storing the FOS result from each trial calculation.
4. Use the many calculated FOS values from the Monte Carlo simulation to estimate (a) the probability, $P(F < F_c)$, (b) the *sample* mean and variance, and (c) the FOS PDF from the histogram.

It should be noted that as each Monte Carlo simulation will use a different sequence of random values, the resulting probabilities, means, variances, and histograms will be slightly different. As the number of trials increases, the differences will become smaller.

Because a Monte Carlo simulation requires a large number of calculations, it often has been viewed as a less desirable option than the Taylor series or point estimation methods. However, with the general availability of faster desktop computers, the Monte Carlo approach has gained wider acceptance due to its inherent simplicity and the ability to use existing computer programs with only minor modifications. Unfortunately, before relying on Monte Carlo output for engineering design, one must decide on the number of trials (i.e., calculations) needed to generate consistent, reliable results.

The minimum number of trials required will depend on the number of component variables and the desired level of confidence. If the results of the Monte Carlo simulation are considered analogous to the binomial distribution, with successes considered as cases where the FOS exceeds a specified critical FOS, we can estimate the required minimum sample size for different confidence levels (Hahn and Shapiro, 1967; Melchers, 1987). Based on this assumption, the number of trial calculations, n, will be given by

$$n = \left(\frac{100\,d}{E}\right)^2 \frac{1 - p_F}{p_F} \times m \qquad \text{(Eq. 6-111)}$$

where n = minimum number of trials for the Monte Carlo simulation
p_F = probability of unsatisfactory performance (or failure)
E = relative percent error in estimating the probability, p_F
m = number of component random variables
d = normal standard deviate according to the following confidence levels:

Confidence Level	Normal Standard Deviate, d
80%	1.282
90%	1.645
95%	1.960
99%	2.576

So for a simulation that requires a confidence level of 95 percent and a relative error of 20 percent, the minimum number of trials for a preliminary estimate will be 90 (using $d = 1.96$, $p_f = 0.5$) for a single random variable. For the same simulation, 384 trials will be required for a relative error of 10 percent.

To apply Equation 6-111, a value of p_F is required *prior* to the simulation. As this is not always available, Hahn and Shapiro (1967) suggest the use of $p_F = 0.5$ for a first estimate of n. In examining Equation 6-111, we can see that n is inversely proportional to the square of the relative error, E, and directly proportional to $(1 - p_f)/p_f$. So in order to assess small probabilities of failure using a Monte Carlo simulation, a larger number of trials is required to satisfy the prescribed relative error and confidence level. Generally, relative percent errors ranging between 5 percent and 20 percent have been found to provide satisfactory results.

After a Monte Carlo simulation consisting of n trials has been completed and the probability of failure, p_F, has been estimated, the relative percent error can be calculated using

$$E = d\sqrt{\frac{m(1 - p_F)}{np_F}} \times 100\% \qquad \text{(Eq. 6-112)}$$

With E computed, the resulting p_F value can be considered to be accurate to within a range of $\pm 0.01E \times p_F$, with the level of confidence associated with the selected value of d.

Once a decision has been made on the minimum number of trials, random values of the component variables are generated, and the FOS values calculated by using the performance function. Most schemes use a function that can generate uniformly distributed random values, $R_1, R_2, \ldots, R_N$, between 0 and 1. With such a sequence, random values for a standardized normal distribution with mean $\mu = 0$ and standard deviation $\sigma = 1$ can be generated using the Box and Muller (1958) method. With this method, pairs of uniformly distributed random variables, R_1 and R_2, are used in the equations

$$\begin{aligned} N_1 &= \sqrt{-2 \ln R_1}\ \cos(2\pi R_2) \\ N_2 &= \sqrt{-2 \ln R_1}\ \sin(2\pi R_2) \end{aligned} \qquad \text{(Eq. 6-113)}$$

to generate two new random values, N_1 and N_2, that are independent, and follow the standard normal distribution. This process is repeated continuously with two fresh uniformly distributed random numbers until the desired number of normally distributed random variables, $N_1, N_2, \ldots, N_i$, have been obtained. These random numbers will have a mean $\mu = 0$ and standard deviation $\sigma = 1$. To convert the standard normal N_is to values of a general normal component variable with mean μ_X and standard deviation σ_X, the following expression is used:

$$x_i = \mu_X + N_i \sigma_X \qquad \text{(Eq. 6-114)}$$

where x_i = value taken on by the normally distributed random variable, which will be used in the calculation of the performance function

N_i = random value taken from the standard normal distribution

The sequences of uniformly and normally distributed random numbers discussed so far are assumed to be *independent* random numbers. If the component variables of a normally distributed random variable are *dependent*, one of a pair of the independent random numbers, say N_1 and N_2, are modified using (Lumb, 1974):

$$N_2^* = N_1 \rho_{XY} + N_2 \sqrt{(1 - \rho_{XY}^2)} \qquad \text{(Eq. 6-115)}$$

where ρ_{XY} = the correlation coefficient between two random variables, X and Y.
N_1 = first value in a pair of two standard normal random numbers
N_2 = second value in a pair of two standard normal random numbers
N_2^* = random number correlated to N_1

Interestingly, if uniformly distributed, dependent random numbers are required, normally distributed, dependent random numbers should be generated first. These are then converted back to uniformly distributed random numbers through the use of the standard normal, cumulative distribution (known as an "inversion process"). Then these pairs of dependent, uniformly distributed numbers can be used to effectively derive new random numbers that satisfy other PDFs such as a triangular or beta distribution. For more information about these and other distributions, the reader should refer to Hahn and Shapiro (1967) or Hall (1992). The worked example at the end of this section will also show this procedure.

Once the dependent random number is generated using Equation 6-115, the correlated values of the desired general normal distribution are obtained by

$$\begin{aligned} x_1 &= \mu_X + N_1 \sigma_X \\ y_1 &= \mu_Y + N_2^* \sigma_Y \end{aligned} \qquad \text{(Eq. 6-116)}$$

With the application of Equation 6-115, the set of values for x and y will be dependent, with a correlation coefficient, ρ_{XY}.

For each set of component variables, the FOS is computed using the performance function. By keeping track of the computed FOS values, all FOS values that are less than the critical FOS value, F_c, can be counted and the probability for unsatisfactory performance reported as

$$P(F < F_c) = \frac{n_c}{n_{\text{total}}} \qquad \text{(Eq. 6-117)}$$

where n_c = number of FOS values less than F_c
n_{total} = total number of trials for the Monte Carlo simulation

The final results of a Monte Carlo simulation can also be presented in the form of a histogram that represents the PDF of the FOS values computed for each set of random values. Although not necessary, the mean and standard deviation of the FOS values can be reported also, as this will allow calculation of the reliability index, β. An example of a Monte Carlo simulation will be presented at the end of this section.

Fourier Analysis If the complete joint PDF of the FOS is required rather than a Monte Carlo histogram of possible FOS values, then one can use the Fourier analysis method, which is based on the convolution of independent PDFs (Feller, 1966). For this procedure to work, the n random variables in the performance function must be transformed into a linear combination of the form

$$F = A_1X_1 + A_2X_2 + \cdots + A_nX_n \qquad \text{(Eq. 6-118)}$$

where A_i = constant coefficient
X_i = random variable (or sometimes a combination of two or more component variables in the performance function)

The intent of the method is to generate the resultant joint PDF of the FOS values, F, knowing that the random variables, X_i, will be defined by their own individual, continuous PDFs. The joint PDF of the FOS may be determined using the following steps:

1. Discretize each of the component PDFs of the random variables into N unique points. So for the variable, X_1, we will have N discrete values: x_{11}, x_{12}, $x_{13}, \ldots, x_{1N}$, and a similar series of values for the other variables, $X_2, X_3, \ldots, X_N$, each discretized at a uniform spacing.

   For convenience, the value of N should be selected such that it is a power of 2, (i.e., 64, 128, 256, 512 and so on) to make use of the Fast Fourier Transform (FFT) method.
2. Multiply each discretized PDF by its constant coefficient, $A_1x_{11}, A_1x_{12}, \ldots, A_1x_{1N}$; $A_2x_{21}, A_2x_{22}, \ldots, A_2x_{2N}$; and so on. This will ensure that the units of each of the component variables of the linearized function are consistent and the discretizing increment, ΔF, is the same for all random variables.
3. Perform a FFT on all discretized PDFs from step 2. This will generate N complex values, $X_{11}, X_{12}, \ldots, X_{1N}$ for each component PDF.
4. Multiply all the corresponding Fourier coefficients such that we have N products of the type

$$\begin{aligned} Y_1 &= X_{11} \times X_{12} \times X_{13} \times \cdots \times X_{1N} \\ Y_2 &= X_{21} \times X_{22} \times X_{23} \times \cdots \times X_{2N} \\ &\vdots \\ Y_N &= X_{N1} \times X_{N2} \times X_{N3} \times \cdots \times X_{NN} \end{aligned} \qquad \text{(Eq. 6-119)}$$

   The products, $Y_1, Y_2, \ldots, Y_N$, will also be complex.
5. Perform an inverse FFT on the resulting N products, Y_i. This will result in N real values $y_1, y_2, y_3, \ldots, y_N$, which will define the desired PDF of the FOS, discretized at the same ΔF increment used in Step 2.

The main difficulty with this Fourier analysis method is the requirement of a linear performance function. However, if a linear performance function can be identified, then an accurate, unique PDF of the FOS will be obtained, without any restrictions on the types of input PDFs that may be used to define the component random variables. This allows the use of highly skewed PDFs, if necessary. For an application of this method to rock slopes, please refer to Miller (1982 a and b).

6.18.6 Quantifying Uncertainty

All of the discussion about the probabilistic analysis of slopes so far has only mentioned the component random variables, without making any effort to identify or quantify these parameters. For each parameter, the uncertainty must address the variance associated with (Christian et al., 1994)

1. *Data Scatter*, which is attributable to
   - spatial variability in the soil profile
   - random testing errors
2. *Systematic Error*, which is influenced by
   - statistical modeling of the mean due to the limited amount of sampling and testing in the laboratory or in the field
   - bias in measurement procedures due to features such as sampling methods and the extent of sample disturbance

Our current knowledge is limited regarding the exact contribution of these two items to the overall uncertainty. But as more data are gathered within this framework, we hope to separate the influence of each category with more confidence.

For most slope stability analyses, the critical parameters include

1. Shear strength—in the form of c and ϕ parameters for various drainage conditions and rates of loading
2. Pore water pressures
3. Unit weight of soils
4. Location of failure surface
5. Thickness of a soft, low-strength zone
6. External environmental effects, such as drawdown and surface loads

If we are to include such parameters into our analysis, we need information about the uncertainty of the above parameters. This uncertainty should be defined by a PDF and by the associated parameters such as the mean value and the standard deviation. Unfortunately, most geotechnical projects will rarely have sufficient data to define a PDF. So we must select PDFs for the component variables based on past experience with similar projects.

Following an extensive review of literature by Lumb (1966), Baecher et al. (1980) and Chowdhury (1984), there is some consensus about selecting the following PDF models:

Material	Property	PDF
Compacted clay cores	Dry unit weight	Lognormal
Silts, sands and clays	Void ratio	Normal
Silty sand, clayey silt	$\tan\phi'$	Normal
Clayey silt	Undrained shear strength	Normal
Various other soils	c', ϕ'	Beta

In general, it appears that the normal distribution may be used for most parameters, unless there is sufficient evidence to suggest otherwise. The true normal distribution, which extends from $-\infty$ to $+\infty$, is truncated for most practical purposes here to include values typically within $\pm 3\sigma$ of the mean. The other significant feature of note is that the parameter $\tan\phi$ is considered to be normally distributed rather than the angle ϕ. For parameters not listed in the above table, the normal probability distribution should be adopted for most cases. If there are concerns that too many unrealistic, erroneous negative random values may overly influence the reliability index, then the lognormal or triangular distribution should be selected as a model for the random variable.

Once the shape of the PDF has been selected, other parameters such as the mean and standard deviation will have to be defined for normal and lognormal distributions. In most instances, the standard deviation is estimated from published values of the coefficient of variation, V_X [for example, see Phoon et al. (1995) and Phoon and Kulhawy (1999a), for a good summary]. Some examples of V_X that have been reported in the literature are shown in Table 6.18. Please note that the ranges given in this table are representative values only and do not cover the entire range of values possible.

TABLE 6.18 Coefficients of Variation

Geotechnical Parameter	Coefficient of Variation, V_X	Source
Unit weight	3–8%	Harr (1977, 1987), Kulhawy (1992)
Effective, ϕ'	2–21%	Baecher et al. (1980), Harr (1977, 1987), Kulhawy (1992), Phoon et al. (1995)
Undrained strength, S_u	13–49%	Bishop (1948), Harr (1977, 1987), Chowdhury (1984), Kulhawy (1992), Lacasse and Nadim (1997), Phoon et al. (1995), Duncan (2000)
Undrained strength, S_u, from vane tests	10–20%	Kulhawy (1992), Christian et al. (1994)
Undrained strength ratio, S_u/σ_v'	5–15%	Lacasse and Nadim (1997), Duncan (2000)

With the availability of such published coefficients and an estimate of the mean value, μ_X, the standard deviation can be calculated using

$$\sigma_X = V_X \times \mu_X \quad \text{(Eq. 6-120)}$$

If there are insufficient data and published values of the coefficient of variation cannot be obtained, then one must resort to the rule of thumb suggested by Dai and Wang (1992). This rule is based on the premise that most data will be normally distributed, and so 99.73 percent of all values will lie within plus or minus three standard deviations from the mean value. So if one can select a possible range of values, which are symmetric about the mean, the standard deviation, σ_X, can be estimated using the following simple equation

$$\sigma_X = \frac{X_{\max} - X_{\min}}{6} \quad \text{(Eq. 6-121)}$$

where $X_{\max}$ = maximum value of random variable X
$X_{\min}$ = minimum value of random variable X

Duncan (2000) believes that there is a tendency to underestimate the range of possible values. In view of this, it would be more realistic to select the standard deviation value on the basis of the equation

$$\sigma_X = \frac{X_{\max} - X_{\min}}{4} \quad \text{(Eq. 6-122)}$$

In this case, we are presuming that the range will account for 95.5 percent of the possible values. If Equation 6-122 is used, a larger, more realistic coefficient of variation will be selected for the analysis.

Instead of presuming that an input variable is distributed according to a normal distribution, the triangular distribution may also be considered due to its overall simplicity and application. Only the minimum, maximum, and the most likely value are needed to define the triangular distribution. With this information, Equations 6-89 and 6-90 may be used to calculate the mean and standard deviation for use with the Taylor series method or Rosenblueth's point estimate method for determining the statistical parameters of the joint PDF of the FOS. For the Monte Carlo procedure, the random values for a triangular distribution can be readily calculated using independent or dependent, uniformly distributed random numbers (Hall, 1992).

6.18.7 Examples

To illustrate probabilistic analysis, the 2H:1V slope of Figure 6.25 will be analyzed using three of the four methods discussed above. The Fourier analysis method will not be used as the FOS equations for the simplified Bishop (Equation 6-48) and simplified Janbu (Equation 6-43) methods cannot be linearized. The slope, analyzed in Section 6.9.6, consists of a single soil type with a unit weight of $\gamma = 16$ kN/m^3 and shear strength parameters $c = 20$ kPa and $\phi = 20°$. For the probabilistic analysis, the same failure surface, as shown in Figure 6.25, will be evaluated with the shear strength parameters, c and $\tan\phi$, considered as random variables and the unit weight held constant.

TABLE 6.19 Soil Parameters Used for Example Slope

Parameter	Mean, μ	Coefficient of Variation, V_x	Standard Deviation, σ_x	Correlation $\rho_{c,\tan\phi}$
c	20 kPa	25%	5 kPa	−0.70 and 0.0
$\tan\phi$	0.3640[a]	10%	0.0364	

[a]Corresponds to $\phi = 20^\circ$

For the three methods that will be used to illustrate the probabilistic analysis, the strength parameters are summarized in Table 6.19.

Taylor Series Method Any available computer method may be used to calculate the expected FOS corresponding to selected values of c and $\tan\phi$. For all results presented in this section, the solution was obtained using the method of slices and the slice data presented in Section 6.9.6. The mean or expected value, $E[F]$, can be easily calculated using the mean values of c and $\tan\phi$. The calculation of the standard deviation requires that the FOS be evaluated for four different combinations of c and $\tan\phi$. This will then allows us to estimate the derivatives required for estimating the variance using Equation 6-98 and 6-101. The sequence of required calculations is shown in Table 6.20.

With the above values, we can now calculate the approximate variance of the FOS, using Equation 6-101. For the simplified Bishop method, the variance is given by

$$\begin{aligned}\text{Var}[F] = \sigma_F^2 &= 0.25 \times [0.21048^2 + 0.21666^2] \\ &\quad + 0.5 \times [0.21048 \times 0.21666 \times (-0.70)] \\ &= 0.00685\end{aligned}$$

Using the same equation for the Janbu method, the variance will be given by

$$\begin{aligned}\text{Var}[F] = \sigma_F^2 &= 0.25 \times [0.21685^2 + 0.20704^2] \\ &\quad + 0.5 \times [0.21685 \times 0.20704 \times (-0.70)] \\ &= 0.00676\end{aligned}$$

TABLE 6.20 Example Calculations for Taylor Series

Variables		Simplified Bishop		Simplified Janbu	
c, kPa	$\tan\phi$	FOS, F	ΔF_i	FOS, F	ΔF_i
20	0.3640	1.544	—	1.469	—
25 15	0.3640	1.652 1.441	0.21048	1.578 1.361	0.21685
20	0.4004 0.3276	1.655 1.438	0.21666	1.573 1.365	0.20704

TABLE 6.21 Summary of Results for Taylor Series

	Simplified Bishop		Simplified Janbu	
Parameter	$\rho_{XY} = -0.7$	$\rho_{XY} = 0$	$\rho_{XY} = -0.7$	$\rho_{XY} = 0$
Mean value	1.544	1.544	1.469	1.469
Variance	0.00685	0.02281	0.00676	0.02247
Standard deviation	0.0828	0.1510	0.0822	0.1499
Coefficient of variation	5.36%	9.79%	5.60%	10.20%

The same calculations can be repeated for the case of independent component variables (i.e., $\rho_{XY} = 0$). The results for this independent case, along with the more realistic dependent case, are summarized in Table 6.21.

With knowledge about the mean and standard deviation, we can also compute the probability of unsuccessful performance. If the prescribed minimum FOS is defined as $F_c = 1.4$, the probability may be computed by assuming that the FOS follows a distribution which is either lognormal, or normal.

For a lognormal distribution, the standard deviation and mean corresponding to the normal distribution must be calculated using Equations 6-87. For the simplified Bishop results, this gives

$$\sigma_N = \sqrt{\ln\left(1 + V_x^2\right)} = \sqrt{\ln(1 + 0.0536^2)}$$

$$= 0.05358$$

$$\mu_N = \ln(\mu_x) - \tfrac{1}{2}\sigma_N^2 = \ln(1.544) - \tfrac{1}{2}0.05358^2$$

$$= 0.43263$$

Next, the probability that the FOS will be less than 1.4 may be computed as

$$p_f = P(F < 1.4) = \Phi\left[\frac{\ln(1.4) - 0.43263}{0.05358}\right]$$

$$= \Phi[-1.7945]$$

$$= 0.0364$$

where $\Phi[\,]$ is the standard normal cumulative distribution function (CDF). Also, the reliability index for the $F_c = 1.4$ case may be calculated using Equation 6-91 to give $\beta = 1.7945$.

If the distribution of the FOS is assumed to be normal, the probability that the FOS will be less than 1.4 may be computed as

$$p_f = P(F < 1.4) = \Phi\left[\frac{1.4 - 1.544}{0.0828}\right]$$

$$= \Phi[-1.7340]$$

$$= 0.0415$$

The reliability index for the $F_c = 1.4$ case may be calculated using Equation 6-93,

TABLE 6.22 Summary of β and p_F Values for Dependent and Independent Cases

		Dependent, $\rho_{XY} = -0.7$		Independent, $\rho_{XY} = 0$	
Method	Distribution	β	$P(F < 1.4)$	β	$P(F < 1.4)$
Bishop	Lognormal	1.7945	0.0364	0.9509	0.1708
	Normal	1.7340	0.0415	0.9502	0.1710
Janbu	Lognormal	0.8326	0.2025	0.4219	0.3365
	Normal	0.8395	0.2006	0.4604	0.3226

to give $\beta = 1.7340$. A summary of the results for the simplified Bishop and Janbu methods, for the case where $F_c = 1.4$, are presented in Table 6.22, above.

In comparing these results, values of the reliability index and the probability of unsuccessful performance are only slightly different for the lognormal and normal distribution assumption.

Rosenblueth's Point Estimate Method For this method, Equations 6-103 to 6-106 will be used to determine the mean and variance of the FOS. For the case of two dependent random variables, c and $\tan\phi$, the FOS is evaluated for the following four sets of parameters:

	Parameters	Factor of Safety, F Bishop	Janbu
1. F_{++}	$c = \mu_c + \sigma_c = 20 + 5 = 25$ kPa $\tan\phi = \mu_{\tan\phi} + \sigma_{\tan\phi} = 0.3640 + 0.0364 = 0.4004$	1.763	1.681
2. F_{+-}	$c = \mu_c + \sigma_c = 20 + 5 = 25$ kPa $\tan\phi = \mu_{\tan\phi} - \sigma_{\tan\phi} = 0.3640 - 0.0364 = 0.3276$	1.541	1.474
3. F_{-+}	$c = \mu_c - \sigma_c = 20 - 5 = 15$ kPa $\tan\phi = \mu_{\tan\phi} + \sigma_{\tan\phi} = 0.3640 + 0.0364 = 0.4004$	1.547	1.464
4. F_{--}	$c = \mu_c - \sigma_c = 20 - 5 = 15$ kPa $\tan\phi = \mu_{\tan\phi} - \sigma_{\tan\phi} = 0.3640 - 0.0364 = 0.3276$	1.339	1.257

Before calculating the mean and variance, the weighting factors, P_{++}, P_{+-}, P_{-+}, and P_{--} are calculated using Equation 6-105

$$\begin{aligned} P_{++} &= P_{--} = \tfrac{1}{4}(1 + \rho_{xy}) \\ &= \tfrac{1}{4}[1 + (-0.70)] = 0.075 \\ P_{+-} &= P_{-+} = \tfrac{1}{4}(1 - \rho_{xy}) \\ &= \tfrac{1}{4}[1 - (-0.70)] = 0.425 \end{aligned}$$

With the above four values of the FOS, for Bishop's method, the expected value is computed using Equation 6-104

$$\begin{aligned} E[F] &= P_{++}F_{++} + P_{+-}(F_{+-} + F_{-+}) + P_{--}F_{--} \\ &= 0.075(1.763) + 0.425(1.541 + 1.547) + 0.075(1.339) \\ &= 1.545 \end{aligned}$$

To calculate the variance for Bishop's method, we must first calculate $E[F^2]$, using

$$\begin{aligned} E[F^2] &= P_{++}F_{++}^2 + P_{+-}(F_{+-}^2 + F_{-+}^2) + P_{--}F_{--}^2 \\ &= 0.075(1.763^2) + 0.425(1.541^2 + 1.547^2) + 0.075(1.339^2) \\ &= 2.393 \end{aligned}$$

Then, Equation 6-106 is used to calculate the variance, Var$[F]$

$$\begin{aligned} \text{Var}[F] &= E[F^2] - (E[F])^2 \\ &= 2.393 - 1.545^2 \\ &= 0.00676 \end{aligned}$$

With this variance, the standard deviation is $\sigma_F = 0.0822$ for $\rho_{XY} = -0.7$, and the coefficient of variation will be $V_F = 5.32\%$. A complete summary of the results for the Bishop and Janbu method is given in Table 6.23.

The reliability index and the probabilities of unsuccessful performance are calculated in the same manner as shown for the Taylor series method. These are summarized in Table 6.24.

These results are very similar to the ones presented for the Taylor series method.

Monte Carlo Method To illustrate the Monte Carlo method for probabilistic analysis, two sets of solutions are presented where the component random variables, c and $\tan\phi$, are presumed to follow (1) a normal distribution, and (2) a triangular distribution. Both of these examples will be completed for *dependent* and *independent* variables. The same parameters are used to illustrate the application of the Taylor series

TABLE 6.23 Summary of Results for the Point Estimate Method

	Simplified Bishop		Simplified Janbu	
Parameter	$\rho_{XY} = -0.7$	$\rho_{XY} = 0$	$\rho_{XY} = -0.7$	$\rho_{XY} = 0$
Mean value	1.545	1.547	1.469	1.469
Variance	0.00676	0.02250	0.00676	0.02247
Standard deviation	0.0822	0.1500	0.0822	0.1499
Coefficient of variation	5.32%	9.69%	5.60%	10.20%

TABLE 6.24 Summary of β and p_F Values for Dependent and Independent Cases

		Dependent, $\rho_{XY} = -0.7$		Independent, $\rho_{XY} = 0$	
Method	Distribution	β	$P(F < 1.4)$	β	$P(F < 1.4)$
Bishop	Lognormal	1.7848	0.0371	0.9169	0.1796
	Normal	1.7591	0.0393	0.9809	0.1633
Janbu	Lognormal	0.8019	0.2113	0.3630	0.3583
	Normal	0.8412	0.2001	0.4610	0.3224

and Rosenblueth's methods will be used for this solution. These parameters are given in Table 6.19.

Normally Distributed Variables To start the simulation, independent, normal random numbers, N_is, are generated using uniformly distributed random variables and the Box-Muller equation (6-113). The first five pairs, N_1 and N_2, of such random numbers from a normal distribution with mean $\mu = 0$ and standard deviation $\sigma = 1.0$ are shown in columns 2 and 3 of Table 6.25. As these values are independent and we require dependent variables, Equation 6-115 is used to modify the N_2 value into a new value, N_2^*. For this example, N_2^* is negatively correlated to N_1, with $\rho_{XY} = -0.7$, and its value is shown in column 4 of Table 6.25. The N_2^* value in the first row is calculated as

$$
\begin{aligned}
N_2^* &= N_1 \rho_{XY} + N_2 \sqrt{1 - \rho_{XY}} \\
&= 1.3145 \times (-0.70) + (-0.3877) \times \sqrt{1 - (-0.7)^2} \\
&= -1.1970
\end{aligned}
$$

Next, Equation 6-114 is used to assign the component random variables, c and $\tan\phi$, according to the distribution parameters given in Table 6.19. The numbers N_1 and N_2^* are used for the dependent set, and N_1 and N_2 will be used for the independent case. The first five sets of numbers are shown in Table 6.26, along with the corresponding values of the angle, ϕ, derived from the $\tan\phi$ value.

TABLE 6.25 Example Normal Random Variables Used for Simulation

Pair	N_1	N_2	N_2^*
1	1.3145	−0.3877	−1.1970
2	2.3329	−1.5977	−2.7740
3	−1.3600	−0.7262	0.4334
4	−0.4366	0.7783	0.8614
5	0.6379	−0.0934	−0.5133

TABLE 6.26 Example Random Component Variables, Normal Distribution

	Dependent, $\rho_{XY} = -0.7$			Independent, $\rho_{XY} = 0.0$		
Pair	c (kPa)	$\tan\phi$	$\phi°$	c (kPa)	$\tan\phi$	$\phi°$
1	26.57	0.3204	17.76	26.57	0.3498	19.28
2	31.66	0.2630	14.73	31.66	0.3058	17.00
3	13.20	0.3797	20.79	13.20	0.3375	18.65
4	17.81	0.3953	21.57	17.81	0.3923	21.42
5	23.19	0.3452	19.04	23.19	0.3605	19.82

The c, $\tan\phi$, and ϕ dependent component variables on the first row were calculated as follows

$$\begin{aligned} c_1 &= \mu_c + N_1\sigma_X = 20.0 + 1.3145 \times 5.0 \\ &= 26.573\text{ kPa} \\ \tan\phi_1 &= \mu_{\tan\phi} + N_2^*\sigma_{\tan\phi} = 0.3640 + (-1.1970) \times 0.0364 \\ &= 0.3204 \\ \phi &= \tan^{-1}(0.3204) = 17.766\text{ degrees} \end{aligned}$$

The independent variables are calculated in the same way, except that the random variable N_2 is used rather than N_2^*. The c and ϕ pairs computed for the independent and dependent case are then used to calculate the FOS using the equations discussed earlier.

However, before we can complete the Monte Carlo simulation, we need to estimate the number of trials required to estimate the probability, p_F, within certain bounds and a level of confidence. Using the approach suggested by Hahn and Shapiro (1967), for a 95 percent confidence level, relative error of 10 percent and for a preliminary estimate of $p_F = 0.5$, Equation 6-111 suggests

$$\begin{aligned} n &= \left(\frac{100\,d}{E}\right)^2 \frac{1 - p_F}{p_F} \times m \\ &= \left(\frac{100 \times 1.96}{0.10}\right)^2 \frac{0.5}{0.5} \times 2 \\ &= 768 \end{aligned}$$

So for a preliminary estimate of p_F, at least 768 trials (or 800, rounded up) must be performed initially. For this simulation, the first 800 pairs of random values, a continuation of the sequence shown in Table 6.26, were used to compute FOS values according to the simplified Bishop (Equation 6-48) and Janbu (Equation 6-43) methods of analysis.

In examining the 800 calculated FOS values, there were 28 and 172 occurrences of values less than the prescribed limit of $F_c = 1.4$ for the Bishop and Janbu methods,

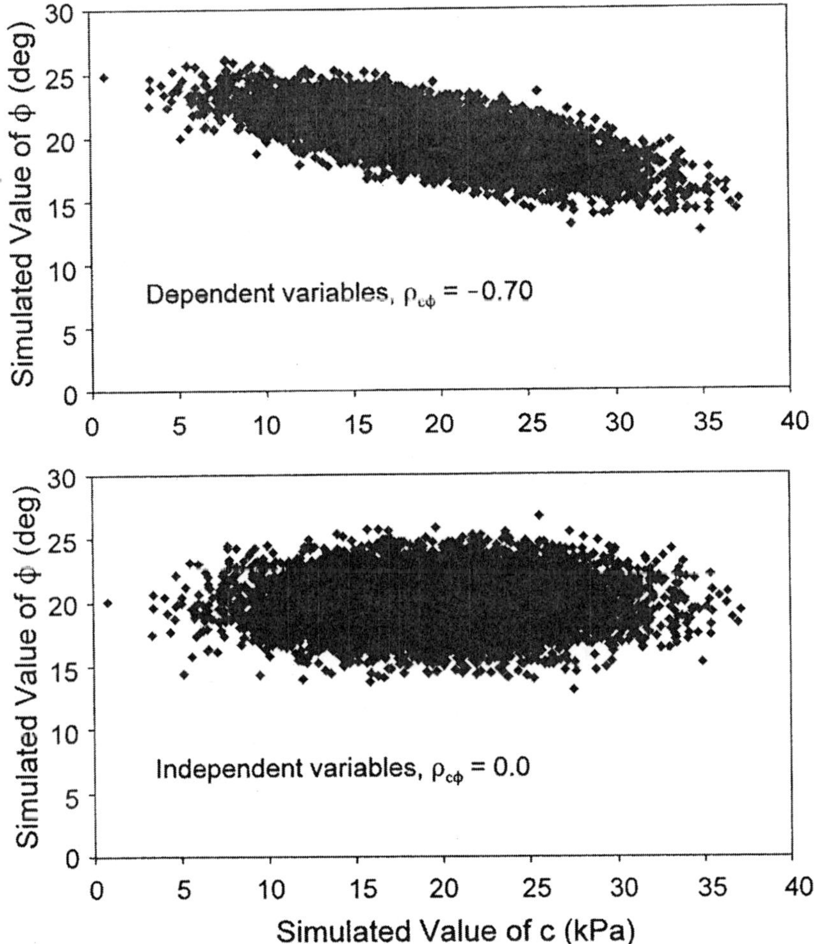

Figure 6.60 Correlation of randomly generated c and ϕ values for dependent and independent cases.

respectively. This indicates an estimated probability, $p_F = 28/800 = 0.035$ for the Bishop method, and $p_F = 172/800 = 0.215$ for the Janbu method.

Now, we can use Equation 6-111 to generate a better estimate of the number of trials required for a 95 percent confidence level and a relative error of 10 percent. For these numbers, Equation 6-111 suggests 21,184 and 2,805 trials for the Bishop and Janbu methods, respectively. For convenience, 10,000 trials will be performed to estimate p_F more accurately, and then Equation 6-112 will be used to calculate the relative error for each case.

Next, 10,000 discrete values of the component variables, c and $\tan\phi$ (or ϕ), were generated for the simulation. Figure 6-60 shows the correlation between c and ϕ. In this figure, note the desirable negative correlation which shows how smaller values of ϕ are generated for high values of c. For independent variables, there is no control

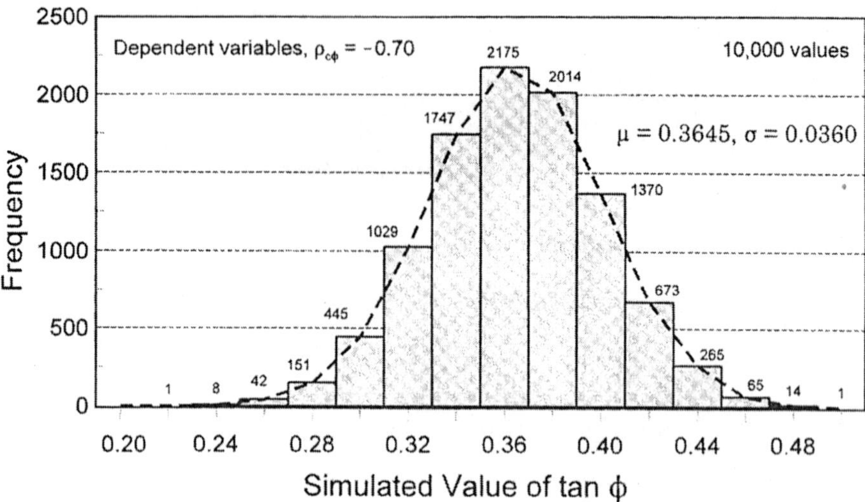

Figure 6.61 Distribution of simulated $\tan\phi$ values for dependent case.

concerning the relative magnitudes of ϕ and c. The distribution of the generated c, $\tan\phi$, and ϕ variables for the dependent case are presented in Figures 6.61 to 6.63 in the form of histograms to show the overall shape of the presumed PDF for the Monte Carlo simulation. For the independent case, the distribution of c is the same, but a slightly different distribution of $\tan\phi$ is used for the simulation. A summary of the statistical parameters for the 10,000 discrete values is given in Table 6.27.

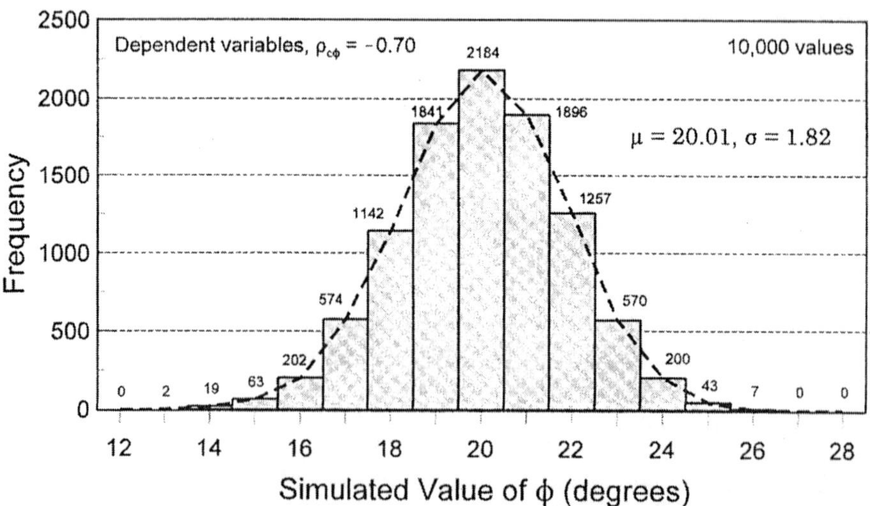

Figure 6.62 Distribution of simulated ϕ values for dependent case.

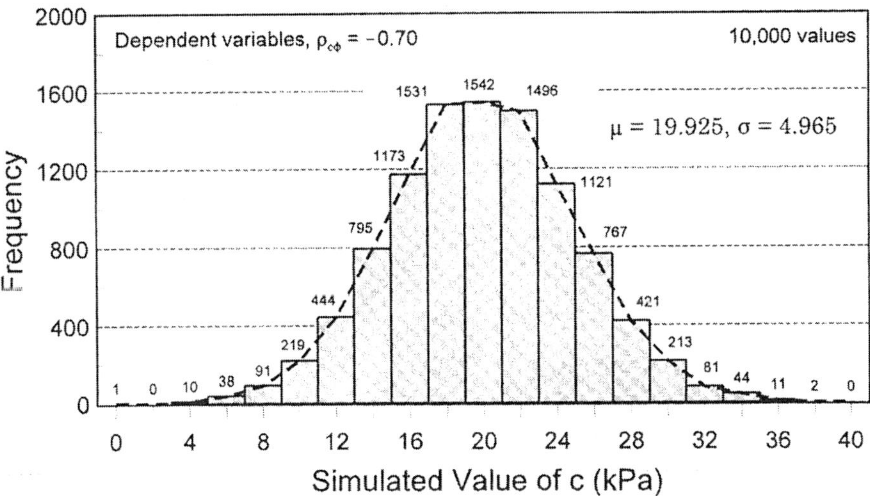

Figure 6.63 Distribution of simulated c values.

In reviewing Table 6.27, one should note that the standard deviation of tan ϕ (or ϕ) for the dependent case is smaller than the independent case. The smaller spread of the dependent data is thus indicative of *less* uncertainty and will lead to a lower probability, p_F.

Once the 10,000 pairs of c and ϕ have been generated, the FOS for the simplified Bishop and Janbu methods may be computed using the method of slices. Once all the FOS values have been computed, the distribution may be presented as a histogram, as shown in Figure 6.64. In this figure, the number of occurrences within a specific range of FOS is indicated next to the bars in the graph. For example, there were 2,123 occurrences of the FOS between the range 1.475 to 1.525 for the simplified Bishop solution. In the case of the Janbu solution, there were 2,182 occurrences of F in the same range.

These results can also be presented as a cumulative distribution function (CDF), indicating the number of occurrences of the FOS below a specific value of F. If this cumulative number is divided by the total number of trial (10,000 in this case), a

TABLE 6.27 Statistical Summary of Component Random Variables, Normal Distribution

	Dependent, $\rho_{XY} = -0.7$			Independent, $\rho_{XY} = 0.0$		
	c (kPa)	tan ϕ	ϕ	c (kPa)	tan ϕ	ϕ
Mean	19.92	0.3645	20.01	19.92	0.3642	19.99
Standard deviation	4.97	0.0359	1.82	4.97	0.0363	1.84
Minimum value	0.79	0.2218	12.51	0.79	0.2325	13.09
Maximum value	37.14	0.4914	26.17	37.14	0.5038	26.74
Coefficient of variation	24.9%	9.8%	9.09%	24.9%	9.9%	9.2%

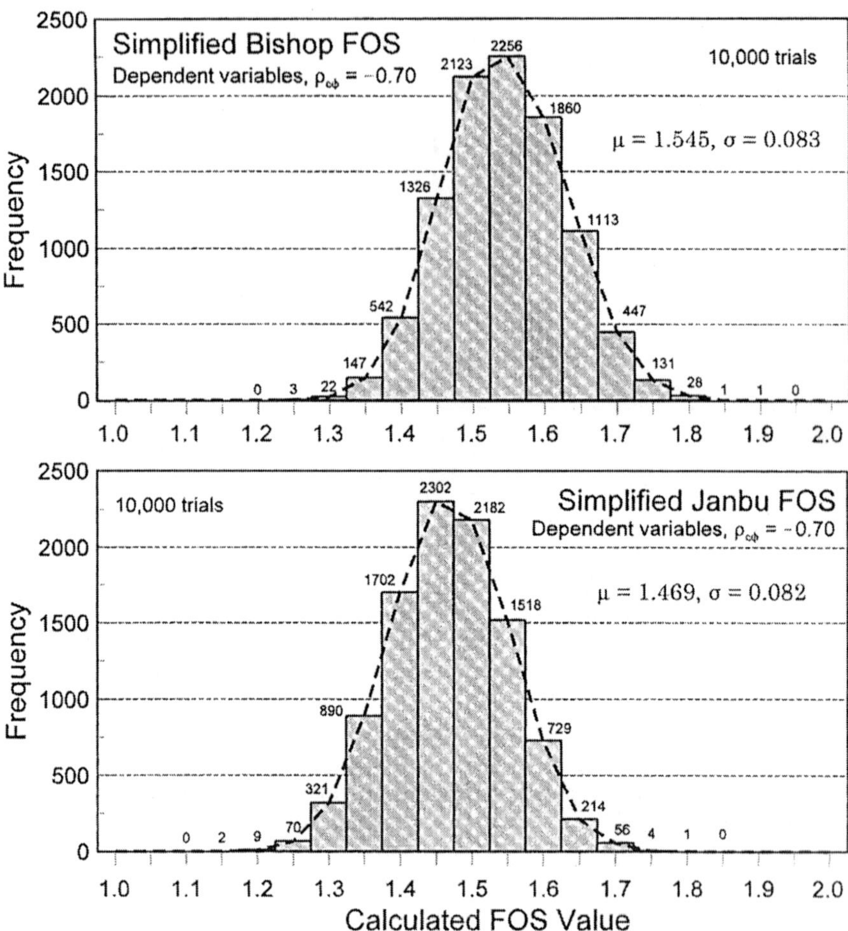

Figure 6.64 Distribution of FOS results using a simulation of 10,000 trials.

normalized CDF may be generated, as shown in Figure 6.65. From such a figure, one can readily estimate the probability, p_F, for any value of the FOS, F. For example, if $F_c = 1.4$ is selected, the probability of unsuccessful performance is estimated as 0.20 and 0.04 for the Janbu and Bishop solutions, respectively.

Table 6.28 presents an overall statistical summary of the 10,000 trials. As expected, the mean and standard deviation of the sample are very close to the values predicted by the Taylor series and point estimate methods. From this summary, we can see that the Bishop FOS values ranged between 0.948 and 2.075 for the independent case, but the range for the dependent case provided a narrower range of 1.245 to 1.882. Calculations for the relative error are also included in this table, now that the desired magnitude of p_F has been determined on the basis of 10,000 simulations. Although the relative error for the dependent Bishop solution appears to be large (14.3%), the

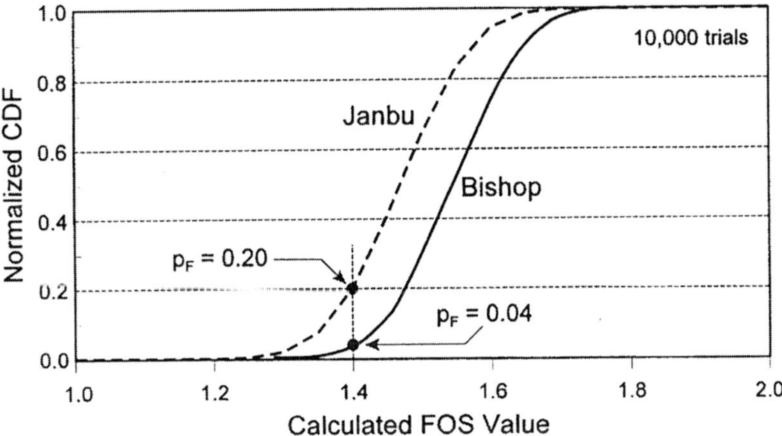

Figure 6.65 Normalized cumulative density function (CFD) for the FOS results.

value of p_F (0.0362) is relatively small in comparison to the other solutions. The 14.3% error indicates that the computed value of p_F is within the range 0.036 ± 0.005, with a confidence level of 95 percent.

As expected, the Bishop FOS was found to be greater than the Janbu FOS due to the adopted performance functions. Also, the standard deviation for the dependent FOS is considerably less than the independent case, which is indicative of the extent of the range of FOS values and the lower value of the probability, p_F. For this particular simulation, the computed FOS was less than 1.0 (failure) only for the case where the component variables are independent. For this case, 1 out 10,000 trials resulted in failure for the Bishop method, and similarly, 8 out of 10,000 cases for the Janbu method. The spread for the dependent variables was too small to generate FOS values less than 1.0, so a probability of failure could not be reported for this simulation.

TABLE 6.28 Statistical Summary of Monte Carlo Simulation, Normal Distribution

	Dependent, $\rho_{XY} = -0.7$		Independent, $\rho_{XY} = 0.0$	
	Bishop	Janbu	Bishop	Janbu
Mean	1.545	1.469	1.546	1.468
Standard deviation	0.0823	0.0823	0.1476	0.1475
Minimum value	1.245	1.145	0.948	0.854
Maximum value	1.882	1.793	2.075	1.985
Coefficient of Variation	5.36%	5.60%	9.54%	10.05%
Number $< F_c = 1.4$	362	2016	1644	3231
Number $< F_c = 1.0$	0	0	1	8
Probability, p_F	0.0362	0.2016	0.1644	0.3231
Relative error[a]	14.3%	5.5%	6.2%	4.0%
Valid p_F range[a]	0.031–0.041	0.191–0.213	0.154–0.175	0.310–0.336

[a]95% Confidence Level

TABLE 6.29 Generating Random Values for a Triangular Distribution

Pair	N_1	N_2	N_2^*	R_1	R_2^*
1	1.3145	−0.3877	−1.1970	0.9057	0.1156
2	2.3329	−1.5977	−2.7740	0.9902	0.0028
3	−1.3600	−0.7262	0.4334	0.0869	0.6676
4	−0.4366	0.7783	0.8614	0.3312	0.8055
5	0.6379	−0.0934	−0.5133	0.7382	0.3039

Triangularly Distributed Variables For this case, the component random variables, c and $\tan\phi$, are presumed to be distributed according to a triangular distribution with the following parameters:

Parameter	Minimum	Maximum	Mode
c	10 kPa	30 kPa	20 kPa
$\tan\phi$	0.3249	0.40404	0.3640
ϕ	18°	22°	20°

The random values for a triangular distribution are generated using standardized, normally distributed, independent random numbers with a mean of zero and a standard deviation of 1.0. To illustrate this approach, the first five sets of random numbers are presented in Table 6.29. Columns 2 and 3 show the normal random numbers generated using the Box-Muller method. In the third column, a dependent N_2^* is computed for the negative correlation, $\rho_{XY} = -0.7$, using Equation 6-115. Next, uniformly distributed random numbers are generated by converting the pair of numbers, N_1 and N_2^*, using the normal distribution "inversion" process. In this way, the converted numbers R_1 and R_2^* will also be correlated, and their values will be within a zero to one range.

With each pair of random numbers, R_1 and R_2, the random variables that follow the specified triangular distribution may be calculated using the equations (Hall, 1992) given below.

$$\text{If} \quad R \leq \frac{b-a}{c-a}, \qquad T = a + \sqrt{R(c-a)(b-a)}$$

$$\text{otherwise}, \qquad T = c - \sqrt{(1-R)(c-a)(c-b)} \qquad \text{(Eq. 6-123)}$$

where R = uniformly distributed random variable
a, b, and c = parameters used to define the triangular distribution
T = random value according the triangular distribution.

For dependent random variables, the values T_1 and T_2 must be calculated using the correlated pairs, R_1 and R_2^*. Table 6.30 shows the computed values of c and $\tan\phi$ (and ϕ), as calculated for the first 5 sets of random values.

A Monte Carlo simulation was performed using 10,000 trial values of c and ϕ. A summary of these 10,000 values is presented in Table 6.31 for the dependent and independent cases, and Figures 6.66 and 6.67 illustrate the actual distribution of

TABLE 6.30 Generating Random Values, c and $\tan\phi$, Triangular Distribution

Pair	R_1	R_2^*	c (kPa)	$\tan\phi$	ϕ°
1	0.9057	0.1156	25.66	0.3438	18.97
2	0.9902	0.0028	28.60	0.3278	18.15
3	0.0869	0.6676	14.17	0.3716	20.38
4	0.3312	0.8055	18.14	0.3792	20.77
5	0.7382	0.3039	22.77	0.3556	19.57

TABLE 6.31 Statistical Summary of Component Random Variables, Triangular Distribution

	Dependent, $\rho_{XY} = -0.7$			Independent, $\rho_{XY} = 0.0$		
	c (kPa)	$\tan\phi$	ϕ°	c (kPa)	$\tan\phi$	ϕ°
Mean	19.94	0.3644	20.02	19.94	0.3646	20.03
Standard deviation	4.06	0.0162	0.82	4.06	0.0160	0.81
Minimum value	10.11	0.3256	18.04	10.11	0.3253	18.02
Maximum value	29.75	0.4036	21.98	29.75	0.4032	21.96
Coefficient of variation	20.3%	4.4%	4.1%	20.3%	4.4%	4.0%

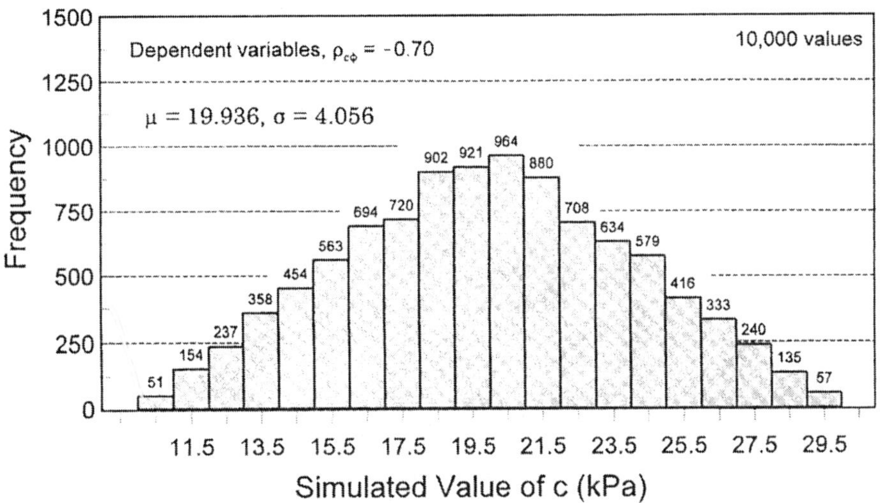

Figure 6.66 Triangular distribution simulated c values.

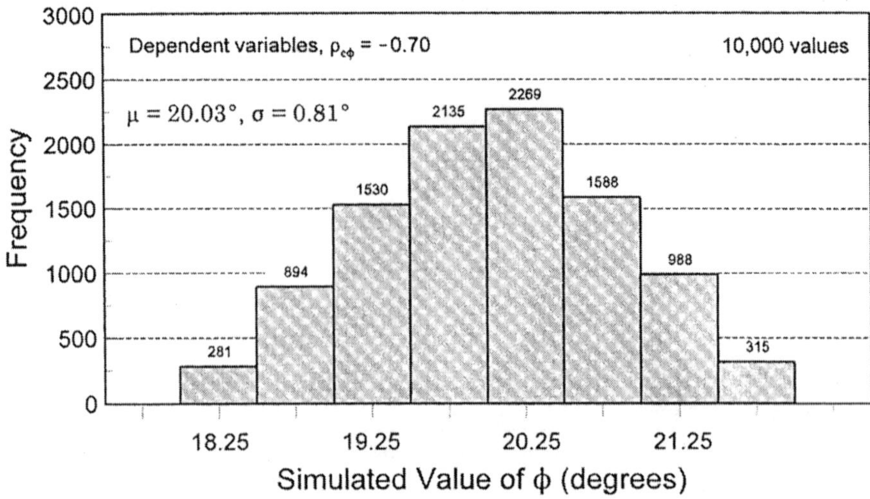

Figure 6.67 Triangular distribution of simulated ϕ values.

the component data. Given these results, the FOS was computed using the Bishop and Janbu methods of analysis. The results of these 10,000 trials are summarized in Table 6.32, and the histograms are presented in Figure 6.68.

6.18.8 Summary

The example probabilistic analyses provided in this section used three different procedures: (1) Taylor series method, (2) Rosenblueth's point estimate method, and (3) Monte Carlo simulation. Overall, all three methods provided very similar results, as summarized in Table 6.33. The presumption whether the component variables are dependent or independent appears to have a significant influence on the results. The standard deviation for the FOS calculated using independent variables is almost

TABLE 6.32 Statistical Summary of Monte Carlo Simulation, Triangular Distribution

	Dependent, $\rho_{XY} = -0.7$		Independent, $\rho_{XY} = 0.0$	
	Bishop	Janbu	Bishop	Janbu
Mean	1.545	1.469	1.545	1.469
Standard deviation	0.063	0.066	0.098	0.100
Minimum value	1.349	1.262	1.249	1.159
Maximum value	1.744	1.663	1.845	1.764
Coefficient of variation	4.1%	4.5%	6.4%	6.8%
Number $< F_c = 1.4$	49	1549	697	2550
Probability, p_F	0.005	0.155	0.070	0.255
Relative error[a]	39.5%	6.5%	10.1%	4.7%
Valid p_F range[a]	0.003–0.007	0.145–0.165	0.063–0.077	0.243–0.267

[a]95% Confidence Level

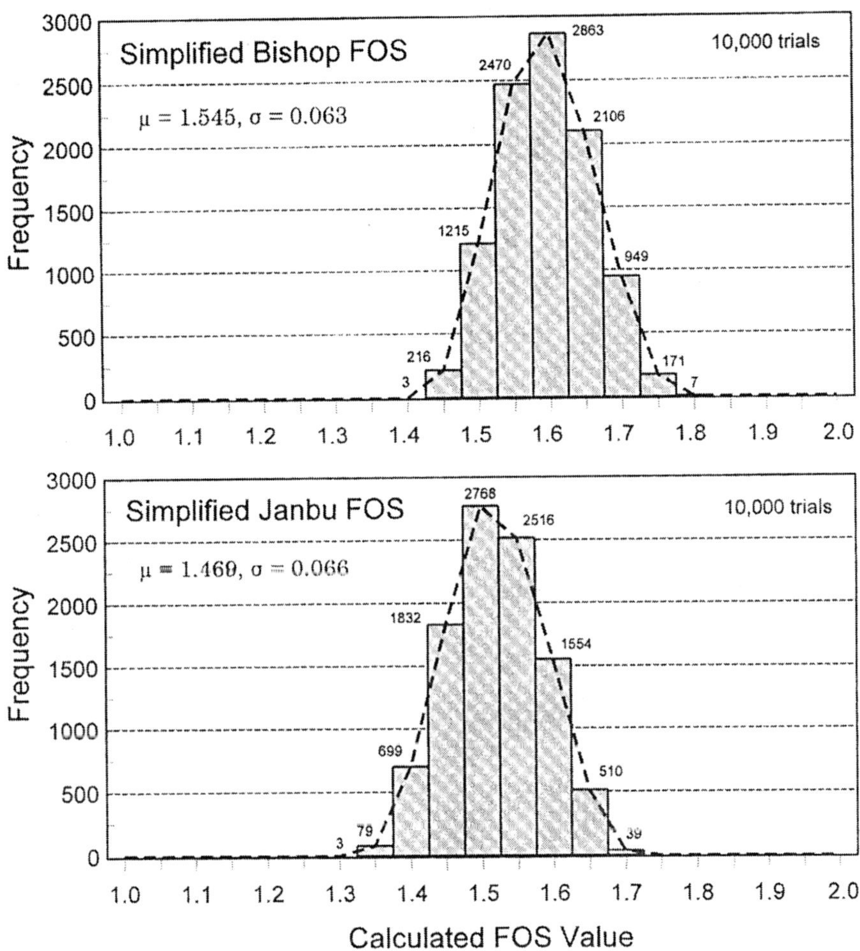

Figure 6.68 Distribution of FOS results for triangularly distributed component variables using a simulation of 10,000 trials.

double the standard deviation for the dependent case. So, if the variables are truly dependent, the assumption of independence will lead to higher probabilities of failure.

For the dependent case, the Monte Carlo simulation estimated a mean Bishop FOS of 1.545, which does not differ much from the results estimated using the Taylor series method (1.544) and the point estimate method (1.545). The estimated standard deviations, which are an indication of the uncertainty, were also very similar. So why use the Monte Carlo simulation method? As mentioned earlier, it provides the opportunity to view histograms of the selected PDFs of the component variables, and the calculated FOS values provide considerably more insight to the probabilistic analyst, even though the procedure requires many more calculations. The availability of faster computers, coupled with visually rich, graphical solutions, is likely to make the elegant, yet simple, Monte Carlo method even more popular.

TABLE 6.33 Summary and Comparison of Results for All Methods

		Dependent case, $\rho_{c,\tan\phi} = -0.7$			Independent case, $\rho_{c,\tan\phi} = 0.0$		
Method	Parameter	Taylor	Point Estimate	Monte Carlo	Taylor	Point Estimate	Monte Carlo
Bishop:	Mean	1.544	1.545	1.545	1.544	1.547	1.546
	Std. Dev.	0.0828	0.0822	0.0823	0.1510	0.1500	0.1476
	c.o.v.	5.37%	5.32%	5.36%	9.79%	9.69%	9.54%
	minimum	—	—	1.245	—	—	0.948
	maximum	—	—	1.882	—	—	2.075
	Lognormal, $p(F < 1.4)$	0.0391	0.0371	—	0.1708	0.1796	—
	Normal, $p(F < 1.4)$	0.0442	0.0393	—	0.1710	0.1633	—
	M-C Simul. $p(F < 1.4)$	—	—	0.0362	—	—	0.1641
Janbu:	Mean	1.469	1.469	1.469	1.469	1.469	1.468
	Std. Dev.	0.0822	0.0822	0.0823	0.1499	0.1499	0.1475
	c.o.v.	5.60%	5.60%	5.60%	10.20%	10.20%	10.05%
	minimum	—	—	1.145	—	—	0.854
	maximum	—	—	1.793	—	—	1.985
	Lognormal, $p(F < 1.4)$	0.2024	0.2113	—	0.3365	0.3583	—
	Normal, $p(F < 1.4)$	0.2004	0.2001	—	0.3226	0.3224	—
	M-C Simul. $p(F < 1.4)$	—	—	0.2016	—	—	0.3231

Of course, the availability of faster computers is also likely to introduce other innovative, analytical procedures that consider uncertainty, such as fuzzy methods (Juang et al., 1998), and even possibly Artificial Neural Networks (ANN) and Genetic Algorithms (GA) for assessing the stability of slopes.

REFERENCES

Abramson, L. W., 1985. "Rock Wedge Stability Analysis on a Personal Computer," *Proceedings of the 26th U.S. Symposium on Rock Mechanics*, Rapid City, South Dakota. Rotterdam: Balkema, pp. 675–682.

Baecher, G. B., M. Chen, T. S. Ingra, T. Lee, and L. R. Nucci, 1980. "Geotechnical Reliability of Offshore Gravity Platforms," Department of Civil Engineering, Massachusetts Institute of Technology, Cambridge, Massachusetts, *Research Report SG80-20*.

Baligh, M. M., and A. S. Azzouz, 1975. "End Effects on Stability of Cohesive Slopes," *Journal of the Geotechnical Engineering Division*, ASCE, Vol. 101, No. GT11, pp. 1105–1117.

Barnes, G. E., 1991. "A Simplified Version of the Bishop and Morgenstern Slope-Stability Charts," *Canadian Geotechnical Journal*, Vol. 28, No. 4, pp. 630–632.

Bishop, A. W., 1948. "Some Factors Involved in the Design of a Large Earth Dam in the

Thames Valley," *Proceedings of the Second International Conference on Soil Mechanics and Foundation Engineering*, Rotterdam, Paper IV b 1, June, pp. 13–18.

Bishop, A. W., 1995. "The Use of the Slip Circle in the Stability Analysis of Slopes," *Geotechnique*, Vol. 5, No. 1, March, pp. 7–17.

Bishop, A. W., and N. R. Morgenstern, 1960. "Stability Coefficients for Earth Slopes," *Geotechnique*, Vol. 10, No. 1, pp. 129–150.

Box, G. E. P., and M. E. Muller, 1958. A Note on the Generation of Random Normal Deviates," *Annals of Mathematical Statistics*, Vol. 29, 1958, pp. 610–611.

Bromhead, E. N., 1992. *The Stability of Slopes*, 2nd ed., Glasgow, Scotland: Blackie Academic & Professional, 411 pp.

California, State of, 1977. "Analysis and Mitigation of Earthquake-Induced Landslide Hazards," Guidelines for Evaluation and Mitigation of Seismic Hazards in California, Division of Mines and Geology, *California Department of Conservation Special Publication 117*, Chapter 5, 15 pp.

Castro, G., S. J. Poulos, and F. D. Leathers, 1985. "Re-examination of Slide of Lower San Fernando Dam," *Journal of the Geotechnical Engineering Division*, ASCE, Vol. 111, No. GT9, pp. 1093–1107.

Chandler, D. S., 1996. "Monte Carlo Simulation to Evaluate Slope Stability," *Proceedings, Uncertainty In the Geologic Environment: From Theory to Practice*, ASCE Geotechnical Special Publication No. 58, Madison, Wisconsin, Vol. 1, pp. 474–493.

Charles, J. A., and M. M. Soares, 1984. "Stability of Compacted Rockfill Slopes," *Geotechnique*, Vol. 34, No. 1, pp. 61–70.

Chen, W. F., and M. W. Giger, 1971. "Limit Analysis of Slopes," *Journal of the Soil Mechanics and Foundation Division*, ASCE, Vol. 97, No. SM-1, January, pp. 19–26.

Ching, R. K. H., and D. G. Fredlund, 1983. "Comparison of Slope Stability Methods of Analysis," *Canadian Geotechnical Journal*, Vol. 14, pp. 429–439.

Chowdhury, R. N., 1978. *Slope Analysis*, Elsevier Scientific.

Chowdhury, R. N., 1981. Discussion of "Stability Analysis of Embankments and Slopes," by S. K. Sarma, *Journal of the Geotechnical Engineering Division*, ASCE, Vol. 107, No. GT-5, May, pp. 691–693.

Chowdhury, R. N., 1984. "Recent Developments in Landslide Studies: Probabilistic Methods—State of the Art Report," *Proceedings of the 4th International Symposium on Landslides*, Toronto, Vol. 1, pp. 209–228.

Chowdhury, R. N., and D. W. Xu, 1994. Slope System Reliability with General Slip Surfaces," *Soils and Foundations*, Japanese Society of Soil Mechanics and Foundation Engineering, Vol. 34, No. 3, pp. 99–105.

Christian, J. T., C. C. Ladd, and G. B. Baecher, 1994. "Reliability Applied to Slope Stability Analysis," *Journal of Geotechnical Engineering Division*, ASCE, Vol. 120, No. GT12, December, pp. 2180–2207.

Chugh, A. K., 1981a. "Multiplicity of Numerical Solutions for Slope Stability Problems," *International Journal for Numerical and Analytical Methods in Geomechanics*, Vol. 5, pp. 313–322.

Chugh, A. K., 1981b. "Pore Water Pressure in Natural Slopes," *International Journal for Numerical and Analytical Methods in Geomechanics*, Vol. 5, pp. 449–454.

Chugh, A. K., 1986. "Variable Interslice Force Inclination in Slope Stability Analysis," *Soils and Foundations, Japanese Society of SMFE*, Vol. 26, No. 1, pp. 115–121.

Clough, R. W., and R. J. Woodward, 1967. "Analysis of Embankment Stresses and Deformations," *Journal of Geotechnical Division*, ASCE, July, pp. 529–549.

Corps of Engineers, 1970, 1982. *Slope Stability Manual EM-1110-2-1902*. Washington, DC: Department of the Army, Office of the Chief of Engineers, Washington, D.C.

Corps of Engineers, 1997. "Engineering and Design Introduction to Probability and Reliability Methods for Use in Geotechnical Engineering," *Engineering Technical Letter No. 1110-2-547*, Department of the Army, Washington, D.C.

Cousins, B. F., 1978. "Stability Charts for Simple Earth Slopes," *Journal of Geotechnical Division*, ASCE, Vol. 104, No. GT-2, February, pp. 267–279.

Dai, S. H., and M. O. Wang, 1992. *Reliability Analysis in Engineering Applications*. New York: Van Nostrand Reinhold.

Dawson, E. M., W. H. Roth, and A. Drescher, 1999. "Slope Stability Analysis by Strength Reduction," *Geotechnique*, Vol. 49, No. 6, pp. 835–840.

Duncan, J. M., 1972. "Finite Element Analysis of Stresses and Movements in Dams, Excavations, and Slopes," *Application of the Finite Element Method in Geotechnical Engineering: A Symposium*, C. S. Desai, Ed., U.S. Army Engineering Waterways Experiment Station, Vicksburg, Mississippi, pp. 267–326.

Duncan, J. M., 1996. "State of the Art: Limit Equilibrium and Finite-Element Analysis of Slopes," *Journal of Geotechnical Engineering Division*, ASCE, Vol. 122, No. GT-7, pp. 577–596.

Duncan, J. M., 2000. "Factors of Safety and Reliability in Geotechnical Engineering," *Journal of Geotechnical and Geoenvironmental Engineering*, ASCE, Vol. 126, No. 4, April, pp. 307–316.

Duncan, J. M., and A. L. Buchignani, 1975. *An Engineering Manual for Slope Stability Studies*, Department of Civil Engineering, University of California, Berkeley, March.

Duncan, J. M., and P. Dunlop, 1969. "Slopes in Stiff-Fissured Clays and Shales," *Journal of the Soil Mechanics and Foundations Division*, ASCE, Vol. 95, No. SM5, May, pp. 467–492.

Duncan, J. M., A. L. Buchignani, and D. W. Marius, 1987. *An Engineering Manual for Slope Stability Studies*, Department of Civil Engineering, Virginia Polytechnic Institute and State University, March.

Fellenius, W., 1927. *Erdstatische Berechnungen mit Reibung and Kohaesion*. Berlin: Ernst.

Fellenius, W., 1936. "Calculation of Stability of Earth Dams," *Transactions*, 2nd Congress Large Dams, Vol. 4, 445 pp. Washington, D.C.

Feller, W., 1966. *An Introduction to Probability Theory and Its Applications: Vol. II*, New York: Wiley.

Finn, W. D., 1988. "Dynamic Analysis in Geotechnical Engineering," *Proceedings of Earthquake Engineering and Soil Dynamics II—Recent Advances in Ground Motion Evaluation*, ASCE Geotechnical Special Publication No. 20, J. L. Von Thun, Ed., Park City, Utah.

Franklin, A. G., and F. K. Chang, 1977. "Earthquake Resistance of Earth and Rock Fill Dams: Permanent Displacements of Earth Embankment by Newmark Sliding Block Analysis," *USAE, Misc. Paper S-71-17*, Report 5, Vicksburg, Mississippi.

Fredlund, D. G., and J. Krahn, 1977. "Comparison of Slope Stability Methods of Analysis," *Canadian Geotechnical Journal*, Vol. 14, pp. 429–439.

Fredlund, D. G., J. Krahn, and D. E. Pufahl, 1981. "The Relationship Between Limit Equilibrium Slope Stability Methods," *Proceedings of the 10th ICSMFE*, Vol. 3, Stockholm, Sweden, pp. 409–416.

Gibson, M., and N. R. Morgenstern, 1962. "A Note on the Stability of Cuttings in Normally Consolidated Clays," *Geotechnique*, Vol. 12, No. 3, pp. 212–216.

Goodman, R. E., 1976. *Methods of Geologic Engineering in Discontinuous Rock.* St. Paul, Minnesota: West Publishing, Chapter 3, pp. 58–90.

Goodman, R. E., 1980. *Introduction to Rock Mechanics*, New York: Wiley.

Gray, D. H., 1978. "Role of Woody Vegetation in Reinforcing Soils and Stabilizing Slopes," *Symposium on Soil Reinforcing and Stabilizing Techniques*, Sydney, Australia, pp. 253–306.

Gray, D. H., and W. F. Megahan, 1981. "Forest Vegetation Removal and Slope Stability in the Idaho Batholith," *USDA Forest Service Research Paper INT–271*, pp. 11–12, May.

Gray, D. H., and A. T. Leiser, 1982. *Biotechnical Slope Protection and Erosion Control.* New York: Van Nostrand Reinhold.

Gray, D. H., and R. B. Sotir, 1992. "Biotechnical Stabilization of Cut & Fill Slopes," *Proceedings: Stability and Performance of Slopes and Embankments-II*, ASCE Geotechnical Special Publication No. 31, Berkeley, California, pp. 1395–1410.

Gray, D. H., and R. B. Sotir, 1996. *Biotechnical and Soil Bioengineering Slope Stabilization.* New York: Wiley.

Griffiths, D. V., and P. A. Lane, 1999. "Slope Stability Analysis by Finite Elements," *Geotechnique*, Vol. 49, No. 3, pp. 387–403.

Hahn, G. J., and S. S. Shapiro, 1967. *Statistical Models in Engineering*, New York: Wiley.

Hall, D. E., 1992. *A Set of Continuous Univariate Pseudo-Random Number Generation Routines for Use in Computer Simulation*, MSc. Thesis, University of Idaho, 96 pages.

Hammond, C. J., R. W. Prellwitz, and S. M. Miller, 1992. "Landslide Hazard Assessment using Monte Carlo Simulation," *Proceedings, Sixth International Symposium on Landslides*, D. H. Bell, Ed., Rotterdam: A. A. Balkema, pp. 959–964.

Harr, M. E., 1977. *Mechanics of Particulate Media*, New York: McGraw-Hill.

Harr, M. E., 1987. *Reliability-Based Design in Civil Engineering*, New York: McGraw-Hill.

Hendron, Jr., A. J., E. J. Cording, and A. K. Aiyer, 1971. "Analytical and Graphical Methods for the Analysis of Slopes in Rock Masses," U.S. Army Engineer Explosive Excavation Research Office, Livermore, California, Technical Report No. 36, 148 p., NTIS No. AD-738–929.

Hoek, E. 1973. "Methods for the Rapid Assessment of the Stability of Three-Dimensional Rock Slopes," *Quarterly Journal of Engineering Geology*, Vol. 6, pp. 243–255.

Hoek, E., and J. W. Bray, 1977. *Rock Slope Engineering*, Institute of Mining and Metallurgical Engineering, London.

Houston, S. L., W. N. Houston, and J. M. Padilla, 1987. "Microcomputer-Aided Evaluation of Earthquake-Induced Permanent Slope Displacements," *Microcomputers in Civil Engineering*, Vol. 2. pp. 207–222.

Hovland, H. J., 1977. "Three-Dimensional Slope Stability Analysis Method," *Journal of the Geotechnical Engineering Division*, ASCE, Vol. 103, No. GT9, pp. 971–986.

Hungr, O., F. M. Salgado, and P. M. Byrne, 1989. "Evaluation of Three-Dimensional Method of Slope Stability Analysis," *Canadian Geotechnical Journal*, Vol. 26, pp. 679–686.

Hunter, J. H., and R. L. Schuster, 1968. "Stability of Simple Cuttings in Normally Consolidated Clays," *Geotechnique*, Vol. 18, No. 3, pp. 372–378.

Hynes-Griffin, M. E., and A. G. Franklin, 1984. "Rationalizing the Seismic Coefficient Method," *Miscellaneous Paper No. G.L. 84-13*, U.S. Army Engineer Waterways Experiment Station, Vicksburg, Mississippi.

I. S. Designs, Inc., 1994. *XSTABL—An Integrated Slope Stability Analysis Program for Personal Computers, Reference Manual*. Interactive Software Designs, Inc., Moscow, Iaho, 83843.

Janbu, N., 1954a. "Stability Analysis of Slopes with Dimensionless Parameters," Thesis for the Doctor of Science in the Field of Civil Engineering, *Harvard University Soil Mechanics Series*, No. 46.

Janbu, N., 1954b. "Application of Composite Slip Surface for Stability Analysis," *European Conference on Stability of Earth Slopes*, Stockholm, Sweden.

Janbu, N., 1968, "Slope Stability Computations," *Soil Mechanics and Foundation Engineering Report*, The Technical University of Norway, Trondheim, Norway.

Janbu, N., 1973. *Slope Stability Computations in Embankment-Dam Engineering*, R. C. Hirschfeld and S. J. Poulos, Eds. New York: Wiley, pp. 47–86.

Jibson, R. W., 1993. "Predicting Earthquake-Induced Landslide Displacements Using Newmark's Sliding Block Analysis," *Transportation Research Record 1411*, TRB, National Research Council, pp. 9–17.

Jibson, R. W., and D. K. Keefer, 1993. "Analysis of the Seismic Origin of Landslides: Examples from the New Madrid Seismic Zone," *Geological Society of America Bulletin*, Vol. 21, No. 4, pp. 521–536.

Juang, C. H., Y. Y. Jhi, and D. H. Lee, 1998. "Stability Analysis of Existing Slopes Considering Uncertainty," *Engineering Geology*, Vol. 49, pp. 111–122.

Keefer, D. K., and R. C. Wilson, 1989. "Prediciting Earthquake-Induced Landslides, with Emphasis on Arid and Semi-Arid Environments," *Proceedings of Landslides in a Semi-Arid Environment*, Vol. 2, Inland Geological Society, Riverside, California, pp. 118–149.

Koerner, R. M., and T.-Y. Soong, 1999. "Assessment of Ten Landfill Failures Using 2-D and 3-D Stability Analysis Procedures," *Proceedings of the 2nd Austrian Geotechnical Conference*, H. Brand, Ed., Vienna, Austria, February, 41 pp.

Kovari, K., and P. Fritz, 1975, "Stability Analysis of Rock Slopes for Plane and Wedge Failure with the Aid of a Programmable Pocket Calculator," *Sixteenth U.S. Symposium on Rock Mechanics*, ASCE, New York, pp. 25–34.

Kulhawy, F. H., 1992. "On the Evaluation of Soil Properties," *Proceedings: Stability and Performance of Slopes and Embankments-II*, ASCE Geotechnical Special Publication No. 31, Berkeley, California, Vol. 1, pp. 95–115.

Lacasse, S., and F. Nadim, 1996. "Uncertainties in Characterizing Soil Properties," *Proceedings, Uncertainty in the Geologic Environment: From Theory to Practice*, ASCE Geotechnical Special Publication No. 58, Madison, Wisconsin, Vol. 1, pp. 49–75.

Lacasse, S., and F. Nadim, 1997. "Uncertainties in Characterizing Soil Properties," *Publication No. 201*, Norwegian Geotechnical Institute, Oslo, Norway, pp. 49–75.

Lam, L., and D. G. Fredlund, 1993. "A General Limit Equilibrium Model for Three-Dimensional Slope Stability Analysis," *Canadian Geotechnical Journal*, Vol. 30, pp. 905–919.

Lambe, T. W., and R. V. Whitman, 1969. *Soil Mechanics*. New York: Wiley.

Li, K. S., and P. Lumb, 1987. "Probabilistic Design of Slopes," *Canadian Geotechnical Journal*, Vol. 24, pp. 520–535.

Lowe, J., and L. Karafiath, 1960. "Stability of Earth Dams Upon Drawdown," *Proceedings of the First Pan American Conference on Soil Mechanics and Foundation Engineering*, Mexico City, pp. 537–552.

Lumb, P., 1974. *Application of Statistics in Soil Mechanics*, in *Soil Mechanics – New Horizons*, London: Butterworth and Co., pp. 44–111.

Lysmer, J., T. Udaka, C. F. Tsai, and H. B. Seed, 1975. "FLUSH—A Computer Program for Approximate Analysis of Soil-Structure Interaction Problems," *Report No. 75-30*, Earthquake Engineering Research Center, University of California, Berkeley, California.

Makdisi, F. I., and H. B. Seed, 1977. "Simplified Procedure for Computing Maximum Acceleration and Natural Period for Embankments," *Report No. UCB/EERC-77/19*, Earthquake Engineering Research Center, University of California, Berkeley, California.

Makdisi, F. I., and H. B. Seed, 1978. "Simplified Procedure for Estimating Dam and Embankment Earthquake Induced Deformations," *Journal of the Geotechnical Engineering Division*, ASCE, Vol. 104, No. GT7, pp. 849–867.

Makdisi, F. I., and H. B. Seed, 1979. "Simplified Procedure for Evaluating Embankment Response," *Journal of the Geotechnical Engineering Division*, ASCE, Vol. 105, No. GT12, pp. 1427–1434.

Marcuson, W. F., and A. G. Franklin, 1983. "Seismic Design, Analysis and Remedial Measures to Improve the Stability of Existing Earth Dams—Corps of Engineers Approach," *Seismic Design of Embankments and Caverns*, T. R. Howard, Ed., New York: ASCE.

McGuffey, V., J. Iori, Z. Kyfor, and D. Athanasiou-Grivas, 1981. "Use of Point Estimates for Probability Moments in Geotechnical Engineering," *Transportation Research Record 809*, TRB, National Research Council, pp. 60–64.

Melchers, R. E., 1987. *Structural Reliability: Analysis and Prediction*, New York: Wiley.

Miller, S. M., 1982a. "Statistical and Fourier Methods for Probabilistic Design of Rock Slopes," PhD Dissertation, Department of Geology and Geophysics, University of Wyoming, Laramie, Wyoming.

Miller, S. M., 1982b. "Fourier Analysis for Estimating Probability of Sliding for the Plane Shear Failure Mode," *Issues in Rock Mechanics: Proceedings of the 23rd U.S. Symposium on Rock Mechanics*, R. E. Goodman and F. E. Hueze, Eds., SME-AIME, New York, pp. 124–131.

Mitchell, J. K, R. B. Seed, and M. Chang, 1993. "The Kettleman Hills Landfill Failure: A Retrospective View of the Failure Investigations and Lessons Learned," *Proceedings of the 3rd International Conference on Case Histories in Geotechnical Engineering*, St. Louis, Missouri, June 1–4, pp. 1379–1392.

Morgenstern, N. R., 1963. "Stability Charts for Earth Slopes During Rapid Drawdown," *Geotechnique*, Vol. 13, No. 1, pp. 121–131.

Morgenstern, N. R., and V. E. Price, 1965. "The Analysis of the Stability of General Slip Surfaces," *Geotechnique*, Vol. 15, No. 1, pp. 77–93.

NAVFAC, 1982. *Soil Mechanics, Soil Dynamics, Foundations and Earth Structures, and Deep Stabilization, and Special Geotechnical Construction*, Design Manual DM-7.1, -7.2 and -7.3, Department of the Navy, Naval Facility Engineering Command, Alexandria, Virginia.

Newmark, N. M., 1965. "Effects of Earthquakes on Dams and Embankments," *Geotechnique*, Vol. 15, No. 2, pp. 129–160.

O'Connor, M. J., and R. J. Mitchell, 1977. "An Extension of the Bishop and Morgenstern Slope Stability Charts," *Canadian Geotechnical Journal*, Vol. 14, No. 1, pp. 144–151.

Phoon, K. K., and F. H. Kulhawy, 1999a. "Characterization of Geotechnical Variability," *Canadian Geotechnical Journal*, Vol. 36, No. 4, pp. 612–624.

Phoon, K. K., and F. H. Kulhawy, 1999b. "Evaluation of Geotechnical Variability," *Canadian Geotechnical Journal*, Vol. 25, No. 3, pp. 625–639.

Phoon, K. K., F. H. Kulhawy, and M. D. Grigoriu, 1995. Reliability Based Design of Foundations for Transmission Line Structures, Electric Power Research Institute, Palo Alto, Report TR-105000.

Prevost, J. H., A. M. Abdel-Ghaffar, and S. J. Lacy, 1985. "Nonlinear Dynamic Analysis of an Earth Dam," *Journal of the Geotechnical Engineering Division*, ASCE, Vol. 111, No. GT7, pp. 882–897.

Rosenblueth, E. 1975. "Point Estimates for Probability Moments," *Proceedings of the National Academy of Science, Mathematics*, Vol. 72, No. 10, pp. 3812–3814.

Sarma, S. K., 1973. "Stability Analysis of Embankments and Slopes," *Geotechnique*, Vol. 23, No. 3, pp. 423–433.

Sarma, S. K., 1979. "Stability Analysis of Embankments and Slopes," *Journal of the Geotechnical Engineering Division*, ASCE, Vol. 105, No. GT-5, December, pp. 1511–1524.

Schmidt, H. G., 1977. "Large Diameter Bore Piles for Abutments," *Proceedings of the Ninth International Conference on Soil Mechanics and Foundation Engineering, Specialty Session on Effect of Horizontal Loads on Piles*, Tokyo, pp. 107–112.

Schnabel, P. B., J. Lysmer, and H. B. Seed, 1972. "SHAKE—A Computer Program for Earthquake Response Analysis of Horizontally Layered Soils," *Report No. EERC 72-12*, Earthquake Engineering Research Center, University of California, Berkeley, California, December.

Seed, H. B., 1979, "Considerations in the Earthquake-Resistant Design of Earth and Rockfill Dams," *Geotechnique*, Vol. 29, No. 3, pp. 215–263.

Seed, H. B., and G. R. Martin, 1966. "The Seismic Coefficient in Earth Dam Design," *Journal of Soil Mechanics and Foundations Division*, ASCE, Vol. 92, No. SM3, May, pp. 25–58.

Sharma, S., 1992. *Slope Analysis with XSTABL*, Technical Report, Intermountain Research Station, U.S. Department of Agriculture—Forest Service, Moscow, Idaho, Contract #INT-89416-RJV, June.

Sharma, S., and C. W. Lovell, 1983. "Strengths and Weaknesses of Slope Stability Analyses," *Proceedings of the 34th Annual Highway Geology Symposium*, Atlanta, Georgia, pp. 215–232.

Sharp, K. D., L. R. Anderson, D. S. Bowles, and R. V. Canfield, 1991. Model for Assessing Slope Reliability, *Transportation Research Record 809*, TRB, National Research Council, pp. 70–78.

Siegel, R. A., 1975a. *Computer Analysis of General Slope Stability Problems*, Report JHRP-75-8, School of Civil Engineering, Purdue University, West Lafayette, Indiana.

Siegel, R. A., 1975b. *STABL User Manual*, Report JHRP-75-9, School of Civil Engineering, Purdue University, West Lafayette, Indiana.

Simons, N. E., and B. K. Menzies, 1974. "A Note on the Principle of Effective Stress," *Geotechnique*, Vol. 24, No. 2, pp. 259–261.

Skempton, A. W., and J. N. J. Hutchinson, 1969. "Stability of Natural Slopes and Embankment Foundations," *Seventh International Conference on Soil Mechanics and Foundation Engineering*, State of the Art Volume, Mexico City, pp. 291–340.

Skempton, A. W., and P. La Rochelle, 1965. "The Bradwell Slip, A Short Term Failure in London Clay," *Geotechnique*, Vol. 15, No. 3, pp. 221–242.

Snitbhan, N., and W. F. Chen, 1976. "Elastic-Plastic Large Deformation Analysis of Soil Slopes," *Computers & Structures*, Vol. 9. New York: Pergamon Press, pp. 567–577.

Sotir, R. B., and D. H. Gray, 1989. "Fill Slope Repair Using Soil Bioengineering Systems," *Proceedings, XX IECA Conference*, Vancouver, British Columbia, pp. 413–429.

Spencer, E., 1967. "A Method of Analysis of the Stability of Embankments Assuming Parallel Inter-Slice Forces," *Geotechnique*, Vol. 17, pp. 11–26.

Spencer, E., 1973. "Thrust Line Criterion in Embankment Stability Analysis," *Geotechnique*, Vol. 23, pp. 85–100.

Stark, T. D., and H. T. Eid, 1998. "Performance of Three-Dimensional Slope Stability Methods," *Journal of the Geotechnical Engineering Division*, ASCE, Vol. 124, No. GT-11, November, pp. 1049–1060.

Taylor, D. W., 1937. "Stability of Earth Slopes," *Journal of the Boston Society of Civil Engineers*, Vol. 24, pp. 197–246.

Taylor, D. W., 1948. *Fundamentals of Soil Mechanics*. New York: Wiley, 700 pp.

Terzaghi, K., and R. B. Peck, 1967. *Soil Mechanics in Engineering Practice*. New York: Wiley, 566 pp.

Tobutt, D. C., 1982. "Monte Carlo Simulation Methods for Slope Stability," *Computers & Geosciences*, Vol. 8, No. 2, pp. 199–208.

Turner, A. K., and R. L. Schuster, Eds., 1996. *Landslides Investigation and Mitigation*, Special Report 247, Transportation Research Board. Washington, DC: National Academy Press.

Vanmarke, E. H., 1977a. "Probabilistic Modeling of Soil Profiles," *Journal of Geotechnical Engineering Division*, ASCE, Vol. 103, No. GT11, November, pp. 1227–1246.

Vanmarke, E. H., 1977b. "Reliability of Earth Slopes," *Journal of Geotechnical Engineering Division*, ASCE, Vol. 103, No. GT11, November, pp. 1247–1265.

Watts, C. F., and T. R. West, 1985, "Electronic Notebook Analysis of Rock Slope Stability at Cedar Bluff, Virginia," *Bulletin of the Association of Engineering Geologists*, Vol. XXII, No. 1, pp. 67–85.

Whitman, R. V., 1996. "Organizing and Evaluating Uncertainty in Geotechnical Engineering," *Proceedings, Uncertainty in the Geologic Environment: From Theory to Practice*, ASCE Geotechnical Special Publication No. 58, Madison, Wisconsin, pp. 1–28.

Wieczorek, G. F., R. C. Wilson, and E. L. Harp, 1985. "Map Showing Slope Stability During Earthquakes in San Mateo County, California," *Miscellaneous Investigations Map I-1257-E*, U.S. Geological Survey.

Wolff, T. F., 1996. "Probabilistic Slope Stability in Theory and Practice," *Proceedings, Uncertainty in the Geologic Environment: From Theory to Practice*, ASCE Geotechnical Special Publication No. 58, Madison, Wisconsin, Vol. 1, pp. 419–433.

Wong, F. S., 1984. "Uncertainties in FE Modelling of Slope Stability," *Computers & Structures*, Vol. 19, No. 5/6, pp. 777–791.

Wright, S. G., 1969. "A Study of Stability and Undrained Shear Strength of Clay Shales," PhD Dissertation, University of California, Berkeley, California.

Wu, T. H., and L. M. Kraft, 1970. "Safety Analysis of Slopes," *Journal of Soil Mechanics and Foundations Division*, ASCE, Vol. 96, No. SM2, pp. 609–632.

Zienkiewicz, O. C., 1971. *The Finite Element Method in Engineering Science*. New York: McGraw-Hill, p. 521.

CHAPTER 7

SLOPE STABILIZATION METHODS

7.1 INTRODUCTION

Slope stabilization methods generally reduce driving forces, increase resisting forces, or both. Driving forces can be reduced by excavation of material from the appropriate part of the unstable ground and drainage of water to reduce the hydrostatic pressures acting on the unstable zone. Resisting forces can be increased by

(1) Drainage that increases the shear strength of the ground
(2) Elimination of weak strata or other potential failure zones
(3) Building of retaining structures or other supports
(4) Provision of in situ reinforcement of the ground
(5) Chemical treatment (hardening of soils) to increase shear strength of the ground

As an alternative to slope stabilization, the unstable slope can be avoided by adjusting the location of construction or selecting a different site all together.

Table 7.1 summarizes the principal methods of slope stabilization. Before the best method can be selected, the actual or potential causes of slope instability must be determined. Quite often there are multiple contributing factors that cause or could cause a landslide or slope instability. Failure to identify contributing causes of failure could render the stabilization work ineffective and slope instability recurrent. Various methods of slope stabilization are discussed in this chapter.

7.2 UNLOADING

Unloading is a technique to reduce the driving forces within a slide mass. The most common type of unloading is excavation of the head of a slide. In the case where the construction of a conventional embankment can lead to slope instability, lightweight fill materials can be used to lessen the driving forces caused by the embankment. These two methods are discussed in further detail below.

7.2.1 Excavation

Excavation is a common method for increasing the stability of a slope by reducing the driving forces that contribute to movements. This can include

(1) Removing weight from the upper part of the slope (also called removal of the head)
(2) Removing all unstable or potentially unstable materials
(3) Flattening slopes
(4) Benching

Some disadvantages associated with excavation are costs linked to accessibility (as the slope must usually be excavated from the top downward), implementing and maintaining safety measures for protection of workers and equipment, disposal of excavated materials unless they can be used for local construction, and right-of-way acquisition that may pose economic and legal problems. The main advantage of excavation is typically low cost.

Removal of the Head of a Slide With this method, relatively large quantities of materials are taken from the head of a slope (Figure 7.1). In practice, this method is usually applied to existing failures. Removal of soil from the head of a landslide reduces the driving forces and tends to balance the failure. This usually leads to fairly permanent solutions so long as the excavation drainage receives careful attention.

Because the quantity of soil or rock to be removed depends on the nature and engineering characteristics of the ground encountered, it is very difficult to estimate accurately prior to construction. As a general guideline, one to two times the quantity originally removed at the toe of the slope during construction should be excavated from the head (Eckel, 1958).

Total Removal of All Unstable or Potentially Unstable Materials There are practical limitations based on the complete removal of a slide depending on the volume of the moving mass and space availability. In most cases, the location of other structures and existing property lines preclude the use of this method. The effect of excavation on area drainage must also be considered.

Flattening of Slopes Flattening of slopes is one of the most widely applied and economical methods for improving slope stability. Often, it is the first option to be

TABLE 7.1 Summary of Methods for Correcting and Preventing Landslides

Method of Treatment and its Effect on Stability of Slide	Treatment	General Use		Frequency of Success[a]			Position of Treatment Related to the Actual or Potential Sliding Mass	Possibilities and Limitations
		Prevention	Correction	Collapse	Slide	Flow		
Avoidance	Relocation	X	X	2	2	2	Outside slide limits	The best method if economical
No effect	Construction of viaduct	X	X	3	3	3	Outside slide limits	Applicable to short stretches of sloping hillsides
Movement of earth Reduction in shear stresses	Removal from the head	X	X	N	1	N	Upper part and head	Large masses of cohesive material
	Slope flattening	X	X	1	1	1	In the cut or embankment slopes	More effective in earth fills on frictional soils
	Terracing in slopes	X	X	1	1	1	In the cut or embankment slopes	—
	Removal of all unstable material	X	X	2	2	2	Throughout the slide	In relatively small superficial masses of moving material
Drainage Reduction in shear stresses and increase in shear strength of the soil	Surface							
	Ditches	X	X	1	1	1	Above the crown	Essential in all types
	Slope treatment	X	X	3	3	3	On surface of sliding mass	Rock covering or permeable apron to control the flow
	Subgrade trimming	X	X	1	1	1	On surface of sliding mass	Beneficial in all types
	Sealing of cracks	X	X	2	2	2	Throughout, crest to toe	Beneficial in all types
	Sealing of joint and fissure planes	X	X	3	3	N	Throughout, crest to toe	Applicable to rock formations
	Subdrainage Horizontal drains	X	X	N	2	2	Located to intercept and divert underground water	Large masses of soil with underground flow
	Stabilizing trenches	X	X	N	1	3		Relatively superficial masses of soil with underground flow
	Drainage galleries	X	X	N	3	3		Deep-seated and large masses of soil with significant permeability
	Vertical drainage wells	X	X	N	3	3		Deep-seated sliding masses, underground water in strata or lenses
	Continuous siphon	X	X	N	2	3		Chiefly used as a ditch or drainage well opening

Retaining structures Sliding resistance increases	Support at the base							
	Rock fill	X	X	N	1	1	Base and toe	Sound rock or firm soil at a reasonable depth
	Earth fill	X	X	N	1	1	Base and toe	As counterweight in the toe gives additional strength
	Common or crib retaining walls	X	X	3	3	3	Base	Relatively small moving masses
	Piles							
	Fixed in the slip surface	—	X	N	3	N	Base	The strength of the failure surface is increased by the amount of stress required to make the piles fail
	Not fixed to slip surface	—	X	N	3	N	Base	
	Anchorages in rock	X	X	3	3	N	Uphill from the highway or structure (cuts)	Stratified rock
	Short anchorages in slopes	X	X	3	3	N	Uphill from highway or structure	Eroding slope protected by screen anchored to a solid underlying formation
Other methods Chiefly an increase in shear strength	Hardening of the sliding mass							
	Cementing or chemical treatment							
	At the base	—	X	3	3	3	Base and toe	Cohesionless soils
	Throughout the mass	—	X	N	3	N	Throughout sliding mass	Cohesionless soils
	Freezing	X	—	N	3	3	Throughout sliding mass	To prevent temporary movement in large masses
	Electro-osmosis	X	—	N	3	3	Throughout sliding mass	Hardens the soil by reducing the water content
	Use of explosives	—	X	N	3	N	In the lower part of the slide	Relatively superficial cohesive mass overlying a mass of rock. Fragmented sliding surface. Explosives may also enable water to drain from the sliding mass

[a] 1: Frequent; 2: occasional; 3: rare; N: not considered applicable.

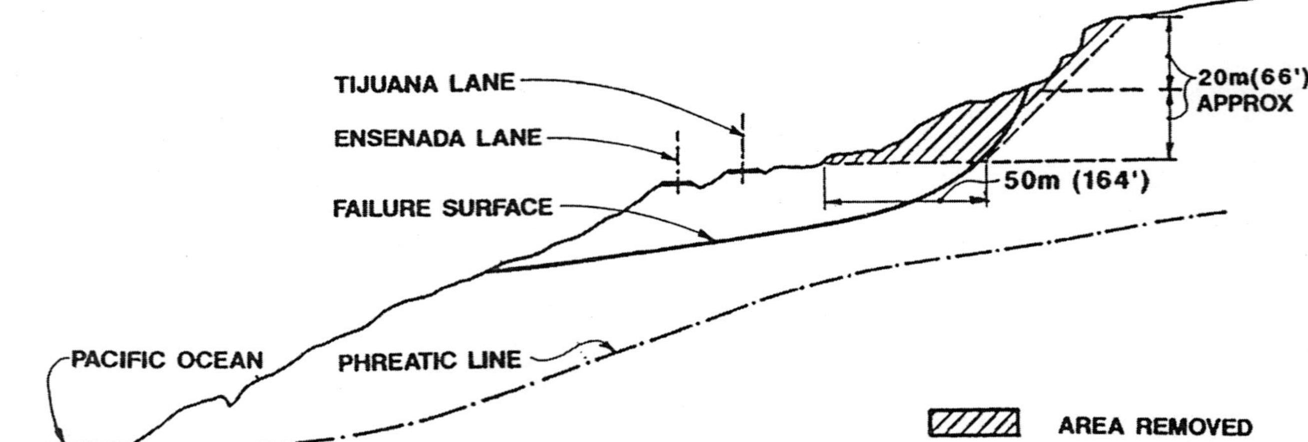

Figure 7.1 Removal of landslide head. (From Rodriguez et al.. 1988, with permission of Trans Tech Publications.)

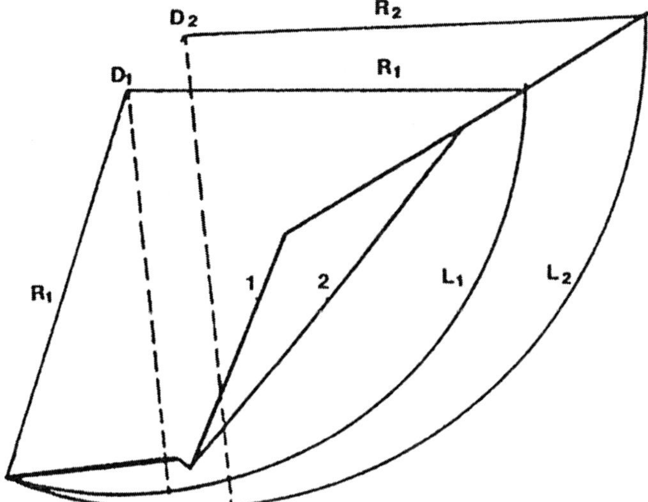

Figure 7.2 Mechanism depicting increase in stability by flattening of cut slope. (From Rodriguez et al., 1988, with permission of Trans Tech Publications.)

considered when stabilizing a slope. As with all methods, it is not universal and its effectiveness can vary from one case to another.

Figure 7.2 illustrates the increase in slope stability that results from flattening a cut slope. Flattening the slope not only reduces the sum of the driving force, but also tends to force the failure surface deeper into the ground. Deeper ground is often firmer because of less weathering, less dissipation of residual stresses through expansion, and higher normal effective pressure if the slide plane has frictional strength characteristics.

The mechanism of flattening an embankment slope is quite different from that of a cut slope (Figure 7.3). Figure 7.3*a* illustrates an embankment where the critical failure circle originally corresponded to a base failure (L_1). By flattening the embankment slope, a new critical circle (L_2) is obtained. In this case, the change lengthens the failure surface and increases the resisting forces because the shear strength is distributed over a wider area. Calculations must be made to assess the extent of improvement gained by flattening the embankment slope.

Figure 7.3*b* illustrates the flattening of an embankment slope where the critical circle passes through the toe of the slope. The new critical circle radius becomes larger, thereby enhancing stability since shearing resistance is proportional to the length of the failure surface.

Figure 7.3*c* illustrates the contributing effect of slope flattening on a translational failure. The slope flattening increases the length of the failure surface within the weak stratum to create more sliding resistance. In a frictional stratum, the weight of fill will cause shear resistance to increase because it is proportional to the normal stress.

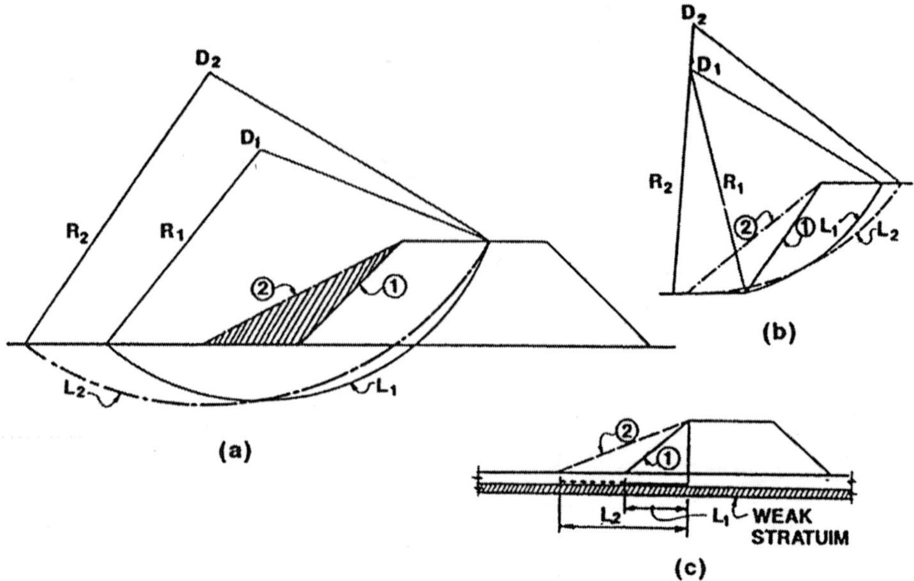

Figure 7.3 Slope flattening in embankment slopes. (From Rodriguez et al., 1988, with permission of Trans Tech Publications.)

Benching of Slopes Figure 7.4*a* and *b* show two typically benched slopes, one in cohesive soils and the other in soils with cohesive and frictional strength. The purpose of benching a slope is to transform the behavior of one high slope into several lower ones. For this reason, the bench should be sufficiently wide. For slopes with cohesive and frictional strengths, the chief objective is to flatten the slope. In sloping ground, benching will result in higher overall slopes and greater excavation quantities overall (Figure 7.4*c*). Benching will reduce subsequent maintenance costs and thereby offset increased construction costs.

Benching of slopes is also used to control erosion and to establish vegetation. The vertical heights of benches are typically about 25 to 30 feet. Each bench should have drainage to convey runoff to a suitable discharge outlet. Typical benching of slopes is shown in Figure 7.5.

7.2.2 Lightweight Fill

In embankment construction, lightweight fill can reduce the driving force of the slope and thereby increase the stability. Lightweight materials, such as slag, encapsulated sawdust, expanded shale, cinders, shredded rubber tires, polystyrene foam, and seashells, have been used successfully. Selection of the type of lightweight material depends on its cost and availability in local areas. For example, wood product waste is available at a reasonable cost in Washington. In one case, sawdust and wood fiber were used to replace soils from a landslide head, as shown in Figure 7.6 (Nelson and

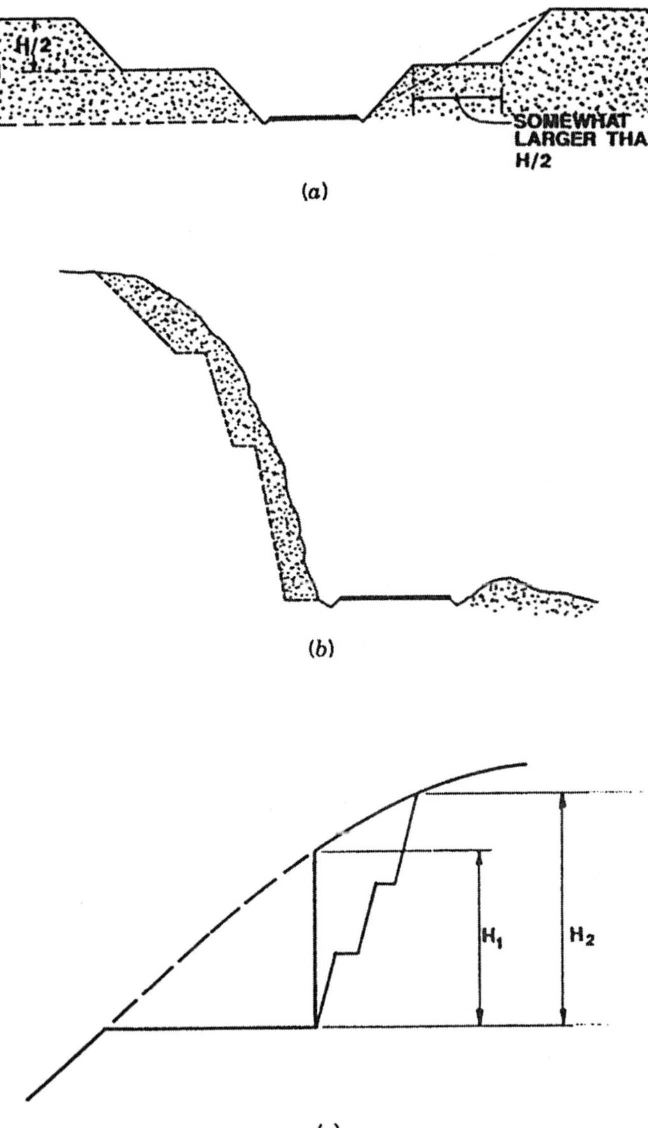

Figure 7.4 Terraces in slopes. (*a*) Terracing in cohesive materials. (*b*) Terracing in competent soils. (*c*) Terracing in sloping ground. (From Rodriguez et al., 1988, with permission of Trans Tech Publications.)

Allen, 1974). Exposed wood fiber and sawdust decay with time. Asphalt encapsulation is commonly applied as a retardant to the decay process.

The use of expanded polystyrene (EPS) for landslide corrections and lightweight embankment fills is growing rapidly. Yeh and Gilmore (1992) report on a case

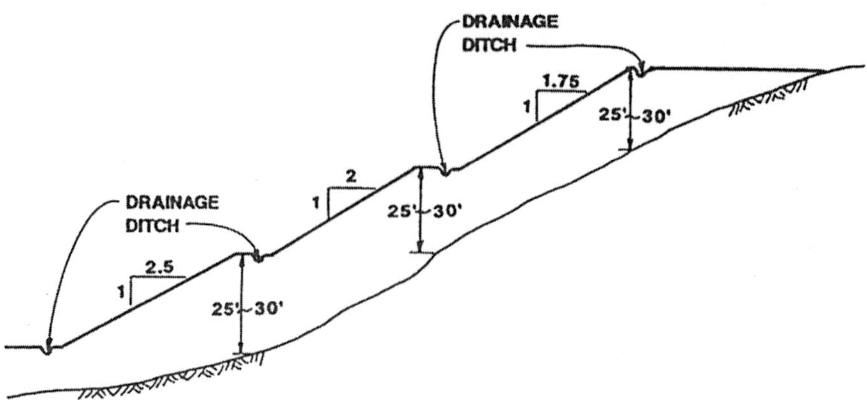

Figure 7.5 Typical benching of slopes.

where EPS was used to correct a landslide along U.S. Highway 160 between Mesa Verde National Park and the town of Durango in southwestern Colorado. The total length of the affected area was around 200 feet and the slide covered an area of about 1 acre. The total amount of slide material involved was around 11,000 cubic yards.

The subsurface conditions and landslide configuration at the U.S. 160 site are shown in Figure 7.7. The slip zone was found to be irregular and the modified Janbu's method of stability analysis was used to back-calculate the soil strength. The assumed soil and rock properties are shown in the figure. It was concluded that a rising water table in the embankment was the primary cause of the slide, and thus an interceptor drain was installed on the uphill side of the facility along the ditch (Figure 7.8). The discharge points were located well beyond the slide area.

Restoration of the embankment consisted of placing about 850 cubic yards of $4 \times 8 \times 2$-foot EPS blocks along 108 feet of facility embankment. The FPS blocks were staggered within each lift and blocks in alternating lifts were placed in a direction perpendicular to those above and below. In order to protect against gasoline spills, reinforced concrete was placed on top of the EPS blocks. To prevent possible floating of the EPS by groundwater, a blanket drain was provided and was connected with the toe drain. All filter drain material was wrapped in nonwoven filter fabric. Ten-inch diameter perforated drain pipe was installed at the bottoms of the trench drains, and 6-inch diameter pipes spaced at 10 feet each were installed in the blanket and interceptor drains.

7.3 BUTTRESSING

Buttressing is a technique used to offset or counter the driving forces of a slope by an externally applied force system that increases the resisting force. Buttresses may

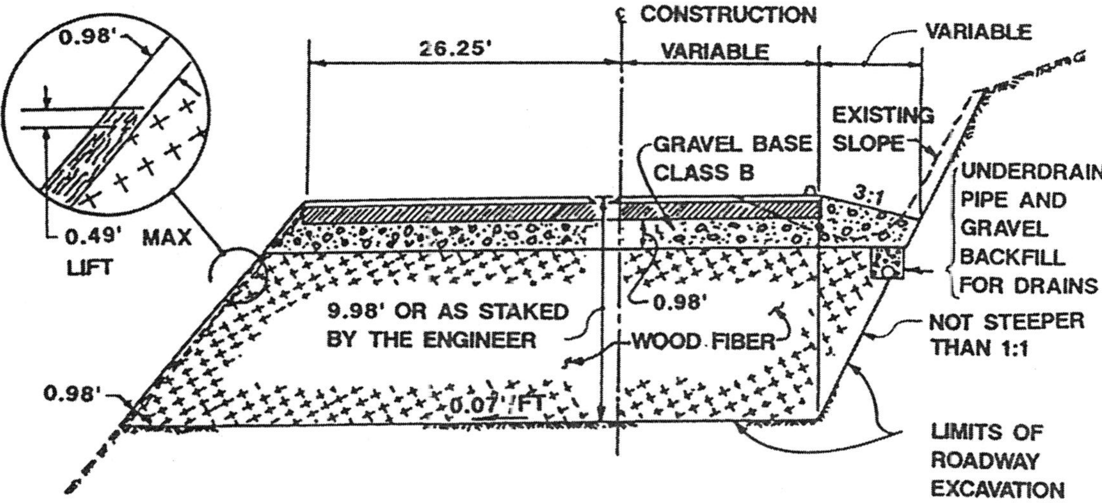

Figure 7.6 Excavation and lightweight fill detail used by Washington Department of Transportation. (Nelson and Allen, 1974.)

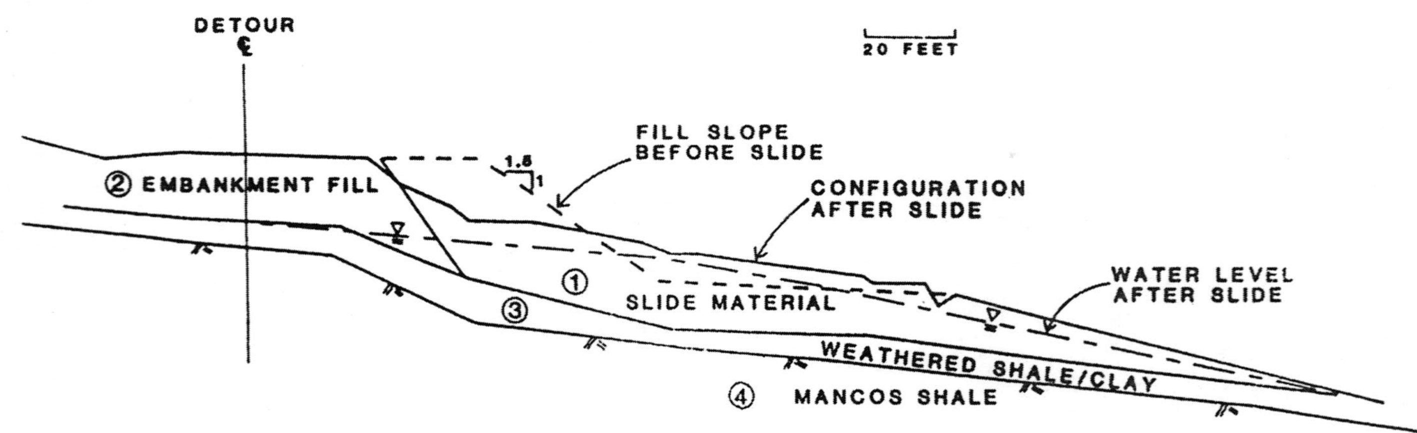

SOIL/ROCK PROPERTIES

SOIL/ROCK TYPE	TOTAL UNIT WEIGHT, PCF	MOISTURE %	L.L.	P.I.	C PSF	Ø
①	115	WET	-	-	50	15
②	115	16	35	14	500	20
③	120	14.1	36	15	1000	0
④	115	-	-	-	9000	0

Figure 7.7 Geometry and cross section through slide area. (From Yeh and Gilmore, 1992, with permission of ASCE.)

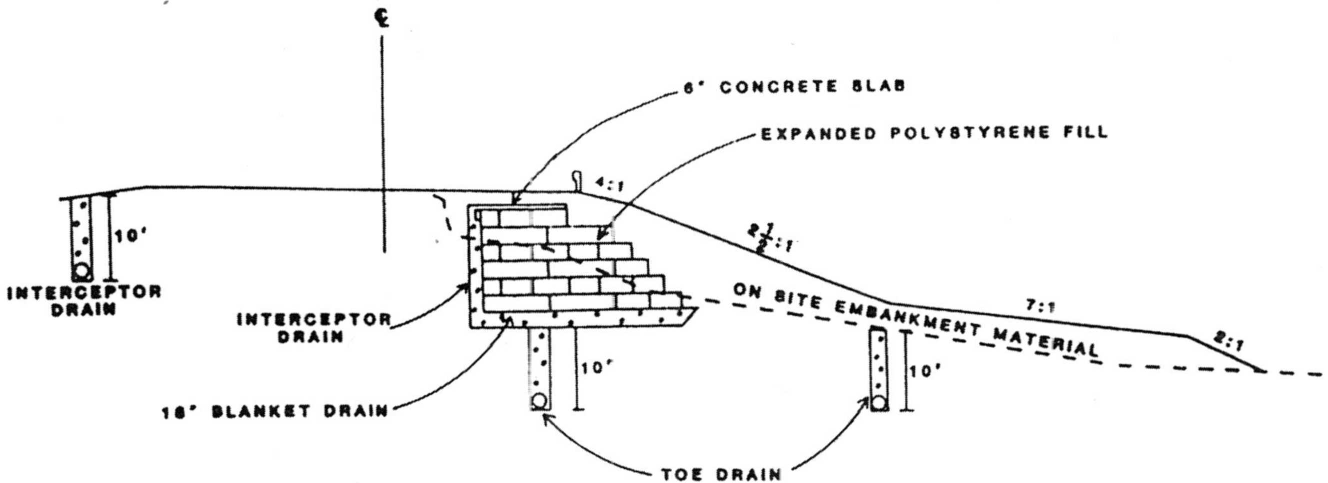

Figure 7.8 Designed embankment and drains. (From Yeh and Gilmore, 1992, with permission of ASCE.)

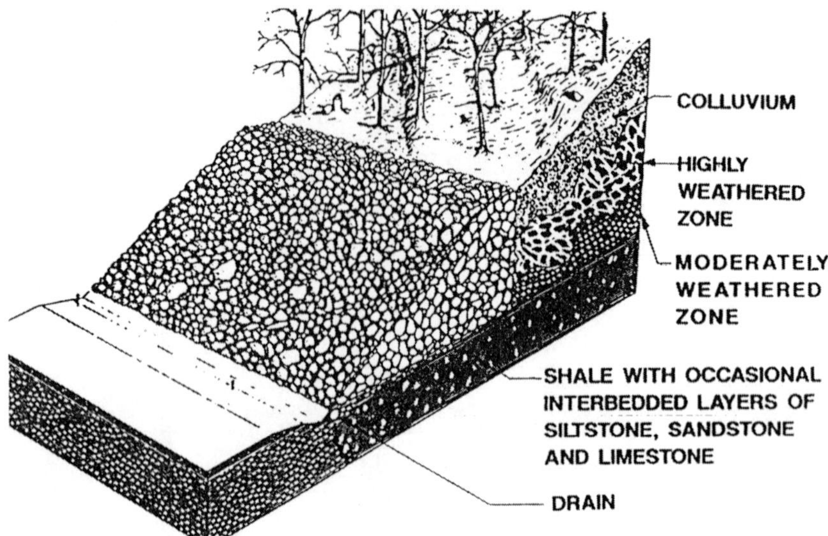

Figure 7.9 Rock buttress used to control unstable slope. (Schuster and Krizek, 1978.)

consist of

(1) Soil and rock fill
(2) Counterberms
(3) Shear keys
(4) Mechanically stabilized embankments (MSE)
(5) Pneusol (tiresoil)

7.3.1 Soil and Rock Fill

Soil and rock fill is used to provide sufficient dead weight near the toe of an unstable slope to prevent movement (Figure 7.9). Where resources are available and where soil and rock fill can be found locally, this method is the most practical way to arrest further movement of an unstable slope.

7.3.2 Counterberms

A counterberm is used to provide weight at the toe of a slope and to increase the shear strength below the toe. This is particularly useful for embankments over soft soils where the ground at the toe can move upward and form a bulge. By locating a counterberm where the upheaval is expected to occur, the resistance against sliding is also increased. The counterberm must be carefully designed in order to utilize the weight most effectively and to assure that it is stable itself. Unless careful investigation and thorough analysis are made, there is a danger that the additional load imposed by

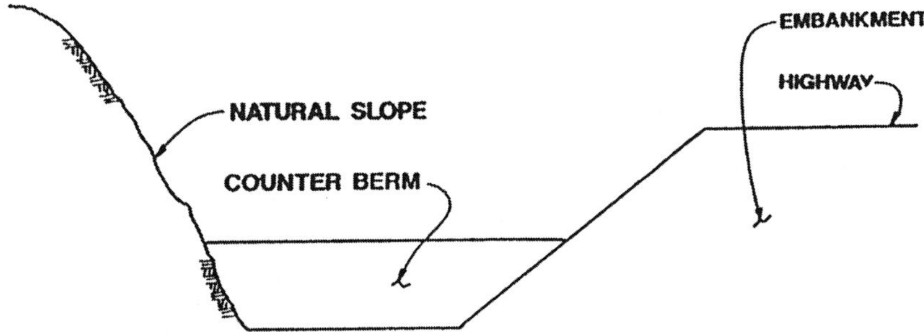

Figure 7.10 Counterberm to provide weight at toe of embankment.

the counterberm may increase the driving force rather than provide added resistance against sliding. The counterberm is safest if it extends between an embankment and a natural bank or hill, as shown in Figure 7.10. Used to stabilize embankments, counterberms increase the length and depth of the potential failure surfaces and increase the resisting moments (Figure 7.11).

7.3.3 Shear Keys

Shear keys are used sometimes to provide additional sliding resistance for a counterberm or rocky/soil buttress. The main purpose of a shear key is to force the critical slip circle deeper into a stronger underlying formation, thereby increasing the resistance along the slip surface (Figure 7.12). This method becomes very practical and cost-effective if the stronger formation is only a few feet below the overlying soft soils. Construction of a shear key requires excavation of a trench at the toe of the slope. In such circumstances, care must be exercised not to further undermine the toe of the slope. This can be done by a careful design of a bracing system for the trench and by excavating only small sections at one time.

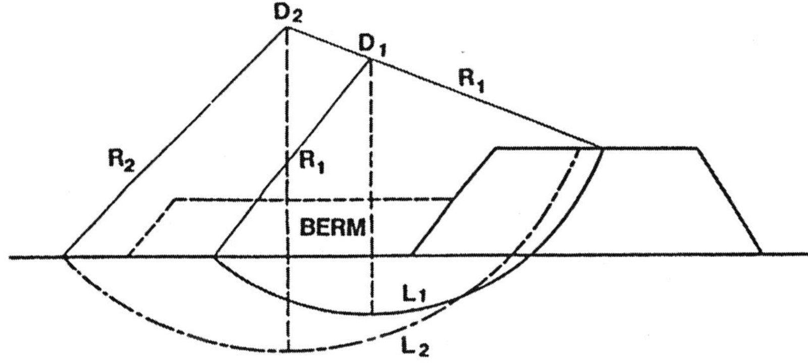

Figure 7.11 Effect of counterberm. (From Rodriguez et al., 1988, with permission of Trans Tech Publications.)

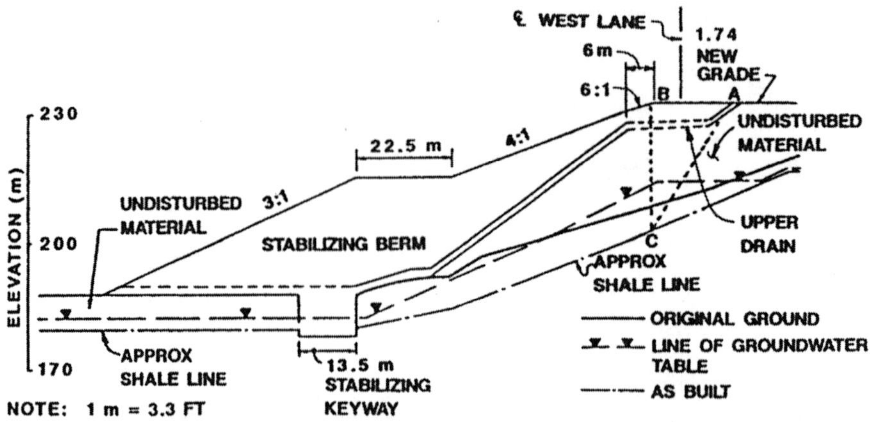

Figure 7.12 Stabilizing effect of a shear key. (Schuster and Krizek, 1978.)

In a somewhat related but different application of shear keys, Sills and Fleming (1992) report on the use of stone-fill trenches to stabilize a levee along the Ouachita River at Rilla, Louisiana, a location 9 miles downstream of Monroe, Louisiana. Prior to 1983, the levee had been set back because of bank caving problems. In the summer of 1983, scouring had caused an instability in the upper clay slope. During low river stages, the slide scarp had propagated into the riverside toe of the previous levee setback. At this time, further setback of the levee was not considered feasible because of existing houses and structures located landside of the levee toe. The recommended repair consisted of a large stone toe dike and a sand berm. The stone dike was intended to serve as a buttress of the sand fill as well as to provide scour protection. Because of problems during construction, such as extended high water and an extremely compressed work area, the slide was not properly stabilized and continued to move slowly. By the summer of 1988, the slide movement had reached a point where it was considered a danger to the flood control system and needed to be repaired prior to the next high water in the spring.

A typical section of the slide area is shown in Figure 7.13. Conclusions from the studies that were subsequently carried out indicated that the initial triggering mechanism of the slide was the steepening of the river bank due to erosion of the silty sand substratum. This in turn caused sliding to occur in the backswamp clays.

A sliding wedge analysis was performed with an assumed slip surface at the base of the clay layer and a phreatic profile assumed to be 5 feet below the ground surface. For a factor of safety (FOS) equal to one, the drained friction angle and cohesion values were back-calculated to be 16° and 0, respectively. It was also determined that an average drained friction angle of 20° would be required for the desired 1.25 FOS. By using a weighted average technique, stone-fill trenches had to be 2.5 feet wide and 15 feet on center to obtain the desired average friction angle (Figure 7.14). The trenches were excavated to the base of the clay layer. This method is also similar to the use of stone columns discussed in Section 7.5.2.

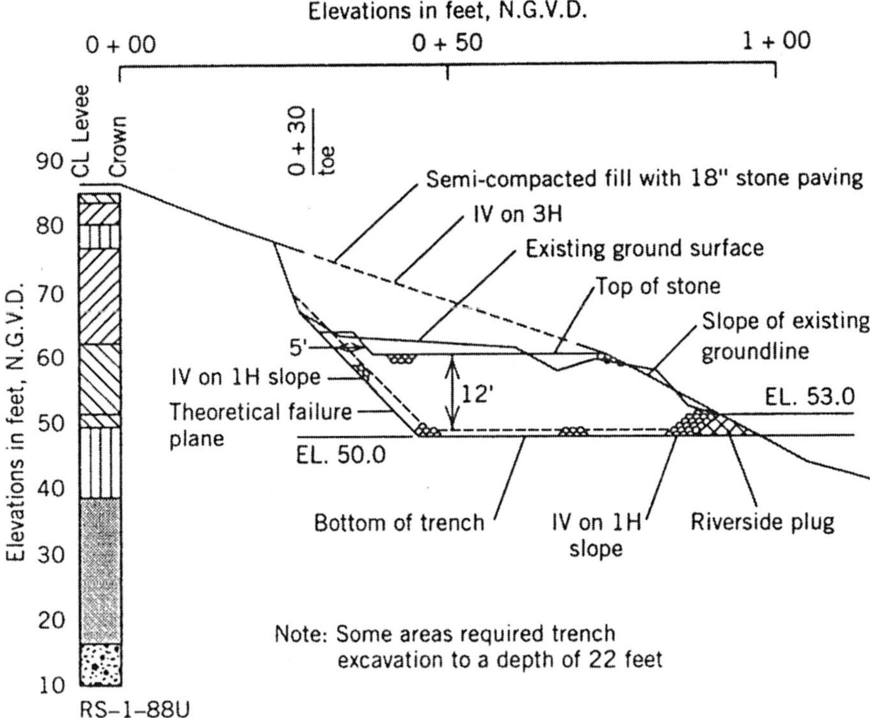

Figure 7.13 Typical section of slide area. (From Sills and Fleming, 1992, reproduced by permission of ASCE.)

7.3.4 Mechanically Stabilized Embankments

Mechanically stabilized embankments (MSE) involve the designed use of backfill soil and thin metallic strips, mesh, or geosynthetic reinforcement mesh to form a gravity mass capable of supporting or restraining large imposed loads (Figure 7.15). The MSE slope face is either vertical or inclined, and the backfill material is typically confined behind metal, reinforced concrete, or shotcrete facing. The mesh or geosynthetic is sometimes wrapped around the soil at the face between reinforcement layers.

MSE slopes must be designed for internal and external stability. Internal stability requires that the reinforced soil structure be coherent and self-supporting under the action of its own weight and any externally applied forces. The reinforcement must be sized and spaced so that it does not fail in tension under the stresses that are applied, and does not pull out of the soil mass. For external stability, an MSE slope must satisfy the same external design criteria as a conventional retaining wall. That is, it must resist forces that can cause overturning, sliding at or below the base, and global instability (i.e., an unsafe failure surface around the entire embankment).

Many numerical analyses, laboratory modeling, and full-scale tests have been conducted to better understand the mechanisms and behavior of MSE slopes. Several

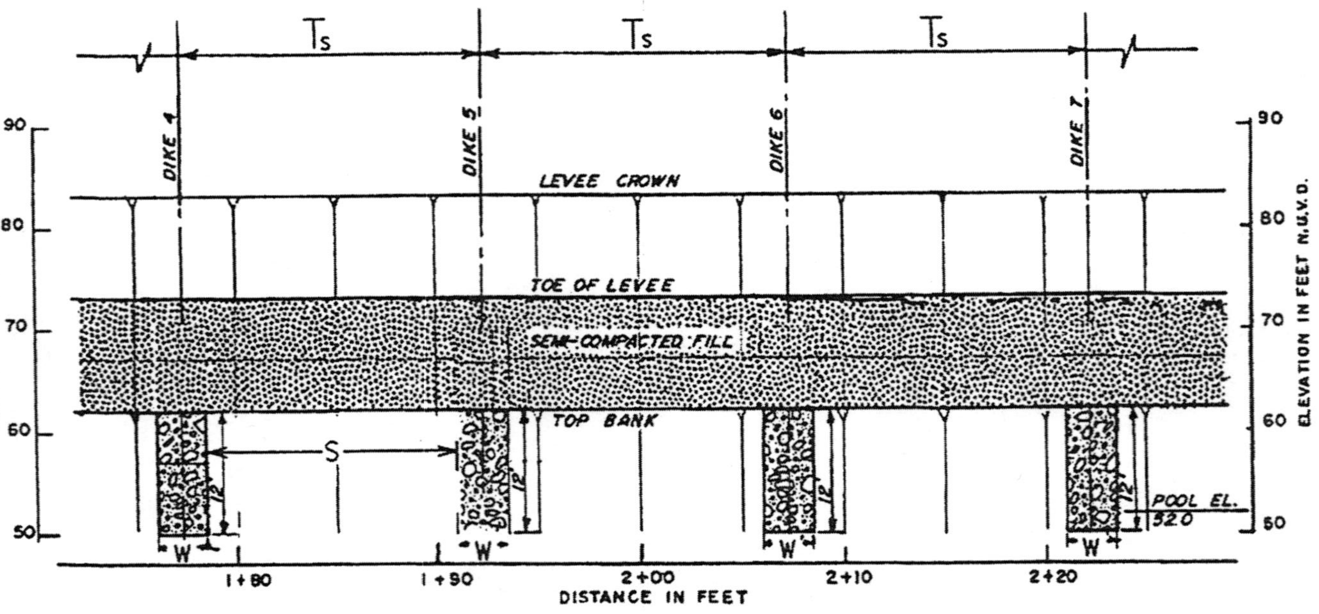

FOR RILLA

T_s = 15 FEET

S = 12.5 FEET

W = 2.5 FEET

Figure 7.14 Typical design section.

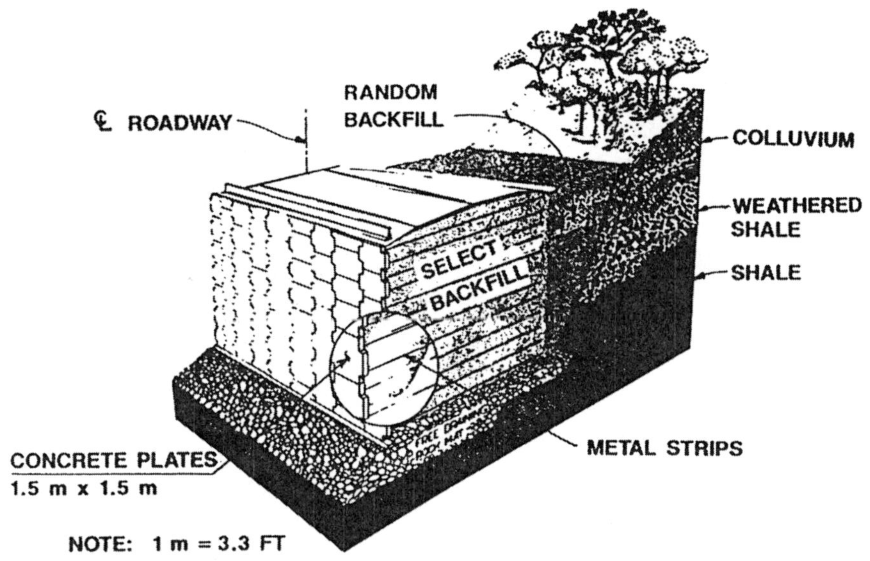

Figure 7.15 MSE used for stabilization. (Schuster and Krizek, 1978.)

different design approaches have been developed; some are based on analysis under failure conditions while others use a working stress approach. The details of these different design approaches can be found in technical papers by Ingold (1982), Juran (1977), Schlosser (1983), Schlosser and Segrestin (1979), Shen et al. (1981), and Stocker et al. (1979).

Granular backfill is commonly used in MSE slopes. Caution should be exercised when using clayey, highly micaceous, and residual soils. This is because significant reduction in pull-out capacity (up to 70 percent reduction) of these soils may occur when saturated (Elias and Swanson, 1983).

MSE walls are often designed by manufacturers or, as a minimum, with guidance given by the manufacturers. For conceptual design, Sowers (1979) illustrated one basic model of MSE systems utilizing galvanized steel strip reinforcement, as shown in Figure 7.16. The design of this type of system requires that the skin resist the soil pressure from the backfill layer between strips, that the strip length, L, be great enough to support the skin and provide a stable mass, and that the strip be strong enough to resist the tension in it.

At a depth of z below the wall top, the force, P, against the skin element defined by S and ΔH, with an earth pressure coefficient of K, is defined as

$$P = K\gamma z S \,\Delta H \qquad \text{(Eq. 7-1a)}$$

The friction force, F, developed by the top and bottom faces of the metal strip of width W, length L, and angle of friction with the backfill of δ is

$$F = 2LW\gamma z \tan\delta \qquad \text{(Eq. 7-1b)}$$

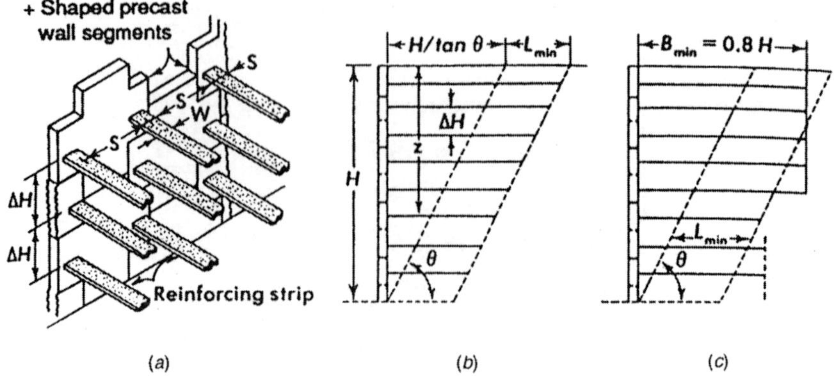

Figure 7.16 Reinforced earth wall. (*a*) Reinforcing strip width and spacing behind segmented precast wall. (*b*) Minimum reinforcing length, L_{min}. (*c*) Reinforcement for overall stability, $L_{\text{avg}} > L_{\text{min}}$. (Reprinted with the permission of Simon and Schuster from the Macmillan College text *Introductory Soil Mechanics and Foundations*, 4th ed., by George F. Sowers. Copyright © 1979 by Macmillan College Publishing Company, Inc.)

The required length, L_{min}, is found by multiplying P by an appropriate safety factor, F_s (usually 1.5 or 2), and equating the expressions

$$F_s P = F$$

$$F_s K \gamma z S \, \Delta H = 2LW\gamma z \tan \delta$$

$$L_{\text{min}} = \frac{F_s K S \, \Delta H}{2W \tan \delta} \qquad \text{(Eq. 7-1c)}$$

Typical dimensions are: for S, 2 feet; for H, 10 to 12 inches; and for W, 3 inches. The length, L, is considered to be measured beyond the zone of Rankine failure. For overall stability, a top width $B = 0.8H$ has been expedient. The uppermost strips may be shorter than L_{min}, but as long as the average L exceeds L_{min}, the wall should be stable. Sometimes the length of the lower strips is made less than $0.8H$, but exceeding L_{min}.

When backfilled with compacted sand, experience has shown that movement is sufficient that $K = K_A$. For other backfills, the data are insufficient; for clays K probably approaches K_0 (although clay backfills are discouraged).

7.3.5 Pneusol (Tiresoil)

Pneusol (tiresoil) uses old automobile tires as inclusions in the soil mass (Figure 7.17) instead of the metal or nonmetal reinforcement in MSE walls. This technique was used to repair a sidehill fill failure along California State Highway 236, north of Santa Cruz (Forsyth and Egan, 1976). After removing the slide debris to below the slip surface, a drainage system was installed and the embankment was rebuilt using tire sidewall mat reinforcement at 2-foot vertical intervals. The individual sidewall

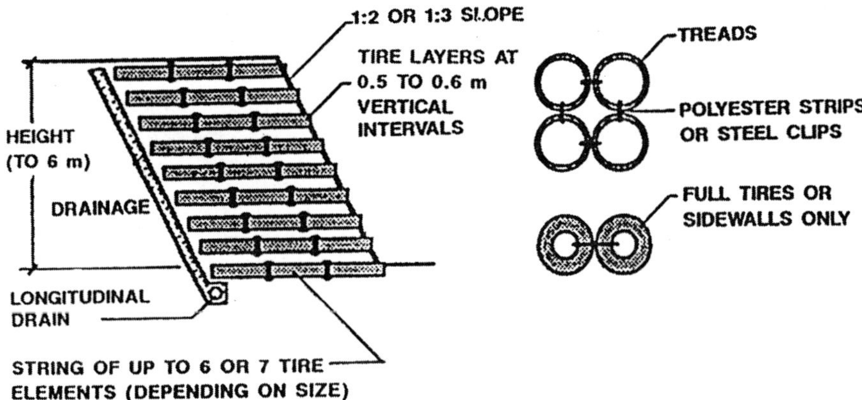

TYPICAL CROSS SECTION (e.g. USING TRUCK TIRES OR TREADS)

PLAN VIEW OF POSSIBLE TIRE, SIDEWALL OR TREAD ARRANGEMENT

Figure 7.17 Pneusol (tiresoil). (From Hausmann, 1992, reproduced by permission of ASCE.)

mats were interconnected by steel clips and lapped over the edge of the embankment, thus providing erosion protection. The reinforced embankment stands with a side slope of 0.5H–1V instead of the conventional 1.5H–1V, thus saving a total of about 92,000 cubic yards of fill (Hausmann, 1992).

This technique is now used widely in France. According to Long (1990), French engineers have built more than 200 structures with tire reinforcement in France and 12 in Algeria. Research on reinforcement in the form of whole tires, sidewalls, or treads placed on edge or flattened has been carried out by the Laboratoire Central des Ponts et Chaussees in France. Some of the major findings were given by Long (1990) and are summarized by Hausmann (1992) as follows:

(1) Pull-out resistance varied from 4.4 kips for a single sidewall mat to 15.3 kips for a group of treads.

(2) Unit weight of trial fills incorporating tires varied from 38 to 50 pounds per cubic foot.

(3) Stress–strain modulus of the reinforced soil is less, but the shear strength is higher than that of an equivalent amount of unreinforced soil.

(4) A variety of connectors can be used including synthetic ropes, straps, or metal hooks.

Design of pneusol is similar to that of MSE slopes where internal, external, and overall stability of the pneusol should be satisfied.

7.4 DRAINAGE

Of all stabilization techniques considered for the correction or prevention of landslides, proper water drainage is the most important. Drainage reduces the destabilizing hydrostatic and seepage forces on a slope as well as the risk of erosion and piping. Various drainage techniques are discussed below.

7.4.1 Surface Drainage

Carefully planned surface drainage is essential for treatment of any slide or potential slide. Every effort should be made to ensure that surface runoff is carried away from and not seeping downwards into the slope. Such considerations should always be made and are extremely important when evaluating a failure. Temporary remedial measures usually considered after a landslide include

(1) Using sandbags to divert water runoff away from the failure zone
(2) Sealing cracks with surface coatings such as shotcrete, lean concrete, or bitumen to reduce water infiltration
(3) Covering the ground surface temporarily with plastic sheets or the like to reduce the risk of movement during construction

Surface runoff usually is collected in permanent facilities such as V- or U-shaped concrete lined or semicircular corrugated steel pipe channels and diverted away from the slide mass. These channels should be placed strategically at the head of the slope and along berms. The detailing of surface water collection systems should provide for minimum maintenance and displacement due to future slide movement.

Catchment Parameters Catchment parameters to be considered in drainage systems design include

(1) Area and shape of the catchment zone
(2) Rainfall intensity
(3) Steepness and length of the slope being drained
(4) Condition of the ground surface and nature of the subsurface soils
(5) Nature and extent of vegetation

These parameters are site-specific and cannot be generalized.

Redirection of Surface Runoff When surface runoff is found to be the cause of a landslide or a potentially unstable zone, it should be redirected to ensure that the stability of the slope is not further worsened. Redirection of surface runoff commonly is the first response to a rainfall-induced failure. The design of the remedial drainage system should consider natural drainage patterns.

7.4.2 Subsurface Drainage

The FOS against failure on any potential slip surface that passes below the phreatic surface can be improved by subsurface drainage. Methods that can be used to accomplish subsurface drainage are

(1) Drain blankets
(2) Trenches
(3) Cut-off drains
(4) Horizontal drains
(5) Relief drains
(6) Drainage tunnels

It is far more cost-efficient to incorporate these methods into initial design and construction than to need them as remedial measures during or following construction.

As compared to engineered embankment slopes, natural slopes are rarely homogeneous enough to allow reliable subsurface drainage design according to simple principles of dewatering (Xanthakos et al., 1994). For a successful dewatering system, the designer must have a good understanding of geological structure and choose a drainage system layout that increases the probability of intersecting the major water-bearing layers (Hausmann, 1992).

Monitoring is important for any subsurface drainage program. Not only should piezometers be installed to measure the preconstruction pore pressure, but they should also be monitored during and after construction to observe the effects of the subsurface drainage systems. In the long run, piezometric readings can indicate reduction of drainage efficiency caused by siltation, deterioration of seals, or breakdown of pumps.

The volume of water flowing out of a drain in a water-bearing zone is determined by formation permeability and hydraulic gradient. When a drain is installed, the groundwater level will be lowered, thereby reducing the head of water and hydraulic gradient. The seepage will gradually reduce from its initial value to a steady-state value. This flow reduction is not necessarily an indication of drain deterioration. Abundant flow out of the drains installed in clayey materials with low permeabilities should not be expected. They can be operating successfully at very low flow rates.

Common slope seepage conditions and potential effects on stability are listed in Table 7.2.

TABLE 7.2 Common Slope Seepage Conditions and Effects on Stability

Slope Seepage Conditions	Effects on Stability
Naturally dry or well-drained	Favorable with little seepage; serves purpose of subdrainage
Subjected to a normal flow that is uncontrolled because of rainfall	Stability reduced as excess pore pressures are produced; flow is generally parallel to slope
Subjected to vertical downward flow, encouraged by subdrainage systems	Vertical downward flow reduces pore pressures and increases stability

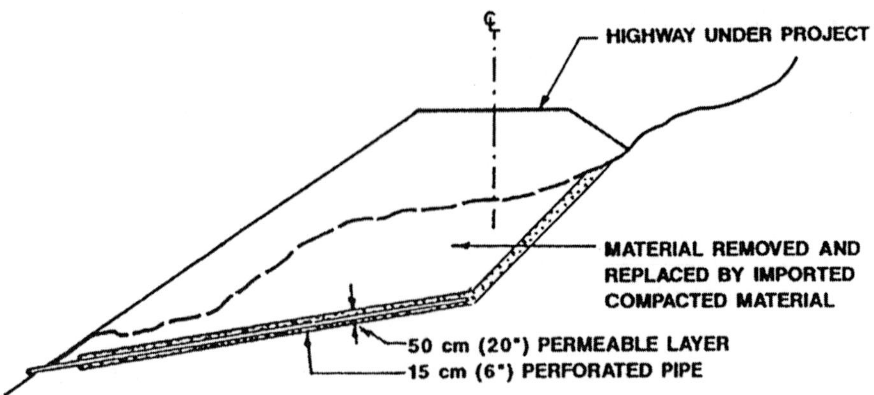

Figure 7.18 Placement of a drain blanket underneath embankments subsequent to removal of poor material. (From Rodriguez et al., 1988, with permission of Trans Tech Publications.)

Subsurface Drainage Blankets When there is a thin layer (not more than 10 feet) of poor-quality saturated soil at a shallow depth (not deeper than 15 feet beneath a proposed embankment), and when there are materials of better quality below that layer, it may be practical to remove the poor quality layer and replace it with a well-draining soil fill (Figure 7.18). The bottom of the excavation should be covered with a layer of filter fabric wrapping a 6- to 24-inch-thick filter stone layer with a perforated pipe embedded in it to capture flow. To avoid blockage of holes by vegetation, the first 5 feet of the outlet end of the perforated pipe should not be perforated. To minimize surface erosion, a drainage ditch should be installed to convey water flow from the outlet of the pipe to a suitable discharge point.

Trenches Deep trenches should be constructed when subsurface water or soils of questionable strength are found at such great depths that stripping of the soils as discussed above is not practically feasible. Trenches usually are excavated at the steepest stable side slopes for the construction period. Shoring may be required. Any trench so excavated should extend below the water-bearing layer. The trench should be backfilled with a layer of pervious material encased in filter fabric that has an underdrain pipe running through it (Figure 7.19).

The number of trenches needed depends on the hydrogeology and geomorphology of the site. If the slope is in a natural depression of limited aerial extent, one trench normal to the centerline of the site may be sufficient. In the case of large areas, an extensive system of trenches, often in the form of a herringbone pattern, may be necessary.

Cut-Off Drains At a site where shallow groundwater is encountered, cut-off drains can be used to intercept the groundwater flow. A typical layout is shown in Figure 7.20.

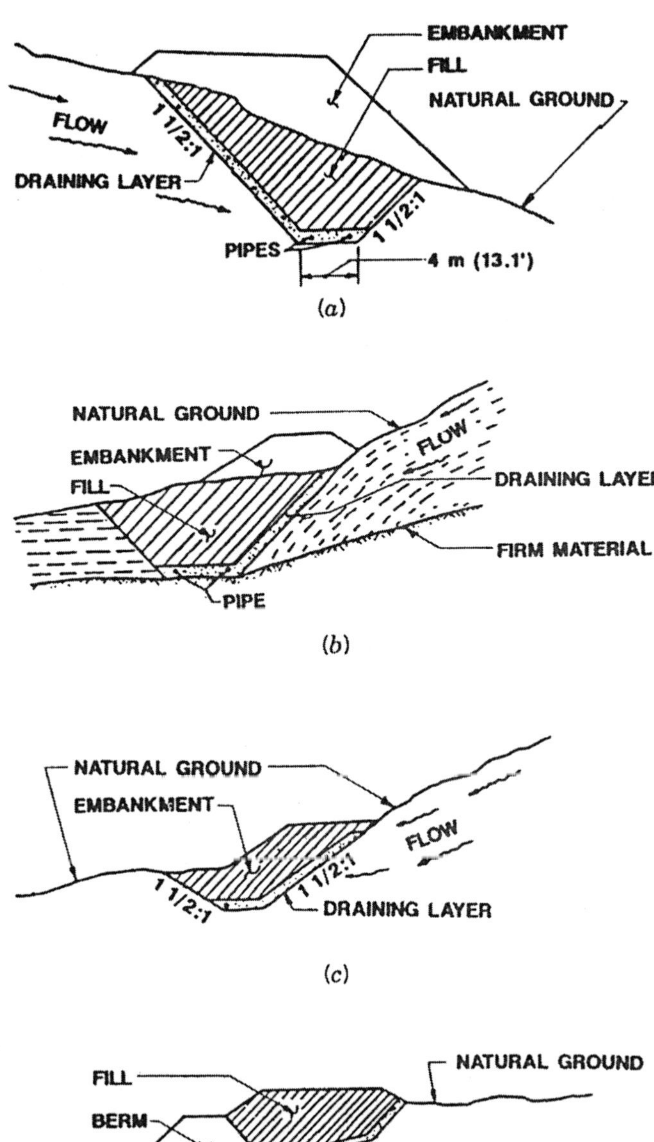

Figure 7.19 Different types of stabilizing trenches. (*a*) Trench under the embankment. (*b*) Trench going down to a firm stratum, providing both drainage and support. (*c*) Trench within the embankment. (*d*) Trench with lateral berm, showing a transverse discharge pipe. (From Rodriguez et al., 1988, with permission of Trans Tech Publications.)

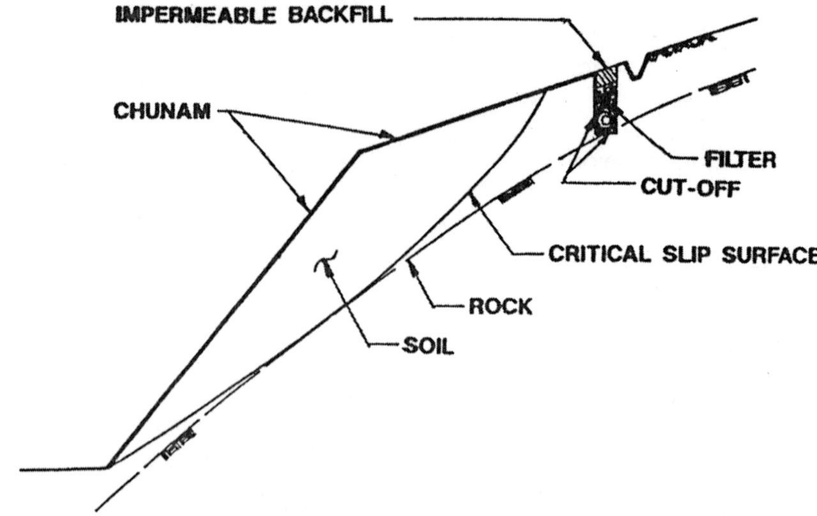

Figure 7.20 Cut-off drain. (Geotechnical Control Office, 1979.)

An impermeable zone or membrane is used as a cut-off downslope of the drain, and the top zone of the trench is backfilled with impermeable material. Runoff from the upper slopes should be collected in drainage channels. The free draining material used to backfill the trenches should be designed to conform with standard filter criteria. The size of perforations in pipes should be compatible with the grain size of the backfill filter material.

Horizontal Drains Horizontal drains can be used where the depth to subsurface groundwater is so great that the cost of stripping or placing drainage trenches is prohibitive. Horizontal drains should be designed specifically to lower the seepage pressures in slopes and to prevent failure. A horizontal drain is a small-diameter (say 3- to 4-inch diameter) hole drilled at a 5 to 10 percent grade and fitted with a 2- to $2\frac{1}{2}$-inch-diameter perforated pipe wrapped in filter fabric (Figure 7.21). Water captured by the drains is often discharged on benches sloped for drainage and the toe of the slope equipped with underdrains. The first 5 feet of drain pipe immediately next to the outlet should not be perforated to avoid the invasion of vegetation and subsequent obstruction.

The length of horizontal drains largely depends on the geometry of the zone in which they are to be installed. They can be made as long as about 300 feet. In general, the length required can be determined by drawing a cross section of the slope with its probable critical circle superimposed on a geologic cross section depicting aquifers. The drains must be installed in such a way as to give thorough protection to any zone that is likely to slide. As for spacing, 5 to 15 feet is needed for low-permeability ground. The effect of any one drain may be relatively small (Figure 7.22).

Installing horizontal drains is difficult in fine silty sands and soils that contain boulders, rock fragments, open cracks, and cavities. Silty sand tends to collapse and form cavities during drilling, as the initial hole is usually not cased for economic reasons.

Figure 7.21 Horizontal drain details.

In the case of bouldery formations, difficulties caused by hardness and heterogeneity must be anticipated.

The success of a horizontal drain system is not necessarily measured by the quantity of water collected by the drains. A very permeable water-bearing stratum with free water may be intercepted by the drains, in which case the volume of flow drained may be impressive. Conversely, drains may be installed in low-permeability clayey formations where they can very efficiently lower pore pressures and greatly contribute to increased stability. The quantities of water they collect may be small. Also, the amount of water collected by drains may vary seasonally.

Horizontal drains can effectively drain materials of lower permeability by using vacuum systems. Vacuum-augmented horizontal drains call increase hydraulic gradient because the drain exhausts to a negative a atmospheric pressure. Vacuum drainage can also redirect seepage forces so that they act perpendicular to the failure surface. Consequently, the normal force acting on the failure surface is increased, with the result that the shear strength of the material is increased. Vacuum drainage may be employed in landslides, rock slopes, waste embankments, and tailings dams (Brawner and Pakalnis, 1982). The maximum practical depth of lowering a groundwater table with one stage of horizontal drains is about 15 feet. Permanent dewatering systems should be avoided at all costs because of the relatively high operation and maintenance costs.

It must be noted that the efficiency of horizontal drains may decrease with time as soil fines and other debris plug the pores of the drains. Thus they should be installed in such a position that they can be cleaned and flushed by pumping water into the drains. Special equipment with wire brushes and water jets mounted on tractors generally is available for this work. The performance of horizontal drains can be checked by

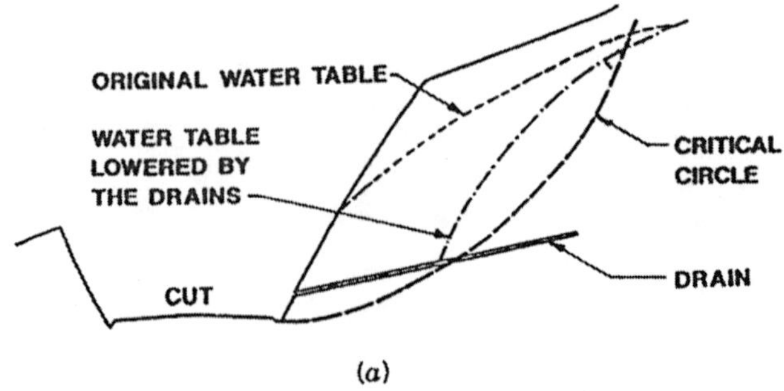

(*a*)

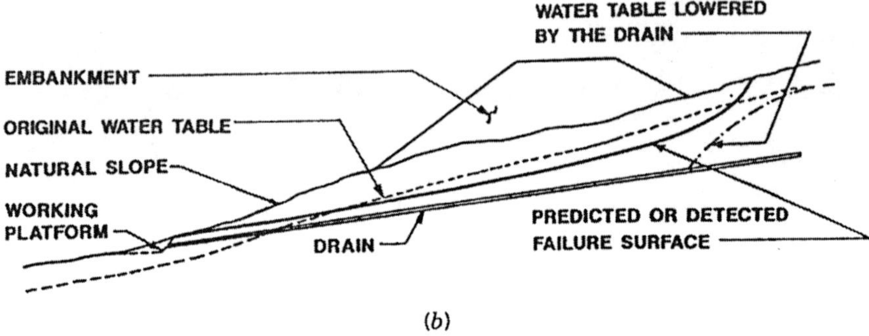

(*b*)

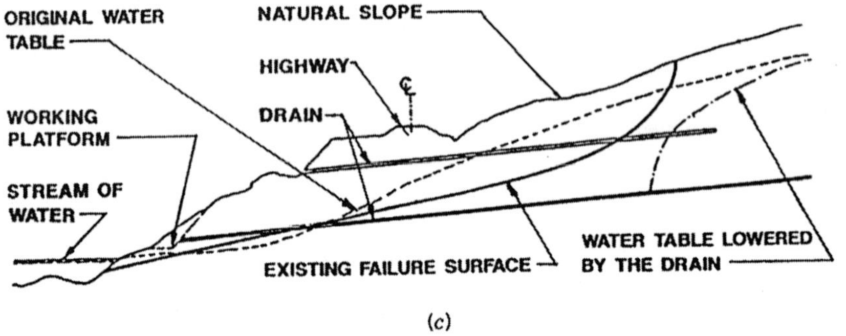

(*c*)

Figure 7.22 Horizontal drains to lower groundwater table in natural slope. (*a*) Cut. (*b*) Embankment on a natural slope. (*c*) Correction of an existing failure. (From Rodriguez et al., 1988, with permission of Trans Tech Publications.)

recording variations in the height of water in the observation wells strategically placed throughout the drained zone.

Roth et al. (1992) report on the use of 32 horizontal drains to enhance the stability of the 780,000 cubic yard Via de Las Olas landslide in Pacific Palisades, California. A plot plan of the slide area is shown in Figure 7.23. A typical cross section is presented

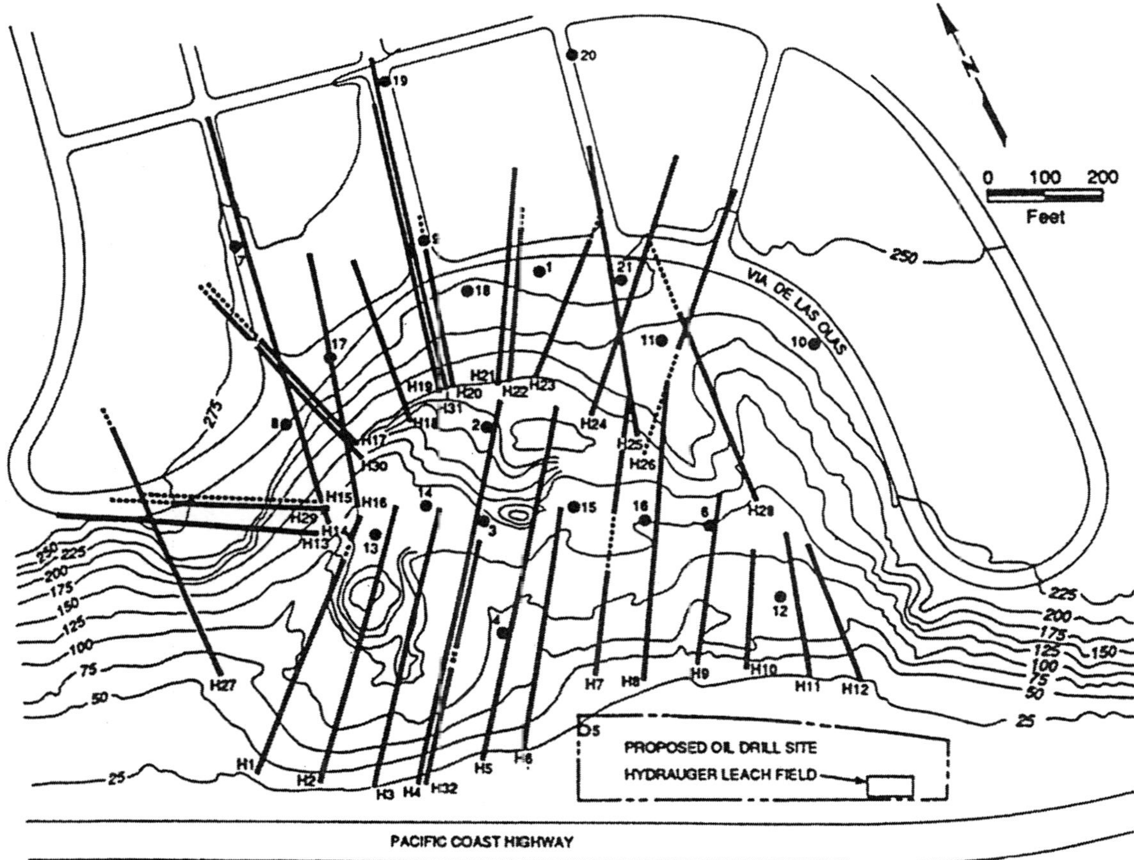

Figure 7.23 Plan view with boring locations and hydraugers. The dotted lines denote sand and gravel zones encountered by the hydraugers. (From Roth et al., 1992, reproduced by permission of ASCE.)

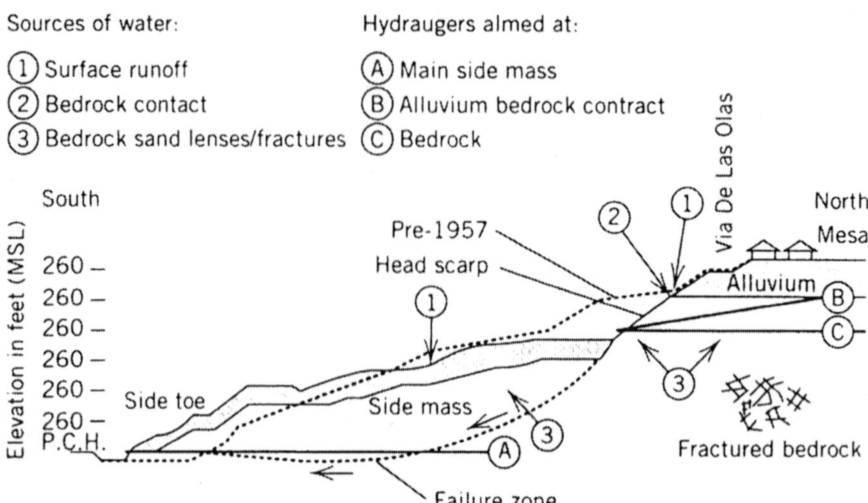

Figure 7.24 Cross section with hydraugers. (From Roth et al., 1992, reproduced by permission of ASCE.)

in Figure 7.24. The slide was about 800 feet square in plan dimensions with a height of approximately 240 feet from the toe to the top of the head scarp. The main head scarp ranged in height from 50 feet in its eastern part to as much as 130 feet to the west. Reported slope movements at the site date back to 1890, and intermittent slides have been noted since the 1920s when residential development of the mesa began. The last major failure occurred in I958, when approximately 780,000 cubic yards of slide debris forced the relocation of the Pacific Coast Highway 200 feet toward the ocean.

Slope stability analyses indicated that water was the primary factor affecting the slide mass. The increase in unit weight and loss in strength experienced by the fractured shale as a result of water saturation made this area particularly susceptible to landsliding. The main slide mass was deemed to be safe from future landsliding because it had assumed a more stable configuration after the 1958 event by shifting weight from the head of the slide mass to the toe. However, the steep head scarp was evaluated and found to be only marginally stable. In addition to possible future damage of adjacent homes on the upper mesa, a slope failure in the scarp area was seen as a potential threat to the stability of the main slide mass. It was therefore recommended to enhance the stability of both the main slide mass and the head scarp through the installation of open surface drains and a subsurface drainage system consisting of hydraugers.

The hydraugers consist of 1.5-inch-diameter slotted PVC pipe inserted in approximately 3-inch-diameter uncased holes. The holes were drilled through a 20-foot long, 5-inch diameter steel casing grouted in place prior to drilling the deep 3-inch holes. Thirty-two hydraugers were installed, varying in length from 250 to 790 feet, at inclinations of 2° to 15° from the horizontal. The total length of installed drainage pipe was about 14,500 feet. Observation wells and pore pressure gages indicated

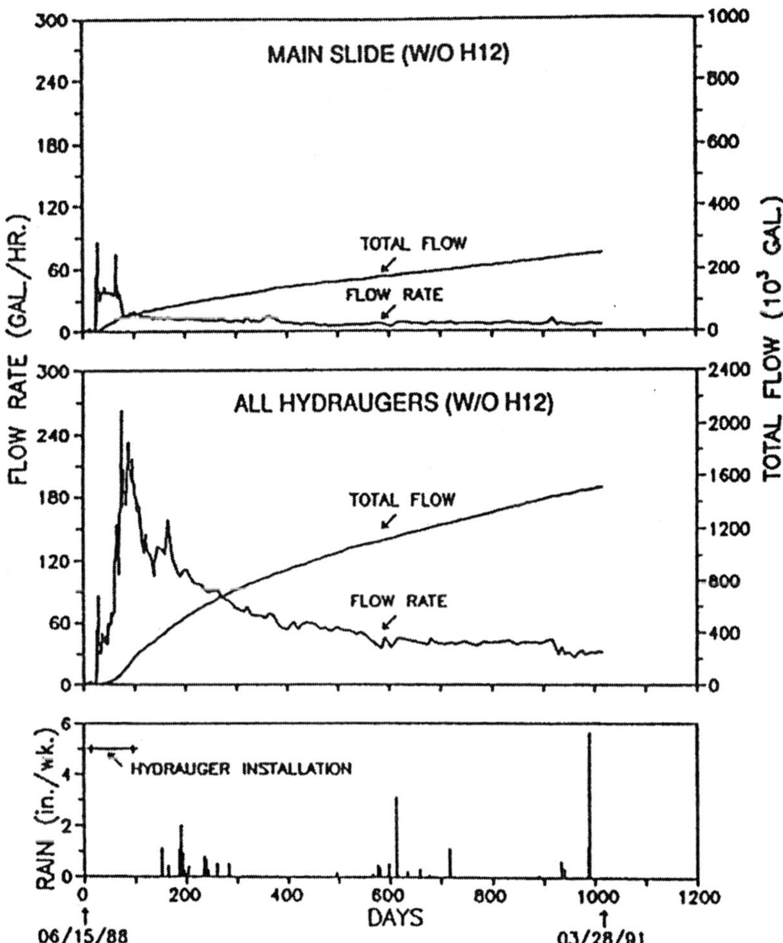

Figure 7.25 Hydrauger flow rates, main slide and total. (From Roth et al., 1992, reproduced by permission of ASCE.)

that the horizontal drains accomplished the purposes for which they were designed (Figure 7.25).

Relief Wells The principal function of relief wells is to lower the water pressures in layers that are deep down in the subsoil, layers that cannot be reached by open excavation methods or horizontal drains because of cost or construction difficulties. Relief wells are vertical holes with a diameter of about 16 to 24 inches. A perforated pipe with a 4- to 8-inch diameter is placed inside the hole. The annular space between the borehole and the pipe should be filled with filter material. A water disposal system using a submersible pump or surface pumping and discharge channels is required to dispose of the water from the wells. Disposal of the water may be very costly, and an effective dewatering system will require frequent maintenance. Alternatively, horizontal drains can be used to tap the relief wells for the disposal of water (Figure 7.26). However, this method requires a careful drilling operation to successfully connect the horizontal drains with the wells.

The spacing between relief wells is very important because it affects the performance and cost of the system. Spacings of 15 to 40 feet are common, and many systems consist of two closely overlapping rows. The depth of relief wells depends on the unstable zone in which stability needs to be improved. Relief wells up to a depth of 160 feet have been built.

To increase the effectiveness of relief wells and to reduce the cost of pumping from the wells, a new technique, called RODREN, which interconnects the toes of the deep wells by 3- to 4-inch diameter discharge pipes, was introduced in Italy (Figure 7.27). This technique has been used successfully in Italy to stabilize landslides affecting urban areas and along highways (Bruce, 1992). Caltrans uses a similar method, consisting of belled caissons and horizontal drains (Woodward Clyde Consultants, 1994).

RODREN is composed of 4- to 7-foot diameter vertical drainage well lines, 15 to 25 feet apart, and connected near their toes by 3- to 4-inch diameter pipes. These connectors use gravity to transfer the groundwater intercepted by each well to a pump-out location. Like the conventional relief wells, the depth of RODREN is normally governed by the hydrogeology of the slope. The depth of the pipe should be located below the slip surface.

The major attraction of RODREN is that it uses gravity flow via connector pipes, and thus requires no active pumping. It is a maintenance-free, gravity-driven interceptor array responding directly to the field groundwater conditions. One practical consideration, however, should be the safety of the operator who connects the pipes from well to well.

Drainage Tunnels or Galleries A drainage tunnel (also called drainage gallery) may be considered when a cut to be dewatered is so large that it requires a substantial number of horizontal drains, when groundwater is at such a depth that it is impossible to reach by open excavation methods, or when the topography makes horizontal drains impractical (Figure 7.28). Tunnels are effective for correcting unstable zones of large proportions, but they have relatively high construction costs.

The stabilization of the Tablachaca Dam landslide was a case where drainage tunnels, as well as other stabilization methods, were warranted. Tablachaca Dam is a 236-foot high concrete gravity-arch dam on the Mantaro River in Peru, located approximately 186 miles from Lima (Millet et al., 1992). In September 1972, the reservoir was first filled from river level at about elevation 8,700 feet to a maximum elevation of 8,840 feet. During the period when the reservoir was rising, movement was observed in part of an ancient slide on the right abutment immediately upstream from the dam (Figure 7.29). Because of this event, the reservoir was lowered and landslide movements were monitored. By April 1982, there was serious concern that a very large active slide mass was in danger of sliding into the reservoir. The active slide surface measured about 1,000 feet along the reservoir bank by 1,150 feet up the slope.

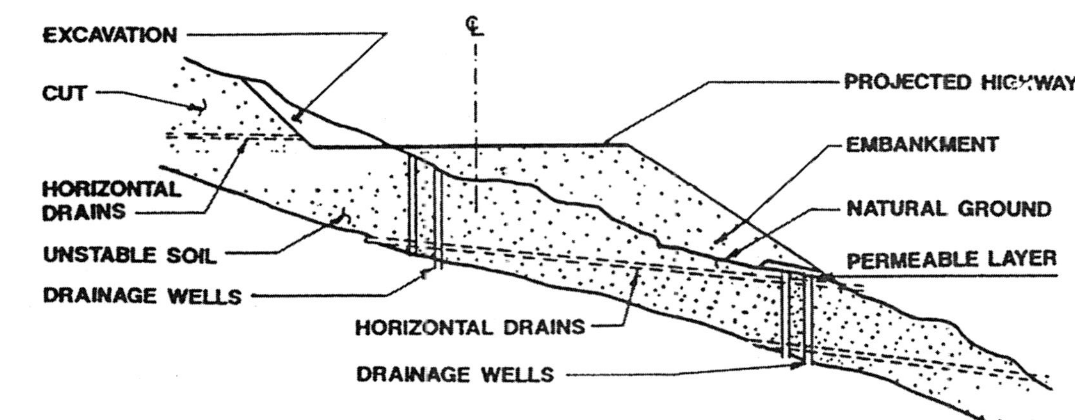

Figure 7.26 Drainage well combined with horizontal drain. (From Rodriguez et al., 1988, with permission of Trans Tech Publications.)

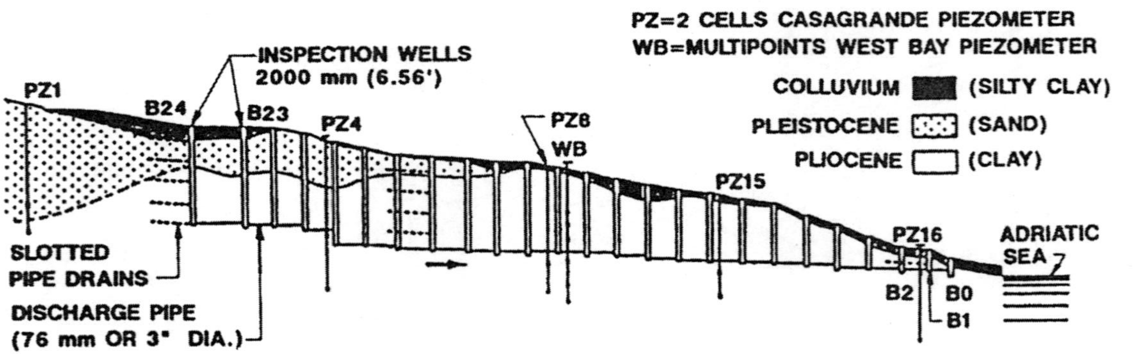

Figure 7.27 Cross section through Rodren alignment B Ancona, Italy. (Bruce and Bianco, 1991.)

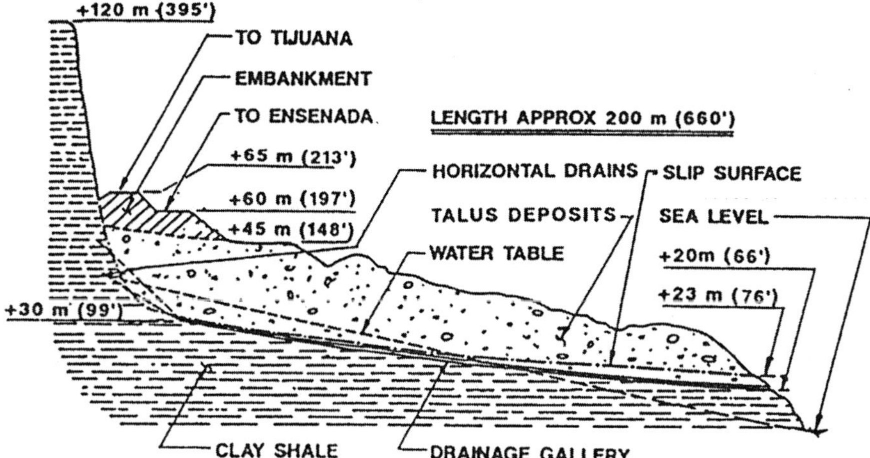

Figure 7.28 Geological cross section of the drainage gallery at the Tijuana-Ensenada highway. (From Rodriguez et al., 1988, with permission of Trans Tech Publications.)

Static slope stability analyses on potential sliding planes (Figure 7.30) indicated a friction angle value of about 37°. Seismic studies indicated that if the reservoir was filled too high and excess pore water pressure built up along a failure plane, significant movement of the slide mass could occur. The emergency remediation

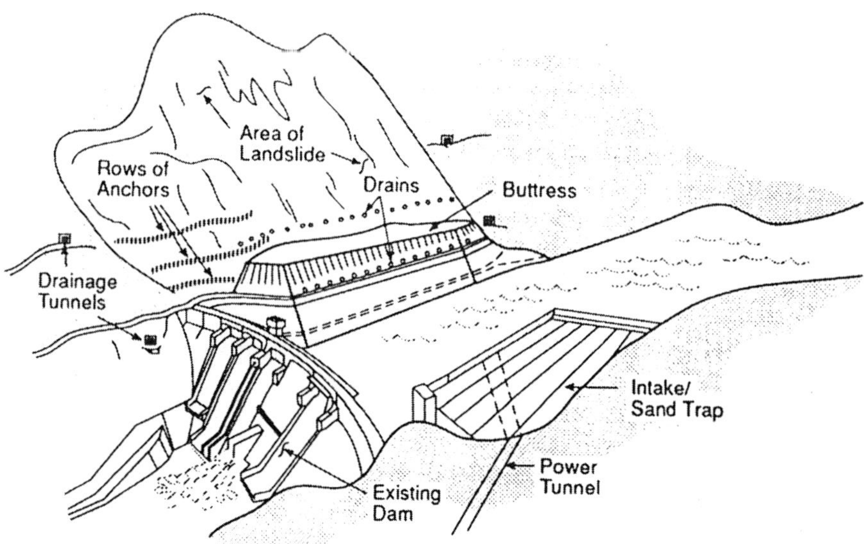

Figure 7.29 Overview of emergency works at Tablachaca Dam. (From Millet et al., 1992, reproduced by permission of ASCE.)

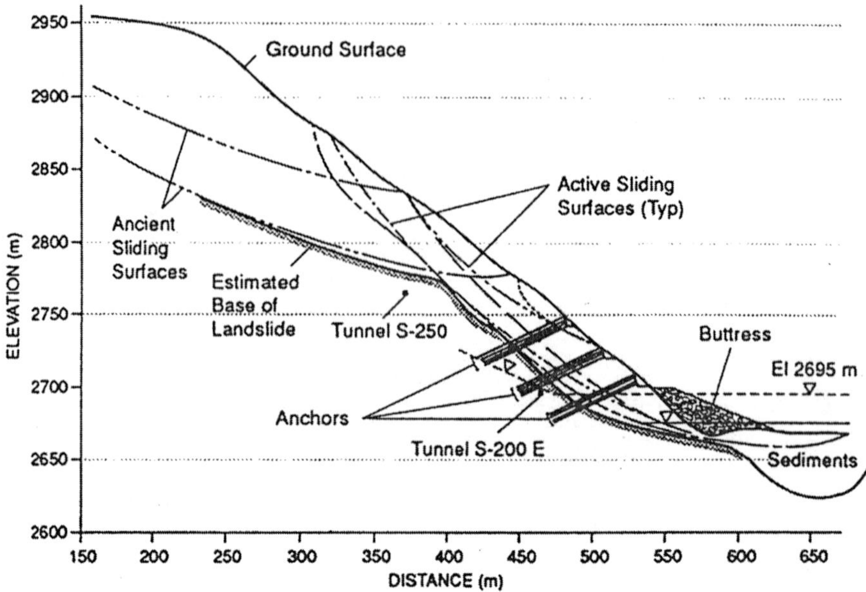

Figure 7.30 Schematic cross section of landslide at Tablachaca Dam. (From Millet et al., 1992, reproduced by permission of ASCE.)

program consisted of the construction of a buttress (Figure 7.31), installation of rock anchors where there was insufficient room for a buttress, construction of two drainage tunnels with radial drains, and installation of horizontal drains. The tunnels and galleries totaled about 5,000 feet in length and were excavated within the bedrock

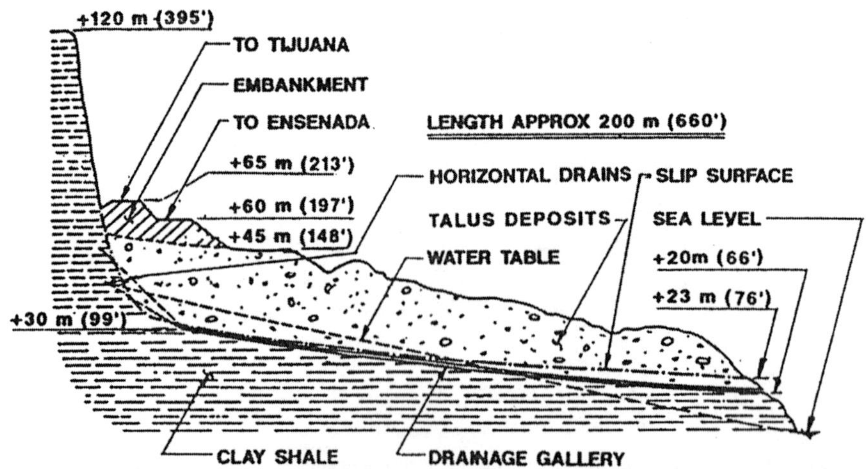

Figure 7.31 Buttress section for remediation at Tablachaca Dam. (From Millet et al., 1992, reproduced by permission of ASCE.)

behind the landslide. The tunnels were about 7 feet wide and 8 feet high and were supported with timber and steel sets and lagging with a precast concrete invert. A total of 190 radial drains totaling about 11,000 feet in length were drilled from the inside of the tunnels. Intercepted water flow was estimated to be between about 20 to 40 gallons per minute. To date, significant movement of the landslide has not occurred since completion of the emergency works.

7.5 REINFORCEMENT

7.5.1 Soil Nailing

Soil nailing is a method of in situ reinforcement utilizing passive inclusions that will be mobilized if movement occurs. It can be used to retain excavations and stabilize slopes by creating in situ, reinforced, soil retaining structures. The main applications are shown schematically in Figure 7.32.

In soil nailed excavations, the reinforcement generally consists of steel bars, metal tubes, or other metal rods that resist tensile stresses, shear stresses, and bending moments imposed by slope movements. The nails generally are not prestressed and are relatively closely spaced. The nails can be installed in the excavation cuts by either

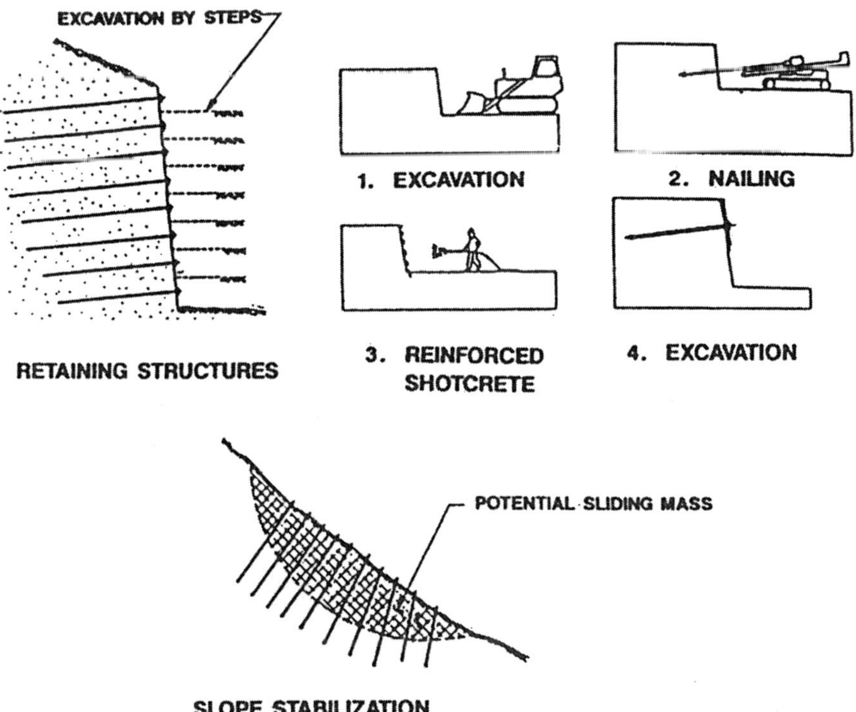

Figure 7.32 Main applications of soil nailing. (Mitchell and Villet, 1987.)

TABLE 7.3 Assumptions of Different Soil Nailing Design Methods

Features	French Method (Schlosser, 1983)	German Method (Stocker et al., 1979)	Davis Method (Shen et al., 1981)	"Modified" Davis (Elias and Juran, 1991)	Kinematical Method (Juran et al., 1989)
Analysis	Limit moment equilibrium Global stability	Limit force equilibrium Global stability	Limit force equilibrium Global stability	Limit force equilibrium Global stability	Working stress analysis Local stability
Input material properties	Soil parameters (c, ϕ') Limit nail forces Bending stiffness	Soil parameters (c, ϕ') Lateral friction	Soil parameters (c, ϕ') Limit nail forces Lateral friction	Soil parameters (c, ϕ') Limit nail forces Lateral friction	Soil parameters ($C/(\gamma H)$, ϕ') Nondimentional bending stiffness parameter (N)
Nail forces	Tension, shear, moments	Tension	Tension	Tension	Tension, shear, moments
Failure surface	Circular, any input shape	Bilinear	Parabolic	Parabolic	Log-spiral
Failure mechanisms	Mixed[a]	Pull-out	Mixed	Mixed	Not applicable
Safety Factors[b]					
Soil Strength, F_c, F_ϕ	1.5	1 (residual shear strength)	1.5	1	1
Pull-out resistance, F_p[c]	1.5	1.5 to 2	1.5	2	2
Tension bending[c]	Yield stress Plastic moment	Yield stress	Yield stress	Yield stress	Yield stress Plastic moment

Design output	GSF[d]	GSF	GSF	GSF	Mobilized nail forces
	CFS[e]	CFS	CFS	CFS	CFS
Groundwater	Yes	No	No	No	Yes
Soil stratification[f]	Yes	No	No	No	Yes
Leading[f]	Slope, any surcharge	Slope surcharge	Uniform surcharge	Slope, uniform surcharge	Slope
Structure geometry[f]	Any input geometry	Inclined facing Vertical facing	Vertical facing	Inclined facing Vertical facing	Inclined facing Vertical facing

Source: Elias and Juran (1991).

[a]Mixed failure mechanisms: Limit-tension force in each nail is governed by either its pull-out resistance factored by the safety factor or the nail yield stress, whichever is smaller. Pull-out failure mechanism: Limit-tension forces in all the nails are governed by their pull-out resistance factored by the safety factor.

[b]Definitions of safety factors used in this analysis:

- For soil strength, $F_c = c/c_m$, $F_\phi = (\tan\phi)/(\tan\phi_m)$; where c and ϕ are the soil cohesion and friction angle, respectively, while c_m and ϕ_m are the soil cohesion and friction angle mobilized along the potential sliding surface.
- For nail pull-out resistance, $F_p = f_l/f_m$; f_l and f_m are the limit interface shear stress and the mobilized interface shear stress, respectively.

[c]Recommended limit nail force.

[d]GSF: global safety factor.

[e]CFS: critical failure surface.

[f]Present design capabilities.

driving or grouting in predrilled boreholes. Stability of the ground surface between nails can be provided by a surface skin—often a thin layer of shotcrete (4 to 6 inches thick) reinforced with wire mesh—or by intermittent rigid elements analogous to large washers. Soil nailing in excavations has been used in granular and cohesive soils and in relatively heterogeneous deposits.

The design of nailed excavations and slopes generally is based on limit equilibrium analyses where critical potential failure surfaces must be assumed. The analyses are predicated on global or partial FOSs. As the hypothetical soil nailed wall being evaluated fails progressively, a global FOS (equal for all nails) does not accurately predict the behavior of nails in different rows during failure. The assumptions of the different design approaches are discussed in Elias and Juran (1991) and summarized in Table 7.3.

The soil nail design methods that emerged in the 1970s and 1980s consist of the Davis method (Shen et al., 1981), the German method (Gassler and Gudehus, 1981; Stocker et al., 1979), and the French method (Schlosser, 1983), which are all limit equilibrium analysis design methods (Elias and Juran, 1991). A more complex and cumbersome method of analysis based on the behavior of MSEs, the kinematical method by Juran (1977), considers kinematically admissible displacement failure modes in a limit analysis framework. The kinematical method places an undue emphasis on nail stiffness and is difficult to use. The detailed design methodologies of these methods are not discussed here but can be found in the respective publications given above.

It must be noted that none of the methods described above solve the soil nailed slope problem without inconsistencies in the input parameters, analytical methods, and comparisons to field behavior. Some of these inconsistencies, according to Walkinshaw (1992), include

(1) Improper cancellation of interslice forces (Davis method)
(2) Lateral earth pressures inconsistent with nail force and facing pressure distribution (all methods)
(3) No redistribution of nail forces according to construction sequence and observed measurements
(4) Complex treatment and impractical emphasis on nail stiffness (kinematical method).

Soil nailing system design involves spacing, size, and length of the nails and design of the wall facing. The spacing, size, and length of the nails must be determined based on global stability and internal stability considerations. Design of the wall facing (materials, thickness, and reinforcement) is dependent on the nail forces assumed. The desired FOSs used in the design analysis are different from case to case and should be compatible with the use of the wall (whether it is a permanent or temporary wall) and economic and risk-to-human-life consequences of the slope. For permanent soil nailed slopes, corrosion protection should always be considered in design. Furthermore, for the design of the soil nail wall system, engineers must choose the method that they feel

TABLE 7.4 Preliminary Design Parameters Used in the Design of Soil Nailed Slopes

	Design Parameters[a]		
Types of Nails/Surrounding Soils	L/H	DL/S	d^2/S
Drilled and grouted nails in granular soils	0.5–0.8	0.5–0.6	$(4–8) \times 10^{-4}$
Driven nails in granular soils	0.5–0.6	0.6–1.1	$(13–19) \times 10^{-4}$
Nails in moraine and marl	0.5–1	0.15–0.2	$(1–2.5) \times 10^{-4}$

Source: Bruce and Jewell (1987).

[a] L = Length of the soil nails; H = height of the wall; D = diameter of the soil nail hole for bond ratio; d = diameter of the nail bar for strength ratio; S = area per soil nail.

most comfortable with and make the appropriate modifications and adjustments based on experience, instrumentation, case histories, engineering judgment, and common sense.

Different researchers and engineers have proposed empirical methods for determining design parameters for soil nailing. Bruce and Jewell (1987) derived parameters based on published case histories that can be used as a first step to design. These parameters are presented in Table 7.4.

In comparison to conventional excavation and retaining systems, such as massive concrete walls, internal bracing systems, and tieback wails, soil nailing presents the following advantages:

(1) *Low Cost* Steel bar reinforcement is inexpensive. The shotcrete between the nails is relatively thin and inexpensive. For example, cost savings of 10 to 30 percent may be realized with soil nailing as compared to tieback walls (Xanthakos et al., 1994).

(2) *Light Construction Equipment* Soil nailing can be done using conventional drilling and grouting equipment. Thus the technique is of particular interest on sites with difficult access and limited space constraints.

(3) *Adaptability to Different Soil Conditions* In heterogeneous ground where boulders or hard rocks may be encountered in softer layers, soil nailing generally is more feasible than other techniques such as soldier piles. This is because it involves only small-diameter drilling for the installation of the inclusions.

(4) *Flexibility* Nailed soil retaining structures are more flexible than classical cast-in-place reinforced concrete retaining structures. Consequently, these structures can conform to deformation of surrounding ground and can withstand larger total and differential settlements. This characteristic of soil nailing can provide economical support for excavations on unstable slopes.

(5) *Reinforcement Redundancy* If one nail becomes overstressed for any reason, it will not cause failure of the entire wall system. Rather, it will redistribute its overstress to the adjoining nails.

Disadvantages are common to other reinforcement systems and are minor as compared to the advantages listed above. The disadvantages of soil nailing are

(Xanthakos et al., 1994):

(1) The ground to be excavated must be strong enough for 3- to 8-foot high cuts to remain stable for at least a few hours to allow time for reinforcement installation.

(2) Free water cannot be flowing out of the face of the excavation to permit the application of shotcrete facing.

(3) Reliable drainage systems are difficult to construct.

(4) Very soft clays are not suitable for this type of reinforcement because of potential creep movements.

(5) Permanent or temporary underground easements may be required; interference with nearby utilities may occur.

The advantages of soil nailing are illustrated in the following examples.

A 50-foot-deep excavation was required at the corner of a sloping site bordered by a city street on one side and residential structures on the other, as depicted in Figure 7.33 (stocker and Riegdinger, 1990). The installation of long tieback anchors underneath these properties was not permitted. The soil consisted of about 2 to 6 feet of silt, sand, and cinder fill, underlain by about 26 feet of medium to stiff clayey, sandy silt and gravel. The overburden soils were underlain by Keuper marl consisting of alternating layers of siltstone and claystone. The unit weights for the fill, silt, and marl were 122, 128, and 134 pounds per cubic foot, respectively. The angle of internal friction for the fill was 30°. The silt had an angle of internal friction of 27.5° and a cohesion value of between about 100 and 200 pounds per square foot. The marl had

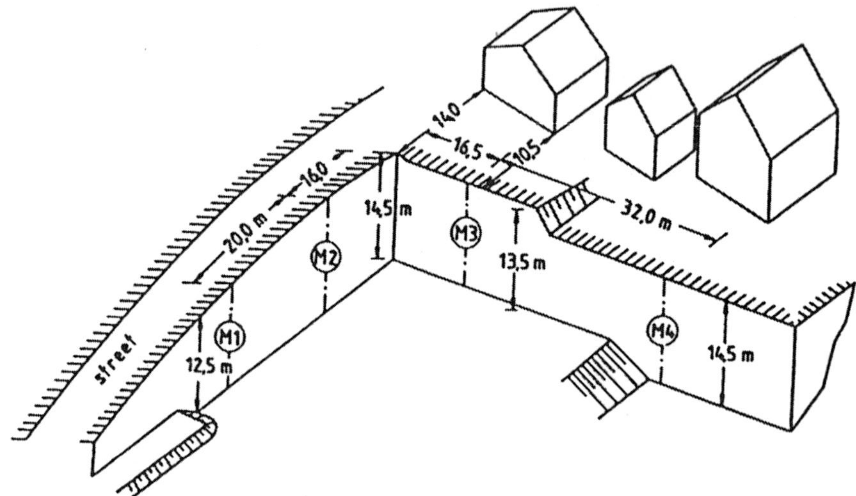

Figure 7.33 Stuttgart building site plan. (From Stocker and Riedinger, 1990, reproduced by permission of ASCE.)

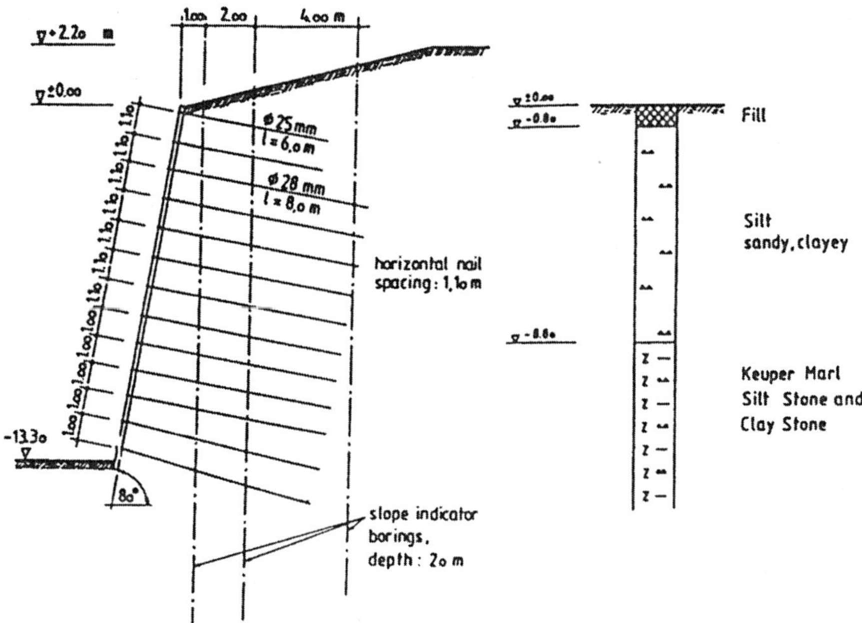

Figure 7.34 Excavation cross section. (From Stocker and Riedinger, 1990, reproduced by permission of ASCE.)

an angle of internal friction of 23° and a cohesion value greater than 1,000 pounds per square foot.

The top two rows of nails consisted of 20-foot-long, $\frac{7}{8}$-inch diameter deformed steel bars (Figure 7.34). The other rows consisted of 26-foot-long, 1-inch-diameter bars. The yield and ultimate strengths of the bars was about 60,000 and 72,000 pounds per square inch, respectively. Corrosion protection consisted of $\frac{1}{25}$-inch-thick PVC sleeves cement grouted to the entire length of the bars. The wall facing consisted of 10-inch-thick, steel mesh reinforced shotcrete. Excavation was carried out in 3- to $3\frac{1}{2}$-foot-high lifts.

The excavation support was instrumented and monitored with slope indicators, load cells, extensometers, and strain gages. Five percent of the nails were load tested to 45,000 pounds. A plot of maximum nail forces after excavation indicated a potential failure zone, as shown in Figure 7.35. Long-term monitoring over a period of 10 years indicated a significant load increase when the daily minimum temperature at ground surface fell below freezing. Maximum horizontal deformation as a function of excavation depth varied between 0.1 and 0.36 percent. Additional amounts of deformation over time ranged between 0.06 and 0.15 percent, probably due to continuing construction and loads from the new building. Deformations generally ceased by the end of the third year of monitoring. Deformations within the ground mass decreased with increasing distance from the wall face (Figure 7.36). The earth

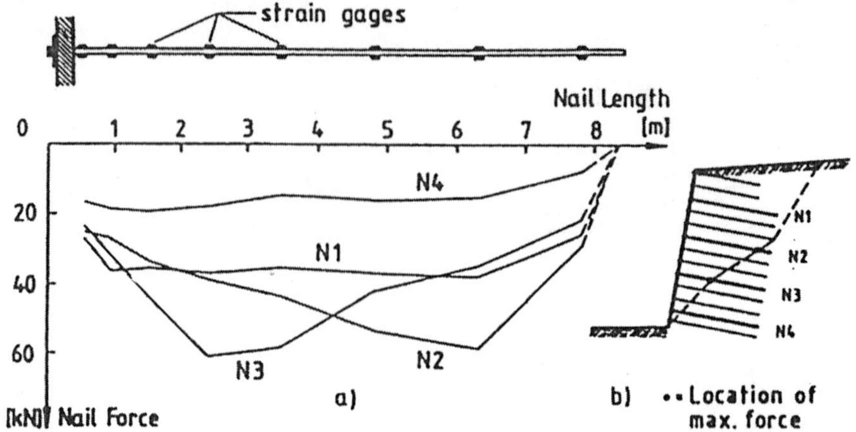

Figure 7.35 (*a*) Observed maximum nail forces. (*b*) Location of maximum force. (From Stocker and Riedinger, 1990, reproduced by permission of ASCE.)

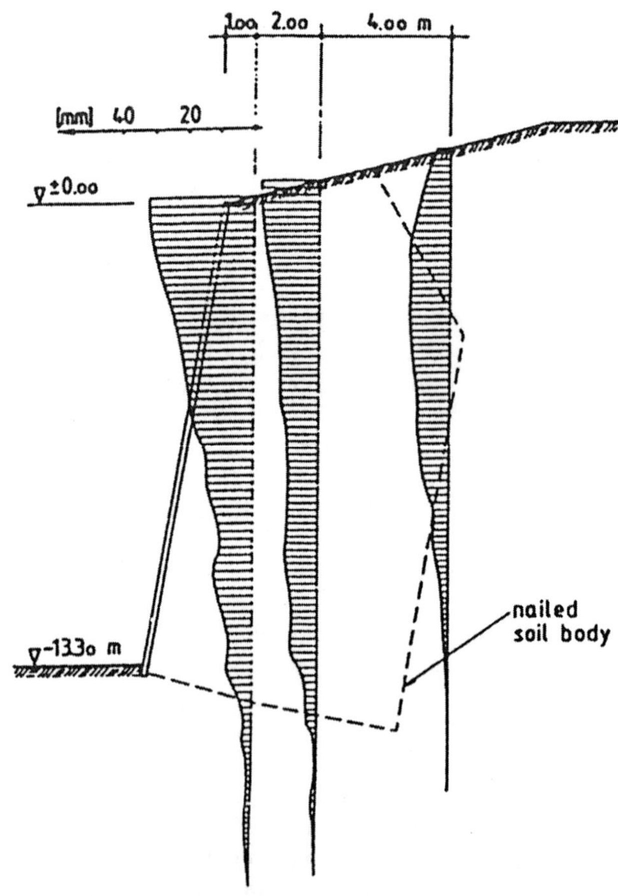

Figure 7.36 Observed ground deformations. (From Stocker and Riedinger, 1990, reproduced by permission of ASCE.)

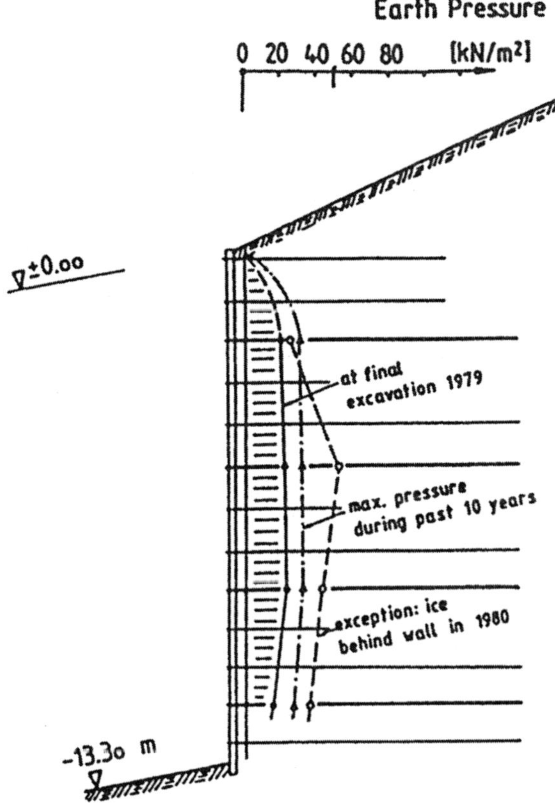

Figure 7.37 Observed earth pressures. (From Stocker and Riedinger, 1990, reproduced by permission of ASCE.)

pressure distribution observed is more or less rectangular with a maximum of about 1,000 pounds per square foot (Figure 7.37).

In another example that used soil nail reinforcement, over 20 miles of State Highway Route 504 (SR 504) was destroyed as a result of the eruption of Mt. St. Helens on May 18, 1980 (Leonard et al., 1988). This road, before it was covered with tons of ash and debris, led to the mountain along the valley of the north fork of the Toutle River. The new road was located on the valley sides above any future debris flows emanating from the mountain. The cuts were on the order of 65 feet high in an area where the natural slopes range up to about 35°. The cut involved compact to dense silty sand and gravel colluvium, talus gravel, cobbles, and boulders, and poor to good quality basalt bedrock. The overburden thicknesses ranged between about 20 and 35 feet. The cuts therefore involved both soil and rock. The wall design included 1H–10V slopes with a bench at the rock surface (Figure 7.38), 25-foot-long nails on a 6-foot pattern, shotcrete initial support, concrete facade final support, drainage fabric, and weeps.

Soil nailing can even be used to stabilize embankments and fills. In one example of this, a highway embankment was stabilized using soil nailing at Mt. White, about

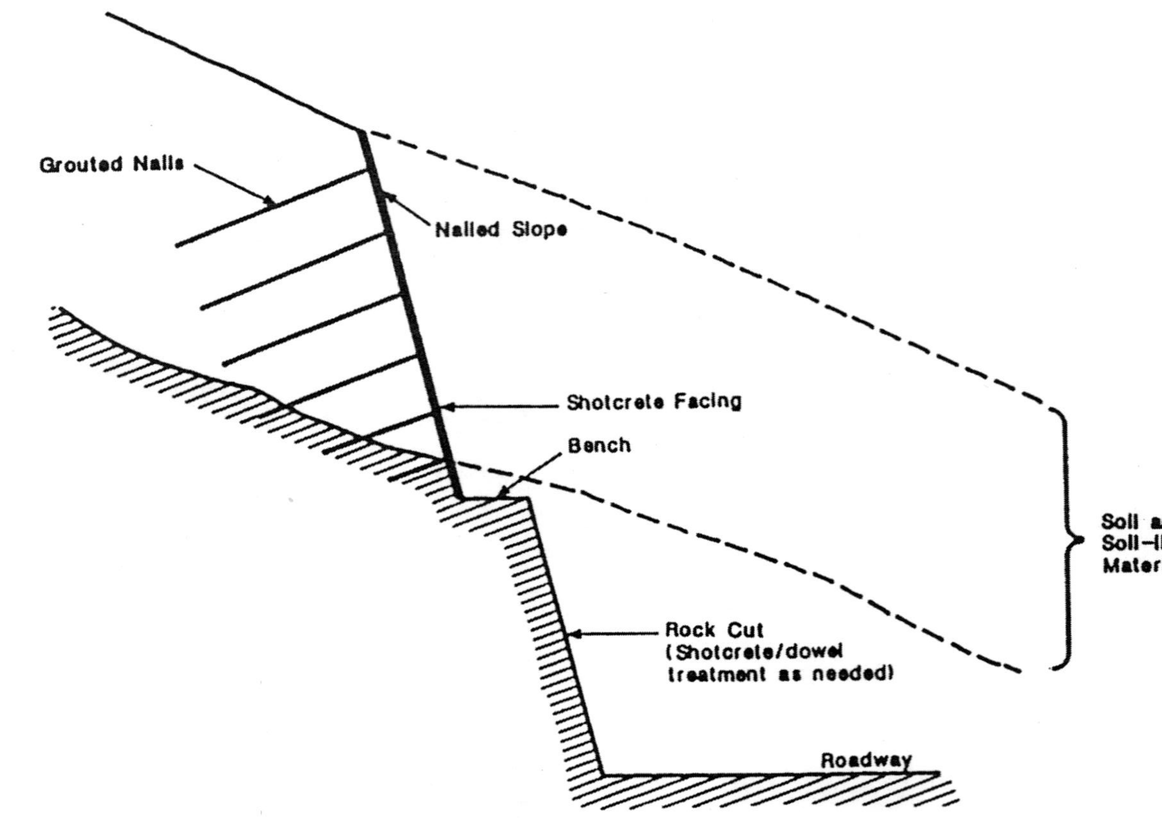

Figure 7.38 State Route 504 highway cut cross section. (From Leonard et al., 1988.)

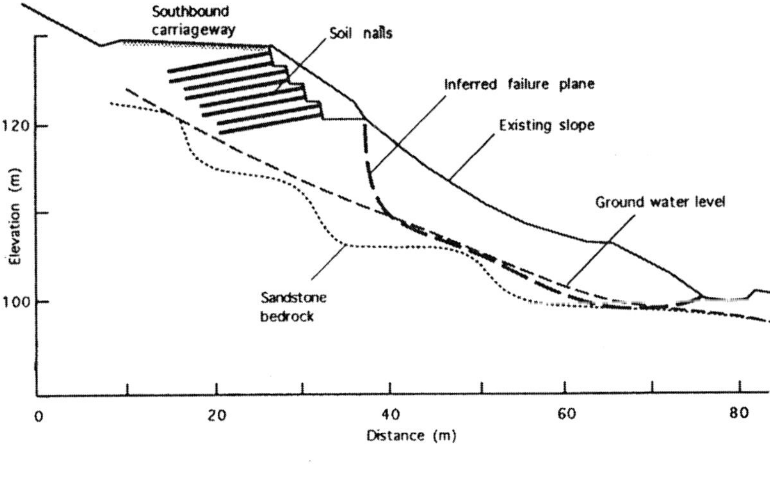

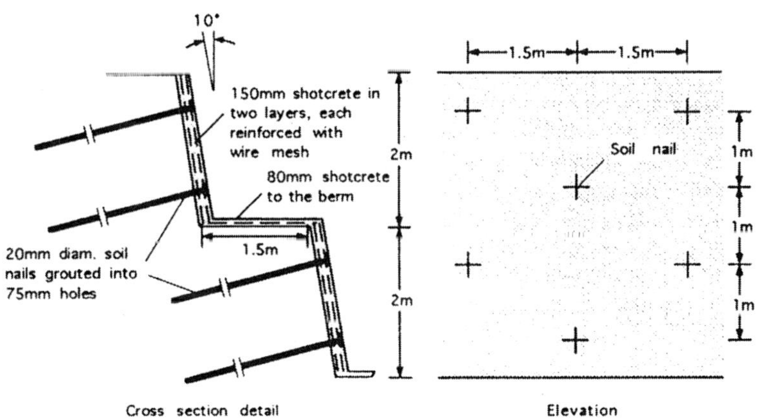

Figure 7.39 Mt. White highway cut cross section. (From Hausmann, 1992, reproduced by permission of ASCE.)

37 miles north of Sydney, Australia (Hausmann, 1992). The embankment consisted of clayey silty sands and sandstone fill, and threatened to fail after heavy rains in June 1989. Four benches were constructed on which two rows of nails were installed (Figure 7.39). A total of 400 nails were installed on a 6.5-foot-pattern horizontally and a 3.3-foot-staggered pattern vertically. The nails were $\frac{3}{4}$ inch in diameter and 39 feet long, installed at an angle of 10° below horizontal. The wall was faced with 6 inches of steel-mesh-reinforced shotcrete.

7.5.2 Stone Columns

Stone columns can be used to stabilize or prevent landslides (Aboshi et al., 1979; Goughnour et al., 1990). This ground improvement technique increases the average

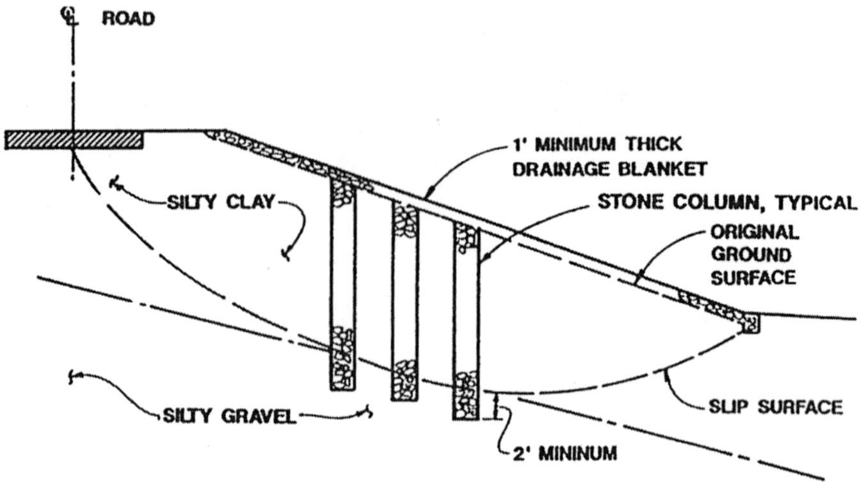

Figure 7.40 Schematic diagram showing stone columns to stabilize an unstable slope.

shear resistance of the soil along a potential slip surface by replacing or displacing the in situ soil with a series of closely spaced, large-diameter columns of compacted stone (Figure 7.40). In addition, stone columns also function as efficient gravel drains by providing a path for relief of pore water pressures, thereby increasing the strength of the surrounding clayey soils.

Construction of stone columns consists of:

(1) Forming a vertical hole in the underlying material, using either the vibro-replacement or vibro-displacement technique.
(2) Placing stone in the preformed hole from the ground surface, as in the vibro-replacement technique, or by means of bottom feed equipment, as in the vibro-displacement technique.
(3) Compacting the stone by repenetration of each lift with the vibroflot, a process that drives the stone laterally to the sidewalls of the hole and thus enlarges the hole.

Review of case histories and discussions with specialty contractors by Bachus and Barksdale (1989) indicate that the stone column technique is more suitable for subsurface soils whose shear strengths range from 200 to 1,000 pounds per square foot. Soils weaker than the lower limit may not provide sufficient lateral support for the stone fill, thus causing large stone consumption and/or excessive deformation. For soil strenght approaching the upper limit, stone columns may not be needed.

Special care should be exercised when stone columns are used in sensitive or organic soils (Bachus and Barksdale, 1989). In sensitive soils, construction should proceed quickly to minimize the amount of vibration to the soils. Peaty soils, in which the layer thickness is greater than one to two vibroflot diameters, should be avoided because of documented poor performance and construction problems.

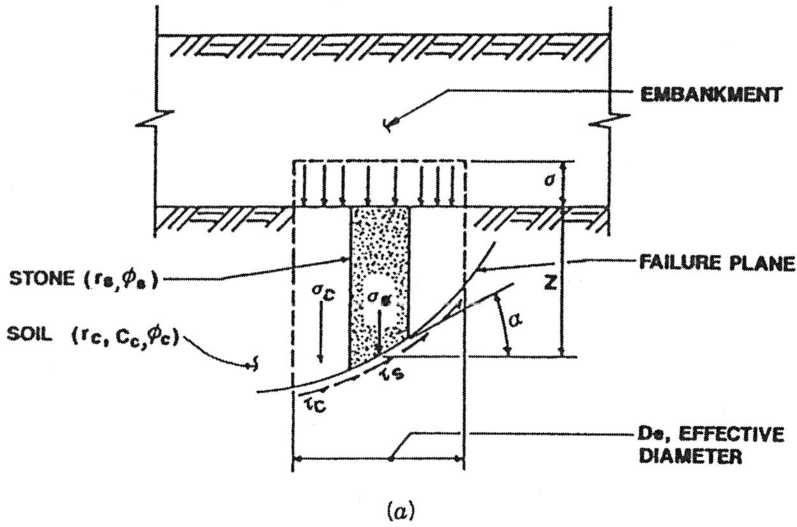

(a)

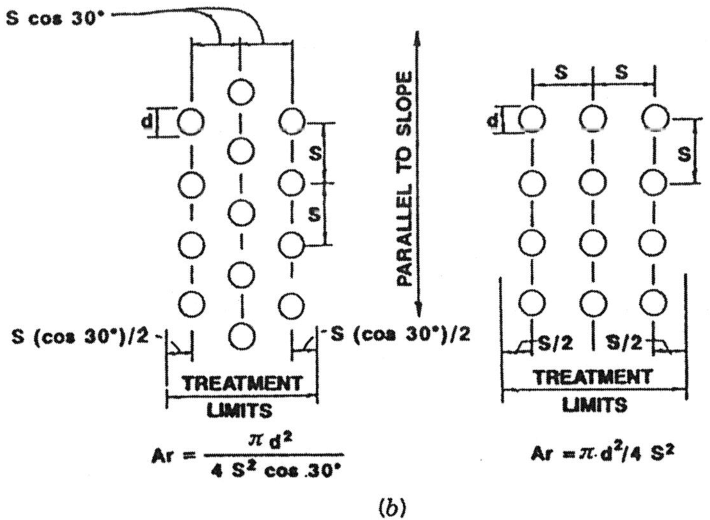

(b)

Figure 7.41 Design methodology of stone columns. (*a*) Definition sketch for Japanese method. (*b*) Definition of stone column treatment limits. (From Goughnour et al., 1990, copyright © ASTM, reprinted with permission.)

Two empirical methods are available for the design of stone columns. The first is based on the technique developed in Japan (Aboshi et al., 1979) for sand columns. The second is called the average strength parameter method (Goughnour et al., 1990). Figure 7.41 illustrates the design methodology for stone-columns by defining an average shear strength that can be applied to the stone-column treated soil. Stability calculations are then carried out by using conventional slope stability analysis methods. The basic equations for developing average shear strength values are

(1) *Japanese Method*

$$\tau_{ave} = (1 - A_r)\tau_c + A_r \tau_s \cos\alpha \qquad \text{(Eq. 7-2)}$$

(2) *Average Strength Parameters Method*

$$C_{ave} = C_c(1 - A_r) + c_s A_r$$

$$= C_c(1 - A_r) \qquad \text{(Eq. 7-3)}$$

$$(c_s = 0 \quad \text{for stone column})$$

$$\tan\phi_{ave} = \frac{(1 - A_r)\tan\phi_c + S_r A_r \tan\phi_s}{1 + A_r(S_r - 1)} \qquad \text{(Eq. 7-4)}$$

$$S_r = 1 + (S_{rv} - 1)\cos\alpha \qquad \text{(Eq. 7-5)}$$

$$\gamma_{ave} = (1 - A_r)\gamma_c + A_r\gamma_s \qquad \text{(Eq. 7-6)}$$

where c_{ave} = average cohesion to be used for the treated soil
c_c = cohesion of the in situ soil
c_s = cohesion of stone
τ_{ave} = average weighted shear strength within the area tributary to the stone column
S_r = stress ratio appropriate to the orientation of the failure surface at that location
$S_{rv} = \sigma_s/\sigma_c$, stress ratio or vertical stress in the stone column divided by that in the in situ soil
$A_r = \pi d^2/4S^2$ for square array and $\pi d^2/(4S^2 \cos 30^\circ)$ for triangular array
τ_c = shear strength of the in situ soil
τ_s = shear strength of the stone column
σ_s = effective vertical stress due to weight of column ($\tau_s z$) and applied loading $\sigma = \tau_s z + s\mu_s$
σ_c = vertical stress in the in situ soil
α = angle of inclination of the failure surface from the horizontal
$\mu_s = S_{rv}/(1 + (S_{rv} - 1)A_r)$
ϕ_s = internal friction angle of the stone
ϕ_c = internal friction angle of the in situ soil
ϕ_{ave} = average internal friction angle of the treated soil
γ_{ave} = average unit weight of the treated soil
γ_c = unit weight of in situ soil
γ_s = unit weight of stone

TABLE 7.5 Case Histories of Stone Columns used for Highway Slope Stabilization

Location	Diameter (feet)	Spacing (feet)	Construction Method
Steel Bayou Bridge, Route 465, Mississippi	3	5	Wet vibro-flotation
Nemadji River Bridge, Superior, Wisconsin	3.5	6 to 7	Preaugering holes
Route 22, New York	3.5	6	Dry displacement method using bottom feed equipment

Source: Goughnour et al. (1990).

Stabilization of highway slopes by means of stone columns has been used in Alaska, California, Florida, Iowa, Kentucky, Mississippi, New York, South Dakota, Texas, Virginia, and Wisconsin. Table 7.5 lists a few case histories of the application of stone columns in highway slope stabilization. See Xanthakos et al. (1994) for more examples.

7.5.3 Reticulated Micropiles

Reticulated micropiles were developed in Italy and are used to create a monolithic rigid block of reinforced soil to a depth below the critical failure surface (Figure 7.42). The piles used in this way are similar to soil nailing systems. The major difference between the reticulated micropiles and soil nailing is that the behavior of micropiles is influenced significantly by their geometric arrangement. Field and model tests (Lizzi, 1985) have demonstrated that the group and network effect of a reticulated micropile system provides higher load bearing and shearing capacities than those of closely spaced vertical piles. Because of the great length of the piles as compared with their

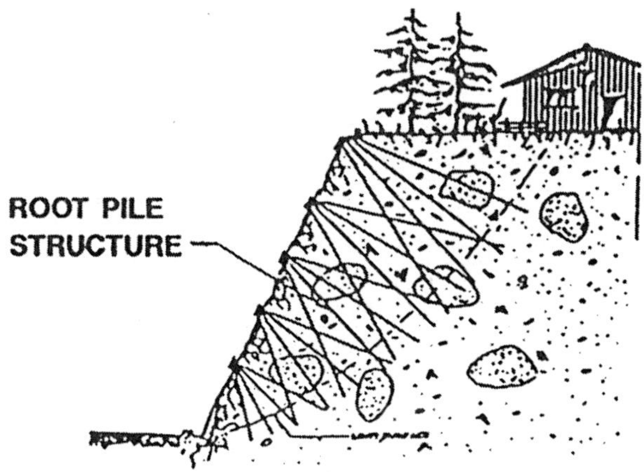

Figure 7.42 Reticulated micropiles used to stabilize slopes. (Lizzi, 1985.)

diameter, the load and shearing resistance are carried not only by the piles but also by the soil they encompass. As a result of the crisscross pattern used, the micropiles are subject to compression and tension forces that provide the required structural stability of the reinforced slope.

The advantages of reticulated micropiles used in slope stabilization are fourfold:

(1) They do not require large soil excavations.
(2) They can be considered in any soils, whatever the permeability and whatever boulders or other obstructions may be present.
(3) They do not prevent water circulation in the subsoil so there is no risk of water accumulating in the system as at the back of conventional walls.
(4) Design can be arranged to counteract many patterns of internal forces.

The drawbacks of reticulated micropiles are that the piles must be founded in stable formations, which can be deep below the slip zone, and provisions must be made to resist long-term corrosion of the steel bars.

Design of micropiles involves the following procedures:

(1) Conduct stability analyses to determine the increase in resistance along a potential or existing slip surface required to provide an adequate FOS, which is expressed as

$$\mathrm{FOS} = \frac{R + R'}{A} \qquad \text{(Eq. 7-7)}$$

where R = total resistance on the critical slip surface
A = driving forces on the same surface
R' = additional shear resistance provided by the micropiles

(2) Check the potential for structural failure of the piles because of loading from the moving soil mass using Figure 7.43. This figure was developed by Pearlman and Withiam (1992) on the basis of the theory devised by Fukuoka (1977). His theory addresses the bending moments developed in a pile oriented perpendicular to a slip plane assuming a uniform velocity distribution of the soil above the slip plane.
(3) Check the potential for plastic failure of soil between the piles. This potential can be analyzed with a procedure developed by Ito and Matsui (1975). Based on Ito and Matsui's theory, the predicted results for various pile spacings and soil conditions can be plotted as shown in Figure 7.44.

7.5.4 Geosynthetically Reinforced Slopes

Geosynthetic soil reinforcement is another technique used to stabilize slopes, particularly after a failure has occurred or if a steeper-than-"safe" unreinforced slope

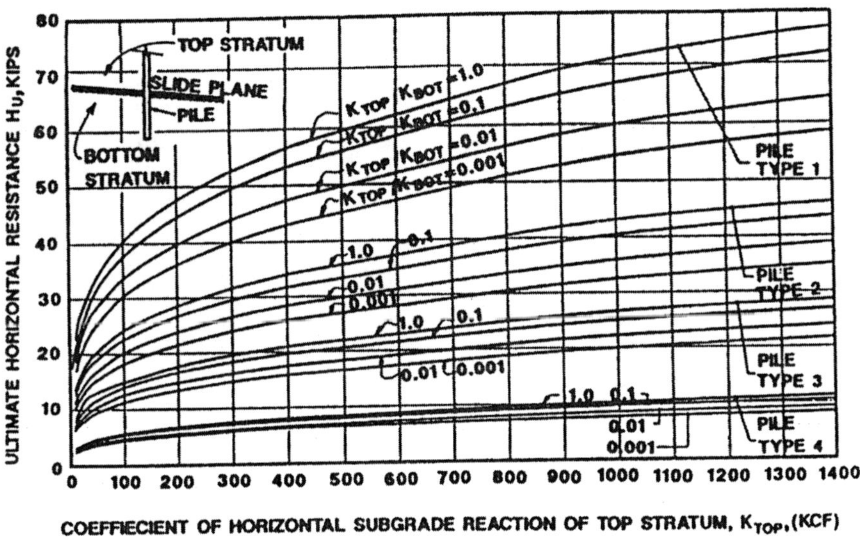

PILE TYPE	HOLE DIA. (in)	PIPE O.D. (in) OR REBAR No.	PIPE WALL THICKNESS (in)	STEEL YIELD STRENGTH Fy (Ksi)	CONCRETE STRENGTH Fc' (Ksi)
1	8	7.0	0.500	80	4
2	6	5.5	0.400	80	4
3	6	4.0	0.375	80	4
4	6	#9	N/A	80	4

Figure 7.43 Preliminary design chart for ultimate horizontal resistance of piles. (From Pearlman and Withiam, 1992, reproduced by permission of ASCE.)

is desirable. In addition, it can improve compaction on the edge of a slope, thus decreasing the tendency for surface sloughing (Figure 7.45).

Similar to MSE slopes, design of geosynthetically reinforced slopes is based on modified versions of classical limit equilibrium slope stability methods. Kinematically, the potential failure surface in a reinforced homogeneous slope is assumed typically to be defined by the same idealized geometry (but not location) as in the unreinforced case (for example, circular, log spiral, bilinear wedge). Statically, the inclination and distribution of the reinforcement tensile force along the failure surface must be postulated. The capacity of reinforcement layers is taken as either the allowable pull-out resistance behind the potential failure surface or as its allowable design strength, whichever is less. The target factor of safety for a reinforced slope is the same as for an unreinforced slope.

Field and numerical model test results (Christopher et al., 1990) indicate that the limit equilibrium approach to design provides a suitable though conservative design approach. A step-by-step design procedure incorporating the circular arc approach may be found in Christopher and Leshchinsky (1991).

Simplified charts for the design of geosynthetically reinforced slopes have been proposed by many researchers and engineers (Christopher and Holtz, 1985, 1989;

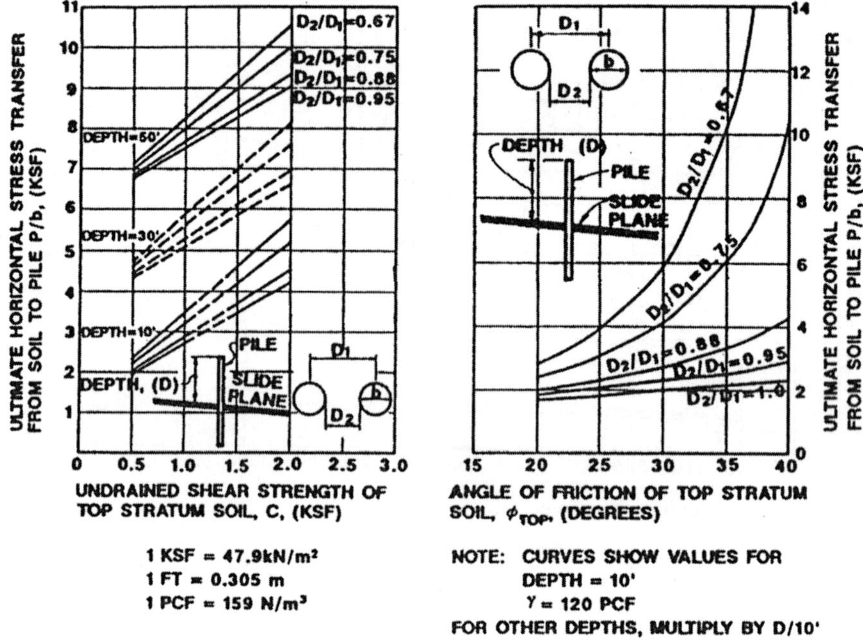

Figure 7.44 Ultimate stress transfer from soil to piles versus shear strength of soil. (From Pearlman and Withiam, 1992, reproduced by permission of ASCE.)

Jewell and Woods, 1984; Jewell et al., 1984). These charts can be used to evaluate the preliminary stability of a geosynthetically reinforced slope before more thorough design procedures proposed by Christopher and Leschinsky (1991) are performed. Figure 7.46*a* to *e* illustrate the simplified design procedure and charts developed by

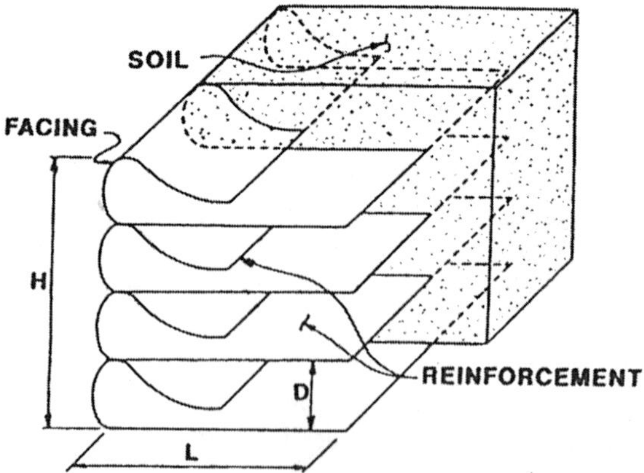

Figure 7.45 Schematic diagram of a reinforced soil using geosynthetics. (Mitchell and Villet, 1987.)

Jewell and Woods (1984), and may be used for preliminary evaluation of geosynthetically reinforced slopes. The following describes the design chart procedure.

Design Chart Procedure (Jewell and Woods, 1984) The steps in the chart procedure and the three pairs of charts for three pore water pressure coefficients, r_u, are as follows:

(1) Select the required slope dimensions and the surcharge loading. The slope geometry is defined by the slope height, H, and slope angle, β (Figure 7.46*a*). A uniform vertical surcharge, W_s, may be used.

(2) Select design values for the soil properties and pore water pressures to represent the loading case to be examined, for example, short-term (end of construction) or long-term loading.

(3) Determine the earth pressure coefficient, K, and the length of the reinforcement, L, from the design charts Figures 7.46*b*,*c*, and *d*.

(4) Choose in situ soil strength properties for the reinforcement and an appropriate partial FOS. Again, the values should be appropriate to the loading case being examined (see Step 2 above) and should anticipate the possibility of site damage to the reinforcement material during construction.

(5) Obtain the factored reinforcement strength, P, by dividing the reinforcement strength from Step 4 by the partial FOS.

(6) Choose a minimum vertical reinforcement spacing, v, that is compatible with the soil compaction layer thickness, and calculate the spacing constant, Q, for the reinforced slope. Q is a function of the soil properties, slope angle, reinforcement strength, and minimum spacing, as defined by

$$Q = \frac{P}{K\gamma v} \qquad \text{(Eq. 7-8)}$$

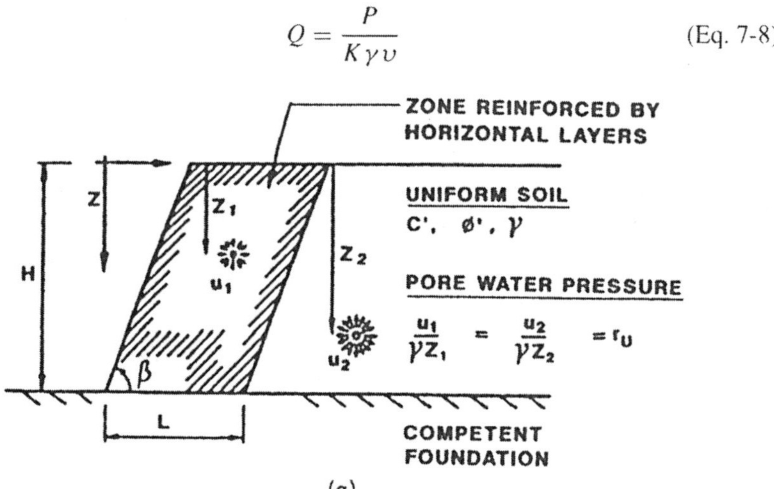

Figure 7.46 (*a*) Steep reinforced slope: definitions of symbols. (*b*) Chart for pore pressure ratio, 0. (*c*) Chart for pore pressure ratio, $r = 0.25$. (*d*) Chart for pore pressure ratio, $r = 0.50$. (*e*) Zones of equal reinforcement spacing.

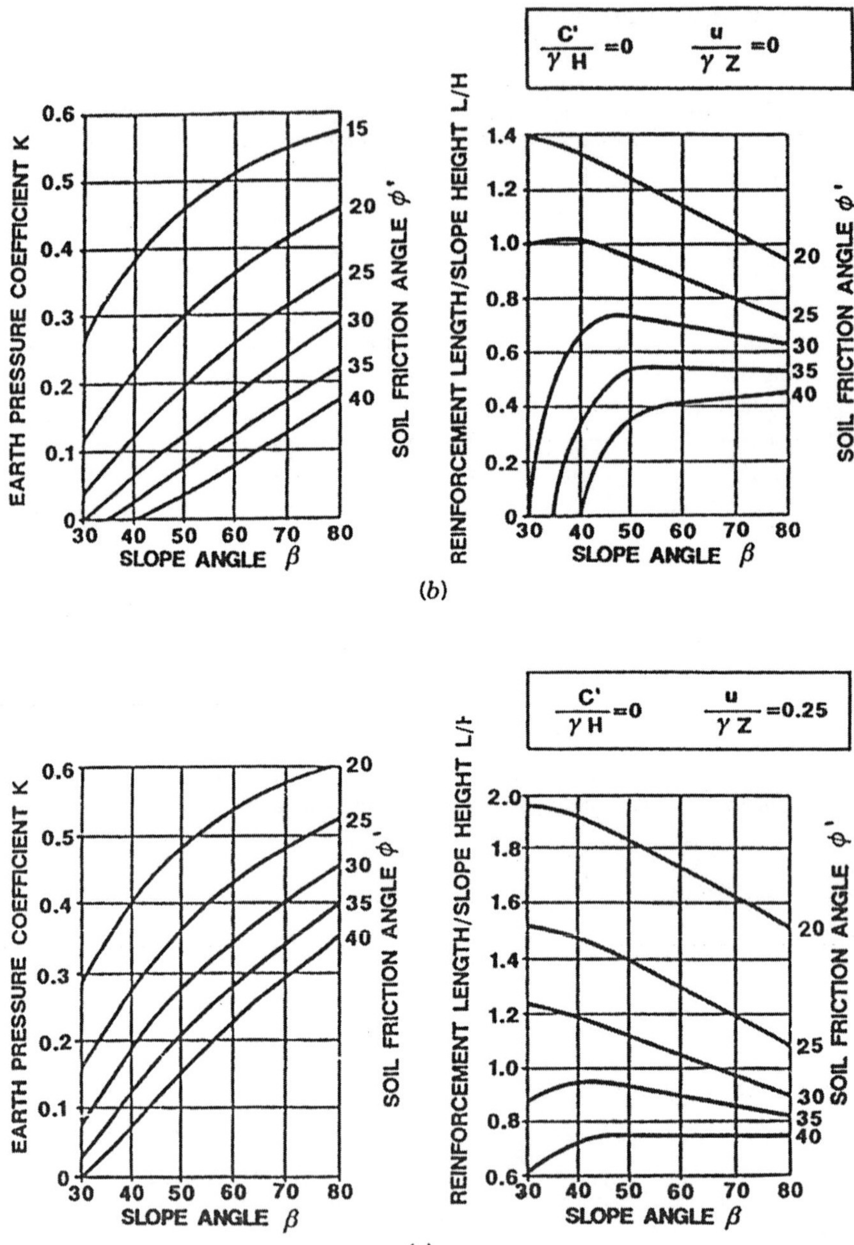

Figure 7.46 (*Continued*)

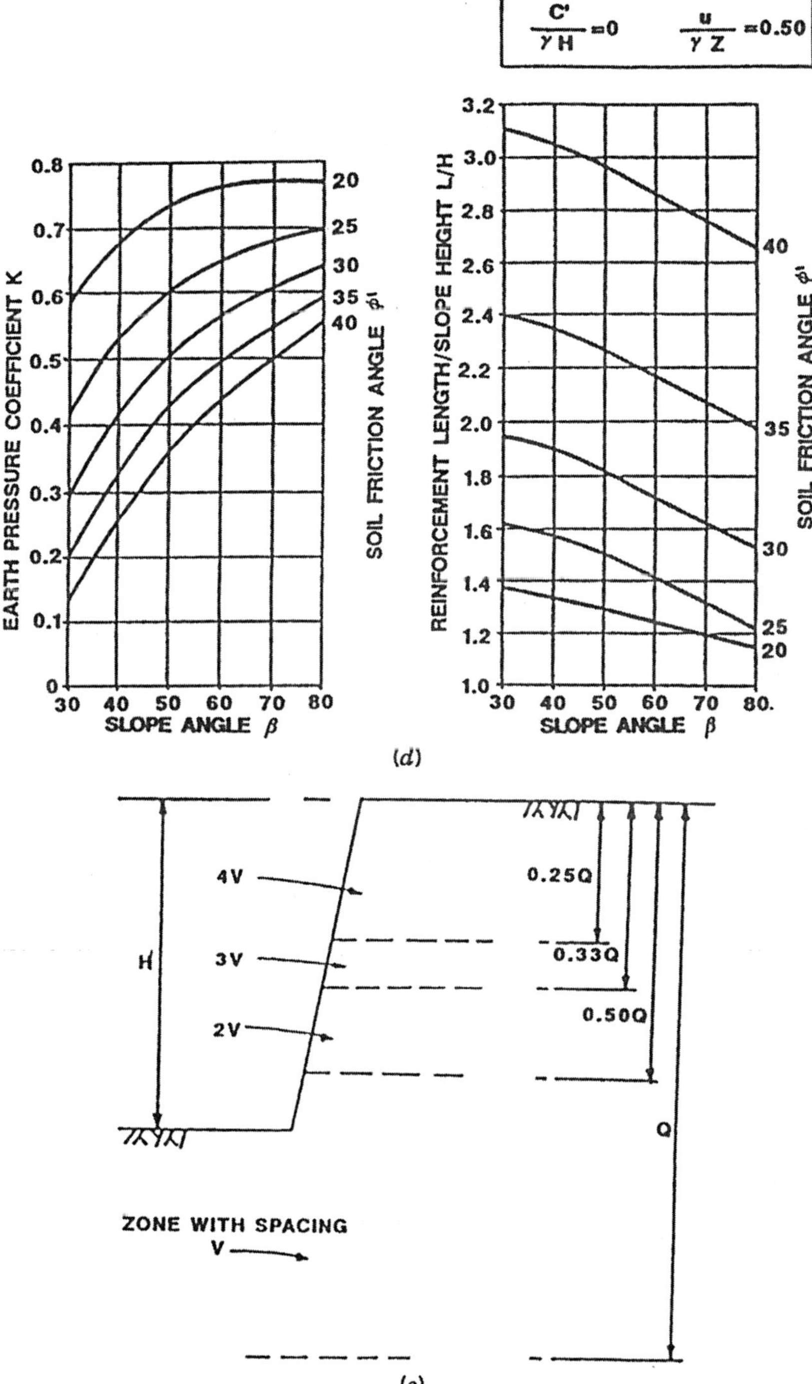

Figure 7.46 (*Continued*)

TABLE 7.6 Zones of Equally Spaced Geosynthetic Reinforcement

Spacing in Zone	Depth to Bottom of Zone	Thickness of Zone
υ	Q	$0.5Q$
2υ	$0.5Q$	$0.17Q$
3υ	$0.33Q$	$0.08Q$
4υ	$0.25Q$	$0.05Q$
5υ	$0.2Q$	$0.03Q$
6υ	$0.17Q$	$0.017Q$

(7) Perform a tabular calculation for the number and spacing of the reinforcement layers. The calculation organizes reinforcement layers into zones of equal multiples of the minimum spacing, υ. The depth from the slope crest to the bottom of any one zone of equally spaced reinforcement (at spacing υ, 2υ, 3υ, etc.) is illustrated in Figure 7.46*e* and is given in Table 7.6.

If the height of the slope $H >$ Q, reinforcement at a closer spacing than υ is required in the lower parts of the slope and special design details will be needed. For cases in which $H < Q$, the thickness and vertical spacing in the deepest reinforced zone depends on the relative values of H and Q.

Geosynthetically reinforced embankments should always have one layer of reinforcement placed on the foundation at the base of the slope to calculate the reinforcement spacings. Working from the base, the number of reinforcement layers in the first zone of equal spacing can be calculated by dividing the thickness of the zone by the spacing of the reinforcement layers in that zone. In Figure 7.46*e*, for example,

$$\text{No. of layers in the deepest zone} = \frac{H - 0.5Q}{\upsilon}$$

The result is unlikely to be an integer, so round down to the nearest whole number of layers and add the remaining thickness to the overlying zone. Repeat the process to the top of the slope.

(8) As a check on the number of reinforcement layers derived above, calculate the total horizontal force required for equilibrium,

$$T = 0.5K\gamma H^2 \qquad \text{(Eq. 7-9)}$$

where T = total horizontal force required for equilibrium
K = earth pressure coefficient
H = unit weight of backfill material
H = height of slope

Check that the following condition is satisfied:

$$\frac{T}{\text{number of layers}} < P$$

where P = factored reinforcement strength. This will ensure that the average load in the reinforcement layers does not exceed the factored design strength.

Two examples that demonstrate the utility and flexibility of geosynthetic reinforced fills for slope stabilization are the Interstate Route H-3 Access Roads and California State Route 4 Willow Pass grade lowering projects. Interstate Route H-3 is a new 11-mile-long highway built northeast of Honolulu on the island of Oahu, Hawaii. Construction involves a series of access roads, viaducts, tunnels, and at-grade sections at this remote site. The first phase of construction consisted of the construction of access roads from existing roads to the tunnel portal sites in heavily vegetated fluted topography. The surficial geology consisted primarily of alluvium, colluvium, and residual soil derived from the host weathered basalt formations.

Geosynthetically reinforced fills were designed for use in locations where cutting and filling of the native slopes alone would not provide the required roadway geometrics for vehicular and construction traffic (Figure 7.47). The length of the fabric layers was determined from analyses of both internal and external stability. The required length of the fabric strips was 14 feet (Riccobono and Hansmire, 1991). Based upon

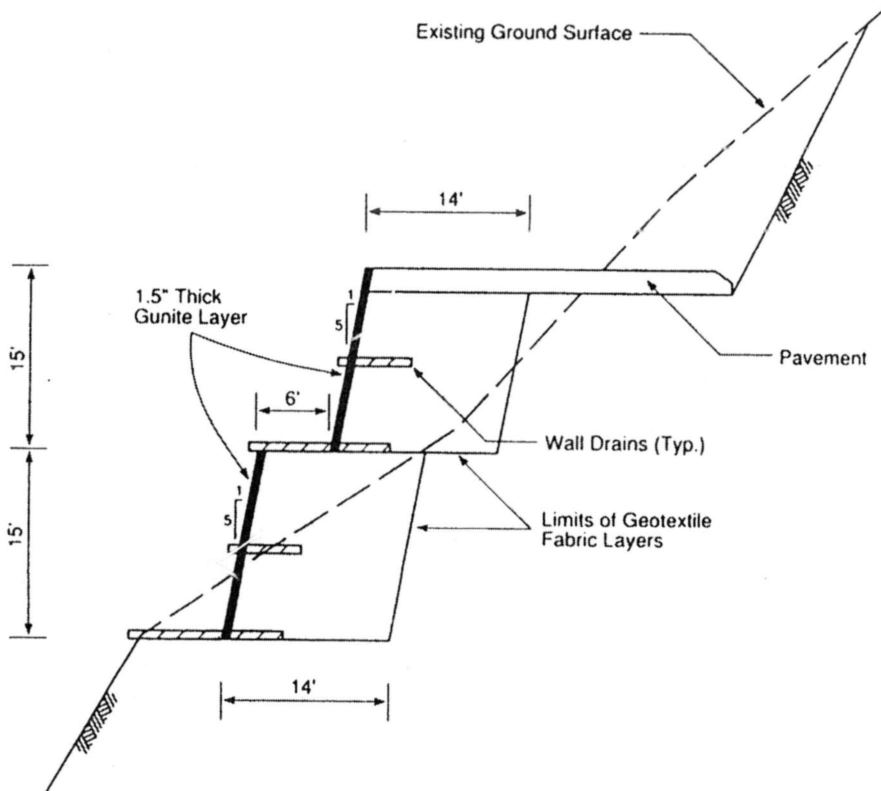

Figure 7.47 Geotextile fabric wall for Interstate Route H-3 access road.

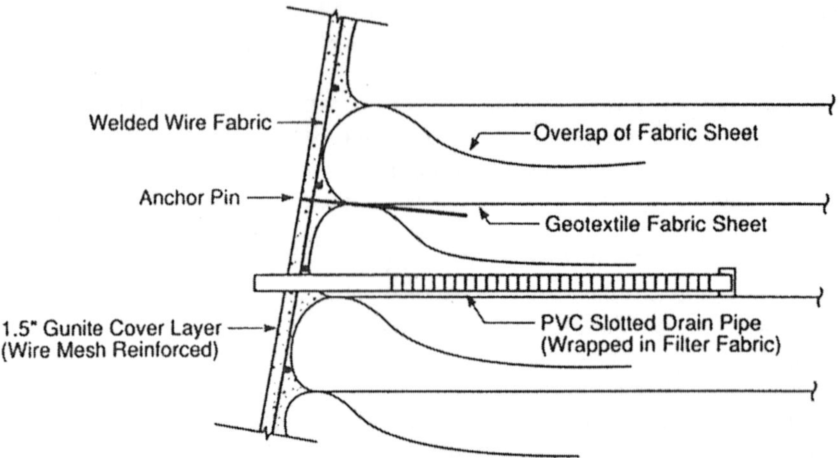

Figure 7.48 Geotextile wall detail.

a minimum wide strip tensile strength of 60 pounds per square inch at 10 percent strain, the required vertical spacing of the fabric layers from 0 to 20 feet below the top of the wall varied between 6 inches and 12 feet. For wall heights greater than 20 feet, the required fabric tensile strength was increased to 90 pounds per square inch. For depths less than 4.5 feet below the top of the wall, an overlap length of 4.5 feet was required (Figure 7.48). For greater depths, the minimum length of 3 feet was required. Shotcrete was used to protect the facing of the fabric wall.

In another example, geosynthetic slopes were used to temporarily retain steep slopes adjacent to traffic lanes as a new embankment was built in stages to expand and improve a busy state road in California. To flatten the grade on State Route 4 in Concord, California, a deep cut was made at Willow Pass and the excavated soil was used to raise the grade adjacent to the pass for the improvement of highway geometrics and to allow for the construction of a transit system extension (Johnston and Abramson, 1994). Staged construction required temporary earth slopes as steep as 0.8H : 1V and up to 36 feet high, reinforced with geosynthetics. No facing elements were used. The slope was designed for heavy truck traffic and a seismic force of $0.15g$. Geotextile filter fabric was installed in layers in the embankment as it was placed (Figure 7.49). Wooden forms were used to wrap the face with fabric (Figure 7.50). The geosynthetic strips were left in place after completion of the embankment. The exposed filter fabric was eventually covered over with soil as the embankment was completed.

7.6 RETAINING WALLS

The most common use of retaining walls for slope stabilization is when a cut or fill is required and there is not sufficient space or right-of-way available for just the slope

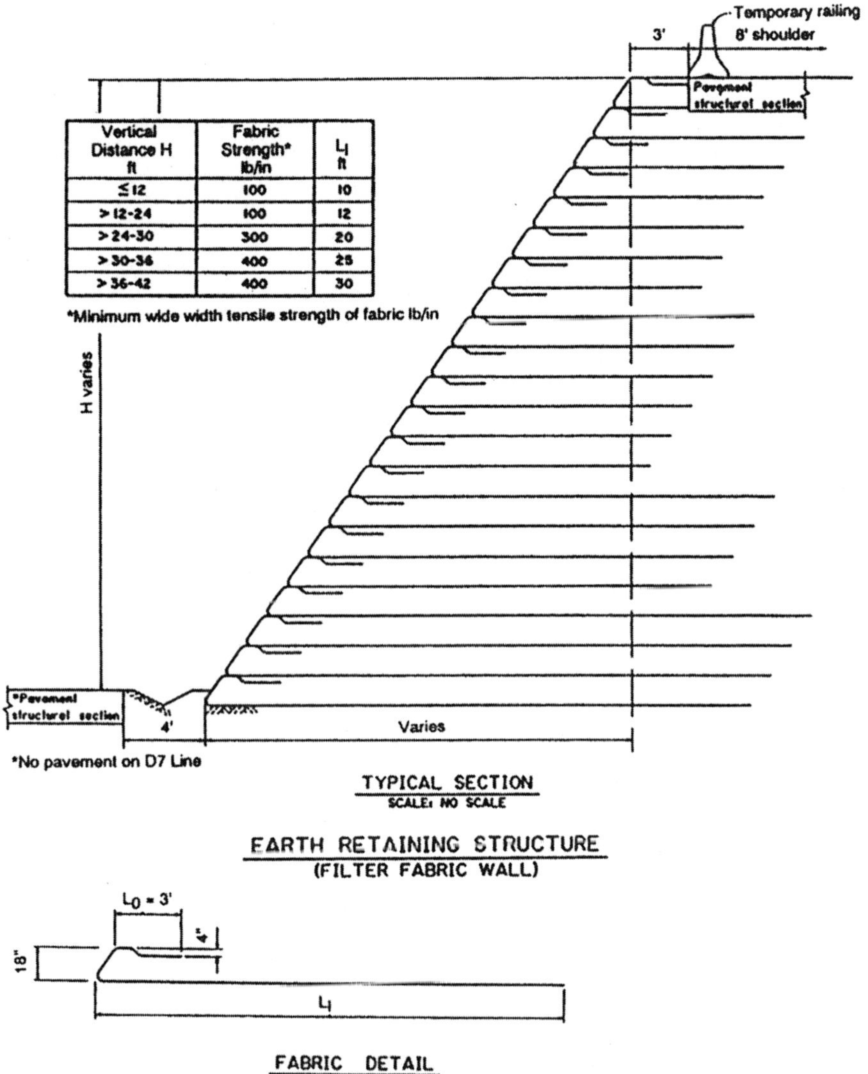

Figure 7.49 Geotextile fabric reinforced slope design for the State Route 4 Willow Pass grade lowering project, California.

itself. The wall should be deep enough so that the critical slip surface passes around it with an adequate FOS, as shown in Figure 7.51. In addition, the ability of the retaining wall to perform as a stabilizing mass is a function of how well it will resist overturning moments, sliding forces at or below its base, and internal shear forces and bending stresses.

Figure 7.50 State Route 4 Willow Pass grade lowering during construction.

Retaining wall types include:

(1) Conventional gravity or cantilever retaining walls
(2) Driven piles
(3) Drilled shaft walls
(4) Tieback walls

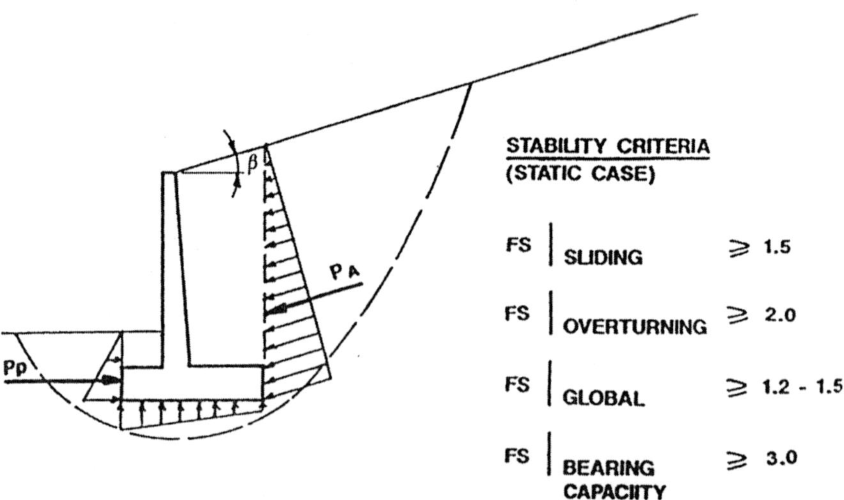

Figure 7.51 Different wall stability criteria (static case).

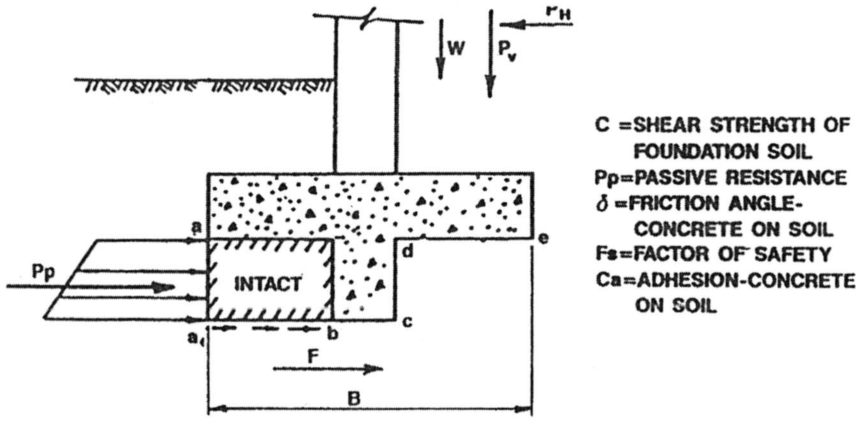

RESISTANCE AGAINST SLIDING ON KEYED FOUNDATIONS

COHESIVE SOILS $F=(W+P_V)\ \mathrm{TAN}\ \delta + C_a(B-\overline{a_1b})+C\ (\overline{a_1b})+P_P$

GRANULAR SOILS $F=(W+P_V)\ \mathrm{TAN}\ \delta + P_P$

$F_S=F/P_H$ (SLIDING)

Figure 7.52 Provision of a shear key to increase sliding resistance. (NAVFAC, 1982.)

Each of these retention systems is discussed in the following subsections. Since it is considerably difficult and costly for the retention system to block groundwater flow, walls should be designed with adequate drainage systems behind or through the walls.

7.6.1 Gravity and Cantilever Retaining Walls

The design of retaining walls is based on classical soil mechanics and should consider the forces that drive overturning and sliding at the wall base. Overall stability of the walls should also be satisfied.

Shear keys are sometimes required to provide adequate sliding resistance (Figure 7.52). Construction of a shear key sometimes requires excavation of a trench at the toe of an unstable slope. In such circumstances, care should be exercised so as not to further undermine the slope by using a bracing and shoring system or only excavating small sections at one time.

7.6.2 Driven Piles

Driven piles are sometimes used to provide stabilization of landslides in natural hillsides and engineered slopes (Figure 7.53). This method is only appropriate for shallow slides and soils that will not tend to flow between the piles. Deep-seated slides often generate very high lateral forces, which cannot easily be resisted by piles.

Piles should be embedded in firm and competent ground to avoid being uprooted and overturned. A slab of reinforced concrete is sometimes placed between the piles

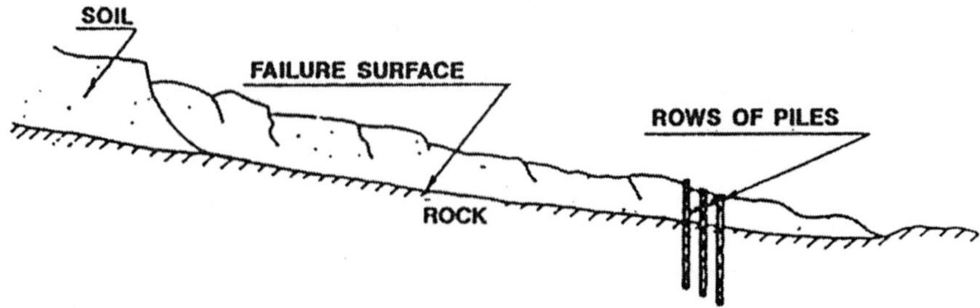

Figure 7.53 Use of driven piles to stabilize slopes. (Zaruba and Mencl, 1982.)

to increase the overall effectiveness of the system and to arrest soil flow between piles. Overall, driven piles are not as effective to arrest the sliding mass of an unstable slope as other stabilization methods.

7.6.3 Drilled Shaft Walls

In many urban locations, flattened slopes or counterweight fills are not a practically feasible solution to slope stability problems. Right-of-way limitations and the presence of existing private and commercial structures limit the types of stabilization methods selected. In such cases, drilled shafts (usually 2 to 5 feet in diameter) can be installed as a restraining system (Figure 7.54*a*).

The drilled shafts must be embedded deeply enough into a bearing stratum to provide resistance against the lateral forces transmitted from the unstable soil mass (Figure 7.54*b*). The depth of the drilled shafts should also pass through the potential critical slip surface. Because of arching effects between drilled shafts, the shafts are usually spaced a distance of three pile diameters apart.

Contiguous drilled shaft Walls (Figure 7.55) are sometimes constructed to retain large open cuts more than 100 feet deep. Because of the contiguous nature of the drilled shafts, adequate drainage systems should be provided behind the walls to facilitate water seepage and to avoid buildup of hydrostatic pressures.

Although drilled shaft walls can correct or prevent soil slope stability problems, they have inherent limitations that must be considered when evaluating their possible use. Driving forces on the wall increase as a function of the height squared. Higher walls necessitate increased depth of penetration below the failure surface, greater drilled shaft diameters, and additional reinforcing to resist the overturning moments. Hence, the construction cost of the walls becomes very expensive with height. The cost of the walls can be reduced when combined with tiebacks or axial post-tensioning.

7.6.4 Tieback Walls

Tieback walls can be used instead of conventional walls when wall location or space constraints limit excavation of the footing (Figures 7.56 and 7.57*a*). Tieback wall designs use the principle of carrying the lateral earth pressure on the wall by a "tie"

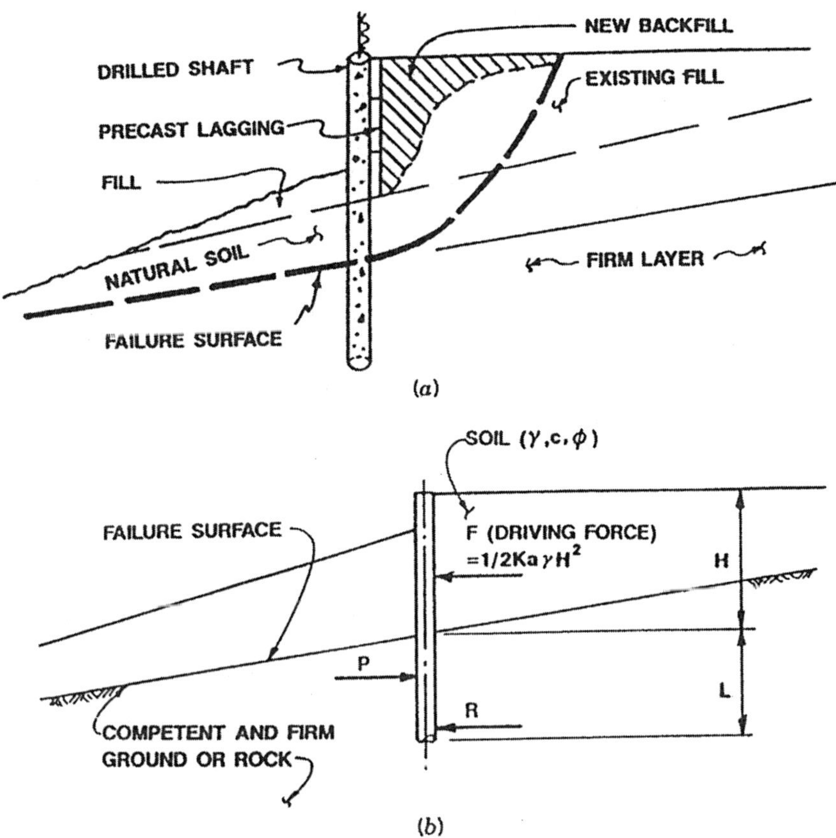

Figure 7.54 (*a*) Drilled shaft wall (cylinder pile wall) to stabilize deep-seated failure. (From Nethero, 1982, reproduced by permission of ASCE.) (*b*) Drilled shaft wall design concept.

system that transfers the imposed load to a zone behind the potential or existing slip plane where satisfactory resistance can be established. The tiebacks consist of post-tensioned steel cables, rods, or wires attached to deadmen (as used in embankments) or grouted to a firm, strong bearing stratum (as used in cuts).

Permanent tiebacks are installed routinely in cohesionless soils but seldom in soft to medium cohesive soils because of concern about long-term load holding ability. According to Weatherby and Nicholson (1982), tiebacks installed in soils with unconfined compressive strengths greater than 1 ton per square foot and a consistency index, I_c, greater than 0.8 do not experience significant loss of load or movement with time. The consistency index is defined as

$$I_c = \frac{LL - w}{LL - PL} \qquad \text{(Eq. 7-10)}$$

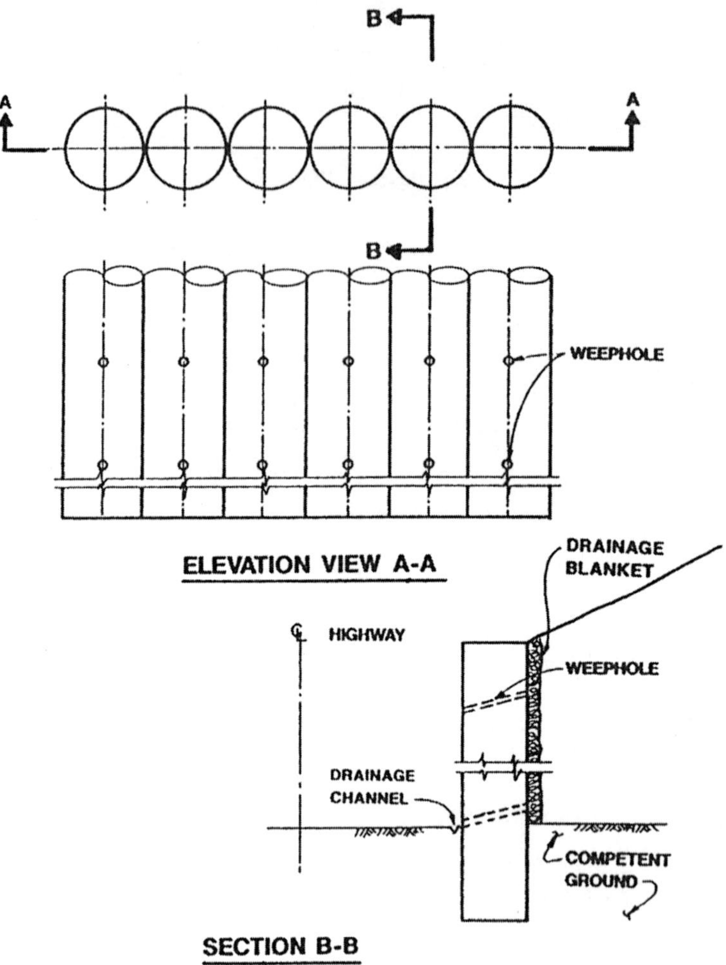

Figure 7.55 Schematic diagrams of a contiguous drilled shaft wall.

where LL = liquid limit
w = natural water content
PL = plastic limit

For the design of a tieback wall, the following items may serve as a guide (Weatherby and Nicholson, 1982):

(1) The design load usually varies between 50 and 130 tons. Tieback tendons of this capacity range can be constructed without heavy equipment and the drill hole need not be larger than 6 inches. In addition, the stressing and testing equipment can be readily handled without using power lifting equipment.

Figure 7.56 Tieback reinforcement for construction of railroad.

(2) The length of a tieback is controlled by stability requirements. The unbonded length of the tieback should be selected so that the anchor is located beyond the potential critical failure surface. The total length of the tieback should be established so that the probable failure surface just behind the tiebacks would have a factor of safety equal to or greater than that on the critical failure surface. Figure 7.57*b* illustrates these two cases.

(3) For shallow failure surfaces and when the wall penetrates the failure Surface, one row of tiebacks is enough to support the wall (Figure 7.57*c*). In this case, the wall is designed to provide shear resistance across the failure surface. At least two rows of tiebacks are required when the failure surface is deep and the wall cannot practically penetrate the failure surface. Figure 7.57*d* illustrates a multitiered tieback wall that does not penetrate the failure surface.

(4) It is desirable that a minimum of 15 feet of overburden be above the anchor bond zone. Most tiebacks are installed at an angle of between 10 and 30° from the horizontal. Special grouting techniques may be required if the angle of inclination is less than 10°. When a suitable anchoring stratum lies at some depth, an angle up to 45° may be chosen. The steeper the angle is, however, the less the tieback force acts horizontally and the longer the tieback needs to be for a given horizontal force required. By increasing the angle of inclination, the vertical load component of the tieback also increases, thus increasing the downward loads on the wall members and the underlying foundation materials. This must be taken into account during design of the soldier piles and other supports.

(5) The drill hole diameter of a tieback is usually between 3 and 6 inches. The majority of soil tiebacks are drilled with augers or casing. Because of the casing weight and the associated handling and drilling problems for larger casings, the largest common size of casing is 6 inches.

(6) Two types of tendons are commonly used:

(a) 0.6-inch-diameter 7-wire strands having an ultimate tensile strength of 270 kips per square inch; and

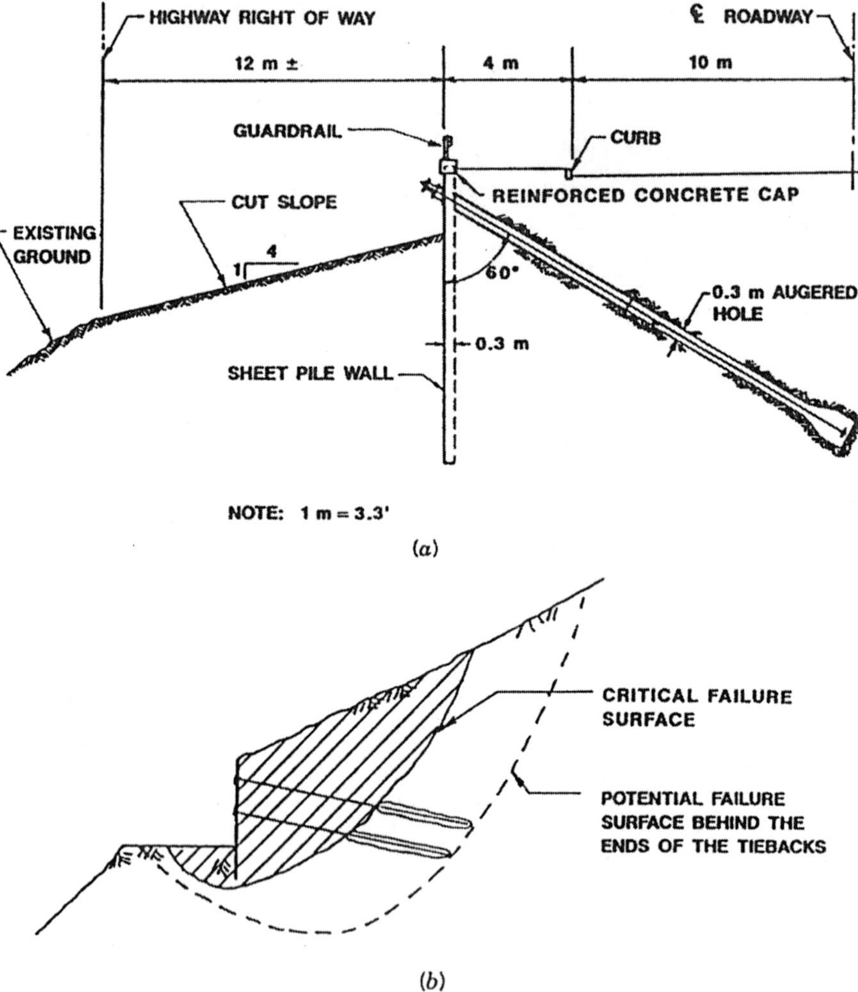

Figure 7.57 (*a*) Section of tieback wall to correct slide condition on New York Avenue in Washington, DC. (*b*) Stability analysis for determining the unbonded and total tieback length. (Weatherby and Nicholson, 1982.)

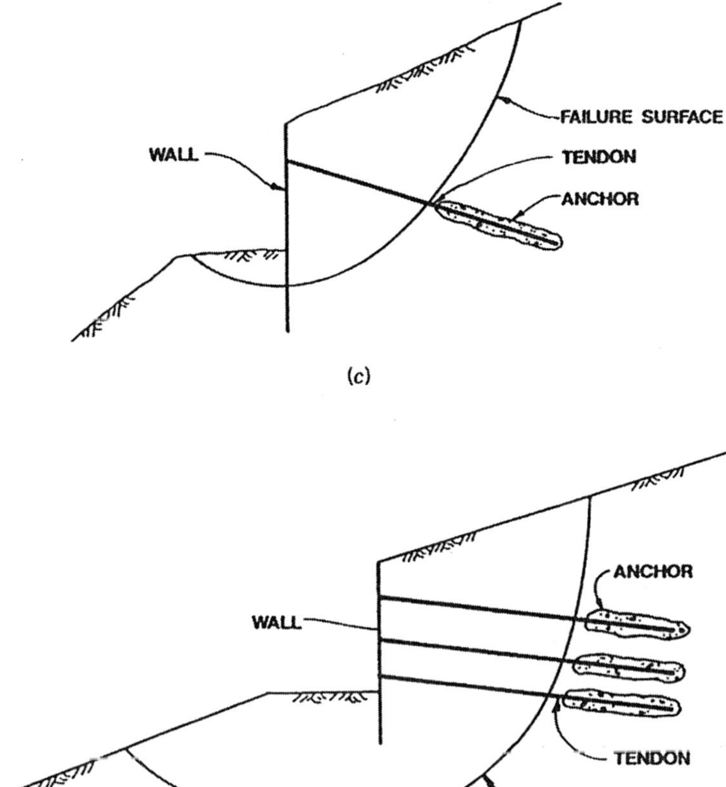

Figure 7.57 (*Continued*)

(b) 1 to 1.375-inches-diameter deformed bars having an ultimate tensile strength of 1.50 kips per square inch.

(7) All permanent tiebacks should be corrosion protected. Most of them are protected by portland cement grout along the anchor length and a grease-filled tube or heat-shrinkage sleeve over the unbonded length (Figure 7.57*e*). If the soil surrounding the anchor length has a pH less than 5, a resistivity less than 2,000 ohm-centimeters, or sulfides present, then the tendon should be completely encapsulated in a plastic or steel tube.

(8) Where necessary, tieback easements from adjacent property owners must be obtained.

The most economical tieback installation will be achieved if the contractor has some degree of flexibility in selecting the tieback type and construction method. The

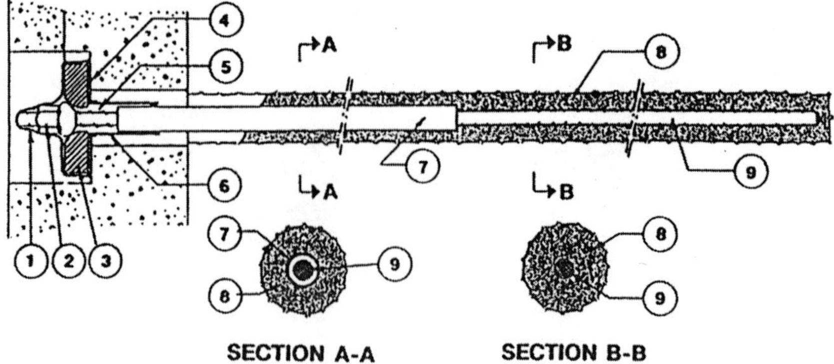

LEGEND:

1) INSULATING COVER OF PREFORMED PLASTIC, HEAT SHRINKABLE SLEEVE, OR MOLDABLE TAPE.
2) NUT.
3) BEARING PLATE
4) BEARING PLATE INSULATION
5) ANTICORROSION GREASE.
6) PLASTIC TRUMPET.
7) GREASE FILLED PVC OR POLYETHYLENE SHEATH
8) ANCHOR GROUT.
9) TENDON.

(*e*)

Figure 7.57 (*Continued*)

designer should specify minimum unbonded length, the minimum total tieback length, and the unit tieback capacity specified or a loading diagram specified at each tieback level. In addition, the type or desired level of corrosion protection must be specified, as well as the method of verifying the long-term load capacity. Finally; each production tieback should be tested to verify that the anchor will carry the design load.

7.7 VEGETATION

Vegetation (grass, shrubs, and trees) is highly effective and advantageous for soil stabilization purposes. Removal of earth to construct cuts and embankments inevitably removes the vegetative covering and the surface soils are left exposed and susceptible to runoff and wind attack. Vegetation stabilizes the soil surface by the intertwining of its roots, minimizes seepage of runoff into the soil by intercepting rainfall, and retards runoff velocity. In addition, vegetation may have an indirect influence on deep-seated stability by depleting soil moisture, attenuating depth of frost penetration, and providing a favorable habitat for the establishment of deeper-rooted vegetation (shrubs and trees).

Vegetation is multifunctional, relatively inexpensive, self-repairing, visually attractive, and does not require heavy or elaborate equipment for its installation. However, there are certain limitations. Vegetation is susceptible to blight and drought. It is difficult to get established on steep slopes. It is unable to resist severe scour or wave action, and it is slow to become established.

7.7.1 General Design Considerations

Vegetation can affect the balance of stresses in a slope due to mechanical reinforcement from the root system of trees, slope surcharge from the weight of trees, modification of soil moisture, reduction of pore pressures by interception and transpiration from the foliage, attenuation of frost depth penetration, and lateral restraint by buttressing and soil-arching action from the trunks or stems.

Soil erodibility can contribute to slope instability and is another design factor to be considered. The susceptibility of soil particles to detachment and transport by rainfall and runoff depends on soil textures, slope lengths, and angles, which are discussed below.

Root Reinforcement Root reinforcement in soils provides apparent cohesion. The results of several root and fiber soil reinforcement studies are shown in Table 7.7

TABLE 7.7 Tensile Strength of Tree Roots

Tree Species	Root Diameter (mm (in.))	Tensile Strength (psi)	Average Tensile Strength, All Size Classes (psi)
Rocky Mountain Douglas fir	2 (0.08)	3,285	
	4 (0.16)	3,226	
	6 (0.23)	2,579	2,653
	8 (0.39)	2,349	
	10 (0.39)	2,152	
Coastal Douglas fir	2 (0.08)	8,214	
	4 (0.16)	8,504	
	6 (0.23)	6,846	7,083
	8 (0.31)	6,482	
	10 (0.39)	6,243	
Spruce–Hemlock	2 (0.08)	1,450	
	4 (0.16)	1,390	1,375
	6 (0.23)	1,380	
Birch	2 (0.08)	6,600	
	2–7 (0.08–0.27)	3,170	5,305
	>15 (0.59)	6,560	

Source: Burroughs and Thomas (1976), Gray (1978), and Turmanina (1965).

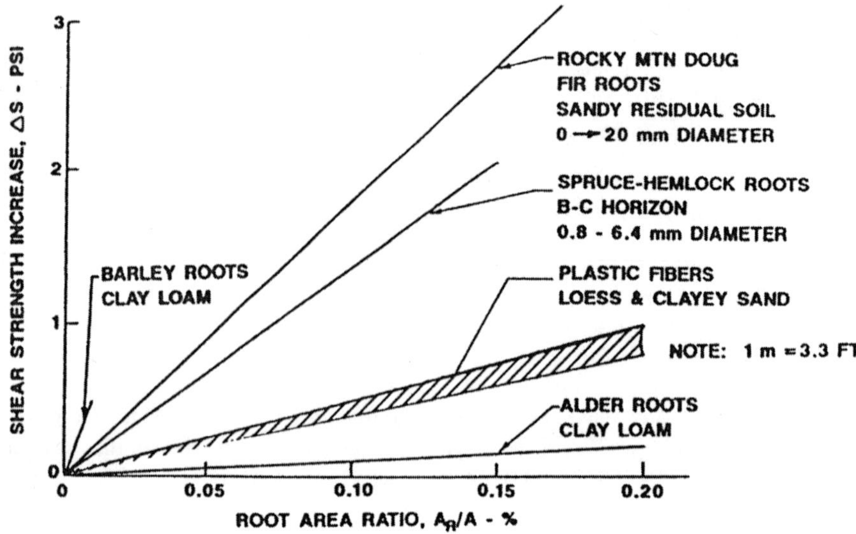

Figure 7.58 Summary of the shear strength increase resulting from fiber and root reinforcement of various soils. (Gray, 1978.)

and summarized graphically in Figure 7.58 (Gray, 1978). The figure shows that for typical root area ratios (0.05 to 0.15 percent), the amount of apparent cohesion or shear increase provided by the live or fresh roots is in the range of 0.5 to 2.5 pounds per square inch. The theory of root reinforcement is discussed in detail by Barker (1986), Gray (1978), and Waldron (1977). Figure 7.59 shows how root strength can be incorporated into stability analyses.

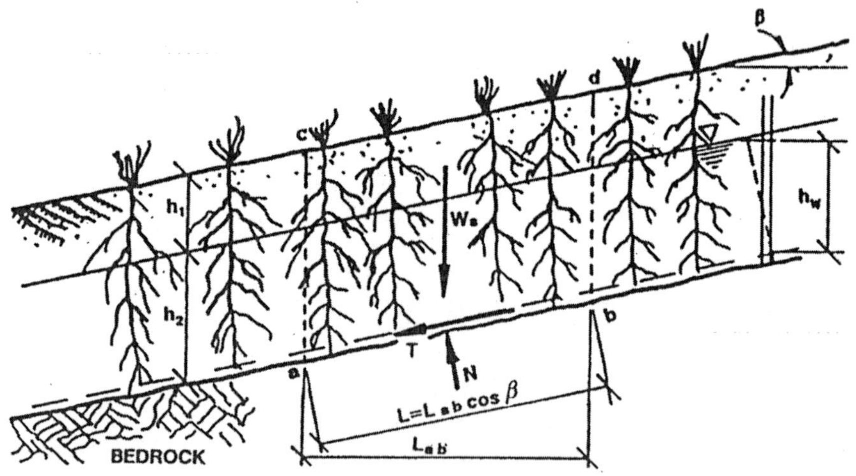

Figure 7.59 Forces on a soil mass about to slide. (Bache and MacAskill, 1984.)

The downslope component of soil weight can be expressed as

$$T = W_s \sin \beta \qquad \text{(Eq. 7-11)}$$

where W_s = bulk unit weight of the soil
β = slope angle

The total shear strength, S, of the soil/root system, reinforced by roots contributing an increase in shear strength of ΔS_R, is given by the modified form of Coulomb's equation for shear strength:

$$S = (S_s + \Delta S_R) + \sigma' \tan \phi' \qquad \text{(Eq. 7-12)}$$

where S_s = shear strength of the root-free soil
σ' = effective normal stress
θ' = effective angle of friction of the soil

Therefore,

$$\text{FOS} = \frac{(c' + \Delta s)L + W_s' \cos \beta \tan \phi'}{W_s \sin \beta} \qquad \text{(Eq. 7-13)}$$

where W_s' = buoyant weight of the soil
$= \gamma_1 h_1 + (\gamma_2 - \gamma_w)h_2$
$W_s = \gamma_1 h_1 + \gamma_2 h_2$
c' = effective cohesion of root-free soil
Δs = shear strength increment per unit area of soil where $\Delta S_R = \Delta s L$
γ_1 = unit weight of soil above groundwater table
γ_2 = unit weight of soil below groundwater table

Another factor contributing to the strength of root-reinforced soil is the density of soil roots per volume of soil. Effective buttressing action of tree roots increases resistance to sliding of soil masses on slopes and is favored by deep penetration of main tap roots and secondary sinker roots. Because of aeration requirements, roots of most trees tend to be concentrated near the slope surface. Some trees, such as ponderosa pine, have an extensive vertical and lateral root system, as illustrated in Figure 7.60.

Hydrologic Effect Vegetation can affect the stability of slopes by modifying the hydrologic regime of the soil. Interception and transpiration of moisture by trees tend to maintain drier soils and mitigate or delay the onset of waterlogged or saturated soil conditions. Such conditions have been known to cause slope failure. Conversely, felling trees tends to produce wetter soils and faster recharge times following intense rainstorms.

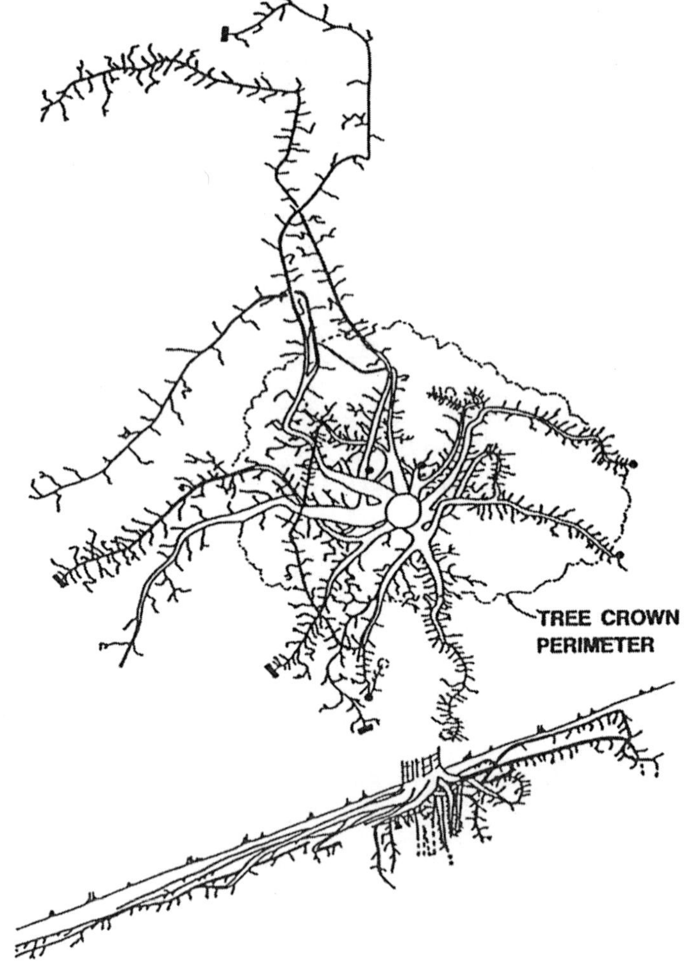

Figure 7.60 Root structure and morphology of a 60-year old ponderosa pine tree. (Curtis, 1964.)

Rice and Krames (1970) suggested that climate determines the relative contribution of transpiration and vegetative modification of soil moisture to prevent landslides. In climates where precipitation greatly exceeds potential evapotranspiration, this contribution is small. However, in more arid climates, where substantial moisture depletion develops each summer, differential use of water by different types of plant covers may significantly affect the occurrence of landslides. The effect of tree cover on soil moisture is shown in Figure 7.61 (Gray, 1978). The figure indicates that the forest cover has little effect on the soil moisture regime once sufficient precipitation falls on the slope and eliminates soil moisture suction. Conversely, the soil moisture suction is definitely higher during the drier season of the year.

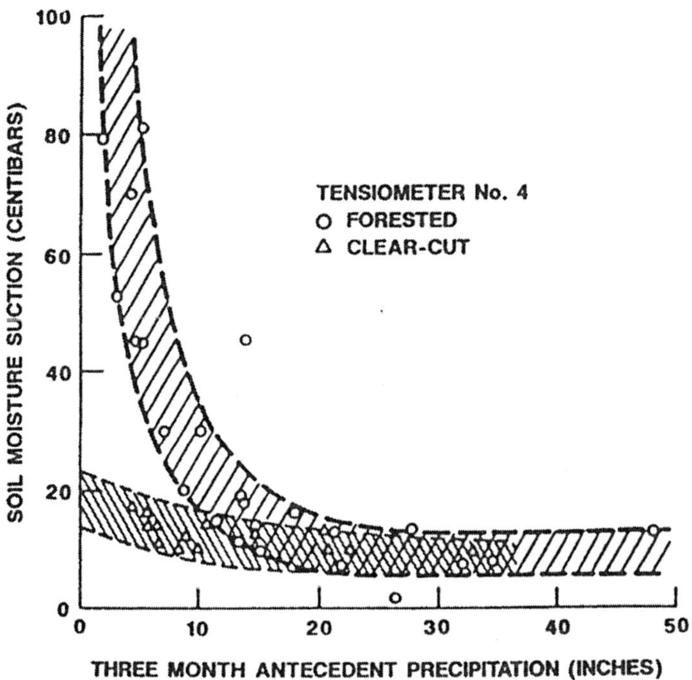

Figure 7.61 Comparison of soil moisture tension versus antecedent precipitation. (Gray, 1978.)

Soil Erodibility Soil erodibility is a measure of the susceptibility of soil particles to being eroded by rainfall and runoff. Erosion, particularly at the toes of slopes, is known to trigger landslides. Soil erodibility is measured by a factor, K, which is affected by soil textures, lengths, and slope gradients. A nomograph that can be used to estimate K values (Figure 7.62) was developed by the Soil Conservation Service of Utah (Erickson, 1977; Wischmeier and Smith, 1965). The K value usually ranges from 0.02 to 0.69.

Slope gradient and slope length strongly influence the transport of soil particles once the particles are dislodged by raindrop impact or by runoff. The combined effect of slope length and slope gradient is measured by a slope length–gradient factor, L.S., which is ratio of soil loss per unit area on a site to the corresponding loss from 72.6-foot-long experimental plot with a 9 percent slope. Values of L.S. factors are listed in Table 7.8 for slopes between 1 and 100 percent (i.e., between 100H : 1V and 1H : 1V) and lengths between 50 and 300 feet. The values are derived from an empirical equation developed by Wischmeier and Smith (1965), given as

$$\text{L.S.} = \left(\frac{65.41s^2}{s^2 + 10{,}000} + \frac{4.56s}{\sqrt{s^2 + 10{,}000}} + 0.065 \right) (l/72.5)^m \qquad \text{(Eq. 7-14)}$$

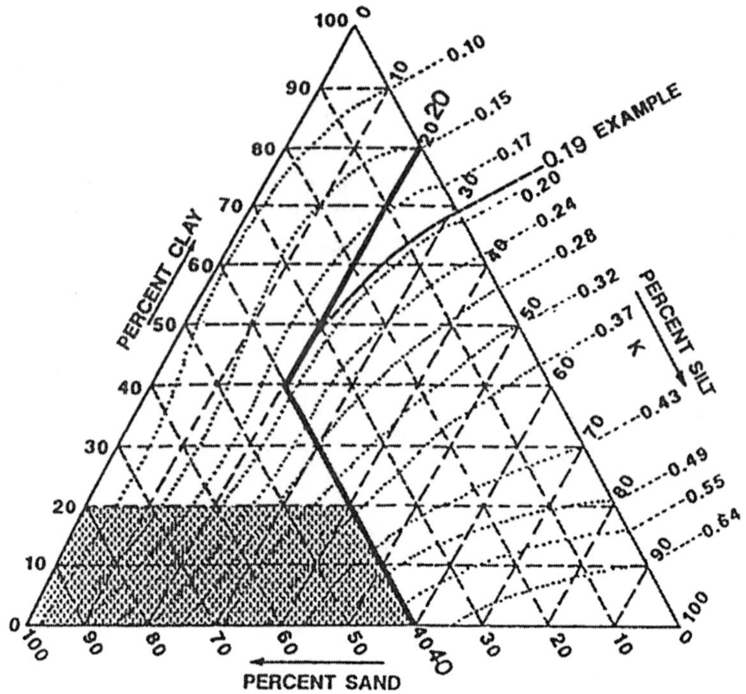

Figure 7.62 Triangular nomograph for estimating K value. (Goldman et al., 1986.)

where L.S. = length slope factor
l = slope length in feet
s = slope steepness
m = exponent dependent upon slope steepness:

TABLE 7.8 Length Slope Factor Values versus Slope and Slope Lengths

Slope (H : V)	Length Slope Factor Values (L.S.) Slope Lengths (feet)					
	50	100	150	200	250	300
100 : 1	0.11	0.12	0.14	0.14	0.15	0.16
20 : 1	0.38	0.53	0.66	0.76	0.85	0.93
10 : 1	0.97	1.37	1.68	1.94	2.16	2.37
5 : 1	2.88	4.08	5	5.77	6.45	7.06
3 : 1	6.67	9.43	11.55	13.34	14.91	16.33
2 : 1	12.6	17.82	21.83	25.21	28.18	30.87
1.5 : 1	18.87	26.68	32.68	37.74	42.19	46.22
1 : 1	29.87	42.24	51.74	59.74	66.79	73.17

Source: Goldman et al. (1986).

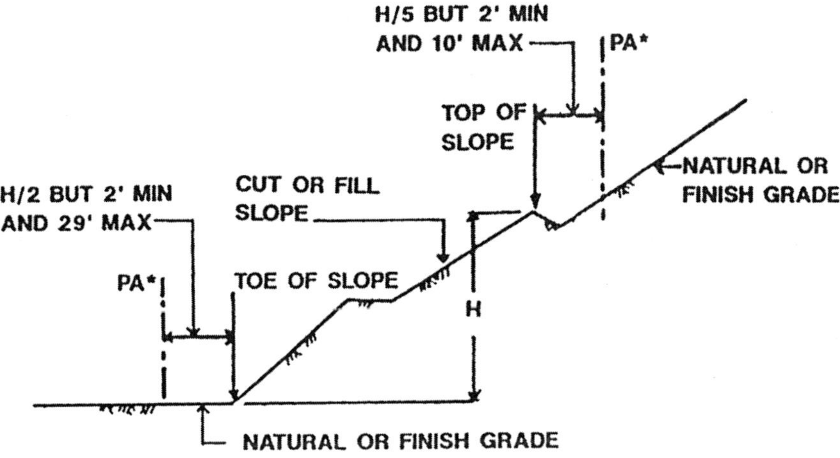

* PERMIT AREA BOUNDARY

Figure 7.63 Terracing of slopes. (Uniform Building Code, 1991.)

0.2 for slopes < 1%
0.3 for slopes 1 to 3%
0.4 for slopes 3.5 to 4.5%
0.5 for slopes > 5%

In flat areas, runoff is slow and soil particles are not moved far from the point of raindrop impact. Thus L.S. is less than 1 for slopes less than 9 percent and lengths less than 50 feet. On steep slopes, soil movement increases dramatically. Doubling the gradient from 3H–1V to 1.5H–1V triples the L.S. factor (that is, triples the soil loss).

The effect of slope length is not as great as the effect of slope angle. Thus very long slopes and especially long steep slopes should be avoided, and those that exist should not be disturbed without arrangements for erosion prevention. Slope length can be shortened by installing midslope diversions. Local building codes often require terraces or drainage ditches at specified intervals. Chapter 70 of the Uniform Building Code (1991) specifies a 30-foot interval (Figure 7.63). Some erosion control manuals (Association of Bay Area Governments, 1981; Geotechnical Control Office, 1984) recommend 15- to 25-foot intervals between terraces.

7.7.2 Vegetation Species

Since plants and grass absorb different amounts of water depending on the type of soil that they grow in, there are several different criteria for the selecton of the most appropriate species. A general rule of thumb is to use local plants and grass that are adaptable to local climate. Deciding exactly what types or species are needed requires the aid of horticulture and landscaping experts. In general, vegetation that absorbs large amounts of water from the soil are best in clayey soils to ensure a drier and stronger soil crust. On the contrary, species that absorb less water would be ideal for sandy soils because intense drying of sandy surface soils makes them more susceptible to erosion.

7.7.3 Erosion Control Mats and Blankets

As discussed above, vegetation can be used to protect slopes from erosion. However, the plant seeds that are placed usually require two to six weeks for germination (a time frame also influenced by the amount of water that occurs). During the period of germination, seeds must be protected. Erosion control mats and blankets are often used to prevent the seeds from washing away.

Erosion control mats and blankets consist of natural (organic) and synthetic mats and blankets, roving, and soil confinement systems, as discussed below.

Natural and Synthetic Mats and Blankets Natural (organic) mats and blankets generally are machine- or hand-woven products composed of wood excelsior, straw, or other natural filaments bound together with a tough biodegradable mesh. They retain soil moisture, control surface temperature fluctuations of the soil, conform to the terrain, protect against sun burnout, and break up rain drops to minimize erosion. These materials disintegrate after the plants grow and become well established.

Synthetic mats and blankets are typically composed of a web of continuous or stapled monofilaments bound by heat fusion or stitched between nettings to provide dimensional stability. They are inert to chemicals and stabilized against ultraviolet degradation and, therefore, can remain intact for years as long-term covers.

Installation of erosion-control mats and blankets should follow the manufacturer-recommended procedures. Figure 7.64 illustrates the general installation procedures. Before installation, the site preparation should be shaped to design specifications and dressed to be free of soil clods, clumps, rocks, or vehicle imprints of any significant size that would prevent the erosion-control material from laying flush with the soil surface.

Roving Roving is a pneumatically applied synthetic fiber covering held in place by emulsified asphalt. There are two types of roving. One is fiberglass roving formed from fibers drawn from molten glass. The other is polypropylene roving formed from continuous strands of fibrillated polypropylene. Erosion-control roving is unique because of the flexibility of application that allows any width or thickness of material to be applied. Like the synthetic mats and blankets, fiberglass and ultraviolet-treated polypropylene are resistant to biodegradation and are a long-lasting erosion control material.

Before installation, the slope should be prepared either mechanically or manually so seed, fertilizer, and lime may be applied. Roving is applied using a special nozzle

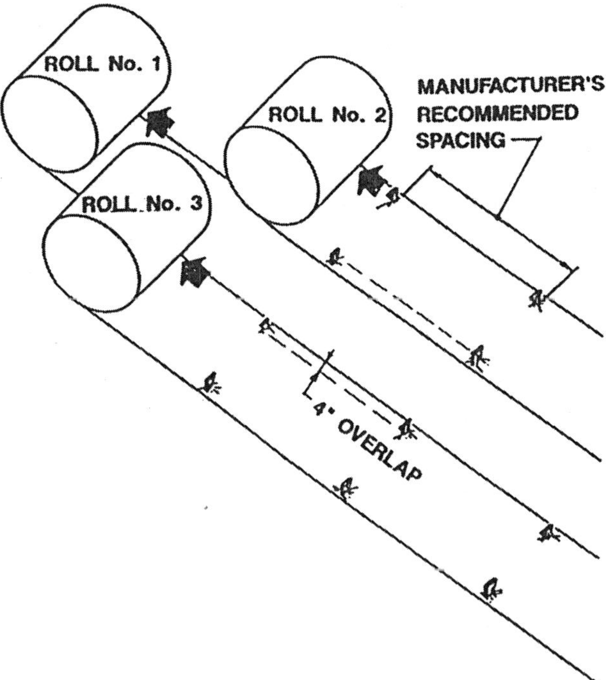

Figure 7.64 Typical installation instruction for erosion-control mats and blankets, follow manufacturer-recommended procedures. (Agnew, 1991.)

connected to an air compressor. As the roving feeds into the nozzle, compressed air, regulated at the nozzle to an approximate pressure of 60 pounds per square inch, propels the roving at a rate of about 0.25 to 0.5 pounds per square yard out of the nozzle, separating the strands into individual fibers. After completing the application, a tack coat of emulsified asphalt is applied at a rate of about 0.25 to 0.33 gallons per square yard to assure adhesion of the fibers to one another and to the soil. Using a coat of black emulsified asphalt makes the roving look somewhat unappealing at first until the vegetation takes over.

Soil-Confinement Systems Soil-confinement systems generally consist of a series of honeycomb-like cells formed into a spreadable sheet or blanket. Sheets of the material are anchored and filled with soil, creating a solid pavementlike surface in areas of poor surface soil stability. The products are generally made from a high-density polyethylene or nonwoven polyester material. Installation of these systems is simple and is illustrated in Figure 7.65. It must be emphasized that, no matter what type of erosion control material is used, it is essential to consult with the manufacturer to ensure proper material selection and installation for specific sites.

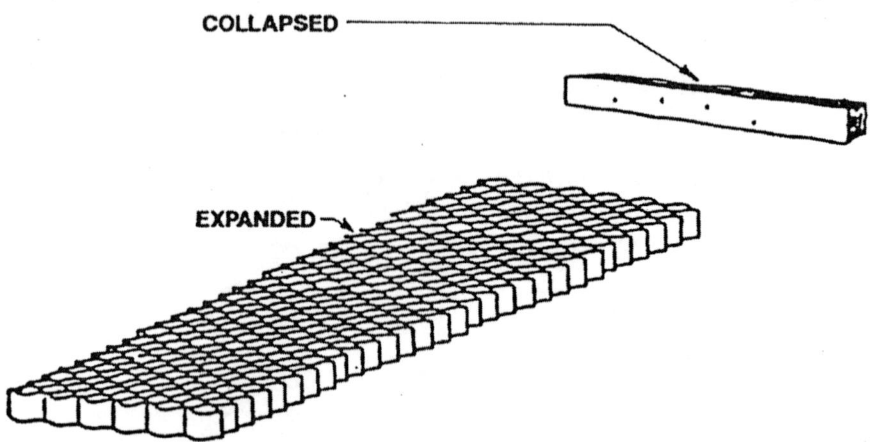

Figure 7.65 Typical installation instructions for soil confinement systems, following manufacturer-recommended procedures. (Agnew, 1991.)

7.7.4 Biotechnical Stabilization

Biotechnical stabilization basically combines the concept of in situ reinforcement by wood stems with surface protection by vegetation. It was pioneered by Gray and Leiser (1982) and entails the use of living vegetation, primarily cut woody-plant material, that is arranged and embedded in the ground to prevent surficial erosion and to arrest shallow mass movement.

The practical application of this technique is confined mostly to fill slopes and embankments, or to soils used as buttresses against a cut. Live cut stems and branches are used as the main soil reinforcement, providing immediate reinforcement. Secondary stabilization occurs as a result of rooting along the length of the buried stems. In addition to its reinforcing role, the woody vegetation serves as a hydraulic drain buttressing element, and a barrier to earth movement. A classification scheme of different biotechnical stabilization techniques is shown in Table 7.9.

The biotechnical stabilization technique employs a method of brushlayering during construction of a fill. A brushlayer fill consists of alternating layers of earth and live, cut branches. The branches are placed in a crisscross pattern between successive lifts of compacted soil, as shown in Figure 7.66. This technique has been used successfully to repair highway slopes in the United States. One example is along State Route 126 in North Carolina. Another example is along State Route 112 in North Massachusetts (Gray and Sotir, 1992).

A brushlayer fill system resembles a buttress of free-draining rock that can be designed as a gravity structure. The brush stems and branches, which reinforce the fill like geosynthetic reinforcement, should be designed for resisting pull-out and tensile failure using the conventional methods (Giroud et al., 1987). The required vertical spacing and length of successive layers of brush reinforcements are determined from the specified FOS, the allowable unit tensile strength, and the interface

TABLE 7.9 Classification of Different Biotechnical Slope Protection and Erosion Control Measures

Category	Examples
Live Construction	
Conventional plantings	Grass seeding Sodding Transplants
Mixed Construction	
Woody plants used as reinforcements and barriers to soil movement	Live staking Contour wattling Brush layering Soft gabions Brush mattress
Plant/structure associations	Breast walls with slope face plantings Revetments with slope face plantings Tiered structures with bench plantings
Woody plants grown in the frontal openings or interstices of retaining structures	Live cribwalls Vegetated rock gabions Vegetated geogrid walls Vegetated breast walls
Woody plants grown in the frontal openings or instices of porous revetments	Joint plantings Staked gabion matresses Vegetated concrete block revetments Vegetated cellular grids "Reinforced" grass
Inert Construction	
Conventional structures	Concrete gravity walls Cylinder pile walls Tieback walls

Source: Gray and Sotir (1992).

friction properties of the reinforcement layer. The allowable unit tensile resistance for a brushlayer can be calculated from the tensile strength of the brush stems, their average diameter, and the number of stems placed per unit width (Sotir and Gray, 1989).

7.8 SURFACE SLOPE PROTECTION

The objective of surface slope protection is to prevent infiltration by rainfall so that the slope can be maintained dry or partially dry. Surface slope protection measures include application of shotcrete or chunam plaster, masonry blocks, or rip-rap. Among these, masonry blocks are the most aesthetically appealing. However, their aesthetic value is not as high as that of vegetation. To improve the appearance of large areas

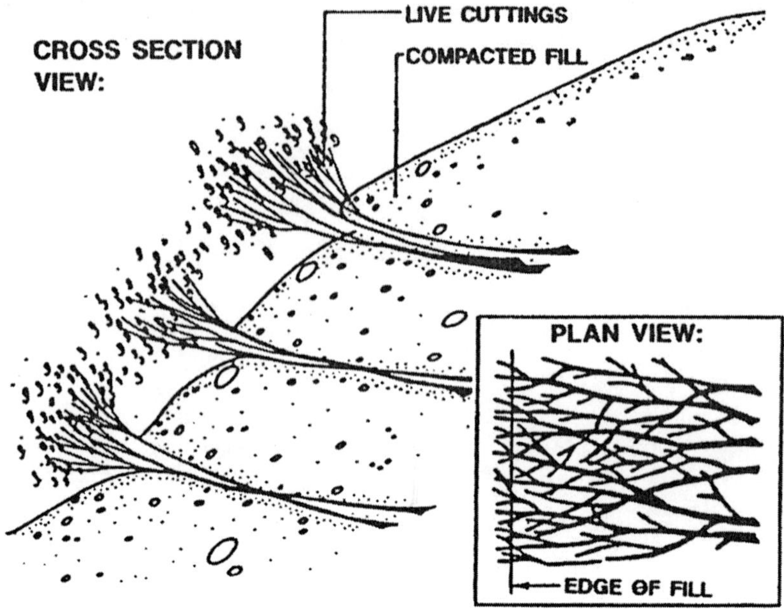

Figure 7.66 Schematic illustration of a brushlayer fill installation. (From Gray and Sotir, 1992, reproduced by permission of ASCE.)

of shotcrete, chunam, plaster, and masonry, selected planting of trees, shrubs, and creeping plants can be used in specially prepared beds (Figure 7.67).

7.8.1 General Design Considerations

The three slope protection measures mentioned above are intended to provide near-impermeable surface protection to slopes. In design, they are not considered to provide resisting forces to the slope like retaining structures, anchoring systems, and buttresses. Because of the near-impermeable nature of the slope protection, consideration should be given to providing drainage of water from the slope. This is normally accomplished with drainage bedding or geosynthetic strips placed on the bare slope face and weep holes penetrating the surface protection on a 5-foot center to center pattern. Cracking of shotcrete and chunam plaster is common due to shrinkage. This can be improved by using a proper mix and by including steel mesh or fibers in the mix.

As for masonry, it is essential to consider the weathering potential and durability of the blocks to be used in construction. Some types of masonry blocks are vulnerable to weathering and subject to losing strength with time under certain types of climatic and environmental conditions.

The selection of slope surface protection measures depends on the overall cost of the work involved. Masonry block pitching is the best choice in terms of strength, durability, and aesthetics. However, it is often the most expensive method because

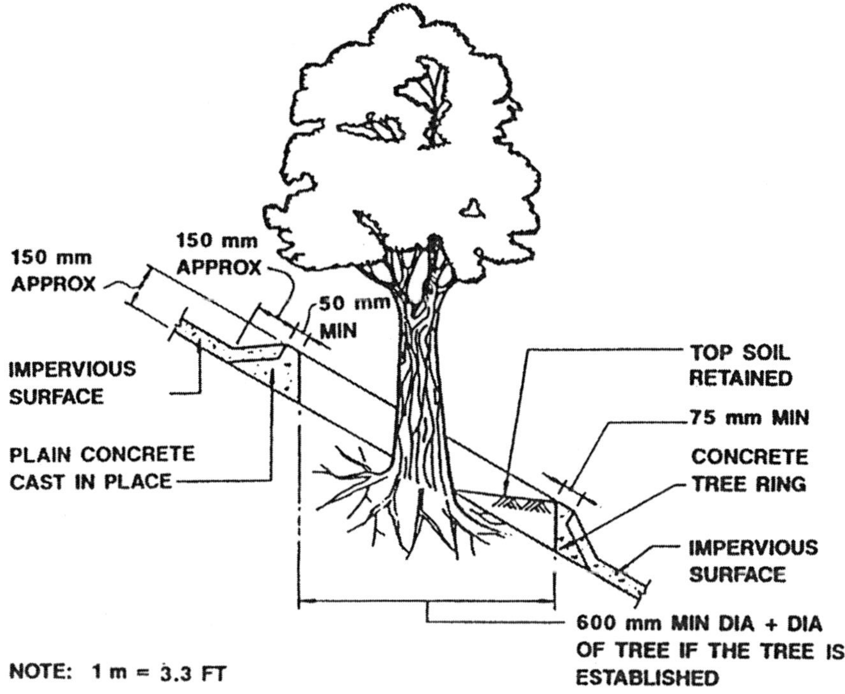

Figure 7.67 Detail of concrete rings around trees on impervious slope surface.

it is very labour-intensive, and masonry blocks could be costly if not found locally. Chunam plaster is the cheapest material, but it also requires labor-intensive installation. Therefore, it could be unaffordable in areas where labor costs are very high.

7.8.2 Shotcrete

The main purpose of shotcrete is to protect the slope from rainfall infiltration. The specifications for materials used are similar to those adopted for conventional concreting, although the aggregates are specially selected not only to meet the requirements of the finished surface but also to prevent segregation while the shotcrete is being pumped and applied. In general, the maximum grain size of the aggregate for slope stabilization should not exceed $\frac{3}{8}$-inch. Steel mesh or fiber reinforcement can be provided where necessary. Admixtures and water : cement ratios need to be fine-tuned for successful shotcrete application. The mix should be tested in the laboratory for compatibility and strength and also tested in the field using test panels and field equipment and personnel. Special training of the nozzle person is required.

Careful consideration must be given to drying and consequent shrinkagecracking that occurs when shotcrete is used for slope surfacing (Figure 7.68). The performance

Figure 7.68 Use of shotcrete for slope stabilization.

of the field test panel, with respect to durability, permeability, and shrinkage, should be assessed and, if necessary, the mix should be modified to meet all requirements. To facilitate drainage, weep holes should be installed in the shotcrete surface, especially where areas of seepage are noted prior to applying the shotcrete (Figure 7.69).

7.8.3 Chunam Plaster

Chunam plaster is a cement-lime stabilized soil used as a plaster to protect the surface of excavations from erosion and infiltration. It was developed in Hong Kong and has been used extensively there and in southeast Asia for slope protection. However, there is no record of its application in the United States, probably because it is so labor-intensive and therefore would be more costly in the United States. However, its application should be considered for slopes less than about 10 feet high.

The recommended mix for chunam plaster is 1 part by weight of portland cement, 3 parts hydrated lime, and 20 parts clayey weathered granitic or volcanic soils (Geotechnical Control Office, 1984). The soils should be free of organic matter, grass, and roots. The cement and lime should be mixed dry before soil is added. The minimum amount of water consistent with the required workability should then be added to the mix in the field. If the water : cement ratio is too high, severe cracking will result.

Prior to applying chunam plaster on a slope, all vegetation, topsoil, and roots must be removed and the slope graded. To hold the chunam in place, 1-inch diameter, 1-foot-long dowels are driven into the surface on about 5-foot centers until about

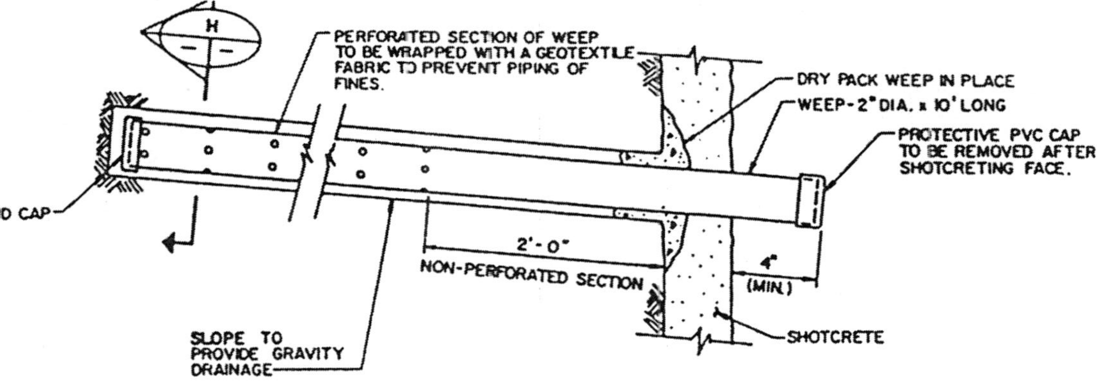

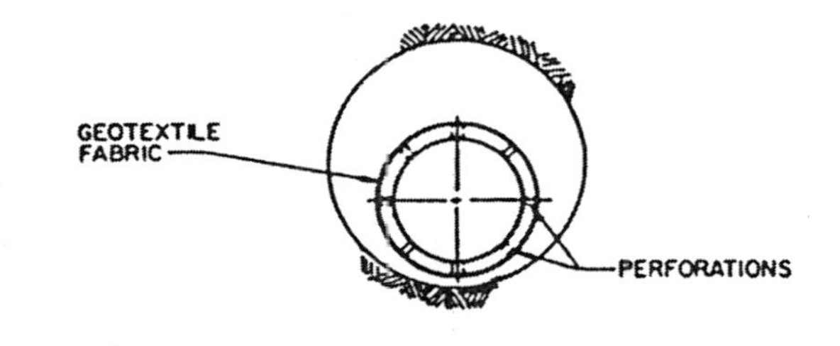

Figure 7.69 Weep hole detail.

1 inch projects from the surface. The chunam should then be applied in two layers, each of a thickness not less than 1 inch. The first layer should be suitably scored with a trowel or left with a rough surface to provide a key for the second layer, which is placed after the first coat has cured, but without an undue delay. The time lapse between the two coatings is usually about 24 hours.

Chunam plaster is normally placed with no regularly formed construction joints. Any joints that occur in the top and bottom layers should not coincide. When regular bays or panels are formed, the joints should be sealed.

7.8.4 Masonry

Slope stabilization by masonry blocks has been used for centuries to protect slope materials against the effects of erosion and weathering. In the past, old masonry walls were dry-packed. Today, masonry blocks are usually bedded on a minimum 3-inch-thick layer of free-draining crushed stone or gravel, conforming to the criteria for the design of filters. Joints between adjacent blocks are usually filled with a 1 part to 3 parts cement-sand mortar to hold the blocks together and prevent infiltration between the blocks. Grass and other vegetation will sometimes establish in the joints. Weep holes draining the bedding material should be provided at the toe of masonry facing to facilitate discharge of groundwater flow. Weeps may also be installed in the wall above the toe.

7.8.5 Rip-Rap

Erosion of toes of slopes by moving water in rivers, streams, and oceans is common and often causes instability if left unattended. The general solution for this problem is to protect the toe of the slope with layers of rip-rap placed at the base of the slope to an elevation of about 3 to 5 feet above the mean high-water level (Figure 7.70).

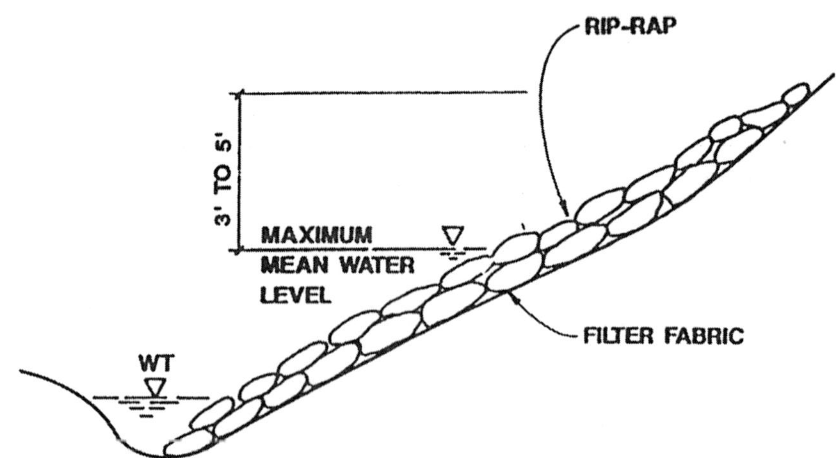

Figure 7.70 Rip-rap to protect erosion at toe of a slope.

TABLE 7.10 Thickness and Gradation Limits of Dumped Rip-Rap

			Gradation, Percentage of Stones of Various Weights[a] (pounds)		
Slope	Nominal Thickness (inches (mm))	Maximum Size	40 to 50 Percent Greater Than	50 to 60 Percent From – To	0 to 10 Percent Less Than[b]
3:1	30 (762)	2,500	1,250	75–1,250	75
2:1	36 (914)	4,500	2,250	100–2,250	100

Source: NAVFAC (1982).

[a] Sand and rock dust shall be less than 5 percent, by weight, of the total rip-rap material.

[b] The percentage of this size material shall not exceed an amount that will fill the voids in larger rock.

Where eroding forces are substantial, a lining of reinforced concrete with hydraulic detailing is constructed to dissipate the eroding forces from the water flow.

Rip-rap can be dumped or hand-placed. Dumped rip-rap consists of large-sized rock dumped on a properly graded filter. Table 7.10 shows design guidelines for dumped rip-rap. Hand-placed rip-rap is laid with a minimum of voids and a relatively smooth top surface. A filter blanket must be provided and enough openings should be left in the rip-rap facing to permit easy flow of water into or out of the rip-rap. The thickness of hand-placed rip-rap layers should be one-half of that for dumped rip-rap but not less than 12 inches. Rip-rap should be hard and durable against weathering and heavy enough to resist displacement by running water or wave action.

7.9 SOIL HARDENING

Stabilization of slopes by drainage may not always be effective in cohesive soils that have low to very low permeability. In these situations, methods borrowed from foundation engineering—known as hardening of soils—can be considered.

7.9.1 Compacted Soil–Cement Fill

Compacted soil–cement fill (cement mixed with local soil material) has been used to construct embankments wherein the embankment sideslopes could be steepened to 1.5H–1V or 1H–1V because of the high shear strength of soil cement. For slope remedial work, soil–cement fill can be used to rebuild a failed slope by forming an earth berm to provide stabilizing resisting forces (Figure 7.71).

By mixing increasing amounts of cement with the soil, the shear strength of the compacted soil cement increases substantially. The mix design is determined in the laboratory for a given soil and desired strength. Review of case histories and laboratory test results indicates that the mix designs used for slope stabilization generally consist of 1 to 10 percent cement by weight of the soil to produce cohesive strengths of 25 to greater than 125 pounds per square inch (Table 7.11).

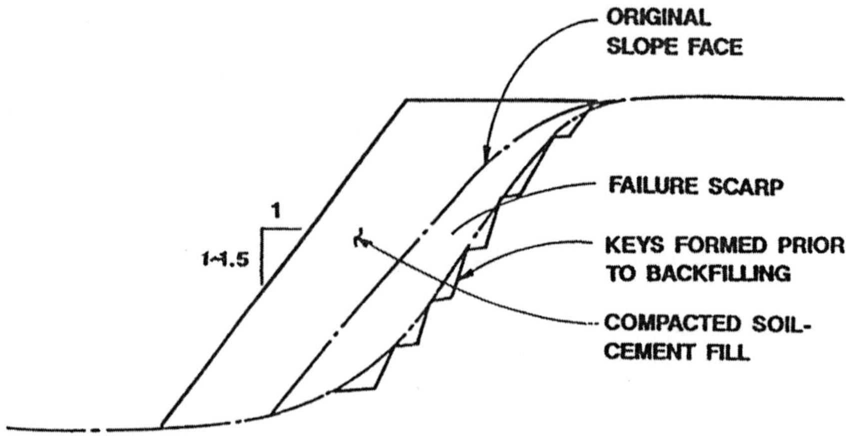

Figure 7.71 Soil–cement fill slope used to stabilize a landslide.

During construction, mixture of the soil–cement fill is controlled at the batch plant for a large project or at the site for a small project. Mix characteristics and properties also must be monitored in the field. Standard compaction equipment is used to achieve compaction of the soil–cement fill. Cement mixes better and more thoroughly with in situ cohesionless soils (sands) than with cohesive soils (clays to silty clays). This method is therefore most often used with cohesionless soils.

Cement mixing not only increases the shear strength of the soil, it also reduces its permeability. This effect poses two concerns about using compacted soil–cement fill. One concern is that it is difficult to establish vegetation on the slope face. A layer of shotcrete or chunam plaster is often required to protect against surface erosion. The other concern is that adequate drainage provisions must be made. That is true for most slopes, however.

7.9.2 Electro-osmosis

Drainage by electro-osmosis has the same overall effect as subdrainage but differs in that the water does not move toward the drainage system under the influence of gravity; rather it is acted on by an imposed electric field. Water migrates toward the cathode under a potential difference between two electrodes, a cathode and an anode. The cathode consists of a perforated pipe from which water is removed by pumping. The method is best suitable to the drainage of silty soils with particles ranging between 0.0002 and 0.002 inch in size. Clay particles contained in the silt harden as water is removed from the soil. This method, however, cannot be applied to fine sand even if the permeability is as low as 3×10^{-4} centimeters per second. This is because gravitational movement of water overwhelms the effect of the electric current. At high voltage, hydrolysis of water may take place. Thus field tests are necessary to establish the parameters of the energy supply that will give the most effective drainage. Electro-osmosis does not have wide application in slope stabilization because of

TABLE 7.11 Triaxial Strength (28 Curing Days) and Cement Content

Soil (AASHTO)	Percent Cement (by Weight)	Cohesion (psi)	Slope Angle (degree)
A-2-4[a]	0	20	29
	2	50	41
	3	58	44
	4	70	44
	6	90	48
	8	100	49
A-1-b[b]	0	10	38
	1	27	45
	2	37	49
	3	50	51
	4	72	52
	5	95	55
A-4[c]	0	5	37
	2.5	30	46
	5.5	65	45
	7.5	85	45
	9.5	125	45

Source: Nusbaum and Colley (1971).

[a] A-2-4 soil: silty or clayey gravel and sand with maximum 35 percent passing No. 200 sieve.
[b] A-1-b soil: gravelly sand or sandy gravel with maximum 25 percent passing No. 200 sieve.
[c] A-4 soil: silty soil with minimum 36 percent passing No. 200 sieve.

high energy costs associated with the method, but nevertheless it is an option for unusual cases.

7.9.3 Thermal Treatment

Thermal treatment of slopes has been used in Romania (Beles and Stanculescu, 1958) and the United States (Hausmann, 1990) to stabilize landslides. The high temperature treatments cause a permanent drying of the embankment or cut slope. Figure 7.72 shows the method used by the Romanian engineers for stabilization of landslides in clay. Two holes are bored and connected to produce a draught exhausting from a chimney tube at the exit of one borehole. Oil is led to a special burner installed in the other borehole, where it is ignited and lowered into the borehole. The thermal treatment technique has limited applications. It is used only in exceptional circumstances, due to high energy requirements and undesirable environmental impacts.

7.9.4 Grouting

Grouting has been used effectively for the stabilization of landslides, especially ones with shallow movement in stiff materials, such as clay shale and stiff clays. Granular soils also can be grouted; however, slaked material cannot be grouted.

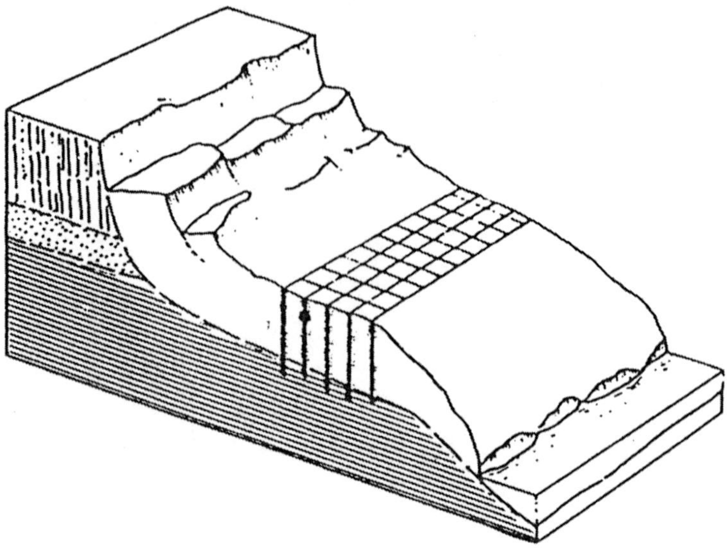

Figure 7.72 Thermal treatment of a slope. (Zaruba and Mencl, 1982.)

The effect of grouting is to displace water from the fissures or pores in the ground as they are filled with cement mortar. The mortar hardens and creates a stable skeleton around the solid soil and rock zones. The grouting pressure is usually greater than that produced by the weight of overburden to allow effective grout penetration into the fissures or voids and along the active slip surface.

For an effective grouting operation, the depth and form of the slip surface must be known. A row of exploratory boreholes can be drilled down to the slip plane and then used for injection pipes. Spacing of the grout holes varies between about 10 and 15 feet. The grouting operation should start with the lowest row in order to increase support near the toe of the endangered slope as grouting progresses. It should be noted that grouting used in slope stabilization may produce a temporary increase in pore pressures that could aggravate slope movements. Pore pressures may increase where drainage paths are blocked by soil infill or healing. Grouting is dealt with in more detail by Xanthakos et al. (1994).

7.9.5 Lime Injection

Slope stability can be improved by injection of lime columns to increase the shear strength of clayey and silty soils. This technique is not effective in sandy soils. Lime injection was first used by Swedish engineers. A rotating disk auger penetrates the ground to a depth below the slip surface, and the stabilizing agent is injected into the resulting kneaded soil column (Figure 7.73). One disadvantage of this method is that at least 80 days must elapse before the columns of stabilized soil can be subjected to loading. However, good results have been obtained with slope movements of limited dimensions.

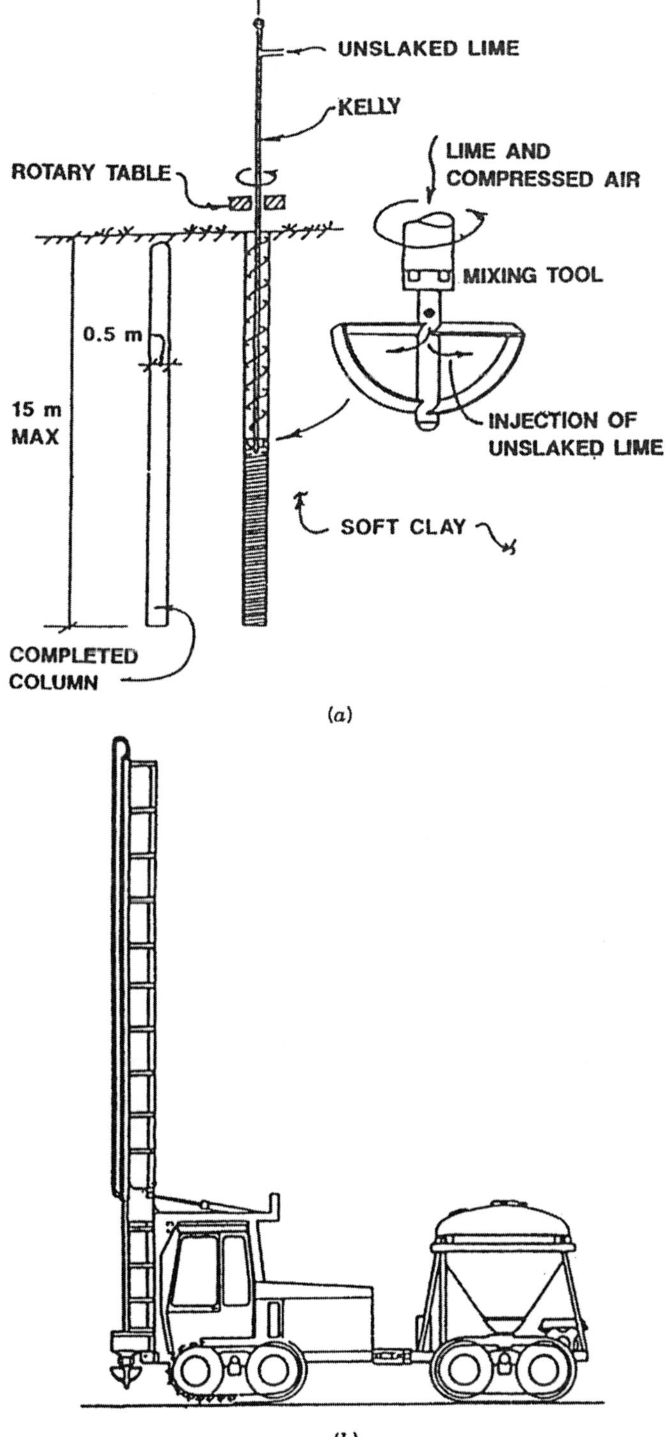

Figure 7.73 (*a*) Lime column installation. (*b*) Lime column machine. (Broms, 1991.)

The lime injection method is an alternative to stabilizing slopes with piles, which in some instances could initiate or worsen slides because of the development of high pore water pressures as a result of pile driving. The dissipation of excess pore water pressure caused by lime columns occurs rapidly, since the columns can also function as vertical drains in the soil.

The increase in soil shear strength due to lime columns can be expressed by estimating the average shear strength along a potential failure or rupture surface in the soil, which is estimated as

$$c_{avg} = c_u(1 - a) + \frac{S_{col}}{a} \qquad \text{(Eq. 7-15)}$$

where c_u = the undrained shear strength of the soil as determined by laboratory tests or field tests

s_{col} = the average shear strength of the stabilized clay within the columns

a = the relative column area (Figure 7.74).

The shear strength s_{col} of the stabilized soil depends on the overburden pressure and the relative stiffness of the columns with respect to the surrounding unstabilized soil. Broms (1991) proposed using the total overburden pressure in the calculation of the shear strength and a friction angle of 30°. The calculated shear strength of the stabilized soil should not exceed $c_{col,max}$, the shear strength of the clay matrix in the columns. The spacing of lime columns and the volume of soft clay requiring

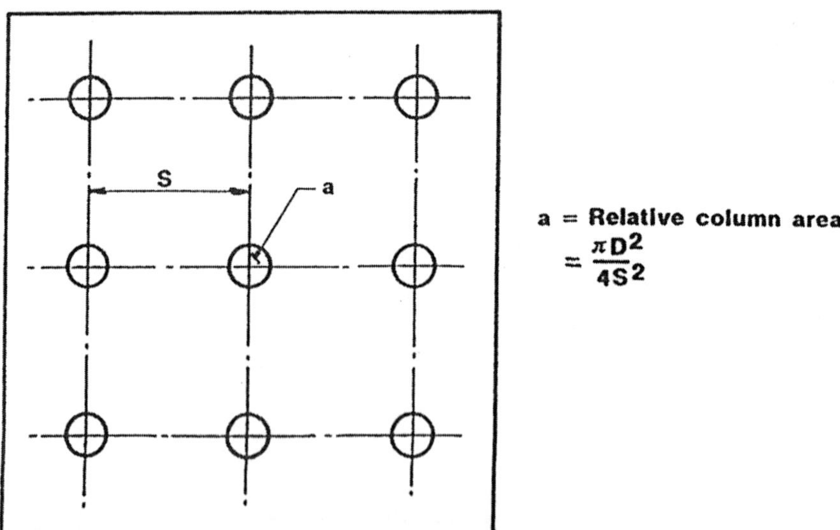

Figure 7.74 Lime column spacing. (Broms, 1991.)

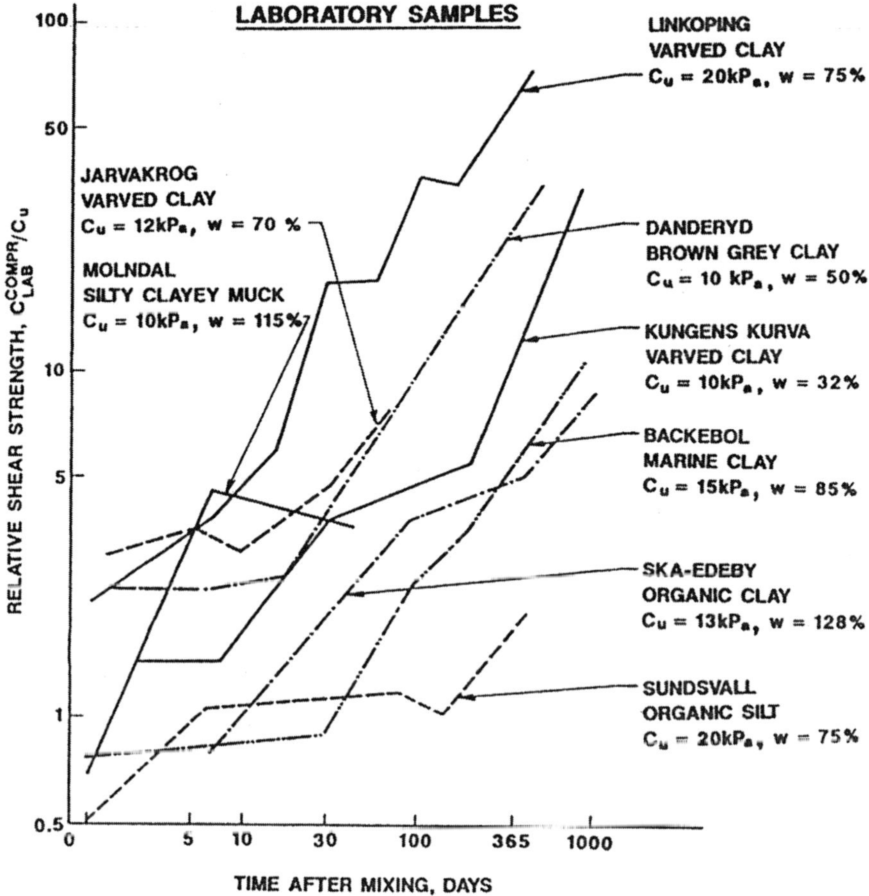

Figure 7.75 Increase of shear strength with lime based on results from unconfined compression tests. (Broms, 1991.)

stabilization can be evaluated by analyzing the stability of the slope along different potential failure surfaces. According to Broms, a safety factor of at least 1.3 to 1.5 should be used.

The shear strength of the stabilized soil, as determined by unconfined compression tests, increases with time, as indicated in Figure 7.75. The undrained shear strength of the stabilized clay can be as high as 5 to 10 tons per square foot after 1 year (Broms, 1991).

7.9.6 Preconsolidation

Another slope stabilization technique is to increase the strength of clayey soils by acceleration of consolidation through the use of a surcharge fill, sometimes in

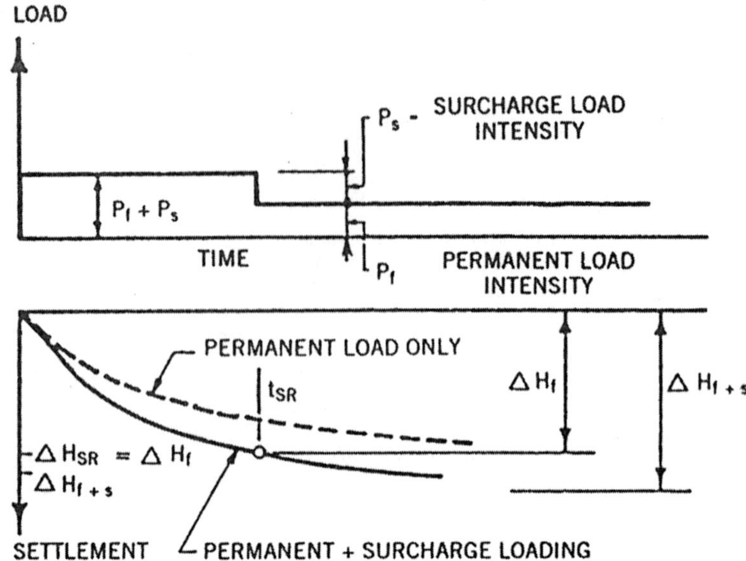

Figure 7.76 Preloading design-compensation for primary settlement by temporary surcharge fill.

combination with wick drains or sand drains (Xanthakos et al., 1994). This technique is suitable for embankment slopes overlying soft foundation soils. The design concept of this technique is to cause a portion of total settlement to occur before paving or other construction takes place. This can be achieved by placing surcharge fill over the embankment to an elevation that would give the same primary consolidation settlement as would have been induced by the embankment fill alone at a certain period (Figure 7.76). If the time to reach the required settlement under the surcharge and embankment fills is long and impractical, wick drains or sand drains commonly are used to expedite the rate of consolidation of the foundation soils by increasing the rate of pore pressure dissipation.

Staged embankment construction is commonly carried out over soft compressible foundation soils in an attempt to minimize global failures and, at the same time, to increase the shear strength of the soft soils. Two to three stages of embankment construction are used with each stage, causing a predetermined amount of consolidation of the underlying materials and dissipation of excess pore pressures to acceptable levels. The rate of pore pressure dissipation usually is monitored by piezometers (see Chapter 3). However, field piezometer measurements in soft soils are sometimes inconsistent and unreliable. To solve this difficulty, it may be necessary to rely on settlement trends to interpret the degree of consolidation. Such a method has been used by the Oregon Department of Transportation to construct an embankment over soft compressible soils (Barrows and Machan, 1991; Machan and Dodson, 1989). The design methodology of embankments over soft foundation soils is well established and can be found in foundation engineering textbooks and manuals.

Design of staged construction must estimate the initial strength of cohesive soil and its rate of increase with time due to consolidation under the applied loading. Methods commonly used for the design of staged constructions include: the total stress analysis (TSA), which is based on unconsolidated undrained (UU) triaxial tests and field vane tests; the effective stress analysis (ESA) (Bishop and Bjerrum, 1960); and Casagrande's method, which requires assumption of an undrained failure when calculating FOSs during staged construction. However, all these methods have their own drawbacks and limitations and may give unsafe FOSs.

Ladd (1991) treats the controversial issue of what type of stability analysis to use for the design of staged construction projects and for checking stability during actual construction. According to Ladd (1991), actual failures of cohesive soils usually occur without significant dissipation of shear-induced pore pressures during staged construction, and most failures during staged construction, and most failures during staged construction occur under essentially undrained conditions. Stability evaluations should include predictions of the initial foundation strength and of any subsequent increases because of consolidation. These predicted strengths will be functions of the initial stress history and of the way stresses change during construction. Neither in situ nor laboratory UU-type tests can provide these predictions, and thus the undrained strength–consolidation stress relationship must be obtained from laboratory consolidated undrained (CU) strength testing. The initial undrained strength anisotropy governs each stage of construction and can be modeled correctly in the laboratory using CK_0U plane strain compression (PSC), direct simple shear (DSS), and plane strain extension tests (PSE).

Ladd (1991) concludes that conventional ESA based on effective strength parameters (c' and ϕ') and measured pore pressures can give highly misleading and unsafe estimates of actual FOSs for staged construction. Ladd recommends using a so-called undrained strength analysis (USA) that treats the in situ effective stresses as the consolidation stresses, uses these consolidation stresses to compute the available undrained strength within the foundation, and defines $\text{FOS}_{\text{USA}} = s_u/s_m$ where s_u is the new undrained strength corresponding to the current in situ consolidation stress history and s_m is the mobilized shear stress required for equilibrium.

Undrained strength analysis requires estimates of the initial undrained shear strength (s_u) of the cohesive foundation soils and the subsequent increases in s_u during staged construction. There is a unique relationship between the in situ undrained strength ratio (s_u/σ'_{uc}) and the overconsolidation ratio ($\text{OCR} = \sigma'_p/\sigma'_{uc}$) of cohesive soils. For preliminary design of staged construction over soft soils, $s_u/\sigma'_{uc} = S(\text{OCR})^m$ can be used with values of S and m obtained from the following (Ladd, 1991).

For homogeneous sedimentary clays plotting above Casagrande's A line,

$$S = 0.22 \pm 0.03$$

For silts and organic clays plotting below A line,

$$S = 0.25 \pm 0.05 \qquad \text{and} \qquad m = 0.88\left(\frac{1 - C_s}{C_c}\right)$$

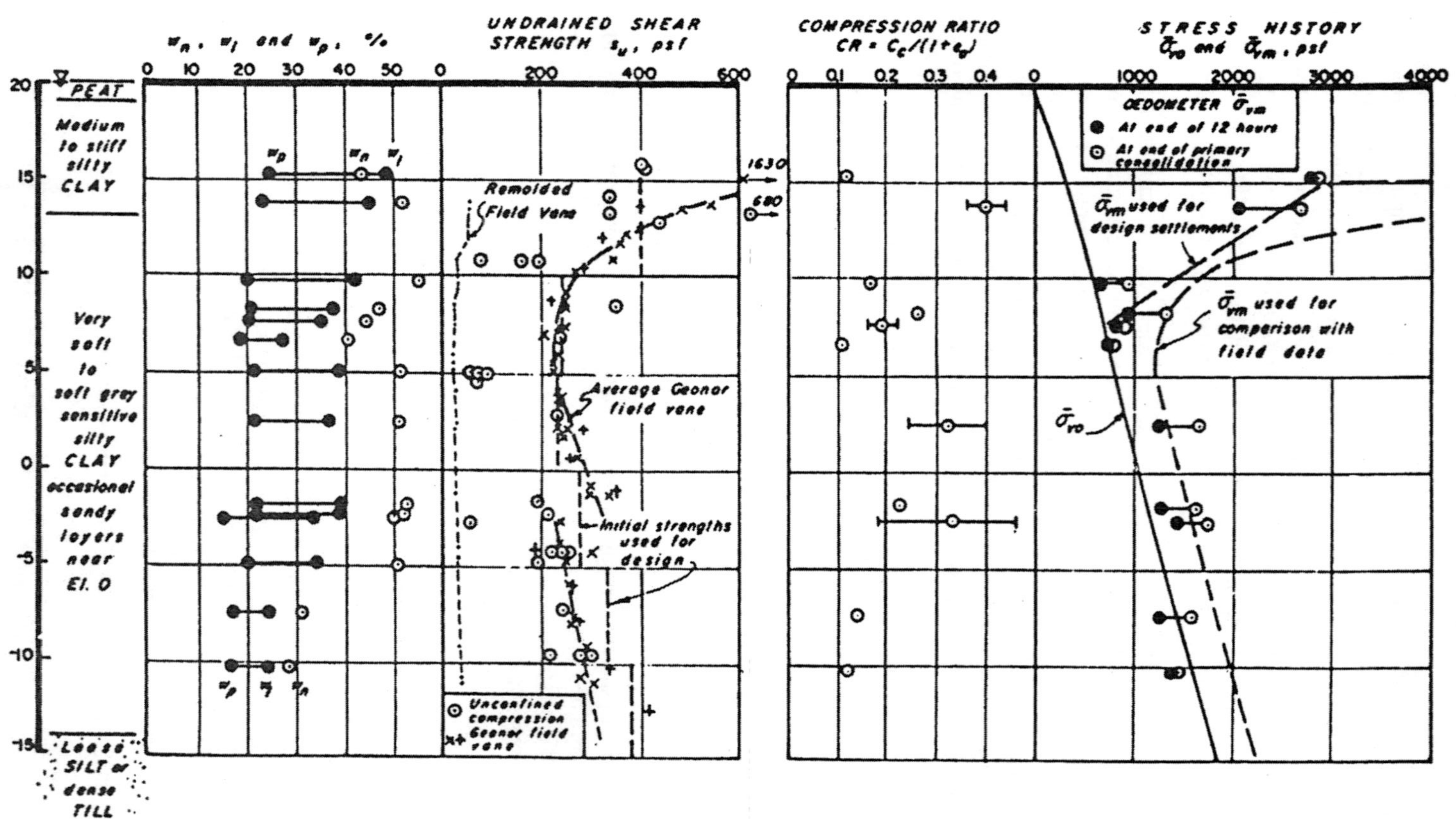

Figure 7.77 Interstate Route 95 interchange subsurface conditions. (From Ladd et al., 1972, reproduced by permission of ASCE.)

where C_s and C_c are equal to the slope of the swelling and virgin compression lines, respectively.

For important (i.e., high risk to life and economic loss) embankments over soft soils, it is advisable to use more sophisticated laboratory testing, such as CK_0UDSS, PSC, and PSE tests to obtain s_u/σ'_{uc} versus OCR relationships.

The following case history illustrates the concepts discussed above. Improvements to the Interstate Route 95 interchange in Portsmouth, New Hampshire, composed of stabilization of the soft clay subsoils beneath the approach embankments requiring several million cubic yards of fill material. A combination of surcharge fills, stabilizing berms, staged construction, and the installation of vertical sand drains was required (Ladd et al., 1972). The interchange project included the construction of five bridges and approach embankments up to 35 feet in height. A major portion of the site is underlain by 35 to 40 feet of highly compressible, sensitive, soft marine clay (Figure 7.77). Typical consolidation curves for two samples tested during the exploration program are shown in Figure 7.78. Typical rates of consolidation determined from the test program were between about 0.1 and 0.15 square feet per day for virgin compression and equaled about 1 square foot per day for recompression. The

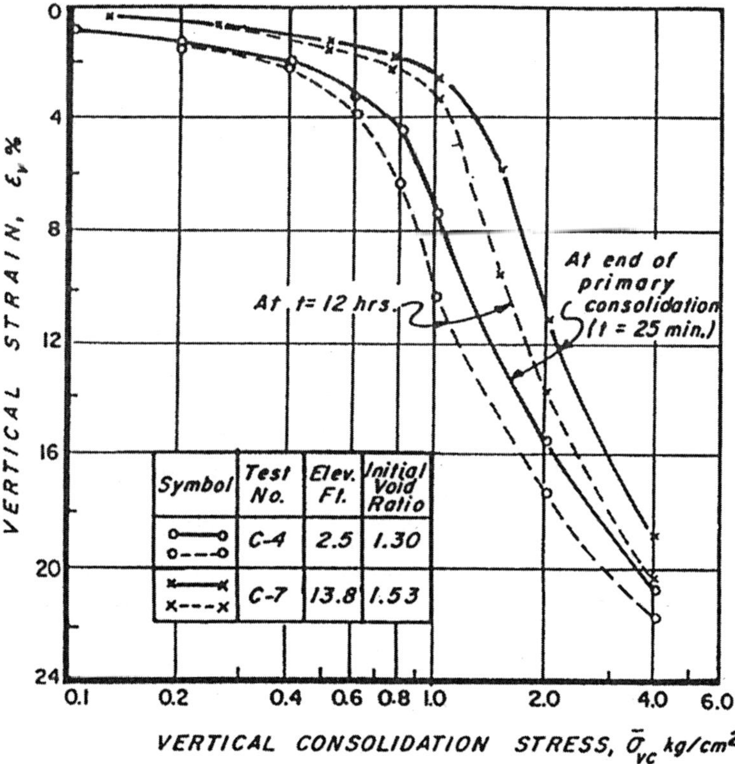

Figure 7.78 Typical consolidation curves. (From Ladd et al., 1972, reproduced by permission of ASCE.)

maximum rate of secondary compression (compression that occurs after 100 percent primary consolidation of the clay) was estimated to be 1.5 percent per log cycle of time at stresses beyond the maximum past pressure. At higher stresses, the secondary compression rate was estimated to be between 0.5 and 1 percent and the rate for overconsolidated clay was estimated to be 0.1 percent.

A system of vertical sand drains was adopted in conjunction with a surcharge scheme in order to achieve sufficient consolidation of the soft clay within the available three year time limit so as to minimize post-pavement settlements. In order to adequately support the required surcharges, extensive stabilizing berms, where possible, were incorporated into the embankment design. In addition, a system of staged construction was adopted to take advantage of the consolidation and resulting anticipated strength increase, which would occur in the sand drain areas beneath the central portions of the embankments.

For design purposes, target limits for post-pavement settlement were set consisting of less than 2 inches for embankments within several hundred feet of the bridge abutments and 6 inches for embankments not in the vicinity of bridge abutments. Based on these criteria, design charts similar to the one shown in Figure 7.79 were developed. In

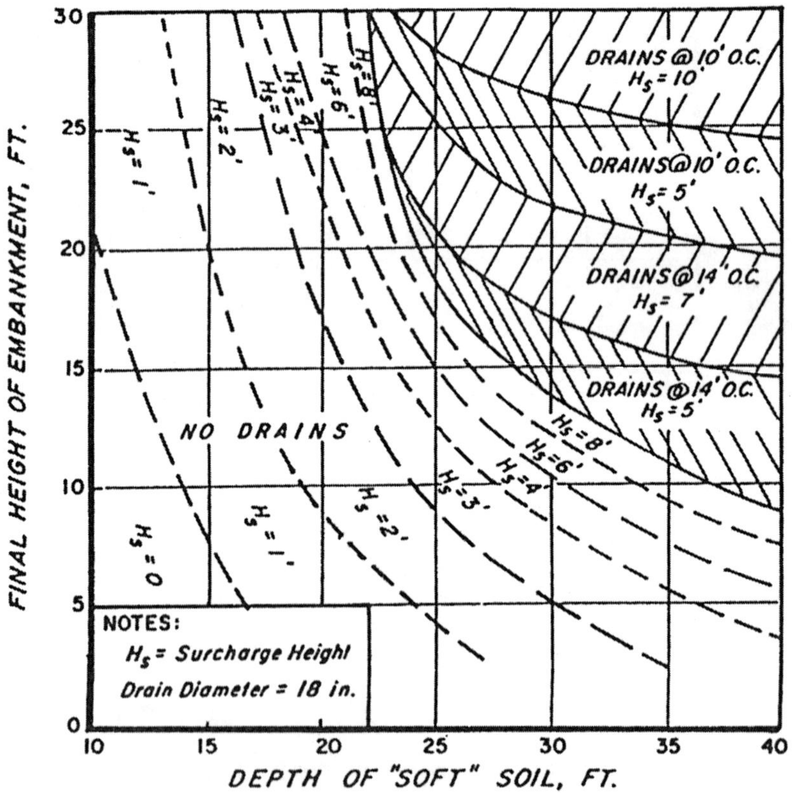

Figure 7.79 Settlement design charts. (From Ladd et al., 1972, reproduced by permission of ASCE.)

order to reduce differential postpavement settlement between differing drain spacings, transition zones with intermediate spacings were utilized. Such zones were particularly important between areas with and without sand drains. Slope stability analyses were performed to determine allowable rates of embankment filling as a function of clay strength increase. A typical instrumented embankment section is shown in Figure 7.80. Typical instrumentation results are shown in Figures 7.81, 7.82, and 7.83.

The equipment used for the nondisplacement-type sand drain installation consisted of a 5-inch-diameter, 27-foot-long "jet-bailer" pipe suspended from a crane. A 12-inch-diameter tooth cutting head was attached to the bottom of the jet-bailer. A 6-inch-diameter high pressure pump supplied water to the jet nozzle at the center of the cutting head. After the sand drain hole was made, concrete sand was placed using shovels to backfill the hole. Sand drain spacings ranged between 9 and 16.2 feet on center.

7.10 ROCK SLOPE STABILIZATION METHODS

There are typically seven repair methods used for stabilizing rock slopes:

(1) Removal of unstable rock
(2) Catchment
(3) Flattening of slope
(4) Buttresses
(5) Surface protection
(6) Reinforcement
(7) Drainage

These methods can be used singly or in combination.

Site-specific conditions normally dictate whether to reinforce the rock or support it. Support methods are most commonly used to stabilize overhangs. Reinforcement is most commonly used to prevent ultimate sliding or rotational failure of potentially unstable rock masses along discontinuities. Also, surface protection using mesh or shotcrete can be used to prevent progressive raveling and attack by sunlight, air, and water.

7.10.1 Removal of Unstable Rock

Removal of potentially unstable rock is typically necessary for slope rehabilitation whether it is to insure long-term performance or simply for worker safety. This may include removal of accumulated rock on benches, surface scaling by hand, and explosive removal of overhangs. Breakage and removal of the rock is normally done using conventional rock excavation equipment. Access to the slope is sometimes limited and may require hand-carried equipment and rappelling expertise. Alternatively, scaling can be done using a crane with specially designed "rake" or demolition ball.

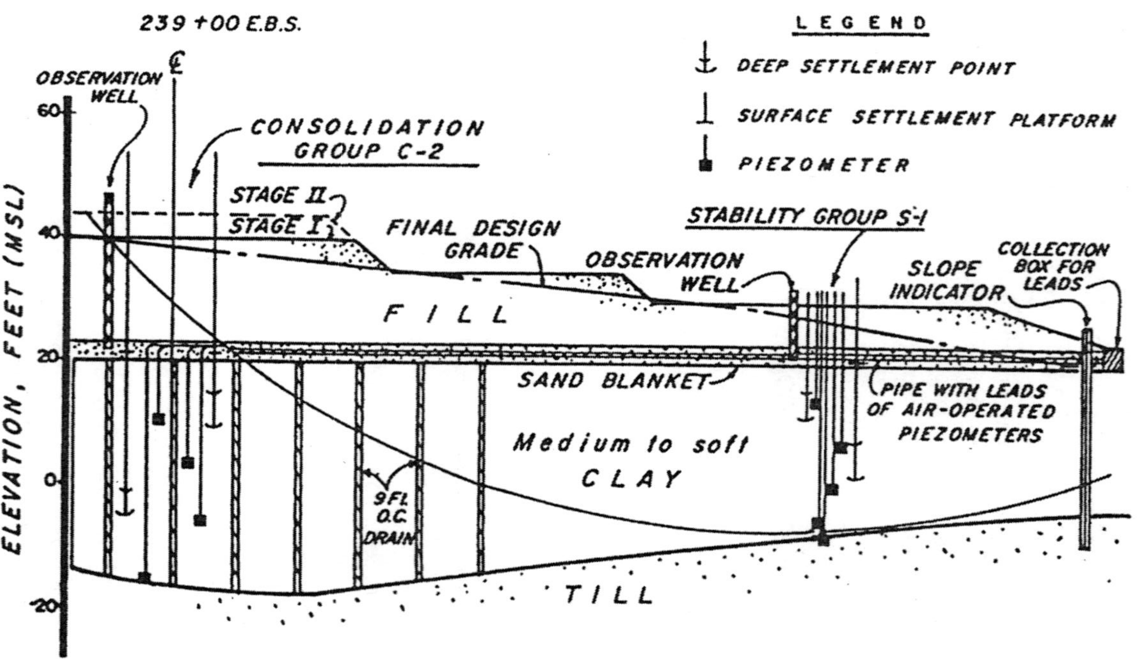

Figure 7.80 Typical instrumentation. (From Ladd et al., 1972, reproduced by permission of ASCE.)

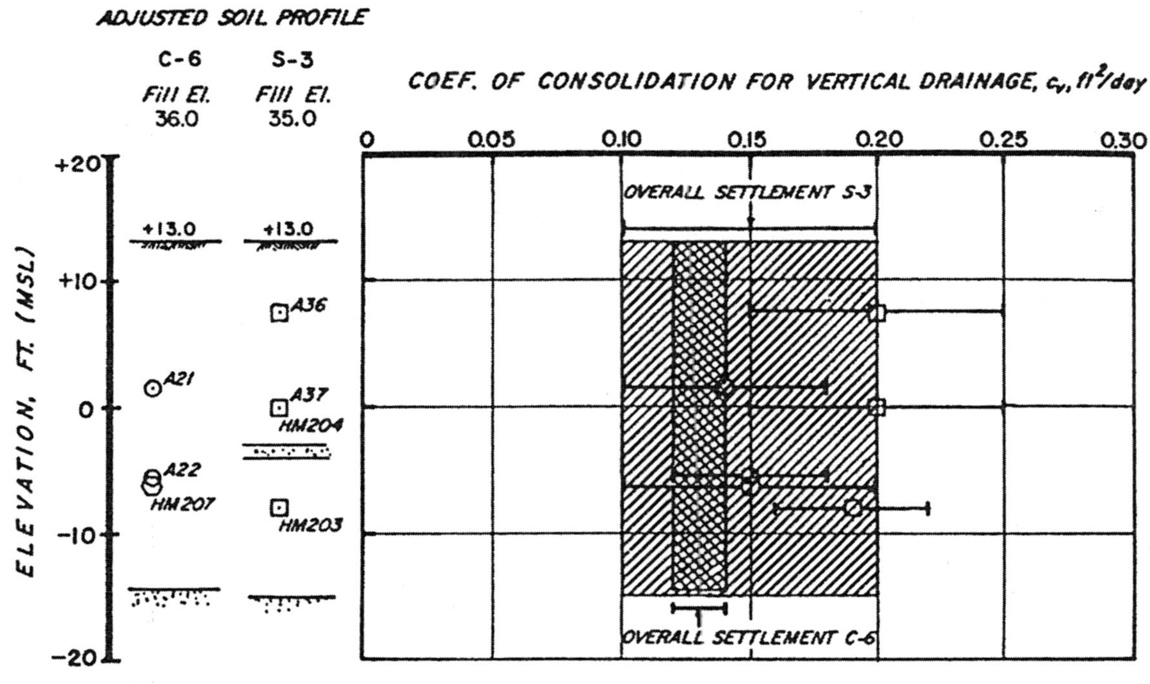

Figure 7.81 Coefficient of consolidation observations. (From Ladd et al., 1972, reproduced by permission of ASCE.)

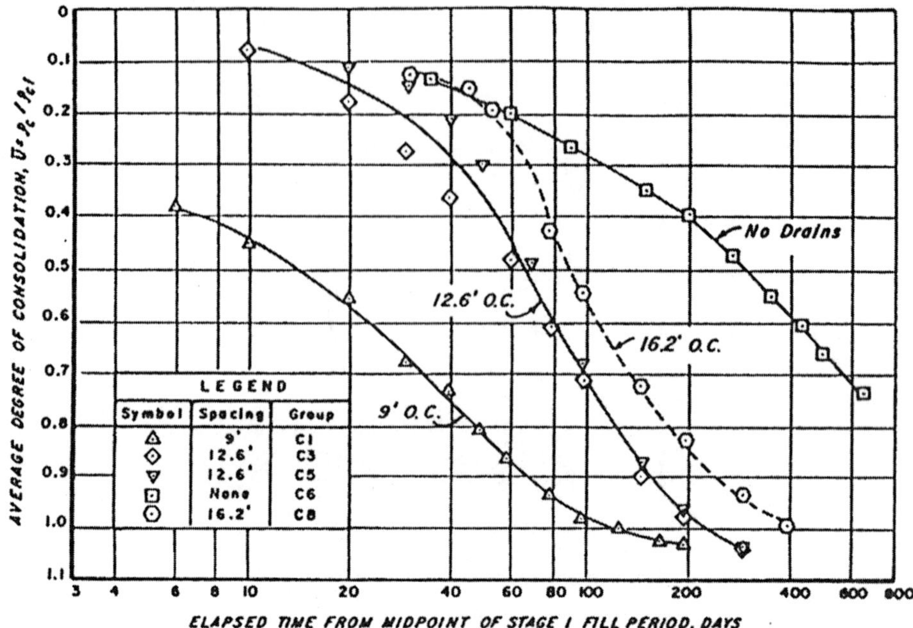

Figure 7.82 Degree of consolidation observations. (From Ladd et al., 1972, reproduced by permission of ASCE.)

Rock removal can be hazardous to the workers doing the work, as well as to workers performing other tasks on-site. Additionally, vehicles and pedestrains passing nearby may be endangered. Therefore, traffic must be stopped or diverted and on-site personnel must be informed about the nature and timing of overhead activities. Protection of existing structures adjacent to the work may also be necessary. All occupational safety regulations governing the facility should be adhered to.

7.10.2 Catchment

Regardless of which of the rehabilitation methods are chosen, there is usually the need for catchment of falling rock. Most slopes contain small pieces of rock that could loosen in the future but do not require extensive removal or reinforcement. Furthermore, just as the original design and construction does not eliminate future rock fall hazards completely, rehabilitation methods will not always be 100 percent effective, due to the continual forces of nature. Catchment can consist of engineered benches, ditches, wide shoulders, berms, steel barriers, nets, fences, and concrete walls (Figure 7.84).

The type of catchment used depends largely on site conditions. Specifically, the height and angle of the slope and clearance between slope and the facility are

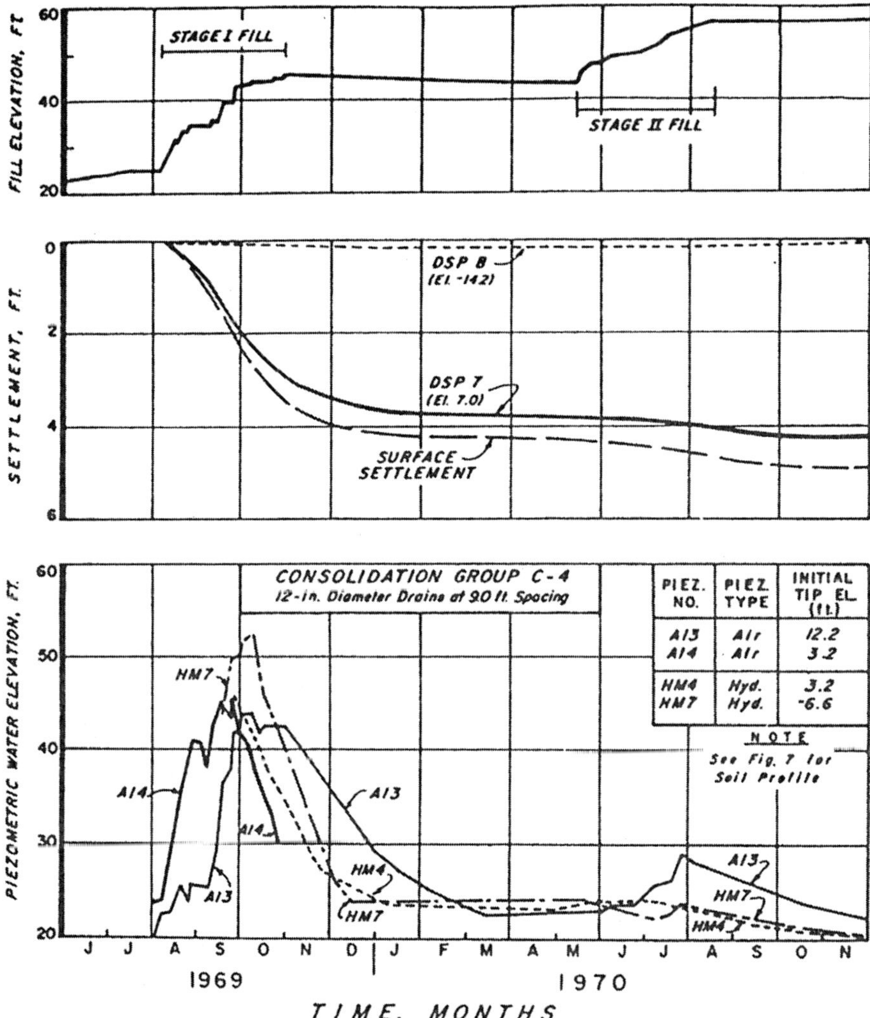

Figure 7.83 Field instrumentation data observations. (From Ladd et al., 1972, reproduced by permission of ASCE.)

important. This not only dictates how much space there is for catchment, but it also relates to anticipated paths of failing rocks. Obviously the catchment area must be somewhere along the paths of the rocks to be effective.

Generally, the flatter the slope is, the more likely falling rocks will bounce and roll. Simple rock falls (i.e., with no bounce or roll) tend to occur on steep slopes and can be kept off of the facility with benches, ditches, and shoulders of appropriate

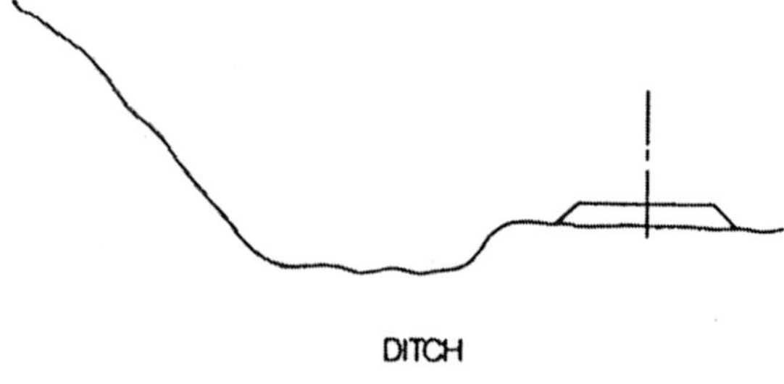

DITCH

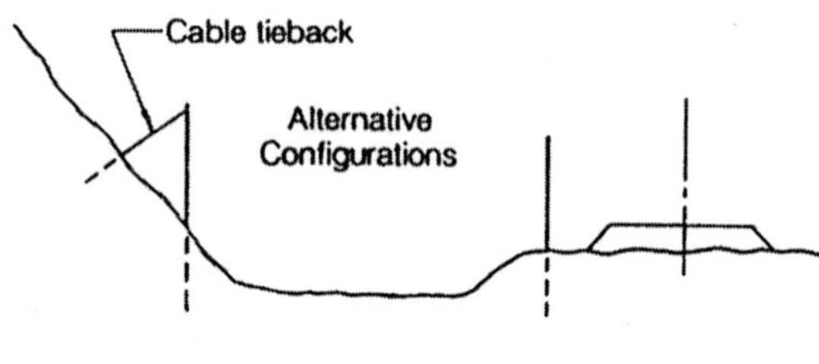

DITCH WITH FENCES

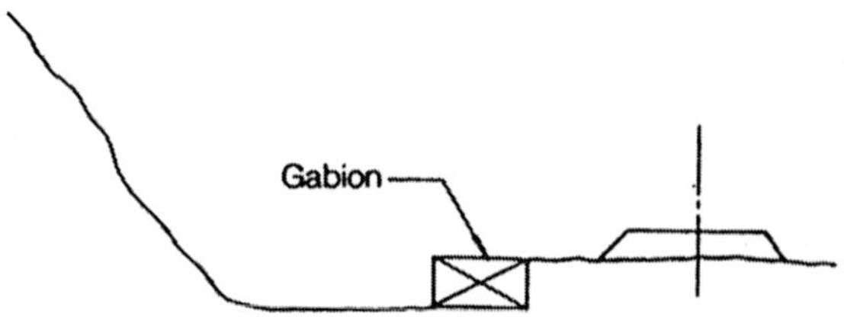

DITCH WITH GABION

Figure 7.84 Catchment methods.

widths and locations. Access should be provided to these areas because the catchment areas sometimes become filled with talus, forming a slope that can eventually direct subsequent rock falls onto the facility. The catchment area then must be accessed periodically to remove this debris, which would otherwise defeat the purpose of the catchment area.

For the flatter slopes where rock falls tend to bounce and roll, a barrier is needed to deflect rocks away from the facility. Key considerations in the design of such barriers are height, location, and strength. Berms, steel barriers, fences, and concrete walls

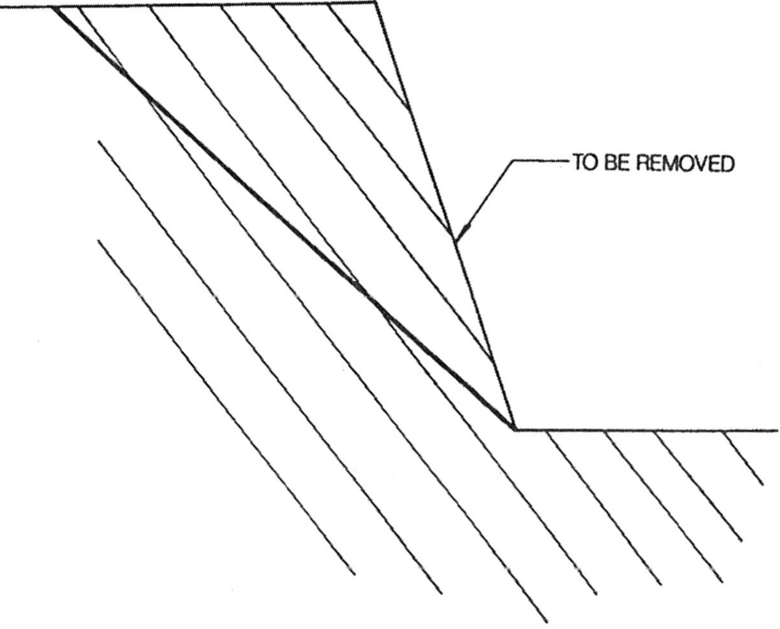

Figure 7.85 Flattening of a steep rock slope.

can be used for this purpose. These are normally used in conjunction with benches, ditches, and shoulders.

7.10.3 Flattening of Slope

Generally speaking, the flatter a slope is, the more stable it is. The potential problems associated with flattening an existing slope are finding a practical place to start the excavation, acquiring additional property or right-of-way, and disposal of the spoil. The goals of flattening a slope are usually to increase the safety factor and eliminate the daylighting of unstable planes or wedges (Figure 7.85).

7.10.4 Buttresses

Support of the rock can be accomplished with buttresses, bulkheads, or retaining walls. These structures are typically constructed with cast-in-place, reinforced concrete, although stone or masonry can be used also. A new technique using the unstable rock as a "shot-in-place buttress" has been used in Tennessee by Moore (1986). Rock reinforcement is often used in combination with support methods.

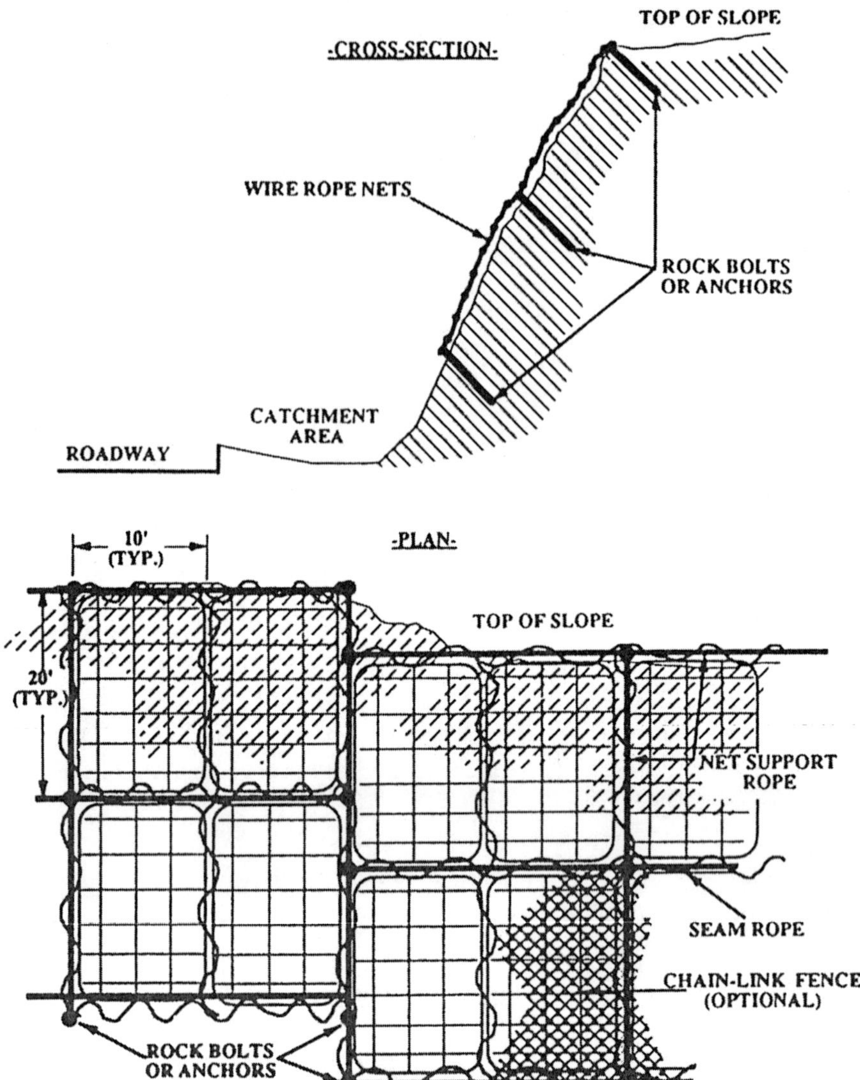

Figure 7.86 Steel mesh rock slope protection. (Courtesy of Brugg Cable Products, Inc., Santa Fe, New Mexico.)

7.10.5 Surface Protection

If space does not permit the construction of adequate catchment areas and barriers, potential rock debris must be held in place directly against the slope face. This can be accomplished using a net that is fastened to or over the slope. The netting commonly consists of chain link fence or gabion wire fabric held in place by rock bolts or cable tendons. Steel fabric, straps, or channels control rock falls between reinforcement elements (dowels, bolts, or tiebacks).

Steel Mesh Steel mesh surface protection usually consists of chain link fence (Figure 7.86). Welded wire fabric can be used, especially if the possibility of using shotcrete also exists. Chain link fence interferes with the proper application of shotcrete. Another type of steel mesh is wire nets, which are discussed further below.

The main design components involved with specifying steel mesh are the wire size and the method and frequency of tie-downs. Standard sizes of chain link fence and welded wire fabric are available. Tie-downs usually consist of short rock dowels with the size, method of grouting, and number per square feet of slope specified. If rock dowels are being used to reinforce the slope in addition to the mesh, the mesh can be attached to the reinforcement dowels without additional tie-downs needed.

Wire Nets Wire nets have recently been introduced to the United States from Europe. Wire nets are similar to chain link fence except they have wider wire spacing and greater energy absorption characteristics.

Shotcrete Shotcrete is forced into open joints, fissures, seams, and irregularities in the rock surface and serves the same binding function as mortar in a stone wall (Figure 7.87). The adhesion of shotcrete to the rock surface, together with the shear

Figure 7.87 Use of shotcrete for slope protection.

strength of the shotcrete layer, provides resistance to the fallout of loose rock blocks as well as confinement of the rock mass. It will also seal rock that is prone to weathering due to exposure to the elements of nature.

7.10.6 Reinforcement

Rock reinforcement dowels, bolts, or tiebacks resist movement along joints and restrict block fallout and loosening. Tensioned reinforcement (bolts or tiebacks) will change the stress state around the slope face by inducing compressive stresses, which provide confinement, thereby improving the strength of the rock mass.

Dowels Dowels are usually made of steel bars grouted in predrilled holes with epoxy resin or cement (Figure 7.88). Dowels are not pretensioned or post-tensioned. The reason for this is that as the rock mass moves it will tend to tension the bar as it resists the movement. Reinforcing steel and high strength bars are most commonly used. The bar diameter size can range between about 1 and 1.5 inches. The hole size depends on the type of grout being used, the length of the bar (whether couplers are needed), and the type of rock.

Tiebacks Tiebacks are similar to dowels except that they are usually longer and are post-tensioned. Steel strand is sometimes used instead of bars (Figure 7.89) Xanthakos (1991) provides a detailed discussion about the design, installation, and testing of tensioned ground anchors. The tensioning of the tieback provides additional confinement to the rock and increased shear strength across joints. Design forces of about 100 kips are not uncommon. The amount of confinement provided can be calculated by dividing the anchor force by the tieback spacing. For example, a 100-kip tieback on a 10 × 10-foot spacing would provide 1 kip per square foot of confinement to the rock mass.

7.10.7 Drainage

To limit the amount of water that accumulates along the rock mass discontinuities, pressure relief drainage systems are used. They tend to prevent hydrostatic pressures from building up along discontinuities, and also limit the volume of water that can freeze in the winter and cause rock block displacement due to expansion. The two general forms of drainage are surface and subsurface. Surface drainage refers to diversion of surface runoff away from tension cracks or open rock mass discontinuities near the slope face. This is usually accomplished with ditches and trenches. Subsurface drainage includes horizontal drains, tunnels, and other methods.

Ditches Ditches can be used to redirect surface runoff away from slopes. Ditches are usually lined with shotcrete or concrete. They can be vee-shaped or rounded. It is important to design the ditches so that they drain into a proper receptacle, have an allowable capacity and grade such that they do not overflow, and are durable such that they do not crack and leak soon after installation.

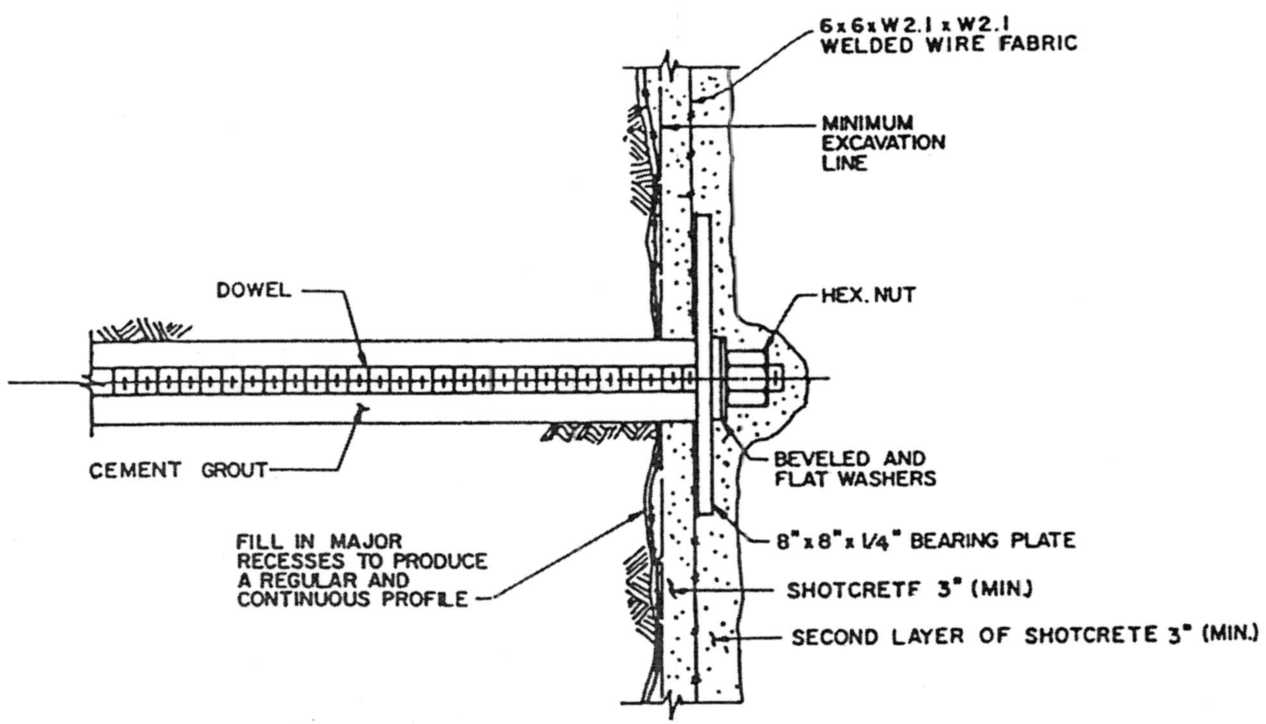

Figure 7.88 Rock dowel detail.

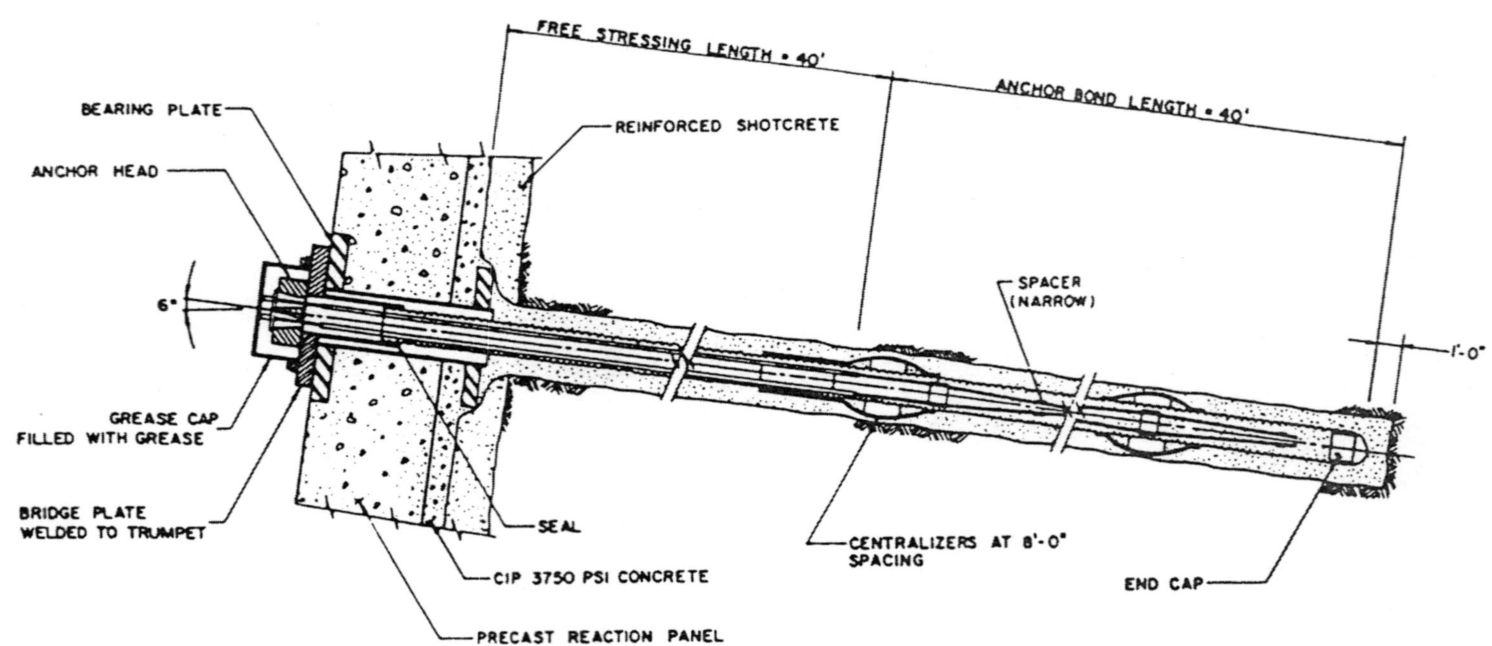

Figure 7.89 Tieback detail.

Trenches Alternative to ditches are trenches, which are lined with geosynthetic fabric and contain previous gravel and drainage pipes. Trenches are more permanent and can intercept large amounts of water before it saturates the area around the slope. The same guidelines for design mentioned above apply to trenches, especially having adequate capacity, durability, and discharge facilities.

Horizontal Drain Pipes Shallow or deep subsurface drainage is usually accomplished with perforated PVC pipes, which are grouted or dry-packed into place near the slope face (Figure 7.90). The holes are normally 2 to 3 inches in diameter and can be drilled to between 100 and 200 feet in length. Most times, not every drain pipe will intercept water-bearing strata or discontinuities. A sufficient number must by installed to compensate for the "dry" ones. Also, the drain pipes must be maintained in operating condition to be effective.

Tunnels and Other Methods Tunnels and other methods have sometimes been used to stabilize rock slopes. Although tunnels are seldom resorted to in stabilizing rock slopes because of the relatively high construction cost, they can be used as drainage galleries to reduce the quantity and rate of water flow at the face of the slope. Other exotic methods might include ground freezing, permanent dewatering systems, and grouted cut-off walls. The high cost of these methods are usually not warranted. Water flow is usually controlled by the methods described above at more reasonable prices.

7.10.8 Use of Explosives

Explosives are sometimes used to break and roughen the failure surface, which is often smooth and polished. For example, slices occur in masses of cohesive soil that overlay rock or much harder soil mantles. The contact surface between the two materials becomes frictional. However, care should be taken to avoid any further slides that can be triggered by the explosion.

Whether correction by means of explosives is permanent is debatable. It is widely believed that for the method to be successful, there must be a hard formation beneath the failure surface. Also, any drainage that may be obtained through the use of explosives is questionable. The resulting fragments may not form a filter. The small spaces between the fragments may become sealed by very fine materials deposited by water flow. This method is not applicable to deep-seated failures because of the large amounts of explosives required.

The greatest advantage of using explosives for slope stabilization is cost. Costs may be so much lower than that of other methods that a program of several successive applications of the procedure over a period of years may still be economically advantageous. The main disadvantages of this method are lack of predictability, noise, and traffic flow requirements.

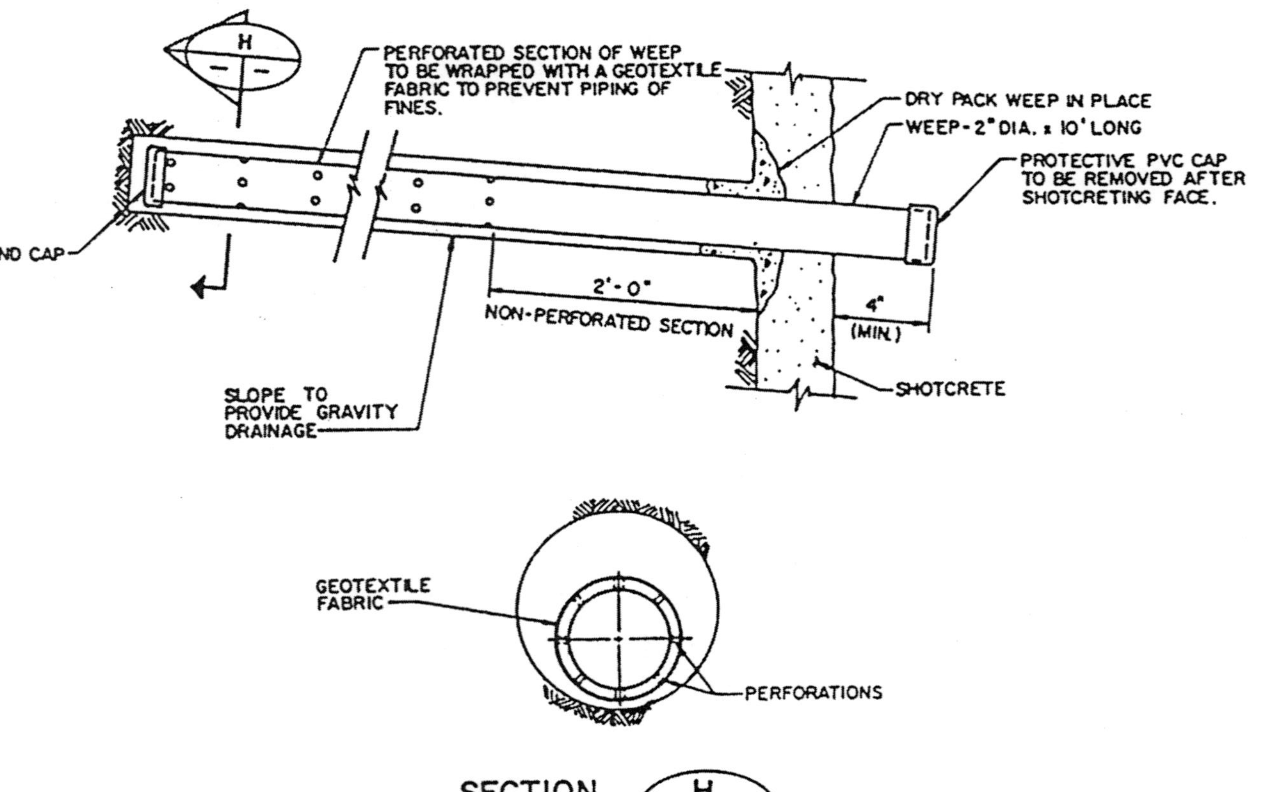

Figure 7.90 Horizontal drain pipe detail.

7.10.9 Rock Slope Stabilization Case Histories

The following case histories illustrate the concepts discussed above.

Interstate Route 40 Rock Slide On March 5, 1985, a massive rock slide involving roughly 14,000 cubic yards of rock debris occurred at the eastern portal area of the twin tunnels that carry Interstate Route 40 through Sterling Mountain in North Carolina near the Tennessee state border (Abramson and Daly, 1986). The 150-foot canopy or tunnel extension, which had been built at the eastern end of the westbound tunnel to protect the interstate from falling rock, was destroyed and the eastbound lane was also completely blocked. A truck and semitrailer just missed getting buried by the rock debris, but luckily, no deaths or private property damage occurred as a result of this failure.

The rock slope adjacent to the tunnel consists of good to excellent quality Longarm quartzite underlain by poor to fair quality Longarm quartzite interbedded with siltstone, slate, and phyllite. The Longarm Formation is one of four formations of the Snowbird Group, which is one of three regional lithologic units of the Ocoee Series. The Ocoee Series was formed from the metasediments that resulted from the erosion of the original Blue Ridge Mountains, which also formed most of the Great Smoky Mountains.

The knob of rock that failed was bounded by a fault trending nearly east-west, a major joint set trending northeast, and bedding striking northwest. Because of the preexisting discontinuities in the rock mass, because of the abundant water flowing over and through the slope, which accelerated weathering along the discontinuities, and because of the freeze/thaw pressures that developed every winter and spring season, the block of rock finally failed, destroying the westbound portal section of the tunnel.

Traffic was reinstated after the failure by diverting westbound traffic through the eastbound tunnel and eastbound traffic around the rock ridge. Existing data, geologic mapping, and core borings were used to characterize the rock mass for selection of remedial measures. These measures included demolition of badly damaged portions of the existing tunnel portal, construction of a new, cast-in-place concrete portal and retaining wall, the repair of damaged portions of the westbound tunnel, construction of a reinforced earth wall along the shoulder as a catchment area, and stabilization of the adjacent rock slopes with a combination of scaling, rock reinforcement, and drainage. Exploration of the site, design of remedial measures, and construction were completed within a total of 8 months after the failure and prior to the approaching winter season, which begins in November.

The construction contract price was for six million dollars and included 17,000 square yards of slope scaling, 93,000 linear feet of rock bolts, 3,400 square yards of steel mesh, 3,000 linear feet of horizontal drains, 10,000 square feet of reinforced earth, portal reconstruction, and tunnel repairs.

Scaling and removal of rock overhangs was carried out first. The contractor piled hay around the existing structure for protection. Explosives were used to remove the

Figure 7.91 Interstate Route 40 rock slope remediation.

rock overhangs. The blaster presplit a back line and then immediately fractured the freed blocks, almost in mid-air. Scalers then rappelled down the side of the slope on ropes with pry bars, knocking loose blocks or wedges that could conceivably be removed by hand. Traffic was stopped during scaling operations.

After scaling was complete, demolition of damaged structures and construction of new ones proceeded. Repair of the concrete tunnel arch began and excavation for new structures was completed. Shoring of the remaining tunnel was also required. The contractor erected reinforcing steel and poured concrete for new footings, retaining walls, and the portal structure while rock bolting was being carried out concurrently (Figure 7.91).

The rock slope was divided into areas according to the type of remediation required. In three areas, selective bolting was required using No. 9 reinforcing bars of varying lengths on an as-needed basis. Locations of these bolts were determined in the field. One area required pattern bolting and wire mesh. The most worrisome area was the one that included large rock blocks that could fail in the same manner as the original failure. Here, 40-foot-long, high strength anchors were specified on a 10 × 15-foot pattern with 20-foot-long bolts to stabilize the surficial slabs in between. The resulting bolt pattern was 5 × 5 feet. All of this work was generally done off of steel platforms hung from cranes.

Additional work included a reinforced earth wall catchment area and rock fence above the new tunnel portal, to collect small-scale rock falls that occur in the future. Also, horizontal drains were installed along the base of the slope to relieve future buildup of excess hydrostatic pressure within the rock mass.

Steel Mesh Used on Highway Projects Three projects in Montana and Nevada utilized double twisted hexagonal heavily galvanized mesh to mitigate rockfall problems (Ciarla, 1986). Along Highway 15 in Montana, the roadway was widened from two lanes to four lanes between Helena and Butte. Between Bernice and Basin, a detour road was required for local traffic during construction. Construction of this road required a 200-foot-high cut in a highly fractured rock formation. Ten thousand square yards of metallic mesh were installed with rock bolts and cables for $177,000 to protect traffic from rock falls. The mesh was delivered in 13-foot-wide × 300-foot-long rolls. It was mounted using a crane by placing the roll at the base of the slope and unrolling it from the bottom to the top of the slope. One-inch-diameter rebar anchors and $\frac{3}{4}$-inch-diameter wire rope were used to fasten the mesh to the slope. These were installed by workers from a crane-supported work cage. The anchors were installed on a 25-foot center to center pattern on the upper portion of the slope and a 50-foot pattern on the lower portion.

State Route 207 in Nevada was lowered, realigned, and widened to accommodate the heavy traffic between the area of Gardnerville and Minden and the Lake Tahoe area. Cuts in decomposed granite were required up to 125 feet in height. To protect the road bed, steel mesh tied to a 50 × 50-foot wire cable grid was used. Falling rocks thereby rolled or slid behind the mesh and were subsequently removed from the toe of the slope. Twenty thousand square yards of mesh were installed on the slope at a cost of $156,000.

The cut slopes along Interstate Route 80 in Nevada, near Sparks and Truckee, require frequent maintenance and rock debris removal. To mitigate the threat of rockfalls onto the travel lanes, steel mesh has been installed. Near Sparks, 37,000 square yards were installed at a cost of $175,000. Near Truckee, 79,000 square yards of mesh were installed at a cost of $458,000.

Tennessee State Route 31 Rock Slide A shot-in-place rock buttress was used to repair a block glide landslide in Tennessee along State Route 31 north of Mooresburg (Moore, 1986). The slide within an existing rock cut damaged the facility shoulder and deformed the roadway pavement. The rock mass consists of medium to thick beds of Clinch Formation quartzose sandstone with thin interbeds of shale and clay (Figure 7.92). The highway cut undercut the dipping beds of rock and exposed the bedding planes dipping into the highway.

As a result of weathering along joint surfaces and the infiltration of groundwater into the joints and bedding planes, movement was allowed to occur along the bedding planes toward the roadway. A 100-foot-long, 200-foot-wide, and 12-foot-thick rock block freed up and slid, forming a 5,200 cubic yard "block glide" type landslide (Figure 7.93). A shot-in-place rock buttress was the chosen method of remediation.

Design parameters for the rock buttress included an angle of internal friction of 38° and a unit weight of 140 pounds per cubic foot. This was estimated to provide an FOS equal to 1.3 against future upslope movements. The base of the buttress was designed to be 25 feet wide with a slope of 23° up to the base of the scarp (Figure 7.94). A

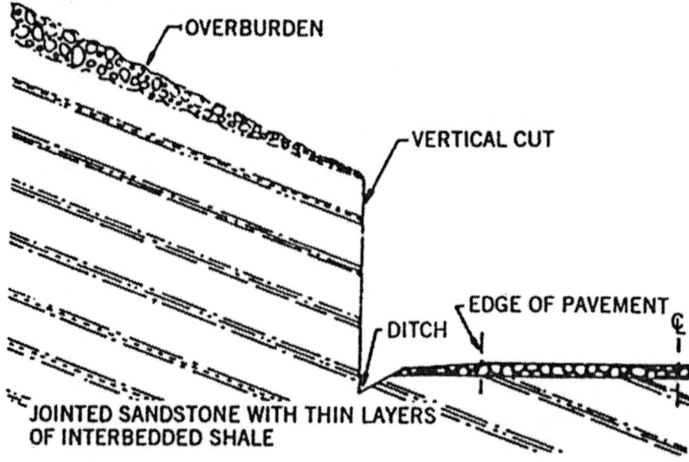

Figure 7.92 State Route 31 rock slope failure subsurface conditions. (From Moore, 1986.)

14-foot-wide berm was left at the top of the buttress. The construction cost for this work was $108,000.

Blasting was carried out in six separate sections. Traffic on the adjacent facility was held during blasting until the rock debris on the road was removed. The final dressed face was graded to 1.25H–1V. There were 2,500 cubic yards of excess rock debris generated by the project. The slope was vegetated and left with no apparent signs of previously having been a landslide.

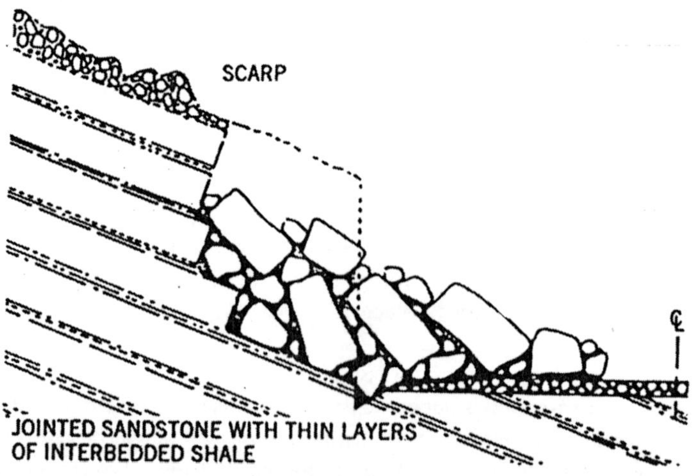

Figure 7.93 State Route 31 block glide failure. (From Moore, 1986.)

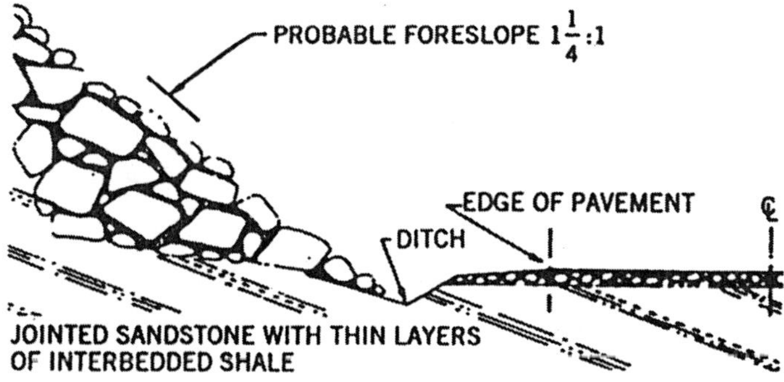

Figure 7.94 State Route 31 buttress cross section. (From Moore, 1986.)

Pittsburgh Light Rail Transit Rock Slope Stabilization Pittsburgh's Light Rail Transit system is utilizing the Mt. Washington Tunnel for access to and from the southern suburbs. This tunnel through Mt. Washington was originally constructed in the 1900s for the trolley system (Voytko et al., 1987). The 150-foot-high face of Mt. Washington above the north portal of the tunnel is composed of horizontally bedded shale, limestone, claystone, and sandstone (Figure 7.95). Differential weathering of the rock has resulted in undercutting of the more resistant beds and overhangs.

Jointing is near-vertical and orthogonal, forming columns of rock (Figure 7.96). Infiltration of groundwater between joint surfaces and subsequent freezing and thawing have caused movement of these rock columns and concern about possible rockfalls that could damage the transit tunnel portal.

The sandstone overhang was repaired with rock bolts and a concrete buttress (Figure 7.97). Fifty 13-foot-long, $\frac{7}{8}$-inch-diameter, Grade 70, expansion anchor rock bolts were installed and postgrouted to stabilize the overhang. Rock bolts and formwork were then installed in preparation for pouring the concrete buttress. For drainage, 4-inch-diameter pipes were installed to convey seepage from the rock to the face of the buttress.

The shale overhang was removed in sections using nonexplosive excavation methods. The overhang was removed to an existing predominant joint located behind the overhang to provide a relatively smooth final surface.

The shale column was intended to be removed in a controlled manner by initially being constrained by a wire mesh and cable restraining system (Figure 7.98) and subsequent removal in small layers. However, the rock column failed before removal operations began and slid to the base of the cut without catastrophic consequences.

After the specialized work described above was completed, the area was scaled and reinforced with welded wire fabric ($3 \times 3-1.4 \times 1.4$) reinforced shotcrete with 2-inch-diameter weep holes spaced 15 feet apart. At the toe of the slope, grouted rip-rap was placed to protect the nonresistant red beds. To further protect the portal structure, a rock fence (Figure 7.99) and earth berm (Figure 7.100) were constructed.

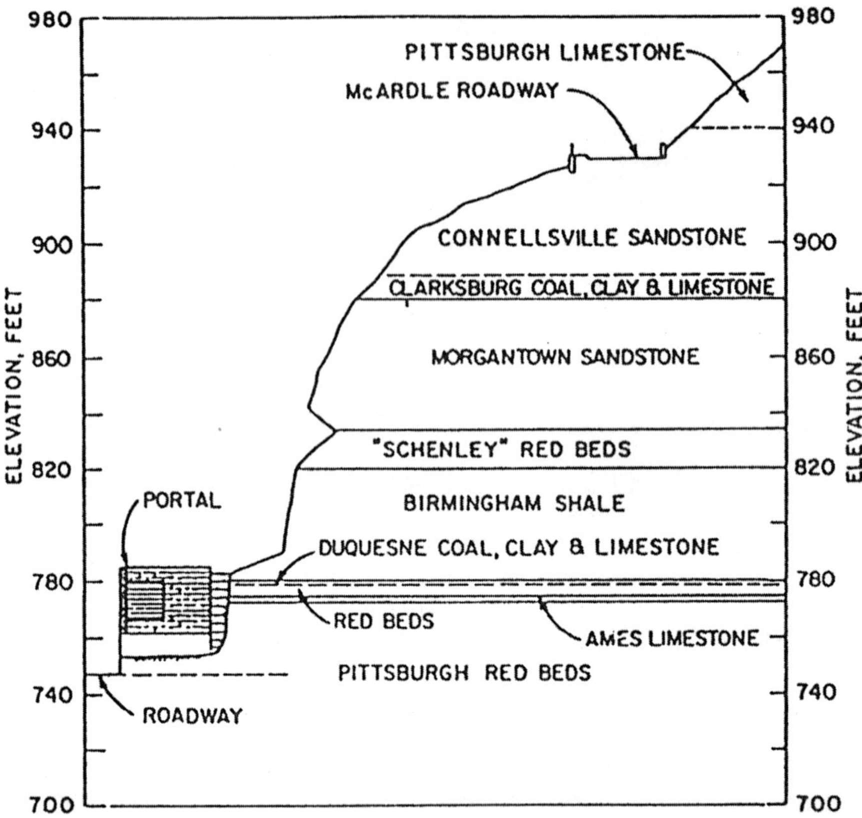

Figure 7.95 Mt. Washington tunnel portal geology. (From Voytko et al., 1987.)

Singapore Rock Slope Failure A rock cut along Bukit Batok Avenue 6 failed in Bukit Batok, a new town in the west-central area of Singapore (Pitts, 1988). The initial failure took place in October 1985 as a result of an unusually high rainfall event, and was further aggravated by additional rainfall in January 1986. The slope was about 100 feet high and 45°. The slope was blasted into quartz sandstone and was coincident with a predominant bedding plane.

The geomorphology of the 5,000 cubic yard slide area is depicted in Figure 7.101. Tropical weathering had taken place along shaly bedding planes and major joint surfaces. A peak shear strength angle of 41° was estimated from laboratory and field tests. A bedding plane angle between 42° and 45° was deduced from joint measurements and stereographic projections of poles. A planar analysis of the slide was carried out with consideration of stress relief and the presence of water along the bedding plane and tension crack. The failure was caused by the shear stress exceeding shear strength due to rain induced pore pressures along the basal bedding

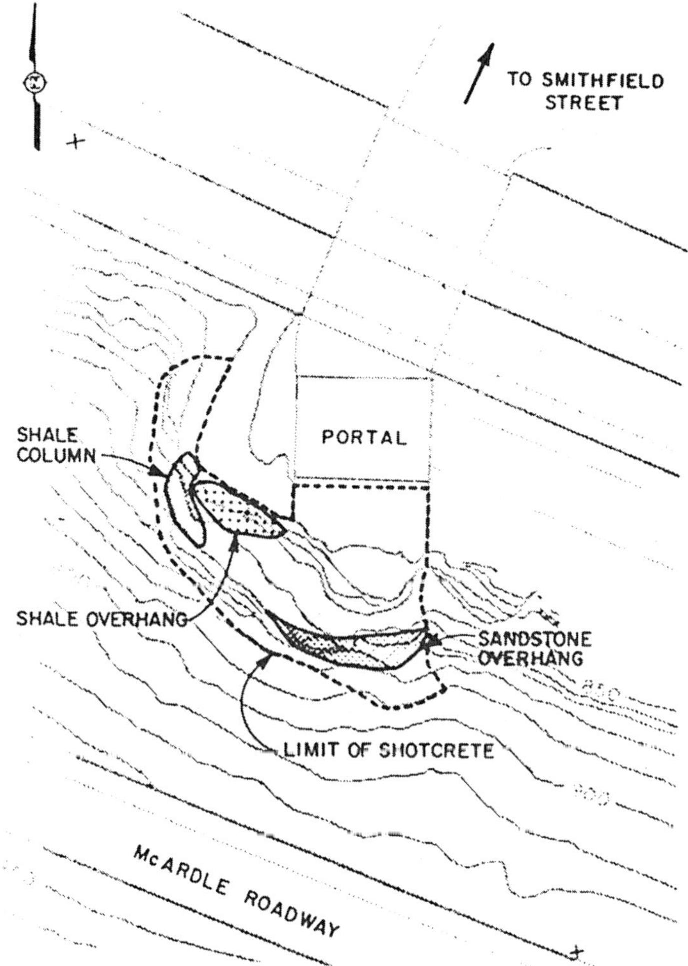

Figure 7.96 Mt. Washington tunnel portal jointing. (From Voytko et al., 1987.)

plane of the marginally stable slope. A sensitivity analysis of these assumptions is depicted in Figure 7.102. Stability was restored by regrading and rock bolting the failed slope.

Interstate Route 93 Rock Slope Failure Approximately 17,000 cubic yards of rock slope failed along an excavated 1H–8V rock cut on Interstate Route 93 in Woodstock, New Hampshire (Fowler, 1976). The facility was being constructed into a mountainside with the northbound lanes founded on the excavated rock and the southbound lanes founded on a viaduct. The failure buried the northbound lane

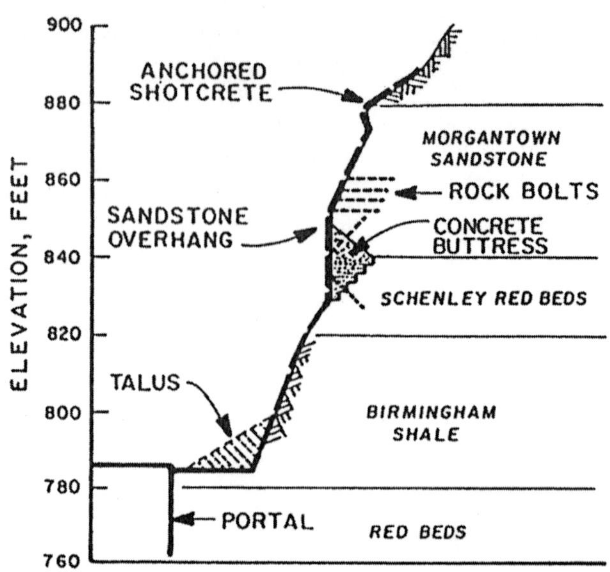

Figure 7.97 Mt. Washington tunnel portal sandstone overhang. (From Voytko et al., 1987.)

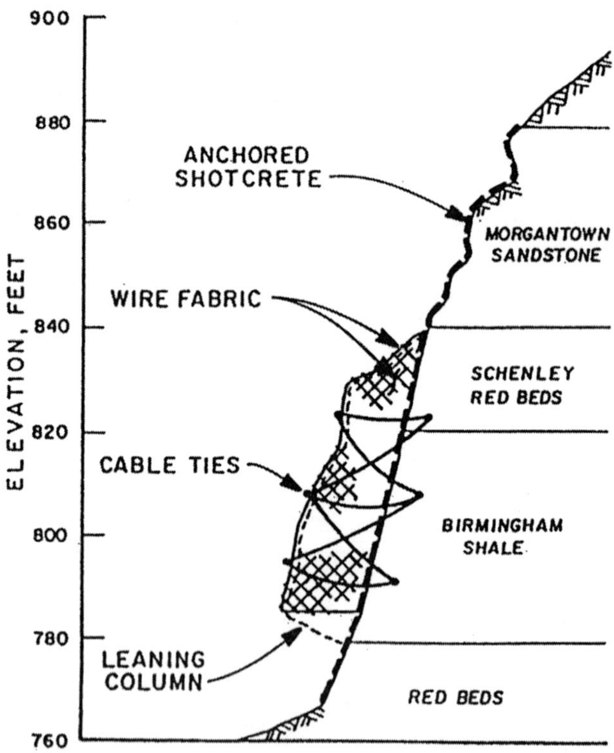

Figure 7.98 Proposed system for removing the Mt. Washington tunnel portal shale column. (From Voytko et al., 1987.)

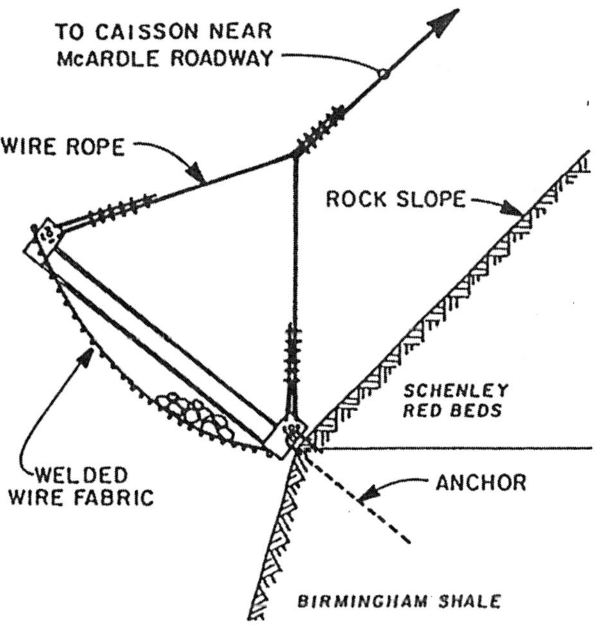

Figure 7.99 Mt. Washington tunnel portal rock fence. (From Voytko et al., 1987.)

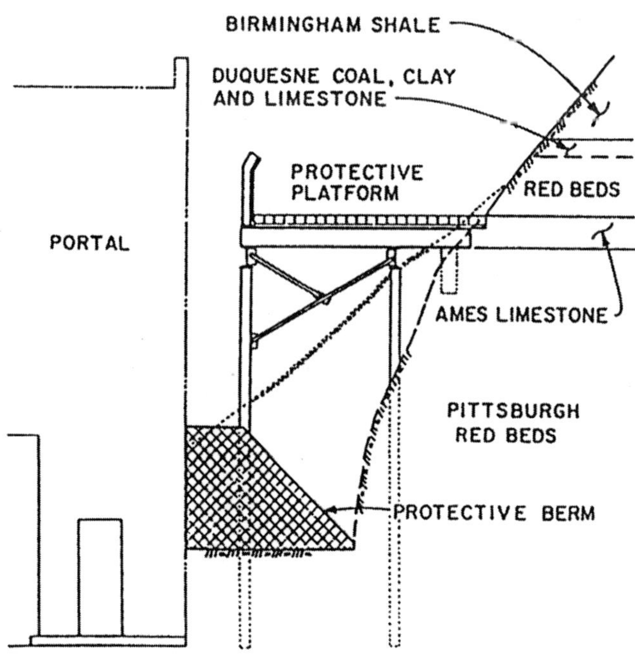

Figure 7.100 Mt. Washington tunnel portal earth berm. (From Voytko et al., 1987.)

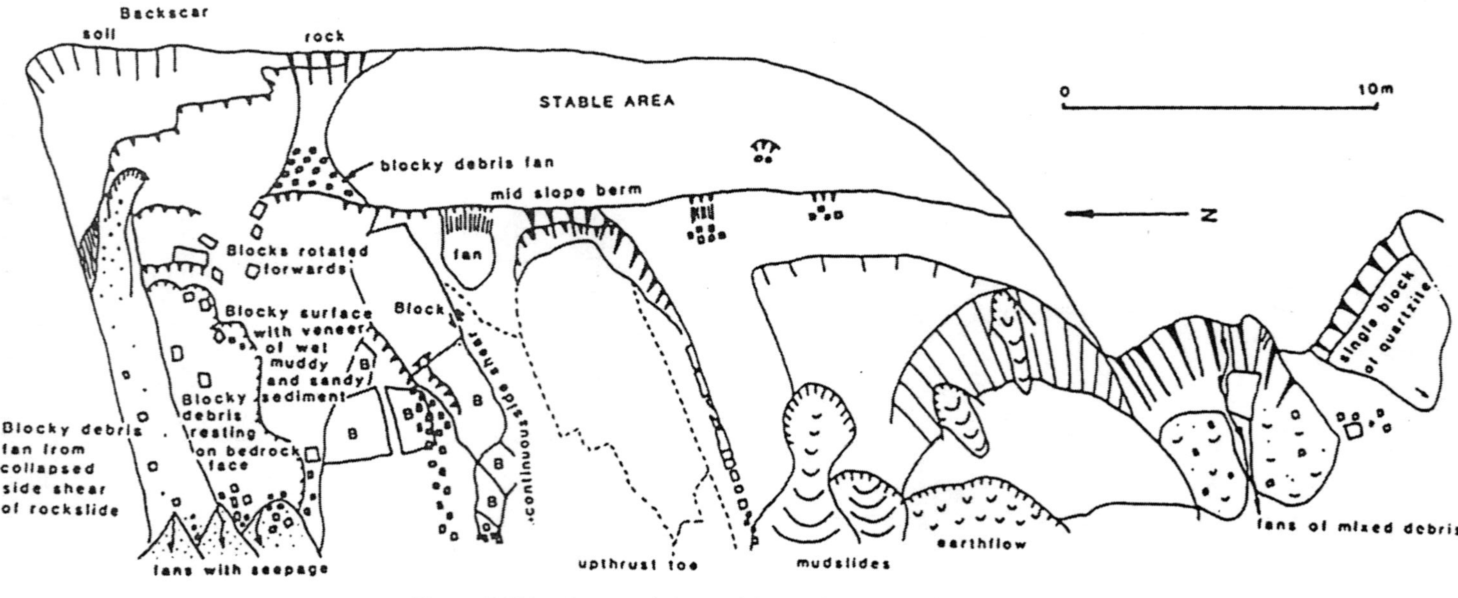

Figure 7.101 Geomorphology of the Bukit Batok rock slide.

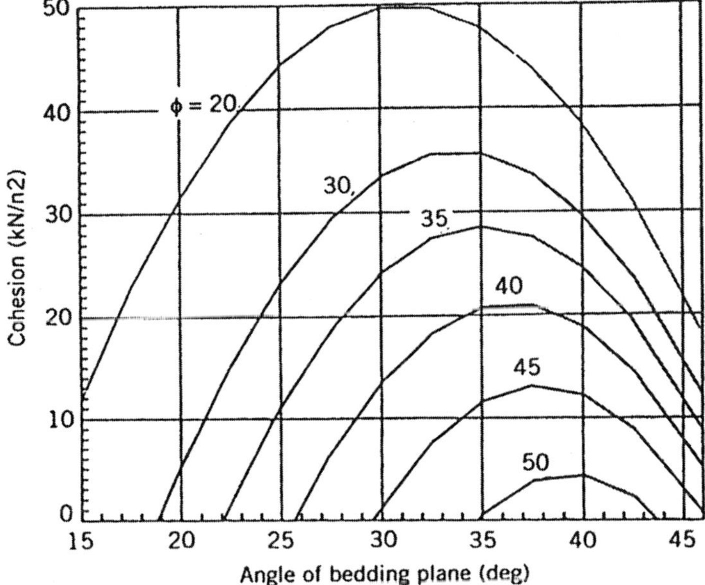

Figure 7.102 Bukit Batok rock slide sensitivity analysis. (From Pitts, 1988.)

under construction and delayed further construction. During construction there had been concern about the adverse direction of jointing, slickensided mylonitic zones, large joint apertures, and high water pressures compounded by overloading of the production and presplit blast holes.

The cut is composed of metamorphic gneissic and schistose rock common to north and central New Hampshire. The schistose foliation is very pronounced in the area of the slide. Rocks in the cut area are also concordantly and discordantly intruded by pegmatite and andesite. Additionally, several layers of mylonitic materials were observed ranging in thickness between $\frac{1}{2}$ inch and 11 feet. Geologic mapping indicated one set of discontinuities parallel to foliation (N50E 75NW) and the other major set perpendicular (Figure 7.103). Some of the mylonitic zones were perpendicular to foliation, although the ones responsible for the failure were parallel to the northbound lane alignment and dipped into the excavation. Back-calculation of shear strength properties assuming an FOS of 1 indicated that the mylonite had an angle of internal friction of 35°. The dip of the failure plane was 38°.

Modifications to the original design precluded further excavation of the southbound lanes and limitations on additional excavation for the northbound lanes (Figure 7.104). The backslope was flattened to $\frac{1}{2}$H–1V and the presplit line spacing was reduced from 36 inches to 18 inches with alternately loaded holes. During further excavation, another mylonitic zone was encountered and necessitated reinforcement of the slope with high-strength steel tendons, rock bolts, and horizontal drains (Figure 7.105).

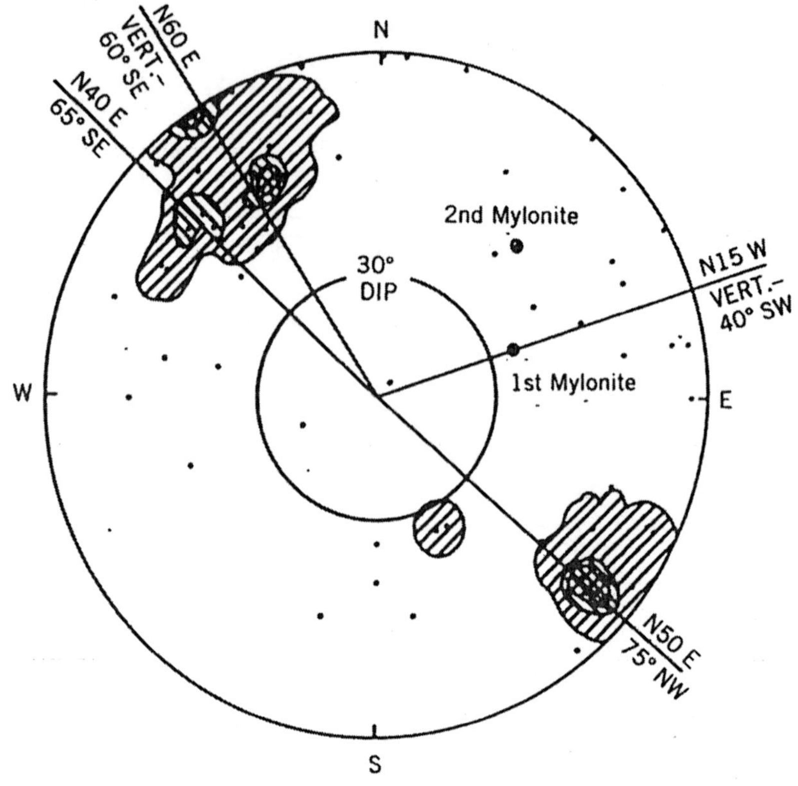

EQUAL AREA NET DATA
FROM STRUCTURAL SURVEY
(LOWER HEMISPHERE)

Figure 7.103 Woodstock rock slide jointing. (From Fowler, 1976, reproduced by permission of ASCE.)

7.11 ALTERNATIVES TO SLOPE STABILIZATION

Stabilization is not necessarily the only answer to slope stability problems. Other options include removal of the unstable mass, relocation of the facility, and bridging over the unstable area. These are discussed in more detail below.

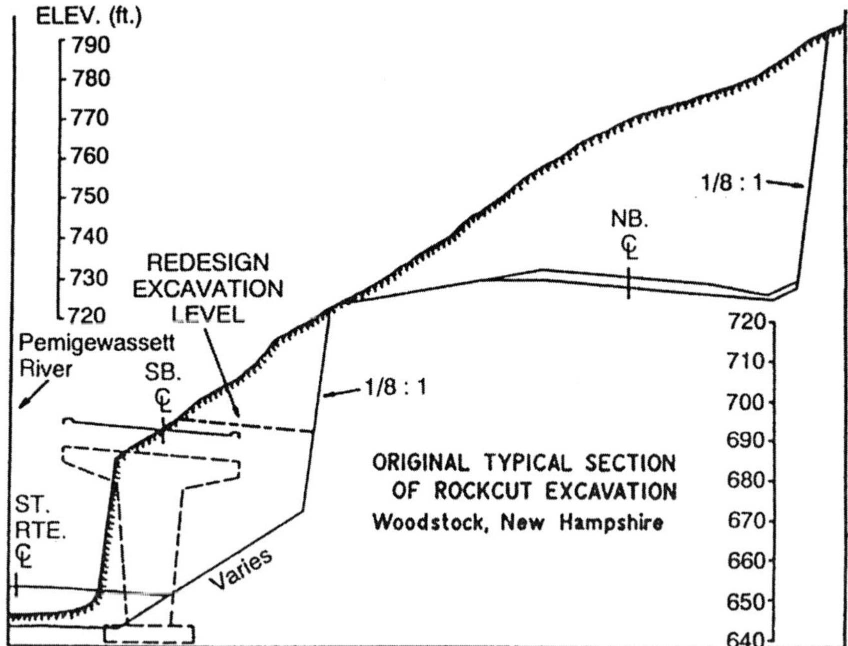

Figure 7.104 Woodstock rock slide excavation limitations. (From Fowler, 1976, reproduced by permission of ASCE.)

7.11.1 Complete Removal of Slide Zone

Complete removal of a slide zone is an alternative to stabilization of the slope (Figure 7.106). The removal of potentially unstable materials can vary from simple stripping of the top few feet of soft material to a more complicated and costly operation of excavating deep soil and rock deposits. Economics and the relative risk of further slope instability, should play an important role in the selection of a final course of action.

7.11.2 Facility Relocation

It is sometimes feasible to avoid a potential slide or a landslide area by changing the location of the facility. Relocation is generally applicable to treatment of potential slides as well as active ones. The main advantage of this method is the high probability of avoiding future instability. No other method, except complete removal, will correct the problem as well. One disadvantage is the physical difficulties that may be produced by a location change. In many cases, this method will represent the most costly correction alternative under consideration.

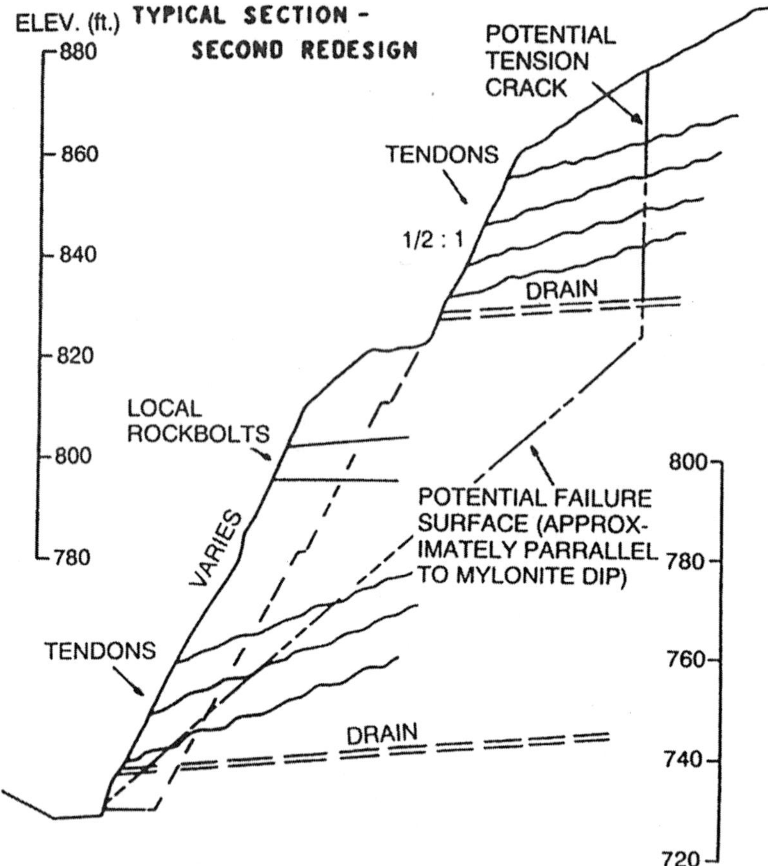

Figure 7.105 Woodstock rock slide remediation. (From Fowler, 1976, reproduced by permission of ASCE.)

One good example of highway relocation is the Hole in the Wall Gulch Slide in Oregon State Facility 86, 30 miles east of Baker, Oregon. After thorough investigation and evaluation of the slide, the Oregon Department of Transportation decided that it would be more economical to build a new road around the landslide ($5 million) than to stabilize it ($10 million). Their solution was to reconstruct several miles of the highway on the other side of the Powder River and to connect it with the undamaged section of the old highway by way of two new bridges, all designed to 55 miles per hour standards.

Another factor that must be considered when deciding whether to relocate a facility is legal liability. If an unstable mass is left in place, further liability to the property owner may develop and stabilization may be necessary simply to avoid the danger of future lawsuits.

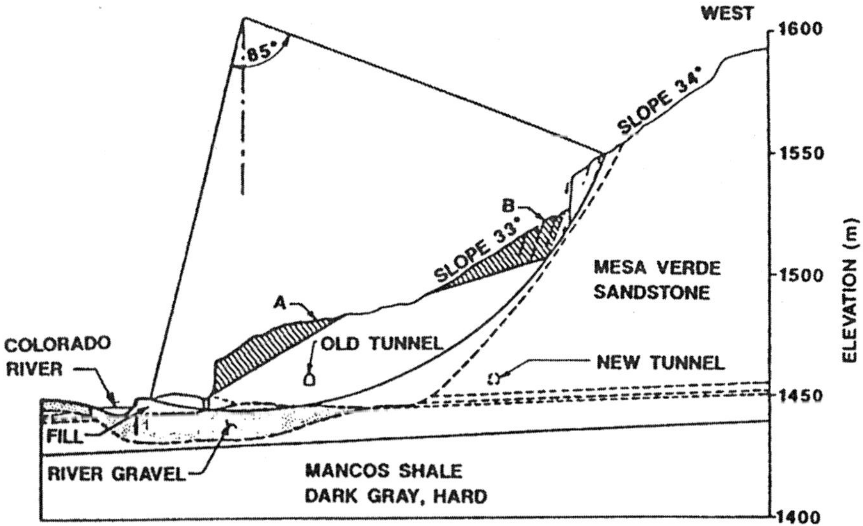

Figure 7.106 Stabilization of a landslide by partial removal of the head (stability analysis determined that removal of volume B was more effective than removal of A). (Eckel, 1958; Schuster and Krizek, 1978.)

7.11.3 Bridging

If there is no way to avoid an area of present or potential landslide, or if removal of steep and long narrow unstable slopes will be too costly, a bridge can be constructed to span the unstable area. This technique retains desired grades and alignments. The support of the bridge should be founded on piles well below the unstable foundation soils. Construction of the bridge may include limited excavations and the use of surface and subsurface drainage.

Bridges are commonly applicable only to small landslides or unstable areas. For large landslides or unstable areas that require bridge spans greater than 100 to 300 feet, it is doubtful whether this method is economically justified when compared with other methods (Eckel, 1958).

7.12 SELECTION OF STABILIZATION METHODS

Not all stabilization methods are appropriate for every type of failure in natural hillsides and engineered slopes. For example, slope flattening and berms, together with surface drainage, are often the first stabilization methods considered. Subdrainage is another very effective and rapid solution to landslides or slope stability problems of any magnitude. Surface protection methods can be used on a small scale to provide protective covering over materials that are prone to weathering. Retaining walls and explosives are most often used on small slides; they very seldom prove to be effective on very large-scale slides. Rock fills, concrete walls, and vegetation have been used successfully to prevent erosion. Where space constraints arise, tieback walls or reinforcement systems are often the only solution to the stability problem.

The number and variety of the stabilization techniques mentioned above are evidence that there can be no rule-of-thumb system for prescribing treatment of a potentially unstable slope or a landslide. Therefore, there is seldom one and only one "correct" method of treatment. The most expensive treatment is not always the most effective, and vice versa. Frequently, the most economical and effective means of treating slopes consists of a combination of two or more of the stabilization techniques described in this chapter.

Sometimes, the final choice between several methods may not belong to the design engineer. In conjunction with selecting the most effective and economical stabilization measure, other factors must also be considered, including safety, construction scheduling, availability of materials, site accessibility, equipment availability, aesthetics, environmental impact of the stabilization work, political issues, and labor considerations.

It is important to emphasize that engineers confronted with stability problems must not allow their minds and imaginations to be limited by those solutions with which they are already familiar. Similarly, standard reference books cannot substitute for local experience. Engineers must therefore use all available knowledge and search for the best solution for the particular case.

7.12.1 Goals

Economy of time and money are frequently the key factors in the selection of stabilization methods. This does not imply that complete or adequate engineering and geologic studies are to be waived in the interest of time and money. It is the engineer's job to accomplish all tasks with the maximum efficiency and economy, and within the time and emergency conditions that exist.

As mentioned earlier, there is usually more than one method of stabilization applicable to a given landslide or potentially unstable slope. The one used should be the most economical, but at the same time, the most effective.

7.12.2 Technical Constraints

Technical constraints include those associated with the type of ground to be stabilized, strain compatibility, creep of in situ ground, soil corrosivity, durability, and constructibility. They are discussed in the following.

The type of ground where the stabilization work will be carried out has a major impact on its selection. Natural ground is treated differently from fill. Granular soils are treated differently from fine-grained soils. Of utmost importance is the location of groundwater. Stabilization work over soft ground requires either highly flexible systems that can tolerate settlement induced by imposed loads, or use of lightweight fill to minimize further shearing of the unstable soil mass.

If movement of the ground during or after construction is anticipated, the stabilization method selected must be able to tolerate these movements without adversely affecting performance. Mechanically stabilized embankments and soil nailing are good examples of systems influenced by the strain compatibility of the system with the soil. The strains required to mobilize the full strength of the reinforcing elements are much smaller than those needed to mobilize the full strength of soil. However, for extensible reinforcing elements such as geosynthetics, the required strains are much larger. Therefore, relatively large internal deformations could occur using these systems and the material properties used for analysis should be measured at large strains (residual strength).

Gravity-type structures are less influenced by the magnitude of internal deformations than systems involving soil reinforcement (Munfakh, 1990). Flexible walls, such as gabions, should be used in lieu of rigid walls for slopes with potentially large vertical and horizontal deformations.

The creep of clayey soil and rock could have a negative impact on the long-term performance of slope stabilization systems including retaining walls and soil nailing. Design of these systems should account for the potential creep effect of clayey soils and rocks.

Corrosivity can adversely affect the long-term performance of steel-reinforced systems and concrete retaining walls. This can be alleviated with adequate drainage provisions that intercept aggressive groundwater that could cause corrosion of metallic reinforcing elements. Drainage elements also need to be protected against corrosion. Methods frequently used are galvanizing and other resistant coatings. Special cement as well as adequate concrete cover over steel reinforcement might be required for concrete and shotcrete elements.

The durability of any system is another factor in slope stabilization selection. When geosynthetics are used as reinforcement, their long-term deterioration caused by chemical attack and exposure to ultraviolet light may be a concern, as the tensile strength and ductility may be reduced, leading to eventual failure.

Factors such as material, equipment, and labor availability must also be considered when selecting types of slope stabilization work. For example, where rock fill is abundant, gabion walls would be suitable and economical, but this may not be the case if the rock requires a long-distance haul. Similarly, stabilization work requiring labor-intensive systems, such as chunam plaster walls, is usually not cost-effective in areas with strong labor union requirements.

7.12.3 Site Constraints

Site constraints for slope stabilization work are usually related to existing right of way and adjacent facilities. For example, a relatively large space is required behind the structure face for mechanically stabilized embankments as compared to that needed for footings of conventional walls. The length of the reinforcing elements is generally 0.5 to 0.7 times the height of the wall. Cantilever retaining wall footings require less than that distance behind the wall, but require room in front of wall for the toe of the footing. MSE walls do not have a footing toe. For very tight spaces, soil nailing or tiebacks in cuts may be the most suitable because little additional excavation is required, although they are influenced by the presence of utilities and buried structures nearby and additional costs for permanent underground easements.

7.12.4 Environmental Constraints

As with most structures, the selection of slope stabilization work should consider its potential environmental impact during and after construction. For example, excavation and disposal of contaminated material at the site and discharge of large quantities of water are of concern. Slope stabilization works that encroach on wetlands or have negative ecological impacts are usually not accepted without considerable companion efforts addressing mitigation. Changes in groundwater flow patterns may have a potentially detrimental effect on the environment.

7.12.5 Aesthetic Constraints

In addition to being functional and economical, slope stabilization work has to be aesthetically pleasing. This aesthetic factor is important when the stabilization work is to be made in natural habitats and forests, as well as in urban environments. Under these circumstances, attractive wall systems with vegetation, architectural treatments, and coloring are usually considered to improve aesthetics.

7.12.6 Schedule Constraints

The stabilization of landslides or slopes prone to sliding must be executed according to a well-thought-out plan, which lists individual measures in order of urgency as well as sequence of implementation. Scheduling must be sensitive to changes, such as existing operation of facilities, time required to order special materials and equipment, and weather. For example, it is counterproductive to begin constructing a stabilization berm before the subsoils have been drained of excess water. A long and often laborious process of investigation, stability analyses, and planning of stabilization work for landslides should not delay the requisite measures that obviously must be taken, as the treatment of active landslides is always a contest with time.

Although the stabilization program may be well-planned, its effectiveness may be affected adversely by delays of implementation because of unexpected schedule constraints, such as legal problems, right of way acquisition, funding, political issues, or union problems. Therefore, contingency plans also must be contemplated in the event that such constraints arise.

All stabilization work must be regularly checked and maintained for proper and efficient performance. Otherwise, the extensive and costly stabilization work may be useless and within a short time new movements may start again. Thus the schedule of inspection and maintenance work should be included in the overall planning of the stabilization works.

Scheduling of stabilization work must focus on weather conditions. In the northern part of the United States, extensive operations are very difficult, or even impossible, in winter when the waterlogged slide areas may be frozen or covered by snow. Therefore, individual operations should be scheduled with caution so that remedial work can be finished at the appropriate time. It may be better to postpone some operations until spring or to more suitable weather conditions. In other cases, winter holidays or spring thaw may represent deadlines by which slope stabilization measures must be completed.

7.12.7 Other Constraints

Other constraints that may influence the selection of stabilization works include those associated with politics and tradition. National or local district policies, trade barriers, and political influences sometimes affect the selection of materials. An example is the use of conventional reinforced concrete retaining walls to support a landslide or unstable area because the local contractors are only equipped for and experienced with that type of construction.

7.12.8 Cost

The total cost of slope stabilization work has many components, including the stabilization system itself, right of way, temporary and permanent easements, disposal of unsuitable materials, and drainage. In general, the construction cost of MSE and nail/tieback walls is less than 60 percent that of conventional cast-in-place concrete walls. According to Mitchell and Villet (1987), concrete cantilever walls become less economical for wall heights greater than 10 feet. Because of intense competition among the various wall systems, substantial savings have been realized on projects when alternate bidding was allowed.

7.13 PROBABLE COST ANALYSIS OF STABILIZATION ALTERNATIVES

7.13.1 Introduction

Often the decision on what slope stabilization methods make the most sense for a project ends up coming down to "the best bang for the buck." The least expensive solution may be fine for the short-term but it might have to be redone periodically. The technically best long-term solution may be prohibitively expensive. So, how does one quantitatively develop an evaluation of "best bang for the buck"? To a certain extent, that depends on how the "best bang" is defined. In some cases, the speed at which a closed facility can be brought back into operation is the primary concern. In other cases, the preferred option might be the one with the least environmental impact. The

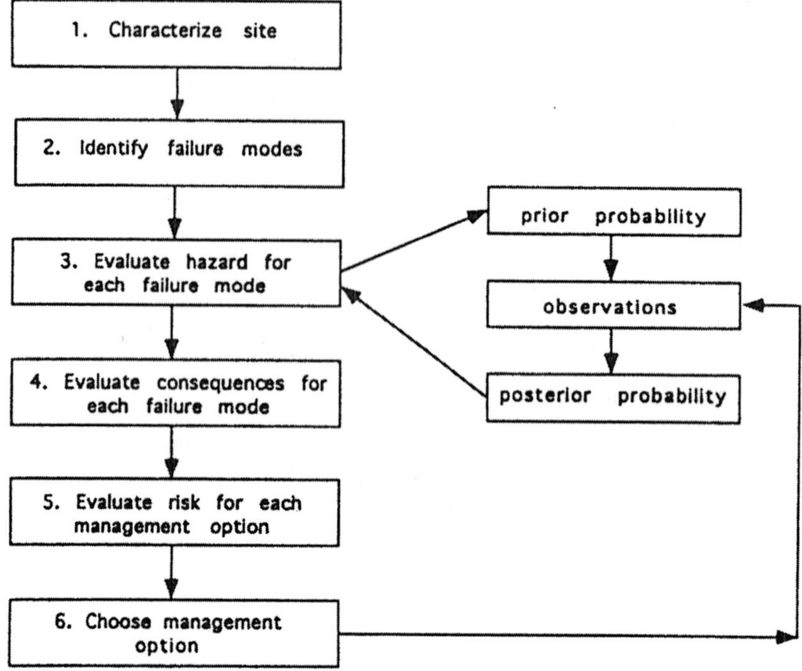

Figure 7.107 Decision-making process for slope stabilization project. (Turner and Schuster, 1996.)

answer lies in analyzing all of the possible costs and outcomes of various options and then making a selection that satisfies the project constraints. A decision-making process is depicted in Figure 7.107.

7.13.2 Developing Alternative Schemes

Typically in the initial stages of a project, a series of alternatives are developed for slope stabilization. The alternatives may consist of individual stabilization methods or combinations of methods. Development of the alternatives may require multiple slope stability analyses and conceptual designs. The conceptual designs should be to a level that a credible cost estimate can be developed (say to a 15 to 30 percent design level). The cost estimates should include sufficient costs for design, construction management, and contingencies. Table 7.12 lists possible options for a hypothetical slope stabilization project.

7.13.3 Estimating Outcomes and Probabilities

Each potential action has a potential outcome. In the simplest sense, one might say for a slope stabilization project, the possible outcomes are that the slope either remains

TABLE 7.12 Alternatives for Example Slope Stabilization Project

Alternative Number	Description	Conceptual Designs	Factor of Safety
1	Do nothing	N.A.[a]	0.87
2a	Flatten slope	1H–1V	1.12
2b		1.5H–1V	1.26
2c		2H–1V	1.48
2d		2.5H–1V	1.74
3a	Install drainage system	One row of 2″ dia., 50′-long horizontal drains at 10′c.c	0.87
3b		Three rows 100′ long	1.54
3c		Five rows 200′long	2.17
4a	Construct rock buttress	20′ wide × 20′ high	1.09
4b		40′ wide × 20′ high	1.15
4c		40′ wide × 40′ high	1.32
5a	Install soil nails	Five rows of no. 10–40′-long soil nails at 4′ c.c	0.87
5b		Ten rows 60′ long	1.34
5c		Twenty rows 60′ long	1.41
5d		Twenty rows 90′ long	1.86

[a]N.A. = not applicable.

stable or it fails. To take it a step further, one could say if the slope fails, the failure could be a minor sloughing or a major landslide. The process could be taken still further by evaluating the potential consequences of a slope failure including loss of use of the facility, property damage, and personal injury. This type of evaluation was carried out for a tunnel and the slopes at each end of the tunnel for a project in Colorado (Parsons Brinckerhoff, 1992). The resulting decision trees are depicted in Figures 7.108, 7.109, and 7.110.

The probabilities of occurrence may be estimated using judgement, inspection and maintenance records, empirical methods, and/or more rigorous methods like those discussed in Section 6.17. If one calculates the probability of failure, the probability of no failure would be 1.0 minus the probability of failure. So for example, if Alternative 2c is chosen from Table 7.13, the probability that the slope will still fail is estimated to be 6 percent. Worded another way, if Alternative 2c is implemented, there is a 94 percent probability that the slope will not fail.

7.13.4 Determining Probable Costs

The probability of slope failure can be calculated as well as the resultant cost that would occur as a result of the failure such as clean-up, lost use of the facility, property damage, personal injury, and so on. Likewise, the probability and cost of no failure as a result of a remediation can also be calculated. In the example above for Alternative 2c,

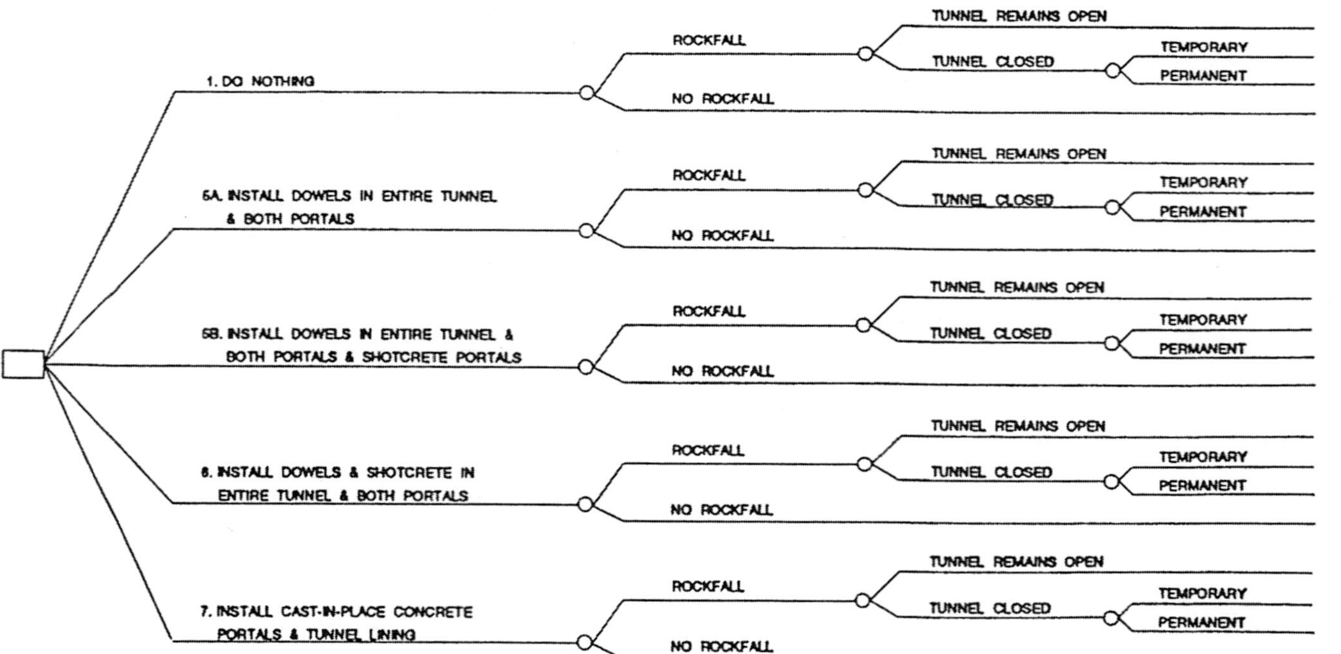

Figure 7.108 Decision tree considering loss of use. (Parsons Brinckerhoff, 1992.)

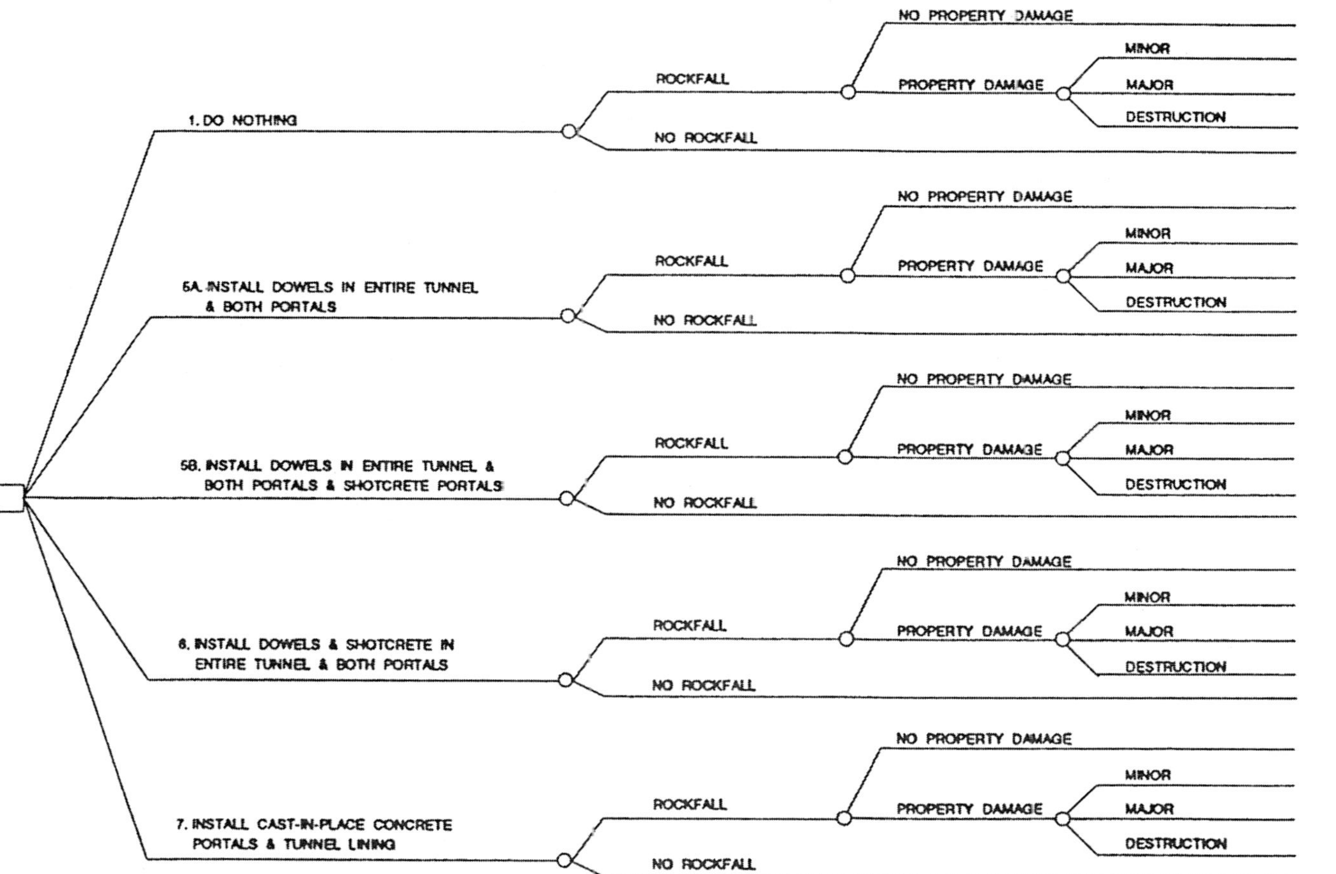

Figure 7.109 Decision tree considering property damage. (Parsons Brinckerhoff, 1992.)

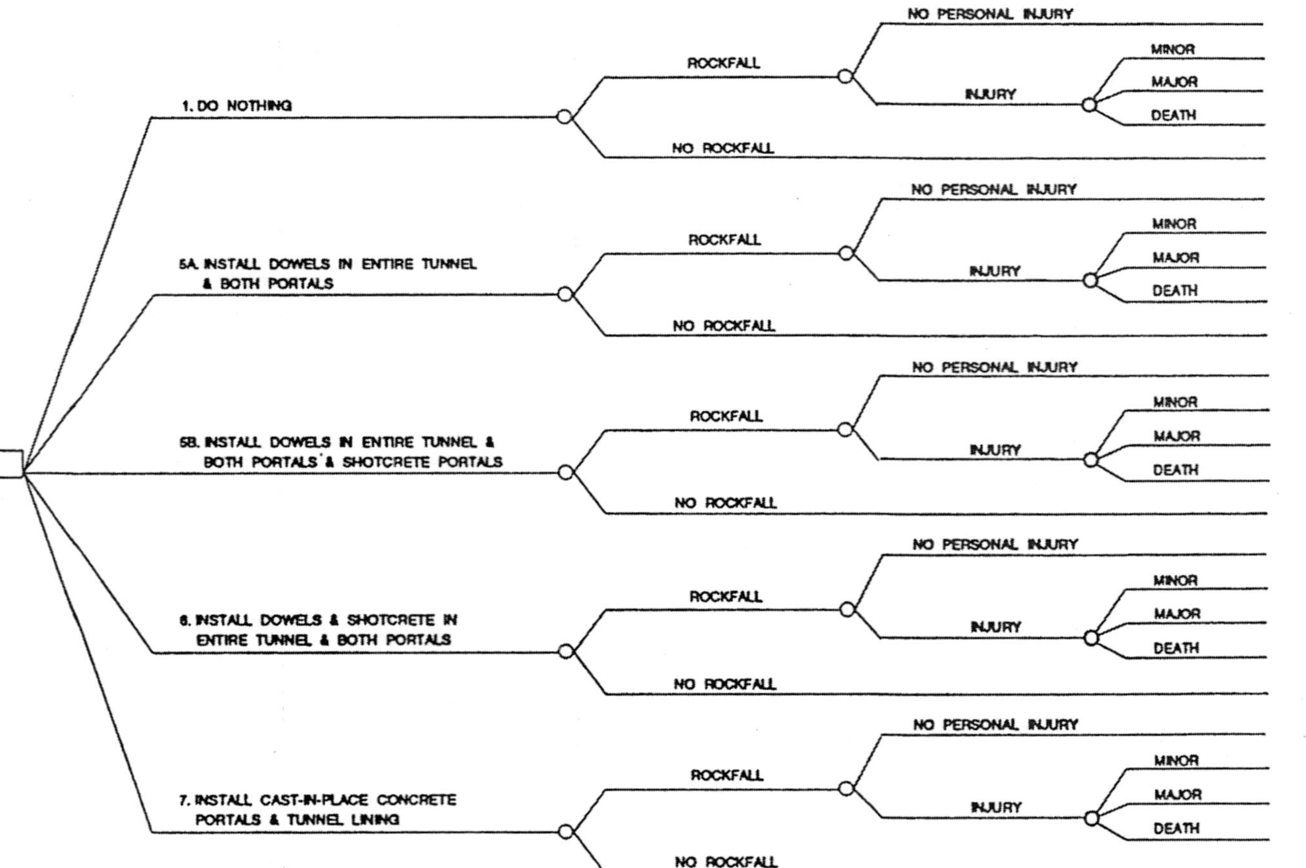

Figure 7.110 Decision tree considering personal injury. (Parsons Brinckerhoff, 1992.)

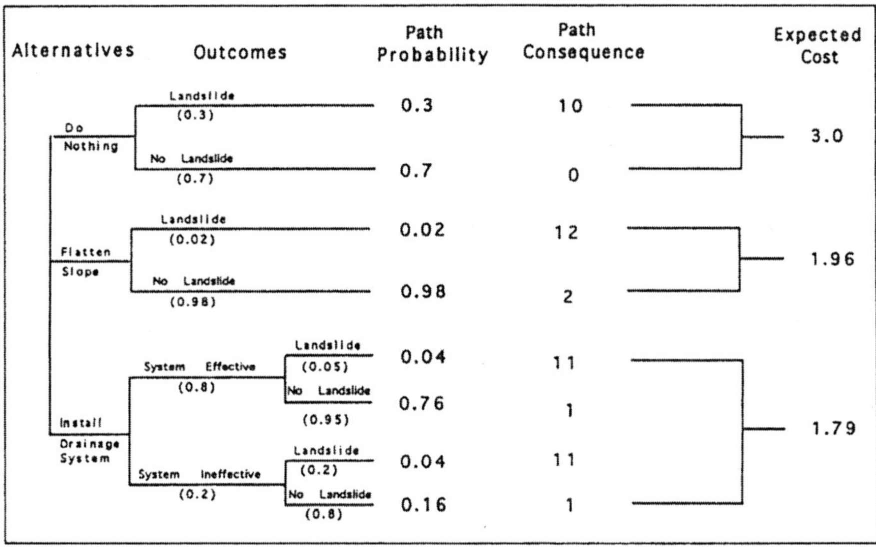

Figure 7.111 Decision tree of slope stabilization project. (Turner and Schuster, 1996.)

the probability of no failure as a result of flattening the slope to 2H–1V is 94 percent and the cost is estimated to be $400 per foot of slope. The probable cost would then be the probability multiplied be the estimated cost or $376 per foot of slope (0.94 × 400).

TABLE 7.13 Alternatives for Example Slope Stabilization Project

Alternative Number	Description	Factor of Safety	Estimated Cost ($ per ft of slope)	Probability of Failure (%)
1	Do nothing	0.87	0	100
2a	Flatten slope	1.12	200	40
2b		1.26	300	25
2c		1.48	400	6
2d		1.74	500	2
3a	Install drainage system	0.87	200	100
3b		1.54	1,200	6
3c		2.17	4,000	0
4a	Construct rock buttress	1.09	50	45
4b		1.15	100	35
4c		1.32	150	18
5a	Install soil nails	0.87	5,000	100
5b		1.34	15,000	20
5c		1.41	30,000	10
5d		1.86	45,000	1

A similar evaluation is depicted in Figure 7.111. This type of analysis can be used to identify alternatives of actions, possible outcomes, and respective consequences or costs for each scenario or path. The probability of each branch of the outcome can be determined either from probabilistic models or subjectively from available information. In this example, the optimal solution for this site is the installation of a drainage system if the expected cost is the criterion for decision. When the probabilities or costs are crude estimates, a sensitivity analysis should be performed to find out if the optimal solution changes with the probabilities and costs.

REFERENCES

Aboshi, H., E. Ichimoto, M. Enoki, and K. Hazaad, 1979. "A Method to Improve Characteristics of Soft Clays by Inclusion of Large Diameter Sand Columns," *International Conference on Soil Reinforcement*, Paris, pp. 211–216, March.

Abramson, L. W., and W. F. Daly, 1986. "Analysis and Rehabilitation of Aging Rock Slopes," *37th Annual Highway Geology Symposium*, Helena, Montana, pp. 56–86.

Agnew, W., 1991. "Erosion Control Product Selection," *Geotechnical Fabric Report*, April, pp. 24–27.

American Concrete Institute, 1984. "State-of-the-Art Report on Fiber Reinforced Shotcrete," Committee 506, ACI 506.1R-84 (89), Detroit.

Association of Bay Area Governments, 1981. *Manual of Standards for Erosion and Sediment Control Measures*, Oakland, California.

Bache, D. H., and I. A. MacAskill, 1984, *Vegetation in Civil and Landscape Engineering*. London: Granada.

Bachus, R. C., and R. D. Barksdale, 1989. "Design Methodology for Foundations on Stone Columns," *Foundation Engineering Congress*, Evanston, Illinois, pp. 244–257, June.

Barker, D. H., 1986. "Enhancement of Slope Stability by Vegetation," *Ground Engineering*, April, pp. 11–15.

Barrows, R., and G. Machan, 1991. "Reinforcement of a Failed Embankment Over Slough Mud," *Proceedings of Geosynthetics Conference*, Atlanta, Georgia, February.

Beles, A. A., and I. I. Stanculescu, 1958. "Thermal Treatment as a Means of Improving the Stability of Earth Masses," *Geotechnique*, Vol. 8, No. 4, pp. 158–165.

Bishop, A. W., and L. Bjerrum, 1960. "The Relevance of the Triaxial Test to the Solution of Stability Problems," *Proceedings, Research Conference on shear Strength of Cohesive Soils*, Boulder, Colorado, ASCE, June.

Brawner, C. O., and R. Pakalnis, 1982. "Vacuum Drainage to Stabilize Rock Slopes on Mining Projects," *1st International Mine Water Conference*, Budapest, Hungary, April.

Broms, B. B., 1991. "Stabilization of Soil with Lime Columns," *Foundation Engineering Handbook*, H. Y. Fang, Ed. New York: Van Nostrand Reinhold, Chapter 24, pp. 833–855.

Bruce, D. A., 1992. "Two New Specialty Geotechnical Processes for Slope Stabilization," *Proceedings: Stability and Performance of Slopes and Embankments—II*, ASCE Geotechnical Special Publication No. 31, Berkeley, California, pp. 1,505–1,519.

Bruce, D. A., and B. Bianco, 1991. "Large Landslide Stabilization by Deep Drainage Wells,"

Proceedings: International Conference on Slope Stabilization Engineering. Shanklin, Isle of Wight, April.

Bruce, D. A., and R. A. Jewell, 1987. "Soil Nailing: Application and Practice—Part 2," *Ground Engineering, The Journal of the British Geotechnical Society*, Vol. 20, No. 1, pp. 21–28.

Bruce, R. L., and J. Scully, 1966. *Manual of Landslide Recognition in Pierre Shale, South Dakota*, South Dakota Highways Department, Research Project 615 (64), December.

Burroughs, E. P., and R. R. Thomas, 1976. *Root Strength of Douglas Fir as a Factor in Slope Stability*, USDA Forest Service Review, Draft, INT 1600-12 (9/66).

Christopher, B. R., and R. D. Holtz, 1985. *Geotextiles Engineering Manual*, National Highway Institute, FHWA, Washington, DC, Contract DTFH61-80-C-00094.

Christopher, B. R., and R. D. Holtz, 1989. *Geotextile Design and Construction Guidelines Manual*, National Highway Institute, FHWA, Washington, DC, under contract to GeoServices, Inc., Contract DTFH61-86-R-00102.

Christopher, B. R., and D. Leschinsky, 1991. "Design of Geosynthetically Reinforced Slopes," *Proceedings: Geotechnical Engineering Congress*, F. G. Mclean, D. A. Campbell, and D. W. Harris, Eds., Boulder, Colorado, pp. 988–1,005, June.

Christopher, B. R., S. A. Gill, J. P. Giroud, I. Juran, J. K. Mitchell, F. Schlosser, and J. D. Dunnicliff, 1990. *Reinforced Soil Structures, Volume 1: Design and Construction*, Federal Highway Administration, Washington, DC, Report FHWA-RO-89-043.

Ciarla, M., 1986. "Wire Netting for Rockfall Protection," *37th Annual Highway Geology Symposium*, Helena, Montana.

Curtis, J. D., 1964. "Roots of a Ponderosa Pine," USDA Forest Service Research Paper INT-9.

Earth Technology Corporation, 1988. "Instability of Landfill Slope, Puente Hills Landfill, Los Angeles County, California," Report submitted to Los Angeles County Sanitation District.

Eckel, E. B., Ed., 1958. *Landslides and Engineering Practice*, Highway Research Board, Special Report 29.

Elias, V., and I. Juran, 1991, "Soil Nailed Structures: Analysis of Case Histories," ASCE Geotechnical Special Publication No. 12, New York, pp. 232–244.

Elias, V., and I. Juran, 1991. "Soil Nailing for Stabilization of Highway Slopes and Excavations," Federal Highway Administration, Publication FHWA-RD-89-193, June.

Elias, V., and P. Swanson, 1983. *Cautions of Reinforced Earth with Residual Soils*, Transportation Research Board 919, Washington, DC.

Erickson, A. J., 1977. *Aids for Estimating Soil Erodibility* $\frac{3}{4}$ *"K" Value Class and Soil Loss Tolerance*, U.S. Department of Agriculture, Soil Conservation Service, Salt Lake City, Utah.

Forsyth, R. A., and J. P. Egan, 1976. "Use of Waste Materials in Embankment Construction," Transportation Research Board 593, pp. 3–8.

Fowler, B. K., 1976. "Construction Redesign—Woodstock Rock Slide, N. H.," *Specialty Conference on Rock Engineering for Foundations and Slopes*, Vol. 1, Boulder, Colorado. New York: ASCE, pp. 386–403.

Fukuoka, M., 1977. "The Effects of Horizontal Loads on piles Due to Landslides," *Proceedings, Specialty Session 10, 9th International Conference on Soil Mechanics and Foundation Engineering*, Tokyo, Japan, pp. 27–42.

Gassler, G., and G. Gudehus, 1981. "Soil Nailing—Some Soil Mechanic Aspects of In-Situ Reinforced Earth," *Proceedings of the 10th International Conference on Soil Mechanics and Foundation Engineering*, Vol. 3, Session 12, Stockholm, pp. 665–670.

Geotechnical Control Office, 1979. *Guide to Retaining Wall Design*. Hong Kong: Engineering Development Department, July.

Geotechnical Control Office, 1984. *Geotechnical Manual for Slopes*, 2nd ed. Hong Kong: Engineering Development Department, May.

Geotechnical Control Office, 1987. *Guide to Site Investigation*. Hong Kong: Engineering Development Department, September.

Giroud, J. P., G. Goldman, and A. Jones, 1987. "Soil Reinforcement Design Using Geotextiles and Geogrids," in *Geotextile Testing and the Design Engineer*, ASTM Specialty Publication 952, pp. 69–116.

Goldman, S. J., K. Jackson, and T. A. Bursztynsky, 1986. *Erosion and Sediment Control Handbook*. New York: McGraw-Hill.

Goughnour, R. R., J. T. Sung, and J. S. Ramsey, 1990. "Slide Correction by Stone Columns," *Deep Foundation Improvements: Design, Construction, and Testing*, M. I. Esrig and R. C. Bachus, Eds., ASTM STP-1089.

Gray, D. H., 1978. "Role of Woody Vegetation in Reinforcing Soils and Stabilizing Slopes," *Symposium on Soil Reinforcing and Stabilizing Techniques*, Sydney, Australia, pp. 253–306.

Gray, D. H., and A. T. Leiser, 1982. *Biotechnical Slope Protection and Erosion Control*. New York: Van Nostrand Reinhold.

Gray, D. H., and R. B. Sotir, 1992. "Biotechnical Stabilization of Cut & Fill Slopes," *Proceedings: Stability and Performance of Slopes and Embankments—II*, ASCE Geotechnical Special Publication No. 31, Berkeley, California, pp. 1,395–1,410.

Hausmann, M. R., 1990. *Engineering Principles of Ground Modification*. New York: McGraw-Hill.

Hausmann, M. R., 1992. "Slope Remediation," *Proceedings: Stability and performance of Slopes and Embankments—II*, ASCE Geotechnical Special Publication No. 31, Berkeley, California, pp. 1,274–1,317.

Ingold, T. S., 1982. "An Analytical Study of Geotextile Reinforced Embankment," *Proceedings 2nd International Conference on Geotextiles*, Las Vegas, Nevada.

Ito, T., and T. Matsui, 1975. "Methods to Estimate Lateral Force Acting on Stabilizing Piles," *Soils and Foundations*, Vol. 15, No. 4, pp. 43–59.

Jewell, R. A., 1988. "The Mechanics of Reinforced Embankments on Soft Soils," *Geotextiles and Geomembrane*, Vol. 7, No. 4, pp. 237–273.

Jewell, R. A., and R. I. Woods, 1984. "Simplified Design Charts for Steep Reinforced Slopes," *Symposium on Reinforced Soil*, University of Mississippi, September.

Jewell, R. A., N. Paine, and R. I. Woods, 1984. "Design Methods for Steep Reinforced Embankments," *Proceedings: Symposium on Polymer Grid Reinforcement*, Institute of Civil Engineering, pp. 18–30.

Johnston, R. E., and L. W. Abramson, 1994. "Geosynthetic Reinforcing of Highway Embankments," *Proceedings of the 45th Annual Highway Geology Symposium*, Portland, Oregon, August.

Juran, I., 1977. "Dimensionnement interne des Ovrages en Terre Armee," Thesis for Doctorate of Engineering, Laboratoire Central des Ponts et Chaussees, Paris.

Juran, I., G. Baudrand, K. Farrag, and V. Elias, 1989. "Kinematical Limit Analysis for Design of Soil Nailed Structures," *Journal of Geotechnical Engineering*, Vol. 116, No. 1.

Ladd, C. C., 1991. "Stability Evaluation During Staged Construction," 22nd Terzaghi Lecture,

Journal of the Geotechnical Engineering Division, ASCE, Vol. 117, No. 4, April, pp. 540–615.

Ladd, C. C., J. J. Rixner, and D. C. Gifford, 1972. "Performance of Embankments with Sand Drains on Sensitive Clay," *Performance of Earth and Earth-Supported Structures*, New York, ASCE, pp. 211–242, June.

Leonard, M., R. Plum, and A. Kilian, 1988, "Considerations Affecting the Choice of Nailed Slopes as a Means of Soil Stabilization," *Proceedings of the 39th Annual Highway Geology Symposium*, Park City, Utah, pp. 288–302, August.

Lizzi, F., 1982. "The Pali Radice (Root Piles)," *Symposium on Rock and Soil Improvement*, Bangkok, AIT, November, Paper D-1.

Lizzi, F., 1985. "Pali Radice" (Root Piles) and "Reticulated Pali Radice, Underpinning," S. Thorburn and J. F. Hutchison, Eds. Glasgrow and London: Surrey University Press, Chapter 4.

Long, N. T., 1990. "The Pneusol," Laboratoire Central des Ponts et Chaussees, Paris.

Machan, G. and T. J. Dodson, 1989. *Coquille Reroute, Coos Bay–Roseburg Highway #35, M.P. 9.6, Coos County, Oregon*, Oregon State Highway Division, Geotechnical Report, December.

Millet, R. A., G. M. Lawton, P. C. Repetto, and V. K. Garga, 1992. "Stabilization of Tablachaca Dam Landslide," *Proceedings of a Specialty Conference on Stability Performance of Slopes and Embankments—II*, ASCE Geotechnical Special Publication 31, Berkeley, California, pp. 1,365–1,381, June.

Mitchell, J. K., and W. C. B. Villet, 1987. *Reinforcement of Earth Slopes and Embankments*, Transportation Research Board, Washington, DC, National Cooperative Highway Research Program Report 290, June.

Mitchell, R. A., and J. K. Mitchell, 1992. "Stability Evaluation of Waste Landfills," *Proceedings of a Specialty Conference on Stability Performance of Slopes and Embankments—II*, ASCE Geotechnical Special Publication No. 31, Berkeley, California, pp. 1,152–1,178, June.

Moore, H. L., 1986. "The Construction of a Shot-in-Place Rock Buttress for Landslide Stabilization," *37th Annual Highway Geology Symposium*, Helena, Montana, pp. 137–157.

Munfakh, G. A., 1990. "Innovative Earth Retaining Structures: Selection, Design and Performance," *Proceedings, ASCE Specialty Conference on Design and Performance of Earth Retaining Structures*, Cornell University, Ithaca, New York, June.

NAVFAC, 1971. *Soil Mechanics, Foundations, and Earth Structures*, Design Manual DM-7, Department of the Navy, Naval Facility Engineering Command, Alexandria, Virginia, March.

NAVFAC, 1982. *Soil Mechanics, Soil Dynamics, Foundations and Earth Structures, and Deep Stabilization, and Special Geotechnical Construction*, Design Manuals DM-7.1, -7.2 and -7.3, Department of the Navy, Naval Facility Engineering Command, Alexandria, Virginia, May.

Nelson, D. S., and W. L. Allen, Jr., 1974. "Sawdust as Lightweight Fill Material," *Highway Focus*, Vol. 6, No. 3, July, pp. 53–66.

Nethero, M. F., 1982. "Slide Control by Drilled Pier Walls," *Proceedings of the Application on Walls to Landslide Control Problems*, R. B. Reeves, Ed., ASCE Convention, Las Vegas, Nevada, pp. 61–76, April.

Nusbaum, P. J., and B. E. Colley, 1971. *Dam Construction and Facing with Soil Cement*, Portland Cement Association.

Parsons Brinckerhoff, 1992. "Repair Alternatives, Risk Analysis, and Remedial Measures for the Needles Eye Tunnel," Consultant Report Prepared for the Department of Public Works, Boulder County, Colorado, April.

Pearlman, S. L., and J. L. Withiam, 1992. "Slope Stabilization Using In-Situ Earth Reinforcements," *Proceedings of a Specialty Conference on Stability Performance of Slopes and Embankments—II*, ASCE Geotechnical Special Publication No. 31, Berkeleys California, pp. 1,333–1,348, June.

Pitts, J., 1988. "Stability of a Rock Slope at Butik Batok New Town, Singapore," *2nd International Conference on Case Histories in Geotechnical Engineering*, Vol. 1, S. Prakash, Ed., University of Missouri—Rolla, Missouri, pp. 115–121.

Riccobono, O., and W. H. Hansmire, 1991. "Geotextile Walls for Hawaii Route H-3 Access Roads," *Geonews*, Parsons Brinckerhoff, November, pp. 8–9.

Rice, R. M., and J. S. Krames, 1970, "Mass-Wasting Process in Watershed Management," *Proceedings on Interdisciplinary Aspects of Watershed Management*, ASCE, pp. 231–260.

Rodriguez, A. R., H. D. Castillo, and G. F. Sowers, 1988. *Soil Mechanics in Highway Engineering*. London: Trans Tech Publications.

Roth, W. H., R. H. Rice, D. T. Liu, and J. Cobarrubias, 1992. "Hydraugers at the Via De Las Olas Landslide," *Proceedings of a Specialty Conference on Stability Performance of Slopes and Embankments—II*, ASCE Geotechnical Special Publication No. 31, Berkeley, California, pp. 1,349–1,364, June.

Schlosser, F., 1983. "Analogies et differences dans le Comportement et le Caicul des Ouvrages de Soutenement en Terre Arme et par Clougae du Sol," *Annales de L'Institut Technique du Batiment et des Travaux Publics*, No. 418.

Schlosser, F., and P. Segrestin, 1979. "Dimensionnement des ouvrages en Terre Armee par la methode de l'Equilibre Local," *Proceedings of International Conference on Soil Reinforcement: Reinforced Earth and Other Techniques*, Paris, Vol. I.

Schuster, R. L., and R. J. Krizek, Eds., 1978. *Landslides Analysis and Control*, Special Report 16, Transportation Research Board, National Academy of Sciences, Washington, DC.

Shen, C. K., L. R. Hermann, K. M. Romstad, S. Bang, Y. S. Kim, and J. S. De Natale, 1981. "An in-Situ Earth Reinforced Lateral Support System," Report 81-03 for the University of California, Davis, California, March.

Sills, G. L., and R. L. Fleming, 1992. "Slide Stabilization with Stone-Fill Trenches," *Proceedings of a Specialty Conference on Stability Performance of Slopes and Embankments—II*, ASCE Geotechnical Special Publication No. 31, Berkeley, California, pp. 1,382–1,394, June.

Sotir, R. B., and D. H. Gray, 1989. "Fill Slope Repair Using Soil Bioengineering Systems," *Proceedings, XX International Erosion Conference*, Vancouver, British Columbia, pp. 413–429.

Sowers, G. F., 1979. *Introductory Soil Mechanics and Foundations*, 4th ed. New York: Macmillan.

Stocker, M. F., and G. Reidinger, 1990. "The Bearing Behavior of Nailed Retaining Structures," *Design and Performance of Earth Retaining Structures*, P. C. Lambe and L. A. Hansen, Eds., ASCE Geotechnical Special Publication No. 25, New York, pp. 612–628.

Stocker, M. F., G. W. Korber, G. Gassler, and G. Gudehus, 1979. "Soil Nailing," *Proceedings International Conference on Soil Reinforcement: Reinforced Earth and Other Techniques*, Paris, Vol. I, pp. 469–474, March.

Turmanina, V. I., 1965. "The Strength of Tree Roots," *Bulletin of the Moscow Society of Naturalists, Biological Section*, Vol. 70, pp. 36–45.

Turner, A. K., and R. L. Schuster, Eds., 1996. *Landslides Investigation and Mitigation*, Transportation Research Board, Special Report 247. Washington, DC: National Academy Press.

Uniform Building Code, 1991. International Conference of Building Officials, Whittier, California, p. 1004.

United States Department of Agriculture, 1977. *Guides for Erosion and Sediment Control in California*, Soil Conservation Service, Davis, California.

Voytko, E. P., V. A. Scovazzo, and N. K. Cope, 1987. "Rock Slope Modifications Above the North Portal of the Mt. Washington Tunnel," *38th Annual Highway Geology Symposium*, Pittsburgh, Pennsylvania, pp. 155–162.

Waldron, L. J., 1977. "Shear Resistance of Root-Permeated Homogeneous and Stratified Soil," *Soil Science Society of American Journal*, Vol. 41, pp. 843–849.

Walkinshaw, J. L., 1992. Personal Communication, United States Federal Highway Administration—Region 9, San Francisco, January.

Weatherby, D. E., and P. J. Nicholson, 1982. "Tiebacks," United States Federal Highway Administration Publication No. FHWA/RD-82/047, July.

Weatherby, D. E., and P. J. Nicholson, 1988. "Tiebacks Used For Landslide Stabilization," *Proceedings of the Application of Walls to Landslide Control Problems*, R. B. Reeves, Ed., Las Vegas, Nevada, ASCE, pp. 44–60, April.

Wischmeier, W. H., and D. D. Smith, 1965. *Predicting Rainfall Erosion Losses from Cropland East of the Rocky Mountains*, Agriculture Handbook No. 282, U.S. Department of Agriculture, Washington, DC.

Woodward Clyde Consultants, 1994. Letter to Parsons Brinckerhoff Quade & Douglas, Inc., concerning remedial measures to repair the landslide at California State Road 4 at Willow Pass.

Xanthakos, P. P., 1991. *Ground Anchors and Anchored Structures*. New York: Wiley.

Xanthakos, P. P., L. W. Abramson, and D. Bruce, 1994. *Ground Control and Improvement*. New York: Wiley.

Yeh, S. T. and J. B. Gilmore, 1992. "Application of EPS for Slide Correction," *Proceedings of a Specialty Conference On Stability Performance of Slopes and Embankments—II*, ASCE Geotechnical Special Publication No. 31, Berkeley, California, pp. 1,444–1,456, June:

Zaruba Q., and V. Mencl, 1982. *Landslides and Their Control*. New York: Elsevier.

CHAPTER 8

DESIGN, CONSTRUCTION, AND MAINTENANCE

8.1 INTRODUCTION

Once the analysis is completed, the slopes must be designed, built, and maintained. The following sections address the preparation of the contract documents, inspection during construction (including some of the geotechnical instrumentation used to verify slope performance), postconstruction inspection, and maintenance. Early and frequent maintenance can help prevent more costly repairs and redesigns in the future.

8.2 CONTRACT DOCUMENTS

Contract documents for new slope construction and slope rehabilitation generally consist of plans, specifications, and geotechnical reports. These documents should fully depict and describe the work to be done and the conditions present and expected. Contractors and other interested parties lacking the extensive knowledge of the site that the engineer has will use these documents. They should relate that knowledge as clearly as possible and should represent a project that is constructible. There is nothing to be gained by withholding information from the contractor and having the attitude that "the contractor's people will have to figure everything out for themselves." This attitude only increases the risk the contractor is accepting and will result in a higher bid price with greater built-in contingency costs.

Design information should be shown in only one place to the extent possible. If an item is covered in the specifications, that item should not be shown on the drawings, and vice versa. Likewise, if something is shown and dimensioned on one drawing, do not duplicate the information on another drawing. It often happens that an item gets changed on one drawing and not on the the other. Conflicts within the drawings must then be resolved in the field, which may lead to a construction claim.

8.2.1 Contract Drawings

The contract drawings should consist of plans, elevations, cross sections, and details sufficient to alert the contractor to site conditions, access, adjacent utilities and facilities, topography, geologic contacts, excavation and fill requirements, slope angles, construction sequences, benches, shoring, reinforcement and stabilization details, borrow and disposal areas, geotechnical instrumentation, traffic control requirements, drainage plan and connections, architectural details, and structural details. Plans should consist of a small-scale plan showing the site location and environs keyed to readily located features such as highways, rivers, and cities, as well as a large-scale detailed plan showing the site itself. Suitable scales are often 1 inch equals 100 or 200 feet for the site plan and 1 inch equals 20 or 40 feet for the detailed plan. If the site is too large to fit on one sheet, multiple sheets should be used with match lines. The detailed site plan should show existing topography as well as all existing features and facilities. Other plans in the contract drawing set should include:

- Utility plans
- Final grading plans
- Traffic control and signing requirements
- Drainage plans and details
- Construction staging and sequence requirements

It is helpful if these plans are drawn to the same scale, so different aspects of the work can easily be related from one plan to the other. A developed elevation should be prepared that depicts existing conditions as well as proposed conditions. Stations, offsets, elevations, and coordinates should be shown as accurately as possible. Different types of treatment can be shown on this elevation, for instance, a retaining wall next to a laid-back slope. Frequently the exact location of geologic contacts and proposed structures and stabilization elements may not be known at every point. It is incumbent on engineers to make their best guess for the purposes of estimating quantities and costs. The exact locations will be determined in the field during construction and should be so noted on the drawing. Any proposed architectural finishes and landscaping should be shown on a elevation separate from the one presenting geometrics.

Cross sections should include typical and extreme conditions existing and proposed. Geologic contacts such as top of rock should also be shown. Slopes, benches, reinforcement, drains, and other pertinent information should be presented in a clear manner. Typical scales for cross sections are 1 inch equals 10 or 20 feet. Often, specific details accompany the cross sections and are drawn to larger scales, typically from 1 inch equals $\frac{1}{2}$ foot to 1 inch equals 5 feet. The horizontal and vertical scales should be the same and the cross section(s) plotted as shown in the plan.

When steel and concrete structures such as retaining walls, drop inlets, and ditches are involved with the work, separate drawings are used to depict layout, dimensions, and reinforcing. Sometimes the structural and architectural drawings are combined.

8.2.2 Specifications

Specifications detail the materials to be used in the work, what submittals are required before the work starts, quality control (QC) requirements, how the work will be executed, how it will be measured for payment, and how it will be paid for. Most states have standard specifications and these should be used to the extent possible, particularly for concrete, earthwork, drainage, and other typical highway appurtenances. Many cities and counties have their own standard specifications. In southern California, for example, the *Standard Specifications for Public Works Construction*, known as the "Green Book," is routinely used. Newer innovations in slope stabilization methods such as soil nailing and mechanically stabilized embankments (MSEs), require special provisions for a particular project. These can be based on Federal Highway Administration (FHWA) generic specifications, FHWA design manual guidelines, and specifications used for similar projects.

8.2.3 Geotechnical Design Reports

Geotechnical design reports are commonly prepared during projects. These may include reports on:

- Available data
- Geologic exploration for the project
- Alternative design analyses
- Slope stability computations
- Investigation of previous failures
- Proposed design
- Interpretation of geologic conditions expected during construction

These reports should be available to contractors bidding on the project. Information contained in geotechnical design reports should include:

- Regional geology
- Project geology
- Hydrology
- Seismology
- Site description
- Description of work proposed
- Project subsurface exploration

- Soil properties
- Depth to bedrock
- Earthwork characteristics
- Cut and fill slopes and stability
- Drainage
- Earth pressures
- Foundation conditions
- Estimated settlements
- Slope stabilization measures
- Construction considerations

Geotechnical reports do become a baseline for the contractor's work and should be written in a clear and concise manner. They should present all of the facts known. The designer should clearly identify any interpretations.

8.3 INSPECTION DURING CONSTRUCTION

Construction inspectors are charged with the responsibility of seeing that the project is constructed according to the plans and specifications. To do so they should be knowledgeable about the intent of the design as well as the contractual requirements of the design documents. Sometimes field conditions will dictate that clarifications and modifications to the design be made. Inspectors should not fear requesting information and should consult the designer on a regular basis regarding clarification and modifications. Contract provisions should never take precedence over worker safety and common sense.

8.3.1 Inspection Guidelines

Standardized inspection guidelines should be in place before the work begins on any construction project. The guidelines should be formalized, bound, and distributed to all personnel. An organization chart should be prepared and displayed, indicating who is in charge and to whom responsibilities are delegated. This should be done for the inspection staff as well as the contractor's staff. Telephone lists that include all involved personnel, as well as police, fire, ambulance, and hospital telephone numbers, should be distributed. Project procedures should include formats for daily inspection reports, reports of workers and equipment on site, payment authorizations, requests for information (RFIs), shop drawing reviews, submittal reviews, schedule tracking, quantity tracking, and field book notes. Other methods of record keeping include notes, coding, and coloring of project plans hung up on the field office walls. These procedures must not only be set up at the beginning of the project, but they must also be known and understood by the entire inspection staff. Weekly on-site coordination meetings with the staff and with the contractor's staff are highly recommended.

Safety briefings should be held at the beginning of the project and on a regular basis thereafter. Occupational Health and Safety Administration (OSHA) regulations should be enforced at all times.

8.3.2 Quality Control/Quality Assurance

Quality control (QC) and quality assurance (QA) are defined as follows:

- Quality control formalizes traditional elements of design and construction including checking, reviewing, examining, and supervising work performed. This is accomplished by having well-qualified professionals performing project tasks, by making certain that the professionals understand and carry out their responsibilities, and by documenting this process.
- Quality assurance formalizes and documents the traditional overview role performed by supervisors and independent reviewers. Items evaluated as part of this process include technical adequacy, clarity, practicality, constructibility, conformance to professional standards and practices, and overall suitability.

On a construction project, all QC procedures that will be used to evaluate the contractor's work should be given in the specifications. Qualified individuals should enforce these specifications and where outside testing is required, laboratories and firms involved in QC verification should have proof of their qualifications and ability to perform the work with the appropriate equipment and according to accepted industry standards.

8.3.3 Instrumentation

During construction, geotechnical instrumention is installed to provide inspectors and designers with information on specific parameters for design and construction of slopes and for long-term performance monitoring. The following steps should be approached during the planning phase of an instrumentation program:

- Predict mechanisms that control behavior
- Define the purpose of the instrumentation
- Define the geotechnical questions to answer
- Select parameters to measure
- Predict magnitudes
- Select instrumentation and location
- Prepare budget, which includes instrument and labor costs
- Write specifications
- Plan for construction phase

Selecting good quality instruments is strongly recommended. Even with the best equipment, a percentage of the instruments will fail or malfunction. Instrumentation programs should be designed with this in mind. When selecting instrument types, try to incorporate cross checks in the system by using different types of instruments rather than duplicating instruments of the same type.

Once the mechanisms have been predicted and the instruments selected, careful planning of the instrumentation monitoring scheme is essential. This includes these steps:

- Define the objectives and develop contingency plans
- Decide on the type, number, and frequency of measurements
- Organize personnel with proper training
- Plan out monitoring, processing, and reporting stages
- Factor in instrument accessibility and impact on construction

Sensitive instruments tend to be expensive. They may also have restricted ranges and can sometimes be less reliable than those that are less sensitive. Sometimes it is more appropriate to install a greater number of cheaper, reliable but less sensitive instruments when the potential behavior of the slope is unknown. Overall accuracy of the instrumentation plan is more important than the quoted sensitivity of each individual instrument.

Once the instrumentation and monitoring plans are established, execution follows. The following steps should be implemented during the construction phase of a project:

- Procure the instruments
- Install the instruments
- Calibrate and maintain the instruments
- Establish factors that influence measurements
- Establish procedures of ensuring data correctness
- Collect data
- Process and present the data
- Interpret data
- Report conclusions
- Implement conclusions

Monitoring Parameters

Visual Clues Tension cracks at the crest of the slope may be the first sign of instability. If cracks appear at the crest of the slope or elsewhere, their widths and vertical offsets should be monitored. Crack measurements give clues to the behavior of the entire slope. Often the direction of movement may be inferred from the pattern of cracking, particularly by matching the irregular edges of the cracks.

Structures such as retaining walls may also show signs of distress. Cracking, rotation, or misalignment of the facade are a few examples of instability.

On natural slopes, bowing of trees is the result of prolonged creep on hillsides. Unusual tilting of trees is a sign of more recent movement. In all cases, these visual cues should be considered as early warning signs of instability. Instrumentation should be installed as soon as possible so that early data can be collected and remediation measures planned and implemented.

Vertical and Horizontal Movement Slope instability in many cases is a slow progressive process. After visual signs are noticed, it becomes important to determine the following information about the ground movement:

- Where?—Locate shear plane
- What is the direction of travel?—Up, down, to the left or right
- How much?—Magnitude of displacement
- How fast?—Rate at which the movement is traveling

This information can be collected by various types of instruments. Simple to sophisticated instruments can be used. Surface and subsurface techniques are available to help answer the where. Of all the parameters, rate is the most important to determine. If movement of a slope is decreasing, the need for an immediate solution is relaxed and time is available to look at various alternatives. If the rate of movement is increasing, then a quick solution is required. The problem in many cases is determining what rate is too high.

Groundwater Level Groundwater in a slope adds a contributing driving force. Determining actual water levels is important for design and stability analysis. Groundwater levels taken in exploration drillholes are not reliable for determining groundwater levels. The flushing action of water during drilling affects the groundwater regime. For reliable observations, the groundwater table should be allowed to stabilize, and this could take several days after drilling is completed. The level of the water table can then be measured in the open borehole, but response time is very slow and infiltration of surface water may cause recharging of the hole.

Alternate methods include installation of an observation well or open standpipe piezometer. Small standpipes will give a shorter response time than does standard borehole casing. Observation wells should be used sparingly as they could intercept more than one water bearing zone, producing meaningless water levels.

Pore Pressure The majority of slope failures occur along distinct geological boundaries. Water infiltration from the ground surface can result in the development of perched water tables at the boundary of two materials with different permeability. Perched water develops pore pressure in the soil and provides an additional driving force. Piezometers are used to measure pore pressure. They consist of an isolated zone within the soil mass. Ideally, the piezometer is located at depths where changes in mass permeability occur. These piezometers will then measure the pore pressure of the perched water table. Piezometers at greater depths are useful for determining the permanent groundwater table elevation.

Types of Instrumentation The type of instruments selected depends on the problem to be monitored. Instrument application is the key factor to consider when selecting the instrument type. Accuracy, cost, environmental conditions, and personnel training are also factors to be considered. In addition to the instruments discussed below, settlement devices, load cells, and strain gages may be of some use in monitoring the behavior of a soil slope.

Optical Surveys Precise surveying techniques can determine the coordinates and elevation (x, y, z) of points anchored in the ground. These measurements are determined from a fixed datum located outside the footprint of the slope. Absolute movements of these points can be obtained with repeated measurements over time. Instrument error of the surveying method must be less than the minimum required accuracy of the movement measurements. Control points should allow for accurate positioning of the survey instruments, since movement of points is often very small.

Surface Extensometers The formation of cracks at the top and base of slopes is an initial sign of instability. Measuring and monitoring the changes in crack width is required. Numerous types of surface extensometers, crack gages, or telltales are available. They all measure or detect further movement between two points spanning the crack. Devices should also show magnitude and direction of the movement. Many of the mechanical devices are crude and simple. Electrical devices are more sophisticated and can provide continuous monitoring capabilities. All devices should be marked with dates of installation.

One problem with placing crack gages or telltales at the top of slopes is that movement has already occurred. Additional cracking will potentially weaken the entire area. Stakes driven in when cracks are first noted may become loose, resulting in inaccurate measurements. Placing stakes further apart is one solution, but the greater span may introduce new errors. Deflections and weights of the measuring device need to be considered.

Crack gages at the top of slopes also introduce safety concerns. Access to the moving portion of the slope may be limited. Limited access must be considered when selecting and placing instruments. All systems must also be protected from adverse weather conditions and animals.

The simplest form of a telltale is two stakes driven into the ground on either side of a crack or scrap. A survey tape, tape extensometer, rod, or ruler measures the separation. Other examples of surface extensometers, gages, and telltales are listed below.

METAL STRIPS A simple telltale consists of metal strip overlying a plate; see Figure 8.1*a*. Two scribed lines on the underlying plate serve as a reference datum. Offsets of the strip to the datum are recorded over time. Movement perpendicular to the face can cause errors in the readings. To overcome this problem, a second telltale may be placed a short distance from the first, with the overlying metal strip bonded to the opposite side of the crack. Movement normal to the face can be determined by measuring the separation of the two plates.

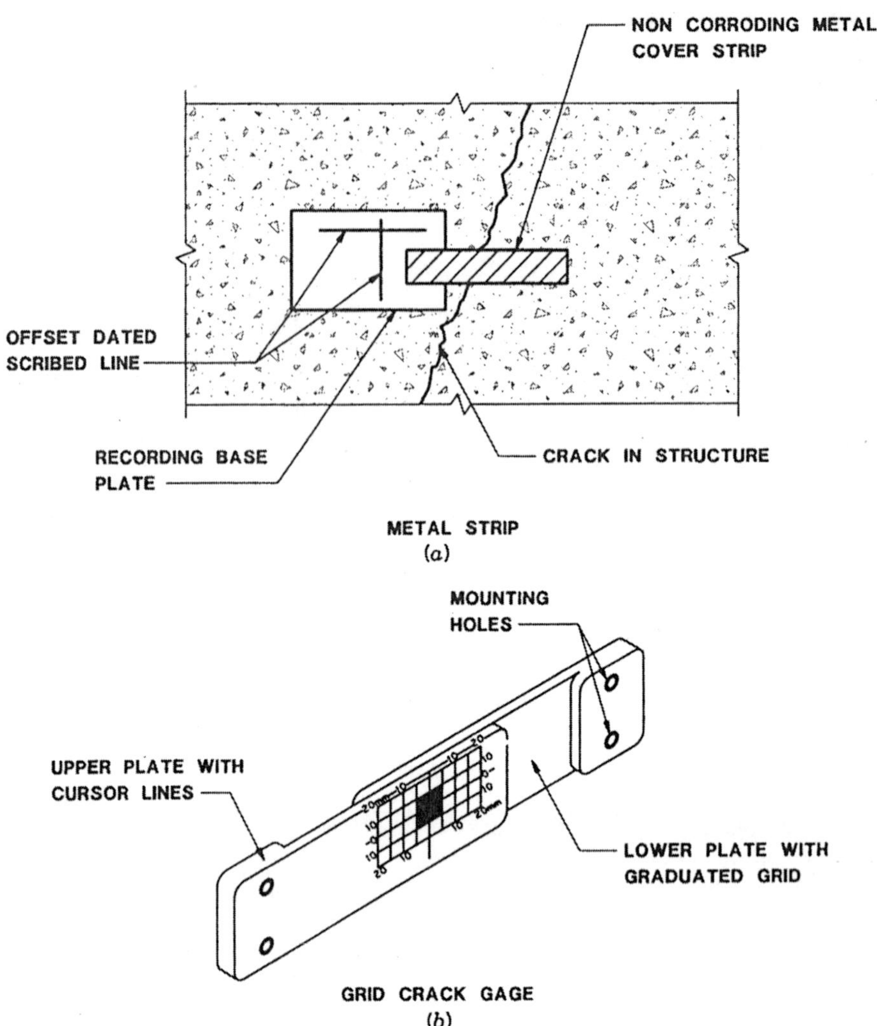

Figure 8.1 Types of surface extensometers. (*a*) Metal strips. (*b*) Grid crack gage. (*c*) Tape Extensometer. (*d*) Electrical crack gage. (Dunnicliff, 1988.)

GRID CRACK GAGES Grid crack gages, also known as calibrated telltales, consist of two overlapping transparent plastic plates. One plate is mounted on each side of the discontinuity. Crossed cursor lines on the upper plate overlay a graduated grid on the lower plate; see Figure 8.1*b*. Movement is determined by observing the position of the cross on the upper plate with respect to the grid. Only displacements up to 1 inch can be measured effectively, since the total grid pattern is small. Grid crack gages are most useful on retaining walls and concrete structures.

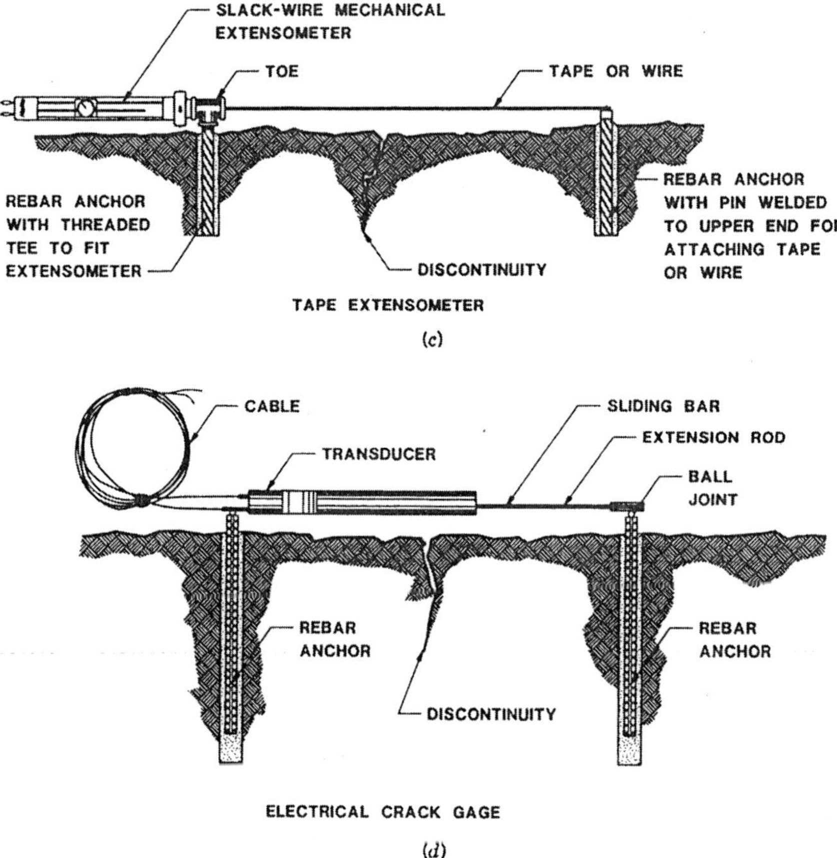

Figure 8.1 (*Continued*)

DEFORMATION GAGES Mechanical deformation gages can accurately measure movement. The gages consist of two anchor points. The distance between the points is spanned with a bar fixed on one end. The other end rests on a knife edge and is free to slide. The bar can be wood, aluminum, or steel. Steel bars can be of invar if temperature variations are a concern. The actual distance measurement between the knife edge and the initial mark established at installation can be measured by a host of different methods. These methods include scales, calipers, micrometers, or tape extensometers (Figure 8.1*c*) and depend on the accuracy required.

ELECTRICAL CRACK GAGES When access to the gage location is not available for monitoring, a remote-read electrical crack gage is required. The displacements can be measured using three general arrangements:

- An electrical linear displacement transducer is attached to a bracket on one side of the crack and arranged to bear against a machined reference surface on the other side.

- A transducer is attached to anchors on each side of the crack via ball joints, as shown in Figure 8.1*d*.
- A transducer is attached to a wire under tension between two anchored points.

These electrical versions measure displacement in one direction. Single gages also can be installed to monitor displacement in the other, perpendicular direction, as required. Electrical crack gages are more expensive than mechanical gages. Their range is also limited. Most electrical gages can be extended by resetting the transducer. Precision is generally between ±0.0001 and 0.005 inch, depending on the transducer. Readings can be affected by lead wire changes, temperature, and other environmental conditions.

Inclinometers An inclinometer measures a casing's inclination from vertical. Successive inclination measurements, when processed, measure lateral deformations with depth. A portable borehole inclinometer is one of the most widely used form of inclinometer systems available. The portable borehole inclinometer system is a composite of three units (Figure 8.2):

- Special casing that deforms under the influence of earth movement
- Sensing unit that measures successive inclinations of the casing
- Electrical readout that supplies voltage to the sensing unit and displays the measured inclinations as numerical readings

INCLINOMETER CASING Special casing is installed in a borehole in the ground. The casing is flexible enough to move with the soil when the soil deforms laterally. The casing generally ranges in sizes from 1.5 to 3.5 inches in internal diameter. It is made of plastic, anodized aluminum, or steel and is provided with continuous longitudinal grooves that are positioned at the quarter points of its inside circumference. See Figure 8.3. During data collection, the groove pairs are referred to as the *A* and *B* directions. Large-diameter casings maximize precision. Rotation of the sensor decreases as casing diameter increases. Casings should be installed well in advance of construction so a set of initial readings can be established and a complete record of all soil movement during construction can be obtained.

The bottom of the inclinometer casing is assumed fixed so deformation can be calculated from the inclination measurements. It is imperative that the casing bottom be anchored firmly. The lower portion of the casing should be installed 10 to 15 feet below the elevation where soil is expected to undergo lateral displacement. Anchoring the casing into rock is recommended if the geologic conditions exist. If any doubt arises as to the fixity of the casing bottom, movement of the casing top should be checked regularly by accurate surveying methods. The difference in movement of the casing top between survey and inclinometer methods should be used to correct the lateral movement profile.

Inclinometer casings usually are installed in 5- or 10-foot lengths that are joined together with couplings. The couplings are either riveted or cemented to ensure a firm

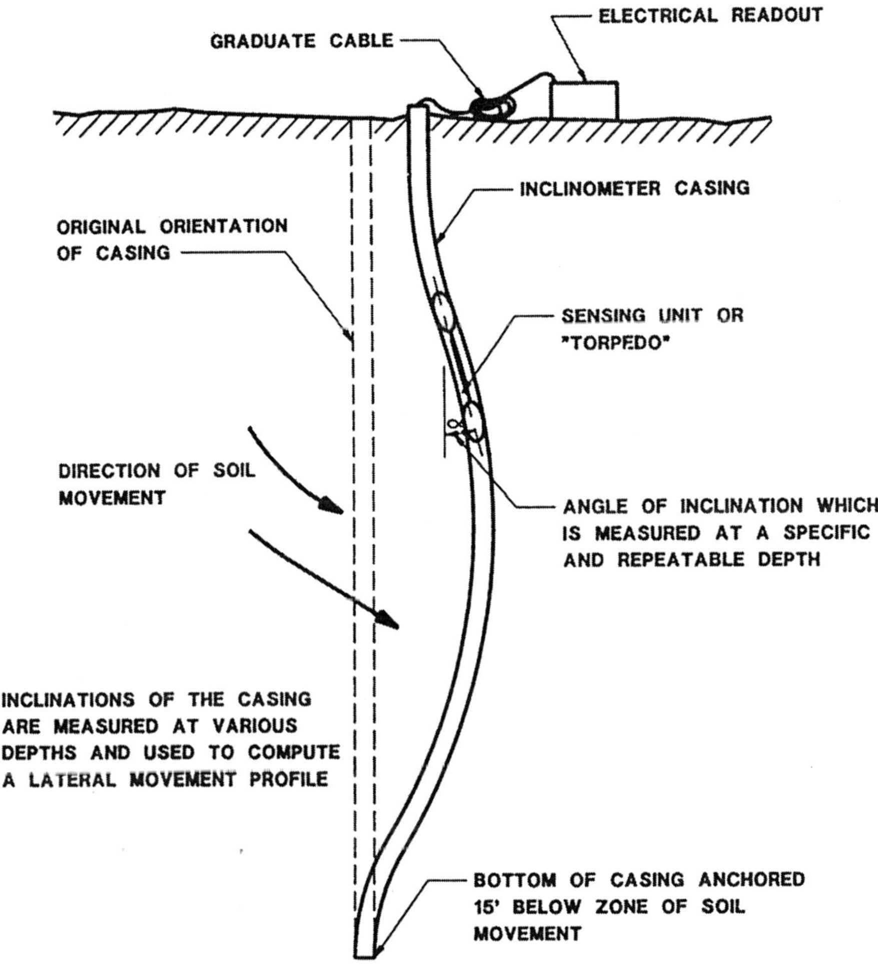

Figure 8.2 Portable borehole inclinometer system.

connection. Each coupling represents a possible source of error. Mud or debris can seep into the casing and be deposited along the internal tracking system. Couplings should be waterproof. *O*-rings or caulking with a waterproof sealant and complete coverage with a waterproof tape are two methods used. The bottom section of the casing should be closed with a watertight cap or plug. If the drill hole is filled with water or drilling mud, the inclinometer casing must be filled with water to overcome buoyancy. The casing should never be filled with drilling mud.

The annular void between the inclinometer casing and the borehole wall must be carefully backfilled to ensure that casing movements will reflect soil movements. Grout is probably the most satisfactory backfill. Grouting can be done through a plastic tube attached at the bottom of the casing on the outside. Sand or pea gravel can also

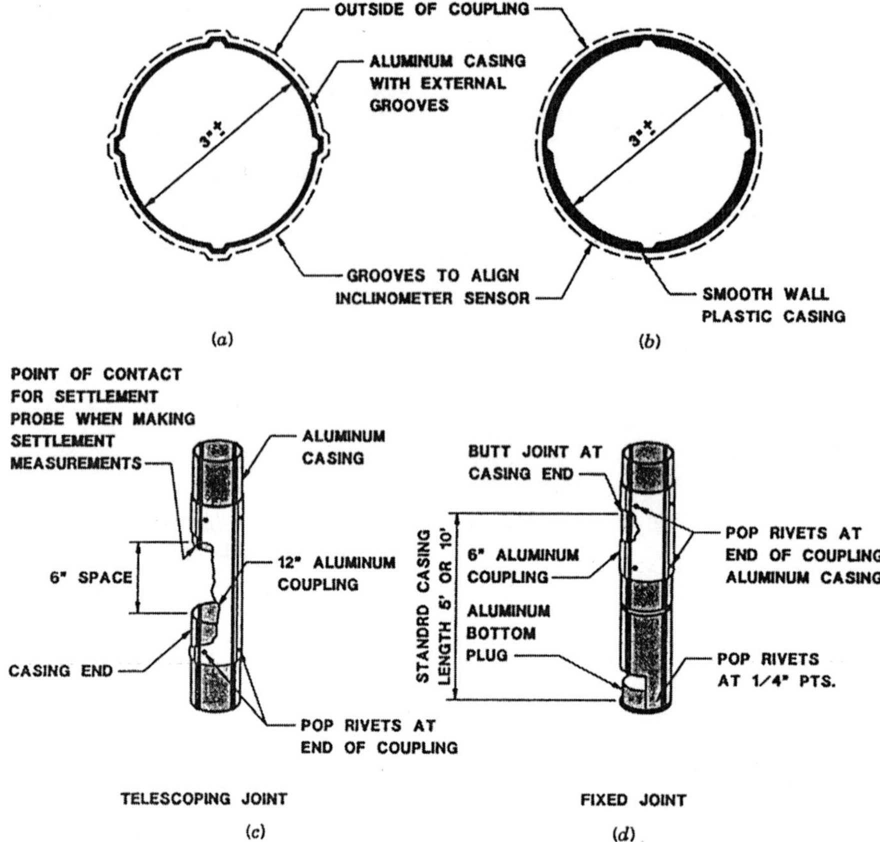

Figure 8.3 Inclinometer casing details. (*a*) Section of aluminum casing. (*b*) Section of plastic casing. (*c*) Telescoping joint. (*d*) Fixed joint.

be used as backfill, but extreme care is required to avoid leaving voids and to obtain satisfactory compaction of the backfill around the casing. Poor quality backfilling around the casing may cause scatter in readings shortly after installation. Readings should stabilize over time as the backfill consolidates.

The casing as manufactured may have some spiral to the grooves, particularly plastic casing. During installation, the casing may become even more twisted or spiralled so that at depth, the casing grooves may not have the same orientation as at the ground surface. Green (1974) reports of spiralling as a high as 18°. One degree per 10-foot section is not uncommon. Deep inclinometer casings should be checked after installation using twist indicators available from the inclinometer manufacturers. Significant errors in the assumed direction of movement may result if the amount of twist is not verified and corrected. Preassembling the casing on the ground surface prior to installation and eliminating excessively spiralled sections minimize casing spiralling. Each casing section and coupling should be numbered and marked. The casing is then installed in the same order, taking care to match the alignment marks. During installation the casing must be lowered straight down and not turned or twisted.

Small irregularities of the casing's tracking surface can lead to errors in observation, especially if careful repetition of the depth to each of the previous readings is not exercised. Plastic casing may warp locally if stored in the sun prior to installation. Aluminum casing may corrode if the placed in contact with alkaline soil, corrosive groundwater, or grout.

Fit the top of the casing with a cap to prevent intrusion of debris and surface water. The cap also protects the reference edge of the casing where the tripod sits. Establish a reference point on the top surface of the casing to allow repeatable optical survey readings. Mark the $A+$ direction on the outside of the casing with a permanent marker. Marking the $A+$ direction allows others to collect data in the proper orientation. Bearing of the $A+$ direction is required to show orientation on plans and interpretive plots. Install inclinometer casing so one set of grooves is parallel to the cut face or slope. This eliminates the need to work with components to determine the maximum amount of displacement. Protect the inclinometer casing at the ground surface. Well-marked small manhole-type covers should be provided to avoid disturbance by local construction activity.

Careful placement of the inclinometer is important to obtain information predicting the behavior of the slope or slide. Figure 8.4 shows a few examples of good and bad inclinometer positions.

SENSING UNIT/TORPEDO The sensing unit or "torpedo" is manufactured as a water-tight housing enclosing an electrical transducer. The torpedo is equipped with spring activated wheels that track within the grooves of the casing and align the torpedo in stable and repeatable positions in a vertical plane. The deflection of the casing, and hence the surrounding soil, is measured by determining the inclination of the casing at various depths with the torpedo. A portable inclinometer sensor is lowered into the casing and read at specific intervals to determine the inclination of the casing. Lateral displacements of the casing at successive times are determined from the changes in inclination occurring since the initial measurements. The total lateral profile of the casing is found by summing the individual lateral displacements from the bottom of the casing to the top.

The procedures for determining the total lateral profile assume that the bottom of the casing remains fixed and does not moves laterally. An additional deformation must be added to the deformation determined from the inclinometer survey if the bottom does move. This additional deformation can be determined by comparing the lateral deformation of the top of the casing as determined by independent optical surveying with that determined by the inclinometer survey. The lateral profile of the casing is usually determined in two mutually perpendicular vertical planes.

The precision of the inclinometer observations generally is limited not by the sensitivity of the inclinometer but by the requirement that successive readings be made with the same orientation of the instrument at the same depth in the casing.

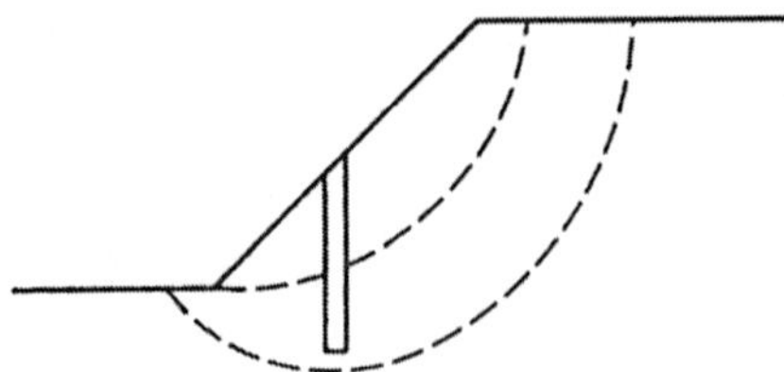

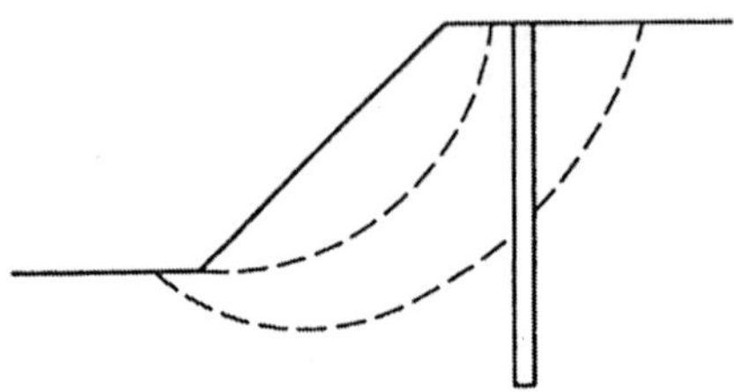

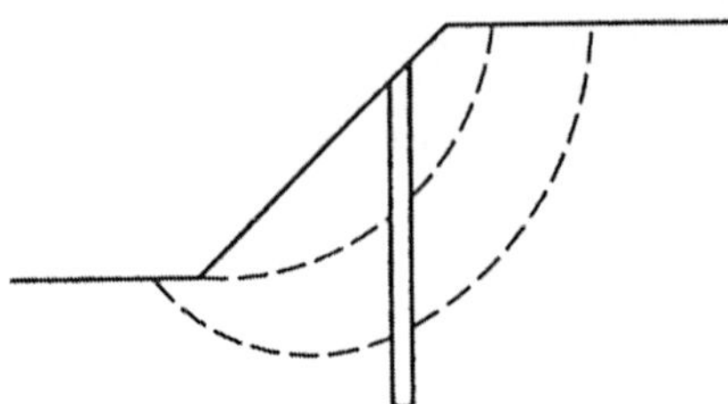

Figure 8.4 Inclinometer locations.

Changes in sensor alignment can occur from damage because of rough handling, wear of the tracking wheels, or realignment of the unit during repair.

The wheels of the torpedo provide the measuring points between which the inclination of the instrument is measured. If the reading interval is greater or less than the torpedo wheel base, the accuracy of the total lateral displacement profile determined will be reduced. Maximum accuracy requires the distance between each reading interval to equal the distance between the upper and lower torpedo wheels. Readings every 2 feet are typical. Readings intervals of 5 feet could result in poor accuracy.

Inclinometer measurements generally are recorded as the algebraic difference between readings 180° apart. Duplicate readings in one vertical plane are taken for each depth by running the instrument up the casing once and then repeating the readings with the instrument turned 180°. Computing the algebraic differences of the readings for each depth eliminates errors resulting from irregularities in the casing and alignment bias within the instrument.

One excellent reliability check of each measurement is to compute the algebraic sum of the 180° readings. The sum of the 180° readings should be approximately constant for all measurements except when readings are made where one set of wheels is influenced by a casing joint. Differences between the sum of two diametrically opposed readings and observed constant sum usually indicate an error or nonparallel casing walls. By summing the 180° readings during observation, errors resulting from mistaken transcription, faulty equipment, or improper technique can be eliminated. When the algebraic sum does not remain constant, the torpedo and readout should be recalibrated before continued use.

ELECTRICAL READOUT The electrical readout may vary in complexity from a simple Wheatstone bridge to more intricate circuitry. The readout and torpedo are connected with a heavy-duty electrical cable. The cable is graduated with external markers that are used to measure the depths of individual readings. A steel wire core is included in the cable to limit stretching. The cable connections are particularly vulnerable to water. The cable should be checked regularly to insure that the cable remains dry. The cable is stored either by winding on a portable reel or by hand coiling. Improper handling of the cable can cause problems. Twists or loops in the line impose a torque on the torpedo. Inclinometer readings can shift a small amount when compared to readings taken with an untwisted cable. A tripod with lockoff clamp can control the position of the cable and torpedo in the casings. Cable markings should be clamped at the same spot during every run so the torpedo will be in the same position every time. Take readings at intervals equal to the wheelbase and in two mutually perpendicular planes for maximum accuracy. These planes are designated *A* and *B*.

DATA PROCESSING The typical inclinometer is a dual-axis unit. Reading are taken at an interval equal to the wheelbase (2 feet) for maximum accuracy in two mutually perpendicular planes. Two sets of readings are taken for each plane with the sensor reversed 180° between readings. A data sheet and processing form is provided in Figure 8.5. Sums of the 180° readings are computed as the measurements are taken to guard against large errors. For the typical inclinometer,

$$\text{reading} = C \sin\theta \qquad \text{(Eq. 8-1)}$$

where θ = angle of inclination
C = a constant provided with the instrument.

To improve the accuracy of the measurements, the algebraic difference between the 180° readings is used rather than a single reading.

$$\text{difference} = 2C \sin\theta \qquad \text{(Eq. 8-2)}$$

INCLINOMETER – I6
INITIAL – DATA SET #1, 06-17-91

DATA SET #14, 10-25-91
GSN, 10:37 AM

Depth (Feet)	Initial A	A0	A180	Difference	Check Sum	Change	Sum Change	Deflec. (in)	Initial B	B0	B180	Difference	Check Sum	Change	Sum Change	Deflec. (in)
2	470	269	-269	538	0	68	1597	0.958	387	172	-195	367	-23	-20	-510	-0.306
4	343	201	-202	403	-1	60	1529	0.917	312	138	-161	299	-23	-13	-490	-0.294
6	181	115	-129	244	-14	63	1469	0.881	214	98	-105	203	-7	-11	-477	-0.286
8	428	242	-235	477	7	49	1406	0.844	404	191	-207	398	-16	-6	-466	-0.280
10	499	263	-265	528	-2	29	1357	0.814	304	135	-155	290	-20	-14	-460	-0.276
12	474	255	-259	514	-4	40	1328	0.797	330	143	-174	317	-31	-13	-446	-0.268
14	436	235	-237	472	-2	36	1288	0.773	259	115	-136	251	-21	-8	-433	-0.260
16	338	185	-194	379	-9	41	1252	0.751	279	119	-142	261	-23	-18	-425	-0.255
18	138	78	-65	143	13	5	1211	0.727	197	107	-95	202	12	5	-407	-0.244
20	272	149	-151	300	-2	28	1206	0.724	0	1	5	-4	6	-4	-412	-0.247
22	435	220	-222	442	-2	7	1178	0.707	-208	-113	87	-200	-26	8	-408	-0.245
24	398	203	-204	407	-1	9	1171	0.703	-231	-124	107	-231	-17	0	-416	-0.250
26	312	166	-179	345	-13	33	1162	0.697	-170	-92	93	-185	1	-15	-416	-0.250
28	343	179	-178	357	1	14	1129	0.677	61	23	-41	64	-18	3	-401	-0.241
30	311	161	-165	326	-4	15	1115	0.669	152	66	-90	156	-24	4	-404	-0.242
32	274	136	-138	274	-2	0	1100	0.660	226	106	-130	236	-24	10	-408	-0.245
34	354	177	-179	356	-2	2	1100	0.660	296	146	-176	322	-30	26	-418	-0.251
36	528	286	-290	576	-4	48	1098	0.659	408	194	-215	409	-21	1	-444	-0.266
38	131	91	-93	184	-2	53	1050	0.630	378	177	-194	371	-17	-7	-445	-0.267
40	89	53	-57	110	-4	21	997	0.598	357	166	-186	352	-20	-5	-438	-0.263
42	148	63	-66	129	-3	-19	976	0.586	250	111	-137	248	-26	-2	-433	-0.260
44	202	110	-114	224	-4	22	995	0.597	245	103	-121	224	-18	-21	-431	-0.259
46	178	122	-125	247	-3	69	973	0.584	279	121	-118	239	3	-40	-410	-0.246
48	192	129	-128	257	1	65	904	0.542	388	184	-186	370	-2	-18	-370	-0.222
50	151	110	-113	223	-3	72	839	0.503	365	156	-178	334	-22	-31	-352	-0.211
52	137	101	-105	206	-4	69	767	0.460	462	204	-230	434	-26	-28	-321	-0.193
54	168	122	-124	246	-2	78	698	0.419	509	226	-243	469	-17	-40	-293	-0.176
56	-16	-33	33	-66	0	-50	620	0.372	404	201	-219	420	-18	16	-253	-0.152
58	130	37	-41	78	-4	-52	670	0.402	307	162	-179	341	-17	34	-269	-0.161
60	127	68	-71	139	-3	12	722	0.433	336	121	-149	270	-28	-66	-303	-0.182
62	-11	22	-25	47	-3	58	710	0.426	352	130	-152	282	-22	-70	-237	-0.142
64	-97	-5	5	-10	0	87	652	0.391	324	142	-162	304	-20	-20	-167	-0.100
66	-44	-6	1	-7	-5	37	565	0.339	303	145	-138	283	7	-20	-147	-0.088
68	176	128	-126	254	2	78	528	0.317	478	231	-227	458	4	-20	-127	-0.076
70	156	103	-107	210	-4	54	450	0.270	315	133	-150	283	-17	-32	-107	-0.064
72	202	111	-115	226	-4	24	396	0.238	212	103	-122	225	-19	13	-75	-0.045
74	157	104	-105	209	-1	52	372	0.223	97	46	-59	105	-13	8	-88	-0.053
76	13	58	-68	126	-10	113	320	0.192	172	75	-73	148	2	-24	-96	-0.058
78	163	121	-115	236	6	73	207	0.124	654	287	-305	592	-18	-62	-72	-0.043
80	80	45	-47	92	-2	12	134	0.080	535	257	-269	526	-12	-9	-10	-0.006
82	39	22	-27	49	-5	10	122	0.073	380	176	-187	363	-11	-17	-1	-0.001
84	-57	-40	38	-78	-2	-21	112	0.067	266	131	-153	284	-22	18	16	0.010
86	-137	-62	67	-129	5	8	133	0.080	202	111	-111	222	0	20	-2	-0.001
88	272	131	-147	278	-16	6	125	0.075	367	179	-188	367	-9	0	-22	-0.013
90	101	54	-57	111	-3	10	119	0.071	353	168	-187	355	-19	2	-22	-0.013
92	8	24	-27	51	-3	43	109	0.065	377	173	-183	356	-10	-21	-24	-0.014
94	-54	-21	19	-40	-2	14	66	0.040	177	79	-95	174	-16	-3	-3	-0.002
96	-149	-89	81	-170	-8	-21	52	0.031	210	118	-106	224	12	14	0	0.000
98	-259	-124	128	-252	4	7	73	0.044	571	282	-289	571	-7	0	-14	-0.008
100	-119	-49	43	-92	-6	27	66	0.040	450	213	-228	441	-15	-9	-14	-0.008
102	-76	-30	21	-51	-9	25	39	0.023	280	139	-138	277	1	-3	-5	-0.003
104	19	19	-14	33	5	14	14	0.008	78	38	-38	76	0	-2	-2	-0.001

Figure 8.5 Inclinometer data processing form.

The lateral displacement of the casing, y, over the length of the inclinometer wheelbase is

$$y = \text{wheelbase} \times \sin\theta = 24\frac{\text{difference}}{2C} \qquad \text{(Eq. 8-3)}$$

The total displacement at any point is the sum of displacements from all the previous displacements starting with the bottom. The displacement is calculated in this way because the bottom point is assumed fixed.

$$\Delta y_n = \sum_{i=1}^{n} \frac{12}{C}(\text{difference})_i \qquad \text{(Eq. 8-4)}$$

$n = 1$ is the bottom reading, with each successive reading as one above. The total lateral displacement at a point from the time of the initial measurement to the time of the present measurement is

$$\Delta y_n = \sum_{i=1}^{n} \frac{12}{C}(\text{difference})_i - \sum_{i=1}^{n} \frac{12}{C}(\text{initial difference})_i \qquad \text{(Eq. 8-5)}$$

Hand reduction of inclinometer data is a tedious and time-consuming operation. The availability of computers has eased the process. Manufacturers provide programs to calculate the lateral displacements and to compare readings. In many cases, comparison of two readings is limiting. Spreadsheets are handy tools for processing data and plotting up to six readings.

Multiple-Point Borehole Extensometers An extensometer consists of tensioned rods anchored at different points in a borehole. Most extensometers have multiple rods and are known as multiple-point borehole extensometers (MPBXs). Anchor types vary from self-boring to groutable. The connecting rods formerly made of steel are now fiberglass. Three-sixteenth-inch diameter rods are enclosed in $\frac{1}{2}$-inch diameter polyethylene tubing that fastens to the anchors. The MPBX measures the change in distance between the anchor and the end of the rod at the reference head. Extensometers are most suited to measuring deformation of and behind retaining structures stressed by anchoring or excavation. Figure 8.6 shows a sketch of a groutable-type MPBX.

Anchor spacing is determined mainly by geological factors. Generally it is best to have one of the anchors positioned in stable ground. This allows the movements of the other anchors to be measured relative to this stationary anchor.

Installation of MPBXs is complex. Proper rod tension is needed to maximize the amount of rod movement in the correct direction. For example, rod/anchor fitting should not be overtightened for fear of cracking the rods.

Relative movement between the reference head and the end of each extensometer rod may be measured mechanically or electrically. The mechanical device consists of a micrometer inserted into the rod casing to measure the distance between the rod and reference head. The electrical system consists of linear potentiometer sensors, which are mounted and fixed on the reference head. Most MPBXs can hold up to five sensors. The sensors are connected to a multiconductor cable for reading by a portable readout device.

Rod displacements should be read after installation is completed. Subsequent measurements are compared to these intial readings. Data should be recorded showing

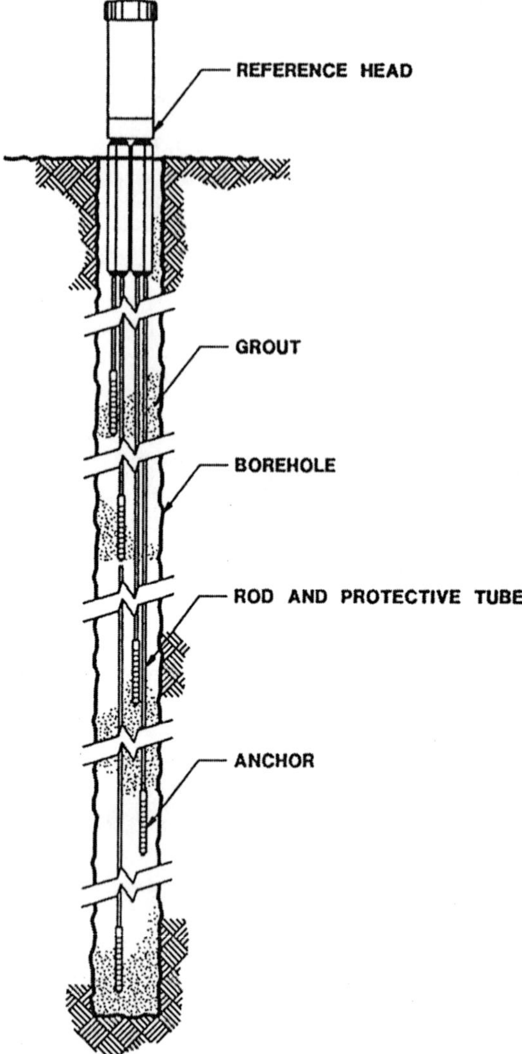

Figure 8.6 Multiple-point borehole extensometer (MPBX).

the time, date, operator, and readings. An example processing form is provided in Figure 8.7.

Data processing may consist of the reduction of field data to include

- Displacement of the instrument head relative to each downhole anchor
- Displacement of each downhole anchor relative to the deepest anchor
- Strain acceleration either between adjacent anchors or between the deepest downhole anchor and the instrument head.

LOCATION - CUT FACE S-T INTERFACE
ELEVATION - 1068 FT.
INSTALLED ON JANUARY 26, 1992
INCLINATION - 20 DEGREES UPWARD   ORIENTATION - S60E

DATE	TIME	TAKEN BY	30 FT READING	30 FT DISPL. (Inches)	30 FT CHANGE (Inches)	50 FT READING	50 FT DISPL. (Inches)	50 FT CHANGE (Inches)	70 FT READING	70 FT DISPL. (Inches)	70 FT CHANGE (Inches)	100 FT READING	100 FT DISPL. (Inches)	100 FT CHANGE (Inches)	NO. OF DAYS	DISPL. AT FACE (Inches)	DISPL. AT 30 FT (Inches)	DISPL. AT 50 FT (Inches)	DISPL. AT 70 FT (Inches)
18-Mar-92	1355	MLF	2.38	0.085	0.076	10.73	0.385	0.225	8.92	0.320	0.274	12.41	0.445	0.334	52	0.334	0.259	0.109	0.060
25-Mar-92	0820	MLF	2.38	0.085	0.076	10.78	0.388	0.227	8.92	0.320	0.274	12.70	0.456	0.345	59	0.345	0.269	0.118	0.071
31-Mar-92	0910	GBM	2.38	0.085	0.076	10.96	0.393	0.234	8.93	0.320	0.274	12.89	0.463	0.352	65	0.352	0.276	0.118	0.077
07-Apr-92	0945	GBM	2.39	0.086	0.076	11.09	0.398	0.238	8.94	0.320	0.275	13.07	0.469	0.358	72	0.358	0.282	0.120	0.083
13-Apr-92	1039	GSN	2.39	0.086	0.076	11.16	0.400	0.241	8.94	0.320	0.275	13.07	0.469	0.358	78	0.358	0.282	0.117	0.083
17-Apr-92	1134	GSN	2.40	0.086	0.077	11.33	0.406	0.247	8.95	0.321	0.275	13.16	0.472	0.361	82	0.361	0.285	0.114	0.086
20-Apr-92	0850	CSM	2.75	0.098	0.089	11.58	0.415	0.256	9.53	0.341	0.296	13.15	0.472	0.361	85	0.361	0.272	0.105	0.065
21-Apr-92	1158	GSN	2.40	0.086	0.077	11.27	0.404	0.245	8.96	0.321	0.276	13.21	0.474	0.363	86	0.363	0.287	0.118	0.088
23-Apr-92	0736	GSN	2.38	0.085	0.076	11.06	0.396	0.237	8.95	0.321	0.275	13.08	0.469	0.358	88	0.358	0.283	0.121	0.083
24-Apr-92	0732	GSN	2.38	0.085	0.076	11.06	0.396	0.237	8.95	0.321	0.275	13.05	0.468	0.357	89	0.357	0.282	0.120	0.082
27-Apr-92	0830	GSN	2.39	0.086	0.076	11.13	0.399	0.240	8.95	0.321	0.275	13.08	0.469	0.358	92	0.358	0.282	0.119	0.083
13-May-92	1325	CSM	2.41	0.086	0.077	11.42	0.409	0.250	8.97	0.321	0.276	13.26	0.476	0.365	108	0.365	0.288	0.115	0.089
08-Jun-92	1000	CSM	2.39	0.086	0.076	11.20	0.401	0.242	8.96	0.321	0.276	13.28	0.477	0.366	134	0.366	0.289	0.123	0.090
01-Jul-92	1100	CSM	2.40	0.086	0.077	11.32	0.406	0.247	8.97	0.321	0.276	13.37	0.480	0.369	157	0.369	0.292	0.122	0.093
13-Jul-92	1417	CSM	2.42	0.087	0.077	11.81	0.423	0.264	8.98	0.322	0.276	13.58	0.487	0.376	169	0.376	0.299	0.112	0.100

Figure 8.7 MPBX data processing form.

Shear Plane Indicators Shear plane indicators are devices installed to measure deformation perpendicular to a borehole or pipe. Typical applications are to determine depth and extent of sliding zones in natural and excavated slopes. Shear plane indicators range from crude and inexpensive rupture stakes to more precise and expensive slope extensometers.

RUPTURE STAKES Rupture stakes can be used in soft clay to examine the areal extent and depth of a slide that is continuing to move. They are wooden stakes pushed or driven into the ground to a depth beyond the anticipated shear plane. Shearing will break the stakes. The depth to the shear plane can be determined by pulling out the upper part of each stake. Stakes can be 2 × 1 inch softwood or hardwood beading without knots.

Rupture stakes placed by hand are an economical solution. If a drill rig is required then shear probes may be the preferred approach. Stakes may break when removed from the ground. A large number should be installed so false data can be discarded and a statistical average established.

SHEAR PROBE The shear probe, also known as a "poor boy," consists of plastic tubing or thin-walled PVC pipe installed in a nominally vertical borehole. The depth to the top of the shear zone is determined by lowering a rigid rod into the piping until it cannot go any further. The depth where the rod stops is assumed to be bent by the slide and hence is the depth of the shear plane.

The depth to the bottom of the shear zone can be measured by leaving a rod with an attached graduated nylon line at the bottom of the piping. Pulling on the line, the rod will stop at the corresponding bend in piping below the shear plane.

The same instrument can also be used as a crude inclinometer. Curvature can be determined by inserting a series of rods of different lengths and noting the depth at which each rod will not pass further down the piping. Curvature is given by

$$R = \frac{L^2}{8(D_1 - D_2)} \qquad \text{(Eq. 8-6)}$$

where R = radius of curvature of the piping
D_1 = inside diameter of the piping
D_2 = outside diameter of the piping
L = rod length

The same setup can also be used in observation wells and open standpipe piezometers, allowing monitoring of both groundwater levels and horizontal displacements.

SHEAR STRIPS The shear strip consists of a parallel electrical circuit made up of resistors mounted on a brittle backing strip and waterproofed. The strip is located in a borehole. The depth at which the shear strip fractures is determined by measurement of the resistance. Figure 8.8*a* is a schematic of a typical setup. Resistors can be

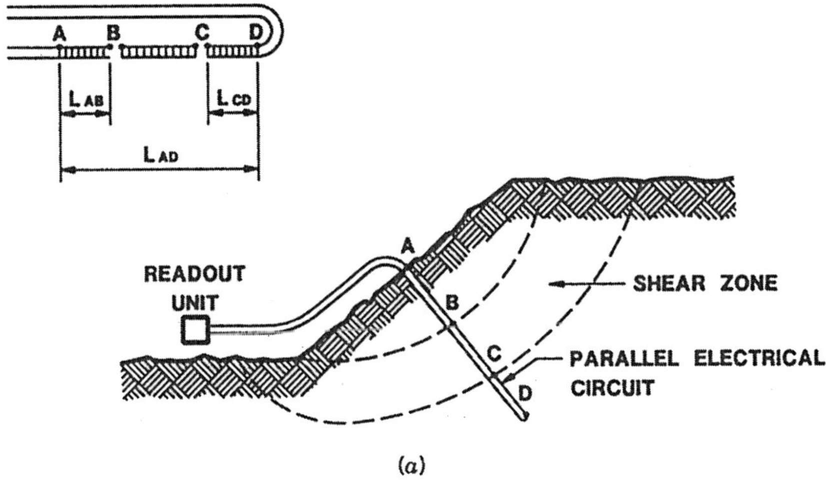

(*a*)

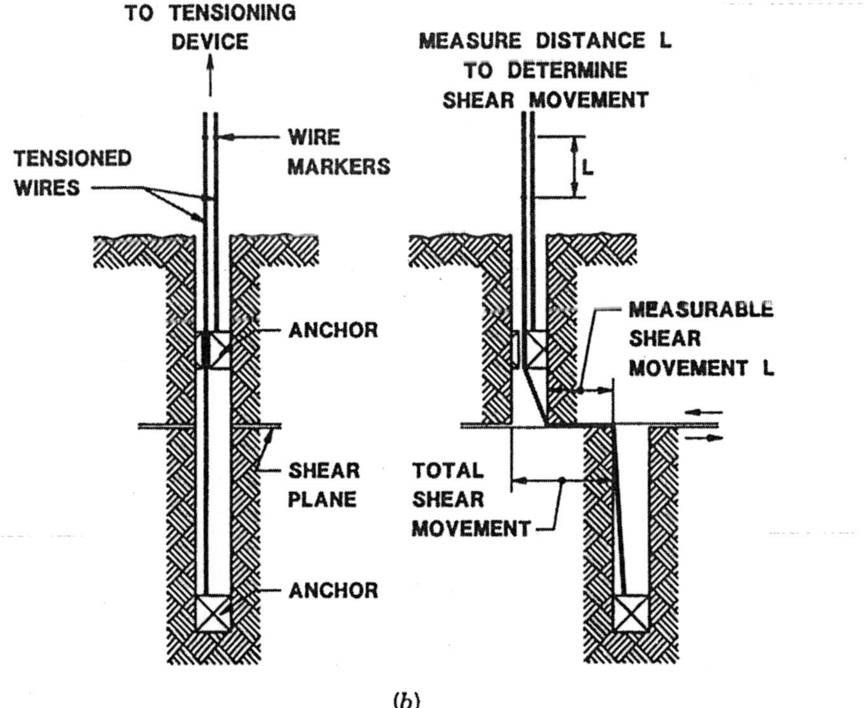

(*b*)

Figure 8.8 Types of shear plane indicators. (*a*) Shear strip (*b*) Slope extensometer. (Dunnicliff, 1988.)

spaced at any interval. Three-foot intervals are a good average spacing unless special conditions exist. The maximum number of resistors per strip is about 100.

A second type consists of a bubbler tube device connected to a thin glass tube within a larger diameter metal pipe. The air pressure required to continually pass bubbles into the ground decreases when shearing breaks the glass tube.

A third type is a pipe with mounted foil strain gages placed inside a PVC pipe. Deformation or shear can be monitored by recording the change of strain in these gages.

SLOPE EXTENSOMETER The slope extensometer is an MPBX with tensioned wires instead of rods. The slope extensometer is arranged for monitoring deformation perpendicular to the axis of the borehole; see Figure 8.8*b*. At the instrument head each wire passes over a pulley and is attached to a weight. If shear deformation occurs between two anchors, no vertical movement of the weights occurs for all anchors above the shear plane. Weights attached to anchors below the shear plane move downward as the wires are dragged across the shear plane. Shear plane deformation can be monitored by measuring movement of the weights relative to the head. A depth band is also established. Up to about 10 anchors and wires can be installed in a borehole.

Initial shear deformation will not cause an equivalent reading change owing to lateral movement of the wires within the borehole. After the borehole has been separated completely, the reading change will equal the shear deformation. This "dead" spot can be reduced by installing each wire within a plastic tube and grouting between the tubes. Measurement precision can be increased by using alternative mechanical or electrical transducers instead of the weights.

The advantages of this system over the conventional inclinometer are

- Simple and rapid reading procedure
- Option to provide an alarm
- Ability to monitor much larger shear deformations.
- Ability to insert it into an open standpipe

Vibration Monitor Vibrations can be caused by construction on or near slopes. Heavy earth moving equipment, pile driving, blasting, and traffic are sources of vibrations. Blasting may occur in a drill and blast operation to excavate slopes or tunnels. Vibrations are usually small in intensity and will not cause a natural slope to fail. Vibration monitoring is suggested if higher than normal vibrations are anticipated in an area. Vibrations can be more of a concern if there are residential areas nearby. Restrictions may be imposed on working times and intensity of tolerable vibration in such cases.

Installation The instruments must be installed properly before, during, or after construction. Poor installation will result in misleading, inaccurate, or false information, and hence waste money. Written specifications should require experienced personnel to install instruments, such as inclinometers and extensometers. A special instrumentation contractor or the manufacturer may be required to install them. It is important to verify any installation experience claims by contractors. The owner or owner's representative should inspect and document the installation. Being available to answer questions and note any difficulties during installation is useful should problems develop during the monitoring and interpretation phases.

Instruments should always be installed by technicians who are fully conversant with the equipment and who have detailed knowledge of the factors influencing the performance of the instruments. Manufacturer's installation instructions are seldom adequate, and installation is best supervised by an experienced instrumentation engineer. Instruments should be installed well in advance of the time they are required to monitor performance so that checks on drift and background noise level can be made and a baseline established for subsequent observations. They should be well protected against corrosion, moisture, other aggressive agents, and vandals.

8.3.4 Instrumentation Monitoring

In all cases, whether as part of construction or remediation, the instruments should be monitored by the owner or owner's representative. The data collected should be processed by the close of business each day and shared with the contractor by the next morning.

Reliability is usually more important than absolute precision. It is essential that instruments be calibrated not only prior to their installation but also from time to time during the life of the monitoring system. Where instruments are expected to remain in use for several years, the possibility of instrument drift cannot be ignored, particularly when electrical instruments are used. The complete system, including the detecting device, associated electronics, and recorder, should be calibrated as one unit. Instruments sensitive to weather and gravity variations should be calibrated at the project site. Probe-type instruments are easier to calibrate, but are unsuitable for situations requiring continuous monitoring.

The instruments should be read systematically by a suitably competent person who has an understanding of their purpose. The frequency of reading will depend upon the situation and the nature of the changes that the instruments are monitoring. Readings are often most critical in inclement weather conditions. For example, piezometers may require reading several times a day during and after heavy rains. Readings should be recorded on standard field sheets that include details of the probable range of readings from the instruments being observed. Any readings that indicate a marked change in conditions should be checked immediately. When possible, the functioning of the instruments should be also checked when unexpected readings occur.

Data Processing All data requires some sort of processing. The easiest way to process instrumentation data is to use spreadsheets. The firs step should therefore always be to make a hard copy of the field data. If data are lost in the computer, you can always reenter the data if a hard copy exists.

The raw data can initially be downloaded or entered by hand. Processing the data can involve converting raw data into strain or stress information, or comparing a new reading to an initial reading. Spreadsheets can be programmed to perform the calculations immediately. The processed data should then be plotted for visual inspection.

All instrument readings should be plotted on a time base so that the significance of the variation can be assessed more easily. Example plots are given in Figures 8.9 and 8.10.

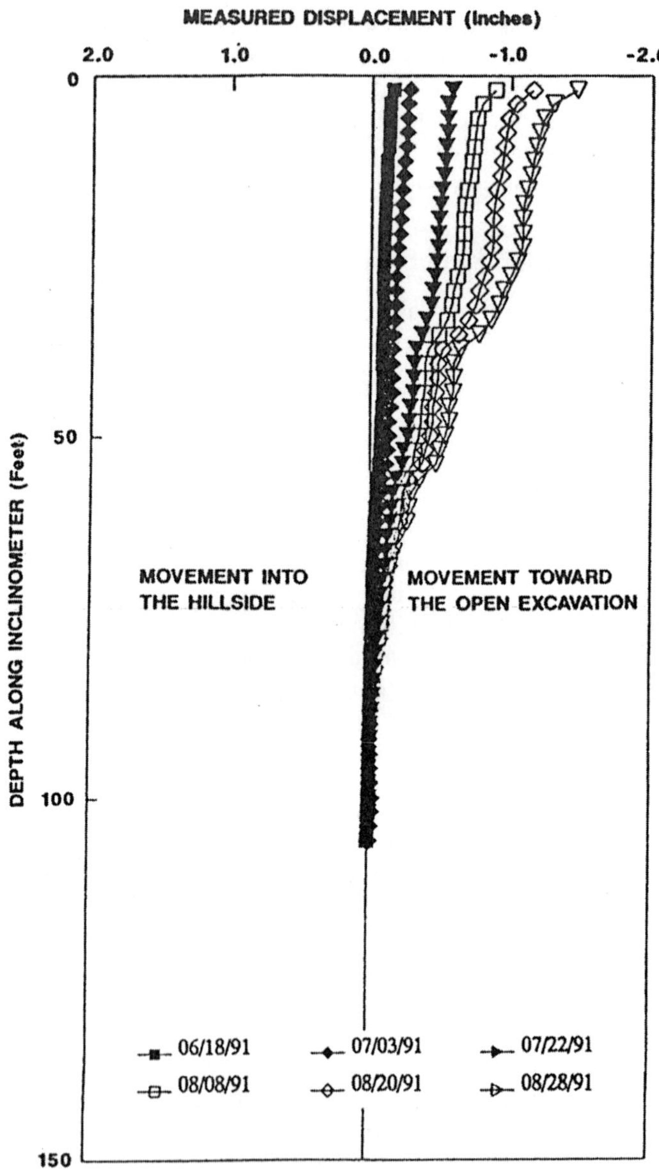

Figure 8.9 Example inclinometer data plot.

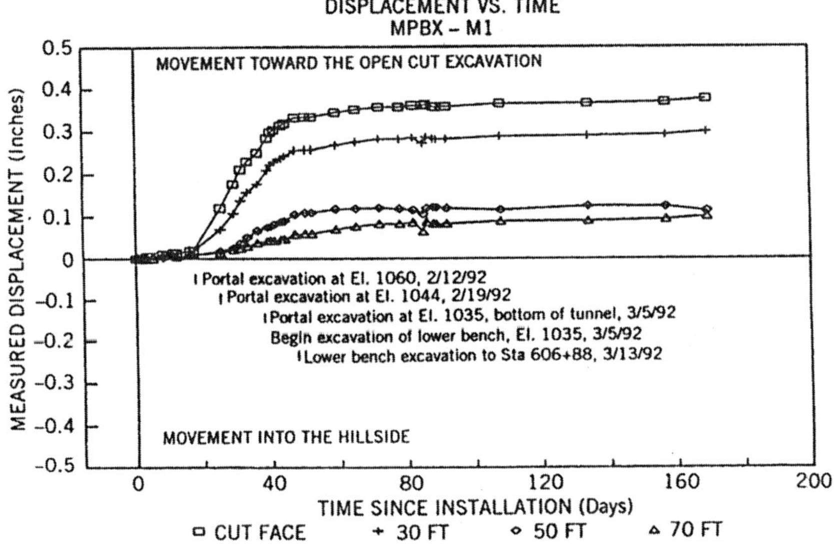

Figure 8.10 Example MPBX data plot.

Interpretation When the data have been collected and processed, the next step is to plot the information. Plots should show displacement versus time. Rates of movement can then be calculated from the slopes of the curves. It is advisable to keep the plots as current as possible. When displacements are occurring at an increasing rate, the interpretative plots are generally needed at a greater frequency.

Interpretative plots require showing all the information available. This additional information includes the ground surface, crack locations, shear planes, groundwater table elevations, pore pressure, and so on. Developing this type of comprehensive plot is time consuming. Interpretative plots should be to scale. All additional information should be clearly labeled. An example is provided as Figure 8.11.

For MPBXs, the most easily visualized and information plot is anchor displacement versus time. If the instrument head is considered to be stable, anchor displacement relative to it should be plotted. For the slope stability application, the deepest anchor is considered to be in the stable ground. In this case, the plot should show anchor displacements and instrument head displacements relative to the deep anchor.

Response Once the information is collected and shown on plots or graphs, remedial measures need to be developed. The contractor must be included in the discussions if problems occur during construction. Remedial measures vary. The corrective action chosen should be sufficient to eliminate the need for future corrective action if at all possible.

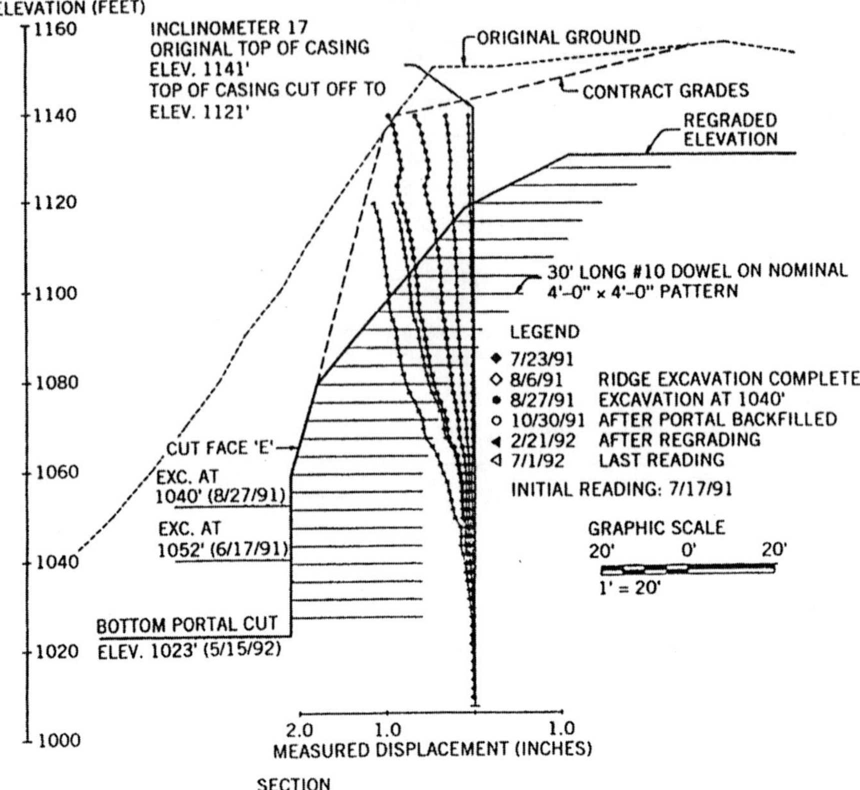

Figure 8.11 Example interpretation plot.

8.4 INSPECTION FOLLOWING CONSTRUCTION

8.4.1 Introduction

Regular inspections and maintenance are essential for the continued stability of highway slopes. Sometimes, recommendations for maintenance of slopes are not followed. Careful design and detailing can reduce both the amount of maintenance required and the physical labor involved. Slopes should be designed to be as maintenance free as possible. However, all slopes require some degree of periodic inspections and maintenance.

It is essential to keep a good construction database for the slope. These data would form the basis for future slope maintenance records. The designer should recommend a slope inspection and maintenance program after construction. Such a program should contain data regarding expected postconstruction behavior of the slope, as well as guidance on maintaining and reading installed instruments, and should indicate the probable range of readings that will be obtained if the structure is functioning as designed. If, during the lifetime of the slope, the observed behavior and readings obtained from instruments indicate that conditions worse than those allowed for in the design are occurring, it may be necessary to undertake remedial measures to ensure slope stability. Instrument readings and inspection records should be kept on standard record sheets, an example of which is shown in Figures 8.12*a*, *b*, and *c*. The maintenance engineer should use these forms when performing regular inspections of the slopes. He or she should be responsible for initiating and scheduling future maintenance inspections and recommending remedial works, if necessary.

It is imperative to establish and maintain a precise record of all indications of slope instability and sections of highway showing distress. Such a record should be periodically reviewed by the geotechnical engineering staff and consulting engineer, if appropriate. This review should be jointly undertaken by the maintenance engineers and the geotechnical engineers.

It should be noted that the maintenance engineers should consult the geotechnical engineers about the geotechnical assumptions used to design the slope and about matters concerning specialty structures, for example, ground anchors or mechanically stabilized embankment walls.

8.4.2 Frequency of Inspections

For slopes that have been in existence for many years, it may be necessary for the maintenance engineer to screen all the slopes under his or her control to decide the frequency of inspections in view of the potential consequences of failure. As a general rule, slopes of high risk category (in terms of loss of human lives and economic loss or loss of highway service) should be more frequently inspected than those of low risk category.

As detailed in Sections 8.4.3 and 8.4.4, inspections should be carried out by the maintenance engineers and the geotechnical engineers at the recommended intervals shown in Table 8.1.

For new slopes, it may be necessary to undertake inspection more frequently for the first year, but less often thereafter if the slope performs as expected (Table 8.2). More frequent inspections in the first year are needed because signs of instability and problems usually arise in the first year, especially after the first wet season. If drought conditions or low rainfall conditions exist during the first year, this "probation period" should be extended until the slope undergoes a substantial saturation and performs well as a result. If problems are detected, they should be repaired and remedied with no

TABLE 8.1 Interval between Maintenance Inspections (Existing Slopes)

	Recommended Interval	
Inspector	High-Risk Slopes	Low-Risk Slopes
Maintenance engineer	6 months	2 years
Geotechnical engineer	2 years	5 years

TABLE 8.2 Interval between Maintenance Inspections (New Slopes)

	Recommended Interval	
Inspector	High-Risk Slopes	Low-Risk Slopes
Maintenance engineer	2 months for the first year and 6 months thereafter	6 months for the first year and 2 years thereafter
Geotechnical engineer	6 months for the first year and 2 years thereafter	Once for the first year and 5 years thereafter

delay. In addition, design assumptions should also be verified and experience gained from the actual performance of the slope.

Tables 8.1 and 8.2 are intended as a recommended guideline regarding the frequency for inspections. The frequency for inspections will vary depending on the geologic complexity, nature of the slope, climatic conditions, and, most of all, the funding and resource availability.

8.4.3 Technical Inspections

The purpose of technical inspections is to ensure that the slope is not deteriorating. These inspections may also be used to identify slopes that need to be upgraded into a higher risk category. It is important that guidance notes are issued to the maintenance engineer. These should highlight such points as the crest, all berms/benches, and the toe of slope, which must be inspected. It is also important that worrisome observations should be brought to the attention of the maintenance supervisor and/or the geotechnical engineer. An action plan of additional repairs should be implemented without delay.

8.4.4 Engineering Inspections

Engineering inspections should be carried out on slopes that are exhibiting signs of instability or questionable behavior. The purposes of engineering inspections are threefold, namely (1) to verify the technical inspections, (2) to identify the cause of slope instability, and (3) to advise on the remedial work requisite for the unstable slope. The engineering inspections should be conducted by a geotechnical engineer who has a broad background in engineering geology and soil mechanics as well as working knowledge of construction and maintenance procedures.

8.4.5 Inspection Reports

It is of utmost importance that all inspections should be properly recorded and that a system should exist to transfer inspection recommendations into remedial works, preventive works, or detailed investigation. The inspection reports should be properly filed and kept for later retrieval. Once the database is established, the reports should be added to produce a historical record of the slope. The maintenance engineer should always check previous inspection records and follow up to see if the last set of recommendations have been fully implemented. Photographs should be used to supplement reports where appropriate.

Figure 8.12*a*,*b*, and *c* are examples of a three-sheet-long slope maintenance inspection record. These are intended for general guidance in recording features that are often observed during inspections, but are not tailored to specific types of slope stabilization work, such as ground anchors, soil nails, or MSE walls.

As a guide, the following information should be collected and included in an inspection report:

- Location (state, county, city, highway number, mile marker)
- Type of slope and general features (cut above road, highway embankment, downhill cut, remote area, populated area)
- Geotechnical conditions (soil, rock, fill)
- Type of instability (rockfall, mud flow, rotational, wedge failure)
- Contributing factors (surface drainage, subdrainage, debris material, rain conditions, blocked drains)
- Inspection data (rate of movement, visual signs of distress, effect on roadway, utility locations, size of slide, dimensions of slide, maintenance activity, property information, amounts of rainfall)
- Sketches and photographs

8.5 MAINTENANCE

Many slides can be prevented or minimized by careful and knowledgeable maintenance practices. If maintenamce personnel are familar with the factors that cause slides, effective maintenance practices will reduce costs.

8.5.1 Access

Access should be provided to all berms, channels, and drainage systems to permit inspection and maintenance. For all new slopes, proper access should be included in the design. Lockable gates should be provided to prevent unauthorized entry and vandalism.

8.5.2 Slope Performance Observations

All slopes should be examined for signs of movement indicative of slope instability. Apart from natural causes such as weathering, water is the major factor that can adversely affect slope stability. Inspection and subsequent maintenance should be principally directed toward preventing the infiltration of water. Forms of surface protection such as shotcrete covering and concrete drainage ditches are generally brittle and therefore are susceptible to cracking. Inspection records should give details of crack positions, lengths, widths, and relative movements. Telltales or gage measurement points should be installed on new cracks where appropriate. During inspections

Slope Maintenance Inspection

Slope Location

Page 1 of 3

Slope Number | Weather
Inspecting Officer | Position | Date

Risk Category Low/Medium/High
Last Inspection Date
Is interval between inspections OK? Yes/No
Previous Risk Category Low/Medium/High
Have past recommendations been carried out? Yes/No

Summary

Major works required	Yes/No
Minor works required	Yes/No
Investigation needed	Yes/No
Slope satisfactory	Yes/No

ACCESS

Is there good maintenance access?	Yes/No
Is it difficult for the public to gain access?	Yes/No
Has the inspecting officer gained access to the crest, the toe, and all berms?	Yes/No

Comments

INSTRUMENTATION

Have all instrumentation systems been checked?	Yes/No
Have all instrumentation results been plotted?	Yes/No
Are all readings acceptable?	Yes/No
Is there a need for new instrumentation?	Yes/No

Comments

P.S. This inspection sheet is to be read together with the slope data sheet/file.

(a)

Figure 8.12 Example of a maintenance inspection record for a slope.

Slope Maintenance Inspection

Slope Location

Page 2 of 3

CONDITION OF SLOPE

	Status of Feature				
	None	Good	Satis-factory	Works Needed	
				Minor	Major
Condition of impermeable surface					
Extent of impermeable surface					
Condition of weepholes					
Capacity of weepholes					
Condition of vegetated surface					
Capacity of surface drainage					
Condition of U-channels & step channels					
Condition of catchpits & sandtraps					
Condition of raking drains					
Condition of associated culverts & nullahs					
Condition of artificial support					
Condition of toe fence/toe barrier					

Comments

		Works Needed	
		Minor	Major
Has there been a recent slope failure?	Yes/No		
Has there been any recent erosion?	Yes/No		
Has there been any recent movement?	Yes/No		
Are there any tension cracks at the crest?	Yes/No		
Is there adequate protection against infiltration above the crest?	Yes/No		
Has there been any recent seepage?	Yes/No		

If seepage give details :-

If movements give details :-

Comments

(b)

Figure 8.12 *(Continued)*

Slope Maintenance Inspection

Slope Location

Page 3 of 3

ASSOCIATED RETAINING WALLS		Works Needed	
		Minor	Major
Have there ever been wall movements?	Yes/No		
Has there been recent wall settlement?	Yes/No		
Has there been recent wall cracking?	Yes/No		
Has there been recent wall titling?	Yes/No		
Has there been recent wall bulging?	Yes/No		
Is the capacity of the weepholes adequate?	Yes/No		
Are the weepholes clear?	Yes/No		
Are the mortar joints/pointing satisfactory?	Yes/No		
Is vegetation adversely affecting the wall?	Yes/No		

Comments

SERVICES & DRAINAGE

Are services adversely affecting the slope?	Yes/No
Do any services need testing?	Yes/No
Have the appropriate authority been informed?	Yes/No

GENERAL & COMMENTS

Does the slope need upgrading to meet the present risk category? Yes/No

RECOMMENDATIONS

Signature :

(c)

Figure 8.12 *(Continued)*

of grass-covered slopes, the positions, depth and extent of erosion scars should be noted, and any cracks monitored with slope stakes as appropriate.

During inspections, seepage traces on and adjacent to all slopes should be recorded. Flow from sources, weepholes, and horizontal drains should be recorded and, where possible, should be examined for signs of migration of soil fines; any such signs probably indicate detrimental effects of internal erosion (piping). Most importantly, the source of the seepage should be investigated and adequate drainage provided.

Rigid surface protections may sometimes be made adjacent to trees. They should be examined for signs of deterioration as a result of root action, and replaced or repaired as needed.

Routine maintenance of soil slopes should include removal of undesirable vegetation. Cracked rigid surfaces should be repaired by cutting a chase along the line of crack and filling it with a suitable caulking material of a suitable mix. On inclined surfaces, it is good practice to cut the chase in a manner that prevents the ingress of water through any shrinkage cracks that may develop between the original surface and the repair.

Eroded grass slopes should be regraded, if necessary with compacted fill or lean concrete where appropriate. Before placing fill or lean concrete, all loose materials found in the eroded area should be removed. If fill is used, the repaired surface should be replanted with grass.

8.5.3 Instruments

The instruments that are most likely to be installed and used for continued monitoring are piezometers, inclinometers, and survey points. The accuracy and serviceability of piezometers rely, among other things, on preventing the ingress of water and foreign matter into the standpipe. The tops of the piezometers (i.e., the protective boxes) should be examined during each inspection. All standpipes should have tight-fitting caps, and all the protective box installations should be drained. If necessary, caps should be renewed and drains cleared.

Other instrument installations, such as inclinometers and survey points, should be checked to ensure that they are functioning under the conditions specified by the manufacturers.

8.5.4 Drainage

Surface Drainage Regular maintenance of surface drainage systems is essential to ensure long-term stability of highway slopes. Inspection and maintenance records should include the following:

(1) Position and extent of broken and cracked channels and catchpits
(2) Position and extent of silted-up section of channels and catchpits
(3) Position and extent of deteriorating channels and catchpits

(4) Details of construction works, possibly outside the boundaries of the development, from which mud and debris could migrate to block the drains of the system being inspected

As mentioned previously, many slope failures are rainfall-induced. It is therefore imperative to pay close attention to surface and subsurface drainage systems in the wet season. Cracks in channels should be repaired with cement mortar or with a suitable plastic sealing compound. If a channel cracks from settlement, the settled section should be removed and reconstructed with no delay. However, the cause of settlement should be investigated and addressed to avoid damage from further settlement. Where sections of channels have to be rebuilt, this work should preferably be done in the dry season when the existing channels may be safely removed without further risk to the slope.

Subsurface Drainage Where horizontal drains are installed to lower groundwater levels, they should be regularly inspected and maintained for full serviceability. This is because the efficiency of horizontal drains will decrease with time. When interpreting the records of flow from horizontal drains, it should be noted that any initially high discharges should decline as the drains lower the groundwater regime and a steady state is achieved. It is therefore important to record flows from individual drains during each inspection, and to correlate these with rainfall records for that area or with readings from piezometers that should be installed as an integral part of any subsurface drainage system.

The efficiency of horizontal drains can be determined by monitoring the piezometric readings. If the piezometric readings indicate a rise in groundwater level, and at the same time, the discharge from the horizontal drains decreases, it should be concluded that the efficiency of the drainage system is decreasing. It may be possible to partially reinstate the system by means of flushing the drains with a suitably designed compressed air and water jet. If flushing of the drains fails to raise the effectiveness of the system to an acceptable level, additional drains should be installed.

The discharge from the drains should also be examined for signs of migrating soil fines, which may be indicative of internal erosion. In some situations, drains are or remain dry after installation, and drains that have had constant discharge in the past turn out to be dry. These conditions may suggest that the drainage system has had the desired effect, that the drains are plugged, or that they do not intercept the groundwater table. If the drains are plugged, they should be flushed and cleaned.

Regular maintenance of horizontal drains includes the removal of obstructions at the outlets, removal of silt from the inside of the pipe, and, if possible, cleaning or replacement of internal filters.

8.5.5 Adjacent Utilities

Sometimes, services such as stormwater drains, sewers, and water mains are located adjacent to highway slopes. Leakage may affect the slope stability. During routine inspections, all services should be inspected for leakage or water flow. When detected, utilities should be repaired or replaced. The inspection records should consist of a drawing showing the position and nature of all services in the vicinity of the slopes. The appropriate organization should be notified to test water mains and sewers where leakage is suspected and where it could lead to instability of the slopes being examined.

8.6 TIME DOMAIN REFLECTOMETRY (TDR)

8.6.1 Introduction

Time domain reflectometry (TDR) is the latest tool used to monitor movements within sliding planes. TDR was originally developed by the power and communications industries to find locations of breaks or strains in cables in 1950s. To test a cable, a voltage pulse waveform is sent down it. If the pulse encounters a change in the characteristic impedance of the cable, it is reflected. This can take the form of a crimp, kink, or a break in a cable. The first geotechnical TDR applications were made by the mining industry in the United States and Canada at sites where coaxial cables embedded in boreholes were installed to monitor unstable ground. The California Department of Transportation (Caltrans) recognized the similarity between mine overburdens and the unstable soil and rock slopes that threaten many of its highways.

8.6.2 Mechanics of TDR

TDR involves embedding a cable in a vertical hole or horizontal trench, then reading the cable "signature" at various times. Ground movement, such as shearing along a failure plane, will deform the cable and change the cable impedance. This change can be monitored to locate the position of failure, and its rate will correspond to the rate of ground movement. Figure 8.13 shows the set up of a TDR system.

Coaxial cables are composed of a center metallic conductor surrounded by an insulating material, an outer metallic conductor surrounding the insulation, and a protective jacket (Figure 8.14). The cables have characteristic impedance determined by the thickness and type of insulating material between the cables. This insulating material is called "dielectric," and may be made of almost any nonconducting material—polyvinyl chloride (PVC) foam, Telflon, or even air. If the cable is crimped or damaged, the distance between the inner and outer conductors changes, as does the impedance at that point. The TDR cable tester can then determine the location of the fault.

The TDR system offers remote and automatic readout capabilities. The instrument is programmed to store the data in a computerized datalogger that automatically turns on a cell phone. The technician can dial that phone at any time to retrieve the data.

The first test ever to look at the capability of the TDR system to monitor landslide application was carried out by Caltrans at the Last Chance Grade landslide on U.S. 101 (Kane and Beck, 1996). This active area, some 520 meters wide and 460 meters long affects about 235 meters of the roadway in Del Norte County in California. One 6-Millimeter diameter, 82-meter long braided coaxial cable was installed along the

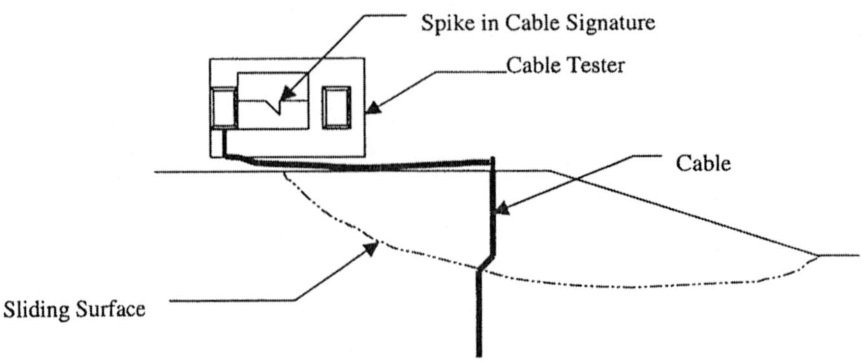

Figure 8.13 Setup of TDR equipment in the field. (Courtesy of KANE GeoTech, Inc.)

outside of an inclinometer casing, and another cable was installed in a groove cut into the pavement and extended beneath a concrete barrier. Inclinometer readings showed that the slide plane was about 40 meters below the roadway at the center of the slide. TDR showed one notable spike at the same depth and location. Traffic control was necessary for the safety of the personnel who read the inclinometer, but the TDR cable could be read from behind the concrete safety barrier.

8.6.3 Advantages and Disadvantages of TDR over Inclinometers

In many instances, there are clear advantages and disadvantages of TDR when compared to inclinometers. Listed below are comparisons between the two technologies:

Cost Cable has an enormous cost advantage over casing. Cables can be commercially purchased at a cost of \$0.20 per foot, as opposed to \$7 to \$10 a foot for

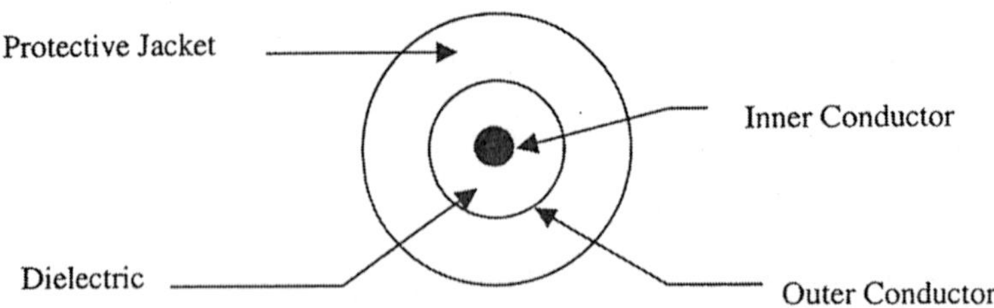

Figure 8.14 Schematic of a coaxial cable. (Courtesy of KANE GeoTech, Inc.)

inclinometer casing. The cost of the electronic equipment is about the same for both TDR and inclinometers, at roughly \$10,000 per unit in 1999. For remote installations, dataloggers and cellular phone equipment run another \$2,000. As for drilling costs, they are equal. One important advantage of TDR over inclinometers is that with TDR, there is no risk of losing an expensive probe due to a kink in the inclinometer casing; only expendable cable is lost.

Time TDR cables can be read more rapidly than inclinometers such that more holes can be read in less time—an additional saving in personnel time. A single cable can be read in about 5 minutes or less, as compared to $\frac{1}{2}$ hour to an hour to read a 100-foot deep inclinometer.

Remote Access Telemetry using a datalogger and cellular phone can be installed for real time readings and/or difficult access and remote areas without the need to visit the slope. In-place inclinometers are available, but they are expensive and must be considered sacrificial.

Safety With TDR, data collection can be carried out from a safe location, even from an office with remote data acquisition equipment. When TDR is installed in highways, there is no need to maintain traffic control as cables can be run under the pavement. In other words, TDR readings can be recorded outside the traffic lane as the cable is buried in a trench underneath the traffic lane. Technicians need not be exposed to dangerous traffic or to landslide and rockfall hazards on unstable slopes.

Data TDR provides instantaneous data as to whether the slope is moving and where is it moving. It is not necessary to download data from a readout box into a computer and then plot the results. With TDR, the screen on the cable tester shows the cable signature immediately. By comparing cable signatures over time, the location of any slip and changes in the rate of movement can be determined. Rates of deformations can be determined by plotting the magnitude of the spike in the cable with time.

Unlike inclinometers, however, TDR cannot determine the direction and magnitude of slope movement. This aspect is considered to be the one of the major disadvantages of TDR as compared to inclinometers. Research is being done in this area in optical time domain reflectometry (OTDR) (Tsang and England, 1995).

REFERENCES

Agnew, W., 1991. "Erosion Control Product Selection,"*Geotechnical Fabric Report*, April, pp. 24–27.

Association of Bay Area Governments, 1981. *Manual of Standards for Erosion and Sediment Control Measures*. Oakland, California.

Dunnicliff, J., 1988. *Geotechnical Instumentation for Monitoring Field Peformance*. New York: Wiley

Geotechnical Control Office, 1984. *Geotechnical Manual for Slopes*, 2nd ed. Hong Kong: Civil Engineering Services Department, May.

Green, G. E., 1974. "Principles and Performance of Two Inclinometers for Measuring Horizontal Ground Movements," *Proceedings of the Symposium on Field Instrumentation in Geotechnical Engineering*, British Geotechnical Society. London: Butterworths, pp. 166–179.

Kane, W. F., and T. J. Beck, 1996. "Rapid Slope Monitoring," *Civil Engineering Magazine*, Vol. 66, No. 6, June, pp. 56–58.

Tsang, C. M., and G. L. England, 1995. "Potential of Fibre Optic Sensing in Geotechnical Applications," *Geotechnical News*, December.

CHAPTER 9

SHALLOW FAILURES

9.1 INTRODUCTION

As discussed in Chapter 3, water has a detrimental impact on slope stability by decreasing shear strength of the soil mass. In 1982–1983 and 1997–1998, major El Niño storms in the United States resulted in thousands of landslides, killing tens of people, and causing tens of millions of dollars worth of damage to public and private facilities. Studies by the National Research Council, an arm of the U.S. Academy of Sciences, cited three main factors attributing to these rainfall-induced landslides. They are:

- Urbanization and development in landslide-prone areas
- Deforestation of landslide-prone areas
- Increased regional precipitation caused by global warming, resulting in change of climate patterns

In the 1990s, the council said that El Niño had caused regional weather changes in western North America that resulted in much heavier than normal precipitation in mountainous areas. One of the results was a tremendous increase in landslide activities in the northwest to western regions of the United States such as California, Colorado, Nevada, Oregon, Utah and Washington. These rainfall-induced landslides are often shallow in nature; that is, the depth to length ratio of the slide (*D/L* ratio) is less than 15 percent, or not deeper than 10 feet, which may include both soil and bedrock.

This chapter focuses on issues related to shallow failures as caused by rainfall and snowmelt and discusses various hydrogeologic conditions that are likely to be responsible for many shallow failures. In addition, the mechanism of rainfall-induced landslides will also be discussed to highlight the reasons why shallow failures occur in the field.

9.2 SEEPAGE FLOW MECHANISM DUE TO INFILTRATION

During heavy rains, water seeps into the ground and travels through unsaturated soils, as shown in Figure 9.1. The results of measurements at various field sites indicate that the unsaturated gradient is generally directed vertically downward during steady rainfall. This water may perch on lower permeability materials or a drainage barrier such as bedrock and highly impermeable clays, and so on, creating a temporary, localized saturated zone. Water then continues downward and downslope, eventually seeping into gullies or reaching the lower groundwater table.

At the drainage barrier, a phreatic surface can develop with flow parallel to the drainage barrier. For intense rainfall, groundwater would not rise gradually, but would rise rapidly to the ground surface because of the near-saturated condition in the unsaturated zone. This would result in a sudden increase in pore pressure, which would, in turn, reduce the shearing resistance of the soil. This condition reduces the factor of safety (FOS) of the slope and may cause surficial failures.

The perched water table is considered to be the key factor in triggering rainfall-induced slope failures, which are often shallow in nature. Studies (Reid et al., 1988)

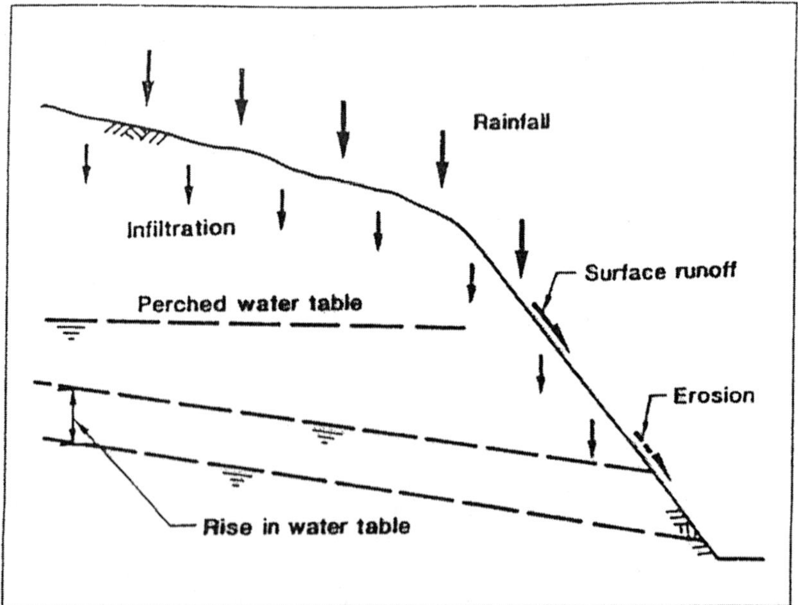

Figure 9.1 Diagrammatic representation of effects of rainfall on slopes. (Brand, 1985.)

have shown that rainfall induces positive pore pressures in hillslope soils, pressures that are accentuated by

- Water perching on low permeability layer,
- A decrease in slope of the perching layer,
- Groundwater flow convergence controlled by surface or bedrock topograph, and
- Spatial variations in soil properties causing localized mounding (Figure 9.2).

All these factors may contribute to the development of a perched water table, resulting in the buildup of pore pressure and subsequent slope failures. The cause of localized mounding, however, is not readily apparent; it could be caused by changes in soil layer geometries or by spatially variable hydraulic properties (Reid et al., 1988). The requirement of a small downslope reduction in hydraulic conductivity with a transition near the upslope end of the failure area may be quite plausible. The decrease in soil hydraulic conductivity serves to "dam" water in the failure area, retarding subsurface flow downslope, resulting in local mounding of perched water table in the failure area and increasing pore pressures. The direction of groundwater flow is rotated upward relative to the hillslope, creating seepage forces that lead to slope instability.

Water can be concentrated at certain locations within a soil slope by various means. The flow of water within a soil slope may be concentrated by factors like the shape of bedrock surface, breaks in the slope, differences in permeability within both soil and bedrock, and human-induced modifications to the slope (e.g., cut slope geometries and installation of ditches and culverts that may modify the patterns of surface flow and concentrate flows in specific locations). Any slopes that cross soil, such as colluvium, must account for the possible effects of the proposed drainage collection and discharge systems on the future stability of these slopes. Many slope failures have been a result of lack of consideration of drainage flow within the slope, thus causing buildup of water within the slopes, which led to subsequent failures.

9.3 MECHANISM OF RAINFALL-INDUCED LANDSLIDES

When failure of a soil slope is caused by rainfall, the mechanism is that water infiltration causes a reduction in soil suction (or negative pore pressure) in the unsaturated soil. Soil suction will be lost upon full saturation. This results in a decrease in the effective stress on the potential failure surface, which is reflected in a decrease in the soil strength to a point where equilibrium cannot be sustained in the slope. Failure occurs under constant total stress and increasing pore pressure.

According to the effective stress principle, the soil shear strength is intimately related to pore pressure. To determine in the laboratory the appropriate shear strength properties for slope design and slope analysis, it is essential to model the field mechanism of failure as closely as possible to what is actually occurring in the field. To determine the shear properties of soils, the common test used is the triaxial test in which the axial stress, σ_1, is increased to failure while the cell pressure, σ_3, is

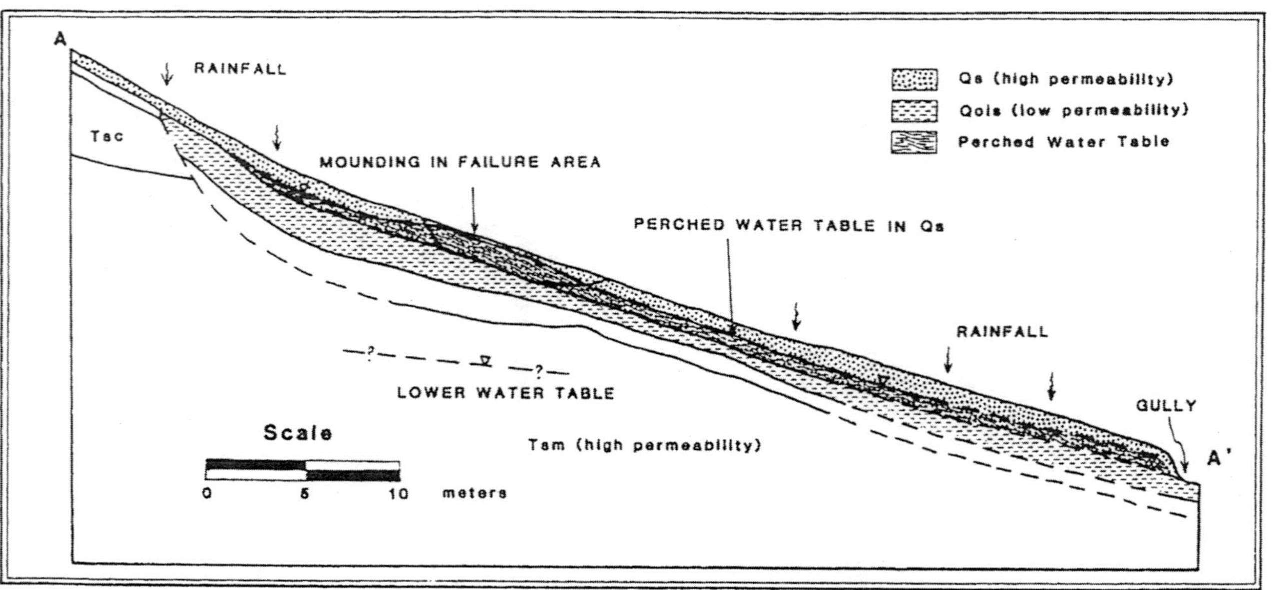

Figure 9.2 Schematic of localized mounding in the failure area with transient, perched water table in the soil overlying the lower permeability old landslide deposit. (Reid et al., 1988.)

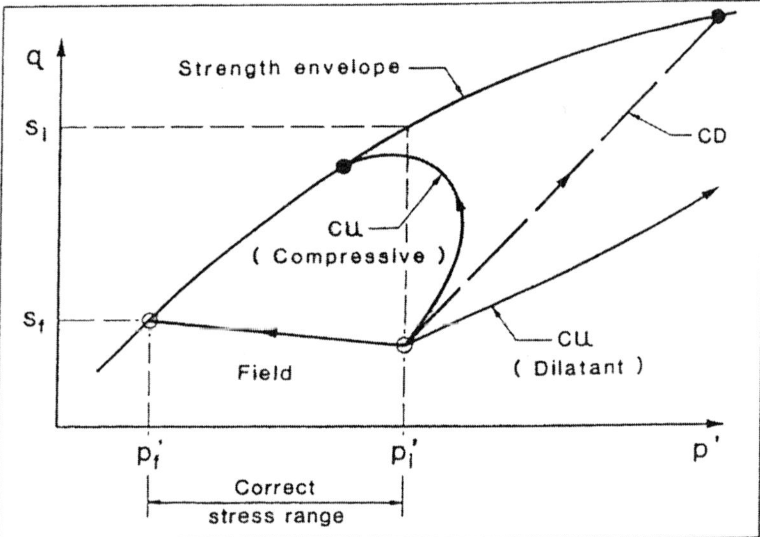

Figure 9.3 Comparison between stress paths for rain-induced slope failure and for triaxial tests. (Brand, 1982.)

maintained constant. Either a drained test is conducted (in which $\Delta u = 0$), or an undrained test (or constant water content test) is carried out during which the pore pressure is measured. However, the stress path followed in the field during a rainfall-induced failure is very different from the stress path in the triaxial test, since both σ_1 and σ_3 are constant and the pore pressure increases until failure occurs.

Figure 9.3 illustrates the difference between the stress paths commonly followed in the conventional triaxial test and that which applies to the field conditions.

The actual mechanism of rainfall-induced failure can only be modeled in the laboratory by means of a constant-load test in which the pore pressure is increased from an initial negative value (i.e., when soil suction exists in the unsaturated soil) until failure occurs. However, in practice, this type of test is not common. Nevertheless, the limited experimental data obtained in this way in Hong Kong has enabled the evaluation of shear strength envelopes in the very low stress range, and this information has proved most valuable.

9.4 FIELD LOADING CONDITIONS

The in situ state of stress within a soil slope is a function of slope angle, the unit weight of the soil, the depth of the groundwater table, and the depth of the soil. During and after periods of prolonged intense rainfall, significant seepage forces may develop and add to the gravity-induced shear stress. In an infinite slope, for example, the shear stresses can easily be resolved on a plane parallel to the slope as shown in Figure 9.4*a*, and the state of stresses is shown in Figure 9.4*b*. When the soil is moist, but unsaturated, the negative pore pressure imposes an isotropic component of effective stress, as shown in Figure 9.4*c* (Anderson and Sitar, 1991). This negative pore pressure (or soil suction) may be very large (as in fine-grained soils) relative to the gravity-induced stresses, as shown in a sketch of representative Mohr circles in Figure 9.4*c*. To model the field situation correctly in the laboratory, these initial stresses must be known, as they determine the laboratory consolidation stresses.

Because of the nature of shallow failures, the initial stresses are typically very small (i.e., 200 to 1,200 pounds per square foot.) Research by other (Brand, 1982; Johnson and Sitar, 1987) has indicated that when the soil is only partly saturated, the pore pressure is negative throughout most of the rainy season, except during and after a prolonged intense rainstorm. In addition, such saturation leading to dissipation of negative pore pressure does not lead to failure and positive pore pressures are required. This suggests that most of the time, the state of stress is similar to that of the right-hand Mohr circle in Figure 9.4*c*. During the early stage of a rainy season, or during the early stage of a long-duration intense rainstorm, the circle migrates left, to the position of the Mohr circle for the saturated soil. This saturated, or nearly saturated, state of stress has been chosen by Anderson and Sitar (1991) as the consolidation state for the anisotropically consolidated triaxial (ACU) test and field stress path triaxial (FSP) test. The stress path to failure in the field starts from this consolidated stress state and moves horizontally to the left as generated by an increase in pore pressures (assuming the seepage force remains constant between the consolidation state and failure) until intersecting the Mohr–Coulomb failure envelope. The field stress path causes an increase in $K_c(\sigma_1'/\sigma_3')$ and, at some critical value, when $K_c = K_f$, shear strain commences. This point of yielding is the initiation of failure and it is reached by a stress change caused by the hydrologic response to rainfall. When the soil is in a loose state, a debris flow is then mobilized and the onset of shear strain causes a rapid reduction in shear strength associated with undrained failure (Anderson and Sitar, 1995).

In summary, the field stress conditions that precede rainfall-induced shallow failures differ from the stress conditions of most laboratory tests. In the field, the stresses are generally lower and the soil is brought to failure by the buildup of positive pore pressure, rather than an increase in shear stress, as occurs in standard triaxial and direct shear tests. In addition, the stress ranges over which triaxial or direct shear tests are usually conducted are not appropriate to the field conditions to accurately model rainfall-induced shallow failures.

9.5 CORRELATIONS BETWEEN LANDSLIDES AND RAINFALL

As mentioned previously, shallow failures often occur during and after periods of intense rainfalls. This leads us to believe that there should be some type of correlation between rainfall and landslides. In fact, the correlation between major rainstorms and the occurrence of landslides has been noted and documented worldwide (Brand, 1982; Lumb, 1975; Pierson, 1977; Slosson and Larson, 1995).

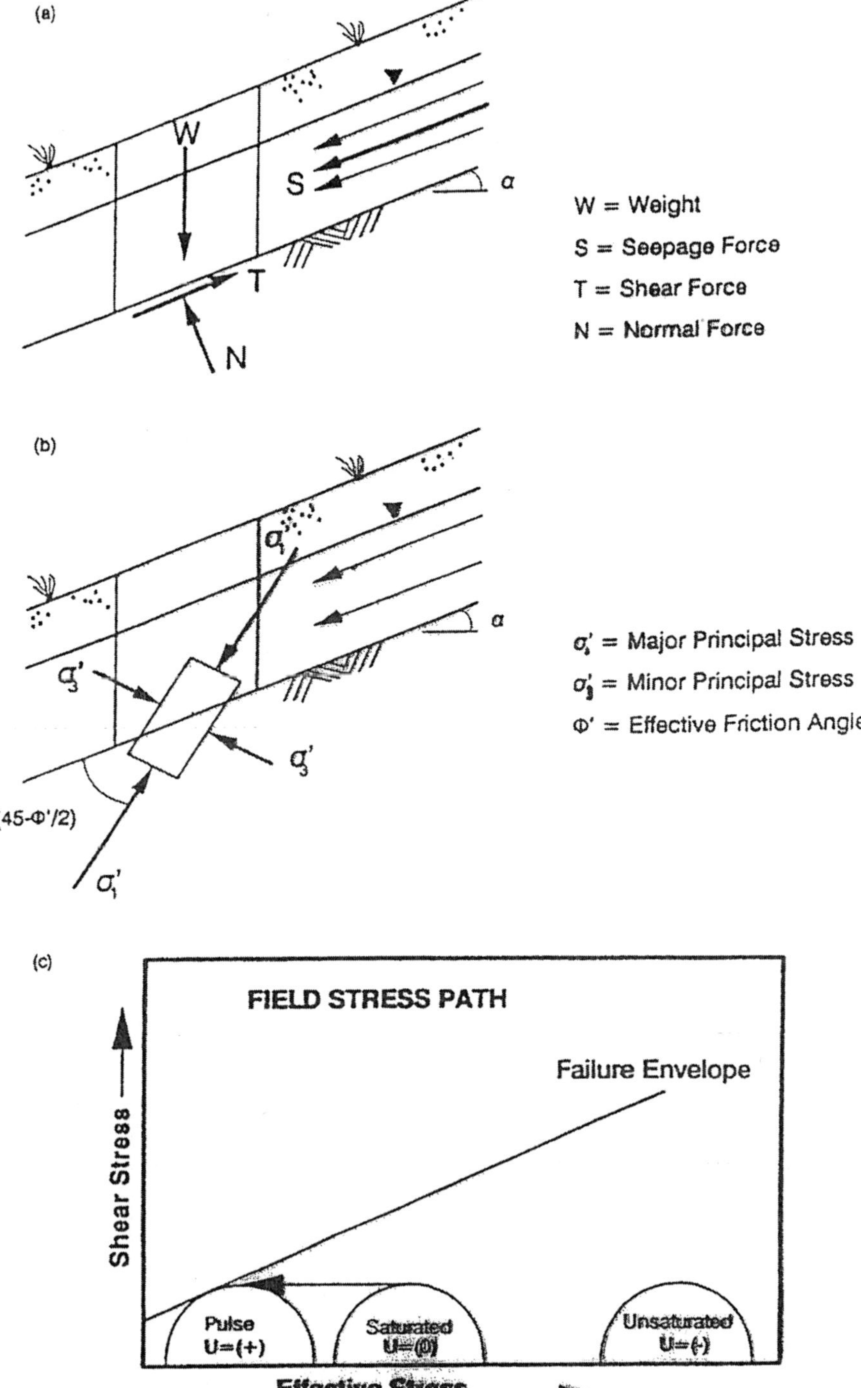

Figure 9.4 Schematic interpretation of stress conditions. (*a*) Forces on an infinite slope in the field. (*b*) Stresses, as resolved for a laboratory triaxial specimen. (*c*) Mohr circles representing the field stress conditions and the field stress path to failure. (Anderson and Sitar, 1991.)

9.5.1 Antecedent Rainfall

In addition to intense rainfall, a certain amount of prestorm rainfall is necessary for the initiation of shallow failures. The antecedent rainfall is necessary to bring the full thickness of the soil mantle to field capacity where the moisture content is such that water will flow out as fast as it flows in. In other words, the soil mantle needs to be fully saturated before subsequent rainstorms can trigger shallow failures or debris flows. The amount of antecedent rainfall that would be necessary for shallow failures will depend upon type of soil mantle, soil conductivity, rate of infiltration, rate of evapotranspiration, and the hydrologic formation of the slope.

In unsaturated soils, moisture is held in soil pores under soil suction or negative pore pressures. Antecedent rainfall has been related to the soil-moisture content and its influence on these pore-water tensions (Johnson and Sitar, 1987). According to Johnson and Sitar (1987), a rainstorm occurring under "wet" conditions (prestorm suctions of less than 500 millimeters of water) is much more likely to generate positive pore pressures than a storm of similar intensity occurring under "dry" conditions (prestorm suctions greater than 1,500 millimeters of water). This is because so much rainfall under "dry" conditions would have to be consumed by wetting the soil.

The influence of antecedent rainfall on landslides has been a subject of debate for many years. Lumb (1975) found that antecedent rainfall had an influence on landslide occurrences. Specifically, it was found that rainstorm events with widespread landslides occurred under high antecedent rainfall conditions. Lumb (1975) examined slope failures in Hong Kong and correlated their occurrence with rainfall intensity (Figure 9.5). He divided rainfall-induced landslide events into four categories:

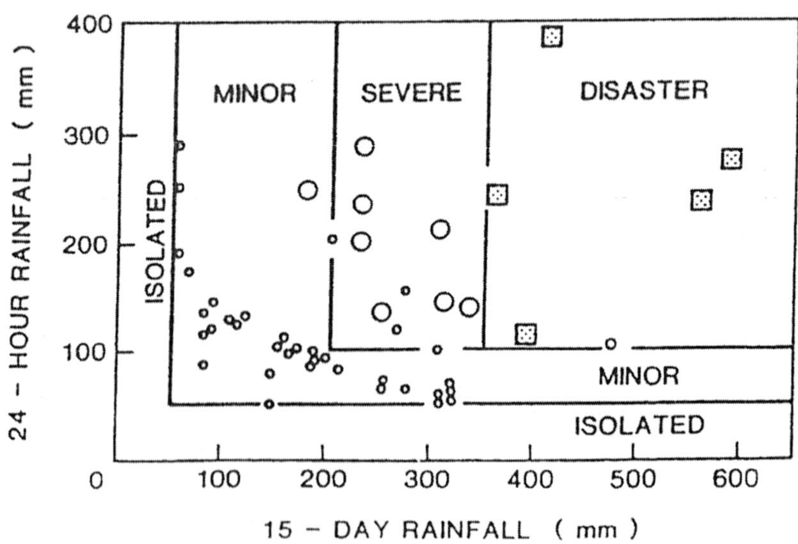

Figure 9.5 Relationship between rainfall and landslides in Hong Kong. (Lumb, 1975.)

- Disastrous events causing more than 50 recorded landslides in one day,
- Severe events causing 10 to 50 landslides in one day,
- Minor events causing less than 10 landslides in a day, and
- Isolated events causing only a single slip.

Using rainfall data obtained from various rain gages, Lumb identified bounds between different levels of event severity and demonstrated that the 15-day antecedent rainfall correlated well with the 24-hour rainfall. In particular, disastrous events occurred when the 24-hour rainfall exceeded 100 millimeters and the 15-day antecedent rainfall exceeded 350 millimeters. Severe events were triggered by rainfall of about 100 millimeters per day after 200 millimeters of antecedent rainfall.

Experience by the Geotechnical Engineering Office in Hong Kong in the operation of the Landslide Warning system for the last two decades has indicated that without high antecedent rainfall, a significant number of landslides are unlikely to be reported even if recorded rainfall has exceeded the warning criteria (70 millimeters per hour) unless the intense rain covers a relatively large area, say 20 square kilometers in an urban area (Pun, 1997).

9.6 RAINFALL THRESHOLDS FOR PREDICTION OF SHALLOW FAILURES

As correlation between rainfall and landslides has been well documented, there have been many studies available to determine whether there is a rainfall threshold for slope failures. Rainfall amounts exceeding 140 percent of normal (mean average over 100 years) in southern California appear to be sufficient to trigger slope failures (Slosson, 1969; Slosson and Krohn, 1982). Wilson et al. (1993) suggested that the threshold for slope instability in the San Francisco Bay Area is approximately 125 percent of normal rainfall, which is about 20 inches per year. Wieczorek (1987) also found that similar thresholds and rainfall patterns caused similar events in the San Francisco Bay Area. These threshold values represent a general rule of thumb in California, indicating when surficial soil slips, shallow slope failures, and debris flows usually begin to occur.

Brand et al. (1984) examined the landslide risk in Hong Kong using three methods:

- Reviewing Lumb (1975) data using rainfall data collected from 46 automatic recording rain gages distributed throughout Hong Kong
- Considering 1-hour rainfalls, 24-hour rainfalls, and antecedent rainfalls for periods varying up to 30 days
- Identifying the precise time of landslide occurrence by using the Fire Services Department reports of calls.

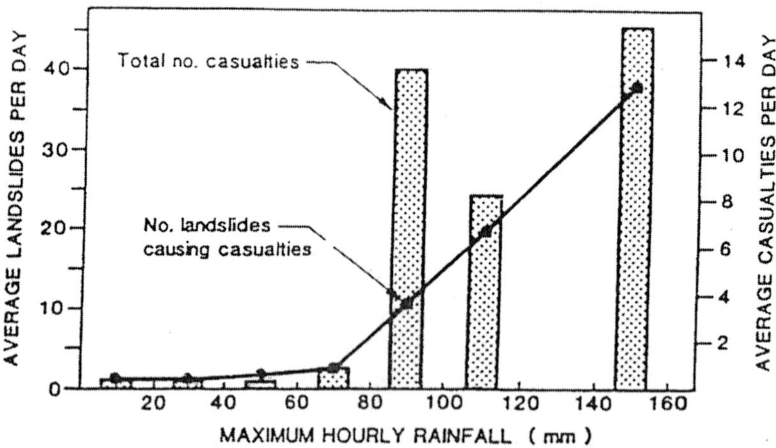

Figure 9.6 Number of landslides causing casualties and number of casualties for period 1963–1983 in Hong Kong compared with maximum hourly rainfall. (Brand 1981; Lumb, 1975.)

Figure 9.6 shows average landslides per day and average casualties per day plotted against maximum hourly rainfall. As can be seen from Figure 9.6, there exists a clear threshold of 70 millimeters per hour of rainfall to trigger multiple landslides with subsequent loss of life. Based on this study, Brand et al. (1984) concluded that

(1) The large majority of landslides in Hong Kong were induced by localized short duration rainfalls, and

(2) A rainfall intensity at about 70 millimeters per hour appeared to be the threshold value above which landslides occurred.

9.7 TYPES OF SOILS, HYDROGEOLOGIC, AND GEOMORPHOLOGIC FEATURES PRONE TO SHALLOW FAILURES

9.7.1 Colluvium

Among various types of soils, colluvium and loess are most likely subject to shallow failures during and after long periods of intense rainfall and snowmelt. Colluvium is formed by the movement and deposition of soil/rock particles by gravity, sometimes with water and wind as secondary transportation agents. It may accumulate as a result of large catastrophic movement of materials, especially debris flows. In rugged terrain, colluvium can be merged and interlayered with alluvium in debris fans and cones associated with minor drainages. Within any given colluvial deposits, particle gradation and density usually vary with depth and in irregular shapes across the entire deposit. This type of geomorphic feature often leads to buildup of perched groundwater on top of an impermeable layer, causing an increase in pore pressures with subsequent slope failures.

Intense rainfall has been cited as the most common triggering mechanism for landslides involving colluvium (Brand et al., 1984; Campbell, 1975; Jones, 1973; Wieczorek, 1987). The soil slope has to be fully saturated before the soil starts to move. In 1943, Kesseli postulated that slope failure of colluvium occurs when the rain is so intense that it causes water to percolate into colluvium at a rate larger than that at which the water percolates into bedrock (Figure 9.7). This condition forms a temporary perched groundwater in the colluvium and a downslope seepage force develops within the soil mass. The Kesseli model has been adopted by others (Campbell, 1975) to explain the initiation of slope failure in colluvium, which is the most commonly observed initial failure mechanism. Campell (1975) suggested that at least 267 millimeters of rainfall is necessary in southern California after a dry season to bring the soils to field capacity (i.e., the water content at which water drains from the soil as fast as it is added). Once the field capacity is reached, it takes 5 to 6 millimeters per hour to trigger debris flow (Campbell, 1975).

9.7.2 Loess

Loess is an eolian deposit composed primarily of silt-sized particles. It covers about 17 percent of the soil deposits of the United States and 11 percent of the land surface of the earth (Turnbull, 1965). Loess is characterized by a loose structure that consists of silt and fine sand particles bound by clay lenses. At low water contents, high negative pore pressures (suction) develop in the clay binder, producing relatively high shear strengths. As a result, as long as it remains unsaturated, loess has the ability to stand in vertical cut slopes and to support relatively large loads as a foundation material. However, upon wetting, the negative pore pressures are eliminated, reducing the effective stresses, and thus the strength, as water content within the clay binder increases to near saturation.

Shallow failures in loess often occur during and after prolonged rainfall and after snowmelt. Shallow failures often come in the form of erosion damage and internal piping primarily in silty loess, although clayey loess is susceptible also in both dry and moist climates. Many large erosion gullies have been initiated by small piping failures originating 5 to 10 feet behind the top of the cut face. These gullies are sometimes initiated by animal burrows that intersect the slope cut.

Erosion damage in loess slopes is often attributed to a lack of good drainage provided at the slopes. Erosion damage to near vertical cuts in Mississippi and Louisiana often occurs where water ponds near the crest of the slope. If natural drainage is directed away from a loess slope, erosion is usually minimal.

Loess slopes are often cut in steep angles and benched to mitigate sheet and rill erosions on the slope faces. In addition, sodded V-ditched channels are often placed on the bench to convey surface runoff. However, drainage channels on the benches can erode severely, which can lead to failure of the entire bench. Paved ditches have been used in Missouri and Illinois to mitigate this effect (Missouri State Highway Department, 1978).

The practice of constructing drainage ditches immediately at the toe of a loessial slope may inadvertently contribute to shallow failures. Periodic highway maintenance

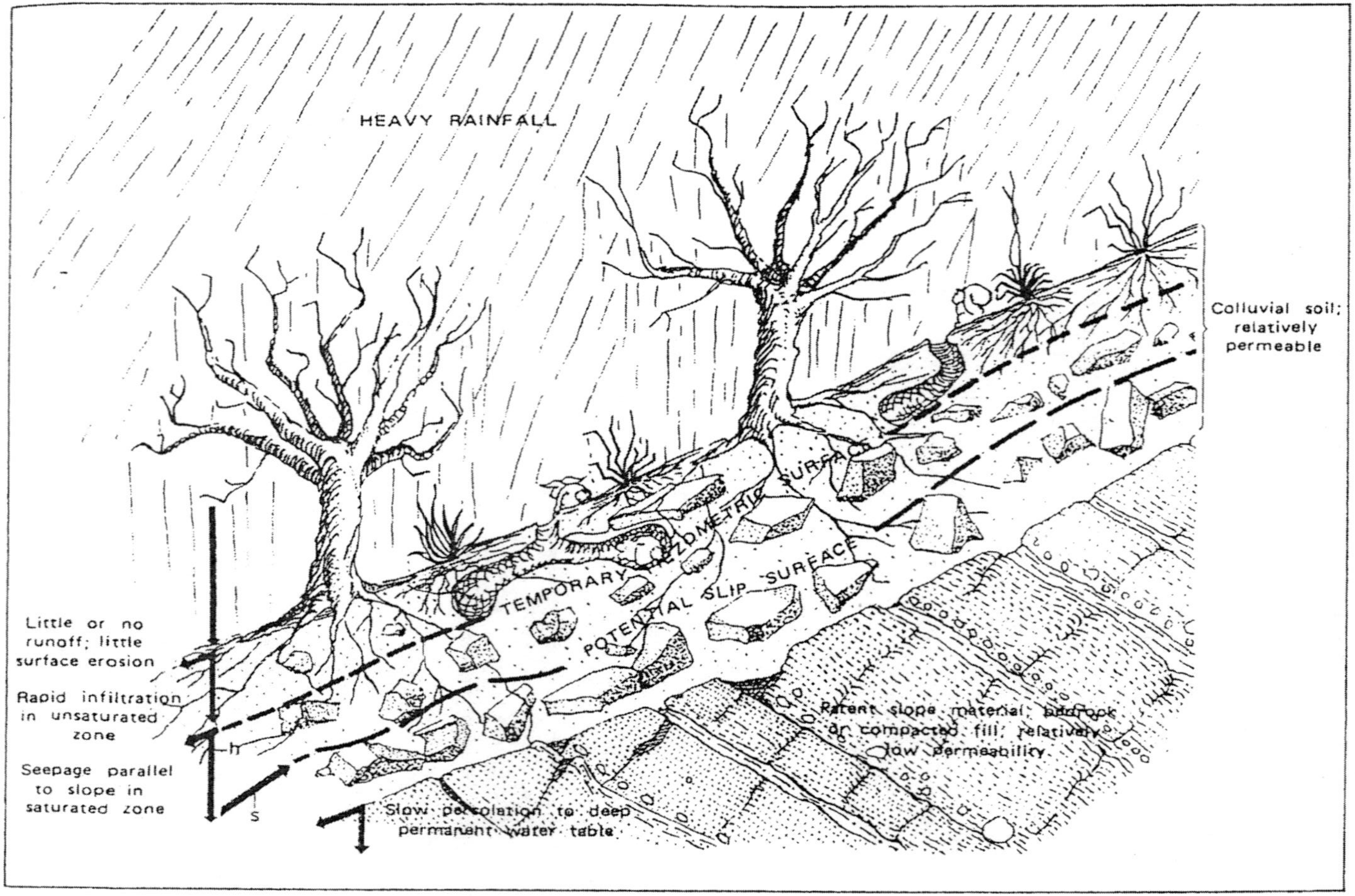

Figure 9.7 Buildup of perched water table in colluvium during heavy rainfall. (Campbell, 1975.)

requires removal of sediment from the ditches, which may result in undercutting of the toe. In addition, drainage ditches, especially the sodded type at the toe, may raise the water content of the toe materials, further exacerbating the condition.

Another serious form of erosion damage in loessial soils, particularly on flattened slopes, is surface erosion during construction and up to the time of establishment of a good vegetation cover. Even with adequate drainage, extensive damage may result from rainfall impact and rill erosion. This problem is mostly critical in silty loess, but clayey loess is also susceptible.

Frost damage is often found in near-vertical slopes in silty loess in Missouri (Missouri State Highway Department, 1978). Failure is often initiated by freeze-thaw action of the ice lenses and often results in slumping along the resulting surface of weakness. The resultant damage leaves soil exposed to erosion and further damage.

9.7.3 Debris Flows

Debris flows are typically shallow in nature. They tend to occur on slopes that are geologically young, steep, and naturally only marginally stable. It follows that the initial stress state is close to the stress state at failure and that a relatively small disturbance can initiate failure.

Debris flows are important geomorphic agents in a wide variety of landscapes. In humid, soil-mantled environments in particular, debris flows often originate in fine-scale valleys in steep, rhythmically dissected terrain (Figure 9.8). In such regions, concave platform contours define topographic swales, referred to as "hollows" in the nomenclature proposed by Hack and Goodlett (1960), that typically contain

Figure 9.8 Ridgecrest north of the entrance to San Francisco Bay in Marin County, California, showing a series of colluvium-filled swales. (Montgomery et al., 1991.)

TABLE 9.1 Characteristics of Debris Flows

- Typically originate in concave, colluvial-filled swales located on steep slopes.
- Topography conducive to debris flow mobilization requires long slopes greater than 27°. Steeper slopes favor acceleration of an initial failure, while gentle slopes inhibit acceleration.
- Induced by prolonged intense rainfall.
- Often happen very quickly.
- Occur in the upper portion of low-order (or first-order) drainages, for example, in valleys and natural ravines.
- Typically limited to granular materials with 8 to 25 percent clays.
- Approximate mobilization index (AMI) = $\omega_{sat}/LL > 1.0$ [Note: soils with $\omega_{sat}/LL < 0.45$ would not suffer debris flows (Ellen and Fleming, 1987).]

colluvial soils significantly thicker than those found on adjacent slopes. Studies in the San Francisco Bay Area indicate that in areas underlain by massive, resistant bedrock, the majority of debris flows originate in colluvium-filled swales (Ellen, 1988; Ellen and Wieczorek, 1988; Ellen et al., 1988; Smith and Hart, 1982). Swales consequently define a mappable debris flow hazard. These colluvium-filled swales should be identified routinely as potential hazards in geotechnical feasibility and design level investigations.

Debris flows generally occur during heavy storms, as pore water pressures build up within the saturated soils. Empirically, the location of many debris flow scars within colluvium-filled swales suggests that topographically driven convergence of runoff into swale axes is a primary factor contributing to debris flow generation. Table 9.1 shows the characteristics of debris flows.

Debris flows tend to occur more often in granular soils than in cohesive soils. Granular soils tend to collapse when sheared, which, under undrained conditions, results in an increase in pore water pressures (Ellen and Fleming, 1987; Johnson and Sitar, 1987). Consequently, surficial failures in contractive soils often evolve into debris flows that may travel great distances because even minor strain may cause liquefaction. Dilatant soils, on the other hand, expand upon shearing, and a continued influx of water is required to sustain their mobilization. Consequently, surficial failures in dilatant soils tend to be relatively slow moving earthflows. Contractive or dilative behavior of a soil can be predicted by the ratio of the saturated water content of undisturbed soil to its Atterberg liquid limit. This approximate mobility index is greater than unity for contractive soils and less than one for dilative soils (Ellen and Fleming, 1987). These indices provide a relatively simple technique for estimating the mobilization potential of a soil.

9.7.4 Residual Soils

Residual soil slopes are subject to continual weathering with time. The weathering profiles of residual soils are often erratic and have an immense influence on both shear strength and groundwater behavior. Residual soils derived from parental rocks often contain relict discontinuities, along which the shear strength may be much lower than the strength of the intact material. The presence of relict discontinuities as planes of weakness in the soil mass is often found to be the culprit for shallow failures. These relict discontinuities affect the stability of slopes in two aspects: failures partly or entirely bounded by adverse oriented relict discontinuities; and reduction in the mass strength of the soil.

Apart from their influence on the mass strength of residual soils, relict discontinuities with clayey infill will interfere with groundwater flow and result in perched water table conditions. Conversely, closely spaced discontinuities can provide preferential drainage path within a relatively less permeable soil mass. Lumb (1975) suggested that for decomposed volcanic soils, field permeability test results that are two orders of magnitude greater than laboratory results for small intact soil samples are due to the influence of jointed planes on the field permeability for these soils.

9.7.5 Rapid Snowmelt

Rapid snowmelt caused by a sudden warming spells or by rain falling on snow can also add water into hillside slopes. Horton (1938) found that snowmelt may provide a continuous cover of moisture over a longer period than that during and after rainfall. Mathewson et al. (1990) discovered that melting of snowpack may also recharge shallow fractured rock and raise pore pressures beneath shallow soils, thus triggering slope failures. In Utah, during an unusually warm 10-day period from late May to early June 1983, a heavy winter snowpack along the Wasatch Front began to melt rapidly and triggered approximately 150 debris flows and other types of landslides (Pack, 1984).

Rain-on-snow events commonly reduce the water content of the snowpack and add sufficient water into soils to trigger slope failures. Snowmelt before rainfall may sometimes augment the piezometric response (Sidle, 1984). In western Oregon, Harr (1981) documented about 85 percent of landslides that could be dated accurately were related to snowmelt during rainfall.

9.8 EFFECT OF PERMEABILITY OF SURFICIAL STABILITY

Before seepage parallel to the face of the slope can occur, the soil must reach saturation. The minimum requirements for achieving saturation are: (1) rainfall intense enough to exceed the infiltration rate of the material; and (2) rainfall lasting long enough to saturate the slope up to a certain depth and create positive pore pressure buildup. Many shallow failures occur when these two requirements are met.

For typical slopes, it is assumed that the material is originally unsaturated and the surface of the slope is continuously being wetted by rain and/or irrigation. The rate at which the unsaturated soil will wet up is critical to the surficial stability of slope. Cohesionless and cohesive would wet up at different rates given the same duration and intensity of rainfall. The process of saturation to a depth of 4 feet, for example, is typically long. It requires a gradual buildup of moisture in the ground over a prolonged period of time. Hence, short heavy rainfall after a prolonged dry period cannot create the conditions leading to saturation. However, a heavy rainfall after a period of moderate rainfalls may saturate the remaining upper few inches of unsaturated soil, and trigger a slough.

Depending on the precipitation intensities at local areas, some types of soil may be more prone to shallow failures. Studies by Pradel and Raad (1993) suggested that, for a return period of 50 years and a saturation depth 1.22 meters (4 feet), $k_{\text{lim}} \approx 10^{-4}$ centimeters per second is the governing threshold permeability of all the locations in Southern California. In other words, in many areas of Southern California, rainfalls with a return period of 50 years or less cannot saturate materials with a permeability greater than approximately 10^{-4} centimeters per second. Hence, clayey and silty soils would be more prone to develop conditions for surficial instability in Southern California. As for slopes made of soils classified as GW, SW, and SP, the chances are small that they would become fully saturated to a depth of 1.22 meters (4 feet) (Pradel and Raad, 1993).

It should noted that the value of k_{lim} suggested by Pradel and Raad (1993) will change if a different depth of saturation is considered and that the stabilizing effect due to plant roots is not accounted for. In particular, k_{lim} increases as the depth of saturation becomes more shallow. This would imply that surficial failures are more likely to occur along shallow planes. However, well-established vegetation can decrease the likelihood of a surficial failure in the upper 1 meter (3 feet).

9.9 STANDARD CODES FOR SHALLOW SLOPE STABILITY

There are few standard codes on shallow slope stability in practice. One well-known set of standard codes is available in the County of Los Angeles. According to the requirements of Los Angeles County's (1978). *Minimum Standards for Slope Stability Analysis*, pertinent aspects of these standards are as follows:

(1) That calculations shall be based on analysis procedures for stability of an infinite slope with seepage parallel to the slope surface or other failure mode, which would yield the minimum FOS against failure.

(2) That calculations of slopes be evaluated using the following equation:

$$\text{FOS} = \frac{c' + (\gamma_t - \gamma_w) z_w \cos^2\alpha \tan\phi'}{\gamma_t z_w \sin\alpha \cos\alpha} \qquad \text{(Eq. 9-1)}$$

where c' = the effective cohesion
ϕ' = the effective friction angle
α = the slope angle
z_w = the vertical depth of saturated soil
γ_t = the total unit weight
γ_w = the unit weight of water

(3) That calculations be performed assuming a vertical depth of saturated soil of at least 4 feet.

(4) That the FOS for surficial stability shall be 1.5.

(5) And that the shear strengths (cohesion and friction angle) determined for use in analysis of gross stability of the same slope shall be considered acceptable for surficial stability analysis without specific justification. The consultant must justify application of higher shear strength values.

Equation 9-1 above was developed by Skempton and Delory (1957), and it analyzes the slope stability of an infinite slope with seepage parallel to the face. This equation was intended for clay slopes, but local codes in Southern California demand that the surficial slopes be evaluated using Equation 9-1.

Considering Equation 9.1 above, if cohesion is neglected, it becomes

$$\text{FOS} = \left(1 - \frac{\gamma_w}{\gamma_T}\right)\frac{\tan\phi'}{\tan\alpha} \qquad \text{(Eq. 9-2)}$$

For cohesionless materials, the friction required by Equation 9.2 to ensure FOS = 1.0 and to satisfy local codes (FOS = 1.5) is listed in Table 9.2 for several slope anlges. In sands, the friction angle typically ranges from 28 to 40. Therefore, the application of the local codes to slopes made of cohesionless materials actually implies that the slopes are made flatter than 3.5 to 1 (horizontal to vertical). Furthermore, Equation 9.2 suggests that slopes steeper than 2.5H–1V would be unstable under heavy rainfalls. This contradicts observations that many 1.5H–1V and 2H–1V slopes made of cohesionless materials remain stable during and after prolonged intense rainfall.

Based on Equation 9.1, an addition of a small cohesion $c' = 100$ pounds per square foot would contribute an additional FOS of about 0.5 to a 2H–1V slope with total unit weight of 130 pounds per cubic foot. Thus cohesion plays a primary role in Equation 9-1, and the frictional angle a secondary role. Therefore, Equation 9-1 suggests that slopes made of cohesive materials, that is, clays and silts, will have a higher factor of safety, and thus will be less susceptible to surficial instability, than those composed of gravelly to sandy materials.

TABLE 9.2 Friction Angle, ϕ', Required by Equation 9-2 for FOS = 1.0 and FOS = 1.5

Slope Ratio (H–V)	ϕ' Required for FOS = 1.0	ϕ' Required for FOS = 1.5
1.5–1	52	62.5
2.0–1	44	55
2.5–1	38	49
3.0–1	33	44
3.5–1	29	39.5

Source: Pradel and Raad, 1993.

Note: Total unit weight of soil = 130 pounds per cubic foot.

Studies by others (Hollingsworth and Kovacs, 1981; Pradel and Raad, 1993) have indicated that clayey slopes in Southern California are more prone to shallow failures than sandy slopes. Hollingsworth and Kovacs (1981) studied approximately 100 sites in the Santa Monica Mountains where surficial failures occurred on natural slopes during heavy rainfalls of February 1980. Their study indicates that the potential for failure is highest for clayey and silty soils and lowest for gravelly to sandy soils. This is in contradiction to Equation 9-1, which suggests that slopes made of cohesive soils are in general less vulnerable to surficial failures.

When evaluating surficial stability, drained direct shear strength tests or triaxial consolidated undrained tests with pore pressure measurements are often conducted to determine the shear strength envelopes. The drained strength envelopes for soils tested at low confining pressures are typically nonlinear (Figure 9.9). The actual effective cohesion at low confining stresses is very small and approaches zero. For higher confining stresses, the drained strength envelopes become more linear with much higher cohesion than that at low confining pressures.

Evaluation procedures using the standards prescribed by the County of Los Angeles are frequently used to evaluate surficial stability. This usually results in an overestimation of the FOS for surficial stability. The effective cohesion intercept for gross stability is not appropriate for surficial stability since it is an extrapolated value that is usually greater than the actual cohesion at low normal stresses (Figure 9.9).

Many 1.5H–1V to 2H–1V slopes have performed well for years with only minor maintenance associated with occasional slump features. This anomaly between analysis and performance may be due to the following factors:

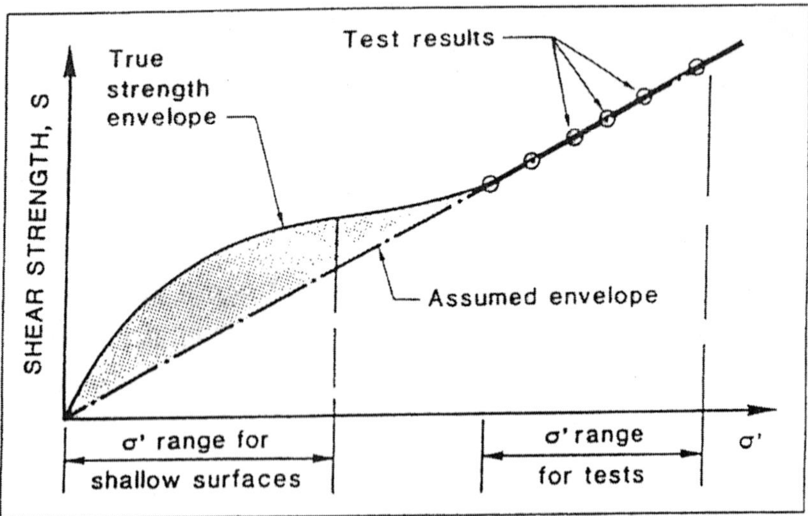

Figure 9.9 Underestimation of shear strengths because of incorrect strength envelope deduced from data obtained from high normal stresses. (Brand, 1985.)

TABLE 9.3 Drained Shear Strengths from Unsaturated and Saturated Tests

	Unsaturated		Saturated	
Location	ϕ'	c'(kPa)	ϕ'	c'(kPa)
Atlanta	30	14	32	0
Hong Kong	36	60	36	25

(1) The soil in the field may not become uniformly saturated as in the laboratory tests conventionally performed. The analysis assumes universal saturation, which may not be the case in the field.

(2) The assumed depth of saturation of 4 feet seems to be excessive for the 1.5H–1V slope. Steeper slopes producing higher runoffs and saturation to 4 feet are very unlikely unless the top soil is highly plastic and cracks to these depths upon drying in summer. Also, the duration of intense rainfall must be very long.

(3) Tests typically done in the conventional laboratory do not follow the actual field stress path that surficial soil would have experienced. Furthermore, the confining pressure should be compatible to the field condition that is typically very low for shallow stability. The effective cohesion and friction angle are sensitive to changes in the very small confining stresses acting in the surficial soil zone as reflected by the nonlinear Mohr–Coloumb envelope in low-stresses. For soils at low effective normal pressures, the shear strength envelope is curved and passes through or near the origin. Using the extrapolated value of the effective cohesion from testing at high confining stress results in an overestimation of c' and a subsequent overestimation of the surficial factor of safety.

(4) For unsaturated clayey to partly clayey soils, negative pore pressure (or soil suction) is present in the soil mass, enhancing the overall slope stability. This negative pore pressure is lost upon full saturation by rainwater or irrigation. An empirical formula, such as that proposed by Skempton and Delory in 1957, which accounts for presence of cohesion, does not truly reflect the actual site condition at the time of slope failure. Table 9.3 shows that higher drained shear strengths can be obtained from unsaturated tests than from saturated tests.

(5) Lack of ability to recover undisturbed samples using conventional drilling methods in weakly cemented sand and uncemented dense gravel may result in lower strengths of these materials. Higher strengths are reportedly obtained through testing of undisturbed block samples in natural gravel deposits (Kort et al., 1979; Sitar, 1990) or reconstituted samples in desert alluvium (Hamel, 1975), as shown in Table 9.4.

9.10 DESIGN PRACTICE FOR SHALLOW SLOPE STABILITY

For years, predicting shallow failures has been a real challenge to practicing engineers. This is because each shallow failure is so unique that it may not be possible to

TABLE 9.4 Strength Characteristics of Natural Gravel Deposits

Sample Type	Type of Test	Density (kN/m^3)	Cohesion (kPa)	ϕ_{peak} (Degrees)
Undisturbed (Kort et al., 1979)	In situ vacuum triaxial	22.5	230	53
Undisturbed (Sitar, 1990)	In situ direct shear	23	100	60
Reconstituted (Hamel, 1975)	Laboratory direct shear	19–20.7	7–165	40–48

generalize the overall failure mechanism. In addition, hydrogeologic features and geologic structures are often unknown prior to failures. These features and structures are often so confined and thin that they cannot be easily recovered using conventional drilling methods.

Brand (1985) evaluated the current ability to predict how a residual slope behaves. He reported that it is possible to predict failure modes based upon observed failures in similar slopes and to estimate shear strength fairly well, but it is not possible to effectively estimate pore pressures at the point of incipient failure. In the light of poor to fair ability to predict slope performance, design practices often depend upon empirical and semiempirical methods. For example, cut and fill slopes in the United States are typically made at 2H–1V or 1.5H–1V. In design of slopes, several approaches have been employed in practice:

- Correlation of slope failures with rainfall pattern
- Terrain evaluation using geomorphological mapping
- Semiempirical methods developed from evaluation of past failures
- Soil mechanics analysis

When evaluating remedial measures for stabilizing slope failures and analyzing cut slope stability near critical structures, soil mechanics analyses are often used. In particular, consideration should be given to alleviating buildup of pore pressures within the soil slopes by application of drainage systems such as horizontal drains, cut-off drains, drainage caissons, and so on. Use of drainage system is essential in mitigating rainfall-induced slope failures. For land use planning and for designing cut and fill slopes along highways, the other three methods listed above are generally utilized.

9.10.1 Debris Flow Hazard Mitigation

In the previous section, we learned that debris flows originating in colluvial deposits within hillside swales cause significant property damage and loss of life in steep, soil-mantled environments. Mitigation measures should be devised to address the debris flow hazards. There are several ways in which the debris flow hazard associated with swales may be mitigated effectively. They are (Montgomery et al., 1991):

- Avoiding locations at risk for debris flow impact,
- Eliminating the causes of initial slope failure, or
- Containing or minimizing the damage caused by a debris flow.

Avoiding locations at risk for debris flow impact is the easiest way to mitigate the debris flows. Building and grading ordinances should address development within and immediately downslope of colluvium-filled swales, along and adjacent to low-order streams, and on debris fans in and proximal to hillside areas. Unfortunately, land-use decisions often are determined by local politics and economics and the real or perceived higher importance of other competing factors, rather than on geologic rationale.

Other approaches to eliminating the causes of initial slope failure include physical removal of colluvium potentially subject to failure and installation of drainage improvements to help stabilize the colluvial deposit.

Other alternatives to reduce the impact of debris flow hazards consist of installation of retaining walls, or other mechanical impediments such as debris basins, diversionary structures, or levees to direct debris flows into predetermined depositional areas or construction of debris fences intended to absorb kinetic energy. All these mitigation measures are intended to protect downslope property.

9.10.2 Design of Loess Slopes

As discussed earlier, the most significant influences on stability of loess slopes are surface drainage, soil water content, and grain-size distribution. For design of the loess slopes, it is essential to understand the following soil properties and to avoid the resulting problems (Higgins and Modeer, 1996):

(1) Loess is highly erodible, and water flow, even at a low to moderate volume and at 5 to 10 percent gradient, can cause severe erosion.

(2) Disturbance of the natural soil structure by construction/farming activities could make the soil more susceptible to erosion.

(3) Saturation of the soil often softens the clay binder and reduces shear strength, leading to slope failure or accelerated erosion.

(4) Silty loess soils are highly susceptible to failure by piping.

With these problems in mind, designers often need to consider ways of preventing water from flowing over the face of loess soil. A good drainage system should be devised to collect water runoff. However, it is necessary that water not be allowed to collect against the toe of the cut in silty or clayey loess. Saturation of the soils in the toe can cause sloughing of the slope. In addition, it is essential not to direct flow into unprotected channels, but line them with concrete or cement mortar.

According to Higgins and Modeer (1996), silty loess soils should perform well in near vertical cuts ($\frac{1}{4}$H–1V) if they are protected by surface drainage structures and the natural moisture contents remain relatively stable (<17 percent can be used as an approximation on the basis of experience in the Midwest). Cuts in clayey loess should perform well if they are not steeper than 2.5H–1V and are protected by a cover of vegetation, and if concentrated runoff flow does not exist on the slope face.

Due to the adverse effect of water on the stability of loess, any cuts in loessial soils should be carried out during the dry season. The drainage structure above the crest of the cut slope should be constructed prior to construction of the cut slope and with as little disturbance to the surrounding vegetation as possible. Once the cut slope is in place, construction equipment should be kept away from the crest.

9.11 ATTENDING LANDSLIDE INCIDENTS

Geologists, engineering geologists, and geotechnical engineers are sometimes required to attend landslide incidents. There are some basic rules that practicing professionals should adhere to while attending landslide incidents. They are as follows:

(1) *Be Prepared* Always be prepared when being called out to attend a landslide. *Always* remember to bring your camera or video camera. A picture is worth a thousand words.

(2) *Don't Panic* Don't panic while attending a landslide. Always stay alert and look for clues and evidence for the cause of the landslide.

(3) *Don't talk to the press* As professionals, we are accountable to our clients for what we find out in the field, but not to the press.

(4) *Don't Talk to the Public* This is to avert any confusion and nervousness that may have resulted from the public who may be directly impacted by the landslide.

Typical Remedial Works Comprise the Following:

1. Cordon off the slide area for public safety.
2. If the slide is significant, consider evacuating people from their homes.
3. Divert surface runoff from the failing slope.
4. Place an impermeable layer, such as plastic sheeting, over the slope.

Typical Permanent Remedial Works May Include:

1. Removing slide debris and placing gravel berm on the slope face.
2. Flattening the slope with a layer of gravel, cobble, or rip-rap.
3. Diverting water from the slope by new drainage systems.
4. Carrying out in-depth studies and designing permanent remedial works using soil mechanics analysis for major failures.

9.12 SUMMARY

Shallow slope failures occur, in general, during and after intense rainfall and/or snowmelt during and after spring. The rapid infiltration of rainfall and rapid melting of snowpack cause soil saturation and a temporary rise in pore-water pressures, thereby reducing shear strength of the soil slope, which subsequently leads to failures. Due to difficulties in determining the maximum pore pressures at the time of incipient failures and the inherent heterogeneity of the soil/rock masses, the variance of geohydrology and the possibility of uncontrolled human activities affecting slopes, prediction of shallow failures in soil and rock may still be a matter of judgement guided by experience.

As illustrated in this chapter, the primary factor triggering shallow failures is water infiltrating into soil/rock slopes, causing a sudden buildup of pore water pressure, and a subsequent reduction in shear strength. Thus preventing water from entering the slopes is an essential part and often the first step to be undertaken in mitigating slope failures. This may come in the form of provisions of adequate drainage systems (e.g., drainage channels, ditches, catchpits, horizontal drains, cut-off trenches, or French drains).

There are inherent uncertainties relating to slope behavior. All slopes undergo a continuous process of deterioration on slope-forming materials, drains, and surface protections, and so on. Accurate prediction of occurrence of shallow failures is still a long way from materializing. As Lumb (1975) pointed out, there can never be an absolute unconditional guarantee of slope safety and the probability of slope failure will always be nonzero.

REFERENCES

Anderson, S. C., and N. Sitar, 1991. "Influence of Stress Path on Soil Strength Parameters and Analysis of Rainfall Induced Slope Failures," Class Notes, University of California, Berkeley, California.

Anderson, S. C., and N. Sitar, 1995. "Analysis of Rainfall-Induced Debris Flows," *Journal of Geotechnical Engineering*, Vol. 121, No. 7, pp. 544–552.

Brand, E. W., 1981. "Some Thoughts on Rainfall Induced Slope Failures." *Proceedings 10th International Conference on Soil Mechanism and Foundation Engineering*, Brookfield, Vermons, A. A. Balkeme, pp. 373–376.

Brand, E. W., 1982. "Analysis and Design in Residual Soils," *Proceedings of theASCE Geotechnical Engineering Division Specialty Conference—Engineering and Construction in Tropical and Residual Soils*, Honolulu, Hawaii, pp. 89–141, January 11–15.

Brand, E. W., J. Premchitt, and H. B. Phillipson, 1984. "Relationship Between Rainfall and Landslides in Hong Kong," *Proceedings 4th International Symposium on Landslides*, Toronto, Canadian Geotechnical Society, Vol. 1, pp. 377–384.

Brand, E. W., 1985. "Geotechnical Engineering in Tropical Residual Soils" (Special Lecture), *Proceedings 1st International Conference on Geomechanics in Tropical Lateritic and Saprolitic Soils*, Brasilia, Vol. 3, pp. 23–91.

Brenner, R. P., H. K. Tam, and E. W. Brand, 1985. "Field Stress Path Simulation of Rain-Induced Slope Failure," *Proceedings of the 11th International Conference on Soil Mechanics and Foundation Engineering*, San Francisco, Rotterdam: A. A. Balkema, Vol. 2, pp. 991–996, August 12–16.

Campbell, R. H., 1975. "Soil Slips, Debris Flows, and Rainstorms in the Santa Monica Mountains and Vicinity, Southern California," USGS Professional Paper 851, U.S. Geological Survey, Washington, DC.

D'Appolonia, E., R. Alperstein, and D. J. D'Appolonia, 1967. "Behavior of a Colluvial Slope," *Soil Mechanics and Foundation Engineering Division*, ASCE, Vol. 93, No. SM4, July, pp. 489–518.

Day, R. W., 1994, "Surficial Stability of Compacted Clay: Case Study," *Journal of Geotechnical Engineering*, ASCE, Vol. 120, No. 11, pp. 1980–1990.

Day, R. W., and G. W. Axten, 1988. "Surficial Stability of Compacted Clay Slopes," *Journal of Geotechnical Engineering*, ASCE, Vol. 115, No. 4, pp. 577–580.

Day, R. W., and D. M. Poland, 1995. "Sea Cliff Erosion Caused by a Storm Drain Failure," *Environmental & Engineering Geoscience*, Vol. 1, No. 4, Winter, pp. 524–529.

Ellen, S. D., and R. W. Fleming, 1987. "Mobilization of Debris Flows from Soil Slips, San Francisco Bay Region, California," *Debris Flows/Avalanches: Process, Recognition, and Mitigation; Geological Society of America Reviews in Engineering Geology*, J. E. Costa and G. F. Wieczorek, Eds. & Vol. 7, pp. 31–40.

Ellen, S. D., 1988a. "Description and Mechanics of Soil Slip/Debris Flows in the Storm," *Landslide Floods, and Marine Effect of the Storm of January 3–5, 1982, in the San Francisco Bay Region, California*, S. D. Ellen and G. F. Wieczorek, Eds. U.S. Geological Survey Professional Paper 1434, pp. 63–112.

Ellen, S. D., and G. F. Wieczorek, 1988. "Landslides, Floods, and Marine Effects of the Storm of January 3–5, 1982, in the San Francisco Bay Region, California," USGS Professional Paper 1434, U.S. Geological Survey, Washington, DC, 310 pp.

Ellen, S. D., S. H. Cannon, and S. L. Renean, 1988. "Distribution of Debris Flows in Marin County," *Landslide Floods, and Marine Effect of the Storm of January 3–5, 1982, in the San Francisco Bay Region, California*, S. D. Ellen and G. F. Wieczorek, Eds. U.S. Geological Survey Professional Paper 1434, pp. 113–132.

Hack, J. T., and J. C. Goodlett, 1960. "Geomorphology and Forest Ecology of a Mountain Region in the Central Appalachians," U.S. Geological Survey Professional Paper 347, 66 pp.

Hamel, J. V., 1975. "Large Scale Laboratory Direct Shear Tests on Desert Alluvium," *Applications of Rock Mechanics, Proceedings of the 15th Symposium on Rock Mechanics*, E. R. Hoskins, Jr. Ed. ASCE, pp. 385–414.

Harr, R. D., 1981. "Some Characteristics and Consequences of Snowmelt during Rainfall in Western Oregon," *Journal of Hydrology*, Vol. 53, pp. 277–304.

Higgins, J. D., and V. A. Modeer, 1996. "Landslides Investigation and Mitigation, Chapter 23, Loess," Transportation Research Board, Special Report 247, pp. 585–606.

Hollingsworth, R., and G. S. Kovacs (1981). "Soil slumps, and Debris Flows, Prediction and Protection," *Bulletin of Association of Engineering Geology*, Vol. 18, No. 1, pp. 17–28.

Horton, R. E., 1938. "Phenomena of the Contact Zone Between the Ground Surface and a Layer of Melting Snow," Association Internationale d'Hydrologie Scientifique, Paris, Bulletin No. 23, pp. 545–561.

Johnson, K. A., and N. Sitar, 1987. "Debris Flow Initiation: An Investigation of Mechanisms," University of California, Department of Civil Engineering Report No. UCB/GT/87–02, University of California, Berkeley, California, 179 pp.

Johnson, K. A., and N. Sitar, 1990. "Hydrologic Conditions Leading to Debris Flow Initiation," *Canadian Geotechnical Journal*, Vol. 27, pp. 789–801.

Jones, F. O., 1973. "Landslides of Rio de Janeiro and the Serras das Araras Escarpment Brazil," USGS Professional Paper 697, U.S. Geological Survey, Washington, DC, 42 pp.

Kesseli, J. E., 1943. "Disintegrating Soil Slips of the Coastal Ranges of Central California," *Journal of Geology*, Vol. 51, No. 5, pp. 342–352.

Kort, I. K., H. H. Musante, and C. Fahrenkrog, 1979. "In Situ Mechanical Properties Measurements of Gravelly Soil Used in an Interaction and Foundation Model for the Santiago Metro," *Proceedings of the 6th Panamerican Conference on Soil Mechanics and Foundation Engineering*, Lima, Peru, Vol. II, pp. 217–224.

Krahn, J., D. G. Fredlund, and M. J. Klassen, 1987. "Effect of Soil Suction on Slope Stability at Notch Hill," *40th Canadian Geotechnical Conference*, Regina, Saskatchewan, Canada, pp. 269–278.

Lambe, P. C., 1996. "Landslides Investigation and Mitigation, Chapter 19, Residual Soils," Transportation Research Board, Special Report 247, pp. 585–606.

Lim, T. T., H. Rahardjo, M. F. Chang, and D. G. Fredlund, 1996. "Effect of Rainfall on Matric Suctions in a Residual Soil Slope," *Canadian Geotechnical Journal*, Vol. 33, pp. 618–628.

Los Angeles County, 1978. *Minimum Standards for Slope Stability Analysis*, Building and Safety Division.

Lumb, P., 1975. "Slope Failures in Hong Kong," *Quarterly Journal of Engineering Geology*, Vol. 8, pp. 31–65.

Mathewson, C. C., J. R. Keaton, and P. M. Santi, 1990. "Role of Bedrock Groundwater in the Initiation of Debris Flows and Sustained Post-Storm Stream Discharge," *Bulletin of the Association of Engineering Geologists*, Vol. 27, No. 1, pp. 73–78.

Missouri State Highway Department, 1978. "Development of Design Criteria for Cut Slopes in Loess, Jefferson City," 78 pp.

Montgomery, D. R., R. H. Wright, and T. Booth, 1991. "Debris Flow Hazard Mitigation for Colluvium-Filled Swales," *Bulletin of the Association of Engineering Geologists*, Vol. XXVIII, No. 3, pp. 303–323.

O'Tousa, J., 1995. "La Conchita Landslide, Ventura County, California," *AEG News*, Fall, pp. 22–24.

Pack, R. T., 1984. "Debris Flow Initiation in Davis County, Utah, During the Spring Snowmelt Period of 1983," *Proceedings of the 21st Engineering Geology and Soils Engineering Symposium*, Moscow, Idaho, pp. 59–77.

Pierson, T. C., 1977. "Factors Controlling Debris Flow Initiation on Forested Hillslopes in the Oregon Coast Ranges," Unpublished PhD Dissertation, Department of Geological Sciences, University of Washington, Seattle, Washington, 166 pp.

Pradel, D., and G. Raad, 1993. "Effect of Permeability on Surficial Stability of Homogeneous Slopes," *Journal of Geotechnical Engineering*, ASCE, Vol. 119, No. 2, pp. 315–332.

Pun, W. K., 1997. "Landslide Warning System in Hong Kong," *30th Session of the Typhoon Committee*, Hong Kong.

Reid, M. E., 1997. "Slope Instability Caused by Small Variations in Hydraulic Conductivity," *Journal of Geotechnical and Geoenvironmental Engineering*, August, pp. 717–725.

Reid, M. E., H. P. Nielsen, and S. J. Dreiss, 1988. "Hydrologic Factors Triggering a Shallow Hillslope Failure," *Bulletin of the Association of Engineering Geologists*, Vol. XXV, No. 3, pp. 349–361.

Sidle, R. C., 1984. "Shallow Groundwater Fluctuations in Unstable Hillslopes of Coastal Alaska." *Zeitschrift fur Gletscherkunde und Glazialgeologie*, Vol. 20, pp. 79–95.

Sitar, N., 1981. "Slope Stability in Coarse Sediments, *Geological Environment & Soil Properites*. ASCE, pp. 82–98.

Sitar, N., 1990. "Seismic Response of Steep Slopes in Weakly Cemented Sands and Gravels," *Harry Bolton Seed Memorial Symposium Proceedings*, J. M. Duncan, Ed., May, pp. 67–82.

Skempton, A. W., and F. A. Delory, 1957. "Stability of Natural Slopes in London Clay," *Proceedings 4th International Conference on Soil Mechanics and Foundation Engineering*, London, England, Butterworths, Vol. 2, pp. 378–381.

Slosson, J. E., 1969. "The Role of Engineering Geology in Urban Planning," *The Governor's Conference on Environmental Geology*, Colorado Geological Survey, Denver, Colorado, pp. 8–15.

Slosson, J. E., and J. P. Krohn, 1977. "Effective Building Codes," *California Geology*, Vol. 30, No. 6, pp. 136–139.

Slosson, J. E., and J. P. Krohn, 1982. "Southern California Landslides of 1978 and 1980," *Storms, Floods, and Debris Flows in Southern California and Arizona 1978 and 1980, Proceedings of a Symposium*, September 17–18. Washington, DC: National Academy Press, pp. 291–319.

Slosson, J. E., and R. A. Larson, 1995. "Slope Failures in Southern California: Rainfall Threshold, Prediction, and Human Causes," *Environmental & Engineering Geoscience*, Vol. I, No. 4, Winter, pp. 393–401.

Smith, T. C., and E. W. Hart, 1982. "Landslides and Related Storm Damage, January 1982, San Francisco Bay Region," *California Geology*, Vol. 36, No. 7, pp. 139–152.

Sowers, G. F., 1993. "Human Factors in Civil and Geotechnical Engineering Failures," *Journal of Geotechnical Engineering*, ASCE, Vol. 119, No. 2, pp. 238–256.

Turnbull, W. J., 1965. "Construction Prolems Experienced with Loss Soils," *Highway Research Record 212*, HRB, National Research Council, Washington, DC, pp. 10–27.

Turner, A. K., 1996. "Investigation and Mitigation, Chapter 20, Colluvium and Talus," Transportation Research Board, Special Report 247, pp. 525–554.

Wieczorek, G. F., 1987. "Effect of Rainfall Intensity and Duration on Debris Flows in Central Santa Cruz Mountains, California," *Debris Flows/Avalanches: Process, Recognition and Mitigation: Geological Society of America Reviews in Engineering Geology*, J. E. Costa and G. F. Wieczorek, Eds. Vol. 7, pp. 93–104.

Wilson, R. C., R. K. Mark, and G. Barbato, 1993. "Operation of a Real-Time Warning System for Debris Flows in the San Francisco Bay Area, California," *American Society of Civil Engineers, Proceedings Volumes*.

Wilson, R. C., and G. F. Wieczorek, 1995. "Rainfall Thresholds for the Initiation of Debris Flows at La Honda, California," *Environmental & Engineering Geoscience*, Vol. 1, No. 1, Spring, pp. 11–27.

CHAPTER 10

STABILITY OF LANDFILL SLOPES

10.1 UNIQUE NATURE OF LANDFILLS

Landfills pose unique slope stability issues chiefly because construction involves different combinations of cuts and fills in a variety of materials including soil, bedrock, water, landfill waste, and geosynthetics. There are three critical periods in the life of the landfill when slope stability is considered:

- Siting and preparation to receive waste
- Waste placement
- Capping and final closure

Initially, when the landfill is being sited and prepared to receive waste, the stability of the host site materials must be analyzed for cut and fill behavior depending on whether the landfill is being placed in a pit, in trenches, against a hillside, in a canyon, or above ground. As the waste is placed in the landfill, temporary fill slopes are constantly being created as different sections of the landfill are filled and then buttressed by additional filling. Finally, when the landfill is full, a cap is placed above the waste and the landfill is closed. Each of these phases presents unique combinations of materials in contact with each other, unique combinations of material and strength properties, and thus, unique slope stability issues.

10.2 TYPICAL LANDFILL CONFIGURATIONS

Five typical landfill configurations are shown in Figure 10.1. The most common landfill configuration in areas of deeper water table is probably the covered pit design

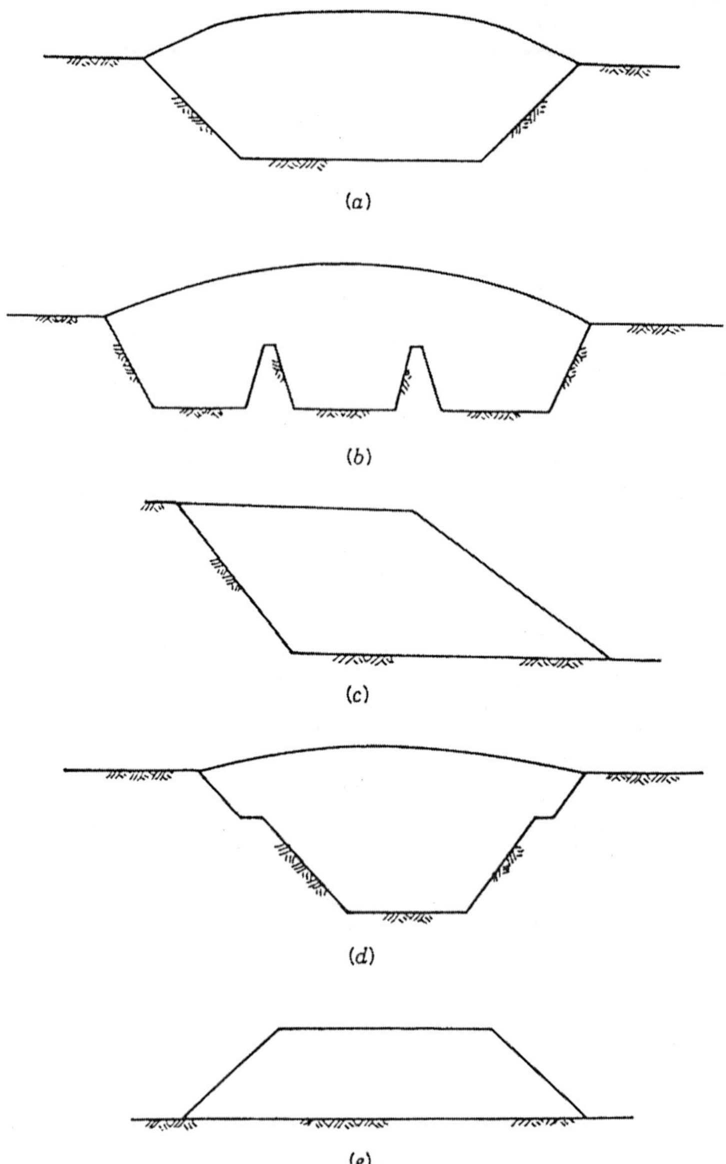

Figure 10.1 Typical landfill configurations. (Sharma and Lewis, 1994.)

shown in Figure 10.1*a*. The advantage of this configuration is that soils excavated from the pit may be used for daily cover. Depending on their nature, these excavated soils may also be suitable for clay liners or granular drainage material. The trenched landfill shown in Figure 10.1*b* has advantages similar to those of the covered pit. It is, however, typical of older landfilling practices, prior to the emphasis in landfill design on maximizing airspace. As is evident in Figure 10.1*b*, significant airspace is lost to the ridges of native soil remaining between cells.

Figure 10.1*c* depicts an upslope landfill. These types of landfills are generally constructed in regions of rolling or hilly terrains. The waste is generally placed on a liner constructed on an excavated natural slope. The canyon or valley landfills (Figure 10.1*d*) are so named because they are typically constructed in natural canyons or valleys. They can be similar to covered pits if the canyon is bowl-shaped: however the boundary side slopes are generally much higher and often steeper. For both upslope and canyon landfills, soil requirements are often satisfied by excavating the natural side slopes, selecting an on-site borrow area, or importing soil. Finally, as shown on Figure 10.1*e*, above-ground landfills are constructed entirely above the natural ground surface. These landfills typically exist in soft subgrade regions, where groundwater is near the ground surface, or where it would be impractical to excavate a pit. Operational soil requirements for these landfills are satisfied by on-site borrow areas or off-site soil import.

A landfill site must meet several locational and geotechnical design criteria and be acceptable to the public. It must be in reasonable proximity to waste generators. The type of site, whether hilly, flat, or whatever, will dictate the configurations that are possible at each site. Data used in the selection of a site often include:

- Topographic maps
- Soil survey maps
- Land use plans
- Transportation maps
- Water use plans
- Flood plain maps
- Geologic maps
- Aerial photographs

Other siting considerations include waste type, waste volume, landfill volume, availability of landfill equipment, recycling and incineration options, existing landfill sites, and funding (Bagchi, 1994). No landfills should be constructed near lakes, ponds, rivers, flood plains, highways, public parks, critical habitat areas, wetlands, airports, or water wells. In fact, a landfill will be an integral part of the landscape and environment (Figure 10.2) and must be evaluated in that context. In analyzing the feasibility of a landfill, the entire life cycle costs of the landfill must be evaluated, including the items shown in Table 10.1.

The landfill waste materials can be placed in a variety of ways, as shown in Figure 10.3. All wastes received by the landfill are spread and compacted in cells and layers within a confined area. At the end of each working day, or more frequently if needed, the area is covered completely with a thin, continuous layer of daily cover material that is then compacted. Daily cover is used to control moisture, control litter, reduce odors, limit rodent and bird contact, provide vehicle access, help prevent fires,

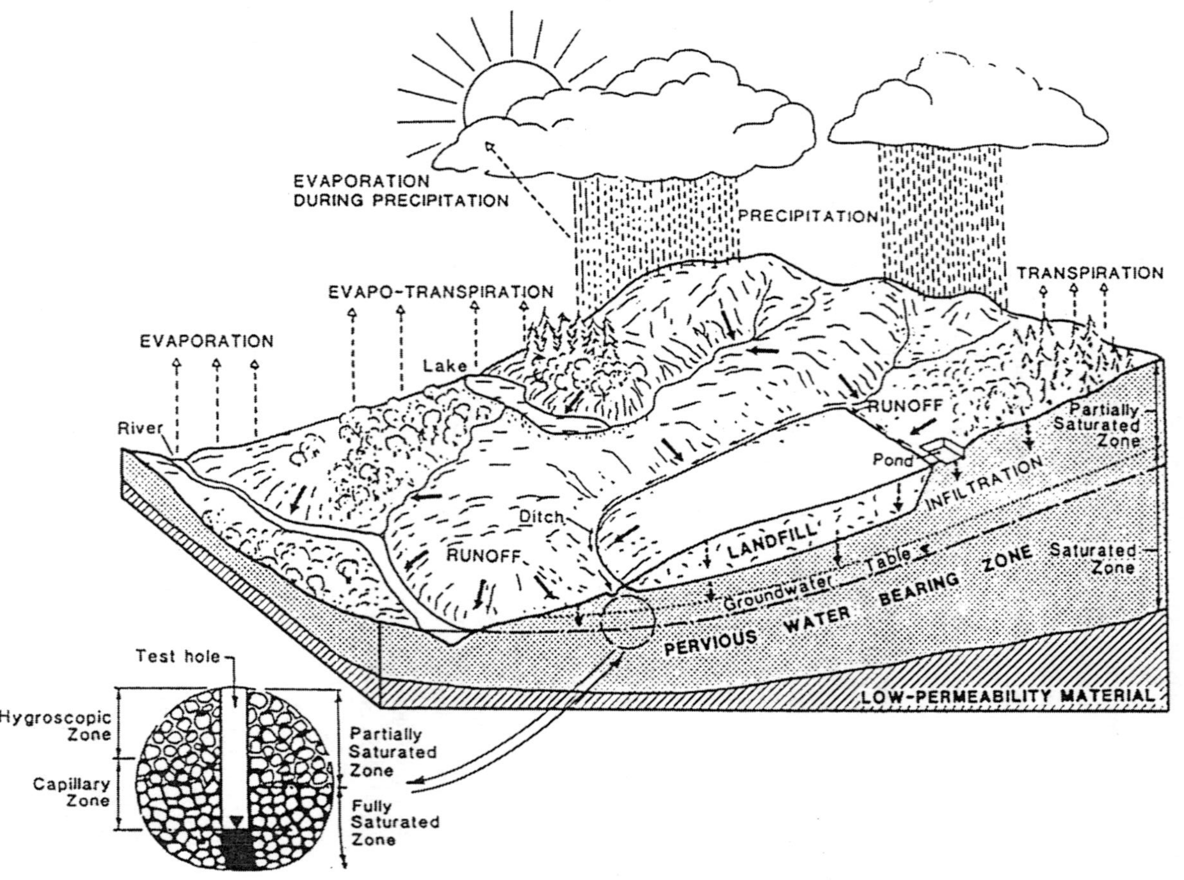

Figure 10.2 Waste storage within the hydrologic cycle. (Sharma and Lewis, 1994.)

TABLE 10.1 Components of Landfill Cost

Predevelopment Costs
Site characterization
Environmental assessment
Engineering design
Hydrogeologic investigation
Professional service fees—design/approvals
Legal consultation
Construction Costs
Land clearing
Excavation
Liner and leachate collection system installation
Leachate management—pumping station and/or treatment systems
Surface water control and final cover construction
Gas management system
Groundwater monitoring systems
Site structures
Operations
Equipment and personnel
Leachate and landfill gas management
Environmental monitoring costs
Community relations
Impact management—dust, odors, and birds
Closure costs
Cap/final cover
Seeding
Runoff control
Long-Term Care Costs
End use plan costs, including trees and shrubs
Site inspections
Land service care
Leachate and gas management
Environmental monitoring
Insurance

Source: McBean et al., 1995.

and improve the appearance of the landfill. The compacted wastes and daily cover material constitute a cell. A series of adjoining cells, all of the same height, make up a lift. The completed landfill consists of one or more lifts, as depicted in Figure 10.4. The two basic methods of operation for landfilling, or sequencing, of the daily cells are the trench method and the area method (Figures 10.5 and 10.6).

After the landfill is completely filled, a cover must be constructed to minimize water infiltration and isolate the waste from the environment, as shown in Figure 10.7. The purposes of the final soil cover over a landfill are to encourage surface water runoff, discourage erosion, retain moisture for vegetation, manage gas migration, provide shaping and contouring, and provide a base for the establishment of a suitable ground cover. Six typical layers of a final landfill cover are shown in Figure 10.8.

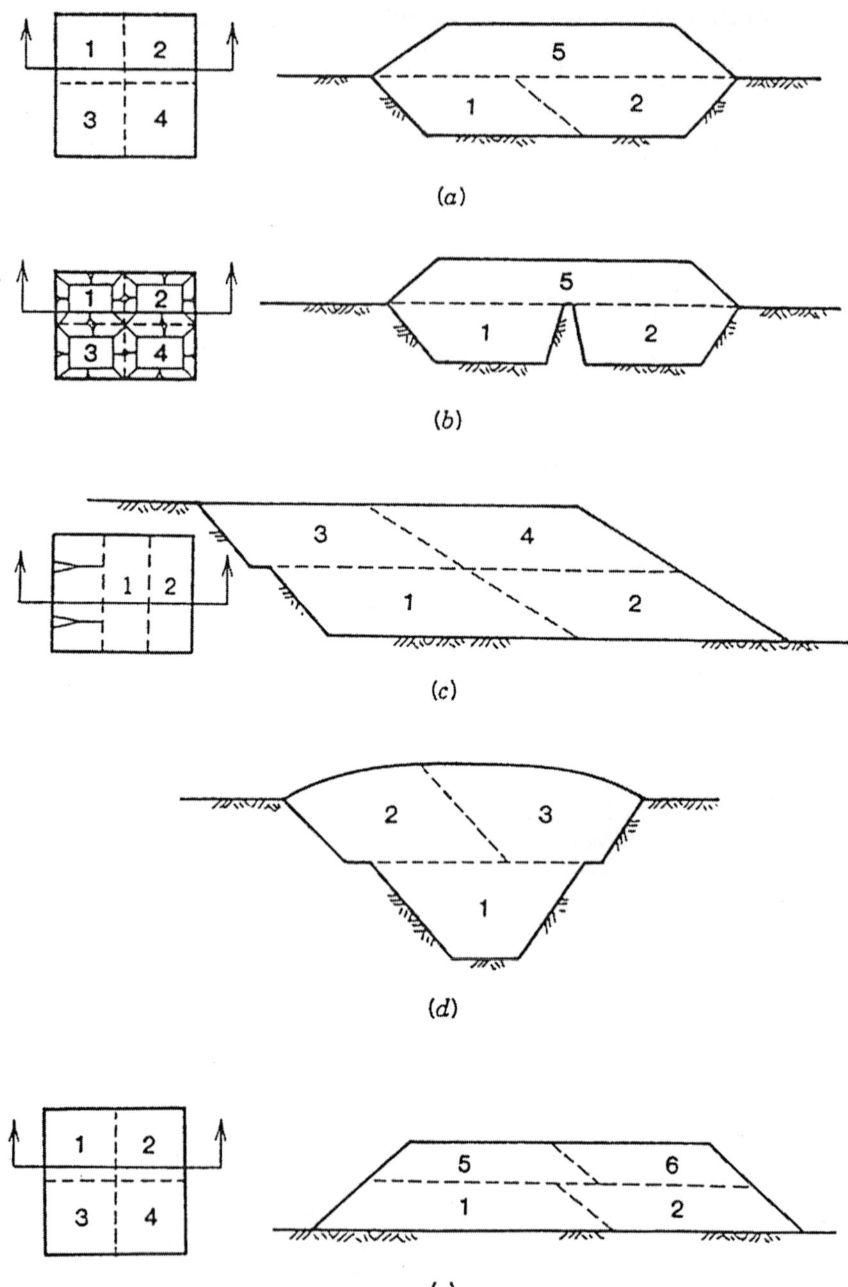

Figure 10.3 Landfill waste placement phases. (Sharma and Lewis, 1994.)

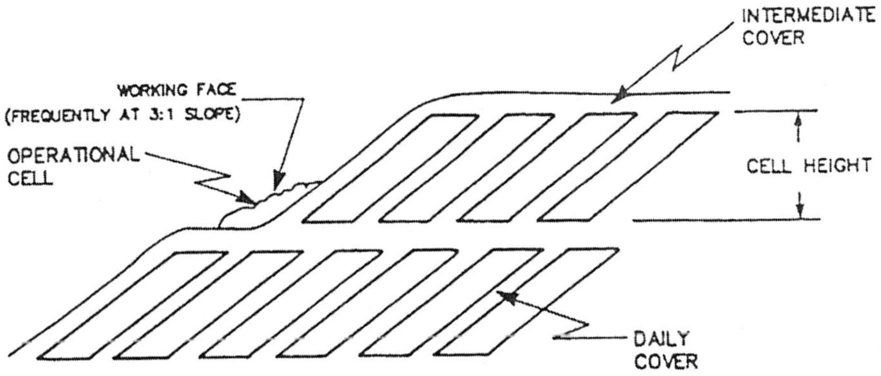

Figure 10.4 Configuration of landfill waste lifts and cells. (McBean et al., 1995.)

10.3 LANDFILL WASTE ENGINEERING PROPERTIES

The engineering properties of landfill waste are quite variable and unpredictable. Landfill waste may contain a multitude of materials including food, rubber, plastic, leather, textiles, wood, furniture, paper, metal, glass, machinery, transportation

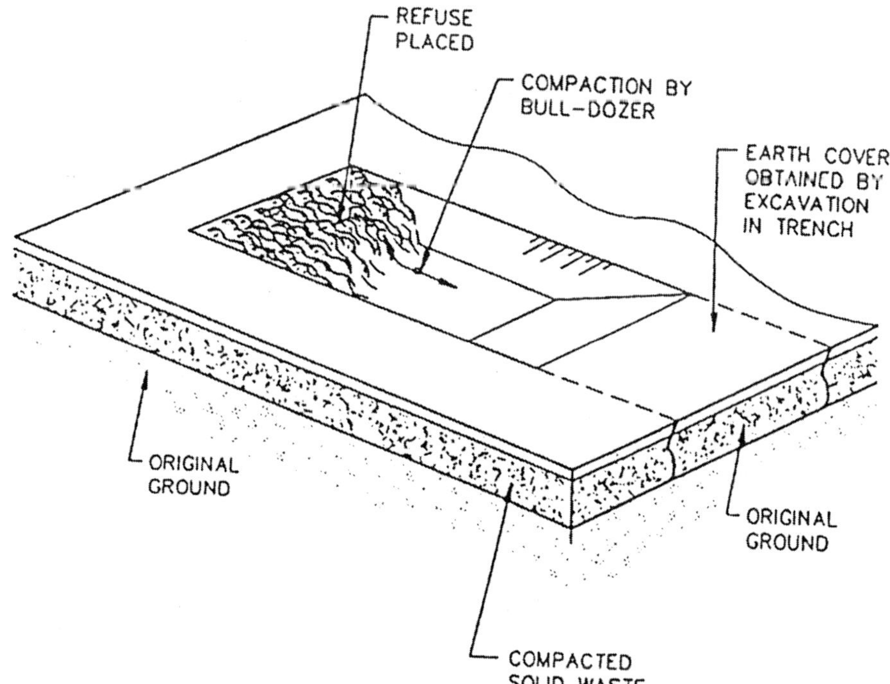

Figure 10.5 Trench method of landfilling. (McBean et al., 1995.)

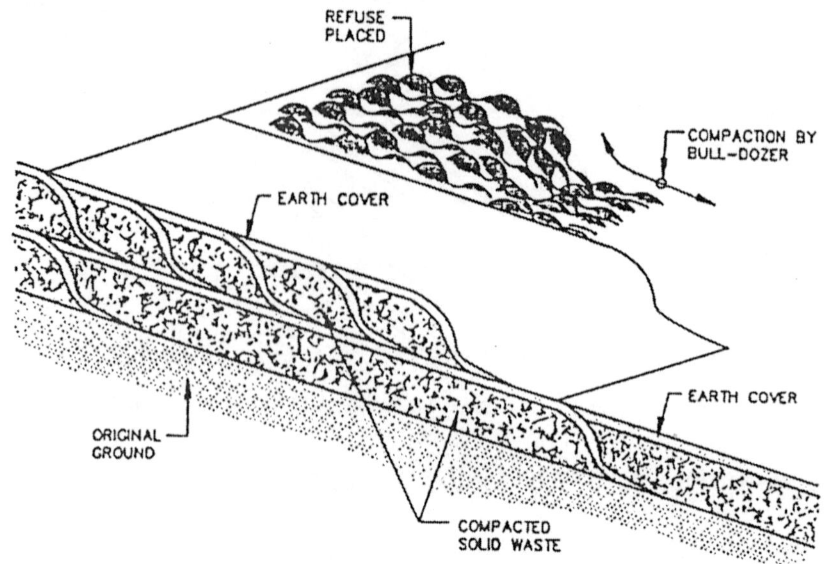

Figure 10.6 Area method of landfilling. (McBean et al., 1995.)

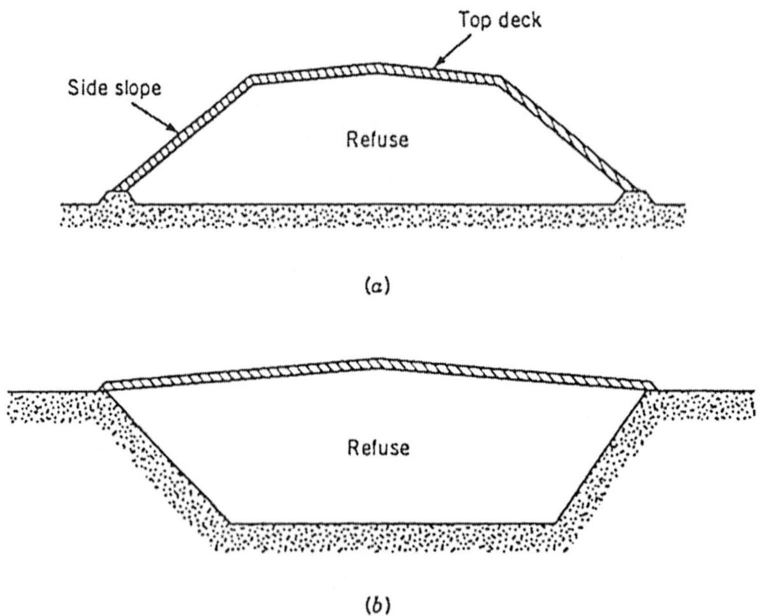

Figure 10.7 Landfill final cover configurations. (Sharma and Lewis, 1994.)

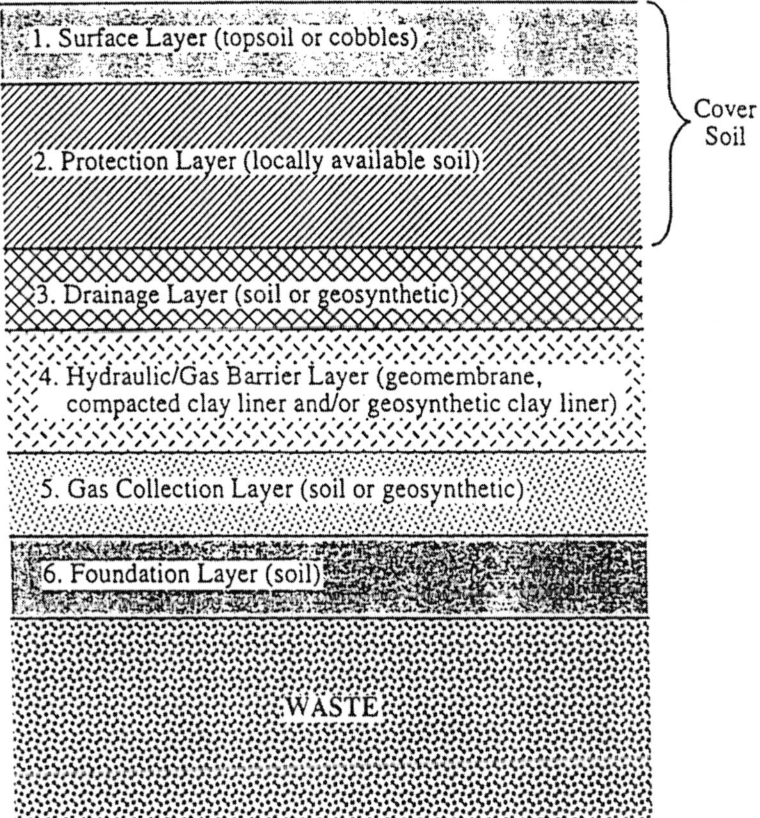

Figure 10.8 Typical layers of landfill final covers. (Koerner and Daniel, 1997.)

equipment, electrical materials, petroleum, coal, chemicals, and other manufacturing products. While conventional geotechnical exploration tools can be used to study and evaluate them, no standard test methods are available. Yet, to carry out slope stability analyses involving these peculiar materials, the engineer must make some assumptions and develop the suite of required material properties and strength values.

There are a multitude of properties related to landfill waste that could be discussed including grain size distribution, porosity, moisture content, hydraulic conductivity, Atterberg limits, unit weight, strength, dynamic properties, and compressibility. Additionally, chemical and leach properties of landfill waste must be known to determine the appropriate disposal locations and methods. The properties most germane to slope stability analyses are unit weight and strength. Some average values for unit weight are given in Table 10.2. Strength properties are often expressed in a graph, as shown in Figure 10.9. The strength of wastes from industrial facilities such as power plants and other manufacturing facilities has received study and some examples are

TABLE 10.2 Average Unit Weights for Landfill Waste

Source	Refuse Placement Conditions	Unit Weight	
		kg/m^3	lb/ft^3
U.S. Department of the Navy (1983)	Sanitary landfill		
	Not shredded		
	Poor compaction	320	20
	Good compaction	641	40
	Best compaction	961	60
	Shredded	881	55
Sowers (1968)	Sanitary refuse: depending on compaction effort	481–961	30–60
NSWMA (1985)	Municipal refuse:		
	In a landfill	705–769	44–49
	After degradation and settlement	1,009–1,121	63–70
Landva and Clark (1986)[a]	Refuse landfill (refuse to soil cover ratio varied from about 2–1 to 10–1)	913–1,346	57–84
EMCON Associates (1989)[b]	For 6–1 refuse to daily cover soil	737	46

Source: Sharma et al. (1990).

[a]These values were obtained from test-pit measurements of refuse at 11 municipal landfills in Canada. Values measured for the Halifax landfill and the August 1983 measurements at the Edmonton and Calgary landfills have not been included, as suggested by the authors.

[b]Based on tonnage records and areal survey maps recorded during the period from April 1988 through April 1989.

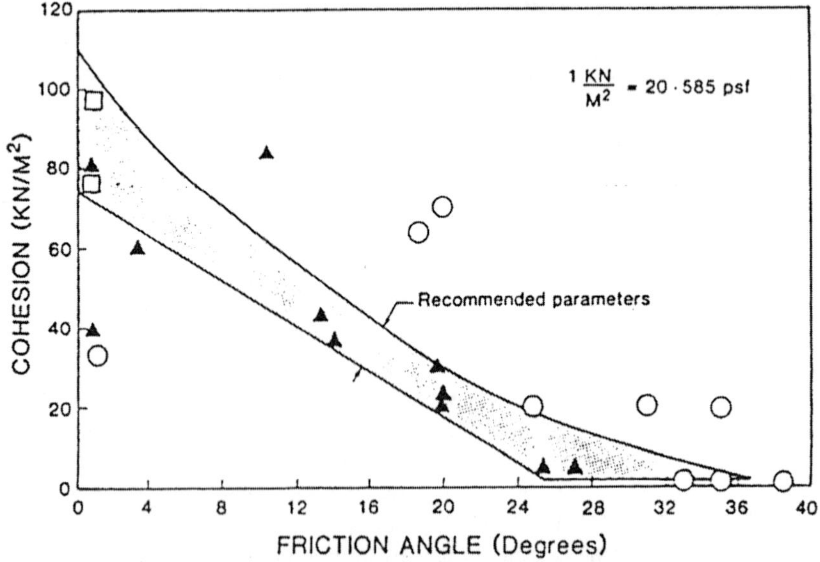

Figure 10.9 Typical shear strength values for landfill waste. (Singh and Murphy, 1990.)

TABLE 10.3 Strength Properties of Mineral Landfill Waste

Description	Cohesion, c (kPa)	Friction Angle, ϕ (degrees)	Undrained Compressive Strength, q_m (kPa)	Water Content, w (%)
Coal refuse				
Undrained condition	10–40	10–28		
Effective stress	0–40	25–43		
Fly ash, Arizona 7-day			223	
Unit wt. 12.6 kN/m^3			331	
Unit wt. 13.4 kN/m^3			587	
Unit wt. 13.8 kN/m^3				
Fly ash (silica 46%, aluminum 34%, calcium 7%)				
Slurry samples	0	37		
Compacted, undrained effective stress	0	41		
Compacted, drained	0	37		
West Virginia fly ash				
Shelby tube samples (consolidated, undrained)		0	34	
Shelby tube samples (consolidated, drained)	0	37.5		
West Virginia bottom ash		38–43		
Flue-gas desulfurization (FGD) sludge				
Consolidated, drained test	0	41.5		
Compacted	0–40	10–40		
Red mud (bauxite residue)				
Unleached			63	52
Leached			0	49
Mud: sand (5:1)			38	41

Source: Oweis and Khera (1990).

presented in Table 10.3. For slope stability analyses, strength values for landfill waste of $c = 400$ pounds per square foot and $\phi = 20°$ is a good starting point (Sharma and Lewis, 1994). Also, of great importance is interface friction angles between waste and other construction materials such as clay caps, geosynthetics, and so on.

10.4 GEOSYNTHETICS IN LANDFILLS AND ENGINEERING PROPERTIES

Fifty to one hundred years ago, landfills did not require linings. Waste was dumped directly onto the native ground and then covered and vegetated as well as possible.

Environmental regulation now requires very specific multiple-layer redundant lining and cover systems (Figure 10.10). Concurrent with stiffer environmental regulation of landfills, the use of geosynthetic materials has forged its way into the geotechnical engineering community including usage at landfills. Geosynthetic materials are materials, mostly plastic, which are commonly used in place of, or to enhance the function of, natural soil materials. Landfill lining and cover systems generally require a variety of geosynthetic materials including geomembranes, geotextiles, geonets, geogrids, and geosynthetic clay liners (GCLs).

10.4.1 Geomembranes

Geomembranes are flexible, polymeric sheets that have extremely low permeability and are typically used as liquid or vapor barriers (Figure 10.11). Table 10.4 lists the major types of geomembranes in current use. Advantages and disadvantages of some commonly used geomembranes are provided in Table 10.5. In landfills, base liners are placed below waste to minimize liquids expelled from and/or filtered through the waste (known as leachate) from contaminating the underlying ground and most important, the groundwater. Cover liners are placed above the final waste configuration to keep water, usually from rain or snow, from entering the waste and producing leachate.

10.4.2 Geotextiles

Geotextiles are synthetic fabrics used in geotechnical engineering for various applications. The majority of geotextiles are composed of polypropylene or polyester fibers; a small percentage are composed of polyamide or polyethylene (Sharma and Lewis, 1994). Geotextiles are manufactured from monofilament, staple, or slit film fibers that are twisted or spun together into yarn. The fibers or yarns are formed into geotextiles using either woven or nonwoven methods. Woven geotextiles are formed using traditional weaving methods and a variety of weave types (Figure 10.12). To create nonwoven geotextiles, the manufactured fibers are placed and oriented on a moving conveyor belt. Needle punching, melt bonding, or resin bonding bonds the fibers. At landfills, nonwoven geotextiles are most commonly used for filtration, separation, cushioning, and drainage. Woven geotextiles are usually used for reinforcement. Both types can be used as an alternative for daily cover.

10.4.3 Geonets

Geonets are used for drainage and consist of two sets of parallel solid or foamed extruded ribs that intersect at a constant angle to form an open net configuration (Figure 10.13). Channels are formed between the ribs to convey either liquid or gases. To prevent intrusion by soils or other adjacent materials in the field, a geocomposite drainage net may be used that consists of a geotextile bonded to the geonet.

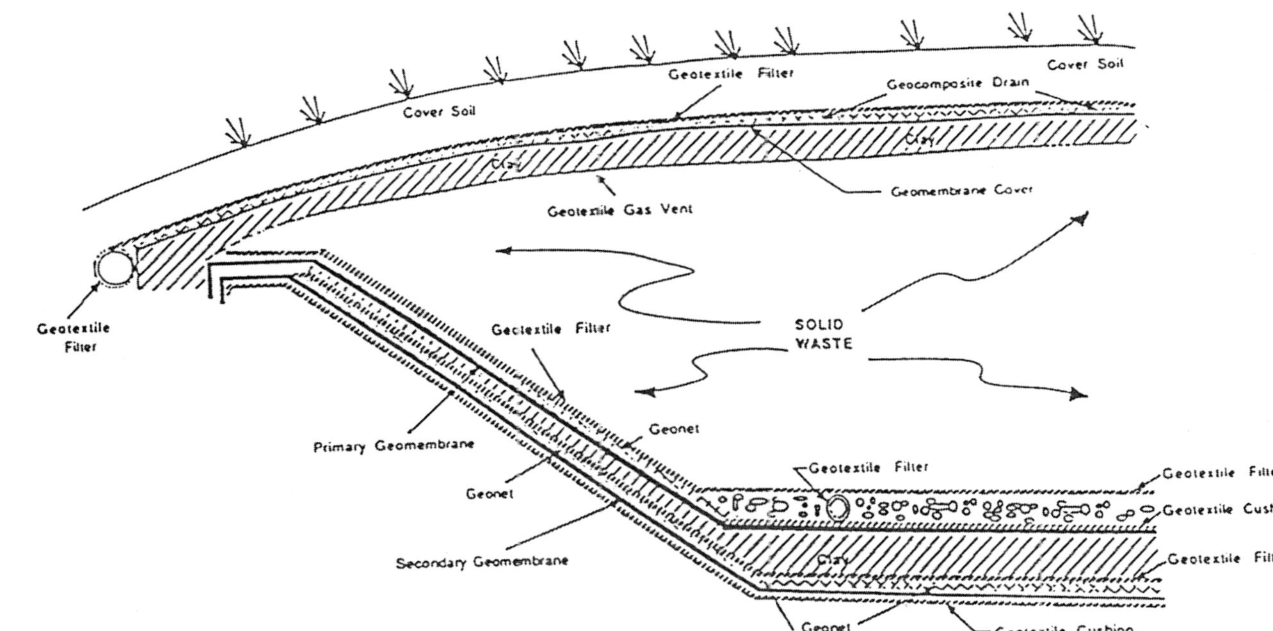

Figure 10.10 Typical landfill liner and cover system. (Koerner and Koerner, 1990.)

Figure 10.11 Geomembrances.

10.4.4 Geogrids

Geogrids are high-strength, soil reinforcement products composed of polypropylene, polyethylene, polyester, or PVC-coated polyester. All geogrids have an open mesh configuration with apertures ranging from $\frac{1}{2}$ to 3 inches (Figures 10.14 and 10.15). They are formed by several different methods. The polyester and polyester PVC-coated geogrids are typically woven or knitted. The polypropylene geogrids are either

TABLE 10.4 Major Types of Geomembranes in Current Use

Thermoplastic Polymers	Thermoset Polymers	Combinations
Polyvinyl chloride (PVC)	Butyl or isoprene-isobutylene (IIR)	PVC–nitrile rubber
Polyethylene (VLDPE, LLDPE, MDPE, HDPE, referring to very low, linear low, medium, and high density)	Epichlorohydin rubber	PE–EPDM
Chlorinated polyethylene (CPE)	Ethylene propylene diene monomer (EPDM)	PVC–ethyl vinyl acetate
Elasticized polyolefin (3110)	Polychloroprene (neoprene)	Cross-linked CPE
Ethylene interpolymer alloy (EIA or XR-5)	Ethylene propylene terpolymer (EPT)	Chlorosulfonated polyethylene (CSPE or Hypalon)
Polyamide	Ethylene vinyl acetate (EVA)	

Source: Koerner (1990).

TABLE 10.5 Advantages and Disadvantages of Commonly Used Synthetic Membranes

Synthetic Membrane	Advantages/Disadvantages
Butyl rubber	Good resistance to ultraviolet (UV) ray, ozone, and weathering elements Good performance at high and low temperatures Low swelling in water Low strength characteristics Low resistance to hydrocarbons Difficult to seam
Chlorinated polyethylene (CPE)	Good resistance to UV, ozone, and weather elements Good performance at low temperatures Good strength characteristics Easy to seam Poor resistance to chemicals, acids, and oils Poor seam quality
Chlorosulfonated polyethylene	Good resistance to UV, ozone, and weather elements Good performance at low temperatures Good resistance to chemicals, acids, and oils Good resistance to bacteria Low strength characteristics Problem during seaming
Ethylene-propylene rubber (EPDM)	Good resistance to UV, ozone, and weather elements High strength characteristics Good performance at low temperatures Low water absorbance Poor resistance to oils, hydrocarbons, and solvents Poor seam quality
Low-density and high-density polyethylene (LDPE and HDPE)	Good resistance to most chemicals Good strength and seam characteristics Good performance at low temperatures Poor puncture resistance
Polyvinyl chloride (PVC)	Good workability High strength characteristics Easy to seam Poor resistance to UV, ozone, sulfide, and weather elements Poor performance at high and low temperatures

Source: Bagchi (1994).

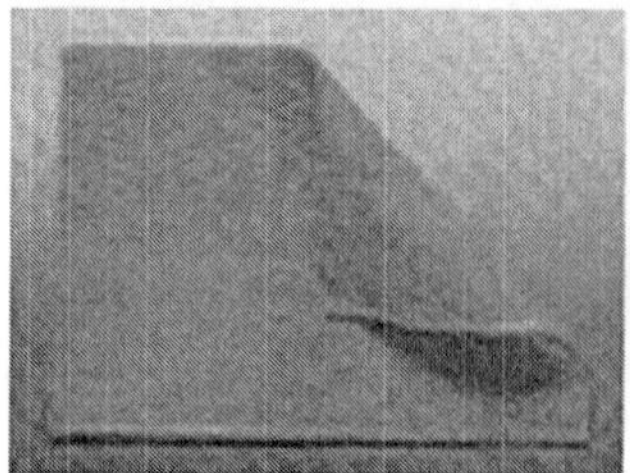
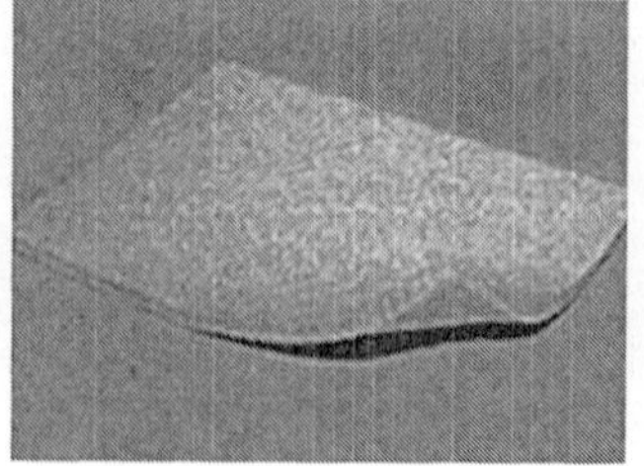
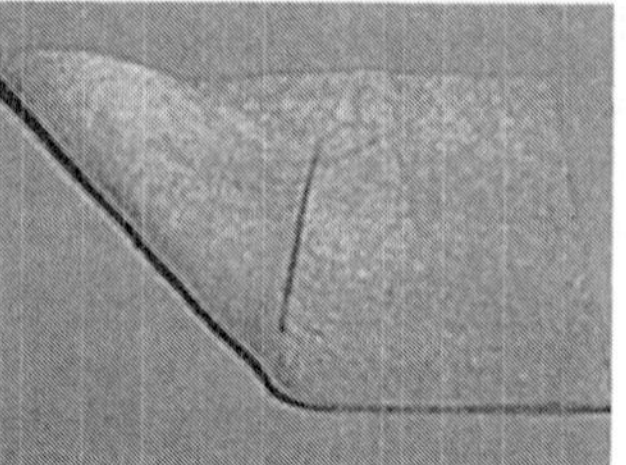

Figure 10.12 Geotextiles.

extruded or punched sheet drawn, and polyethylene geogrids are exclusively punched sheet drawn. At landfills, geogrids may be used to support a lining system over a weak subgrade or to support final landfill cover soils on steep refuse slopes. They are also used sometimes in "piggyback" landfills between the old landfill and the new one. Typical ranges of tensile strengths for geogrids made of different materials are listed in Table 10.6.

10.4.5 Geosynthetic Clay Liners (GCLs)

Geosynthetic clay liners (GCLs) are very low permeability barriers consisting of a layer of unhydrated, loose granular or powdered bentonite that is chemically or mechanically adhered to a geotextile or geomembrane. They are generally used as an alternative to compacted clay liners. Differences between GCLs and compacted clay liners are presented in Table 10.7.

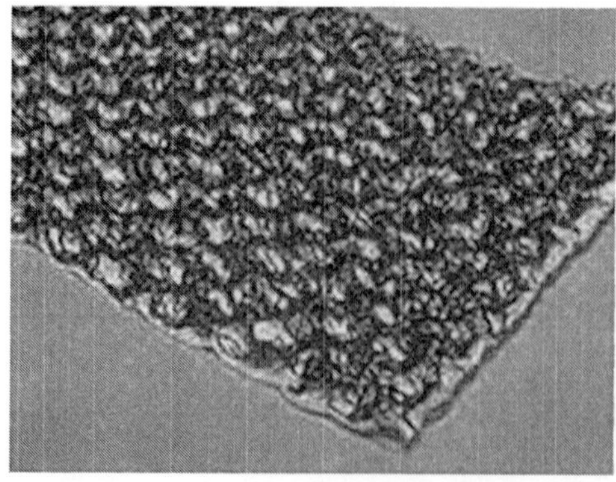
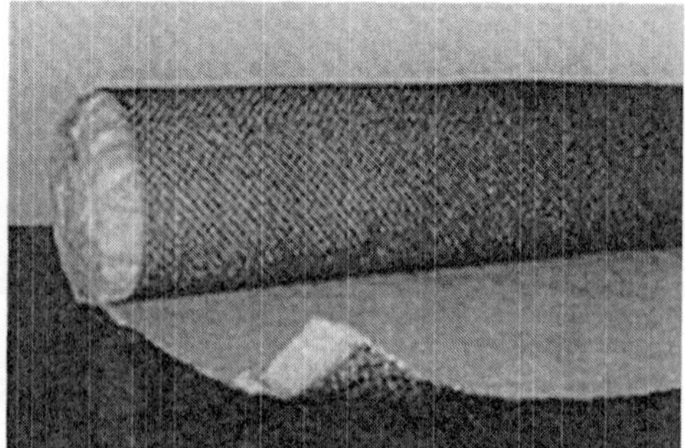

Figure 10.13 Geonets.

Figure 10.14 Geogrid.

10.4.6 Engineering Properties of Geosynthetics

There are a multitude of engineering properties that are of interest when using geosynthetics in landfills. Many of these deal with the ability of geosynthetics to perform their intended function as a barrier or pathway for liquid and gas migration. Many of the material properties of concern are listed in Table 10.8. However, from a slope stability standpoint, the frictional behavior of these materials and between these materials and other materials is of paramount importance. This property is known as interface friction. The possible interfaces that may be of interest in a landfill slope stability analysis are:

- Soil/geotextile
- Soil/geomembrane

Figure 10.15 Geogrid. (Martin, 1998.)

TABLE 10.6 Typical Range of Wide Width Geogrid Strengths in Maximum Strength Direction

Geogrid	Tensile Strength, 5% Strain (lb/in.)	Ultimate Tensile Strength (lb/in.)
Polyester	750–2,300	2,600–8,500
Polyester, PVC coated	600–7,600	1,500–25,400
Polypropylene	50–110	70–190
Polyethylene	140–460	300–810

Source: Sharma and Lewis (1994).

- Geosynthetic/geosynthetic
- Geosynthetic/clay liner

Interface friction angles can be measured in the laboratory using ASTM D 5321. Using this method, each of the combinations of materials can be tested. There is also an abundant amount of information available in the literature, as shown in Tables 10.9, 10.10, 10.11, and 10.12.

10.4.7 Anchor Trenches

In addition to selecting the appropriate types and combinations of geosynthetics, anchoring details must be specified to hold the geosynthetics in place and prevent

TABLE 10.7 Differences between GCLs and Compacted Clay Liners

Characteristic	Geosynthetic Clay Liner	Compacted Clay Liner
Materials	Bentonite clay, adhesives, geotextiles, and geomembranes	Native soils or blend of soil and bentonite
Construction	Manufactured and then installed in the field	Constructed in the field
Thickness	Approximately 10 mm	Approximately 0.5 to 1.0 m
Hydraulic conductivity of clay	10^{-10} to 10^{-8} cm/s (typical)	10^{-8} to 10^{-7} cm/s (typical)
Speed and ease of construction	Rapid, simple installation	Slow, complicated construction
Water content at time of construction	Essentially dry; cannot desiccate during construction and produces no consolidation water	Nearly saturated; can desiccate and can produce consolidation water
Cost	\$5 to \$11 per square meter	Highly variable (estimated range: \$8 to \$32 per square meter)
Experience level	Limited due to newness	Has been used for many decades

Source: USEPA (1993).

TABLE 10.8 Importance of Geosynthetic Material Properties

Geosynthetic	Main Material Properties of Concern	Common Test Methods
Geomembranes	Thickness	ASTM D 374, D 751, D 1593
	Tensile behavior	ASTM D 412, D 638, D 882, D 4885
	Tear resistance	ASTM D 1004
	Punch resistance	FTMS 101C
	Chemical resistance	USEPA 9090
	Seam strength	ASTM D 4437
Geotextiles	Thickness	ASTM D 177, D 5199
	Tensile behavior	ASTM D 4632, D 4595
	Bursting behavior	ASTM D 3786
	Tear behavior	ASTM D 4533
	Apparent opening size	ASTM D 4751
	Permittivity	ASTM D 4491
	Transmissivity	ASTM D 4716
	Seam strength	ASTM D 1683, D 4884
Geonets	Thickness	ASTM D 1777, D 5199
	Crush strength	ASTM D 1621
	Transmissivity	ASTM D 4716
Geogrids	Tensile strength (Also see geomembranes and geotextiles)	ASTM D 4595
Geosynthetic clay liners (GCLs)	(Address each composite material separately including standard tests on bentonite materials)	

tearing. As shown in Figure 10.16, anchor trenches can generally be classified as flat, rectangular, or V-shaped. Selection of the appropriate anchor trench configuration for any particular site depends on the required holding capacity, access considerations, dimensional constraints, and available construction equipment. The holding capacity of anchor trenches is developed by the applied normal load of the soil placed above the geosynthetics that creates frictional resistance between the geosynthetics and the

TABLE 10.9 Typical Range of Reported Soil Geotextile Friction Angles

Geotextile	Sand Friction Angle (degree) (Efficiency)	Clay Friction Angle (degree) (Efficiency)
Woven	23–42 (0.68–1.0)	16–26 (0.61–0.93)
Nonwoven, needle-punched	25–44 (0.67–1.0)	15–28 (0.62–0.99)
Nonwoven, resin- or heat-bonded	22–40 (0.56–0.91)	17–33 (0.60–0.85)

Source: Sharma and Lewis (1994).

TABLE 10.10 Typical Range of Reported and Recommended Soil Geomembrane Friction Angles

Geomembrane	Reported Sand Friction Angles (degree) (Efficiency)	Recommended Sand Friction Angles, δ (degree)	Reported Clay Friction Angles (degree) (Efficiency)	Recommended Clay Friction Angles, δ (degree)
PVC	21–33 (0.62–0.93)	20–30	6–39 (0.53–1.0)	6–15
HDPE	17–28 (0.45–0.81)	17–25	5–29 (0.47–0.88)	5–10
Textured HDPE	30–45 (0.86–1.0)	30–40	7–35 (0.70–1.0)	9–15
VLDPE[a]	21–28 (0.62–0.67)	—	—	—

Source: Sharma and Lewis (1994).

[a] Since VLDPE is a relatively new product, limited results were reported in the literature. It is anticipated that the range of efficiencies for VLDPE to sand interfaces is broader than shown. Blank (—) means insufficient data at this time.

underlying soil. There is minimal friction resistance developed between the upper soil and the geosynthetic since the soil above the geosynthetic is likely to move with the geosynthetic. The soil depth, type of soil or other material underlying the geosynthetics, and geosynthetic anchorage length are therefore the key factors in developing the required anchor trench holding capacity. Sharma and Lewis (1994) provide additional information and design formulas for anchor trench design.

10.5 LANDFILL CONSTRUCTION

The construction of a landfill is a carefully planned, methodical, highly regulated, and dynamic process. Substantial planning, study, and design must take place before a landfill can be sited, opened, and/or modified. Additionally, constant monitoring, testing, and reporting are required throughout the life of the landfill. More relevant

TABLE 10.11 Typical Range of Reported Geosynthetic to Geosynthetic Friction Angles in Degrees

	PVC	HDPE Smooth	HDPE Textured	Geonet
Woven geotextile	10–28	7–11	9–17	9–18
Nonwoven, needle-punched geotextile	16–26	8–12	15–33	10–27
Nonwoven, resin/heat-bonded geotextile	18–21	9–11	15–16	17–21
Geonet	11–24	5–19	7–25	—

Source: Sharma and Lewis (1994).

TABLE 10.12 Direct Shear Test Results under Variable Hydrating Conditions

Hydrating Fluid	GCL Type[a]	Measured Property	Dry[b]	Constrained Swell[c]	Free Swell[d]
Distilled water	Claymax	Ø (deg)	37	16	0
		c (kPa)	6.9	3	4
	Gundseal	Ø (deg)	26	19	0
		c (kPa)	50	5	3
	Bentomat	Ø (deg)	42	37	23
		c (kPa)	14	6	5
	Bentofix	Ø (deg)	36	31	10
		c (kPa)	68	7	9.0
Tap water	Claymax	Ø (deg)	37	18	0
		c (kPa)	6.9	3	3
	Gundseal	Ø (deg)	26	18	0
		c (kPa)	50	5	3
	Bentomat	Ø (deg)	42	43	26
		c (kPa)	14	6	10
	Bentofix	Ø (deg)	36	34	15
		c (kPa)	68	6.9	7
Mild leachate	Claymax	Ø (deg)	37	24	4
		c (kPa)	6.9	6	3
	Gundseal	Ø (deg)	26	18	13
		c (kPa)	50	5	4
	Bentomat	Ø (deg)	42	39	25
		c (kPa)	14	8.3	14
	Bentofix	Ø (deg)	36	43	20
		c (kPa)	68	5	12
Harsh leachate	Claymax	Ø (deg)	37	19	0
		c (kPa)	6.9	6	3
	Gundseal	Ø (deg)	26	13	0
		c (kPa)	50	7.6	3
	Bentomat	Ø (deg)	42	45	32
		c (kPa)	14	5	12
	Bentofix	Ø (deg)	36	39	30
		c (kPa)	68	4	8.3
Diesel fuel	Claymax	Ø (deg)	37	44	38
		c (kPa)	6.9	4	6
	Gundseal	Ø (deg)	26	24	29
		c (kPa)	50	4	6
	Bentomat	Ø (deg)	42	42	40
		c (kPa)	14	6	5
	Bentofix	Ø (deg)	36	51	46
		c (kPa)	68	4	5

Source: USEPA (1993).

[a]Claymax and Gundseal are unreinforced GCLs; Bentomat and Bentofix are reinforced. Claymax also manufactures a reinforced GCL, but it was not used in this testing program.

[b]Dry refers to product as-received, placed under desired normal stress, than sheared at midplane.

[c]Constrained swell refers to product hydrated under desired normal stress, (i.e., constrained swell) then sheared at midplane.

[d]Free swell refers to product hydrated under zero normal stress, then placed under desired normal stress, and then sheared at midplane.

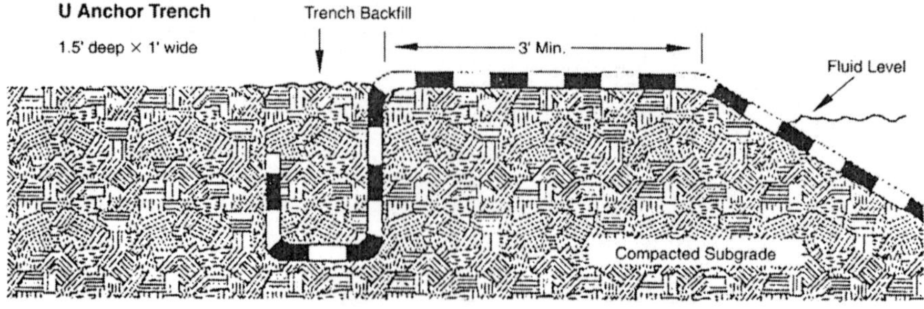

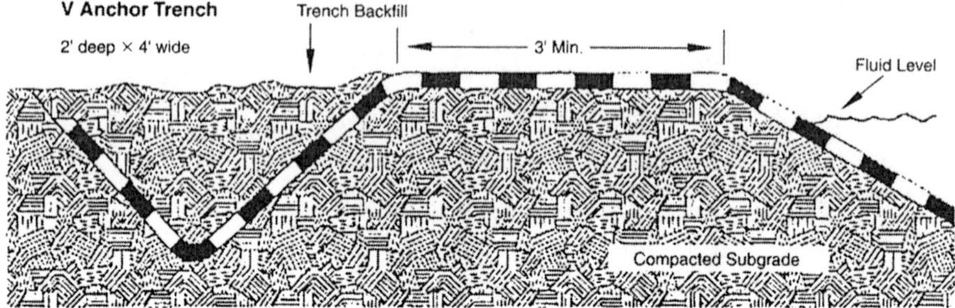

Figure 10.16 Typical anchor trench configurations. (Martin, 1998.)

to slope stability issues is the fact that numerous activities are going on at one time in multiple locations with a variety of materials and equipment. Also, precipitation, wind, and other environmental factors further contribute to the variability of site conditions.

The basic sequence of construction at a landfill includes:

- Subbase preparation
- Liner construction
- Waste placement
- Landfill closure

The subbase for a landfill refers to the ground surface on which the liner is constructed. Compaction and grading of the subbase are necessary so that the actual liner can be constructed easily. The subbase may be developed on existing ground surface or may be constructed on fill or cut. Great care must be taken so that the subbase provides a sound foundation for the landfill. Conventional earthwork equipment and methods are used to prepare the subbase for a landfill.

Liner construction often consists of multiple layers of soil and geosynthetics (Figure 10.17). Placement of the soil follows conventional earthwork practices. Construction of geosynthetic liners requires specialized methods particularly related to connection of adjacent sheets/layers and prevention of damage (i.e., puncturing, tearing, blocking of drainage, etc.) during construction. As each new layer is placed,

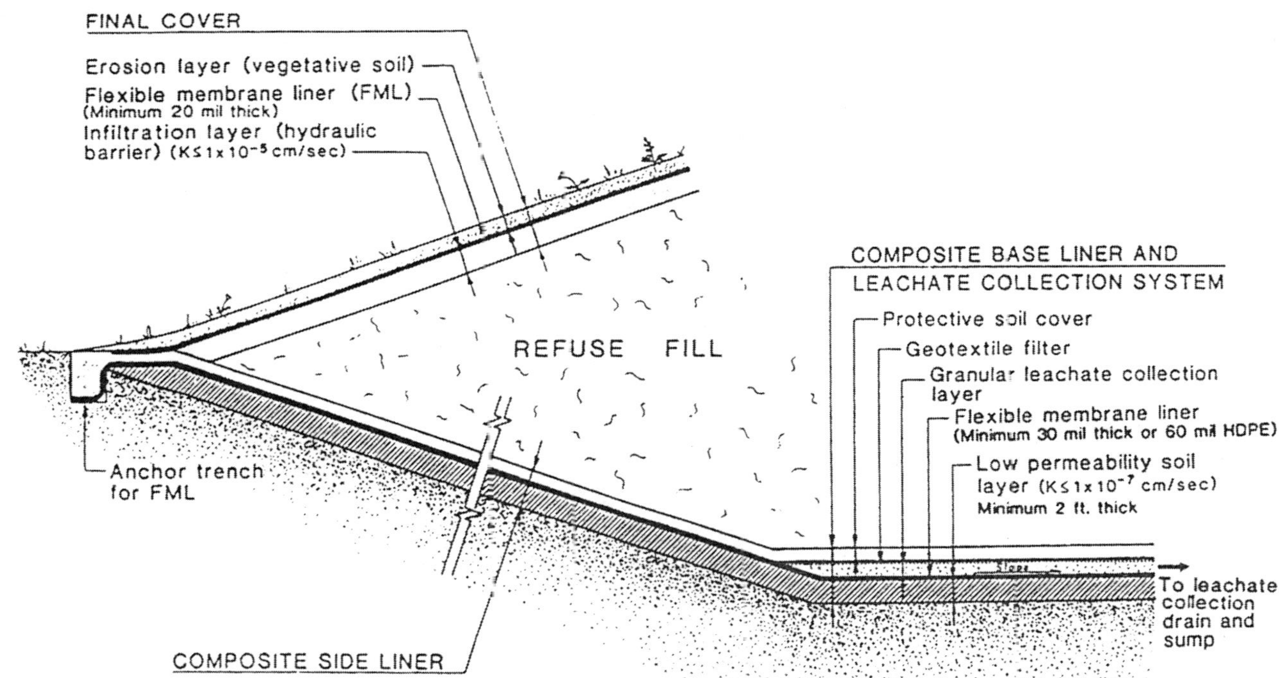

Figure 10.17 Subtitle D liner and final cover for a municipal solid waste landfill. (Sharma and Lewis, 1994.)

the shear strength between layers could change significantly. On a slope, the layer that possesses the lowest shear strength governs stability. It is conceivable therefore that as the liner layers are built up and then piled with waste, a slope could go from being stable to being unstable due to the introduction of a weaker layer.

Waste placement is a unique process during the life of the landfill because a variety of waste materials are being placed in a variety of configurations. Temporary slopes are commonly made and modified on a regular basis. Some of these slopes may be in contact with the liner system; some may not. Also, slopes made during this phase will be traversed several times by a variety of equipment and vehicles going a multitude of different directions. Since a slope stability analysis cannot be carried out every time a new load of waste arrives at the landfill, there must be governing restrictions on the steepness and location of temporary and permanent slopes. These restrictions must take into account precipitation and groundwater conditions.

After the landfill has received its maximum amount of waste, it must be closed such that it performs well and does not impact the environment in the long term. It receives a final cover or cap that is similar to the underlying liner system. In other words, there are a series of layers that get placed over the waste to control infiltration of water into the landfill, thus minimizing leachate, controlling the release of gases from the landfill, and providing for a physical separation between the waste and environment for protection of public health. A typical cross section of a landfill cap is shown in Figure 10.18.

Of the documented case histories of failures, most occur during liner construction and waste placement. Often, when failures have occurred, they were due to underestimation of interface friction angles between landfill, soil, and geosynthetic materials or the dynamic effects of equipment sequencing or loading. For instance, Martin (1998) reports on a failure in Washington that was due to bulldozer braking acceleration and deceleration loads. From a slope stability standpoint, all materials that are placed at some angle from horizontal must be analyzed for stability relative to the adjacent materials as well as all construction effects including sequencing, layering, equipment, weather, and so on.

10.6 SLOPE STABILITY CONSIDERATIONS

Slope stability considerations for a landfill must include all phases of development including subbase preparation, liner construction, waste placement, and landfill closure. Slope stability issues during subbase preparation are the same for any conventional cuts or fills as discussed previously in Chapter 6. Landfill slope stability is unique because after subbase preparation, multiple layers of soil, geosynthetics, and waste get piled on top and underneath each other, creating a unique combination of materials that behave in a composite fashion with uniquely different shear strength properties (Figure 10.19).

All of the methods of slope stability analysis discussed in this book apply to landfills. However, it is important to note the potential modes of failure in a landfill and the unique materials and material properties found at a landfill.

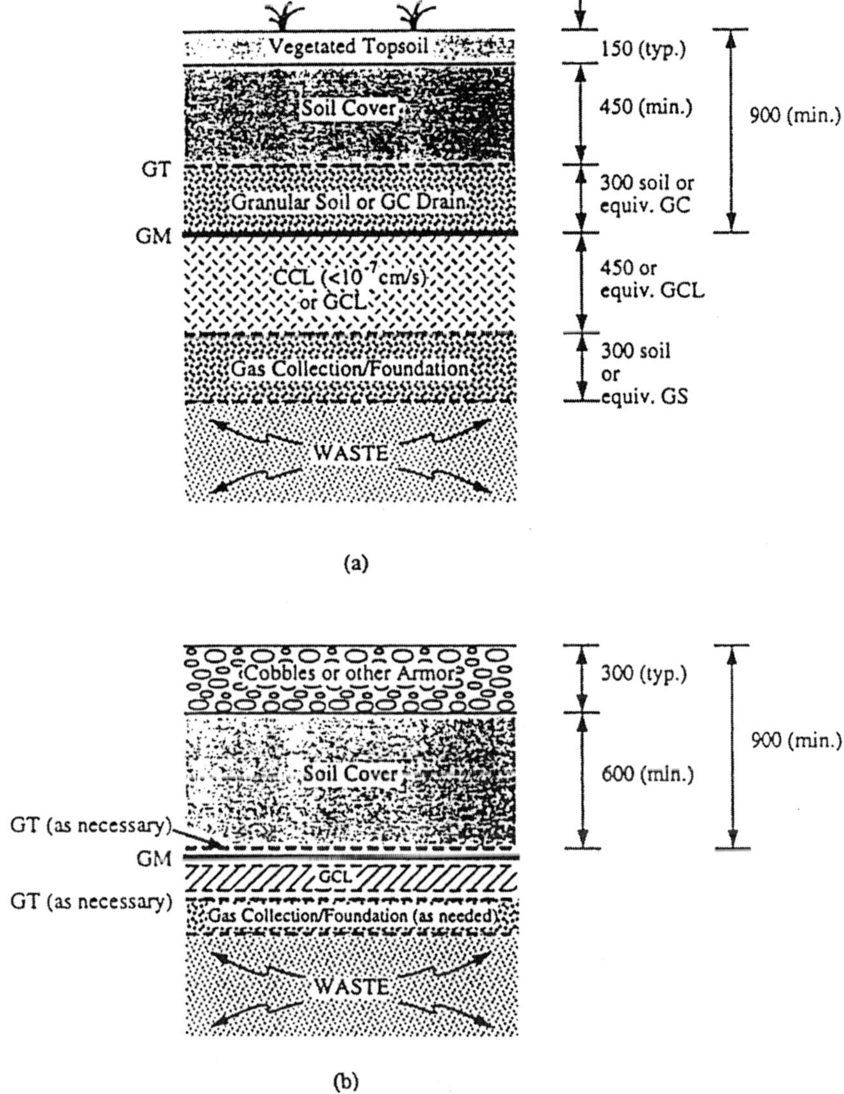

Figure 10.18 Examples of municipal solid waste final covers. (Koerner and Daniel, 1997.)

10.6.1 Excavation Slope Stability

Figure 10.20 illustrates some of the potential slip surfaces that may occur in landfill excavation slopes (Sharma and Lewis, 1994). For covered pit and trenched landfills, it is often simplest to adjust the excavation slope to meet stability and constructability requirements. Since the crest of the landfill excavation slope is typically limited by property boundaries, flattening an excavation slope from 2H–1V to 3H–1V can have a significant effect on landfill airspace and revenue. Most landfill owners therefore

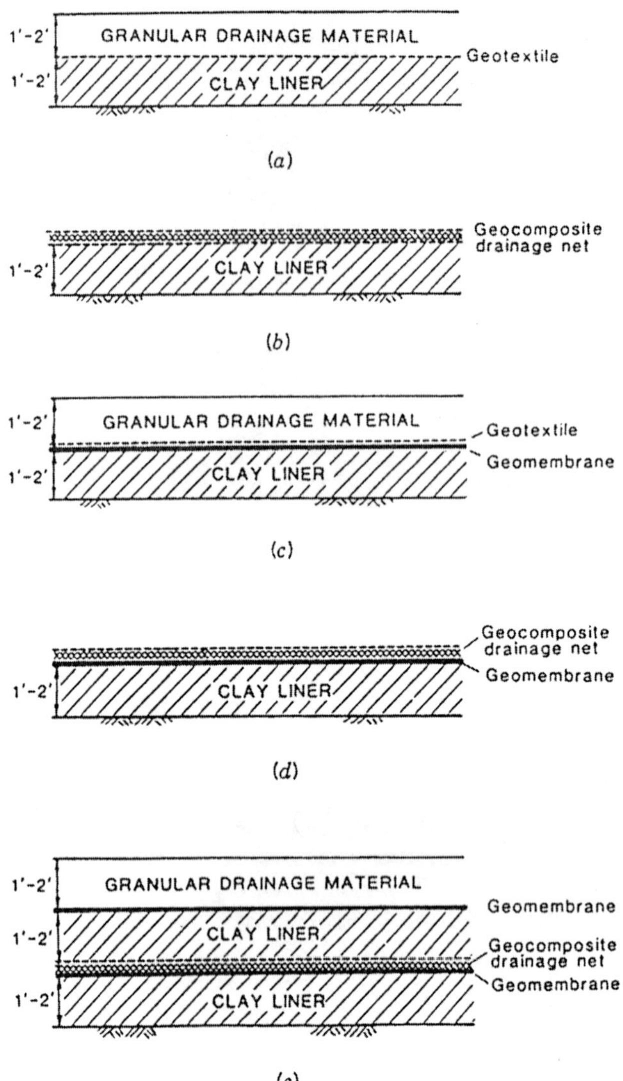

Figure 10.19 Types of landfill containment lining systems. (Sharma and Lewis, 1994.)

attempt to excavate slopes to the maximum grade achievable. Typical excavation slopes for covered pit and trenched landfills are approximately 3H–1V. This is because most soil slopes are stable at 3H–1V inclination and clay liners may be constructed on the side slopes at this grade.

Upslope and canyon landfills are often constructed at grades steeper than 3H–1V, due to the steep inclination of the natural slopes. Although these slopes may be trimmed back to meet a 3H–1V grade, a significant volume of excavation may be required due to the typically large slope heights. A groundwater table or weak soil

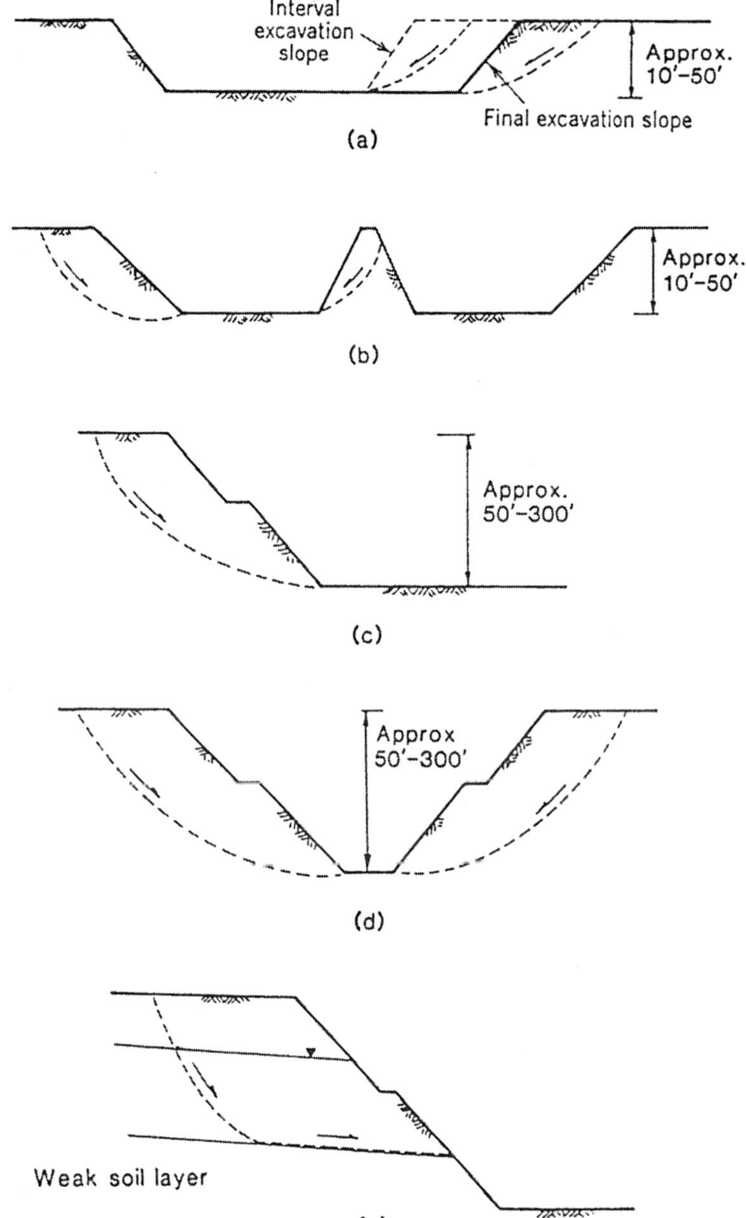

Figure 10.20 Potential slip surfaces in landfill excavation slopes. (Sharma and Lewis, 1994.)

layer also often influences the stability of upslope and canyon landfills. These items must be considered in the stability analyses.

For many excavation slopes, the worst-case conditions for analysis are either during construction or immediately following construction. Often, waste placed against the excavation slope acts as a buttress and prevents movement.

10.6.2 Waste Fill Stability

Potential slip surfaces through landfill waste generally occur in one of three ways, as illustrated in Figure 10.21:

- Through the waste alone
- Along the liner system
- A composite surface through waste and along the liner

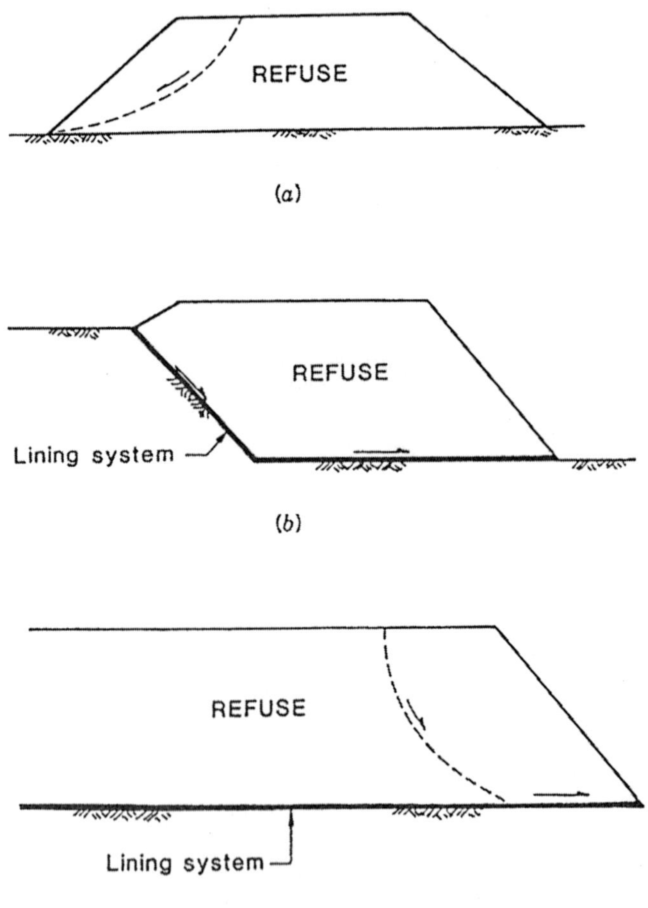

Figure 10.21 Potential slip surfaces in landfill waste. (Sharma and Lewis, 1994.)

Since waste is generally a heterogeneous material that may contain isolated masses of high-strength materials, such as discarded refrigerators or demolition debris, it is unlikely that slip surfaces through waste materials are circular. However, since the current state of technology affords no better solutions, waste fill stability analyses are typically performed using circular surfaces through waste materials. If a geosynthetic lining system is placed in a landfill, a translational block surface along the liner system or a composite slip surface through waste and along the liner system may be critical. There are several interface surfaces within a geosynthetic lining system where a weak plane may exist.

Figure 10.22 depicts typical slip surfaces that may occur through covered pit and trench-type landfills. For these types of landfills, the critical stability condition typically occurs in interim waste fills, prior to the pit or trench being filled to the ground surface (Figure 10.22*a*). Once the pit or trench is filled to the ground surface it is inherently stable (Figure 10.22*b*). If waste is placed above the ground surface, either a circular surface through the waste itself, or a block-type surface extending to the base lining system may occur (Figure 10.22*c*).

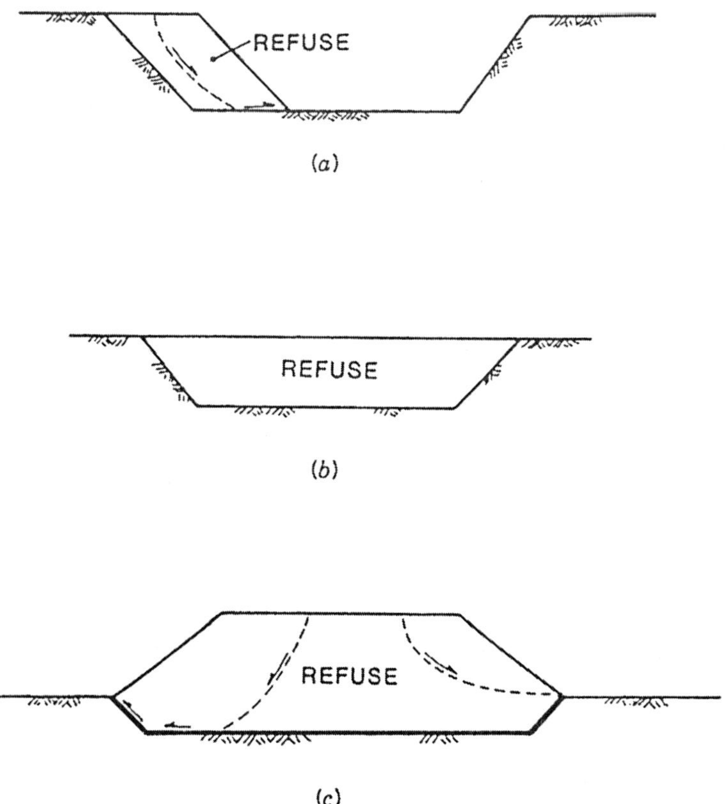

Figure 10.22 Potential slip surfaces in waste for pit and trench landfill configurations. (Sharma and Lewis, 1994.)

Figure 10.23 illustrates potential waste fill stability concerns in upslope or canyon-type landfills. From a stability standpoint, upslope landfills are perhaps the least desired waste fill configuration. This is because both interim slopes (Figure 10.23*a*) and final slopes (Figure 10.23*b*) have a significant potential for instability, especially

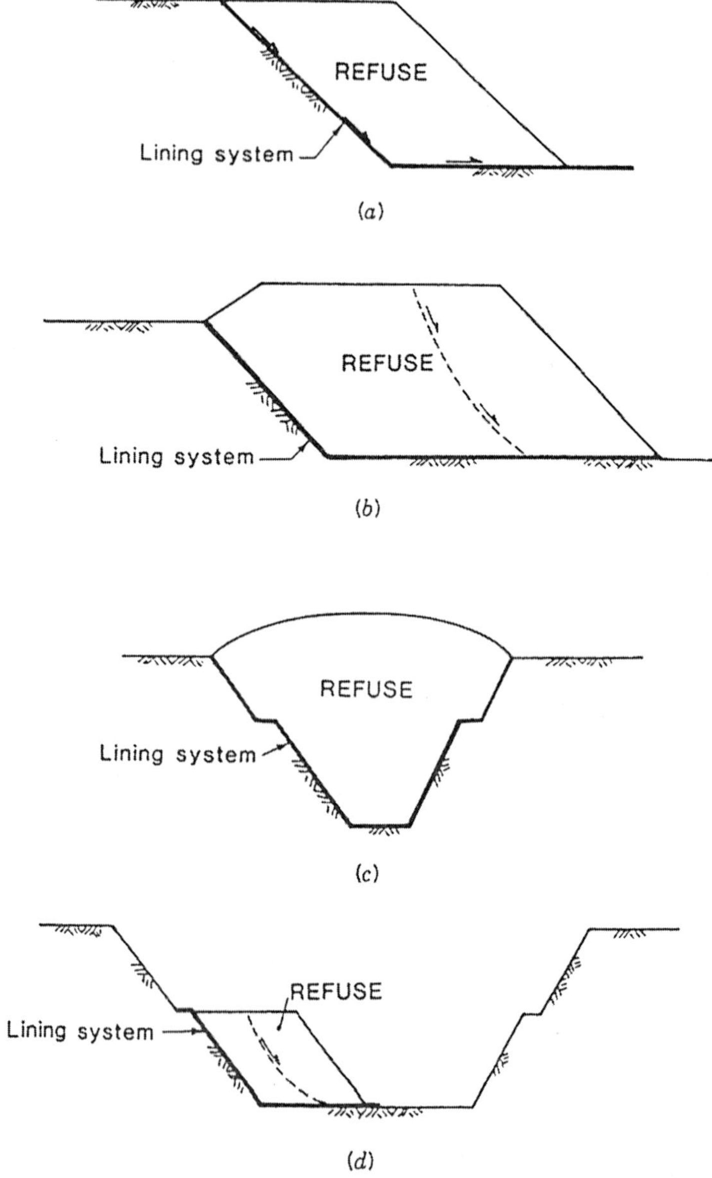

Figure 10.23 Potential slip surfaces in waste for upslope and canyon landfill configurations. (Sharma and Lewis, 1994.)

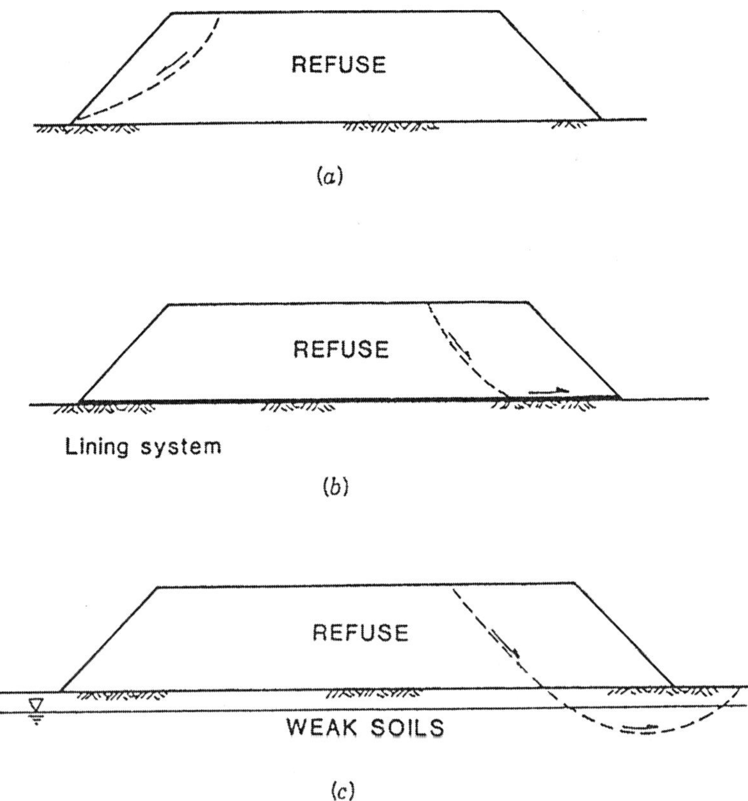

Figure 10.24 Potential slip surfaces in waste for aboveground landfill configurations. (Sharma and Lewis, 1994.)

when overlying a geosynthetic lining system. Canyon-type landfills are generally stable in their final configuration (Figure 10.23*c*), but may be unstable during interim fill conditions (Figure 10.23*d*).

Finally, aboveground landfills are generally stable when founded over firm ground, with some potential for a circular slip surface occurring if the side slopes are too steep (Figure 10.24*a*). If a geosynthetic lining system is employed, a composite-type slip surface may extend through the waste mass (Figure 10.24*b*). Aboveground landfills are also often constructed over soft subgrades or in zones of high groundwater. Bearing capacity subgrade failures (Figure 10.24*c*) should therefore be checked in these types of landfills.

In all of the landfill configurations illustrated, proper fill sequencing can alleviate many of the stability concerns. The key is to maximize the "resisting" portion of the potential slip surface prior to increasing the "active" portion. Figure 10.25 illustrates how this is achieved for covered pit and upslope landfills.

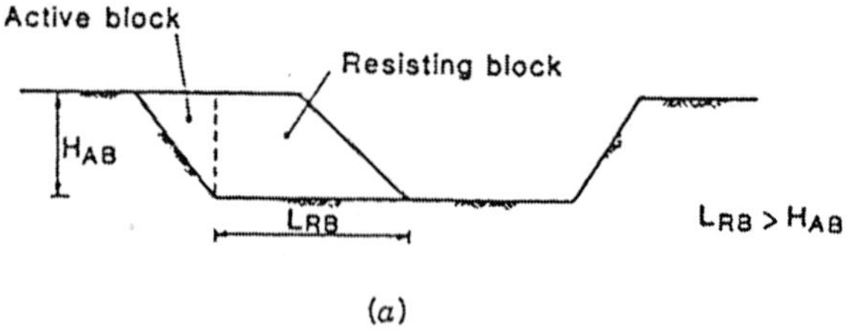

(a)

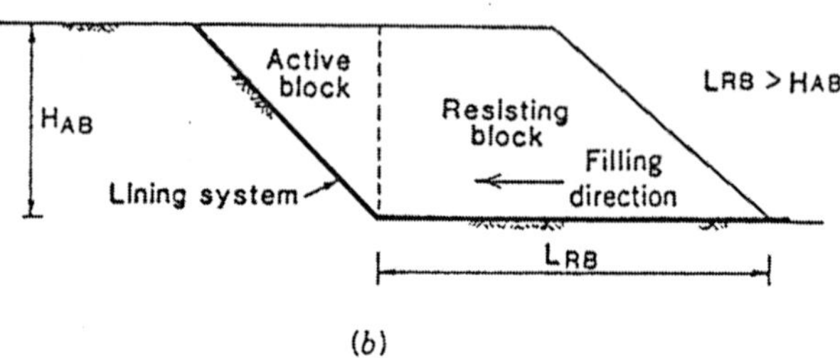

(b)

Figure 10.25 Active and passive stability blocks in covered pit and upslope landfills. (Sharma and Lewis, 1994.)

10.6.3 Cover System Stability

During preparation of landfill site development plans, consideration should be given to the stability of the landfill final cover system so that appropriate final landfill grades may be selected. Most landfills are constructed with final grades of approximately 3H–1V, which are generally stable for soil cover systems. There is some concern, however, regarding the ability of soil cover systems to deform with the underlying

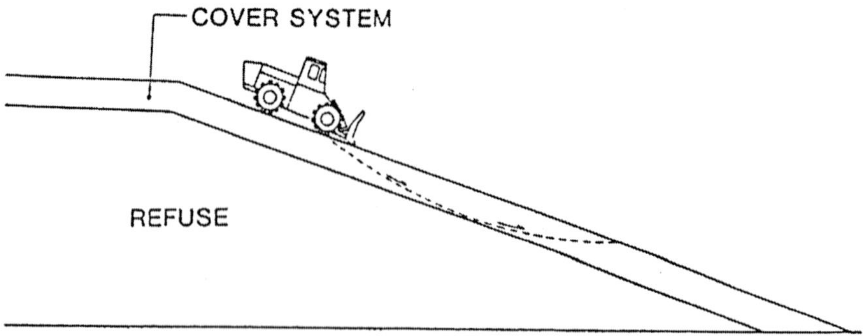

Figure 10.26 Shallow failure surface in final landfill cover due to construction equipment. (Sharma and Lewis, 1994.)

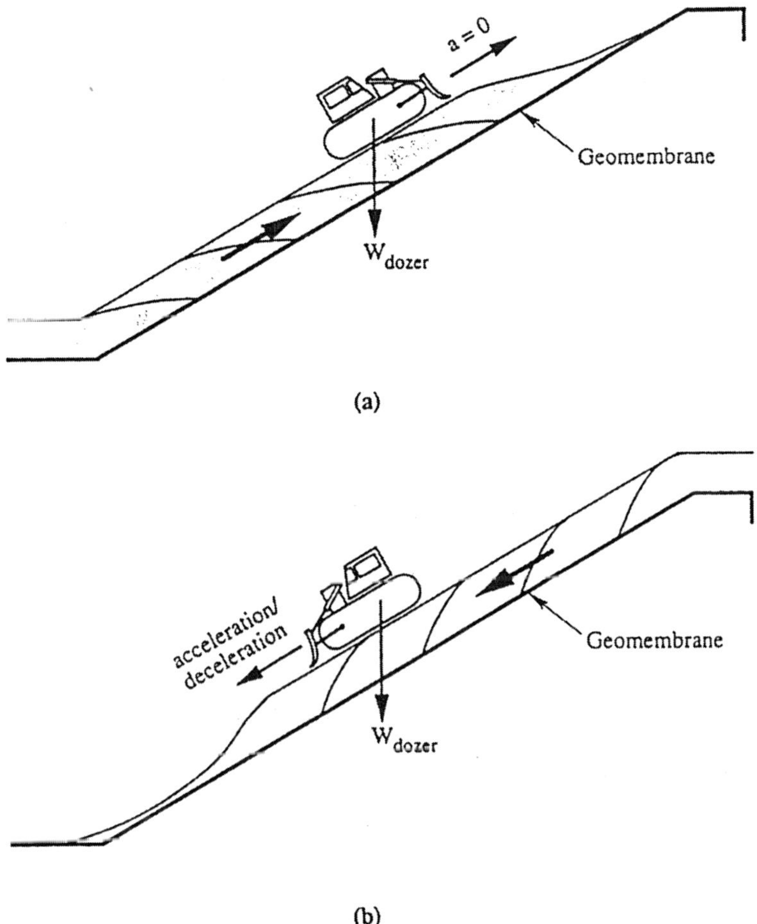

Figure 10.27 Effect of construction equipment loads on final landfill cover stability. (Koerner and Daniel, 1997.)

waste settlements. Geosynthetics are therefore often used in final cover systems since they are inherently more flexible. They may, however, create stability concerns.

Potential slip mechanisms in final covers are generally planar, occurring between material interfaces or through the material itself. Potential slip surfaces could occur along the interfaces or materials shown on Figure 10.26; however, there are some interfaces that are more likely than others to be critical weak planes. In some situations, the worst-case condition for a final cover system is during construction. This is due to the large equipment loads applied during construction. Failure due to equipment loads would probably result in a localized shallow circular slip surface especially when the final cover is being placed from the top of the landfill down to the base as illustrated in Figure 10.27.

REFERENCES

Bagchi, A., 1994. *Design, Construction, and Monitoring of Landfills*. New York: Wiley.

EMCON Associates, personal communications between Larry Burch and Hari Sharma, July 1989.

Koerner, R. M. 1990. *Designing with Geosynthetics*. Englewood Cliffs, New Jersey: Prentice Hall PTR.

Koerner, R. M., and D. E. Daniel, 1997. *Final Covers for Solid Waste Landfills and Abandoned Dumps*. Reston, Virginia: American Society of Civil Engineers.

Koerner, G. R., and Koerner, R. M., "Biological Activity and Potential Remediation Involving Geotextile Landfill Leachate Filters," in *Geosynthetics Testing for Waste Containment Applications*, ASTMP STP 1081, Robert M. Koerner, Ed. ASTM, Philadelphia, 1990a.

Landva, A. O., and Clark, J. I., "Geotechnical Testing of Wastefill," in *39th Canadian Geotechnical Conference*, Ottawa, Canada, 1986, pp. 371–385.

Martin, S., 1998. "Stability and Constructability of Geosynthetics," University of Idaho, Moscow.

McBean, E. A., F. A. Rovers, and G. J. Farquhar, 1995. *Solid Waste Landfill Engineering and Design*. Englewood Cliffs, New Jersey: Prentice Hall PTR.

National Solid Waste Management Association (NSWMA). *Basic Data: Solid Waste Amounts, Composition and Management Systems*. Technical Bulletin 85–6, National Solid Waste Management Association, October 1985, p. 8.

Oweis, I. S., and R. P. Khera, 1990. *Geotechnology of Waste Management*. Sevenoaks, Kent, England: Butterworth and Company (Publishers) Ltd., 273 pp.

Sharma, H. D., and S. P. Lewis, 1994. *Waste Containment Systems, Waste Stabilization, and Landfills—Design and Evaluation*. New York: Wiley.

Sharma, H. D., M. T. Dukes, and D. M. Olsen, 1990. "Field Measurements of Dynamic Moduli and Poisson's Ratio of Refuse and Underlying Soils at a Landfill Site," *Geotechnics of Waste Fills: Theory and Practice*, ASTM STP 1070, A. Landva and G. D. Knowles, Eds. Philadelphia: ASTM, pp. 57–70.

Singh, S., and B. J. Murphy, 1990. "Evaluation of the Stability of Sanitary Landfills," *Geotechnics of Waste Fills: Theory and Practice*, ASTM STP 1070, A. Landva and G. D. Knowles, Eds. Philadelphia: ASTM, pp. 240–258.

Sowers, G. F., "Foundation Problems in Sanitary Landfill," *Journal of the Sanitary Engineering Division, ASCE*, Vol. 94, No. SA1, February 1968, pp. 103–116.

U. S. Environmental Protection Agency (USEPA), 1993. *Report of Workshop on Geosynthetic Clay Liners*, USEPA, Washington, DC, August.

U.S. Department of the Navy, *Soil Dynamics, Deep Stabilization, and Special Geotechnical Construction*, Design Manual 7.3, NAVFAC DM 7.3M, Naval Facilities Engineering Command, Alexandria, Va., April 1983, p. 7.3–79.

INDEX